MODERN INDUSTRIAL HYGIENE

Volume 2

Biological Aspects

Jimmy L. Perkins
University of Texas Health Science Center – Houston
School of Public Health
San Antonio Regional Campus

ACGIH®
American Conference of Governmental Industrial Hygienists, Inc.
Cincinnati, Ohio, U.S.A.

ISBN: 1-882417-48-8

Printed in the United States of America

ACGIH®
1330 Kemper Meadow Drive
Cincinnati, Ohio 45240-1634
U.S.A.
Telephone: (513) 742-2020
Fax: (513) 742-3355
E-mail: publishing@acgih.org
http://www.acgih.org

To the people who formalized the information in this field and turned it into a discipline.

CONTENTS

Contributors xvii
Foreword xix
Preface xxi
Acknowledgments xxiii

1. The Respiratory System **1**
Arch I. Carson, MD and George L. Delclos, MD
Functions of the Respiratory System 1
Gas Exchange 1
Acid-base Regulation 2
Metabolism of Drugs, Toxins, and Endogenous Substances 2
Cardiovascular Pressure Regulation 3
Anatomy and Physiology of the Respiratory System 3
Cardiopulmonary Function 3
Upper Airways 3
Lower Airways 5
Respiratory Zone 6
Important Differences Among Mammalian Species 8
Tests of Respiratory Function 10
Spirometry and Peak Flow 10
Subdivisions of Lung Volume 12
Diffusing Capacity 13
Arterial Blood Gases 14
Cardiopulmonary Stress Testing 14
Bronchial Challenge Testing 14
The Respiratory System as a Portal of Entry for Toxic Substances 15
Determinants of Respiratory System Injury 15

Diagnosis of Occupational Respiratory Disease 19
Management of Occupational Respiratory Disease 21
Epidemiological Trends in Occupational Respiratory Disease 22
Questions 23
References 24

2. Respiratory Tract Deposition and Penetration 27
Lung Chi Chen, MD
Introduction 27
Deposition Mechanisms 27
Factors Affecting Deposition 31
Penetration of Particles into the Respiratory System 35
Deposition of Particles Within the Human Respiratory Tract 35
Experimental Deposition Data 35
Localized Patterns of Deposition 39
Predictive Deposition Models 40
ICRP Deposition Model 40
NCRP Deposition Model 42
Summary and Conclusion 42
Questions 42
References 43

3. Occupational Dermatotoxicology: Significance of Skin Exposure in the Workplace 47
Mark Boeniger, MS
Introduction 47
Skin as a Target Organ 49
Types of Skin Disease 49
Diagnosis of Occupational Contact Dermatitis 54
Prevalence and Incidence of Workplace Dermatoses 56
Cost of Occupational Dermatitis 60
Causes of Allergic and Irritant Dermatitis 61
Infectious Diseases 63
Management and Prevention of Occupational Dermatitis 64

Skin as a Route of Systemic Exposure 65
Skin as a Contributor to Oral Ingestion 70
Skin Route Contributing to Respiratory Sensitization 71
Skin Anatomy and Physiology 73
Physicochemical Characteristics Favoring Permeability 75
Mathematical Modeling of Permeation 77
Factors Affecting Dermal Absorption and Effects on the Skin 82
Diagnostic and Experimental Test Methods Related to the Skin 104
Allergens 104
Irritants 107
Permeation 109
Barrier Function 111
Questions 115
References 116

4. The Practice and Management of Occupational Ergonomics 135
Miriam Joffe, MS, PT and David C. Alexander, MSIE
A Brief History of Occupational Ergonomics 135
Background 135
Ergonomics Versus Human Factors 137
History 137
Integrating the Industrial Hygienist into Occupational Ergonomics 139
Understanding Ergonomics in Industry 139
The Background Survey and the Structure of Industry 139
Understanding the Production and Assembly Process 142
The On-Site Visit 147
A Classification System 149
The Size of People (Anthropometry) 150
The Strength of People (Musculoskeletal) 152
The Control of Work by People 155
The Endurance of People (Physiology) 156
The Comfort of People (Environment) 157
The Memory of People (Cognition) 157
What Are Musculoskeletal Disorders or MSDs? 159
Preventing Workplace Problems by Design 161

Designing the Workplace 164
- Using Anthropometry for Workplace Design 164
- Manual Materials Handling 172
- Determining Push and Pull Limits in MMH 184

Detecting Musculoskeletal Disorders (MSDs) 185
- Data Review 185
- Symptom Identification 186
- Observing Jobs and Reviewing Job Descriptions 186

Correcting Workplace Design Problems 189
- Administrative Controls 194

The Ergonomics Problem Solving Process 195
- Introduction 195
- Step 1: Identification 196
- Step 2: Investigation 199
- Step 3: Development of Alternate Solutions 205
- Step 4: Selection of the Most Appropriate Solution(s) 208
- Step 5: Implementation 208
- Step 6: Following Up 213

The Certification Process for Professional Ergonomists 214
- Introduction 214
- Certification Levels 214

Questions 215

Further Readings 216

References 217

5. Biological Agents – Recognition **219**

Jeroen Douwes, PhD, Peter S. Thorne, PhD, Neil Pearce, PhD, and Dick Heederik, PhD

Definition of Biological Agents 219

Infectious Diseases 220
- Agents 221
- Transmission 222
- Workers at Risk 223
- Incidence 231

Respiratory Diseases 232

Background Immunology 234
Allergic Respiratory Diseases 236
Non-allergic Respiratory Diseases 252
Agents Causing Allergic and Non-allergic Respiratory Diseases 264
Allergenic Agents 264
Pro-inflammatory and Toxic Agents 267
Cancer 276
Mycotoxins 276
Farming-related Occupations 278
Wood-based Industries 280
Other Manufacturing Industries Using Biological Materials 280
Acknowledgments 282
Questions 282
References 282

6. Biological Agents – Monitoring and Evaluation of Bioaerosol Exposure **293**
Dick Heederik, PhD, Peter S. Thorne, PhD, and Jeroen Douwes, PhD
Introduction 293
Assessment Methods for Microorganisms 294
Culture-based Methods 294
Growth Media 298
Non-culture Based Methods 298
Special Issues in the Evaluation of Microbial Exposure 300
New Molecular Biological Approaches for Analysis of Microorganisms 303
Assessment Methods for Constituents of Microorganisms 305
Endotoxins 306
Glucans 309
Mycotoxins 310
Extracellular Polysaccharides (EPS) 310
Ergosterol 311
Fungal Volatile Components 311
Assessment Methods for Bio-allergens (High Molecular Weight Sensitizers) 312
Immunoassays for Measurement of Allergens 312
Evaluation of Bioaerosol Data for Health Hazard Evaluations 314

How to Sample 314
Sampling Duration and Sampling Frequency 314
Exposure Standards 316
Practical Evaluation of Bioaerosol Levels 318
Sources for Bioaerosol Samplers and Other Information 319
Suppliers for LAL Based Endotoxin or Glucan Assays 320
Questions 320
References 320

7. Biological Agents – Control in the Occupational Environment 329
Kenneth F. Martinez, MSEE, CIH and Peter S. Thorne, PhD
Introduction 329
The Basics 330
Clinical Endpoints (Health Effects) 334
Infectious Disease 334
Immunologic Response 335
Disease Incidence 337
Prevention 337
Microbial Agents in Health Care Facilities 340
Airborne Pathogens 341
Bloodborne Pathogens 347
Monitoring Environmental Contaminants in the Healthcare Setting 349
Microbiology Laboratories and Biosafety Level 350
Primary Barriers 353
Secondary Barriers 359
Non-industrial Indoor Environments 361
Control 363
Remediation 36
General Industry 370
Agriculture 370
Biotechnology 372
Summary 374
Questions 375
References 376

8. Biological Monitoring in Occupational Health **383**
Glenn Talaska, PhD
Introduction 383
Air Sampling: Strengths and Weaknesses 383
Interactions Between the Person, the Agent, and the Environment 384
Exposure by All Routes 387
Chronic Effects and Self Selection in the Workplace 388
Biological Monitoring and Biomarkers 389
Biomarkers of Internal Dose 392
Effective Dose Biomarkers 394
Early Effect Biomarkers 396
Sampling Matrices and Sample Collection Timing 397
Biological Monitoring Matrices 401
Breath Analysis 402
Blood Analysis 407
Urine 409
Multiple Exposures 412
Effects of Unusual Work Schedules and "Moonlighting" (Outside Employment) 415
What Should be Considered Before Adding Biological Monitoring to an Occupational Health Program? 417
Why is this Sampling Being Done? 418
Costs Versus Benefits 418
What Markers Should be Sampled, and in What Matrix? 418
Who Gets the Results? 420
Biological Monitoring and the Interaction Between Occupational Health Professionals 421
Biological Monitoring Data and the Health Outcome 422
Standards, Advisory and Legal 423
BEIs® As Standards 423
German Biological Tolerance Values for Working Materials, BATs 425
Biological Monitoring: Applications and Examples 425
Metals 425
Solvents 427

Genetic Susceptibility and Biological Monitoring 428
- Introduction 428
- Screening to Identify Fitness for Duty 432
- Screening to Identify Fitness for Exposure 432

Frontiers of Biological Monitoring 438
- Carcinogens 438

Biological Monitoring in Hygiene Practice 443
- Weaknesses of Biological Monitoring 444

Areas of Further Research 448
Conclusions 448
Acknowledgment 449
Questions 449
Appendices 450
References 454

9. Dermal Exposure Assessment **463**
Richard A. Fenske, PhD

The Dermal Route of Exposure 463
Rationale for Dermal Exposure Assessment 464
Determinants of Dermal Exposure 465
- Theoretical Considerations 465
- Physical States and Exposure Pathways 467
- Behavioral Aspects of Dermal Exposure 470

Personal Monitoring of Dermal Exposure 472
- Surrogate Skin Techniques 472
- Chemical Removal Techniques 477
- Fluorescent Tracer Techniques 478
- Workplace Area Monitoring 482
- Recent Advances in Dermal Exposure Assessment Methods 483

Other Industrial Hygiene Aspects of Dermal Exposure 483
- Sampling Strategies 483
- Biological Monitoring and Dermal Exposure Assessment 486

Dermal Exposure Control Strategies 486
Regulatory Aspects of Dermal Exposure Assessment 487
Questions 488
References 488

10. Epidemiology **495**
Elizabeth Delzell, DSc
Introduction 495
Definition, Scope, and Uses 495
Measures of Disease Frequency 497
Incidence and Mortality Rates 498
Incidence Proportion 499
Prevalence Proportion 500
Estimation of the Effect of a Factor on a Disease: Measures of Association 501
Point Estimates of Measures of Association 501
Absolute Measures of Association 501
Relative Measures of Association 501
Validity 505
Selection Bias 506
Information Bias 506
Confounding 511
Precision 513
Confidence Interval 514
P-Value 514
Power 515
Major Epidemiologic Study Designs 515
Follow-up Studies 515
Case-Control Studies 529
Advantages and Disadvantages of Follow-up and Case-Control Studies 541
Follow-up Studies 542
Nested Case-Control Studies 542
Registry-Based Case-Control Studies 543

Other Study Designs 544
Interpretation of Epidemiologic Research 546
Assessing the Validity of a Single Study (Context 1) 546
Evaluating Causality (Context 2) 547
Concluding Remarks 548
Questions 548
References 549

11. Health Risk Assessment and the Practice of Industrial Hygiene 553
Dennis J. Paustenbach, PhD
Introduction 553
Fundamentals of Health Risk Assessment 554
Hazard Identification 555
Dose-Response Assessment 559
Exposure Assessment 571
Risk Characterization 575
Cost Benefit Analyses 584
Applying Risk Assessment to the Occupational Setting 585
Occupational Exposure to Airborne Chemicals 586
Uptake Via the Skin 587
Interpreting Wipe Samples 606
Retrospective Exposure Assessment 608
Closing Thoughts 609
Questions 613
References 614

12. The History and Biological Basis of Occupational Exposure Limits 631
Dennis J. Paustenbach, PhD
Introduction 631
Intended Use of OELs 635
Philosophical Underpinnings of TLVs® and Other OELs 637
Occupational Exposure Limits in the United States 638
Approaches Used to Set OELs 639

- Uncertainty Factors 642
- Setting Limits for Systemic Toxicants 647
- Setting OELs for Sensory Irritants 648
- Setting OELs for Irritants Using Models 651
- Setting Limits for Reproductive Toxicants 654
- Setting Limits for Developmental Toxicants 656
- Setting Limits for Neurotoxic Agents 657
- Setting Limits for Aesthetically Displeasing Agents and Odors 658
- Setting Limits for Persistent Chemicals 659
- Setting Limits for Respiratory Sensitizers 661
- Setting Limits for Chemical Carcinogens 663
- Setting Limits for Mixtures 669

Do the TLVs® Protect Enough Workers? 670

Where the TLV® Process is Headed 675

Corporate OELs 676

Models for Adjusting OELs 679

- Brief and Scala Model 679
- Haber's Law Model 683
- Pharmacokinetic Models 684
- Hickey and Reist Model (1977) 685
- A Physiologically-based Pharmacokinetic (PB-PK) Approach to Adjusting OELs 689

Occupational Exposure Limits Outside the United States 689

- Argentina 689
- Commonwealth of Australia 690
- Austria 691
- Belgium 691
- Brazil 691
- Canada 691
- Chile 693
- Denmark 693
- Ecuador 693
- Finland 693
- Germany 694

Ireland 694
Japan 694
Netherlands 694
Philippines 695
Organization of Russian States (Former USSR) 695
Conclusions 696
Questions 697
References 698
Supplemental References 710
Appendix **727**
Answers to Study Questions 727
Index **759**

CONTRIBUTORS

David C. Alexander, MSIE
Auburn Engineers
Auburn, AL
Chapter 4

Mark Boeniger, MS
National Institute for Occupational Safety and Health
Cincinnati, OH
Chapter 3

Arch I. Carson, MD
University of Texas Health Science Center – Houston
School of Public Health
Houston, TX
Chapter 1

Lung Chi Chen, MD
New York University
School of Medicine
New York, NY
Chapter 2

George L. Delclos, MD
University of Texas Health Science Center – Houston
School of Public Health
Houston, TX
Chapter 1

Elizabeth Delzell, DSc
University of Alabama at Birmingham
School of Public Health
Birmingham, AL
Chapter 10

Jeroen Douwes, PhD
Institute for Risk Assessment Sciences
Utrecht University
The Netherlands
and
Centre for Public Health Research
Massey University
Wellington, New Zealand
Chapters 5, 6

Richard A. Fenske, PhD
University of Washington
School of Public Health and Community Medicine
Seattle, WA
Chapter 9

Dick Heederik, PhD
Institute for Risk Assessment Sciences
Utrecht University
The Netherlands
Chapters 5, 6

Miriam Joffe, MS, PT
Auburn Engineers
Auburn, AL
Chapter 4

Neil Pearce, PhD
Centre for Public Health Research
Massey University
Wellington, New Zealand
Chapter 5

Kenneth F. Martinez, MSEE, CIH
National Institute for Occupational Safety and Health
Cincinnati, OH
Chapter 7

Glenn Talaska, PhD
University of Cincinnati
Department of Environmental Health
Cincinnati, OH
Chapter 8

Dennis J. Paustenbach, PhD
ChemRisk
Woodside, CA
Chapters 11, 12

Peter S. Thorne, PhD
University of Iowa
College of Public Health
Iowa City, IA
Chapters 5, 6, 7

FOREWORD

In his Foreword to Volume 1 of Modern Industrial Hygiene, my colleague Bob Harris wrote: "Students, educators, and practitioners of our profession are greatly indebted to Dr. Perkins for taking on the huge and daunting task of preparing a planned four–volume textbook on industrial hygiene." Over the 40 years of my practice as an industrial hygienist, with half in the educational setting, I have had the opportunity to review and evaluate most, if not all, the textbooks on industrial hygiene designed to comprehensively cover the practice of our profession. However, before I discuss this Volume of Dr. Perkins' *quadrumvirate*, I think the potential readers of this and the other three volumes should know something about the person who made the commitment to undertake this "huge and daunting task", especially his commitment to professionalism, education and learning.

I first met Jimmy (as he was and still is known to me) in the first class session of advanced biostatistics being taught at the University of Texas School of Public Health. It was the beginning of the 1974-75 academic year and I had taken leave from NIOSH to study for the Doctor of Pubic Health. Jimmy was a young master's student with a background in biology and an interest in environmental health. At the end of the academic year, I invited Jimmy to join me at NIOSH where I directed the Criteria and Recommended Standards program. For the next several years, Jimmy worked on several criteria documents with primary responsibility for the document on Pesticides, our first to deal with multiple chemicals and emphasize control methods beyond exposure limits and air monitoring. At the end of his three-year commitment, Jimmy returned to the UTSPH to work on his Ph.D. in environmental health sciences. About the same time, I also left NIOSH and initiated the graduate program in industrial hygiene at the UAB School of Public Health. As Jimmy finished his PhD, we were in a position to hire additional faculty for the program. Dr. Perkins was my first choice and, although he was reluctant to leave his native Texas, he joined us as an Assistant Professor. For the next 15 years he and I, along with the additional faculty we recruited, developed the University of Alabama at Birmingham industrial hygiene program as well as programs in occupational medicine, occupational health nursing and, in cooperation with Auburn University, occupational safety and ergonomics. It was during this period I saw Dr. Perkins' dedication to teaching reach full maturity. He and I taught a number of courses together, with the most intensive being Case Studies (which later became a textbook). The concept behind Case Studies was to have the students learn by going beyond textbooks and scientific papers to find additional sources including the experiences of industrial hygiene professionals, i.e., networking. I also observed Dr. Perkins' desire to see students go beyond assigned class reading material to find and explore subjects more in depth, with curiosity and imagination being their guide. This is the vision Dr. Perkins brought to *Modern Industrial Hygiene*. Dr. Perkins was recognized for his commitment and dedication in 1996 when he received the UAB President's Excellence in Teaching Award.

In Volume 1 of Modern Industrial Hygiene, Dr. Perkins presented the tools for recognition and evaluation of workplace chemical hazards in a manner emphasizing his vision of the learning process. In this, Volume 2, the reader will explore the biological basis of industrial hygiene. While Volume 1 was entirely authored by Dr. Perkins, he enters into more specific scientific areas in Volume 2 and has decided to draw on the knowledge of scientists who have specialized in the biological areas of interest to hygienists. While the biological basis for the practice of our profession has always been of critical importance, the awareness of the threat posed by biological hazards has grown considerably in the past decade. Beginning with the concern for bloodborne pathogens and growing with industrial hygienists evaluating and controlling exposure to bioterrorism weapons such as anthrax, the industrial hygienist's need for a critical core of knowledge in applied biology takes its rightful place along with chemistry, physics and engineering.

Our profession is indebted to Dr. Perkins for undertaking this "huge and daunting task." I am especially pleased to again be involved in his effort by contributing a few thoughts to one of the final products of what has been a long and winding road. I say involved because I was there when Dr. Perkins began this task many years ago. Dr. Perkins was the first faculty member in our school to be rewarded with a sabbatical. The purpose of a sabbatical (which is not a guaranteed right of a professor) is to take time off, usually six to twelve months, from his day-to-day teaching and research activities and explore new ideas and opportunities for career development. This is usually done in an academic or research setting away from the faculty's place of work. For one year, Dr. Perkins joined the faculty at Griffith University, Queensland, Australia as a Visiting Professor. It was during this period that the textbook and later books, found fertile ground, were nourished and flourished. Those of us, except his secretary, back in Birmingham enjoyed hearing his dictation tapes for Volume 1 so very much, especially the sounds of the tropical birds living around his patio. But the seeds grew and matured and our profession is enriched by the bountiful harvest resulting from Dr. Perkins' time "down under" and all of the time and effort he has devoted since.

Jimmy, my friend and colleague, thank you for staying the course, for it is a job well done.

Vernon E. Rose, DrPH, CIH
Professor of Public Health Emeritus
The University of Alabama at Birmingham

PREFACE

The preface to Volume 1 will not be repeated here. However, since virtually all of it is applicable, it should be summarized. The purpose of these volumes is to try to make the learning experience a more meaningful one. The classroom experience is, of course, the primary focus, but it is hoped and expected that this volume, like Volume 1, will bridge the gap between the classroom and the practice setting. As in Volume 1, we have done several things to try to accomplish this.

First, we have supplied questions at the end of every chapter that should give the reader a better idea of the important concepts of the chapter. In addition, the questions should cause the reader to probe a little deeper into the true meaning of what is being written. In other words, it's one thing to believe that one understands what has been written and another thing to put it into practice. We attempt to help the reader put the knowledge into practice. The answers for each of these questions, placed at the end of the textbook, should provide the reader with not only a sense for what is right (or thought to be right), but a better understanding of why we currently find that certain things (sometimes commonly held beliefs) are wrong.

There was an attempt to instill in each author an appreciation for thoroughness. While some of the chapters, for example epidemiology, can only be introductory, since the subjects are covered elsewhere in one or more volumes of textbooks, they are nonetheless meant to be inclusive of the key information that hygienists need to practice successfully. Other chapters, for example those on biological agents, are meant to be detailed as this is likely to be the only reference for the reader.

The information in each chapter should not merely suffice to give an introduction to the undergraduate student. It should also be challenging to the graduate student and supply answers to the practitioner or at least alert her to the need for further research beyond the information contained here. The reference lists, as a result, should be adequate to tame the curiosity of most students and practicing hygienists.

By its nature, this volume is an eclectic mix of subjects. However, all of them deal with the biology of industrial hygiene. Exposure routes are addressed in the first three chapters. The important subject of ergonomics, the interface between workers in their environment, is addressed in Chapter 4.

The recognition and evaluation of chemical agents was addressed in Volume 1; the control aspects of chemical agents are to be addressed in Volume 3. In this volume, the recognition, evaluation, and control of biological agents are addressed in three chapters. In Volume 1, exposure assessment for chemicals was addressed but two related subjects were not. First, biological monitoring, a subject of increasing importance and the likely endpoint for subsequent attempts to address total exposure assessment, is addressed. Second, Volume 1 only addressed sampling strategies for the inhalation route of exposure; this volume addresses the important dermal exposure route.

The volume ends with three loosely affiliated subjects: epidemiology, risk assessment, and biological basis for standards. Risk assessment is dependent upon epidemiology for input to the risk assessment process, and setting of standards, regardless of the technique used, is a step in the risk assessment process.

ACKNOWLEDGMENTS

The preparation of this volume began just before the publication of Volume 1 more than five years ago. Much has happened in the interim. The original publisher went out of business. A subsequent publisher sold the copyright and all existing copies of Volume 1 to me. ACGIH® agreed to distribute Volume 1 and agreed to publish subsequent volumes. In the first few years of this century, ACGIH® went through some very difficult financial times relative to a series of lawsuits. Although all turned out well for ACGIH®, the financial difficulties led to staffing shortages which led to delays in the publication of this volume. The Volume has been through three publishers and eight editors. Thus, the most important acknowledgments for Volume 2 are to the authors and ACGIH®.

Several of the authors submitted chapters on time for the original deadline (1997) and consequently, because of delays, had to diligently update their chapters. Their patience and perseverance is greatly appreciated, and I will forever be indebted to them. At ACGIH®, Tony Rizzuto, current Executive Director, took an early interest in the project and nursed it through some trying times. Amy Bloomhuff used her excellent organizational skills to get the project back on track. When Amy was needed elsewhere and promoted, Rita Williams took on the project as editor and worked diligently to see it through. Their efforts are greatly appreciated. Thanks is also offered to the numerous reviewers of these chapters. There were at least two for each, often ACGIH® members, who added considerably to the effort.

Dr. Vernon E. Rose prepared the Foreword for this volume. I consider him to be one of my greatest mentors and friends.

ACKNOWLEDGMENTS

1

THE RESPIRATORY SYSTEM

Arch I. Carson, MD and George L. Delclos, MD

The respiratory system of the human body plays a vital role in the maintenance of life and health on a moment-by-moment basis. Thus, it is a critical target of effects that impair its function. Further, with every breath, air from the environment, along with its contaminants, is pulled into the lungs through an intricate set of passageways. The respiratory system structures contacted by environmental air may, therefore, be the site of deposit and absorption or toxic effects of these contaminants. An understanding of the physiology of the respiratory system and its sensitivities to toxic exposures is, therefore, important to the professional evaluation and control of the workplace environment.

FUNCTIONS OF THE RESPIRATORY SYSTEM

Gas Exchange

The most important role of the respiratory system is to provide a means for the equilibration of the blood-bound gases with those in air. Of course, the most obvious benefit of this **gas exchange** function is to allow the bloodstream to gather oxygen from the external environment, which can then be transported to all parts of the body for support of cellular respiration and various biochemical functions. A second benefit is to allow the excretion of gases from the body, such as carbon dioxide and volatile substances, which are metabolic waste products or blood contaminants.

This equilibration process is accomplished by bringing the circulating blood into close proximity with air without the two actually mixing. The resulting passive diffusion process of individual gas molecules is driven by concentration gradients of the concerned substances. Each breath of inhaled air is spread out thinly over a thin membrane, the **pulmonary alveolar membrane**. In the average adult, the surface area of this membrane is very large, about 143 m^2, or roughly the size of a tennis court (Pinkerton, 1991; Weibel, 1973). This large surface area available for gas exchange is critical for the efficient and continuous transfer of large quantities of oxygen from the air into the blood. Any shrinking of this surface area, or thickening of the membrane across which oxygen must diffuse, represents important pathologic changes associated with disease of the respiratory system.

In frogs, other amphibians, and other lower animals, the exchange of oxygen and other gases between the air and the bloodstream occurs fairly efficiently through the skin, which is in direct contact with air or water rich in dissolved oxygen. In humans, however, the skin serves an important protective function, preventing the loss of moisture and heat from the body. Therefore, it is necessary for humans to have an internal system for gas exchange.

None of the gas exchange functions of the respiratory system mentioned above could be accomplished in humans without the ability to bring fresh air into the body from the outside, and to exhaust stale air to the environment. This ability to "breathe" is called **ventilation**, and involves the distribution of the breath through a branching system of air passages to microscopic bubbles of air where the bloodstream lies waiting. That same air is then expelled in reverse through the same distribution system, as an exhaled breath. The breakdown of this ventilation system is an important pathologic change associated with disease of the respiratory system.

Acid-base Regulation

The acid-base status of the blood is determined by its pH and the relative concentrations of electrolyte ions that circulate in it. These various parameters must be maintained within narrow physiological limits for normal functioning of body systems and maintenance of health, and are tightly controlled by metabolic and ventilatory influences. Metabolic processes add ions to the bloodstream through breakdown of foods, body tissues, or drugs and toxins, or remove them through further metabolism or excretion, i.e., in the kidneys or the digestive tract. These alterations affect the blood pH and the influences of normal electrolytes. Ventilation is an important compensation mechanism that is capable of balancing metabolic alterations in blood pH by dumping or retaining acid in the form of carbon dioxide.

Carbon dioxide, a common product of metabolism of organic fuels, is removed from cells by its dissolution in the blood where, with water, it forms carbonic acid. When the blood flows through the lungs, some of this carbonic acid leaves the blood and enters the air as carbon dioxide, which is then exhaled. The quantity and efficiency of ventilation regulate the amount of this acid leaving the blood. Deeper, faster breathing will result in more acid leaving the blood, causing the blood pH to increase. Conversely, slower, shallower breathing results in greater retention of acid and a general lowering of the blood pH. This relationship is regulated by a complex feedback system of monitors and control points and, during states of good health, is usually not perceived. This process of maintenance of controlled conditions within the body is referred to as **homeostasis**.

Metabolism of Drugs, Toxins, and Endogenous Substances

The lungs, specifically various specialized tissues within the lungs, are important sites of rapid metabolic activity. They become important points for the metabolism and detoxification of drugs and toxins (Dahl and Lewis, 1993), and because of their high metabolic enzyme content, can sometimes incur injury in the course of that process. Much of this injury may be the result of local oxidative stress and the production of free radicals and reactive oxidant molecules during biochemical reactions (Crapo, 1992).

Cardiovascular Pressure Regulation

An important hormone that controls blood pressure by modulating the diameter of the peripheral arterial blood vessels, **angiotensin**, is produced in an inactive form by the kidney and released into the bloodstream, but is activated by enzymes in the lungs. One major class of blood pressure-reducing drugs, the "ACE inhibitors," act by inhibiting the angiotensin convertase enzyme (ACE) activity in the lungs, thus reducing the levels of active angiotensin hormone in the circulation and reducing the systemic blood pressure.

ANATOMY AND PHYSIOLOGY OF THE RESPIRATORY SYSTEM

Cardiopulmonary Function

The respiratory system and the cardiovascular system work in concert to obtain oxygen from the environment, and to transfer it to the tissues of the body. The respiratory system provides ventilation of air and the necessary apparatus for gas exchange. The cardiovascular system provides a double circuit circulatory system with an integral pump. It also provides blood having the capability to carry large quantities of oxygen on the hemoglobin of its red blood cells and the mechanism for delivering oxygen to the tissues and carrying waste products away. These functions are under the control of various feedback systems that regulate throughout based on demand.

Since the respiratory and cardiovascular systems are so intimately linked, a functional deficiency in either may result in similar symptoms. For example, if the lungs cannot adequately ventilate air, or if they cannot adequately permit oxygen diffusion into the blood, or if the heart does not sufficiently circulate the blood, or if the blood itself does not possess adequate oxygen carrying capacity, the resulting effect will be inadequate delivery of oxygen to the tissues and a perception of shortness of breath. The body attempts to compensate for this deficiency by increasing the depth and frequency of respiration, relaxing the airways to achieve their maximum diameter, speeding up the heart to pump more blood with each beat and increase flow to the lungs and body tissues. These compensations may also be perceived as symptoms, such as breathlessness or heart palpitations. Thus, deficiencies in any part of the cardio-respiratory chain might be mistaken for disease at another location (see Figure 1-1).

Upper Airways

The upper airways include the air passages of the head and neck that, in addition to their other specialized functions, act as a conduit for air between the external environment and the interior of the chest. Principal structures are the oropharynx (mouth and throat), nasal passages,

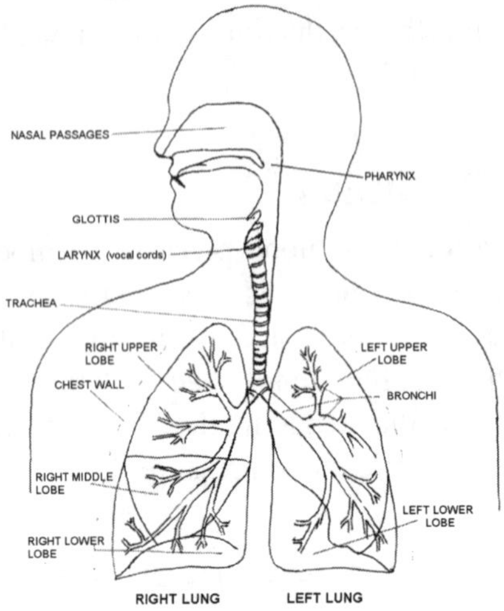

Figure 1-1. Schematic diagram of respiratory tract.

and larynx (voice box). Most experts also include the paired sinuses (ethmoid, sphenoid, frontal, and maxillary) and the middle ear in this category.

Most normal breathing occurs through the nose, nasal passages, posterior pharynx, and larynx. Under these conditions, each breath of environmental air is warmed to body temperature, filtered by sieving and elutriation, humidified to water vapor saturation, and passed over the smell sense organ. During periods of high airflow demand, during speech, and at times of nasal passage dysfunction, the usual route can be bypassed, resulting in mouth breathing. During mouth breathing, air is detoured through the mouth and anterior pharynx directly into the posterior pharynx and larynx. This is a low resistance pathway, but provides less efficient warming, filtration, and humidification of air.

Most of the surface of the upper airways is lined by a mucous membrane that continually secretes respiratory mucus onto its surface. Mucus is composed mostly of water, but also contains small amounts of proteins and carbohydrates that lend to it its sticky, semi-gel characteristics. It also contains secretory antibodies (immunoglobulin A or IgA), which are important defenses against invasion of pathogens into the respiratory system. The water content of the mucus provides the source of humidification for the air passing through the upper airways. Mucus that is secreted to the surface of the membrane is usually swept slowly along the surface by the action of ciliated cells in the underlying membrane. This action steadily sweeps the mucus from the anterior border of the smell sense organ (cribiform plate) forward toward the anterior tip of the nose, or from the posterior border toward the throat (glottis) (see Figure 1-2).

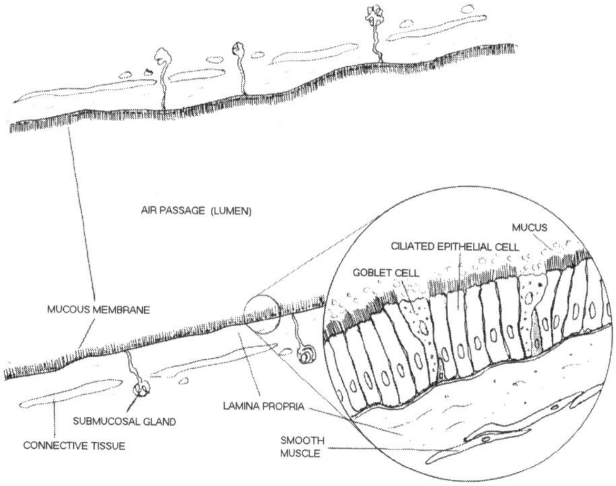

Figure 1-2. Structures of a mucus membrane.

Lower Airways

The term "lower airways" refers to the branching, tree-like, structure of air tubes that extends from the single large airway in the neck, the **trachea**, to the deepest parts of the lungs where gas exchange occurs. The primary purpose of this system is to conduct air from the outside of the body to a place where it can come into close proximity with the bloodstream, thereby allowing oxygen to be transferred into the blood. These airways can be thought of as a system of roadways stretching from a single heavily-traveled superhighway (**trachea**) onto smaller, yet still major, highways (**bronchi**), then to city streets (**bronchioles**), and finally to neighborhood streets (**terminal bronchioles**) and driveways (**respiratory bronchioles**) leading to individual businesses (**alveoli**) or groups of homes on a *cul de sac* (alveolar sacs). The conducting airways are generally divided into two classifications based on their size. The **large airways** are those lower airways that are larger than about 2 mm in internal diameter. In general, surrounding rings of cartilage support their walls, they have a relatively abundant ciliary carpeting of the mucous membrane, and the rate of both airflow and mucus clearance is fast. The **small airways** are those lower airways that are smaller than about 2 mm in internal diameter. In general, their walls are tethered to, and supported by, the surrounding lung tissue through which they travel, they have a relatively sparse ciliary carpeting of the mucous membrane, and the rate of both airflow and mucus clearance is slower.

In humans, airways travel for a short distance and then branch into slightly smaller daughter airways (usually two), which diverge from one another at angles from the axis of the parent airway (dichotomous branching). As in a family tree, each level of branching in the tracheobronchial tree is termed "a generation." This branching process repeats itself as many as 21 to 23 times in humans before the airways reach microscopic size and approach the gas exchange membrane (Weibel, 1973) (see Figure 1-3). Each time a branch occurs, even though the individual resulting airways are smaller than their parent, the total cross-sectional area of the daughter airway generation is greater than that of the parent generation. This phenomenon is important because it relates to the speed of airflow through the airways. Briefly, a single volume of air flowing into the respiratory system must flow through the trachea in its entirety because there is only one trachea. However, as that same volume of air moves deeper into the airway system, it is divided among more and more airways, each successive generation with a larger total cross-sectional area, and therefore moves slower and slower. For this reason, the most widely employed test of pulmonary function, **spirometry**, is very sensitive to changes that affect airflow in the large airways where airflow is fast and turbulent, but is much less sensitive to changes occurring in the small airways where airflow is slow and laminar.

Respiratory Zone

The purely conducting airways give way, in the **respiratory zone,** to blind sacs of air surrounded by blood capillaries, the **alveoli.** An alveolus is the anatomical unit of the respiratory zone. Each alveolus is formed during embryological development. A wispy net of connective tissue, collagen and elastin is laid down. Blood capillaries begin to grow into this system and form around and between the alveolar nets (see Figure 1-4).

Extremely thin epithelial cells that layer out and carpet the inner surface of the developing alveolus, line the interior of the alveolar connective tissue nets. These cells are called **Type I alveolar epithelial cells, Type I pneumocytes,** or just **Type I cells**, and have two important functions in the physiology of the respiratory system. First, they must be as thin as possible because oxygen and other gases that transfer between the air and the bloodstream must pass through them on that route. Secondly, they provide a barrier between the air and the fluid compartments of the blood and extracellular space. If this barrier is removed or breaks down, fluid may begin moving into the air space of the alveolus, severely degrading the efficiency of gas exchange.

A second cell type forms within the lining layer of the alveolus. This is the **Type II alveolar epithelial cell**, thicker and more compact than Type I cells, and containing granules and a prominent Golgi apparatus. This cell serves two major functions. First, it manufactures an important substance called **pulmonary surfactant** which, when secreted onto the inner surface of the alveolus, reduces the surface tension associated with the moisture in the respiratory system, and prevents the alveoli from collapsing as adjacent surfaces come into contact with one another. Second, as the Type I cells reach the end of their life spans and leave the membrane, the Type II cells can divide, spread out thinly, and fill in the bare spots left by their deceased neighbors, becoming new Type I cells.

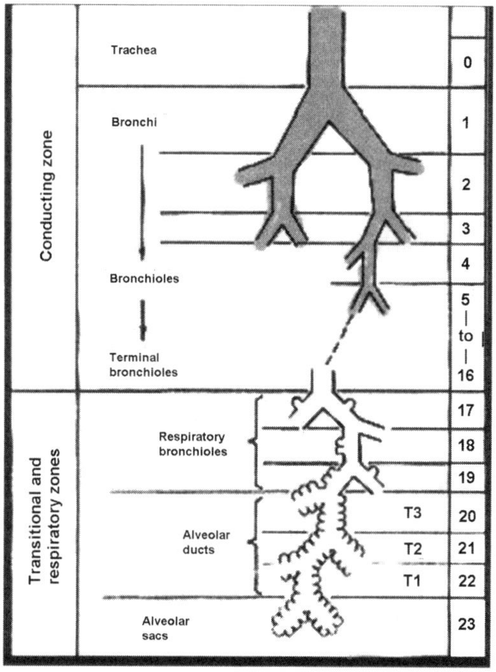

Figure 1-3. Schematic diagram of airway branching (with large and small airways and generations numbered).

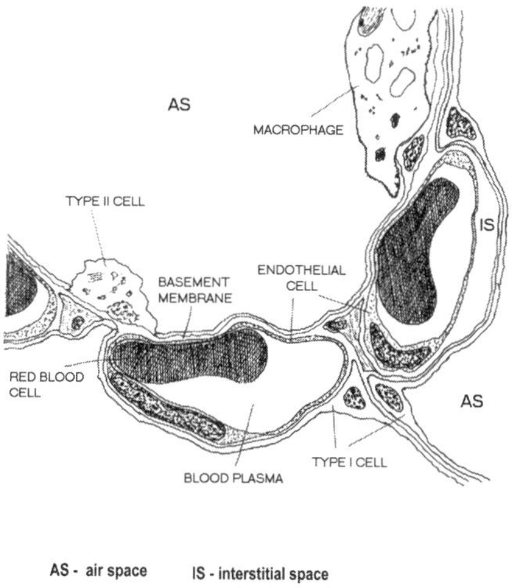

Figure 1-4. Structures of the alveolar septum.

A third cell type has a prominent role in the function of the alveolus. This is the **alveolar macrophage**, a large, mobile, amoeboid cell that moves freely about the inner surface of the alveolus. Its function is to engulf (phagocytize), devour, and digest extraneous materials that it encounters in the alveolus. These could be remnants of the disintegration of old Type I cells, residue of used surfactant, dust or bacteria brought in with the air from the external environment, etc. The alveolar macrophage contains digestive enzymes, which it uses for these purposes.

The movement of oxygen from the environment into the bloodstream involves multiple steps that are influenced by the anatomy and physiology of the respiratory system (see Figure 1-5). A change in the characteristics of any of these stages (inflammation, scarring, dilution, swelling, etc.) may influence the ease of oxygen movement into the blood. The amount of oxygen in air is nominally 21% and varies only by a few tenths of a percent in the open air at the earth's surface. Human made environments and natural enclosed spaces may present large variations in the amount of available oxygen. Changes in the inspired oxygen concentration will directly affect diffusion gas exchange rate.

Important Differences Among Mammalian Species

Much of what we know of respiratory function is learned through research performed using laboratory animals of various species as surrogates for humans. This is often a useful approach, lending researchers much information that would not otherwise be available. Difficulties arise, however, when this information is interpreted as if the animals involved are nothing more than little humans.

DICHOTOMOUS VS. MONOPODIAL BRANCHING OF THE AIRWAYS

Human airways, as mentioned earlier, present important airflow dynamic properties that affect the limits and characteristics of airflow velocities, and the deposition of substances carried by the air in the airways themselves or in the deeper lung. Human airways undergo **dichotomous** branching. This means that, when an airway branches into more numerous smaller airways, the resulting airways are each about the same size as one another and they diverge geometrically at angles away from the line of flow of the parent airway and away from each other (see Figure 1-6). In many animal species this is not the characteristic form of airway branching. Most carnivores, e.g., cats, dogs, ferrets, etc., sometimes used in respiratory research, have a monopodial airway branching scheme. In **monopodial** branching, the parent airway continues in the same general direction beyond branch points, but gives off radially spaced smaller airways at angles to it as it travels. This difference between humans and experimental animals can lead to serious interpretation problems with regard to substance deposition and airflow velocity limitations (Phalen and Oldham, 1983).

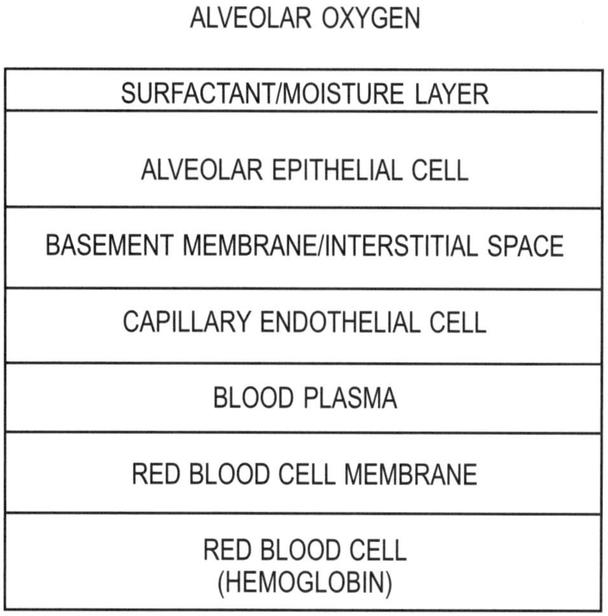

Figure 1-5. Pathway of oxygen from the alveolar air into the red blood cell.

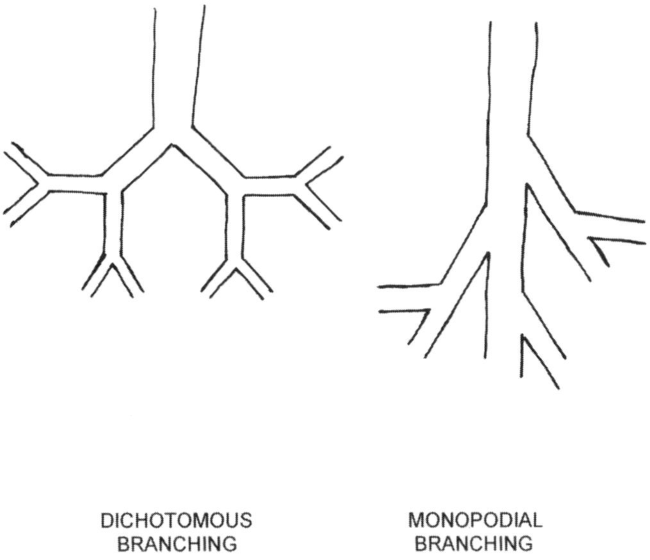

Figure 1-6. Modes of airway branching in humans (dichotomous) and some common laboratory animals.

THE PRESENCE OR ABSENCE OF RESPIRATORY BRONCHIOLES

The human respiratory system contains a transitional structure (respiratory bronchioles) between the purely conducting terminal bronchioles and the purely diffusion-oriented alveolar sacs. Many animal species, e.g., rats, mice, hamsters, rabbits, and guinea pigs, lack these transitional structures (see Figure 1-7) (Phalen and Oldham, 1983). As a result, research performed using these common, inexpensive test animals may be difficult to interpret owing to the participation of respiratory bronchioles as a site of pathological change in much human respiratory injury and disease.

TESTS OF RESPIRATORY FUNCTION

Spirometry and Peak Flow

During normal respiration at rest, air is inhaled then exhaled in a repeating cyclical pattern like the ebb and flow of ocean tides. This analogy is so strong that normal breathing has been called **tidal breathing**. Measurement of the volumes of air that are ventilated through the respiratory system can be accomplished in several ways. However, the most common are by simple volume displacement or airflow integration **spirometry.** The device used to make such mea-

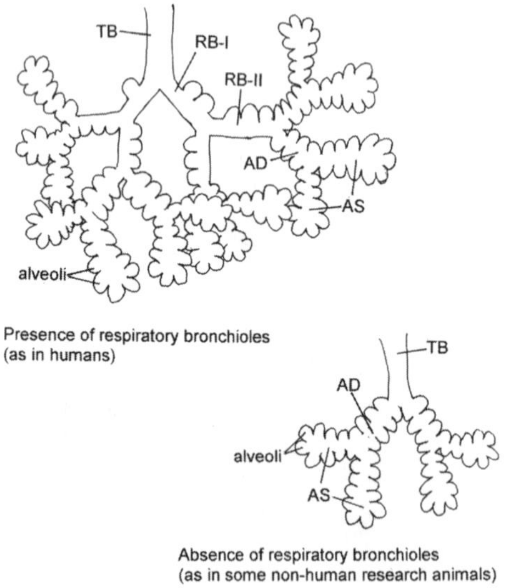

Figure 1-7. Schematic diagram of airway branching (presence vs. absence of respiratory bronchioles). TB (terminal bronchiole), RB-I (first order respiratory bronchiole), RB-II (second order respiratory bronchiole), AD (alveolar duct), AS (alveolar sac)

surements is called a **spirometer**. It measures change in contained volume with respect to the passage of time. Volumes are usually expressed in liters of air at body temperature, ambient pressure, saturated with water vapor (BTPS). Time is expressed in seconds.

The subject is requested to take a full deep breath and then to exhale as forcefully and completely as possible into a spirometer. The breath is recorded as exhaled volume with respect to time (volume–time curve), or expiratory flow rate with respect to exhaled volume (flow-volume curve). This maneuver is repeated several times to assure validity and consistency of results. A valid test provides information regarding the limits of ventilatory capability of the airways, and the amount of air that can be contained and then expelled with respect to time.

Several measurements can be derived from this forced expiratory maneuver. Three that have fundamental usefulness in workplace-related testing will be described here. 1) **Forced Vital Capacity (FVC)** is equal to the maximum volume of air that can be expired forcibly following a deep breath. It is expressed in liters at BTPS for each trial, and is determined graphically from the volume-time curve or the flow-volume curve at the greatest volume excursion shown. It is used as a reliable estimate of respiratory volume capacity. 2) **Forced Expiratory Volume at One Second (FEV_1)** is equal to the volume of air, at BTPS, that is expired during the first one second of each trial. It is determined graphically from the volume-time curve as the volume at the intersection of the one-second time line with the curve. It is used as a reliable indicator of expiration airflow rate and airway caliber. 3) **FVC/ FEV_1 %** is simply a ratio of the FEV_1 to the FVC, expressed as a percentage (normalized airflow measurement). Additional measurements can be made, i.e., volumes at various time points ($FEV_{0.5}$), expiratory flow rates at various levels of lung volume (FEF_{50}), standardized averaged flow rates ($FEF_{25\text{-}75}$), and expiratory times to reach various volumetric milestones (FET), but are less widely utilized. The information obtained can be compared with standardized normal values or can be used in longitudinal comparisons over time for an individual (see Figure 1-8). Spirometry is both simple to perform and inexpensive. It requires the participation of a trained, experienced technician during the testing procedure to assure valid results, and is a frequent component of routine workplace pre-placement and periodic surveillance exams. Standards for performance and interpretation of this test in the work-related setting have been established, and have, in some cases, been incorporated into governmental safety and health regulations (ATS, 1995; Townsend, 2000).

Peak flow measurement utilizes very inexpensive equipment (peak flow meter) to assess one aspect of the subject's respiratory status, i.e., the maximum expiratory flow rate that can be produced during exhalation. Because this depends upon the internal diameter of the airways, it can be used to effectively follow changes that occur over time that are associated with an episodic obstructive component, as in asthma and other forms of reactive airways disease. Subjects can perform this test while at work or at home, or to document functional changes associated with symptoms. Some peak flow meters are now equipped with automatic data logging capabilities, providing a ready record of peak flows with respect to time-of-day, date, and certain other designated events.

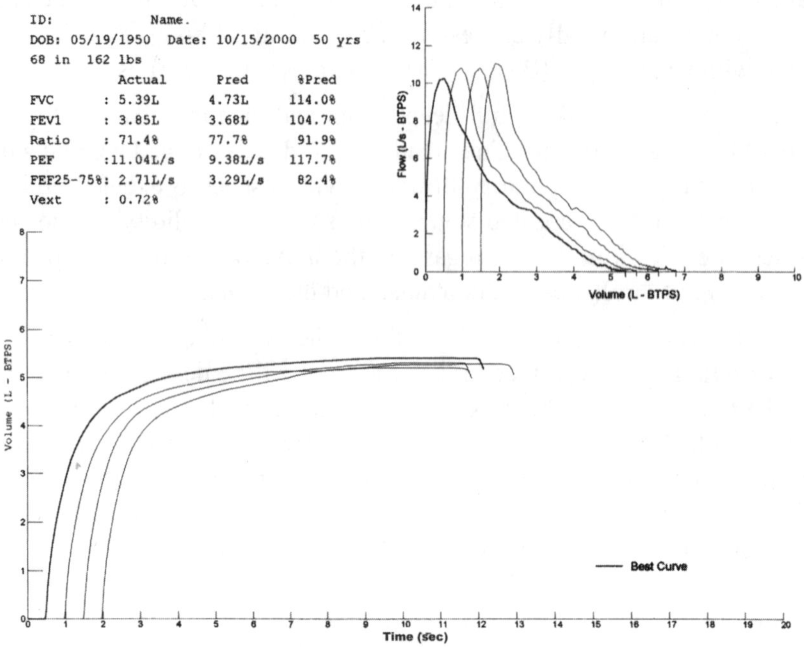

Figure 1-8. Spirogram. A maximal expiratory test (single breath) performed by a healthy subject, expressed both as a volume-time curve and flow-volume curve. Several (4) trials are displayed.

The following tests of respiratory function can be performed in a specialized laboratory under the supervision of a physician.

Subdivisions of Lung Volume

Volumes of air in the respiratory system range from a state of completely empty (a theoretical absolute zero) to one of completely full **(total lung capacity or TLC)**. Several points can be identified within this range, which demarcate physiologically useful subdivisions of lung volume (see Figure 1-9). At the end of a normal tidal exhalation in humans, the forces producing airflow into or out of the body are in equilibrium and airflow stops. This point is called **functional reserve capacity (FRC)** and represents a balance between the force of the lungs to collapse and the force of the chest wall to resist collapse. This is the point of relaxation during respiratory muscle paralysis. The amount of air that is ventilated in and out of the body during a normal tidal breath is termed **tidal volume.** This cyclic process has FRC as its starting point. If, after a subject exhales a normal tidal breath, he inhales deeply, his respiratory system will fill to a point of maximum inflation, or TLC. If he then maximally exhales as far as he can, a minimum inflation level is reached. At the minimum inflation level, the lungs continue to maintain a content of air that can never be voluntarily exhaled. This remaining volume of air

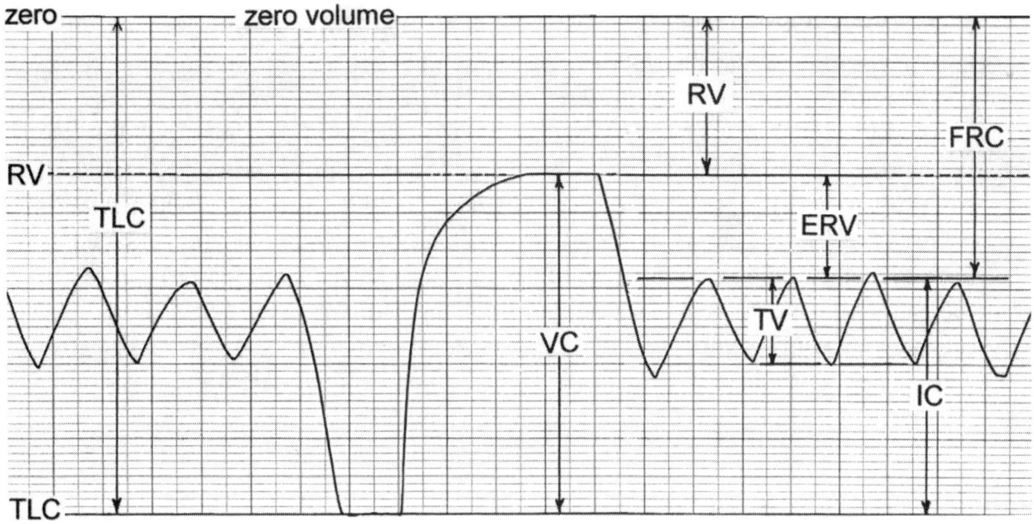

Figure 1-9. Subdivisions of lung volume. TLC = total lung capacity; VC = vital capacity; RV = residual volume; TV = tidal volume; ERV = expiratory reserve volme; IC = inspiratory capacity; and FRC = functional residual capacity.

and the point of ventilation reached are both termed the **residual volume** (**RV**). The volume of air that was exhaled from the point of TLC to the point of RV is termed the **vital capacity** (**VC**), and represents the total amount of air that can be exhaled following a maximal inhalation. These basic divisions of lung volume are very constant and reproducible in a healthy individual and can be assessed in screening for dysfunction. Several methods are available for doing this, including inert gas dilution techniques, whole body plethysmography employing Boyle's Law, and imaging studies with volume calculation.

Diffusing Capacity

Adequate respiratory function depends upon the ability of oxygen to diffuse in large quantities through the pulmonary membrane into the bloodstream. A model of this process can be tested using carbon monoxide as a surrogate for oxygen. The resulting test is called D_LCO or Diffusing Capacity of the Lung for Carbon Monoxide. In the most common form of this test, a single breath of a breathing gas mixture containing a small amount of carbon monoxide (100-500 ppm) and an inert gas tracer (10% helium) is inhaled rapidly and held for 10-15 seconds, and is then rapidly exhaled. A sample of the exhaled breath is captured by a valving system, and the concentrations of CO and He are determined by analyzers. The amount of CO lost across the pulmonary membrane can be determined per unit time and partial pressure, thus quantifying gas diffusion potential. The result is expressed as milliliters of CO per second per Torr (D_LCO), or as this figure adjusted for the estimated surface area of the pulmonary membrane. ($D_LCO_{specific}$).

Arterial Blood Gases

Blood traveling through the arteries represents fully oxygenated blood coming from the lungs prior to reaching the systemic capillary beds. It is, therefore, a useful medium by which to assess respiratory function in individuals who have evidence of respiratory illness. A small sample of blood (1-3 ml.) is withdrawn from an artery, usually in the wrist, and is immediately analyzed for oxygen partial pressure, carbon dioxide partial pressure, and blood pH. Additional factors can be calculated. Often the percentage of carboxyhemoglobin (an indicator of CO exposure) and methemoglobin (a measure of irreversibly reduced hemoglobin) can be obtained simultaneously with the other measurements from the same sample. The arterial blood gas analysis can be a useful tool in assessing the ability of the respiratory system to oxygenate the blood and excrete waste carbon dioxide. It can also help differentiate among various physiological states that can augment or mimic respiratory disease.

Cardiopulmonary Stress Testing

Oxygen in the bloodstream is necessary for cellular respiration, including that which supplies energy for muscular activity. A cardiopulmonary stress test imposes a physical exercise load on the subject, via treadmill walking or stationary bicycle riding, and measures: energy output, heart rate, respiratory rate and tidal volume, oxygen consumption, and carbon dioxide production in response. Parameters can be altered according to standard protocols to test the integrated cardiopulmonary system.

Bronchial Challenge Testing

Persons who experience episodic respiratory illness, such as asthma, in response to exposure to specific substances or classes of materials may have normal respiratory function at other times. In these individuals, specific or non-specific bronchial challenge testing may provide useful information.

SPECIFIC CHALLENGE TESTING

Persons who are allergic to specific substances that produce respiratory symptoms are likely to show physiological changes in pulmonary function when exposed to the offending agent. These tests may have significant associated risks, and should only be performed by experienced personnel at a center where they are routinely performed under rigorous protocols. The advantages of such testing are: they can identify serious worker/substance incompatibilities; they can help establish a cause and effect relationship; and they can lead to economical administrative, engineering, or work practice changes, which can prevent future illness in the individual or in others who might become sensitized. The disadvantages of such testing include cost, time, and the potential for serious side effects. Testing must be performed in a special laboratory and can require several days to complete for a single substance. Often, it is not possible to accurately identify a specific offending agent from a multitude of contaminants present in a workplace exposure, making specific challenge testing not always feasible.

NON-SPECIFIC BRONCHIAL CHALLENGE TESTING

Persons who have respiratory symptoms in response to exposure to multiple individual substances or to classes of substances, e.g., organic proteins, acid gases, or solvent vapors, may be especially sensitive to the non-specific irritant actions of these substances. The **non-specific bronchial challenge test** employs a short-acting pharmacologic agent (such as methacholine or histamine) administered by water mist, to alter the normal neurological balance affecting airway tone, thus unmasking individuals who have **hyperresponsiveness** of the airways and providing a quantitative assessment (ATS, 2000). This test is useful for identifying those with a tendency toward asthma-like symptoms when exposed. It can be performed in a routine pulmonary clinical laboratory, and is not associated with the serious side-effect potential of specific challenge testing. The endpoint of the test is generally accepted to be the dose producing a 20% decrement in the forced expiratory volume at one second (FEV_1), a spirometric measurement, or a 100% increase in the airway resistance (R_{AW}), a plethysmographic measurement.

THE RESPIRATORY SYSTEM AS A PORTAL OF ENTRY FOR TOXIC SUBSTANCES

The respiratory system is a nearly direct conduit for substances present in the environmental air into the bloodstream. Therefore, any inhaled substance that can diffuse across the alveolar membrane can gain access to the blood. This process is controlled by the magnitude of the concentration gradient between the alveolar air and the blood and the air/water partition properties of the substance. Small molecular weight of the substance, electrical neutrality, and lipid solubility are factors associated with enhancement of this process. Substances that combine with the blood or react chemically with it to form different substances, e.g., carbon monoxide, may make this passage very rapidly and efficiently.

Some substances may also be absorbed into the bloodstream through mucous membranes of the airways, although much less efficiently due to the relatively lower surface area and longer pathway for diffusion. This route can predominate, however, in the case of substances deposited primarily in the conducting airway system.

The respiratory system can also act as a collecting filter for aerosol substances present in the inhaled air. These substances may later be distributed into the body by way of the gastrointestinal system as they are cleared from the airway by mucociliary activity.

DETERMINANTS OF RESPIRATORY SYSTEM INJURY

Several factors determine the likelihood that an occupational or environmental toxicant exposure will lead to the development of respiratory disease.

First, exposure to the offending agent must be of sufficient ***intensity*** to cause injury to the respiratory system. Exposure intensity, in turn, is dependent on the concentration that a person is exposed to and the duration of that exposure. It is essential that investigators always take the necessary time to characterize an individual's exposure in as much detail as possible.

A second determinant of injury to the respiratory system relates to the ***physical and chemical characteristics*** of the agent. In general, airborne respiratory toxicants can be classified into *particulates* and *gases or vapors* (see Volume 1, Chapters 5, 9, and 10). In the case of particulates, particle size is a critical determinant of the site of deposition within the respiratory tract. Particulates with a mass median aerodynamic diameter (MMAD) greater than 10 micrometers tend to be predominately deposited in the nasopharyngeal region, where (depending on their chemical characteristics) they may cause irritation of the nose, paranasal sinuses, and/or pharynx, producing congestion and drainage of these structures. Alternatively, they may exert no effect, and be eliminated. Particulates with an MMAD in the range of 5 to 10 micrometers are preferentially deposited in the tracheobronchial region, and can lead to the development of an irritative cough, phlegm production and/or bronchospasm. When the MMAD is less than 5 micrometers, particulates are predominately deposited in the respiratory zone. Depending on the toxicity of the agent and the intensity of exposure, these smaller particulates may ultimately damage these delicate structures, which have limited regenerative capacity, and lead to respiratory pathology. More detail on deposition is found in Chapter 2.

When the respiratory toxin is a gas or vapor, other physical and chemical characteristics of these toxins become important. Certain gases injure the respiratory system by competing directly with oxygen in the process of respiration. These gases are termed **asphyxiants**. **Irritant** gases cause injury through direct contact with the lining of the respiratory system where they cause tissue damage and trigger an inflammation reaction. If these irritant gases are highly water-soluble, they tend to affect those parts of the respiratory system rich in water content, such as the lining of the nose, throat, and paranasal sinuses. This level will experience most of the injury, including swelling of the cell lining and secretion of fluid. On the other hand, poorly water-soluble irritant gases (also known as deep lung irritants) tend to involve lower, more distal parts of the respiratory system, including the small airways, alveolar structures and lung interstitium. This effect of water-solubility is so predictive that models have been developed from animal research that can be used to predict health effects of exposures in humans (Overton, 1987; Kimbell, 1993).

A third determinant of injury to the respiratory system relates to the ability of ***host defense mechanisms*** to control the majority of undesirable exposures that occur on a daily basis. There is a balance between exposure intensity and physical-chemical characteristics of the toxin on one hand, and the host defense mechanisms on the other, that will ultimately determine whether injury to the respiratory system will occur. The main defense mechanisms are: *distance* between the external environment and the gas exchanging portion of the respiratory system; *cough and sneeze*, which rid the upper and large airways of many large particulates; the *mucociliary escalator*, which lines approximately two-thirds of the airways and provides an efficient clearance mechanism for particulates, as well as a scrubber system for many water-soluble gases;

and *alveolar macrophages*, which act as "scavenger cells" that attack and phagocytize many particulates and ultimately detoxify them.

The fourth determinant of injury to the respiratory system relates to the ***anatomical site*** affected by the toxicant exposure. In a practical sense, the respiratory system can be divided into three areas: *upper airways* (nasal passages, paranasal sinuses, oral pharynx, and larynx); *lower airways* (trachea, bronchi, and bronchioles), and the region of the *lung parenchyma and pleura* (respiratory zone, capillary endothelium, lung interstitium, and pleura). Different exposures can affect each of these areas in a different manner.

The fifth determinant of respiratory system injury is the ***mechanism of injury***. In general, there are four main types of injury: *irritation*, *sensitization*, *cytotoxicity,* and *carcinogenesis*.

Irritant exposures cause injury through direct contact between the toxicant and the lining of the respiratory system, with initiation of an inflammation reaction. The clinical manifestation of the inflammation varies, depending on the anatomical site. Exposure of the upper airways to a water-soluble irritant can produce symptoms of nasal congestion (**irritant or chemical rhinitis**), congestion and drainage of the paranasal sinuses (**sinusitis**), sore throat (**pharyngitis**), and/or hoarseness (**laryngitis**). Irritative symptoms of the lower airways (**tracheitis, chemical bronchitis**) include cough and phlegm production. In some cases, massive irritant exposures can also lead to the development of **irritant-induced occupational asthma** (Brooks et al. 1985; Brooks et al., 1998) (see Table 1-1). Irritant injury of the lung parenchyma by deep lung irritants (e.g., oxides of nitrogen, ozone, phosgene) can produce an intense inflammation of these structures (**bronchiolitis/pneumonitis**), with copious secretions flooding the alveoli **(pulmonary edema)**, impeding effective gas exchange and causing severe shortness of breath. This can be immediately life threatening and its resolution can leave permanent anatomical defects and physiological impairment.

Sensitization (also known as "allergy") occurs when certain susceptible individuals (probably due to genetic factors), after an initial exposure to an agent capable of triggering an allergic reaction ("allergen" or "antigen"), develop symptoms when re-exposed to the allergen. These allergic reactions, which present as itchy eyes, nasal congestion, chest tightness, wheezing or (in rare cases) systemic anaphylaxis, occur via an immunological mechanism. Non-susceptible individuals do not develop these reactions. Depending on the anatomical site involved, clinical manifestations will vary. Allergic reactions involving the upper airways will present as **allergic rhinitis, sinusitis, pharyngitis,** and/or **laryngitis**. Involvement of the lower airways, such as in **allergy-mediated occupational asthma**, produces episodic chest tightness, shortness of breath, and/or wheezing caused by acute narrowing of the lower airways in response to allergen exposure. Allergy-mediated occupational asthma can be caused by exposure to high molecular weight antigens (e.g., animal proteins) and low molecular weight compounds (e.g., isocyanates, plicatic acid). Repeated exposure to certain thermophilic microorganisms or molds, capable of penetrating to the deep lung, or some low molecular weight antigens, can trigger an allergic reaction of the lung parenchyma known as **hypersensitivity pneumonitis**. Because this latter reaction involves the alveoli and interferes with gas exchange, the main respiratory symptom in

Table 1-1
Substances associated with irritant-induced airway hyperresponsiveness

Class	Agent
Acids	glacial acetic acid muriatic (hydrochloric) acid phosgene sulfuric acid sulfur dioxide
Bases	ammonia calcium oxide caustic soda aerosol
Volatile organics	cement sealant mixed solvent vapors and mist perchloroethylene volatile organic vapors from poor indoor air quality
Specific toxins	acrylates – methacrylate and cyanoacrylate ester chlorine gas diphenylmethane diisocyanate hexamethylene diisocyanate household bleach hydrazine titanium tetrachloride mist toluene diisocyanate uranium hexafluoride
Mixtures	burned freon fumes diesel exhaust fire retardant aerosol flour dust metal coat remover pesticide vapors spray paint welding fumes

persons with this condition is shortness of breath. This condition is difficult to diagnose because symptoms are often insidious and non-specific. Exposure to beryllium salts, a very potent sensitizer, can also cause an allergic reaction of the lung parenchyma, **chronic beryllium disease** that mimics the more common disease, pulmonary sarcoidosis (an idiopathic granulomatous disease involving the lungs and sometimes the skin and other organs).

A third type of injury to the respiratory system is via ***cytotoxicity***. This phenomenon has been studied best in the pathophysiology of the **pneumoconioses**, a group of lung diseases that are caused by chronic exposure to certain dusts, including silica, asbestos and coal dust. These dust particulates can be directly toxic to alveolar macrophages, causing the death of this cell once it has phagocytized the particulate. As it dies, the macrophage releases a variety of digestive enzymes and strongly reactive chemical mediators that, in turn, can damage the delicate respiratory bronchiole and alveolus. Alveoli have a limited ability to regenerate and, if this damage is sufficiently extensive, alveolar tissue is ultimately replaced by scar tissue, in a process known as *fibrosis*. Depending on the type of dust exposure, this fibrosis is referred to as **silicosis, asbestosis, coal worker's pneumoconiosis**, etc.

A fourth mechanism of injury to the respiratory system occurs when there is malignant transformation of certain respiratory cells leading to the development of cancer (***carcinogenesis***). Depending on the anatomical site, various types of cancers have been described in association with occupational exposures. For example, **cancer of the nose and pharynx** has been causally associated with chronic exposure to nickel, chromium, and probably certain types of wood dust. **Bronchogenic carcinoma** (lung cancer), although primarily caused by chronic cigarette smoking, has also been linked to exposures to uranium, radon daughters, asbestos, and certain chemicals, such as the chloromethyl ethers. **Mesothelioma**, a rare and invariably fatal cancer of the pleura, has been strongly related to both occupational and non-occupational asbestos exposure.

By combining the anatomical site and mechanism of injury described above, a simple and clinically useful table can be constructed that allows the classification of the most common occupational respiratory diseases. Table 1-2 presents a summary of this classification scheme.

DIAGNOSIS OF OCCUPATIONAL RESPIRATORY DISEASE

Although many different tests are available to clinicians for the diagnosis of occupational respiratory disease, the basic "tools of the trade" consist of the *occupational and medical history*, a focused *physical examination* of the patient, *chest radiography*, and simple *tests of lung function*. The accurate diagnosis of *any* occupational respiratory disease depends heavily on detailed characterization of workplace exposures, through a carefully obtained occupational history, as the cornerstone of the clinical evaluation. When the history is ignored, correct occupational diagnoses are missed! For this reason, it is imperative that health care professionals who are knowledgeable and experienced in occupational medicine practice make the diagnosis. In general, their activities will be greatly enhanced by specific information relating to the physical and chemical nature of the exposure, its intensity and time-course, characteristics of the workplace environment, and job activities. Once a good occupational and medical history is obtained, the value of each of the remaining tools will vary depending on the specific respiratory disease. Some illustrative examples follow.

For most irritative syndromes, the diagnosis is based mainly on a compatible history of exposure to an irritant, and physical findings on inspection of the affected part of the respiratory anatomy (Delclos, 1995). For example, in the case of exposure to a water-soluble irritant, a compatible history would be one where symptoms developed abruptly at the time of the exposure. Coupled with findings of irritation (redness, swelling and/or increased secretions) of the nose and throat, this is often sufficient for an accurate diagnosis. Occupational irritative syndromes are very common, yet very under-diagnosed, usually because of failure to take an adequate occupational history. Common misdiagnoses are allergic rhinitis/sinusitis or upper respiratory infection.

The diagnosis of occupational asthma requires use of other tools, in addition to a compatible history of exposure to an agent (allergen or irritant) suspected of causing asthma and symptoms of chest tightness, shortness of breath and/or wheezing. Pulmonary function tests that demonstrate the characteristic changes of asthma (i.e., reversible decreases in airflows) are essential for diagnosing asthma. These changes can be documented by performing spirometry before and after the use of bronchodilators, measuring serial changes in peak flow, or by nonspecific bronchial challenge testing with agents such as methacholine or histamine. Once the diagnosis of asthma is established, the next step is to demonstrate the relationship between this

Table 1-2
Classification of occupational respiratory disease by anatomical site and mechanisms of injury

	Upper Airways	Lower Airways	Lung Parenchyma and Pleura
Irritation	• Rhinitis • Sinusitis • Pharyngitis • Laryngitis	• Tracheitis • Chemical bronchitis • Irritant-induced asthma	• Pulmonary edema • Pneumonitis
Sensitization	• Allergic rhinitis • Sinusitis • Pharyngitis • Laryngitis	• Allergy-mediated asthma	• Hypersensitivity pneumonitis • Chronic beryllium disease
Cytotoxicity			• Pneumoconioses
Carcinogenicity	• Nasopharyngeal cancer	• Bronchogenic carcinoma	• Mesothelioma

reversible airflow obstruction and workplace exposure. This can often be done through serial measurements of lung function both at work and away from the workplace; in some cases, specific bronchial challenges are used (Lombardo et al., 2000; NAEPP). Occupational asthma is also very under-reported, not because asthma is difficult to diagnose, but rather because the connection of asthma to a workplace exposure is often missed.

The diagnosis of any pneumoconiosis (for example silicosis, asbestosis, or coal worker's pneumoconiosis) is based, once again, on a compatible history of dust exposure. Pneumoconioses typically appear long after the onset of exposure to these dusts, with a latency period that is often measured in decades. In addition to the exposure history, virtually all pneumoconioses will cause an abnormal appearance to the lungs when these are examined with a chest radiograph (chest x-ray). Different pneumoconioses produce different (and often characteristic) patterns of abnormality on a chest x-ray. When there is no history of chronic dust exposure, the predictive value of these radiographic abnormalities is low. However, it improves greatly in the setting of a compatible occupational exposure history (Murphy et al., 1986).

Abnormal chest x-ray changes are often supported by abnormal findings on auscultation of the lungs with a stethoscope and by characteristic spirometric abnormalities (loss of volume or "restriction"). Pneumoconiosis can also be diagnosed by examining biopsy obtained lung samples under a microscope for characteristic pathological changes.

The diagnosis of any respiratory disease is the job of a clinician. However, information leading to the correct diagnosis and ultimately supporting appropriate treatment of the patient or guiding institutional response is often provided by the careful investigation of the environment and documentation of exposures occurring there.

MANAGEMENT OF OCCUPATIONAL RESPIRATORY DISEASE

For most occupational respiratory diseases, the most important element of managing the disease rests on *control or removal of the inciting exposure.* Often, this can cure the disease (e.g., irritative syndromes) or prevent further progression (e.g., early stages of some pneumoconioses). Exposures are best managed through the application of the traditional hierarchy of controls, i.e., engineering controls, administrative controls, effective personal protective equipment, and worker education. At times, however, control of the exposure does little to limit the course of an occupational respiratory disease. Such can be the case with advanced pneumoconioses or occupational asthma. In these cases, it is important to address *worker placement*, in the event that the worker is still capable of gainful employment. *Pharmacological therapy* with medications plays a major role in treatment of irritative syndromes and asthma, but rarely in and of itself leads to a cure. Instead, the aim of most pharmacological therapy is to control symptoms. When the disease does not resolve, or a worker is not capable of returning to his/her previous employment, it becomes necessary, as part of the clinical evaluation, to measure the extent of functional loss of the respiratory system that occurred as a consequence of the disease. This is performed through an *impairment evaluation*, which is an essential component in determining the degree of *disability* caused by an occupational respiratory disease. Several semi-quantitative systems are available to assist physicians in assigning a level of impairment for a given patient (ATS, 1993; AMA, 2000).

In addition to controlling harmful exposures, ongoing *medical surveillance* of workers at risk of future development of occupational respiratory disease is a critical element of disease *prevention*. The tool or test that forms the basis of a medical surveillance program should be selected based on knowledge of the natural history of the occupational respiratory disease. Thus, when surveying for irritative syndromes related to irritant gas exposures, a surveillance system based on the detection of early symptoms of irritation may be very useful, since most irritative syndromes tend to reverse following removal of the irritant. For other diseases, such as the pneumoconioses, surveillance based on symptoms is not useful, since symptoms tend to appear at an advanced stage of the disease. Rather, it would make more sense to implement medical surveillance based on the detection of early changes on a serial chest x-ray or pulmonary function, since these abnormalities often precede the development of symptoms by months to years. Moreover, if pneumoconiosis is detected at an early stage, and the worker is removed from exposure, disease progression can be slowed or even ended (though never reversed).

EPIDEMIOLOGICAL TRENDS IN OCCUPATIONAL RESPIRATORY DISEASE

Work-related diseases of the respiratory system make up approximately 6% of all reported occupational illness in the United States. Of these, approximately one sixth comprise the "classic" chronic dust diseases of the lungs (pneumoconioses), and the remainder is made up of toxic inhalation injuries. However, these statistics probably underestimate the true magnitude of occupational lung disease, since they depend on recognition by clinicians of the work-related nature of the illness. Although diagnosis is often straightforward when diagnosing a pneumoconiosis (a lung disease which can only be caused by chronic exposure to certain dusts), workplace associations are less likely to be recognized in the case of respiratory diseases that may have multiple causes (e.g., asthma), including certain occupational exposures.

Overall mortality from pneumoconioses has been gradually declining in the United States from a peak of over 5,000 deaths in 1972 to approximately 3,200 in 1992 U.S. DHHS, 1996). Of these, approximately 60% were deaths due to coal workers pneumoconioses ("black lung"), 15% were attributable to silicosis and 10% to asbestosis. There is a distinct geographic distribution of these pneumoconioses deaths, with clustering in those areas close to the source of workplace exposure. In this regard, for example, most black lung deaths are found in the coal mining regions of the United States, whereas deaths from asbestos-related disease tend to cluster in the coastal areas where shipyards would likely be the source of occupational asbestos exposure.

Asthma prevalence and severity have been increasing over the past 15 to 20 years, not only in the United States, but also worldwide (NAEPP; U.S. DHHS, 1996). The causes of this worsening trend are felt to be multi-factorial, and one factor is likely to be an increase in occupational and environmental exposures. It is estimated that over 20 million workers may potentially be exposed to asthma-causing substances; of these, approximately 9 million workers have the potential for exposure to established sensitizers and irritants. The proportion of all asthma that can be attributed to workplace exposures varies markedly in various studies that have addressed this issue, from 5% to 21% (Contreras et al., 1994; Kogevinas et al., 1996; Meredith et al., 1991; Milton et al., 1998). This wide range in estimates is partly explained by differences in the level of recognition of occupational asthma, as well as by variations in the definition of asthma used in these studies. In contrast to the decline in pneumoconioses, occupational asthma is now the most frequently reported occupational respiratory disease diagnosis in the United States among patients visiting occupational medicine clinics and in workers' compensation claims for respiratory disease (Chan-Yeung et al., 1995). Similar observations have been made in Canada (Contreras et al., 1994) and the United Kingdom (Meredith et al., 1991). Deaths from asthma and chronic obstructive pulmonary disease (COPD) are now the fourth leading cause of death in the United States (Feinlieb et al., 1989).

QUESTIONS

1.1. List four main functions of the respiratory system.

1.2. Assume a group of workers is acutely exposed to a monodisperse dry particulate of a slightly water-soluble strongly irritant chemical substance, having a MMAD of 8 µm and σ_g 1.1.

a. What sort of respiratory symptoms will you expect to be predominant in these workers?

b. In which of three respiratory tract regions will you expect most injury to occur (upper airways, lower airways, or respiratory zone)?

c. What health status characteristics could put some workers at greater risk than others?

1.3. List five determinant factors for respiratory tract injury.

1.4. Discuss the value of spirometry, assessment of lung volumes, assessment of D_LCO, peak flow self tests, cardiopulmonary stress testing, non-specific bronchoprovocation testing, and chest x-ray in:

a. asbestosis

b. occupational asthma

c. non-respiratory heart disease

REFERENCES

American Medical Association, Guides to the Evaluation of Permanent Impairment, Fifth Edition, Cicchiarella, L. and Andersson, G.B.J., Eds., AMA Press, Chicago (2000).

American Thoracic Society, Guidelines for the evaluation of impairment/disability in patients with asthma. Am Rev Respir Dis 147: 1056-61 (1993).

American Thoracic Society, Standardization of Spirometry: 1994 Update, Amer J Respir Critical Care Med 152: 1107-1136 (1995).

American Thoracic Society, Guidelines for Methacholine and Exercise Challenge Testing-1999, Amer J Respir Critical Care Med 161: 309-329 (2000).

Brooks, S.M., Weiss, M.A., Bernstein, I.L.: Reactive airways dysfunction syndrome (RADS): persistent asthma syndrome after high level irritant exposure. Chest 88:376-84 (1985).

Brooks, S.M., Hammad, Y., Richards, I., Giovinco-Barbas, J., Jenkins, K.: The spectrum of irritant-induced asthma: sudden and not-so-sudden onset and the role of allergy. Chest 113:42-49 (1998).

Chan-Yeung, M., Malo, J.L.: Current concepts: occupational asthma. N Engl J Med 333:107-112 (1995).

Contreras, G.R., Rousseau, R., Chan-Yeung, M.: Occupational respiratory diseases in British Columbia, Canada in 1991. Occup Environ Med 51: 710-712 (1994).

Crapo, J., Miller, F.J., Mossman, B., Pryor, W.A., Kiley, J.P.: Relationship between acute inflammatory responses to air pollution and chronic lung disease. Am Rev Respir Dis 145:1506-1512 (1992).

Dahl, A.R., Lewis, J.L.: Respiratory tract uptake of inhalants and metabolism of xenobiotics. Ann Rev Pharmacol Toxicol 32: 383-407 (1993).

Delclos, G.L., Carson, A.: Acute Gaseous Exposures. Occupational and Environmental Respiratory Diseases, Harber, P., Schenker, M., Balmes, J., Eds. 1995 Mosby Yearbook Pub., St. Louis.

Feinlieb, M., Rosenberg, H.M., Collins, J.G., Delozier, J.E., Pokras, R., Chevarley, F.M.: Trends in COPD morbidity and mortality in the United States. Am Rev Respir Dis 140:S9-S18 (1989).

Kimbell, J.S., Gross, E.R., Joyner, D.R.: Application of computational fluid dynamics to regional dosimetry of inhaled chemicals in the upper respiratory tract of the rat. Toxicol Appl Pharmacol 121: 253-263 (1993).

Kogevinas, M., Antó, J.M., Soriano, J.B., Tobías, A., Burney, P., and the Spanish Group of the European Asthma Study: The risk of asthma attributable to occupational exposures. A population-based study in Spain. Am J Respir Crit Care Med 154:137-143 (1996).

Lombardo, L.J., Balmes, J.R.: Occupational asthma: a review. Envir Health Perspect 108 Suppl 4:697-704 (2000).

Meredith, S.K., Taylor, V.M., McDonald, J.C.: Occupational respiratory disease in the United Kingdom 1989: A report to the British Thoracic Society and the Society of Occupational Medicine by the SWORD project group. Br J Ind Med 48:292-298 (1991).

Milton, D.K., Solomon, G.M., Rosiello, R.A., Herrick, R.F.: Risk and incidence of asthma attributable to occupational exposure among HMO members. Am J Ind Med 33:1-10 (1998).

Murphy, R.L., Becklake, M.R., Brooks, S.M., Gaensler, E.A., Gee, B.L., Goldman, A.M., Kleinerman, J.I., Lewinsohn, H.C., Mitchell, R.S., Utell, M.J., Weill, H.: The diagnosis of nonmalignant diseases related to asbestos. Am Rev Respir Dis 134:363-368 (1986).

National Asthma Education and Prevention Program. Expert Panel Report: Guidelines for the diagnosis and management of asthma, NIH Publ. No. 91-3642. National Heart, Lung, and Blood Institute, National Institutes of Health, Bethesda, MD.

Overton, J.H., Graham, R.C., Miller, F.J.: A model of regional uptake of gaseous pollutants in the lung II: The sensitivity of ozone uptake in laboratory animal lungs to anatomical and ventilatory parameters. Toxicol Appl Pharmacol 88: 418-432 (1987).

Phalen, R.F., Oldham, M.J.: Tracheobronchial airway structure as revealed by casting techniques. Am Rev Respir Dis 128:S1-S4 (1983).

Pinkerton, K.E., Gehr, P., Crapo, J.D.: Architecture and cellular composition of the air-blood barrier. In Treatise on Pulmonary Toxicology: Comparative Biology of the Normal Lung. Parent, R.A., Ed., pp. 121-128, CRC Press, Boca Raton (1991).

Townsend, M.C., Lockey, J.E., Velez, H., Carson, A., Cowl, C.T., Delclos, G.L., Gerstenhaber, B.J., Harber, P.I., Horvath, E.P., Jolly, A.T., Jones, S.H., Knackmuhs, G.G., Lindesmith, L.A., Markham, T.N., Raymond, L.W., Rosenberg, D.M., Sherson, D., Smith, D.D., Wintermeyer, S.F.: ACOEM Position Statement – "Spirometry in the Occupational Setting" JOEM 42:228-245 (2000).

U.S. Department of Health and Human Services, National Institute for Occupational Safety and Health. Work-related lung disease surveillance report: 1996. DHHS (NIOSH) Publ. No. 96-134. U.S. Government Printing Office, Washington, DC.

U.S. Department of Health and Human Services, Centers for Disease Control and Prevention. Asthma mortality and hospitalization among children and young adults: United States, 1980-1993. MMWR 45:350-353 (1996).

Weibel, E.R.: Morphological basis of alveolar capillary gas exchange. Physiol Rev 53:419 (1973).

2

RESPIRATORY TRACT DEPOSITION AND PENETRATION

Lung Chi Chen, MD

INTRODUCTION

Particles encountered in ambient air and the work environment have a variety of physical, chemical, and biological properties. These properties, acting alone or in combination, may determine the way particles interact with the respiratory system and the subsequent toxic response. For a given particulate matter, it must penetrate and deposit within the respiratory system in order to produce a biological response. The respiratory tract is a complex and unique organ system. It has a number of subdivisions with markedly different structure and function. To establish reasonable dose-response assessment, comprehensive characterization of material deposited is required.

An aerosol is a suspension of finely dispersed solids or liquids in air (see Volume 1, page 82). Deposition is defined as the removal of particles from inhaled air through contact with the airway surface. As depicted schematically in Figure 2-1 (Schlesinger, 1995), there are five significant mechanisms by which airborne particles may deposit in the respiratory tract (for an introduction, see Volume 1, page 130). These are impaction, sedimentation, Brownian diffusion, electrostatic precipitation, and interception.

Deposition Mechanisms

When a particle with sufficient mass (between 1 and 100 μm d_{ae}) is moving in an air stream, a resistive force of air proportional to μvd is developed, where μ is the viscosity of air, v is the velocity of the particle relative to the air, and d is the particle diameter. When the viscous resistance of air matches the force(s) on the sphere responsible for its motion (i.e., gravity), a constant velocity develops for particles in this size range. This constant velocity is termed the terminal velocity of the particle. For particles less than 1 μm, the viscous resistance of air on the particle begins to be overestimated and the particle's terminal velocity underesti-

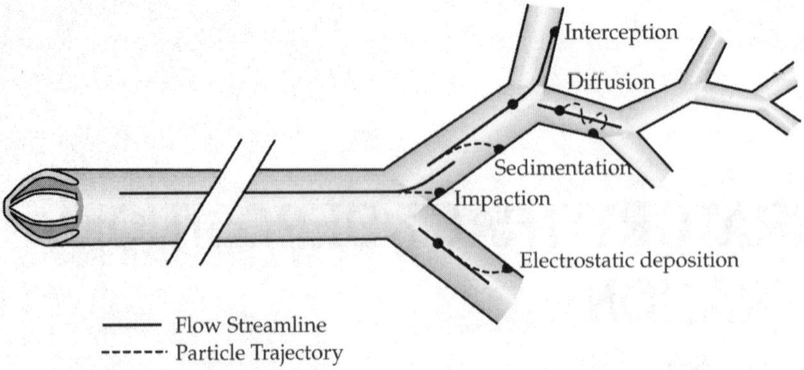

Figure 2-1. Schematic diagram of particle deposition mechanisms (from Schlesinger, 1995).

mated, and a Slip Correction factor is developed to correct these errors (see Volume 1, page 185). These slip corrections become more important as the particle diameter nears, or is less than, the mean free path of air molecules (approximately 0.068 μm at 25°C and 760 mm Hg air pressure).

IMPACTION

Impaction occurs when a particle fails to follow the streamlines of airflow caused by sudden changes in airstream direction and velocity. The resulting inertia causes a particle to deposit onto an airway surface. Impaction occurs in the upper airways where there are high air velocities and sharp directional changes. The probability that a particle with an aerodynamic diameter (see Volume 1, page 186), d_{ae}, moving in an airstream with an average velocity, U, will impact at a bifurcation is related to a parameter called the Stokes number, Stk; defined as

$$\mathrm{Stk} = \frac{\rho d_{ae}^2 U}{9 \mu D_a} \qquad \text{[Equation 2-1]}$$

where ρ is the density of air, μ is the viscosity of air, and D_a is the diameter of the airway.

The probability of impaction, therefore, increases with increasing air velocity, particle size, and density. Impaction is the main mechanism by which particles having diameters ≥0.5 μm deposit in the upper respiratory tract and at or near tracheobronchial tree branching points. In Landahl's lung deposition model (1950a) of impaction in the tracheobronchial region, impaction efficiency was proportional to

$$\frac{\rho d_{ae}^{2} U_{i} \sin\theta_{i}}{D_{ai} S_{i-1}} \qquad \text{[Equation 2-2]}$$

where U_i is the air velocity in the airway generation i, θ_i is the branching angle between generations i and i-1, D_{ai} is the diameter of the airway generation i, and S_{i-1} is the total cross-sectional area of airway generation i-1.

SEDIMENTATION

All airborne particles are constantly subject to gravitational force. When a balance is achieved between acceleration of gravity and the viscous resistance of the air, the particle will fall out of the air stream at a constant rate, which is the terminal velocity. The terminal velocity v_t is defined as:

$$v_t = \frac{d_{ae}^{2} \rho g K_s}{18\mu} \qquad \text{[Equation 2-3]}$$

where d_{ae} is the aerodynamic size, r the particle density (g/cm^3), g the acceleration due to gravity, K_s the slip correction factor, and μ the viscosity of air. The probability of sedimentation is proportional to the particle's residence time in the airway, particle size, and density, and decreases with increasing breathing rate. Sedimentation is an important mechanism of deposition for particles ≥ 0.5 µm in airways with relatively low air velocity, e.g., mid to small bronchi and bronchioles. It should be noted that the particles between the size range of 0.1 and 1.0 µm are too small to have sufficient inertial momentum or terminal velocity to be deposited by impaction or sedimentation and too large to diffuse. Thus, particles of this size range are the most persistent and stable in aerosols and are the least to be deposited in the respiratory tract.

DIFFUSION

Sedimentation becomes insignificant relative to diffusion as the particles become smaller. The terminal velocity of a 0.5 µm unit density spherical particle is about 0.001 cm/s so that gravitational forces become negligible for smaller particles. These small particles acquire random motion due to collision with gas molecules. In small airways, the distances between the particles and airway epithelium are small, resulting in contact of particles with the airway wall. The flux of particles deposited by diffusion is governed by the diffusion coefficient defined as

$$D = \frac{\kappa T K_s}{3\pi\mu d} \qquad \text{[Equation 2-4]}$$

where κ is the Boltzmann constant, T is the absolute temperature, μ is the viscosity of air, and K_s is the slip correction factor. Diffusion is inversely proportional to particle size (specifically the cross-sectional area) but is independent of particle density. Deposition due to diffusion increases in the bronchioles and the pulmonary alveolar region where bulk flow is low or absent. For very small ultrafine particles (paticles less than 0.1 μm in diameter), diffusion has also been shown to be an important deposition mechanism in the extrathoracic regions such as the trachea and larger bronchi (Cheng et al., 1988, 1990).

ELECTROSTATIC PRECIPITATION

All airborne particles are randomly charged by the omnipresent air ions. As particles are produced, they acquire charges from these ions by collisions with them due to their random thermal motion. Eventually, a Boltzman equilibrium is achieved when charge distribution of an aerosol is in charge equilibrium with bipolar ions. Particles that are electrically charged may exhibit enhanced deposition over that expected from size alone. This is due to image charges (charges with opposite polarity) induced on the surface of the airway by these particles, or to space-charge effects whereby repulsion of particles containing like charges results in increased migration toward the airway wall (Schlesinger, 1995). The effect of charge on deposition is inversely proportional to particle size and air flow rate. Compared to the other mechanisms, contribution by electric force for overall deposition is small. Many freshly generated particles are electrically charged. Experimental studies conducted using lung casts and in humans and laboratory animals have shown that deposition increases with charged particles (Cohen et al, 1996). Above a threshold level, deposition increases linearly with an increasing number of electric charges on the particles (Yu, 1985).

INTERCEPTION

Interception is defined as the process when an edge contacts, or intercepts, an airway wall. This process is a significant deposition process for airborne fibers, which are those particles having a length/diameter ratio ≥ 3. Deposition of fibers in the respiratory tract is a complex process involving a combination of inertial, gravitational, diffusional forces as well as the interception process (Asgharian and Yu, 1988, 1989). The deposition efficiencies of fibers is influenced by their orientations, which are strongly related to the direction of airflow. An equivalent mass diameter, d_{em}, can be used to estimate the orientational effects on deposition of fibers in the respiratory tract (Asgharian and Yu, 1988, 1989). The equivalent mass diameter, d_{em}, is defined as

$$d_{em} = d_f b^{1/3} \qquad \text{[Equation 2-5]}$$

where d_f is the fiber diameter and b is its aspect ratio (length/diameter). The probability of interception increases with increasing fiber length, while the probability of impaction/sedimentation is influenced by fiber diameter. Interception also increases with decreasing airway diameter, but it can also be a significant process of deposition of fiber in the upper respiratory tract.

Factors Affecting Deposition

When considering particle dosimetry, it is essential to examine the factors that generally affect particle dose to the entire respiratory tract.

PHYSICAL/CHEMICAL PROPERTIES OF INHALED PARTICLES

Airborne particles can consist of almost any material. To describe an aerosol, one has to include such important factors as size, shape, surface area, density, and its chemical composition. For aerosols found in ambient air and the work environment, their size distribution can usually be described by a lognormal distribution in which the distribution will resemble the bell-shape Gaussian error curve when plotting the frequency distribution against the logarithms of the particle size (see Volume 1, Chapters 7 and 9). Generally, the median diameter (logarithmic mean of the diameter distribution) and the geometric standard deviation, σ_g, (ratio of the log 84 percentile/log 50 percentile size cut or log 50 percentile/log 16 percentile size cut, where the 50 percentile size cut is the median) are used to describe the size distribution of an aerosol.

From the discussion above, it is obvious that the major particle characteristic that influences deposition is size. When particles are subject to impaction and sedimentation, i.e., for particles larger than 0.5 μm in physical diameter, their equivalent aerodynamic diameter (d_{ae}) determines the deposition probability. Since, in terms of many dose-response relationships, the mass of the particles is of concern, the mass median aerodynamic diameter (MMAD) is generally used to evaluate the particle deposition pattern. For radioactivity or biological activity, activity median aerodynamic diameter (AMAD) is used to described the median of the distribution of these particles. For particles smaller than 0.1 μm, diffusion is the dominant deposition process. Therefore, it is their actual physical diameter that governs the particles' deposition probability.

The variation in the distribution of particle size in an aerosol, which is generally described by its σ_g, is also important in terms of ultimate deposition pattern. If σ_g of an aerosol is less than 2, the total amount of deposition within the respiratory tract will probably not differ substantially from that of a monodisperse aerosol ($\sigma_g \geq 1.2$) having the same median size (Diu and Yu, 1983). However, for an aerosol with a larger σ_g, the regional distribution of particle deposition in the lung may differ substantially from that of a monodisperse aerosol having the same median size. For example, when the deposition (in hamsters) of a monodisperse and a polydisperse aerosol having similar median diameter and total respiratory deposition (percentage of amount inhaled) was compared, the latter was found to deposit to a greater extent in the upper respiratory tract due to the presence of large particles that were removed by impaction (Thomas and Raabe, 1978).

HYGROSCOPICITY

Particles entering the respiratory tract encounter an environment that is radically different from the ambient air. The inspired air is rapidly warmed and humidified. Particles with high hygroscopicity, which is the ability of an aerosol particle to grow in size upon entering a more moist environment, will dynamically change their size upon inhalation. In ambient air, hygroscopic particles are predominantly inorganic salts, although a small amount of organic acids are also present. The growth rate of the hygroscopic particles is dependent on the particles' size, i.e., the smallest particles being the fastest to reach an equilibrium size.

The deposition of hygroscopic particles will depend on their hydrated (rather than their initial dry) size. Using sulfuric acid aerosols, Dahl and Griffith (1983) have shown greater total and regional deposition of these aerosols in rats compared to nonhygroscopic aerosols having the same MMAD. Deposition of sulfuric acid aerosols generated at 20% RH was also higher than those generated at 80% RH, indicating that the increased deposition was caused by the growth of the particle in the highly humid environment of the respiratory tract. The deposition of sulfuric acid particles in different regions of the lungs may be higher or lower depending upon the initial size. The regional deposition patterns of inhaled hygroscopic aerosols can be reasonably predicted by computational models in populations of normal subjects. In humans, theoretical model studies have shown that the influence of hygroscopicity of sulfuric acid aerosols is to increase total lung deposition for larger particles, whereas the opposite occurs for smaller particles (Martonen and Zhang, 1993). In this model, enhanced inertial impaction of larger sulfuric acid particles due to hygroscopic growth increases their deposition efficiency, while the growth of the smaller particles loses their diffusional force and reduces the deposition.

RESPIRATORY TRACT GEOMETRY

The respiratory tract is a complex structure and affects particle deposition in many ways. There is a considerable variability in the structure of the respiratory system between different individuals. This may account for the large inter-subject differences in particle deposition pattern observed experimentally (Heyder et al., 1982). The diameter of the airway determines the distance a particle has to travel before it contacts the surface. The regional length and cross-sectional area determine the velocity and flow pattern for a given flow rate. Furthermore, the branching angle and branching pattern are also important in determining the flow characteristics within a given airway. In addition, differences in path lengths within different lung lobes may affect regional deposition. The effect of path length is more obvious for large particles that deposit in the lung through impaction and sedimentation. The shorter the average path length between the trachea and terminal bronchioles, the greater pulmonary deposition for these particles will occur. For smaller particles, especially ultrafine particles, the deposition tends to be more evenly distributed throughout the lung regardless of path length.

Particle penetration into lung airways during normal respiration is also affected by the exchange of inspired air and residual gas. Using excised human lungs, Fang et al., (1993) found that monodisperse particles (0.70, 0.90, 0.96, and 1.44 microns) suspended in $He\text{-}O_2$ penetrated deeper than particles suspended in air. These particles penetrated least in $SF_6\text{-}O_2$. In contrast, dog lungs, which have more asymmetrical airway branching patterns than human lungs, had no significant particle penetration differences associated with carrier gas composition. Thus, particle penetration during the inspiratory phase is dependent on factors that determine flow profile development, such as branching pattern and the Reynolds number of the carrier gas.

VENTILATION PATTERN AND MODE OF BREATHING

The respiratory tract is also a dynamic structure. The caliber, length, and the branching angles of the airway change during respiration. The structural changes occurring during inspiration differ from those occurring during expiration. In addition, even if airways were rigid structures, the behavior of air flow into divergent airways during inspiration would differ from that produced during expiration in a converging direction. As discussed earlier, flow characteristics in the airway influence the probability of particle deposition. It is not surprising that changes in ventilation pattern and mode of breathing influence the sites and relative amount of regional deposition. During exercise or other enhanced activity, the increase in the respiratory rate and tidal volume causes an increase in linear air velocity and the development of turbulence within the conducting airways. Consequently, particle deposition due to impaction is enhanced, while deposition due to sedimentation and diffusion is decreased. While total deposition within the respiratory tract may increase with exercise for particles larger than 0.2 - 0.5 μm in diameter (Harbison and Brain, 1983; Zeltner et al., 1991), a shift in the deposition pattern towards the upper respiratory tract and central region can occur (Bennett et al., 1985; Morgan et al., 1984). In contrast, the deposition of ultrafine particles within the upper respiratory tract and tracheobronchial tree decreases as flow rate increases, and exercise may not increase total respiratory tract deposition of these particles (Hesseltine et al 1986).

The volume of air inhaled during a single breath, i.e., tidal volume, may determine the depth of the inspired air penetrating into the lungs. An increase in tidal volume may result in deeper penetration of inhaled particles, with a potential increase in deposition in the smaller conducting airways and pulmonary region.

By use of a serial bolus aerosol delivery technique, Kim, et al., (1996) had shown that, at a constant flow rate (at 150, 250, and 500 ml/s), deposition efficiency is greater with smaller flow rate in all lung regions. Deposition was distributed fairly evenly throughout the lung regions with a tendency for an enhancement in the distal lung regions for monodispersed particles of 1 micron in diameter. Deposition distribution was highly uneven for monodispersed particles of 3 and 5 microns in diameter, and the region of the peak deposition shifted toward the proximal regions with increasing particle size. Surface dose was 1-5 times greater in the small airway regions and 2-17 times greater in the large airway regions than in the alveolar regions. The results suggest that local or regional enhancement of deposition occurs in healthy subjects, and that the local enhancement can be an important factor in health risk assessment of inhaled particles.

As the activity level increases, humans often switch from nasal breathing to oronasal breathing (combined oral and nasal breathing). This switch in the mode of breathing would bypass the nasal passages, which are more efficient than the oral passage in removing inhaled particles, and increase particle penetration into the lungs.

OTHER FACTORS MODIFYING DEPOSITION

Previous or co-exposure to airborne irritants and lung disease may change the airway geometry and ventilation pattern of an individual resulting in alterations in deposition patterns. For example, irritant exposure may induce bronchoconstriction which would increase deposition in the upper respiratory tree via impaction. Similarly, airway narrowing or obstruction due to diseases such as chronic bronchitis tends to increase total respiratory tract deposition of large particles due to enhanced deposition within the upper respiratory tract and tracheobronchial tree, even though peripheral deposition may be reduced. In contrast, ventilation impairment may prevent penetration of particles into certain portions of the lungs (Thomson and Short, 1969; Thomson and Pavia, 1974; Lourenco et al., 1972).

Alterations in the pulmonary structure in diseases such as emphysema may also affect particle deposition. The increase in alveolar size results in greater distances to deposit on a surface and a concomitant reduction in pulmonary region deposition efficiency (Brain and Valberg, 1979). Similar reduced deposition was observed in rodents with enzyme-induced emphysema (Damon et al., 1983; Hahn and Hobbs, 1979). In contrast, in rats with fibrotic disease such as pneumoconiosis induced by silica or coal, particles tend to deposit more distally than in normal animals (Heppelston, 1963).

Anatomical changes with aging may account for increased pulmonary region deposition of particles larger than 1 μm in older adults compared to younger adults (Phalen et al., 1991). However, the deposition of ultrafine particles may not show dramatic differences between children and adults, nor with aging (Phalen et al., 1991; Swift et al., 1992). More recently, Bennett, et al. (1996) found that deposition fraction of inhaled 2 μm (MMAD) particles was independent of age. There was a tendency toward greater deposition in female than in male subjects. However, because the males had 45% higher minute ventilations than the females, the deposition rate, or particles depositing per unit of time, was 30% greater in males than in females. Multiple regression analysis showed that among all subjects, the variability in deposition was best predicted by variability in the breathing period associated with the pattern used to breathe the particles, and by the subject's specific airway resistance.

Kim and Kang (1997) measured lung deposition in normal healthy control subjects (N) and in subjects with varying levels of airway obstruction: smokers (S), smokers with small airways disease (SAD), asthmatics (A), and patients with chronic obstructive airway disease (COPD), and found a marked increase in particle deposition in patients with obstructive lung disease. Particle deposition was highest in COPD followed by A, SAD, and S. In addition, when all of the subject data were combined, lung deposition fraction correlated well with percent of predicted FEV_1 and $FEF_{25\text{-}75}$ (see Chapter 1).

PENETRATION OF PARTICLES INTO THE RESPIRATORY SYSTEM

Toxicologically, only those particles that can penetrate into the nose or mouth and that deposit on respiratory tract surfaces are of interest. Inhalability of an aerosol is defined as the ratio of the number concentration of particles of a certain aerodynamic diameter, d_{ae}, that are inspired through the nose or mouth to the number concentration of the same d_{ae} present in the inspired volume of ambient air (ICRP66, 1994). Swift (1976) estimated that particles > 61 μm d_{ae} have a negligible probability of entering the nasal passages due to the high impaction efficiency of the external nares (based on a nasal entrance velocity of 2.3 meters/second at rest and a nasal entrance width of 0.5 cm). The upper limit for inhalability is approximately 40 μm d_{ae} for individuals breathing at 15 breaths per minute at rest (Breysse and Swift, 1990). In ICRP's model, the intake efficiency of the head, h_I, i.e., the particle inhalability, is represented by

$$h_I = 1 - 0.5\,(1-[7.6 \times 10^{-4}\,(d_{ae})^{2.8} + 1]^{-1}) + 1 \times 10^{-5}\,U^{2.75}\,\exp(0.055\,d_{ae}) \qquad \text{[Equation 2-6]}$$

where d_{ae} is in μm and U is the wind speed (m s^{-1}) (for 0 d" U d" 10 m s^{-1}).

In contrast to inspirable fraction, the respirable fraction is defined as that portion of the inhaled dust that penetrates to the nonciliated portions (alveolar or pulmonary region) of the lung (Hatch and Gross, 1964).

DEPOSITION OF PARTICLES WITHIN THE HUMAN RESPIRATORY TRACT

Experimental Deposition Data

The respiratory tract can be divided into three functional regions that differ from one another in retention time at the deposition site, the elimination pathway, or both (Schlesinger, 1995). These regions are: pulmonary region, tracheobronchial region, and the upper respiratory tract. The upper respiratory tract includes regions such as nose, oral cavity, nasopharynx, pharynx, and larynx.

As discussed earlier, the fractional deposition in each of these regions is dependent on the aerodynamic particle size, airway geometry, and ventilatory characteristics (flow rate, breathing frequency, tidal volume). The deposition of particles within the human respiratory tract as a function of particle size and respiratory parameters has been assessed by numerous investigators (Schlesinger, 1995, Lippmann, 1995). Figure 2-2 shows the pattern of overall respiratory tract deposition as a function of particle size from studies done with both oral and nasal breathing (Schlesinger, 1995). All values are expressed as deposition efficiency, i.e. the percentage deposition of the total amount inhaled. All appear to show the deposition minima over the 0.2 to 0.5 μm size range, with increasing deposition and increasing size for larger particles, and with decreasing size for smaller ones. As discussed above, particles between these sizes are minimally influenced by impaction, sedimentation, and diffusion and are exhaled.

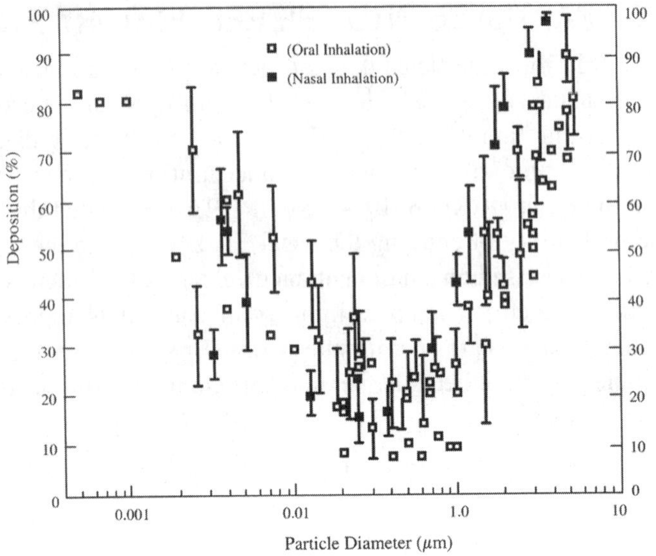

Figure 2-2. Total respiratory deposition as a function of particle size (from Schlesinger, 1995, with permission).

Due to variations in the experimental protocols such as the use of different test particles, different breathing frequency, and flow rate employed, the results of these studies vary quite considerably. When these parameters are tightly controlled, the amount of scatter for total deposition data diminish (Stahlhofen et al., 1989). There is also considerable individual variation among subjects due to intrinsic variability of airway and air space size among individuals in a population. For example, the coefficient variation of bronchial airway sizes has been found to be between 0.21 and 0.25 in healthy young nonsmokers based on aerosol deposition data (Chan and Lippmann, 1980; Lapp et al., 1975) or measurements of lung section taken at autopsy (Matsuba and Thurlbeck, 1971).

As shown in Figure 2-2, it is evident that the mode of breathing, i.e., breathing through the nose or the mouth, has significant influence on particle deposition. For particles larger than 0.5 μm, nasal breathing enhances the deposition of these particles in the upper respiratory tract and results in greater total deposition than does oral breathing. Breathing mode has little influence on the deposition for particles between 0.02 and 0.5 μm. For very small particles, a small increase in deposition was observed for nose breathing as compared to mouth breathing (Schiller et al., 1988).

Figure 2-3 shows the pattern of deposition in the upper respiratory tract. Similar to that observed for the total respiratory tract deposition, deposition is higher during nasal inhalation compared to oral. For $d_{ae} > 0.2$ mm, deposition at the upper respiratory tract is usually expressed

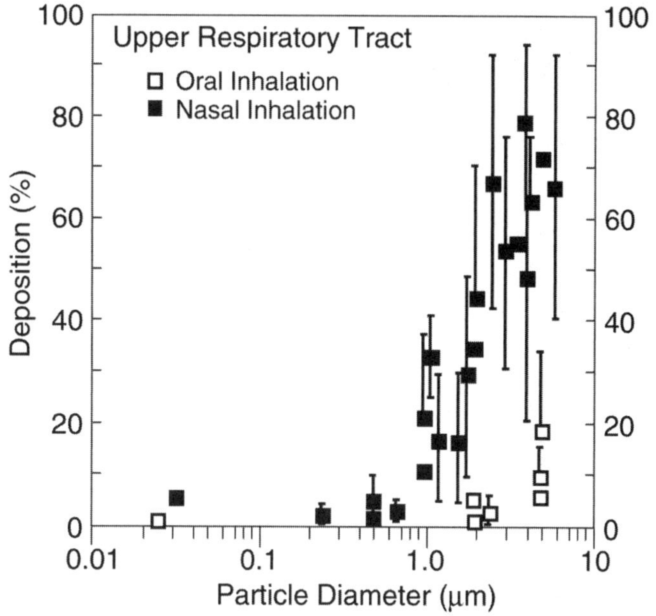

Figure 2-3. Upper respiratory deposition as a function of particle diameter (from Schlesinger, 1995, with permission).

as a function of an impaction parameter, $d_{ae}^2 Q$, where Q is the flow rate, thus normalizing impaction-dominated deposition with actual flow rates used in the experiments. In ICRP's (ICRP66, 1994) dosimetry model, the deposition efficiency via the nose (η_N) or mouth (η_M) is expressed as

$$\eta_N = 1 - [3.0 \times 10^{-4} (d_{ae}^2 Q) + 1]^{-1} \qquad \text{[Equation 2-7]}$$

$$\eta_M = 1 - [1.1\ 10^{-4} (d_{ae} Q^{0.6} V_t^{-0.2})^{1.4} + 1]^{-1}$$

where V_t is the tidal volume.

Even with this correction impaction factor, deposition data for the upper respiratory tract exhibit a very large amount of scatter (Stahlhofen et al., 1989).

Figure 2-4 shows the deposition in the tracheobronchial tree as a function of aerodynamic diameter for oral breathing (Schlesinger, 1995). There is minimal deposition for particles less than 1 μm. For larger particles, the relationship between fraction of deposition and particle size is not as clear as that observed in the other regions. When compared to Figure 2-3, it is also evident that the greater the deposition in the upper respiratory tract, the less the amount available for deposition in the lung. The removal of large particles by the upper respiratory tract may be the reason that the fractional tracheobronchial deposition is relatively constant over a wide particle size range.

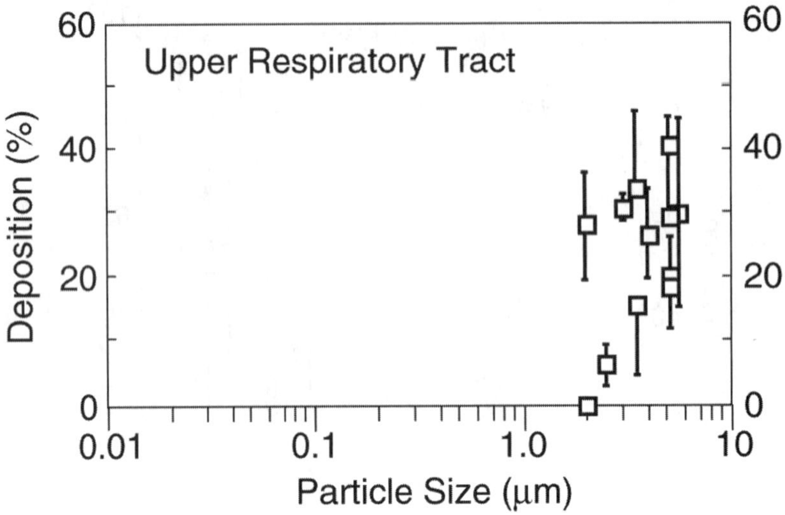

Figure 2-4. Tracheobronchial tree deposition as a function of particle size (from Schlesinger, 1995, with permission).

Figure 2-5 shows the deposition in the pulmonary region (Schlesinger, 1995). With oral breathing, deposition increases with decreasing particle size after a minimum at approximately 0.5 µm. For particles larger than 0.5 µm, however, the removal of particles by the upper respiratory tract during nasal breathing causes a decrease in deposition in the pulmonary region as compared to oral. It can be seen that for mouth breathing, the size for maximum deposition is approximately 3 µm, and approximately 50% of the inhaled particles deposit in this region. For nose breathing, the maximum deposition occurs at 2.5 µm and is only one-half of what occurs during mouth breathing. There is a nearly constant pulmonary deposition of about 20% for particles between 0.1 and 4 µm.

Recently, there is an increasing interest in investigating the toxic response of ultrafine particles (those particles smaller than 0.1 µm). Toxicological studies have shown that ultrafine particles with low solubility appear to be significantly more inflammatory in the lung than are larger size particles of the same composition (Driscoll et al., 1994). This may be due to a number of factors: 1) Ultrafine particles are more rapidly transferred to the interstitium than are fine particles of the same composition, and ultrafine particles exhibit a greater accumulation in the regional lymph nodes and a greater retention in the lung (Oberdorster et al., 1992; Oberdorster, 1995); and/or 2) On a theoretical basis, the greater the number of particles deposited in the lung, the greater the number of toxic 'hits.' In assessing effects of acidic particles, for example, Hattis and colleagues (1987) have proposed an irritation-signaling model which hypothesizes that, because the particle number per unit mass concentration declines dramatically with increasing particle size, larger particles will deliver relatively fewer localized signals (or 'hits') of a toxi-

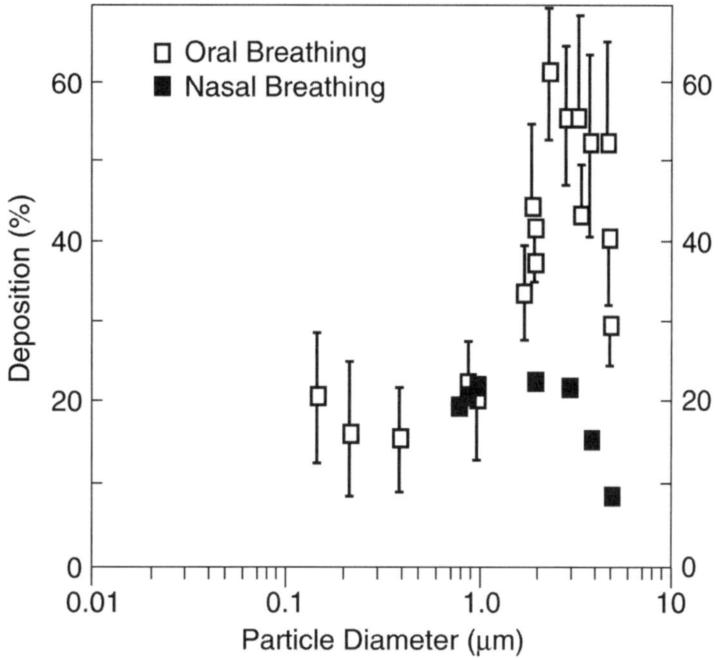

Figure 2-5. Particle deposition in the pulmonary region as a function of particle size (from Schlesinger, 1995, with permission).

cant per unit of mass than would smaller particles. In addition to their potentially inherent toxicity, ultrafine particles also have a large surface area to volume ratio on which gases and vapors may be adsorbed, absorbed, or react with the core constituent to form additional, and perhaps more reactive, chemical species.

Localized Patterns of Deposition

Deposition of particles on the airway surface is not distributed uniformly. Specific patterns of enhanced local deposition are important in determining the dose, which may be a factor in the site selectivity of certain diseases, such as bronchogenic carcinoma (Schlesinger and Lippmann, 1978). As discussed earlier, abrupt changes in flow direction, air turbulence, and high air velocity in the airway may result in enhanced localized particle deposition. In the upper respiratory tract, the larynx, oropharyngeal bend, and nasal turbinates are the likely places where enhanced localized deposition occurs (Swift, 1981; Swift and Proctor, 1988). In humans, air turbulence produced by the larynx results in enhanced localized deposition in the upper trachea and larger bronchi, while deposition is also greatly enhanced at bronchial bifurcations, especially along the carinal ridges relative to the tubular airway segments (Schlesinger et al., 1982). This occurs for spherical particles > 0.5 μm in diameter due to impaction, and for fibers due to both impaction and interception (Asgharian and Yu, 1989). Enhanced deposition at bifurcations is also seen with submicrometer particles having diameters down to about 0.1 μm due to turbulent diffusion (Cohen et al., 1988). As particle size decreases further, deposition along airway surfaces is more uniform since particle deposition is less likely to be affected by localized flow pattern (Gradon and Orlicki, 1990). For example, no enhanced deposition of 0.04 μm particle was found at bifurcations in a cast of the human upper bronchial tree (Cohen et al., 1988). However, for particles down to the molecular size range such as radon progeny, deposition maxima were observed at the carina (a ridge near the bifurcation of the trachea) and along the secondary tubes downstream from the carina of an aluminum model of a lung bifurcation (Kinsara et al., 1995).

There are few data on localized deposition for the pulmonary region. Fibers show nonuniform deposition in distal airways of animals, preferentially depositing on bifurcations of alveolar ducts near the bronchioalveolar junction (Brody and Roe, 1983, Warheit and Hartsky 1990). The presence of early fiber-induced lesions in the bronchioalveolar junction suggests that fibers may also deposit preferentially in these regions of the human respiratory tract (Brody and Yu, 1989).

PREDICTIVE DEPOSITION MODELS

In 1973, the International Commission on Radiological Protection (ICRP66, 1994) adopted a model using a single aerosol parameter, the activity median aerodynamic diameter (AMAD), to estimate the regional deposition within the respiratory tract. For tidal volume of 1450 ml, there are relatively small differences in estimated deposition over a very wide range of s_g (Lippmann, 1995). However, the prediction for total and alveolar deposition by this model and earlier models developed by other investigators differs substantially from the best available experimental data for normal healthy adults. Furthermore, these models do not take into account the very large inter-individual variability in deposition efficiency, nor of the changes produced by cigarette smoking and lung diseases (Lippmann, 1995).

ICRP Deposition Model

In 1984, both the ICRP and the National Council on Radiation Protection (NCRP125, 1997) appointed task groups to review and develop a new dosimetric model of the respiratory tract. The objective of the model developed by ICRP is 1) to facilitate calculation of biologically meaningful doses; 2) to be consistent with the morphological, physiological, and radiological characteristics of the respiratory tract; 3) to incorporate current knowledge; 4) to meet all radiation protection needs; 5) to be more sophisticated than necessary to meet dosimetric objectives; 6) to be adaptable to development of computer software for calculation of relevant radiation doses from knowledge for a few readily measured exposure parameters; 7) to be equally useful for assessment purposes as for calculating recommended values for limits on intake; 8) to be applicable to all members of the world population; 9) to allow for use of information on the deposition and clearance of smoking, air pollutants, and diseases on the inhalation, deposition, and clearance of radioactive particles from the respiratory tract.

The extrathoracic airways are partitioned into two distinct clearance and dosimetric regions: The anterior nasal passage (ET_1) and all other extrathoracic airways (ET_2), i.e., the posterior nasal passages, the naso- and oropharynx, and the larynx. Activity deposited in the thorax is divided into bronchial (BB), bronchiolar (bb), and alveolar-interstitial (AI) regions. Figure 2-6A depicts the fractional deposition in each region of the respiratory tract for reference light

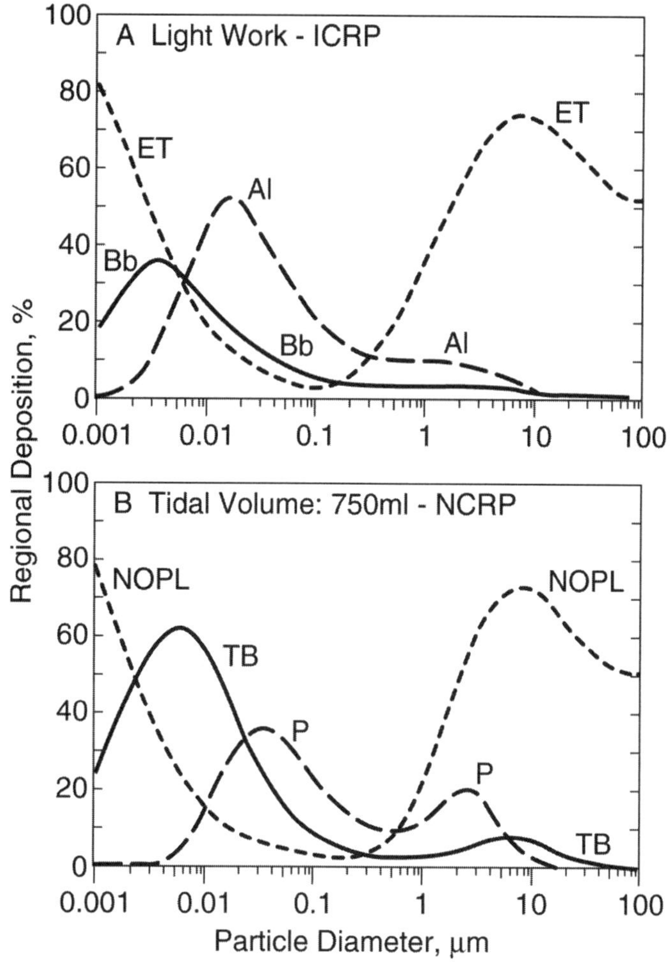

Figure 2-6. Fractional deposition in each region of respiratory tract for reference light worker (normal nose breather). A: 1994 ICRP model, B: 1994 NCRP Model. The relation of particle diameter to calculated regional deposition is for spherical particles of density 1 g/cm (from Lippmann, 1995, with permission).

workers (normal nose breather). In this model, regions ET_1 and ET_2 are combined as ET and regions BB and bb are combined as Bb.

NCRP Deposition Model

The model developed by the National Council on Radiation Protection (NCRP125, 1997) separates the respiratory tract into Naso-or-pharyngo-laryngeal (NOPL, equivalent to ICRP's ET_1+ET_2), tracheobronchial (TB, equivalent to ICRP's BB+bb), and the pulmonary (P, equivalent to ICRP's AI) regions. Figure 2-6B shows the fractional deposition in each region of the respiratory tract predicted by NCRP's model (750 ml tidal volume). Even though both models used the same databases in formulating their predictions, it can be seen that the inclusion of the inspirability factor, i.e., the aspiration efficiency for entry of ambient particles into the human nose or mouth (NCRP), strongly affects the predicted deposition in the head airways. There are also substantial differences in the modal particle size and maximal efficiencies for AI versus P and for TB versus BB+bb (Lippmann, 1995).

SUMMARY AND CONCLUSION

For a given airborne material, the potential hazards arising from inhalation are clearly related to its ability to penetrate and deposit in the respiratory system. Since deposition and penetration of particles are heavily related to particle size, inhalation hazard evaluation should be based on their size rather than the gross air concentration. In addition, there are anatomic and physiologic alterations in individuals that may influence particle deposition in the lung.

QUESTIONS

2.1. Which of the following are correct in regard to inhalation exposure?

a. Deposition by sedimentation occurs mainly with particles having a mass median diameter of less than 1 µm.

b. Deposition by diffusion occurs principally in the deep lung where the directional changes are minimal.

c. Impaction at bifurcations occurs where flow rates are highest.

2.2. Which of the following factors is least likely to influence deposition of particles in the conducting airways?

a. particle size

b. turbulent air flow

c. particle solubility

d. airway bifurcations

e. particle density

2.3. Name at least four mechanisms of particle deposition in the respiratory tract. Identify one that is associated with the deposition of very large particles and one that is associated with the deposition of very small (nanometer size) particles.

REFERENCES

Asgharian, B.; Yu, C.P.: Deposition of fibers in the rat lung. J Aerosol Sci 20:355-366 (1989).

Asgharian, B.; Yu, C.P.: Deposition of inhaled fibrous particles in the human lung. J Aerosol Med 1:37-50 (1988).

Bennett, W.D.; Messina, M.S.; Smaldone, G.C.: Effect of exercise on deposition and subsequent retention of inhaled particles. J Appl Physiol 59:1046-1054 (1985).

Bennett, W.D.; Zeman, K.L.; Kim, C.: Variability of fine particle deposition in healthy adults: effect of age and gender. Amer J Resp & Crit Care Med 153(5):1641-7 (1996).

Brain, J.D.; Valberg, P.A.: Deposition of aerosol in the respiratory tract. Am Rev Respir Dis 120:1325-1373 (1979).

Breysse, P.N.; Swift, D.L.: Inhalability of large particles into the human nasal passage: In vivo studies in still air. Aerosol Sci Technol 13:459-464 (1990).

Brody, A.R.; Roe, M.W.: Deposition pattern of inorganic particles at the alveolar level in the lungs of rats and mice. Am Rev Respir Dis 128:724-729 (1983).

Brody, A.R.; Yu, C.P.: Particle deposition at the alveolar duct bifurcations. In: Extrapolation of dosimetric relationships for inhaled particles and gases, J.D. Crapo, F.J. Miller, E.D. Smolko, J.A. Graham, A. Wallace Hayes, Eds., pp 91-99. Academic Press, San Diego (1989).

Brown, J.H.; Cook, K.M.; Ney, F.G.; Hatch, T.: Influence of particle size upon the retention of particulate matter in the human lung. Am J Public Health 40:450 (1950).

Chan, T.L.; Lippmann, M.: Experimental measurements and empirical modeling of the regional deposition of inhaled particles in humans. Am Ind Hyg Assoc J 41:399-409 (1980).

Cheng, Y.S.; Yamada, Y.; Yeh, H.C.; Swift, D.L.: Deposition of ultrafine aerosols in a human oral cast. Aerosol Sci Technol 12:1075-1081 (1990).

Cheng, Y.S.; Yamada, Y.; Yeh, H.C.; Swift, D.L.: Diffusional deposition of ultrafine aerosols in a human nasal cast. J Aerosol Sci 19:741-751 (1988).

Cohen B.S.; Xiong, J.Q.; Li, W.: The influence of charge on the deposition behavior of aerosol particles with emphasis on singly charged nanometer size particles. In: J.C.N. Morynissen and L. Gradon, Eds. Aerosol Inhalation: Recent Research Frontiers. Kluwer Academic Publishers, Printed in the Netherlands, pp 153-164 (1996).

Cohen et al (charged particles)

Cohen, B.S.; Harley, N.H.; Schlesinger, R.B.; Lippmann, M.: Nonuniform particle deposition on tracheobronchial airways: Implication for lung dosimetry. Ann Occup Hyg 32(Suppl 1): 1045-1052 (1988).

Dahl, A.R.; Griffith, W.C.: Deposition of sulfuric acid mist in the respiratory tracts of guinea pigs and rats. J Toxicol Environ Health 12:371-383 (1983).

Damon, E.G.; Mokler, B.V.; Jones, R.K.: Influence of elastase-induced emphysema and the inhalation of an irritant aerosol on deposition and retention of an inhaled insoluble aerosol in fischer-344 rats. Toxicol Appl Pharmacol 67:322-330 (1983).

Diu, C.K.; Yu, C.P.: Respiratory tract deposition of polydisperse aerosols in human. Am Ind Hyg Assoc J 44:62-65 (1983).

Fang, C.P.; Wilson, J.E.; Spektor, D.M.; Lippmann, M.: Effect of lung airway branching pattern and gas composition on particle deposition in bronchial airways: III. Experimental studies with radioactively tagged aerosol in human and canine lungs. Experimental Lung Research 19(3):377-396 (1993).

Gradon, L.; Orlicki, D.: Deposition of inhaled aerosol particles in a generation of the tracheobronchial tree. J Aerosol Sci 21:3-19 (1990).

Hahn, F.F.; Hobbs, C.H.: The effect of enzyme-induced pulmonary emphysema in syrian hamsters on the deposition and long-term retention of inhaled particles. Arch Environ Health 34:203-211 (1979).

Harbison, M.L.; Brain, J.D.: Effects of exercise on particle deposition in Syrian golden hamsters. Am Rev Respir Dis 128:904-908 (1983).

Hatch, T.F.; Gross, P.: Pulmonary deposition and retention of inhaled aerosols. Academic Press, Inc., New York, NY (1964).

Hattis, D.; Wasson, J.M.; Page, G.S.; Stern, B.; Franklin, C.A.: Acid particles and the tracheobronchial Region of the respiratory system - an "irritation-signaling" model for possible health effects. J Air Pollut Control Assoc 37:1060-1066 (1987).

Heppelston, A.G.: Deposition and disposal of inhaled dust. Arch Environ Health 7:548-555 (1963).

Hesseltine, G.R.; Wolff, R.K.; Mauderly, J.L.; Cheng, Y.S.: Deposition of ultrafine aggregate particles in exercising rats. J Appl Toxicol 6:21-24 (1986).

Heyder, J.; Gebhart, J.; Stahlhofen, W.; Stuck, B.: Biological variability of particle deposition in the human respriatory tract during controlled and spontaneous mouth-breathing. In: Walton, W.H., Ed. Inhaled Particles V: Proceedings of an international symposium, September 1980, Cardiff, Wales. Ann Occup Hyg 26:137-147 (1982).

ICRP Publication 66: Human Respiratory Tract Model for Radiological Protection. Annals of the ICRP, Vol. 2413 (1994).

Kim, C.S.; Hu, S.C.; DeWitt, P.; Gerrity, T.R.: Assessment of regional deposition of inhaled particles in human lungs by serial bolus delivery method. J App Physiol 81(5):2203-13 (1996).

Kim, C.S.; Kang, T.C.: Comparative measurement of lung deposition of inhaled fine particles in normal subjects and patients with obstructive airway disease. Amer J Resp & Crit Care Med 155(3):899-905 (1997).

Kinsara, A.A.; Loyalka, S.K.; Tompson, R.V.; Miller, W.H.; Holub, R.F.: Deposition patterns of molecular phase radon progeny (218Po) in lung bifurcations. Health Physics 68(3):371-82 (1995).

Landahl, H.D.: On the removal of air-borne droplets by the human respiratory tract: I. The lung. Bull Math Biophys 12:43-56 (1950a).

Lapp, N.L.; Hankinson, J.L.; Amandus, H.; Palmes, E.D.: Variability in the size of airspaces in normal human lungs as estimated by aerosols. Thorax 30:293 (1975).

Lippmann, M.: Size-selective health hazard sampling. In Air Sampling Instruments, Chapter 5, American Conference of Governmental Industrial Hygienists, Inc., Cincinnati, OH (1995).

Lourenco, R.V.; Loddenkemper, R.; Cargan, R.W.: Patterns of distribution and clearance of aerosols in patients with bronchiectasis. Am Rev Respir Dis 106:857-866 (1972).

Martonen, T.B.; Zhang, Z.: Deposition of sulfate acid aerosols in the developing human lung. Inhal Toxicol 5:165-187 (1993).

Matsub, K.; Thurlbeck, W.M.: The number and dimensions of small airways in non-emphysematous lungs. Am Rev Respir Dis 104:516 (1971).

Morgan, W.K.C.; Ahmad, D.; Chamberlain, M.J.; Clague, H.W.; Pearson, M.G.; Vinitski, S.: The effect of exercise on the deposition of an inhaled aerosol. Respir Physiol 56:327-338 (1984).

Oberdorster, G.; Gelein, R.; Corson, N.; Mercer, P.: Association of particulate air pollution and acute mortality: Involvement of ultrafine particles. Inhal Toxicol 7:111-124 (1995).

Oberdorster, G.: Lung dosimetry: Pulmonary clearance of inhaled particles. Aerosol Sci Technol 18:279-289 (1993).

Oberdorster, G.; Ferin, J.; Gelein, R.; Soderholdm, S.C.; Finkelstein, J.: Role of the alveolar macrophage in lung injury: Studies with ultrafine particles. Environ Health Perspect 97:193-199 (1992).

Phalen, R.F.; Oldham, M.J.; Schum, G.M.: Growth and aging of the bronchial tree: Implications for particle deposition calculations. Radiat Prot Dosim 38;15-21 (1991).

Schlesinger, R.B.: Deposition and clearance of inhaled particles. In "Concepts in Inhalation Toxicology," 2nd Edition, Edited by McClellan, R.O. and Henderson, R.F., p. 192. Taylor and Francis, Washington, DC (1995).

Schlesinger, R.B.; Lippmann, M. Selective particle deposition and bronchogenic carcinoma. Environ. Res. 15: 424-431, 1978.

Schlesinger,. R. B., Gurman, J.L.; Lippmann, M.: Particle deposition within bronchial airways: Comparisons using constant and cyclic inspiratory flow. Ann Occup Hyg 26:47-64 (1982).

Stahlhofen, W.; Rudolf, G.; James, A.C.: Intercomparison of experimental regional aerosol deposition data. J Aerosol Med 2:285-308 (1989).

Swift, D.L.: Aerosol deposition and clearance in human upper airways. Ann Biomed Env 9:593-604 (1981).

Swift, D.L.: Design of the human respriatory tract to facility removal of particulates and gases. AIChE Sym Ser 72(156):137-144 (1976).

Swift, D.L.; Montassier, N.; Hopke, P.K.; Karpen-Hayes, K.; Cheng, Y.S.; Su, Y.F.; Yeh, H.C.; Strong, J.C.: Inspriatory deposition of ultrafine particles in human nasal replicates cast. J Aerosol Sci 23:65-72 (1992).

Swift, D.L.; Proctor, D.F.: A dosimetric model for particles in the respiratory tract above the trachea. Ann Occup Hyg 32(suppl 1):1035-1044 (1988).

Thomson, M.L.; Pavia, D.: Particle penetration and clearance in the human lung. Arch Environ Health 29:214-219 (1974).

Thomson, M.L.; Short, M.D.: Mucociliary function in health, chronic obstructive airway disease and asbestosis. J Appl Physiol 26:535-539 (1969).

Warheit, D.B.; Hartsky, M.A.: Species comparisons of alveolar deposition pattern of inhaled particles. Exp Lung Res 16:83-99 (1990).

Yu, C.P.: Theories of electrostatic lung deposition of inhaled aerosols. Ann Occup Hyg 29:219-227 (1985).

Zeltner, T.B.; Sweeney, T.D.; Skornick, W.A.; Feldman, H.A.; Brain, J.D.: Retention and clearance of 0.9 μm particles inhaled by hamsters during rest and exercise. J Appl Physiol 70:1137-1145 (1991).

3

OCCUPATIONAL DERMATOTOXICOLOGY: SIGNIFICANCE OF SKIN EXPOSURE IN THE WORKPLACE

Mark Boeniger, MS

Disclaimer

The views expressed in this chapter are those of the author and do not necessarily reflect the views or policies of the National Institute for Occupational Safety and Health.

INTRODUCTION

In Volume 1, Chapter 6, occupational exposure to toxic chemicals was shown to occur by three possible routes. These include inhalation, cutaneous contact, and ingestion, and each may be significant in the workplace. This chapter will focus on the cutaneous route, with reference to ingestion only when hand to mouth transfer may occur.

Chemicals that contact the skin can interact with it in two ways. First, and most obvious, is when the skin itself is affected and there are pathological changes. The most likely effects include allergic and irritant contact dermatitis. Another way skin contact can affect the worker, but is often much more obscure, is when potentially toxic chemicals are absorbed through the skin, adding to the systemic body burden and toxicity in internal organs.

The preponderance of occupational hygiene measurements have historically been conducted to evaluate inhalation exposures. There are relatively few published studies characterizing skin exposure, and in practice it is rare for occupational hygienists to actually measure skin exposures. This should not be inferred to mean that occupational skin exposures are of little importance as a potential hazard. To the contrary, it has been estimated that 42 percent of the

U.S. workforce is at risk of dermal exposure to hazardous chemicals (NIOSH, 1993). Rather, the lack of attention most likely stems from the sometimes crude and non-validated measurement techniques, a lack of guidance criteria, and lack of government compliance emphasis on skin exposures relative to inhalable exposures. What may have become commonplace is a dangerous philosophy that, if one doesn't look for problems, they won't be found and one will not have to try to figure out how to deal with these exposures as potential problems. The result of this course of inaction is that both employees and employers may suffer the consequences of this ignorance. Fortunately, there seems to be a growing number of professionals in a wide cross-section from government, industry, and labor that believe that more attention to documenting and reducing skin exposures is necessary if we truly intend to protect workers. It is the purpose of this chapter to convince the reader that skin exposures that potentially result in illness are important. Chapter 9 in this volume further demonstrates how dermal exposures are identified and evaluated.

In past years, a subtle process of substituting more volatile chemicals with less volatile ones may have inadvertently shifted what were primarily hazards from inhalation of chemicals, to hazards that primarily contact the skin (see Volume 1, page 295). Rather than having a volatile compound that can easily leave the process or be captured by local ventilation, new non-volatile substitutes persist and may accumulate on surfaces throughout the facility. The choice of low volatility chemicals, combined with a general trend towards lowering occupational air exposure limits, has tended to increase the hazard towards dermal contact. This has created, in some cases, exposure situations that many occupational health and safety staff are not adept at controlling. A case in point has been the substitution of toluene diisocyanate (TDI) with a vapor pressure of 0.05 mm Hg, with methylene diphenyldiisocyanate with a vapor pressure of 5×10^{-6} mm Hg. This substitution apparently has *not* led to a reduction of occupational asthma in recent years, possibly because the dermal route is quite efficient in causing systemic sensitization to chemicals (Kimber, 1996). The number of disability benefit cases due to asthma in the U.K., for instance, has increased since widespread substitution around 1990 of TDI for MDI, with isocyanates being by far the leading chemical class attributed to this disease (Health and Safety Commission, 1997). In Japan, workers exposed to MDI – the low air hazard isocyanate – were found to be far more likely to develop asthma than comparable TDI-exposed workers (Jang et al., 2000).

Actually, about 80% of chemicals for which occupational exposure limit criteria exist are relatively non-volatile (≤5 mm Hg). These compounds might become inhalation hazards if they are heated, sprayed, or aerosolized, but otherwise remain in place for extended periods of time. Over time, low volatility compounds that have become temporarily airborne through heating or physical dispersal can be distributed throughout the workplace. Such compounds may also affect the skin if the aerosolized chemicals impinge upon bare skin. The other principal way low volatility compounds contaminate the skin is when they are physically transferred, which can occur when contacting contaminated surfaces. In one analysis of this issue where a subsample of 176 TLV® compounds were selected, 60% of these were considered non-volatile, of which two-thirds of those compounds could be appreciably absorbed through the skin. For the volatile compounds, half of those could appreciably be absorbed through the skin (Fiserova-Bergerova, 1990). From an occupational risk perspective, the hygienist may appreciably underestimate the total risk for chemical exposures if only the inhalation route is considered and only air samples are taken. Collecting air sampling results alone could give a false sense of safety if additional non-inhalation routes of exposure exist but are not measured.

Volatilized compounds are less likely to affect the skin in a vapor state because, generally, the mass concentration in contact with the skin is so low. The exception to this is if workers were to enter a highly contaminated environment with respiratory protection but without skin protection (Susten et al., 1990; McDougal et al., 1990; Jacobs and Phanprasit, 1993). A few organic vapors, such as 2-butoxyethanol, can be appreciably absorbed through the skin even at air concentrations that are equivalent to the occupational exposure limit (Johanson and Boman, 1991; Corely et al., 1997). When high solvent vapors are present and only normal work clothes are worn, protection of the skin is insignificant. Therefore, whole body exposure should be assumed in the risk estimation (Piotrowski, 1971).

SKIN AS A TARGET ORGAN

Types of Skin Disease

Occupational skin disease includes any abnormality of the skin induced or aggravated by the work environment. The term dermatitis relates only to skin conditions with an inflammatory component to their pathogenesis, while dermatosis relates to skin disease from any cause and with any pathologic outcome (Tucker and Key, 1992).

Causes of occupational dermatoses include (1) mechanical, caused by friction, pressure, and mechanical disruption, (2) chemical, (3) physical, caused by extremes in temperature and radiation (principally ultraviolet) , and (4) biological, caused by microbiological and parasitic organisms (Tucker and Key, 1992; Harvey and Hogan, 1995). One estimate is that about 75% of occupationally related skin disease seen in the infirmaries of industrial plants was attributed to mechanical trauma. It was noted that while this type of injury is usually relatively minor, it can predispose the skin to more serious dermatoses due to the skin's compromised mechanical barrier (Tucker and Key, 1992).

About 90-95% of all work-related dermatoses, not including those caused by mechanical trauma, are referred to as occupational contact dermatitis (OCD) or eczema (Lushniak, 1995). Occupational contact dermatitis is typically characterized by inflammation and erythema (reddening), itching, or the formation of scales as a result of contact with external chemicals or substances. The occurrence of pustules (small pus-containing superficial lesions) is rare and occurs only with secondary infection. Contact dermatitis can be further divided into two aetiological classes: allergic and irritant.

Allergic contact dermatitis is a delayed-type immunological reaction in response to contact with an allergen in sensitized individuals. This reaction is also referred to as Type 4, or cell-mediated, since there is a procession of cellular events within the body leading up to the inflammatory response. Allergenic chemicals penetrate the intact skin as small molecules (usually <400 MW), and they are incompletely allergenic (haptens) until they bind to protein and form a complete allergen (Marzulli and Maibach, 1996). Langerhans cells are specialized cutaneous immune effector cells that direct the allergen to a regional lymph node where interaction with T lymphocytes is followed by replication of sensitized T lymphocytes to complete the induction phase. Sensitization can occur after a single exposure, but requires a lag period of a few days to a couple of weeks for induction to be complete. Once sensitized, it normally takes from 12 to 96 hours for a reaction to occur, but more usually 48 to 72 hours after exposure (Magnusson and Kligman, 1970). Table 3-1 lists some common chemicals known to cause allergic contact dermatitis in industry. Allergic contact dermatitis accounts for about 30 to 50% of the cases of contact dermatitis in the workplace (Holness, 1994).

Induction of allergic contact dermatitis is known to depend on the concentration of the allergen on the skin surface. If a sensitizing dose of the allergen is spread over a larger surface area, the likelihood of sensitization declines appreciably. It is believed that sensitization is dependent on the number of allergen molecules per Langerhans cell, a small number of cells bearing many molecules being more effective than having many cells bearing a few molecules (Upadhye and Maibach, 1992). Table 3-2 shows the effect of varying the concentration of dinitrochlorobenzene, a very potent sensitizer, and surface area, upon sensitization. There appears to be a threshold surface concentration for induction of all sensitizers, and the range of induction concentrations is quite large. Some caution is warranted in strictly interpreting experimental laboratory data, since factors in the workplace which might increase percutaneous absorption of chemicals could theoretically reduce the surface concentration that is necessary to cause sensitization. For instance, repeating small exposures over a period of time seems more effective in inducing sensitization than a single large dose. Genetic disposition plays a prominent role in determining individual susceptibility (Magnusson and Kligman, 1983). Although there appears to be a fairly linear dose-response to sensitizing compounds, once sensitized, there is wide variability in the provocation threshold. The concentration necessary for a response can span at least a 100-fold range (Basketter et al., 1997).

It has been reported that 90% of all occupational allergic contact dermatitis was found on the back of the hands and the forearms (Meneghini and Angelini, 1984). However, contact dermatitis among housewives occurred in almost 50% of cases on the palms, whereas 15% of the time it affected the back of the hands and fingers (Cronin, 1985). In another study of dental laboratory technicians, the fingertips were primarily involved in allergic contact dermatitis (93%), whereas in irritant contact dermatitis, the dorsum of the fingers were affected (80%) (Rustemeyer and Frosch, 1996). Figure 3-1 depicts the locations on the body for occupational diagnosed contact dermatitis based on 879 recent Oregon workmen's compensation cases for the period 1988-1992 (NIOSH, 1997).

Contact urticaria (Type I) is an immediate immunological response in the skin resulting from circulating chemical-specific antibodies coming into contact, most commonly, with exogenous proteinaceous molecules (e.g., animal dander, latex proteins, foodstuffs, industrial enzymes). Its appearance is usually pruritic (i.e., wheal and flare response) and the reaction rarely

Table 3-1
Some chemicals causing allergic dermatitis among workers

Chemical	Occurrence
acrylates	paint plasticizer, plastics
bacampicillin	pharmaceutical
benzocaine	pharmaceutical
chloracetamide	water-base preservative in paints, glues, cosmetics
colophony	electronic solder flux, adhesives
cobalt metal, fume and dust	metal smelting
diglycidyl ether of bisphenol A	epoxy, product fabrication with resin
ethylenediamine	solvent and chemical intermediate
formalin	textiles, embalming
hydrazine	soft solder flux, chemical intermediate, metal cleaning
d-limonene	cleansers, degreasers
mercaptobenzothiazole	rubber, PPC, antimicrobial agent
methacrylate compounds	dental laboratory denture technicians
methylene diisocyanate	rigid polyurethane
neomycin sulfate	pharmaceutical antibacterial
nickel	stainless steel, metal products
p-phenylenediamine	oxidative hair dyes, cosmetology
parabens mixture	preservative in skin medication, cosmetics, cleansers
phenyl glycidyl ether	epoxy resin
picric acid	battery manufacture, colored glass, explosives
poison ivy	outdoor work
potassium dichromate	histology, leather, matches, spackle cpd., photography
substilisins	detergent manufacture
thirams	rubber manufacture, PPC, food disinfectant, lub oils
toluene 2,4-diisocyanate	polyurethane foam manufacture

Table 3-2
DNCB skin sensitizing dose-response with changing concentration on human subjects

Concentration (ug/cm^2)	Application Area (cm^2)	Total DNCB Applied (ug)	Percent Sensitized
8	7.1	62.5	8
16.4	3.5	58.0	54
16.4	7.1	116	50
16.4	14.2	232	73
17.7	7.1	125	62
35.4	1.8	62.5	85
35.4	7.1	250	83
71.0	7.1	500	100

Adapted from Upadhye and Maibach (1992).

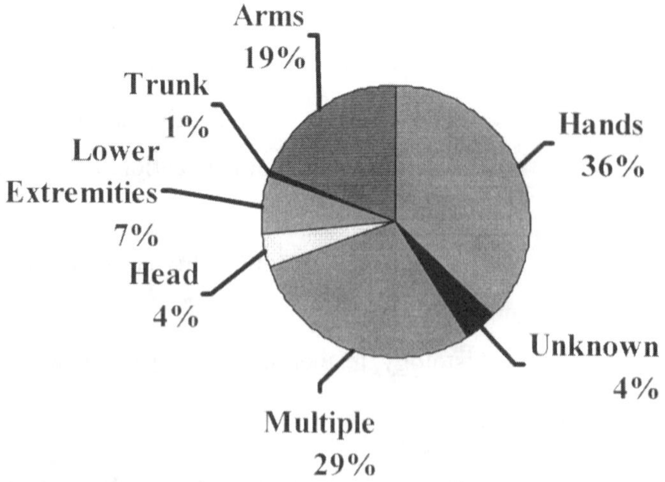

Figure 3-1. Body sites affected by contact dermatitis, as reported on 879 Oregon workmen's compensation cases for the period 1988-1992 (NIOSH SENSOR Dermatitis Program, 1997).

lasts longer than 24 hours. A wheal, or hive, is a firm rounded or flat-topped elevated lesion that results from edema (swelling) of the dermis. Wheals are often pink in color.

In addition to the skin, the respiratory and gastrointestinal tracts, as well as the cardiovascular system, may respond after cutaneous exposure to allergens. Much less frequently, contact urticaria can result from exposure to low molecular weight chemicals forming chemical-protein conjugates (e.g., 2-ethylhexyl acrylate) in the epidermis. Response is mediated by allergen-provoked release of histamine from cutaneous mast cells. In Finland, a recent survey of OCD cases from 1990-1994 found that almost 30% of all occupational immunologically mediated dermatoses were due to contact urticaria, while the remainder were allergic contact dermatitis (Kanerva et al., 1996).

Contact dermatitis from irritants constitutes about 50 to 80% of all OCD cases (Holness, 1994). There are several forms of response common to irritant exposures that are dependent on the chemical substance, the concentration, and the individual exposed. The first type is caused by a single application of a strong compound that results in a toxic, acute reaction. The second type results from repeated exposure that results in erythema, chapping, and fissures in the skin. The third type also results from repeated exposures, but develops into a chronic dermatitis that is characterized by erythema and scaling, with frequent fissuring of the stratum corneum (Weltfriend et al., 1996). A compensatory process of tolerance for irritants (sometimes called "hardening") can occur resulting in lichenified (thickened) skin. A subcategory of irritant dermatitis manifests only after a lag time of 8 to 24 hours or longer, and is thus referred to as being a delayed type (Weltfriend et al., 1996). Some industrial chemicals known to cause delayed effects include epichlorohydrin, ethylene oxide, hydrofluoric acid, some acrylates like hexanediol and butanediol diacrylate, and propane sulfone. In agriculture, the pesticide triphenyl tin hydroxide can cause delayed skin effects. Compounds that cause a delayed irritant reaction characteristically penetrate the stratum corneum slowly, and are cytotoxic to the viable epidermis.

Usually acute irritant response is rapid and begins to subside in 24 to 72 hours (Bjornberg, 1987). However, in experimental studies with sodium lauryl sulfate (SLS), complete functional skin recovery after a single 24 hour exposure had not completely occurred 12 days following exposure (Patil, 1994). The prognosis for complete resolution of both the first and second type of acute irritation is good if exposure is quickly discontinued. The prognosis for chronic irritation of the skin is variable.

Substantial individual range of susceptibility has been found to exist between individuals when exposed in the same way to a model irritant, such as the detergent SLS. The range of threshold concentrations necessary to induce an irritant response to SLS is indicated by the data in Figure 3-2, which cover a range of about 100-fold in concentration. In this study, 110 individuals representing all skin types, ranging from very fair skin (burns easily, never tans) to deep normal pigmentation, were challenged with a 4-hour occluded patch test. There seemed to be little relationship between skin pigmentation type, ability to sunburn, or gender as predictors of skin sensitivity to irritants (McFadden et al., 1998).

The probability of a given concentration to induce irritation is also dependent on the season. This is because of seasonal differences in ambient humidity, and hence skin hydration. In

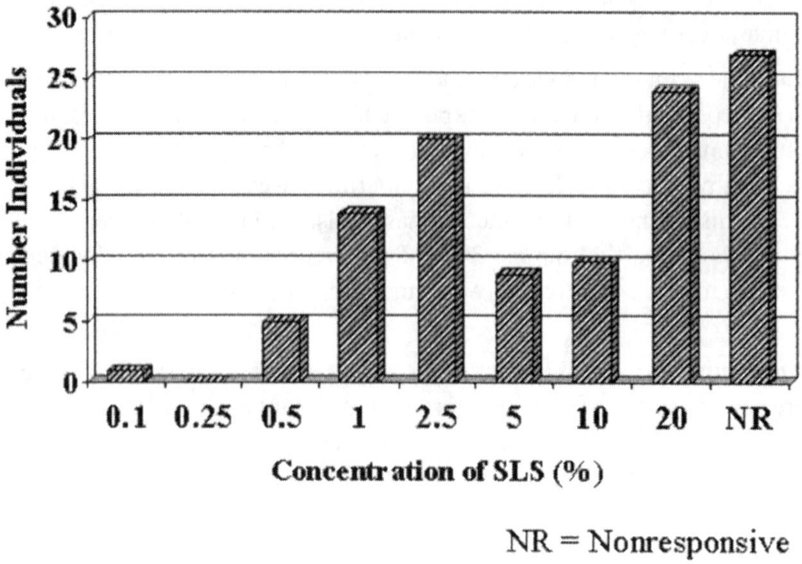

Figure 3-2. Variation in the threshold concentrations necessary for response in "normal" human skin using the irritant sodium lauryl sulfate. Adapted from McFadden et al., 1998.

one test group, 45% reacted to 20% SLS in summer, whereas 91% reacted to this concentration in the winter (Basketter et al., 1996). For other compounds, like alkalies and powders, irritant response seems more likely in the summer (Bjornberg, 1987).

In addition to the primary allergic and irritant types of dermatitis, there are several non-eczematous occupational skin diseases, including fungal and bacterial infections, furuncles (boils), acne, folliculitis, changes in pigmentation, nail diseases, and skin cancer. For information on the potential of various chemical, plant, and biological agents known to cause occupational skin disease, several reference texts are recommended (Adams, 1990; Marks and Vincent, 1992; Lovell, 1993; Hogan, 1994a; Rietchel and Fowler, 1995).

DIAGNOSIS OF OCCUPATIONAL CONTACT DERMATITIS

To the naked eye, irritant and allergic contact dermatitis are virtually indistinguishable. In acute stages the clinical signs of contact dermatitis include erythema, papules, vesicles, and exudation. The affected person may experience itching or a burning sensation. In chronic cases, as a result of hyper-proliferation in the epidermis from chemical injury, fissuring, scaling and lichenification (thickening) develop (Leung et al., 1997).

Patch testing and a thorough history are currently the best tools for distinguishing irritant from allergic causes. Difficulties can arise when the patient does not remember or does not know the composition of a complex product. Compounds that form in-situ after the primary ingredients are mixed, and are often not identified by material safety data sheets. The MSDS identification of a compound when the concentration is less than 1 percent (10,000 ppm) is not required in the U.S. or Europe. However, there are many sensitizers that will illicit a response at or below this concentration (deGroot, 1994).

The diagnosis of occupational allergic contact dermatitis could be made if there was a history of previous work exposure, and if there was a positive skin patch test result to that compound. Conversely, a diagnosis of irritant contact dermatitis is made in the absence of a patch test response when the test concentration is regarded to be non-irritating at the concentration applied. Complications arise when the chemical being tested is applied at a concentration that is irritating to some individuals. This is most likely to occur when the chemical is a weak or moderate sensitizer and higher concentrations are needed in order to avoid false negative results (Rietschel and Fowler, 1995). An ideal patch test concentration for allergen response will be low enough not to cause irritation, but concentrated enough to illicit a response in truly sensitized persons.

It is very difficult to distinguish between an allergic and irritant response on a skin patch test, although the allergic response typically tends to itch more, and has the general appearance of having raised, palpable, vesicular surfaces with borders that spread beyond the patch test site; the spread is not typically observed with irritation (Rycroft, 1996). Irritant responses have glazed-looking surfaces with sharp borders determined by the patch dimensions. It is important to apply the test concentrations to normal skin, and typically the human back fits this requirement. A reaction to an irritant patch test tends to develop quickly and disappear quickly (within 7 days); whereas allergic patch test reactions often develop after 48 hours and become more intense with time (Tucker and Key, 1992). A practical overview of patch testing and the technical and ethical problems associated with it is covered by Storrs (1996).

Recent research suggests that cytokine profiles might be used to distinguish between irritant and allergic contact dermatitis, but the utility of this approach has not yet been proven. This idea represents a potentially important area for future research (Enk and Katz, 1992; Paludan and Thestrup-Pederson, 1992).

Sometimes both irritants and allergens are involved in the dermatoses, since the damage achieved by irritants to the skin can facilitate the passage of sensitizing compounds through the skin (Angelini et al., 1996). Although irritant contact dermatitis is usually far more prevalent in the workplace, persons with allergic contact dermatitis are more likely to seek medical treatment because the symptoms are generally more severe and more persistent (Meding, 1989; Lantinga, 1984).

Skin testing can also be used to diagnose contact urticaria. The test compound may be applied directly, without covering, usually to dermatitic skin, and the response can be read in 40 to 45 minutes. A comprehensive list of industrial chemicals that have been identified to cause allergic contact urticaria are available in Rietschel and Fowler (1995).

Skin patch testing should only be performed by a qualified dermatologist. The patch test protocol typically consists of applying a number of compounds, at a concentration that is generally non-irritating, under occlusion, for a period of 48 to 96 hours. After that period, the test patch is removed and visually graded for signs of erythema and vesicles (Rietschel and Fowler, 1995). The test site is read at 48 hours and again at 96 hours after initial application. Usually, the response in allergic contact dermatitis becomes evident 24 to 48 hours after exposure. Distinguishing between allergic and irritant dermatitis, and contact urticaria is important. The diagnosis will influence the degree of exposure avoidance necessary (prevention), the treatment, and long-term prognosis.

PREVALENCE AND INCIDENCE OF WORKPLACE DERMATOSES

In epidemiologic reports of disease cases, the terms prevalence and incidence are often used. These are both expressed as rates. Prevalence is the number of existing cases during a defined time period per number of persons. Incidence indicates the number of new cases per defined time period and number of persons. Because prevalence includes all existing cases, including ones that may have first become apparent years earlier and continue to persist, this number is usually appreciably higher than the incidence rate.

Statistics on the incidence of occupational skin disease in the U.S. are derived from annual surveys conducted by the Bureau of Labor Statistics (BLS). The statistics indicate a general decline from the early 1970s until 1990. However, it has been suggested that this may not be a true decline, but reflects reporting changes where skin disease is treated as first-aid rather than as a lost-time incident (Cooley and Nethercott, 1994). After 1990, the rate of new cases increased slightly, and more recently appears to remain stable (see Figure 3-3).

Dermatoses caused or aggravated by work have been estimated to be under-reported by between 10 and 50-fold (Discher et al.,1975; Mathias, 1985). In the early days of required OSHA reporting of occupational illness, there was evidence of significant under-reporting of dermatitis.

The Discher study, commissioned by NIOSH and completed in 1975, indicated that only 1 out of 13 identified dermatitic conditions due to work were entered into the OSHA-200 log (Discher et al.,1975). Employers were most likely to record when compensation claims had been filed or acute conditions occurred, and claims where a date of onset was known. Among 600 manufacturing industry workers that were examined, occupational dermatitis had affected 8.3%, which increased to 13.3% in those workplaces where only marginal or inadequate controls were in place.

Even with recognized under-reporting, occupational dermatitis has been a major cause of illness and lost work days ever since records first began being collected. Some estimates of the true prevalence of occupational skin disease cases put the figure in the U.S. between 0.5 to 2.9 million cases per year (Lushniak, 1995). The 1988 National Health Interview Survey, Occupational Health Supplement, interviewed 30,074 persons nationwide (Nat. Ctr. Health Stat., 1989; Behrens et al., 1994). A requirement for recording dermatitis was that it occurred for at least three consecutive days during the past year. Dermatitis prevalence was found among 11.2% of

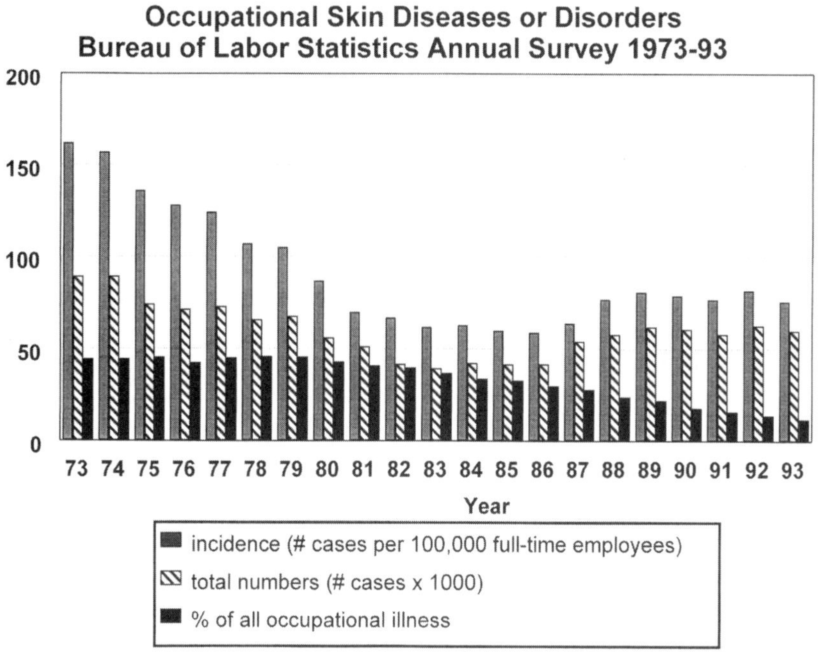

Figure 3-3. Incidence of new reported cases of dermatitis in the United States per year. Data from the U.S. Bureau of Labor Statistics.

the survey, while the prevalence of OCD was indicated in 1.7%, or an estimated 1.87 million cases nationally. This is equivalent to 1700 cases per 100,000 workers. This self-reported prevalence rate is in sharp contrast to the 76/100,000 cases of occupational skin disease or disorders reported to the U.S. Bureau of Labor Statistics in 1993. Interestingly, only 16/100,000 workers' cases were severe enough to result in one or more days away from work (Burnett et al., 1998).

In 1984, the National Labour Inspection Service (NLIS) in Denmark established the Register of Occupational Diseases. In Denmark, all doctors and dentists are required to notify the NLIS of all work-related diseases, even suspected cases. During the period 1979-1989, 17,746 cases of skin disease were reported, which ranked third behind musculoskeletal disorders and hearing damage. This is equivalent to an annual incidence rate of about 85/100,000 workers. Respiratory diseases, by comparison, excluding allergies, contributed 8,107 cases, with an additional 4,600 cases attributed to respiratory allergies (Haklier-Sorensen, 1996a, 1996b).

In Finland, the Act on the Supervision of Labour Protection of 1974 has obligated doctors to report every case of occupational disease. In 1993, 16% of the occupational diseases were attributed to dermatoses, ranking fourth behind musculoskeletal diseases, hearing loss, and diseases caused by asbestos. Eighty-four percent of all occupational skin disease was classified as contact dermatitis. Irritant dermatitis constituted 50%, allergic contact dermatitis 36%, and contact urticaria 14% of these cases. The most common causes of irritant contact dermatitis were detergents (33%), wet work (10%), oils, greases, cutting fluids, organic solvents, and dirty work (18%), and foods (8%). The most common causes of allergic contact dermatitis were rubber chemicals (26%), synthetic resins, plastics, glues, and paints (19%), and metals (16%). The source of the rubber chemical allergies were most often attributed to protective gloves. On the other hand, no cases of occupational allergic dermatoses were detected in the rubber industry during 1993 (Kanerva et al., 1995).

Some industries exhibit a greater risk of contact dermatitis than others. For instance, the agricultural workforce, which represents 1% of the private sector workforce, had 4% of the cases, while 65% of occupational skin disease cases were found in manufacturing, where 30% of the workforce is employed (Wang, 1978). An incidence of 12.6 % of occupational skin disease was found in one factory using phenol/formaldehyde resins in Sweden (Bruze and Almgren, 1988). Health-screening for occupational skin disease in construction workers indicated a current prevalence of 16%, although only 8% was attributed to occupation (Wahlberg, 1969). A questionnaire survey and follow-up exam of painting trade workers in Sweden indicated that 16.7% currently had skin disease, while 30% reported current or past skin disease. Upon detailed examination, 34% were classified as being from occupational causes, 22% were doubtful, and 44% were not occupationally related (Hogsberg and Wahlberg, 1980). The prevalence of irritant skin changes in third year hairdresser apprentices was 55% in one study in Germany (Wolfgang et al., 1998), while 83% of professional hairdressers in a city in Taiwan had contracted occupational dermatitis (Goh, 1994). Table 3-3 lists the major industries in the U.S. with the highest incidence of occupational dermatitis. More detailed analyses of the occurrence of occupational dermatitis in other countries can be found elsewhere (Smit et al., 1993; Smit et al., 1993a; Smit et al., 1995; Kanerva et al., 1995).

Although there is a reasonable concern that work-related skin conditions are under-diagnosed, there is also danger of possible over-diagnosis. This stems from the fact that only about 65% of all skin disease seen among workers is attributed to occupation (HSE, 1993). One critical study of 250 workers from 14 different industries indicated that only 51% of study participants had skin findings consistent with workplace exposure (Plotnick, 1990). Thus, non-work causes could confound the proper attribution of cause and effect in regard to bona-fide occupational dermatitis.

The prognosis for contact dermatitis in chronic cases is surprisingly poor (Burrows, 1972). This seems true for both allergic or irritant dermatitis, as neither seems to clearly have a better prognosis (Hogan et al., 1990; Fitzgerald and English, 1995; Cooley and Nethercott, 1994). Fregert (1975) reported that only one quarter of the patients with hand eczema were symptom-free two to three years after diagnosis. In another study of chemical workers, recurring symptoms of contact dermatitis varied between 35% and 80%, depending on the severity of the symptoms, the period of follow-up, and the intensity of exposure (Williamson, 1967). Driessen et al. (1982) prospectively studied eczema patients for four to seven years after they had initially seen a dermatologist, and found that 56% of the patients with irritant contact dermatitis

Table 3-3
Occupations in U.S. with occupational skin diseases or disorders, numbers and incidence rates by major industry

Major Industry	Total No. Cases	Incidence (Cases per 100,000 Full-Time Employees)
Agriculture/forestry/fish	3600	345
Manufacturing	31,800	179
Services	12,600	60
Construction	2200	56
Transportation/utilities	2800	53
Mining	200	29
Wholesale/retail trade	4400	23
Finance/insurance/realty	600	10
Total	58,200	77

Data from U.S. DOL, 1993.

were cured, versus 37% of the patients with allergic contact dermatitis. Wall and Gebauer (1991) noted that over one quarter of workers who changed jobs because of occupational dermatitis chose occupations in which the new work environment further added to their occupational skin disease. On the other hand, Rosen and Freeman (1993) reported an improvement in occupational contact dermatitis in patients who were able to change their working pattern, and an even greater improvement in those who left the original industry altogether. Others have not detected such an improvement in outcome (Keczkes et al., 1983). The possible reasons for the poor prognosis of occupational contact dermatitis have been succinctly summarized by Hogan (1994) and Birmingham (1986). These include misdiagnosis, continued exposure, misuse of topical medications, atopy, chronicity of condition, insufficient advice to patients, non-dermatological factors, poor wash facilities, improper cleaning agents, cross-sensitivity, and multifactorial causes.

Irritant contact dermatitis is more prevalent in occupations involving wet work such as cleaning, hairdressing, nursing and health care, and food handling. Water itself is damaging to the skin if it is in contact with the skin for prolonged periods. Furthermore, resistance to chemical irritants and physical damage is reduced. This is probably reflected in the high prevalence of skin problems in such groups as professional cleaners (Nielsen, 1996), and among those who wash their hands frequently with soap and water, such as food handlers and health care workers (Mathias, 1986; Grunewald et al., 1995).

Individual risk factors that have been determined to be most likely to be associated with contact dermatitis are a history of atopy (especially atopic eczema) and dryness of skin (Smit et al., 1994; Leung, 1995). Atopic individuals have a 13-fold increased susceptibility to irritants, probably because of the poorer integrity of their stratum corneum (Shmunes and Keil, 1983). Surprisingly, and not completely understood, but believed to be related to a T-lymphocyte deficit, is that atopy has a protective effect on the risk of allergic contact dermatitis (Smit and Coenraads, 1993; Rees et al., 1990). However, atopic individuals are more susceptible to immediate, contact urticaria than normal persons due to a significantly higher presentation of IgE on their B-lymphocytes (Ward et al., 1991). Dry skin, probably as a result of overuse of soaps and other exogenous factors, is indicative of mild damage to the skin. Damaged skin is more prone to assault from a number of external agents capable of causing dermatitis.

COST OF OCCUPATIONAL DERMATITIS

Costs associated with any type of occupational illness include lost or reduced productivity, medical diagnoses and treatment, administrative costs, and when the worker is unable to work due to the illness, workmen's compensation for lost wages.

Although allergic dermatitis accounts for about 20% of all occupational contact dermatitis, it is responsible for a disproportionate number of lost time cases. In one study, 37% of patients with allergic dermatitis took sick leave, versus 14% with irritant contact dermatitis (Meding and Swanbeck, 1990). The proportional distribution of the number of days taken as sick leave for dermatitis in one state are shown in the pie diagram in Figure 3-4.

An earlier study, using average lost workday duration and a wage of $8.25, calculated total lost productivity at $11 million annually. However, accounting for the under reporting and indirect costs of skin disease, the estimated total cost of dermatitis might be between $222 million and $1 billion annually (Mathias, 1985). A recent survey estimated that only between 9% and 45% of workers with all types of compensable conditions actually filed claims (Biddle et al., 1998). In 1985-1986, dermatitis in the U.S. ranked sixth for permanent partial disability compensation, following non-chemical injuries such as hearing loss and musculoskeletal damage (Leigh and Miller, 1998).

In Denmark, contact dermatitis accounted for 41% of all recognized cases of occupational disease between 1979-1989. Of the total cases of dermatitis, 64% resulted in compensation for permanent injury and 11% resulted in compensation for loss of earning capacity. Nearly 17% of the total payments for compensation were for skin diseases. Compensation paid by the insurance companies was 32 million Danish Crowns (D.C.) or about $4,726,735, which was followed by respiratory diseases and musculoskeletal disorders. The average payment for com-

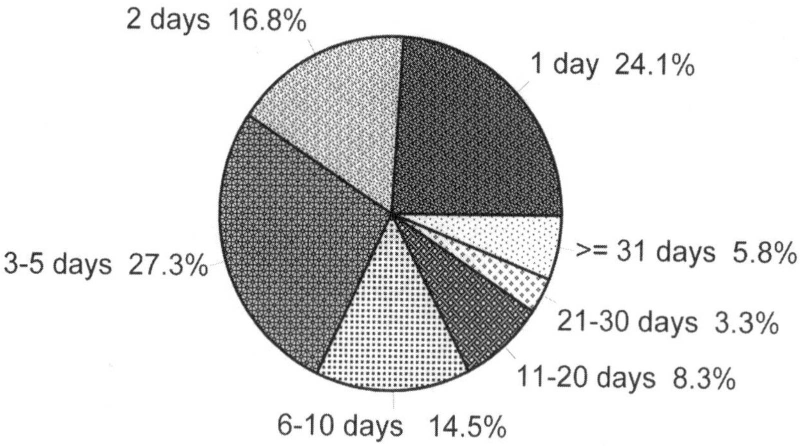

Figure 3-4. Days away from work per case for skin diseases reported to the U.S. Bureau of Labor Statistics, 1993. The total number of cases reported was 12,613.

pensation for permanent injury due to skin disease was 26,000 D.C. ($3,840) and for compensation of loss of earning capacity, 345,000 D.C. ($50,960) (Halkier-Sorensen, 1996a; 1996b).

There seems to be a direct association between higher medical and compensation costs and delayed referral to physicians (Shmunes and Keil, 1983; Gallant, 1986). Not only is the medical prognosis poor, but chronic dermatitis can have severe detrimental social and economic consequences for the affected worker (Breit and Turk, 1976). Each of these reasons points to the importance of early diagnosis and intervention.

Additional costs result when a change of occupation results from disease associated with specific types of work. Data from the former Federal Republic of Germany show that there are workers in many occupations that change professions because of occupational dermatitis (see Figure 3-5). Dermatitis in many occupations far exceeds the number of cases of workers who are forced to change occupations because of respiratory problems (Fed. Rep. Ger., 1990). This change of occupation results in the loss of skilled labor in one profession, and requires time, reduced wages, and training expenses while mastering new skills when new employment is obtained in another occupation.

CAUSES OF ALLERGIC AND IRRITANT DERMATITIS

About 300 potential sensitizers of occupational relevance have been identified and approximately 3,000 chemicals are known to act as possible contact allergens (Menne and Nilesen, 1994). de Groot (1994) provides suggested concentrations for skin patch testing on 3,700 chemicals, although some of these may not be true allergens because they are irritants at low concen-

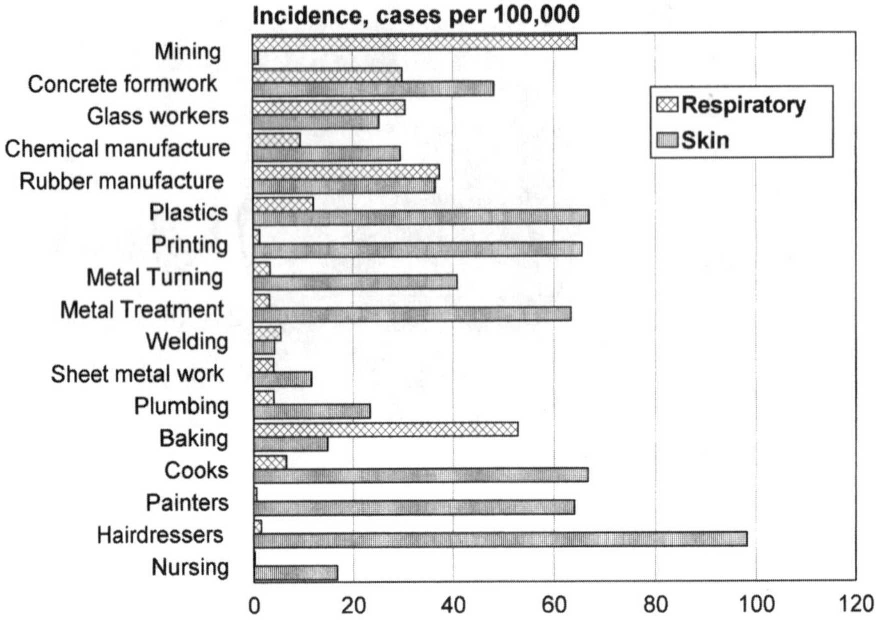

Figure 3-5. Incidence of reported occupational illness in former Federal Republic of Germany requiring change or loss of occupation. Only cases for dermatitis and respiratory disease are reported to compare the relative occurrence of these two classes of disease. Adapted from Berufskankheitenrisiken (Bundesgestaltfur Arbeitsschutzund Arbeitforschung, Federal Republic of Germany (1990)).

trations. The NIOSH Registry of Toxic Effects of Chemical Substances (RTECS) identifies 2,650 chemicals with the potential to irritate the skin (SilverPlatter, 1997).

In Finland, during 1991, the eight most common types of chemicals responsible for allergic contact dermatitis were rubber chemicals, nickel, epoxy resins, formaldehyde, thiuram sulfides, chromate (six valence), isothiazolinones, and colophony. It is likely that the principal contact with rubber chemicals, thiuram sulfides, and isothiazolinones occurred from wearing chemical protective gloves. Sensitization to nickel was reported often for hairdressers who work with stainless-steel scissors, but it is believed that the initial sensitization may occur from wearing inexpensive nickel-plated jewelry. Epoxy resins are encountered in industry, as well as colophony, which constitutes the flux in many solders. Formaldehyde is used as a preservative, and is found in cleaning products and in the textile industry. It is also a principal component in most embalming fluids (Kanerva, et al., 1994).

The most common cause by far of the occupational cases of contact dermatitis, reported in Denmark between 1984 and 1991, was water and detergents (25%). Other significant single causes of contact dermatitis were nickel compounds, hand cleansers (soap), solvents, rubber, metal (unspecified), cutting oils, gloves, and self-copying (carbonless) paper, accounting for the exposure in over 50% of the reported cases (Halkier-Sorensen et al., 1995). It is interesting that hand cleansers and gloves are near the top of the list when paradoxically the intent of these products is either to protect the skin of the employee, or to protect the health of the recipient of the product or service (e.g., health care) that is provided. Because some workers wash up to 100 or more times per day in some occupations, it is understandable how skin damage might occur (Rustemeyer and Frosch, 1996).

The most common occupations in Finland to experience cutaneous urticaria in 1991 were farmers, bakers, domestic animal husbandry, food preparation, and nurses (Kanerva et al., 1994). In each of these occupations, the allergen is principally proteinaceous. These allergens typically have molecular weights that are in excess of 10 kilodaltons. Currently under investigation is how these HMW proteins penetrate the skin, although if the skin is damaged penetration certainly seems plausible. Natural rubber latex (NRL) protein allergy resulting from glove use has grown significantly among nurses and other health care professionals due to concerns about HIV and hepatitis, and NRL is a significant cause of contact urticaria and systemic sensitization (Turjanmaa, 1996).

INFECTIOUS DISEASES

Occupational activities may potentially expose workers to infectious agents, as might occur in occupations such as solid waste sanitation, livestock rearing, sewage treatment, and the health professions. The skin, especially if damaged by physical or chemical exposure, may be a target site for such infections. In England and Wales, the incidence of new cases of occupational dermatitis associated with infective agents as diagnosed by dermatologists and occupational physicians was 1.6% and 8.4%, respectively (Health and Safety Commission, 1997). In the U.S., approximately 19% of all reported cases of occupational disease were attributed to infections of the skin and subcutaneous tissue (Burnett et al., 1998). One might also predict that damaged skin is more likely to present a portal of entry for systemic infectious disease agents. In general, the literature supports this belief (Abrams and Warr, 1951; Levin and Behrman, 1938; McCulloch, 1962; Watt, 1987; Nield, 1990), and the skin has been suggested as a potential portal of entry for specific infectious agents including viral hepatitis, bacterial infections such as tuberculosis, anthrax, and brucellosis, as well as fungal infections (Gantz, 1995; Veraldi et al., 1992; Meneghini, 1982; Adams, 1990; Ancona, 1990), as examples. Thus, a worker with dermatitis or mechanically damaged skin might be at much greater risk of infection than a worker with healthy, intact skin. Such workers should be adequately protected or removed from such activities.

Management and Prevention of Occupational Dermatitis

The techniques for minimizing occupational dermatitis are very similar to minimizing any other occupational hazard. They include a thorough audit of products and processes in the workplace that might cause or aggravate existing dermatitis, and an on-going medical surveillance program that can be related directly to job activities in which affected cases are involved. Often a worker afflicted by dermatitis first visits a personal or company medical doctor who might then refer the patient to a dermatologist. It is usually rare for the dermatologist to seek information from or provide information to the hygienist at the workplace. Thus, important information needed to find the cause and solution to the problem are often not shared. If through the medical surveillance program, it appears that there might be a problem with dermatitis, seeking professional, competent assistance may be the most expedient means of finding solutions.

Studies have found direct associations between the duration of dermatitis prior to treatment and a reduction in the favorable long-term prognosis. Because dermatitis is not generally life-threatening, affected workers often tend to "tough it out." In one survey of workers with allergic reactions to epoxy resins, it was found that almost 30% had never consulted a physician about that condition (Nixon and Frowen, 1991). In one chemical manufacturing site where over 1,000 employees worked, additional training in prevention and early reporting of skin conditions resulted in a new dermatitis rate that was only 39% of the former incidence (Heron, 1997). In conclusion, cost savings and a better prognosis of dermatitis might be achieved by encouraging employees to seek early medical consultation, and/or by implementing periodical medical screening that includes dermatological exams (Halbert et al.,1992; Cooley and Nethercott, 1994).

When considering potential new hires for work that might potentially expose them to contaminants that could affect the skin, prior history of dermatitis alone cannot be used as a reason for not hiring. Provisions of the Americans with Disabilities Act of 1990, which came into full effect on July 26, 1992, places a general duty on the employer to allow a prospective worker employment in a job provided he or she can perform the "essential components" of the job without putting other workers at risk or placing him or herself at "material risk of harm." In addition, the law requires that the employer make "reasonable accommodations" in the work tasks and conditions to allow a prospective employee to perform the essential components of the job. This might include provision of simple personal protective equipment. Neither the doctor nor the employer has the right to deny a person a job unless they are able to quantify the risk and establish that the job may result in material harm. Furthermore, an employer is no longer able to require a medical examination of a prospective employee prior to offering employment, with the exception of a drug test. The practical outcome of this would be to strive to provide a safe work environment even for more susceptible workers, or depend on the employee's determination to seek employment elsewhere where it is relatively free of such risks (Nethercott, 1994).

Protecting and restoring the barrier function of the skin are crucial measures in the prevention and management of contact dermatitis. Reducing the causes of contact dermatitis by minimizing contact with offending agents and conditions is the obvious solution. This can be performed through safer chemical substitutions, automation and closed systems, using milder soaps for cleaning the skin, not washing with hot water, keeping the skin as dry as possible, avoiding prolonged skin occlusion (e.g., under gloves), protecting the skin from mechanical and chemical exposure, and using hypo-allergenic gloves when needed. The prevalence rate of work-related skin rashes was 3.4% among solderers when they typically wore cotton gloves, compared to 15.4% when they did not, demonstrating the efficacy of this simple precaution (Koh, 1994). To date, the efficacy of barrier creams is insufficiently documented (Wahlberg, 1986).

Skin emollients applied to skin may help restore the barrier function of the skin, but studies have detected marked differences in the efficacy of various creams (Halkier-Sorensen, 1996c). Lotions applied to contaminated skin may possibly facilitate percutaneous bioavailability (Davies et al., 1991). A possible deterrent to the use of skin emollients at work is that they must be compatible with the work operations. In the U.S., for instance, food handlers cannot apply skin lotions or creams that are not approved by the Food and Drug Administration. If wearing gloves, some petrolatum or oil-based emollients may help transfer allergens from the gloves to the skin or even degrade the glove (Beezhold et al., 1994; Baur et al., 1998). It is, therefore, important that the efficacy of skin care products is evaluated under conditions of use before being recommended to workers.

Unfortunately, a common mistake made by occupational health management is to over prescribe the use of gloves when intending to protect against skin exposures. As alluded to previously, there is a significant association between the use of gloves as the cause or exacerbation of allergic OCD. Chronic use of gloves can certainly damage the skin through occlusion and by irritation caused by the friction from the glove and glove powders (see Occlusion below and Boeniger, 2002; Taylor, 1994; Wrangsjo et al, 1994; Burke et al., 1995). Use of gloves should be limited when possible to short durations, combined with good work practices and frequent changing.

Self-treatment using topical anesthetics, antibiotics, and antihistamines should be avoided in the treatment of irritant dermatitis, as the skin is more likely to become sensitized to potential allergens in these products. Allergic contact dermatitis to topical corticosteroids and antibiotics, such as neomycin and thiomersal, may occur (Hogan et al., 1990; Wilkinson and English, 1992).

Skin as a Route of Systemic Exposure

The human skin is exquisitely well suited to perform the function that evolutionary demands have required of it. Those main requirements were to help retain internal water, exclude external water soluble compounds, and regulate body temperature through the release of eccrine sweat. Along with the development of our modern industrialized society, the skin is often exposed to a wide variety of chemicals to which it did not evolve to come into contact with.

Molecular diffusion through healthy skin rapidly diminishes for compounds with a molecular weight above 500, but much larger molecules and particles might penetrate the stratum corneum through physical channels (e.g., sebaceous glands, hair follicles). Penetration and diffusion through the appendageal route is also referred to as "shunt" diffusion. Physically or chemically damaged skin might also offer physical pathways of least resistance.

Hair follicles and the associated sebaceous gland (pilosebaceous unit) as well as eccrine sweat pores, generally constitute less than a 0.1% to 0.2% cross-sectional area of the human skin, and for most chemicals that can normally permeate the intact skin, this additional area is insignificant. However, the hair follicle is an invagination of the epidermis extending deep into the dermis, providing a much greater actual area for potential absorption below the skin surface. Penetration through these portals can be considerable for chemicals that do not permeate intact skin, such as larger molecules and even small particles. For instance, 7 μm particles were detected primarily deep within the hair follicles, while 3 μm particles were detected in both the intact stratum corneum and hair follicles (Lauer et al., 1995). Previous studies on particle fate suggest that particles up to 25 μm may penetrate deep into the stratum corneum where they might later dissolve (Schaefer et al., 1982). Most persons consider only liquid chemicals to be available for skin absorption. However, solid materials may become readily mobile by dissolution into the surface sweat and lipids present on the skin, or by transfer from contaminated clothing (Quan, 1994; Wester and Maibach, 1996). Surface dissolution readily occurs, even for dry powders in contact with dry synthetic gloves (Fricker and Hardy, 1994).

Dermal absorption of chemicals can appreciably increase the overall body burden. The cumulative systemic dose will depend on the amount of skin area exposed, duration of exposure, and absorption rate of the chemical exposure. Figure 3-6 shows the relative dermal contribution to systemic dose from limited skin contact with neat chemical in comparison to an 8-hour inhalation exposure at the TLV®. It is clear that even a small area, exposed for intermittent periods, over as little as 15 minutes out of the work day, can equal the systemic dose from a high inhalation concentration.

Systemic absorption of organic chemicals can contribute to a wide variety of adverse health effects. The literature is replete with recorded instances of acute episodes of poisoning from dermal exposures. Table 3-4 lists some agents responsible for acute poisonings with references for further reading. In addition to workplace exposures, acute and subacute toxicity resulting from the topical application of pharmaceutical and cosmetic products have been documented (Freeman and Maibach, 1992). Not reported as frequently are illnesses and functional impairment resulting from chronic exposure by topical contact. A few examples are listed in Table 3-5. Such exposures may be more common than acute episodes, but cause and effect is often far less certain. For example, skin exposure to solvents is often likely, but its contribution to chronic diseases, such as neurological and neuropsychiatric disorders, are unknown (Bos et al., 1991; Hogstedt, 1994; O'Donaghue, 1985; Hanninen et al., 1976; Spencer et al., 1987). Several compounds have been associated with reproductive injury, such as acrylamide, carbon disulfide, dibromochloropropane, glycol ethers, and inorganic mercury. For these compounds, percutaneous absorption is believed to be the principal means of exposure (Tyl et al., 1993).

Metallic compounds in aqueous solutions are less well absorbed, due principally to ionization, which tends to make chemicals less skin permeable than organic compounds (see Volume 1, page 141). Two generalizations appear to be correct: the metal must be water soluble to be ionized, and cations seem more permeable than anions. The mean absorption is less than 1% in five hours when metal compounds such as cobalt chloride, zinc chloride, and silver nitrate

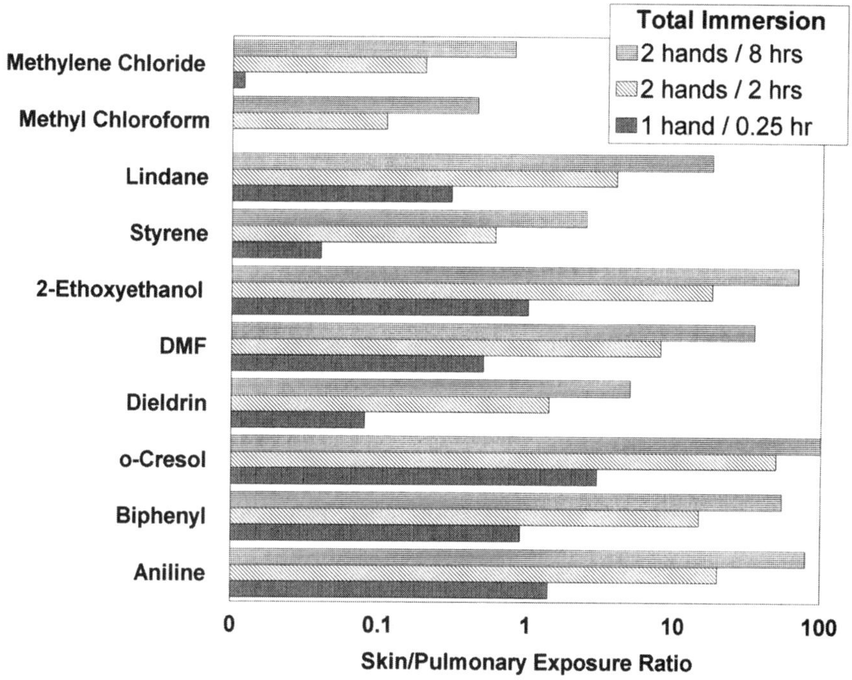

Figure 3-6. Relative absorption of some chemicals from exposure to the hands or by inhalation to TLV® air concentration for 8 hours. Legend indicates relative proportion compared to how much skin area and time the skin would be in contact. Data adapted from Droz-PO, et al., 1990.

are applied to the skin. For mercuric chloride, potassium mercuric iodide, and methyl mercury dicyanidiamide, the absorption increases with higher concentrations between 3.2% and 4.5% in five hours (Subcommittee on the Toxicology of Metals, 1971). Matrix effects appear important. Whether the metallic salt is applied in water or in soil appears to dramatically affect absorption. In *in vitro* experiments with mouse skin, 62% of sodium arsenate was absorbed from water, while <0.3% of the applied dose was absorbed when applied in soil (Rahman et al., 1994). Thus, soil binding appears substantial for some metallic compounds.

Covalently bound organo-metallic compounds are far more permeable since they do not ionize (Siegers and Sullivan, 1992). They also have profound detrimental effects on the central nervous system. The tragic death in 1997 of a Dartmouth College professor conducting environmental toxicity testing was believed to be the result of a single drop, perhaps two, of dimethyl mercury on her permeable latex glove (Lewis, 1997). This is a dramatic example of the potency of some organo-metallic compounds. Organo-metallic compounds, and inorganic metallic cations, which are water soluble or readily dissociate in water, are usually good permeants. Lead acetate and lead nitrate are examples of good permeants (Lilley, S.G. et al., 1988; Stauber

Table 3-4
Some examples of acute and subacute poisoning from transcutaneous exposure

Compound	Industry	Severity	Reference
acrylamide	chemical manufacturing	fatigue, peripheral neuropathy, muscle weakness	He et al., 1989
arsenic trichloride	chemical manufacturing	death	Delepine, 1923
2,4-dichlorophenol	manufacturing	death	Kintz et al., 1992
dinitrocresol	agricultural spraying	death	Herman et al., 1956
ethylene glycol monomethyl ether	textile printing	encephlopathy, bone marrow damage	Ohi and Wegman, 1978
ethylene chlorhydrin	industr. cleanup	death	Middleton, 1930
hydrofluoric acid	petrochemical refinery	ventricular fibrillation, pulmonary edema, death	Tepperman, 1980
hydrocyanic acid	chemical manufacturing	dizziness, weakness throbbing pulse	Drinker, 1932 Potter, 1950
methanol	equipment cleaning	vision impairment dizziness, dyspepsia	Downie, 1992
4,4'-methylene-dianiline	epoxy resin manufacturing	hepatitis	McGill and Motto, 1974
nicotine	tobacco harvesting	nausea, vomiting, dizziness, headache	Ghosh et al., 1991
paraquat	manufacturing and agriculture	pulmonary failure death	Smith, 1988
phenol	disposal of industrial waste	central nervous system toxicity, death	Soares and Tift, 1982
phenol-formaldehyde	manufacturing	hypertension respiratory distress renal impairment	Cohen et al., 1989
thallium	rodenticide	sensory changes	Glamme and Sjostrom, 1955
	manufacturing	parathesia, hair loss, neurological pain	
trichloroethylene	manufacturing	death after 11 mo.	Lockey, 1987

Table 3-5
Some examples of chronic poisoning from skin exposure

Compound	Occupation	Severity	Reference
nitroglycerin	explosives manufacturing	ischemic heart disease	Hogstedt and Axelson, 1977 Hodstedt and Stahl, 1980
triarylphosphate	mechanic, hydraulic fluid	polyneuropathy	Jarvholm et al., 1986
monomethyl ether	eyeglass frame factory	reversible hematological	Larese et al., 1992
methymethacrylate	dental technicians	neurotoxic	Rajaniemi, 1986 Rajaniemi et al., 1989 Sappalainen and Rajaniemi, 1984
polychlorinated	electrical utility	cancer	Loomis et al., 1997

et al., 1994). On the other hand, inorganic lead oxide, which does not dissociate, does not appear to traverse the skin according to Bress and Bidanset (1991), even after prolonged occlusion, although Florence et al. (1988) and Stauber et al. (1994) suggested that there is rapid absorption of both lead oxide and lead metal through the skin. Still, transcutaneous absorption of strontium chloride, with only 0.26% absorbed through intact skin in six hours, increased to 57.4% (200-fold increase) when the skin was abraded (Ilyin, 1975). Similar findings were reported for cobalt chloride, cerium chloride and cesium chloride radionucleotides (Inaba et al., 1979). Thus, the skin can play an important role as a route of exposure for almost any compound if the necessary conditions exist that will facilitate uptake.

Skin as a Contributor to Oral Ingestion

Not only can chemical contaminants be transferred to the skin, but skin contamination can be subsequently transferred to the mouth. It is generally well accepted that smoking and eating in the workplace should not be allowed because workers who have "dirty" hands might transfer this contamination to their cigarettes or food. How much contamination is of concern is a matter of the toxicity of the compound. For the purposes of illustration, lead is a good example.

Lead is a particularly dense compound. A drop of lead weighs about 570,000 μg (1/20th mL × 11.34 grams per mL × 1 million μg/gram). If only 1/1000th of a drop of lead were dispersed over the surface of two hands, this would be equivalent to about 570 μg.

The OSHA PEL for lead is 50 $\mu g/m^3$ in air. If a worker performing light work activity inhales 10 cubic meters of air during the work shift, 500 μg of lead would be inhaled. Hand wipe samples of battery manufacturing plant workers indicate lead contamination levels of up to 20,000 μg per pair of hands. After washing with soap and water, these levels may be reduced to an average 530 μg (Esswein et al., 1996). Even though these workers washed and apparently had clean hands, the amount of lead on the hands was equivalent to the full work shift inhalation exposure at the PEL. Hand contamination could be effectively transferred to food that is prepared or eaten by hand. Furthermore, surface contamination of the eating facility, that is invisible to the eye, could be transferred to recently washed hands and be a significant source of oral ingestion (Esswein et al., 1996).

For simplicity, this theoretical example did not take into account retention and bioavailability in the respiratory and oral routes. An analysis of this would suggest that this oversight may not be very significant. The penetration and retention of airborne particles into the respiratory tract is generally less than 20%, whereas the absorption of inhaled lead into the blood for the average adult was estimated to be only 40%, or about 8% total absorption (Drill et al., 1979). On the other hand, absorption of lead in the gut is between 8% and 10% (Tola et al., 1973; WHO, 1977). The net absorbed lead by either route are approximately the same.

Several workplace studies support the above calculation that personal hygiene can be important to the total contribution of workplace exposures. These studies have focused on exposures to toxic metals such as lead, cadmium, chromium, and arsenic, and also measured the internal biological levels of these contaminants. Using multiple regression analysis and observational or questionnaire data and personal hand contamination measurements, investigators have found that personal hygienic factors and working methods explain biological levels at least as well as do air monitoring data. In one instance, the correlation between exposure and internal dose at least doubled when information on the oral route was included (Lumens et al., 1994).

Eating and smoking with contaminated hands are viewed as poor work practices that might lead to ingestion of the contaminants. Studies on the amount of transfer of contaminants from hands to food is not available, but Wolfe et al. devised a protocol to determine the amount of contaminants transferred to cigarettes (Wolfe et al., 1975). In that study, parathion transferred from hands to cigarettes was calculated to be a potentially significant contributor to exposure. However, several assumptions were made that might over-estimate the amount inhaled, including that all volatilized compound was inhaled, none was thermally decomposed, and none would be trapped in the butt end or filter of the cigarette. If any of these assumptions were not true, less exposure would occur. Although extensive data are lacking, the reasoning for prohibiting eating and smoking with contaminated hands seems valid.

In one study, the average worker's hand contamination with cadmium was up to 1,200 µg/hand during the workday, and up to 300 µg/hand before lunch or before leaving the factory (e.g., after washing). These samples were collected simply by rinsing (no scrubbing) the hand with 500 mL of NaOH 0.1N, thus, this represents readily transferable material (Roels et al., 1982). In another study, the mean hand lead increased 33-fold from the pre-shift levels on Monday morning (33.5 µg/500 mL) to 1121 µg/500 mL on Thursday afternoon. Mouth lead contents were measured by rinsing the inside of the mouth with deionized water. Over the same period of time, these concentrations increased 16-fold (Far et al., 1993). These studies, and others, support stressing the impact of hygienic behavior and work practices as important to decreasing the uptake of toxic agents in the workplace. Furthermore, the potential contribution of this route of exposure can be evaluated by measuring the amount of contaminants on the skin (Askin and Volkmann, 1997; Ulenbelt et al., 1990).

A good rule of thumb is that if the occupational air concentration exposure criteria specify less than 10 mg/m^3 of air, the compound is fairly toxic, and the potential for biologically significant amounts of material to be present on the hands for transfer to the mouth exists. At that air concentration, and assuming that 100 mg/day is inhaled during light work, this mass would have a volume of roughly 100 µL. This volume is equivalent to about two drops of water, a mass that could easily be present on unwashed hands (Kissel, 1996). The amount of contaminant present on washed skin is dependent on many variables, and is best determined by objective measurements. However, in one experimental study it was concluded that the average adult (non-working) ingests 10 mg of soil and dust per day from non-food sources (Stanek et al., 1997). This estimate agrees with other experimental measurements of the amount of fine soil transferred from fingers (three fingers above the first knuckle) to the mouth during mouthing. It was found that the geometric mean amount was 11.6 mg, representing 16% of the amount originally on the skin surface (Kissel et al., 1998). Based on the above information, it would seem plausible that substantial amounts of material could be transferred to the mouth from contaminated hands, especially while eating.

Skin Route Contributing to Respiratory Sensitization

In experimental immunotoxicology, an efficient means of sensitizing animals is by topical or intradermal injection (Magnusson and Kligman, 1970). Intradermal injection is most effective for compounds that do not permeate the skin well, but *penetration* of topically applied compounds through damaged skin, as may occur in the workplace, is a real possibility. Egg albumen, for example, when injected intradermally, is very effective as a means of creating systemic humoral sensitization (Arakawa et al., 1995). It is not known to what extent, if any, dermal contact with egg protein has in contributing to the high prevalence of occupational asthma often seen in the egg breaking industry, which has not been studied (Smith et al., 1990).

Compounds that can permeate the skin may cause systemic sensitization as well. Isocyanates are a case in point.

Isocyanatc compounds are clearly capable of causing dermal sensitization when applied topically to various animal species. Most of the major isocyanate compounds have been shown to possess this capability, including 2,4- and 2,6-toluene diisocyanate (TDI) (Duprat et al., 1956;) 4,4'-diphenylmethane diisocyanate (MDI) (Tanaka, 1987), isophorone diisocyanate (IPDI) (Stern, 1989), and dicyclohexylmethane-4,4'-diisocyanate (HMDI) (Stadler and Karol, 1985). Dermal sensitization to these compounds in animals parallels reported cases of human dermal sensitization to HMDI (Emmett, 1976; Malten, 1977), TDI (Pham, 1978; Huang, 1991), MDI (Linden, 1980), and 1,6-hexamethylene diisocyanate (HDI) (Wilkinson, 1991).

It would seem illogical to believe that individual organs could be isolated from interaction, and that effecting chemical sensitization in one organ would have no effect on another. In fact, there is increasing evidence that compounds that enter the skin as allergens may be *more* likely to induce respiratory sensitization than compounds entering only through the respiratory system (Kimber, 1996).

Isocyanates are a good example of chemicals possessing the potential for topical exposure to cause respiratory sensitization. There are at least four instances where this has been shown in animal species (Karol et al.,1981; Erjefalt and Persson, 1992; Rattray et al., 1994 Bickis, 1994). One of the most dramatic of these studies is summarized below.

Rattray et al. (1994) tested the influence of the route of exposure on the development of respiratory sensitization in guinea pigs. The test animals received either an epicutaneous, intradermal, or inhalation exposure to MDI, and were challenged with various airborne concentrations of MDI 21 days post exposure. The epicutaneous exposure consisted of a single topical application of either 10%, 30%, or 100% MDI solutions to the shaved scapular region of the guinea pigs. The application sites were occluded for 6 hours. Twenty-five percent of the animals exposed to the 10% MDI solution (2 of 8) and 30% MDI solution (2 of 8) displayed respiratory sensitization upon inhalation challenge; this effect was also observed in 3 of 7 animals in the 100% MDI solution exposure group. IgG_1 anti-hapten antibodies were detected in 5 of 8 and 7 of 8 guinea pigs dermally exposed to the 30% and 100% MDI solutions, respectively. The investigators were unable to sensitize animals to MDI via the inhalation exposure route. In addition, an epicutaneous MDI challenge exposure 22 days following the initial MDI treatment induced dermatologic reactions (redness and swelling) in 71% (17 of 24) of the animals in the 10%, 30%, and 100% exposure groups. These data suggest that dermal exposures are important in MDI-induced respiratory sensitization, and that these exposures may be more effective than inhalation exposures in inducing sensitization.

To date, the only evidence of respiratory sensitization following skin contact in humans consist of anecdotal cases of skin splashes followed by respiratory sensitization, known skin contact with possibly incompletely cured polyurethane products, and workplace studies where respiratory symptoms are prevalent without existence of measurable air concentrations (NIOSH, 1994; NIOSH, 1996; DOW, 1996; Nemert and Lenaerts, 1993; Petsonk et al., 2000). Recently the ACGIH® TLV® Committee began assigning unique notations for sensitization to compounds, including chemicals capable of causing sensitization by skin or inhalation exposures (ACGIH®, 1999). Historically, the ACGIH® criteria applying to the skin notation purposely excluded such compounds, limiting it only to chemicals capable of causing systemic toxicity. Also, in Germany, compounds that are known to sensitize either the skin or respiratory tract are provided unique notations (MAK and BAT Values, 1997).

SKIN ANATOMY AND PHYSIOLOGY

In Volume 1, page 138, a basic description of the skin anatomy and physiology is presented. For review, the stratum corneum is the uppermost layer exposed to the outside world. It is composed of flattened denucleated cells called corneocytes, which contain mainly highly cross-linked fibrous keratin proteins. The corneocytes are non-respiring cells that are tightly connected, forming a physically rugged membrane. Each corneocyte is typically 30-40 μm (micrometers) in diameter and only 0.2 to 0.5 μm thick. These are randomly stacked 15 to 25-layers thick over most human skin surfaces (Flynn and Stewart, 1988).

The thickness of the stratum corneum varies by anatomical location on the human body. Table 3-6 shows the average skin thickness for several locations. Overall, the stratum corneum is surprisingly thin. Most of the body surface stratum corneum is only 10 to 16 μm thick, with limited areas of increased thickening related to the need for abrasion resistance, such as on the hands and feet. To help put the thickness of the stratum corneum in perspective, the average human hair has a diameter of about 50 to 70 μm. Polyester adhesive tape (e.g., 3M Scotch Tape®) is about 25 μm thick. The thinnest synthetic glove is about 7 μm thick.

The intercellular space between the corneocytes consists of well organized lipophilic and hydrophilic domains that represent channels of least resistance to diffusion of either water insoluble (lipophilic) or water soluble (hydrophilic) compounds, respectively (Elias et al., 1977; Swartzendruber et al., 1989; Forslind, 1994). About 10% of the stratum corneum mass consists of lipids. The lipid domain constitutes an important, but tortuous pathway for molecular transport through the stratum corneum. A simplified schematic of the stratum corneum construction has been referred to as a brick and mortar model (Michaels et al., 1975). The surface of the skin also naturally accumulates some of these intercellular lipids, as well as sebaceous gland oils, making the stratum corneum generally water resistant.

It is clear that the stratum corneum provides the main barrier protection to the skin. Because chemicals permeating the skin must traverse the intercellular spaces between the flattened corneocytes, the permeability will be reduced by about 1,000 times relative to a pure lipid phase (Michaels et al., 1975). As such, the apparent effective thickness of the stratum corneum has been calculated to be about 500 to 750 μm (Potts and Francoeur, 1991; Potts and Guy, 1992). Another way to look at this is that it would require a homogeneous film at least 500 μm thick, impregnated with the same lipids as the skin, to provide equivalent protection from chemical permeation. Table 3-7 shows how physically removing the thin stratum corneum by cellophane

Table 3-6
Human stratum corneum thickness by anatomical site[1]

Skin Area	Stratum Corneum Thickness, mm
Abdomen	15
Volar Forearm	16
Back	10.5
Forehead	13
Scrotum	5
Back of Hand	49
Palm	400
Sole	600

[1]Adopted from Scheuplein and Blank, 1971

tape stripping increases the permeation of methanol and phenol through mouse skin (Behl et al.,1983).

The viable epidermis is beneath the stratum corneum. This tissue is considerably different physically from the stratum corneum, in that it is primarily aqueous. The germinal stratum granulosum, where new corneocytes are formed, is located within this layer. A complete replacement of the skin corneocytes that make up the stratum corneum occurs every 14 to 28 days in humans (Treherne, 1965). The epidermis provides an important and rather unique immune capability by providing both specific and nonspecific protection against pathogenic microorganisms and environmental antigens. The major cellular constituents are keratinocytes, Langerhan's cells, skin-infiltrating T-lymphocytes, and post-capillary venule endothelial cells. Regional lymph nodes link the skin with the systemic immune system (Leung et al., 1997). The viable epidermis is about 50 to 100 μm in thickness, depending on location.

The thickest layer of the skin is the dermis. It is a richly vascularized area. The appendageal structures, like the hair follicles, eccrine and sebaceous glands, originate in this layer. Within a cubic centimeter of human skin are 11 to 100 philosebacious glands with hair follicles and 100 to 400 eccrine glands. The total cross-sectional area of the appendages is probably 0.1% to 1% of the surface area of the skin, and the total volume available for percutaneous absorption

Table 3-7
Permeation as a function of tape stripping mouse skin, in vitro at 37°C

Number of Tape Strippings	Permeation Coefficient, cm/h	
	Methanol	Phenol
0	3	22
5	48	120
10	260	277
25	291	275
Dermis	395	301

Adapted from Behl, 1983

(excluding hair diameter) is only about one-tenth of that (Blank and Scheuplein, 1969). Although these appendageal structures represent less than 1/100th of the skin surface area, they may represent an important shunt through the stratum corneum for large hydrophobic compounds and particles (Lauer et al., 1995).

Physicochemical Characteristics Favoring Permeability

It has long been recognized that certain chemicals permeate the skin much more readily than others. Knowing the extent of dermal uptake can be used to predict the relative potential of absorption of chemicals in the workplace. This, coupled with the toxicity of the chemical once absorbed, can be used to estimate the potential hazard of exposure if appropriate steps of control are not taken. It is, therefore, desirable to have accurate skin permeation data.

It has been determined from extensive study of the experimental permeation data for humans that the two principal factors that determine a chemical's likelihood to permeate are its solubility and molecular size. Solubility is typically expressed as water solubility or as the octanol-water partition coefficient, the latter being a representation of the chemical's solubility in a non-polar phase versus a polar phase (Surber et al., 1990; Mannhold and Dross, 1996). The log of the P_{ow} is usually reported because of the wide numerical range of octanol-water, or P_{ow} values. Chemicals with a log P_{ow} between -0.5 and 3 usually permeate best, because the intercellular channels contain bipolar moieties. Chemicals that are very lipophilic will rapidly enter the lipophilic stratum corneum but not readily diffuse through the hydrophilic viable epidermis. Such compounds often remain in the stratum corneum for a considerable amount of time, eventually permeating through the viable epidermis or being sloughed off by normal cell desquamation. Similarly, very hydrophilic compounds are also typically poor penetrants of the lipophilic stratum corneum, possibly residing within the upper stratum corneum for a week (Bucks, 1993). In addition, some compounds appear to have a strong affinity for proteins, reversibly interacting with specific chemical sites, and P_{ow} is a strong predictor of protein binding affinity. Taken together, protein binding affinity and P_{ow} favor long residence. Some compounds have low solubility in both water and oil, which further preclude transcutaneous penetration. Finally, ionization of weak acids that have a pKa within the physiological pH range of the stratum corneum also favors retention (Miselnicky et al., 1988; Artuc et al., 1980). Prolonged residence in the stratum corneum is referred to as substantivity.

The second chemical-specific property of importance to percutaneous penetration is the molecular size of the compound. Above a molecular weight of 500, flux through the intact skin is usually negligible for most chemicals. Within this range, however, are a large number of chemicals of commercial and toxicological importance.

Although large compounds may not permeate the skin rapidly, they apparently can penetrate by some (at present poorly understood) physical means. For example, contact urticaria results from cutaneous contact with high molecular weight (HMW) protein allergens, usually exceeding 10,000 MW (>10 kilodalton). Contact urticaria is a common problem among workers that are exposed to animal products, enzymes, and food proteins (Kanerva et al., 1996). It is not known how these proteins traverse the stratum corneum to first induce sensitization, and then later to allow absorption sufficient for elicitation of a response, but there is increasing evidence that such large compounds can cross the intact stratum corneum. Those professions experiencing high contact urticaria prevalence may also experience physical damage to the skin from cuts and abrasions, and penetration of the protein may be enhanced because of this. However, mice have been sensitized through topical application on intact skin by ovalbumin (Wang et al., 1996). The molecular weight of ovalbumin (OVA) is 45 kilodalton. In the experiment, OVA was left on the intact skin under occluded patches for seven continuous days, and occlusion could affect the barrier integrity of the stratum corneum. This resulted in statistically significant elevations in OVA-IgE specific antibody in the dosed group. It was also determined that the concentration required to induce sensitization decreased with repeated epicutaneous exposures. In addition, natural rubber latex proteins were determined to penetrate intact human skin in small amounts, and exposing the proteins to damaged skin resulted in significantly increased penetration (Hayes et al., 2000). As with OVA, mice have been sensitized to natural latex protein allergens by topical exposure, and show respiratory response upon later inhalation exposure (Woolhiser et al., 2000). Additional studies are presently under way at various laboratories to better understand the mechanisms by which HMW proteins and particles reach the viable epidermis and induce an antibody response.

Mathematical Modeling of Permeation

Once solubilization of low molecular weight compounds occurs in the surface of the stratum corneum, transit by molecular diffusion takes over. In solution, percutaneous diffusion of compounds through the skin is predicted by Fick's First Law, which simply states that the flux (J_s) of the compound through the skin is a function of the concentration gradient and the permeation rate constant (cm/hr) so that:

$$J_s = C_s \times K_p \qquad \text{[Equation 3-1]}$$

where C_s is the concentration of the compound (mg/cm^3) and K_p is the permeation constant (cm/hr). Thus, flux is expressed in units something like mg/cm^2/hr. By reorganizing the equation so that K_p equals J_s/C_s, it can be realized that K_p can be experimentally determined by dividing the equilibrium mass absorbed by the concentration at the exposed surface. Ideally, this is performed at several concentrations, and the relationship should be linear if saturation of the skin has not occurred. Thus, K_p is theoretically concentration independent up to saturation and linearly related to concentration differences across a membrane at steady-state. Skin saturation is an important limitation in that Fick's Law applies only to dilute solutions where the capacity of the chemical to solubilize into the limited permeation pathways within the stratum corneum is not exceeded. Often, the concentration range in which K_p is constant is quite narrow, above which the flux rate proportionally diminishes with increasing concentration. Once saturation occurs, as when high substantivity of the chemical in the stratum corneum forms a concentrated skin depot, the concentration gradient across the stratum corneum membrane will diminish. As a result, the flux rate for pure compounds is often no greater than for saturated aqueous solutions, unless chemical damage of the stratum corneum occurs (e.g., methanol, carbon tetrachloride).

In a bit more complex modeling of skin flux, it is realized that the chemical's diffusivity (Dskin) through the skin, as well as partitioning of the chemical in the vehicle in which a chemical exists and in the skin ($R_{skin/veh}$), which are due to respective solubilities, and the thickness of the skin (l_{skin}) are inherent components of K_p. Thus,

$$J = K_p C_{veh} \cong \frac{D_{skin} R_{skin/veh}}{l_{skin}} C_{veh} \qquad \text{[Equation 3-2]}$$

It helps to understand the above components of K_p, and in some instances K_p can be calculated from experimentally measuring these components, however, the latter is rare. It must be realized that vehicle effects can be very influential with regard to $R_{skin/veh}$, and if the compound is in a solvent other than water, the K_p must be empirically determined for the compound in that matrix. It is not appropriate to use a K_p for water to estimate flux from another solvent (Flynn, 1990).

When experimental K_p is not available, mathematical approaches might be used to predict this. One approach is to use statistical algorithms based on best fit of equations to empirical data. One example is the Potts and Guy (1992) equation, which incorporates solubility as measured by P_{ow} and molecular weight such that:

$$\log K_p = -2.72 + 0.71 \log P_{ow} - 0.0061\,MW \qquad \text{[Equation 3-3]}$$

This equation has an r^2 of 0.67 based on the empirical data set of approximately 100 compounds compiled by Flynn (1990). There are many other correlation models determined by other researchers that have attempted to predict K_p from physico-chemical data. In fact, Vecchia and Bunge (2001) have recently compiled 22 different skin permeation correlation models taken from the literature.

Among the more advanced models for predicting K_p are that of Cleek and Bunge (1993) and Wilschut et al. (1995), which attempt to take into account the two phase differences between the lipid-rich horny stratum corneum and aqueous viable epidermis or the theoretical existence of polar and lipophilic pathways, respectively. This approach has increased the prediction accuracy for the more hydrophilic compounds compared to other models. Researchers have considered multiple physical-chemical descriptors in addition to log P_{ow} and molecular weight, including topological indices, hydrogen bonding, solvation free energy, and molecular orbital descriptors, in an attempt to create better quantitative structure-activity relationship (QSAR) models (Sitkoff et al., 1994; Barratt, 1995; Roberts et al., 1995; Cronin et al., 1999). To date these attempts have been limited by small data sets.

Once a permeation coefficient is obtained, it can be used to calculate the approximate amount of compound absorbed through a given skin exposed area, and duration of exposure for a saturated aqueous concentration. Adopting the Fick's Law equation to include exposed skin area and duration of exposure, one can estimate the internally absorbed mass (M) by:

$$M = K_p \times C_{veh} \times A \times t \qquad \text{[Equation 3-4]}$$

where A = area (cm^2) and t = time.

Since flux is the product of the permeation coefficient and the concentration in aqueous solution, it is intuitive that the maximum flux ($MaxJ_s$) of a compound in an aqueous solution can be estimated by

$$MaxJ_s\ (mg/cm^2/hour) = K_p \times WS\ (S_w) \qquad \text{[Equation 3-5]}$$

where S_w = maximum solubility of the compound in water (in mg/cm^3 or in $moles/cm^3$).

Using the aqueous saturation concentration to calculate the maximum flux is a rough approximation of a compound's maximum flux rate, because it only considers the physical limitation of aqueous saturation in computing $MaxJ_s$, and does not necessarily consider skin saturation aspects.

Note that in aqueous solution, the maximum flux *decreases* with increasing octanol-water partition coefficient, while the permeation coefficient *increases*. They are *inversely* related because in an aqueous solution, the concentration needed to saturate the solution by lipophilic compounds decreases with increasing lipophilicity, and the flux decreases since flux is dependent on concentration. However, if you compared two chemicals in an aqueous solution that were of equal concentration, the one with the higher K_p would permeate faster.

The modeling approach selected by Wilschut et al. (1995) sufficiently estimates K_p, at least for some compounds. For example, the calculated K_p for aqueous phenol is 0.0066, whereas experimentally determined K_p was 0.0082 (Roberts et al., 1977). The calculated maximum flux using equation 5 is 0.0383 mg/cm^2/hr, whereas the experimentally determined flux for aqueous phenol was 0.008 to 0.02. However, for pure phenol the experimental flux is only 0.004 mg/cm^2/hr. As the aqueous dilutions have higher empirical fluxes than neat phenol, this probably exemplifies the importance of the strong partitioning effect of water in this instance (see discussion under Factors in next section on variables affecting percutaneous absorption). Although these models are useful for initial risk assessments concerning skin exposures, they are far from infallible. Calculated K_p values can differ from experimental K_p values by up to three orders of magnitude (Vecchia and Bunge, 2001). The underlying question is which data, be it empirically determined or mathematically computed, is more accurate for estimating human absorption. Obtaining large amounts of empirical skin permeation data using a standard laboratory protocol and a wide range of log P_{ow} and molecular weights, is critical to improving these models.

Substance P_{ow} values are available from a number of sources, including the Hazardous Substance DataBank through Medlars Database System (National Library of Medicine, Bethesda, MD), the Silver Platter "Chembank" CD-ROM, or on the internet by contacting mjollnir@daylight.com, subject: help (the last printed version of this database was in Hansch, 1979). Many P_{ow} values for chemical compounds, along with calculated or experimental flux rates, are provided in a report from the U.S. EPA (1992).

Permeation models for chemicals through the skin that are based on Fick's Law assume that the permeant be, strictly speaking, in a dilute aqueous solution. These models predict fairly well for compounds of moderate lipophilicity. For the extremes in aqueous solubility (very hydrophilic or lipophilic compounds), the partitioning effect due to the presence of water plays an increasingly important role. For instance, hydrophilic compounds prefer remaining in water rather than the lipophilic stratum corneum, thus less chemical tends to be absorbed from the aqueous solution compared to the pure compound (Bunge et al., 1995).

For neat highly water soluble or lipophilic compounds, the permeation rate cannot be easily estimated, and direct experimental determination is required, as the flux may be higher or lower than the flux for saturated aqueous solutions. The pure chemical may also directly damage the stratum corneum, which will dramatically increase absorption, e.g., methanol.

Several cautions must be kept in mind regarding all such model equations and the inherent limitations of the empirical data from which they are derived. The models do not take into account various abnormal physiological and pathological conditions, and physico-chemical factors that could greatly influence the dermal penetration rate (Barber et al., 1992; Vecchia and Bunge, 2001).

- General models may predict empirical results better for certain classes of compounds than others that may be scarcely represented in the empirical data set. Typically, compiled empirical data sets have not included many highly hydrophilic or highly lipophilic compounds. *In vitro* experimental permeation systems are especially prone to errors with highly lipophilic compounds, because the compounds may not be as readily removed from the viable epidermis due to lack of a functioning microcapillary blood perfusion.
- The models have been developed from experiments using normal skin. As will be discussed below, physically damaged skin, or skin previously exposed to chemicals, will likely allow much more penetration of chemicals than healthy, intact skin. Further, empirical data from aqueous solutions may not account for possible damaging effects that higher concentrations of chemicals might have on the skin and its barrier function.
- The empirical data using aqueous solutions do not reflect non-aqueous matrix effects of the vehicle, including whether the compound is in soil.
- Some of the compounds in the empirical databases are prone to ionization, which is pH and concentration dependent, and a factor that can affect skin penetration rate by one to two orders of magnitude.
- The empirical data is often performed at temperatures ranging from room temperature (i.e., ~25ºC) up to body temperature (i.e., ~37ºC). Permeability coefficients roughly double with a temperature increase of 5º – 7ºC (Vecchia and Bunge, 2001).

These are all potentially important limitations of the present models in prediction chemical penetration during workplace exposures.

Table 3-8 provides some examples of the flux through human and animal skin for some compounds in saturated aqueous solution and neat concentrations. The experimental flux from an air concentration at the TLV® for some of the compounds is also shown for comparison, when available. It should be noted that experimental fluxes often differ when reported by different experimenters. Since experimental K_p can be imprecise due to many experimental variables, the calculated fluxes may provide a practical alternative when sound experimental data are not available.

Table 3-8
Absorption flux of neat compounds, aqueous saturated solutions, and some vapors through human and animal skin

Compound	Neat Compound, In Vivo ($mg/cm^2/hr$)[1]	Saturated Aqueous Calc. Flux from Experimental K_p ($mg/cm^2/hr$)	Saturated Aqueous Calc. Flux from Calculated K_p[5]	Vapor Uptake, In Vivo at TLV® ($mg/cm^2/hr$)
Aniline	0.2 - 0.7	0.75	0.11	1.5×10^{-4} [6]
Benzene	0.24 - 0.4; 0.22 [3]	0.2	0.04	3.2×10^{-4} [6]
2-butoxyethanol	0.05 - 0.68 [2]; 0.9 [3]		1.6	2.4×10^{-4} [7,8]
n-butanol	0.53 [4]	0.31	0.40	
dimethylformamide	9.4		0.34	
ethylbenzene	22- 33		0.009	
2-ethyoxyethanol	0.8	0.23	0.43	
methanol	11.5 [2]	1.3	1.25	
2-methoxyethanol	2.8		1.27	
methyl ethyl ketone	5.3 [2]; 5.8 [2]		0.53	
styrene	9 - 15; 0.06 [2, 9]	0.01	0.01	2.6×10^{-4} [6] 0.04 [8] 0.016; 0.05
toluene	14 - 23; 0.32 [3] [10] 0.08 [2] [11]		0.02	0.014 [6]
m-xylene	0.12 - 0.15; 0.06 [9]		0.009	0.011 [6]

[1] See Leung and Pastenbach (1994) for references
[2] Recent data. [3] Used Pig skin, generally 2-3 times greater flux than human;
[4] DiVincenzo & Hamilton (1978) dog in vivo
[5] Used Robinson Model, see Wilschut et al. (1995); [6] Riihimaki & Pfaffli (1978); [7] Corley et al.(1997)
[8] Calculated from experimental K_p; [9] Tsuruta, rat skin (1982); [10] Jacobs & Phanprasit (1993)
[11] Ursin et al. (1995)

Factors Affecting Dermal Absorption and Effects on the Skin

ANATOMICAL DIFFERENCES

Permeation of chemicals through the stratum corneum occurs through the intercellular bipolar lipid channels. However, it has been determined that there are regional variations in skin permeability that correspond to the differing amounts of intercellular lipids. Table 3-9 shows the relative difference in permeation of three compounds and water through various regions of the human body. Parathion is the least water soluble, while hydrocortisone is the most. The increase in hydrocortisone permeation in some regions of the body appears to correspond to eccrine sweat production. Note how the palm, even though about 27 times thicker than the forearm (refer to Table 3-6), is almost equivalent in its barrier function. The planar stratum corneum (palms and soles of the feet), although thicker and able to resist physical abrasion better, has less extracellular lipid and poor diffusion barriers to chemicals, especially those that are more hydrophilic. The percentage weight of lipids in the soles, for instance, is 1.3%, whereas lipids account for 7.2% by weight in the forehead (Elias et al., 1981). Comparison of the water

Table 3-9
Comparative relative permeability of human skin to topical ^{14}C Hydrocortisone, Parathion, Malathion, and Water

Regional Variation	Parathion[1]	Malathion[1]	Hydrocortisone[1]	Water[2]
Forearm (ventral)	1	1	1	1
Palm	1.3	0.9	0.8	3.7
Ball of foot	1.6	1	—	12.6
Abdomen	2.1	1.4	—	1.1
Back of hand	2.4	1.8	—	1.8
Scalp	3.7	—	3.5	—
Angle of jaw	3.9	—	13	—
Forehead	4.2	3.4	6	2.7
Axilla	7.4	4.2	3.6	—
Scrotum	11.8	—	42	5.5

[1]data from Maibach et al., 1971
[2]data from Schueplein and Blank, 1971

loss through each site (20 - 40 $g/m^2/hr.$, versus 4-7 $g/m^2/hr.$, respectively) correlates with the difference in extracellular lipid (Lotte, 1987).

The greater permeability of some body regions can be important if that is the location of contact with chemicals. For instance, the head region, notably the jaw angle and behind the ear, are particularly permeable. Aerosol deposition of chemicals to these typically exposed areas, or transfer of chemicals through direct contact with contaminated hands or safety equipment (e.g., eye glasses, respirators), is certainly possible. Workplace studies have produced evidence that indicates that, not only surface contact concentration, but also the anatomical site, along with its unique skin permeability, should be taken into account when interpreting skin deposition data. A study of pesticide exposure found that patch data from some select locations, such as the neck and ankles, correlate better with biological monitoring results than whole body measurements (de Cock, 1995). In a study of skin permeability to polyaromatic hydrocarbons, average absorption rate constants at different skin sites were found to range from 0.036/hr to 0.135/hr (Van Rooij et al., 1993). Thus, the amount absorbed per skin site from equivalent skin concentrations will vary.

Just as there are regional differences in skin permeability, there are also differences in susceptibility to skin irritants. Permeability and irritant response are probably partially related, although biological mediators of response also play a role (Harvell and Maibach, 1994). Cua et al., 1990, found the thigh to be the most sensitive to the irritant sodium lauryl sulfate, while the palm and ankle were least responsive to this chemical.

INDIVIDUAL DIFFERENCES

Inter-individual differences in persons with apparently healthy skin appear common and can be appreciable. According to Feldman and Maibach (1974), the standard deviation expected is one-third to one-half the mean value. Assuming a normal distribution, one person in 10 will absorb twice the mean value, while one in 20 will absorb three times this amount (Maibach et al.). Up to 10-fold differences in interpersonal skin absorption rate have been seen within small groups exposed to such compounds as hydrocortisone and parathion (Feldmann and Maibach, 1965; Maibach, 1976; Dary, 1994). The apparent transdermal absorption rate of nitroglycerin in healthy volunteers, using the same site, but three different topical delivery systems, resulted in variations from 21% to 78% with six subjects (Wester and Maibach, 1987). Individual variation in alveolar air concentrations following a 30-minute skin exposure to five different chlorinated solvents is shown in Table 3-10. A wide range in the apparent uptake was seen, assuming minor confounding from differences in *in vivo* metabolism of the absorbed compound (Stewart and Dodd, 1964). Similar large inter-individual differences were seen by Lauwerys et al. (1978) when 11 male volunteers were asked to immerse both hands into pure m-xylene for 20 minutes. The skin of all volunteers was free of lesions. The total amount of m-xylene absorbed ranged from 16 to 110 mg. Same person differences tested twice over at least a one week period were two-fold or less. The range of skin permeability seen in healthy

Table 3-10
Individual range in alveolar air concentration following a 30-minute dermal exposure

Solvent	Subjects	End of Exposure	30-min. Post-exposure
carbon tetrachloride	3	0.11– 0.83	0.45 – 0.79
trichloroethylene	3	0.033 – 0.76	0.10 – 0.40
tetrachloroethylene	5	0.17 – 0.17	0.26 – 0.35
methylene chloride	3	2.3 – 3.6	1.1 – 6.6
1,1,1- trichloroethane	6	0.19 – 1.02	0.54 – 0.77

Adapted from Stewart and Dodd (1964)

individuals may also account for the wide range of susceptibilities of response to irritants (Judge et al., 1996; McFadden et al., 1998).

Gender and race, as these relate to the skin, have only been marginally studied to date. However, significant gender-related differences have not been found after repeated, daily application of an irritant (Lammintausta et al., 1987; Goh and Chia, 1988). Black skin was found to respond to irritation more than white skin, using objective techniques such as transepidermal water loss and increased cutaneous blood flow. However, no differences in erythema or in diagnosed cases of dermatitis have been noted (Basketter et al., 1996; Anonymous, 1973; Behrens et al., 1994). Age decreases the thickness, lipid content, and transepidermal water loss, but responsiveness to both irritants and allergens appears to decline with age (Cua et al., 1990; Harvell et al., 1994).

In occupational situations, where the skin is regarded as the principal route of exposure, individual variation in skin absorption may contribute to the differences often seen in biological monitoring results when the extent of exposure appears comparable. Controlling for these differences in skin absorption, probably due to differences in skin integrity, would be an advance in skin exposure monitoring. It may be possible, in the future, to use the various noninvasive instrumental techniques that are presently available to check individual skin characteristics, however, these have not yet been utilized in studies simultaneously measuring internal dose. Skin measurement techniques have been used mainly in laboratory settings, but are increasingly being used to study individual differences in skin integrity in the workplace, which may be indirectly related to susceptibility to irritants, allergens, and systemically absorbed chemicals.

PHYSICAL DAMAGE

The barrier properties of the stratum corneum, given its thinness, are quite unique. The important practical aspect of this knowledge is the realization that when the stratum corneum is healthy, it can perform an outstanding job of resisting chemical insults. However, if this thin barrier is physically damaged, or the intercellular lipids are altered, the stratum corneum becomes much less of a barrier. Table 3-11 presents some examples of damage by abrasion and ultraviolet light irradiation to the stratum corneum, and the effect that different types of damage might have on absorption.

Abrasions and cuts are probably the most common insults to workers' skin. As shown in Table 3-11, a non-intact stratum corneum offers little protection against permeation. There have been numerous documented cases where exposure to chemicals, not normally absorbed through the skin in sufficient amounts to cause even mild effects, have actually resulted in death when a few scratches were present. In one case, a woman pruning orchard trees that had been sprayed earlier with paraquat, developed scratches on her unprotected arms and hands during her work. Normally, paraquat is not a dermal exposure hazard due to its poor permeation through healthy skin. In this case, death ensued from respiratory failure a few days later (Newhouse, 1978). Experimentally removing the stratum corneum by tape stripping resulted in 2.6 to 8.5-fold increases in absorption of seven different pesticides (Maibach and Feldman, 1974). Some dangerous radionuclides, like cobalt-chloride (which are poorly absorbed through intact skin (<0.1%)) will rapidly penetrate abraded skin (52%) after 60 minutes (Kusama et al., 1986).

Table 3-11
Effect of type of physical damage on skin absorption of nicotinic acid human skin

	% Absorbed	
Condition of Skin	**In Vivo**	**In Vitro**
Normal	7	5
Abraded	47	51
Tape Stripped		58
UV Irradiated[1]		
1.5 minutes	22	7
6 minutes	51	13

[1] application was made 3 days after irradiation
Adapted from Bronaugh and Stewart, 1985a

Workplace risk assessments should consider skin condition. In one study of chlorophenol exposed workers in a timber mill, the urine concentrations exceeded what was predicted based on skin contact measurements by 70%. It was subsequently determined that the workers' forearms typically became abraded from unprotected contact with the wood, probably allowing more percutaneous absorption than would occur through intact skin (Fenske et al., 1987).

Ultraviolet irradiation from natural sunlight has been shown to increase the permeability of chemicals through the skin. Studies in both animals and humans have shown increases of two- to three-fold after acute UV irradiation for compounds such as hydrocortisone, nicotinic acid, and ethanol (Solomon and Lowe, 1979; Bronaugh and Stewart, 1985; McAuliffe et al., 1991). However, after prolonged exposure to UV irradiation, the skin seems to become more resistant to permeation. This is presumably due to an increased amount of stratum corneum lipids in irradiated skin (Lehmann et al., 1992). More research is needed to determine if skin permeation is affected in workers chronically exposed to sunlight.

Cold-induced damage to the skin has been studied and found to disrupt the percutaneous penetration of model compounds, and to inhibit the repair of the barrier. This type of damage could help to explain the high prevalence of occupational dermatitis in some jobs like fish processing (Halkier-Sorensen et al., 1995).

In addition to physical causes of damage to the stratum corneum, chemical exposures can alter the skin as well. In the presence of organic solvents or surfactants, the surface as well as the intercellular lipids can be removed or disorganized, changing the partitioning and permeation of chemicals into the skin (Surber et al., 1990). After acute to subacute exposures, normal barrier function may be restored fairly quickly (~2 days) or take two to three weeks for recovery, depending on the type of solvent or surfactant and the chronicity of exposure (Grove, 1985; Patil et al., 1994; Malten, 1968; Effendy et al., 1995; Wilhelm et. al., 1994; Lamaud et al., 1984).

DERMATOSES

Dermatoses constitute a broad range of skin conditions that can result from a variety of chemical and physical traumas to the skin, and, as pointed out before, damaged skin presents a less effective barrier. Dermatitis is a form of damage, especially when the dermatitic skin is in the acute stage of response. For instance, percutaneous absorption of the pesticide lindane in patients with severe scabies was 10 to 40-fold greater than in persons without scabies (Lange et al., 1981). In other studies, measurable differences were seen in metabolite excretion of carbon disulfide exposed workers and in dimethylformamide exposed workers who had skin irritation or skin disease (Drexler et al., 1995; Wrbitzky and Angerer, 1998). Percutaneous absorption of xylene vapor appeared to be about three times greater in a volunteer with atopic dermatitis (Riihimaki and Pfaffli, 1978). Finally, a higher blood concentration of a pesticide was seen in a formulator when compared to his co-workers, which was attributed to the presence of scleroderma in the worker with high blood levels (Starr and Clifford, 1971). Hyperproliferation of the stratum corneum, as in psoriasis, exfoliative dermatitis, or ichtyosis, also results in increased permeability of the skin (Solomon and Lowe, 1979; Scheuplein and Bronaugh, 1983). Rapidly generated stratum corneum, while thicker, is a poorer barrier.

Irritation of the skin can also result in increased percutaneous absorption. The data in Table 3-12 show an enhancement in percutaneous absorption that is related to the water solubility of four model compounds (Wilhelm et al., 1991). These compounds were applied to the upper back of hairless guinea pigs after 0.5% sodium lauryl sulfate in water was applied in an occlusive chamber for 24 hours. Because the diffusional resistance of the stratum corneum is greater to polar compounds than to non-polar compounds, any disruption of the barrier should enhance penetration of hydrophilic compounds to a greater extent than would occur for lipophilic compounds. The data support this in that enhancement was 260% for the most hydrophilic compound but only 130% for the most lipophilic.

In another experiment with sodium lauryl sulfate, which induces irritation, response to nickel salts in sensitized persons, as well as animals, was appreciably increased with the addition of the irritant (Samitz, 1958). Irritation, as well as several other factors, which can influence sensitization, is reviewed elsewhere (Kligman,1966).

OCCLUSION

Occlusion is one of the most effective means of increasing chemical permeation through the skin. Covering the skin with a moisture impenetrable barrier not only prohibits evaporative loss of any chemical that may have contaminated the surface of the skin, but rapid skin hydration and increased temperature increases the permeation rate of the contaminant. This effect has long been known, as demonstrated by Burckhardt (1939), who successfully used occlusion to promote sensitization to turpentine. With occlusion, nearly all animals were sensitized; with-

Table 3-12
Effect of chemical irritation on percutaneous penetration

Compound	Log Partition Coefficient	Percent Enhancement
HC	1.6	260
IM	3.1	160
IB	3.5	190
AC	6	140

Adapted from Wilhelm et al., 1991

out it none were sensitized. In this experiment, the site of exposure was occluded for eight to 12 hours.

The stratum corneum normally contains between 5% and 15% water, but this can be increased to as much as 50% by external factors (Bird, 1981). Occlusion is an effective way to hyper-hydrate the skin. The mechanisms by which hydration can influence percutaneous absorption include altering the partitioning between the surface chemical and the skin due to the increasing presence of water, swelling the corneocytes and possibly altering the intercellular lipid phase organization, increasing the skin surface temperature, and increasing the blood flow. It has been postulated that with hyper-hydration of the stratum corneum, the effective partition coefficient of the penetrant between the stratum corneum and viable epidermis is reduced, because the two tissue phases now appear more similar (Bucks et al., 1991).

Occlusion does not appear to influence the percutaneous absorption of all compounds equally. Rather, the impact of occlusion is influenced by the polarity of the chemical. It most increases the absorption of moderately lipophilic molecules, but is less effective on the absorption rate of highly lipophilic or highly hydrophilic compounds. Figure 3-7 illustrates this process for four steroids with and without occlusion. There may be some chemical-specific factor that also affects occlusive penetration, since the same degree of enhancement at each value of

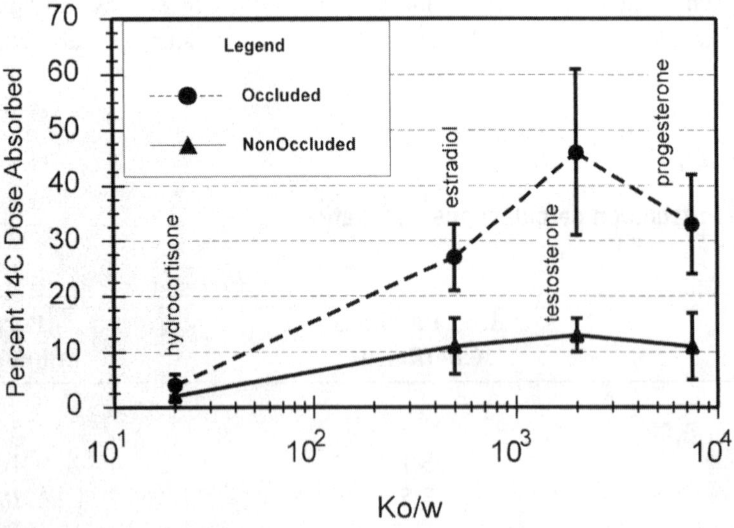

Figure 3-7. Effect of occlusion on four steroids as a function of different water solubilities. The results are from a single topical application of 4μg/cm² to the ventral forearm of human volunteers. Occlusion was continuous for 24 hours, after which the skin site was washed. Adapted from Bucks et al., 1991.

K_{ow} is not always seen with all compounds (Makki et al., 1996; Behl, 1980). Nevertheless, occlusion appears to enhance percutaneous absorption to some extent for all compounds tested.

Just as the permeability of human skin was shown to vary from location to location, it seems that there are anatomical differences in the extent of enhancement of percutaneous absorption due to occlusion (see Figure 3-8). In an experiment by Qiao et al. (1993) using weanling pigs, occlusion greatly increased parathion absorption in most sites tested, but to a lesser degree in the skin of the shoulder region. In terms of human relevance, the pig abdominal skin absorption value matches the absorption value for human forearm skin in non-occluded tests with parathion. This data also underscores the importance of skin site selection when conducting animal experimentation and comparing results between laboratories.

It should be realized that skin permeation effects of occlusion in the workplace scenario are likely to occur when the contaminant gets underneath personal protective clothing. Gloves or a rubber respirator provide excellent occlusive coverings. There have been reports where increased personal clothing protection has increased exposure because the clothing became contaminated on the inside (Kusters et al, 1992).

The above studies demonstrate the increase in permeation due to a one-time occlusion of contaminants on normal skin. In addition, prolonged occlusion has been found to produce chronic impairment of the barrier function of the skin. Figure 3-9 shows the results of an

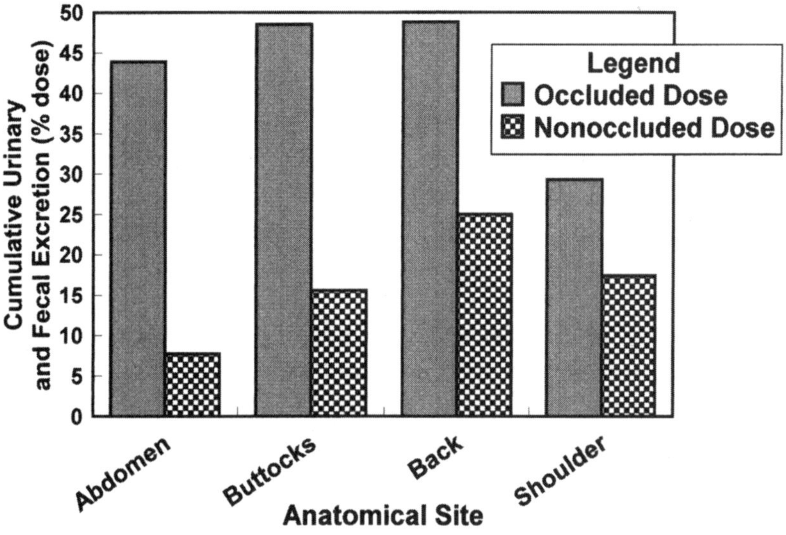

Figure 3-8. Effects of anatomical site and occlusion on the percutaneous absorption and excretion of 2,6-parathion in vivo in pigs. Skin sites were occluded and urine and feces were collected for 168 hours. Maximal excretion occurred for the occluded and nonoccluded groups in 24-36 and 36-48 hours, respectively. While occlusion significantly increased absorption for all sites, regional differences by skin site were found.

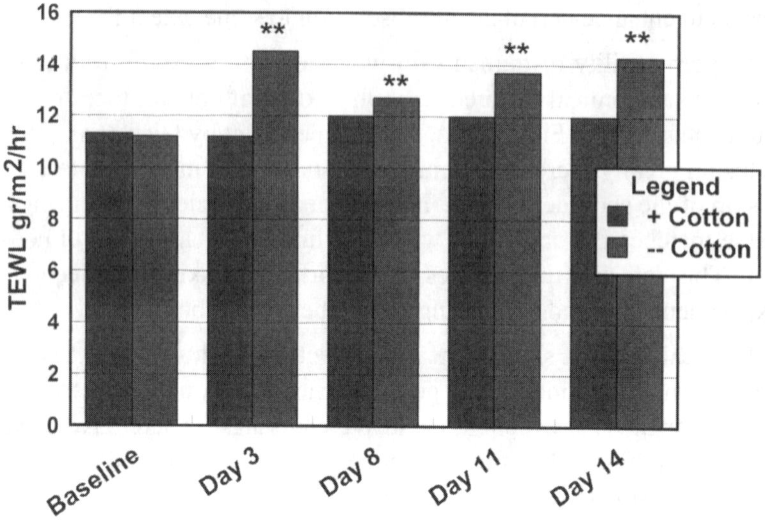

Figure 3-9. Effect of long term experimental occlusion on human skin. Wearing of occlusive glove for 6 hours per day for 14 days had a significant detrimental effect on skin barrier function, as measured by transepidermal water loss measurements. When a cotton glove was worn underneath, the effect was not detected. It was concluded that gloves may be a substantial factor in the pathogenesis of cumulative irritant contact dermatitis, but that with proper use the risk might be minimized. Adapted from Ramsing and Agner, 1996.

experiment to determine the long-term effect on barrier function by prolonged use of an occlusive glove (Ramsing and Agner, 1996). The protocol involved 18 volunteers who wore non-latex hypo-allergenic gloves for a minimum of six hours while sleeping at night. During sleep, the skin temperature and friction were reduced (a best case situation). Half of the volunteers wore cotton gloves underneath the occlusive glove. The results indicated increased trans-epidermal water loss for each of the measurement days up to day 14 for the skin with the occlusive glove only. However, use of the cotton glove prevented this damage. The authors concluded that prolonged occlusion can damage the barrier function of the skin, and might enhance susceptibility to irritants and sensitizing agents. Minimizing the use of gloves and using a cotton glove underneath are two ways to reduce the occlusive effects on the skin.

Susceptibility to irritants was demonstrated by Graves et al. (1995) after the skin was occluded. In their experiment, the skin was occluded for four hours or eight hours. After the covering was removed, skin permeability was evaluated by measuring the time to onset of hyperthermia due to topical application of the irritant hexyl nicotinate. After four hours of occlusion, the time to onset was reduced to 59% of its pre-occlusion value. After eight hours of occlusion, this was further reduced to 38% of the pre-occlusion value.

Within two days of occlusion, marked cytotoxic damage to Langerhans cells, melanocytes, and keratinocytes occurred (Klingman, 1996). Intercellular edema and marked swelling of corneocytes was also prominent. The damaging effects of prolonged hydration can quickly lead to dermatitis.

Interesting questions asked by Graves et al. (1995) were: (1) At what duration of wearing a glove continuously would there be a significant impairment of the stratum corneum barrier function? (2) How frequent and how long should non-occluded periods be to prevent a cumulative effect? (3) Is the impairment of barrier function seen in these experiments sufficient to increase the incidence of irritant contact dermatitis? and (4) Does this impairment correspond to an increased susceptibility to mechanical trauma? These are all relevant and practical questions, and further research will be needed to provide answers.

TEMPERATURE AND RELATIVE HUMIDITY

Molecular diffusion and solution will increase within matrices as the temperature increases. It should be no surprise that as temperature increases, the permeation of chemicals through membranes such as the skin will also increase. Normally, the skin surface temperature is between 32° and 35° C, but this can increase or decrease with heightened or lowered ambient temperature, or if the temperature of a liquid contacting the skin is different.

Figure 3-10 illustrates how the ambient temperature affected the excretion of p-nitrophenol in the urine of human volunteers after 5 grams of 2% parathion dust had been applied topically

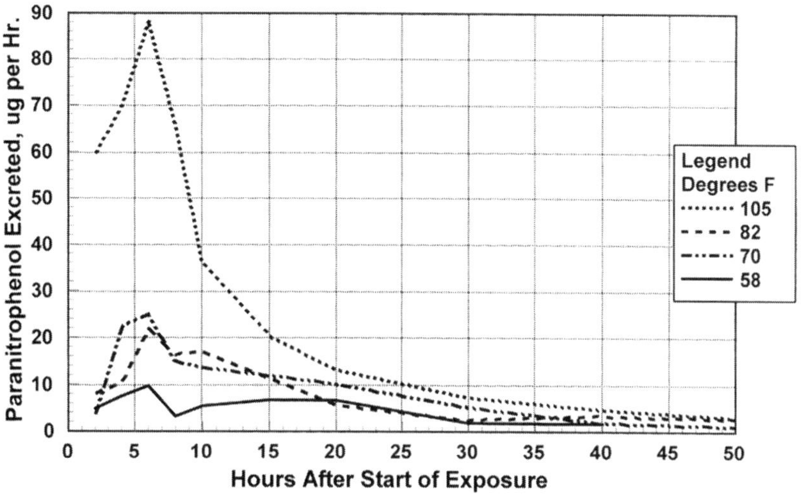

Figure 3-10. Para-nitrophenol excreted by human volunteers following dermal exposure to 2% parathion dust at different ambient temperatures. There were two replicate tests at 58°F, one at 70°F, three at 82°F and nine at 105°F. Adapted from Funckes, et al., 1963.

for two hours. Only one hand and forearm were exposed and placed into a temperature-controlled chamber. Following the two hour exposure, the skin surface area was decontaminated by scrubbing with soap and water for five minutes, followed by two washes with ethyl alcohol. The average excretion rate increased from 4.9 μg/hr. at 58° F (14° C) to 19.6 μg/hr. at 105° F (40° C) over 41 hours (Funckes et al., 1963). Chemical vapor penetration has also been empirically determined to increase with ambient temperature. Percutaneous absorption of aniline vapor, for example, increases about 20% for each 5° C increase in air temperature (Dutkiewicz, 1960). Percutaneous uptake was also significantly increased for 2-butoxyethanol vapor when the air temperature (but also humidity) was increased from 23° C to 33° C (Johanson and Boman, 1991).

Increasing relative humidity has been shown to increase percutaneous absorption. Chang and Reviere (1991) found that the percutaneous absorption of parathion increased by two- to four-fold when the humidity was increased from 20% to 90%, while the temperature remained the same.

DURATION AND FREQUENCY OF SKIN CONTACT

In the workplace, contact with chemicals may be intermittent and sporadic. A key characteristic of the chemical that will determine the residence time of a single contact is the volatility of the chemical. For uncovered skin, highly volatile chemicals will evaporate before appreciable percutaneous absorption occurs (Riefenrath, 1989). Volatility inversely correlates with the persistence of a chemical on the skin surface and the amount of chemical ultimately absorbed systemically. Figure 3-11 presents the results of continuous dermal exposure to n-butanol (vapor pressure [v.p.] = 6 mm Hg) and 1-minute exposure every 30 minutes. The periodic exposure results in a blood concentration that is lower than for continuous exposure. For more volatile compounds, such as toluene (v.p. = 21 mm Hg) and 1,1,1-trichloroethane (v.p. = 100 mm Hg), no such cumulative effect occurs (Boman et al., 1995). The absorbed dose of highly volatile compounds from repeated short exposures is appreciably less when compared to continuous immersion (Stewart and Dodd, 1964). Presumably, the systemic absorption of compounds less volatile than n-butanol would eventually approach continuous contact conditions as evaporative loss decreases further.

Typically, experimental percutaneous absorption studies test naive skin where the chemical agent is applied only once. Is this indicative of repeated exposures that are more likely to occur in the workplace? Some studies have attempted to answer this. Briefly, no cumulative effect of repeated exposures has been observed concerning the permeability of the skin, *unless* the skin is damaged by the chemical exposure (Bucks et al., 1990). Increased absorption was observed in skin damaged by chemical exposure when salicylic acid or hydrocortisone were repeatedly applied to the skin (Roberts and Horlock, 1978; Wester et al., 1977). Both chemicals are known to affect the integrity of the stratum corneum. With other compounds, such as malathion and benzoyl peroxide, no difference in absorption was seen after repeated exposures (Wester et al., 1983; Courtheoux et al., 1986). Washing with soap and water between dose

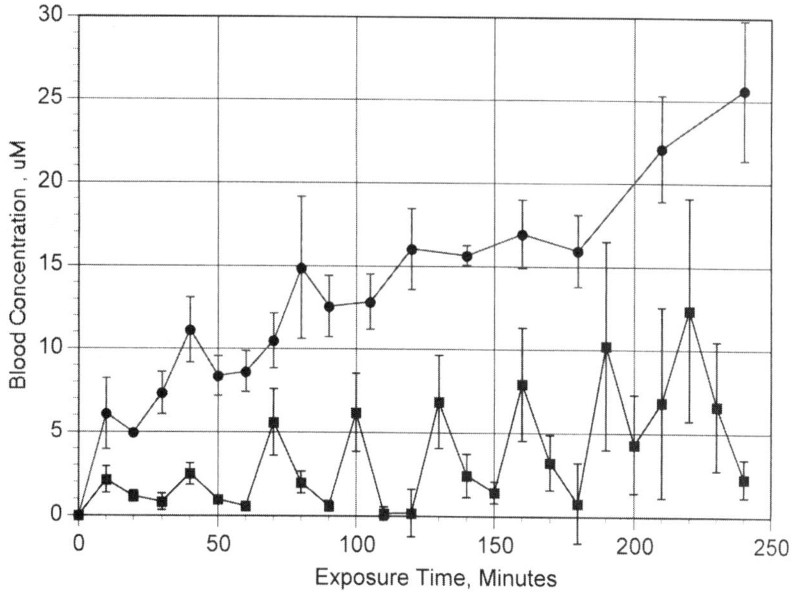

Figure 3-11. Concentration of n-butanol in blood in guinea pigs during continuous and intermittent skin exposure. There is generally an increasing blood concentration with repeated exposures over time. Compounds that are more volatile do not exhibit this trend, and it is rational to expect the converse, that compounds that are less volatile will remain on the skin longer and better mimic continuous exposure. Adapted from Boman et al., 1995.

applications contributed to increased absorption of most compounds, but this has only been demonstrated with animals. Human skin may be more resistant to damage from soaps, but this question will require further study (Bucks et al., 1989a).

VEHICLE

A vehicle in this context is commonly a liquid in which another contaminant, which may be more toxic, is contained. Vehicles are important to percutaneous absorption because they may enhance the absorption rate by either disrupting the stratum corneum barrier function, or encouraging partitioning towards the skin, or a combination of these processes. As will be shown in the next examples, vehicles, or the co-components of a mixture, can have significant effects on how much of a toxic chemical enters the skin. Thus, the rate of uptake can differ appreciably from neat (pure) chemical absorption.

A dramatic example of how the vehicle can influence the uptake of contaminants in a solution is depicted in Figure 3-12. Three dilutions of alachlor in the water soluble commercial formulation were tested for permeation using *in vitro* human skin (Bucks et al., 1989b). One

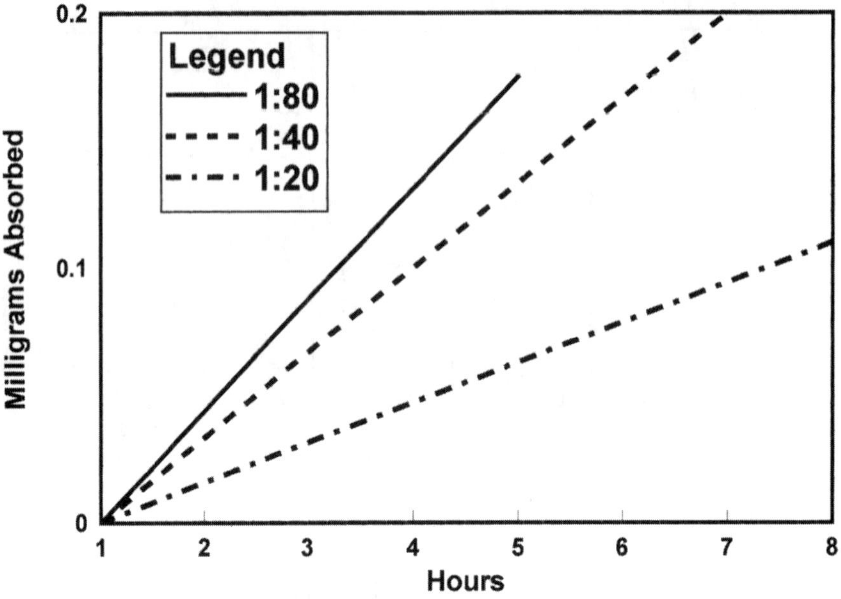

Figure 3-12. Effect of formulation dilution on the in vitro percutaneous absorption of alachlor. Increasing the dilution of the alachlor formulation resulted in significant enhancement ($p<0.01$) in the rate and extent of alachlor penetration. The reason for this might be due to a favorable shifting of the partition coefficient with dilution in water. Adapted form Bucks et al. (1989).

might assume increasing absorption with more concentrated dilutions would occur because diffusional mass going into the stratum corneum is dependent on the concentration of the chemical on the surface. However, the opposite was seen in these experiments. The 1:80 dilution led to twice the absorption as the 1:20 dilution. The authors hypothesized that, in this experiment, the more concentrated dilution contained more of the non-aqueous solubilizing material, which is able to "hold" the alachlor in the solution. When the concentration is decreased in the most dilute mix, most of the solution is aqueous and the highly hydrophobic alachlor preferentially migrates towards the skin. The skin:vehicle interface concentration may, therefore, be greater in the 1:80 dilution than in the more concentrated dilutions.

Another example of vehicle effects on percutaneous absorption, particularly where the vehicle is a mixture of solvents, is demonstrated in the data in Table 3-13. In this experiment, xylene absorption as measured by methylhippuric acid excretion, was compared to exposures of pure xylene, 1:1 mixture of xylene plus isobutanol, or 10:10:1.5 mixture of xylene, isobutanol, and water to saturation (Riihimaki, 1979). In each case, volunteers immersed both hands up to the wrists in the solvents for 15 minutes at room temperature. Interestingly, the addition of a small amount of water to the mix dramatically increased the xylene permeation rate so that it was almost three times greater than the xylene-isobutanol mix, and slightly more than pure xylene alone. With only xylene and isobutanol, there appeared to be a conspicuous dehydration

Table 3-13
Percutaneous absorption of m-xylene and mixed solvent effects in *in vivo* human skin

Solvent	Symptoms	Methylhippuric Acid Excretion
Pure Xylene	Erythema No "tightness"	1.0
1:1 Xylene + Isobutanol	Erythema No "tightness"	0.4
10:10:1.5	Mild wrinkling; skin dry and oily	1.1

Adapted from Riihimaki, 1979

of the skin by isobutanol and a delay in the absorption of xylene evident in the methylhippuric acid excretion (data not shown). The result seen by adding a small amount of water might be due to the increased hydration of the stratum corneum provided by the addition of water, or increasing the partitioning of xylene out of the saturated aqueous mixture towards the skin.

More typically, the effect of the vehicle is compared among different pure solvents. The rule of thumb that "likes dissolve likes" applies to permeation in that hydrophilic solutes will tend not to partition to the skin if in a hydrophilic solvent, and lipophilic solutes will tend not to partition as well into the skin if present in a lipophilic solvent. The importance of the lipophilicity of the vehicle on maximum permeability for a wide range of compounds was studied by Lien and Gao (1995), who found that the ideal lipophilicity of the permeants in a lipophobic vehicle was between log P_{ow} 2.5 to 6, but in a lipophilic vehicle the ideal lipophilicity of the permeants was shifted to 0.4 – 0.6. This principle is exemplified by the 10-fold difference in the skin flux rate of benzocaine when in a water (lipophobic) vehicle or in polyethylene glycol (PEG) 400 (lipophilic). The solubility of benzocaine in water is 1.26 g/L whereas the solubility is 435 g/L in PEG 400. The skin flux rates were 0.1 mg/cm2/hr and 0.01 mg/cm2/hr, respectively (CTFA, 1983).

A summary of many of the above-mentioned factors that can affect skin permeation rate are included in the results shown in Figure 3-13. These experiments assessed the impact of adding water to saturation, to a variety of alcohols with carbon numbers 1 through 10 (Scheuplein and Blank, 1973). By adding carbon length to a simple alcohol reduces its polarity (increases lipophilicity). These experiments with aqueous and neat alcohol solutions demonstrate the importance of the aqueous contribution, its effect on key physical parameters, and the interaction of these measurement parameters with percutaneous flux (J_s) through the epidermis. The partitioning (K_m) effect from the aqueous solution is strong, but the concentration reduction of the alcohols with increasing carbon number causes the flux to decrease since flux is dependent on concentration. Notice how the permeation rate (K_p) increases in the aqueous solution as the carbon number increases. The opposite is seen for the pure alcohol exposures. This is because the stratum corneum/chemical partition coefficient is low for pure lipophilic compounds. To illustrate this point, liquid hexanol (concentration 8.2 moles) is approximately 150 times more concentrated than saturated aqueous hexanol (0.55 moles), yet the K_p of aqueous hexanol is almost twice as great. The large measured flux for the pure alcohols with low carbon number (especially methanol) is attributed to the damaging effects on the stratum corneum. This damaging effect increases the theoretical flux about 10,000 times.

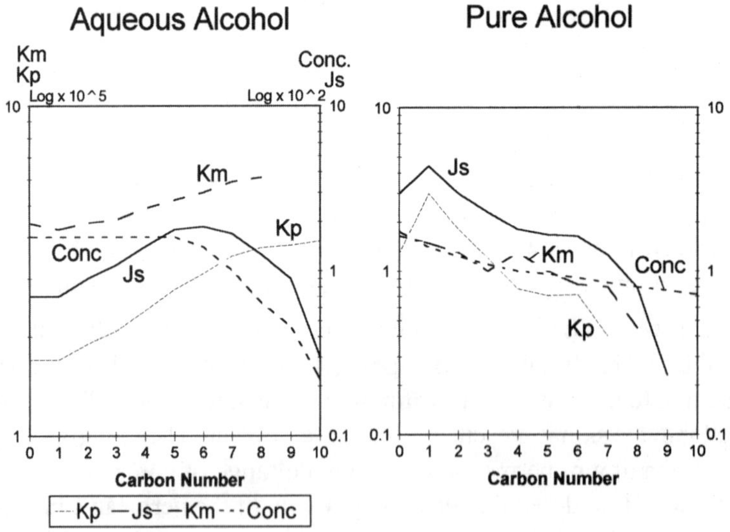

Figure 3-13. Effect of adding water to alcohols with 1 - 10 carbon numbers. Refer to text for complete explanation.

Key: K_p = permeability constant
J_s = flux of solute
K_m = tissue: solvent partition coefficient
Conc. = solute concentration (moles/liter)

Adapted from Scheuplein et al., 1973.

MATRIX

Often workplace exposures to chemicals occur through a particle matrix. Thus, toxic compounds may adhere to soil particles, dust, or other "dirt-like" matrix and come into contact with the skin. Unfortunately, a limited amount of research has been published on this subject.

The absorption of eight compounds *in vivo* in rhesus monkeys over a 24-hour period, when the compound was applied in either soil or one of three solvent vehicles, is shown in Figure 3-14. The application sites were covered with a non-occlusive patch during exposure to prevent loss of particles from the skin surface. Although soil reduced the overall absorption of these compounds to about 60% of absorption compared to when applied in solvent, there were some compounds for which the absorption was similar regardless of matrix (Wester and Maibach, 1996).

In the same study, an *in vivo* percutaneous absorption study of the pesticide 2,4-D was performed with rhesus monkeys with the objective of comparing the kinetic rate of absorption (Wester and Maibach, 1996). The pesticide was applied to the skin in either acetone (and left to quickly evaporate) or soil (1 mg/cm^2). The mean percentage absorbed at 8 hours was 3.2 ± 1.0 when applied in acetone, but only 0.05 ± 0.04 when applied in soil (see Figure 3-15). Absorption increased for the soil vehicle at 16 hours to 2.2 ± 1.2, and at 24 hours 9.8 ± 4.0 percent had been absorbed. The increase of absorption when applied in acetone appeared linear over time and at 24 hours peaked at 8.6 ± 2.1%. Thus, over a long period (24 hours), absorption from either vehicle appeared about equal for 2,4-D, regardless of whether it was in a soil matrix or applied in acetone. However, a marked delay in absorption presumably indicates the time it must take for 2,4-D to get from the soil into and through the skin.

Different findings were reported in a prior study. Shu et al. (1988) found that the dermal penetration of TCDD in soil after just 4 hours of contact with *in vivo* rat skin was approximately 60% of the amount absorbed following 24 hours of contact. The skin absorption kinetics appeared faster for TCDD than the findings described above for 2,4-D. Despite the rapid initial flux, at 24 hours, only approximately 1% of the TCDD was absorbed. As in the study of 2,4-D, the soil was kept in place while on the skin with a non-occlusive covering. Other variables, like TCDD concentration in soil (10 vs 100 ppb), and the presence of used crankcase oil (0, 0.5, 2.0%) had no significant influence on dermal bioavailability.

Turkall et al. (1994) performed a study on pure and soil-absorbed naphthalene in dermally exposed male rats. Within 12 hours after application, it was found that approximately 50% of the naphthalene dose was excreted in the urine and was equivalent when applied in the pure form or as the clay soil-absorbed mix. The corresponding result was 33% for sandy soil-bound TCDD. The study protocol directed that the administered doses be sealed in a vapor tight chamber attached to the skin.

On the other hand, Yang (1989) compared the percutaneous absorption of benzo[a]pyrene (B[a]P) from petroleum crude or fortified soil, using rats as the experimental animal, and found

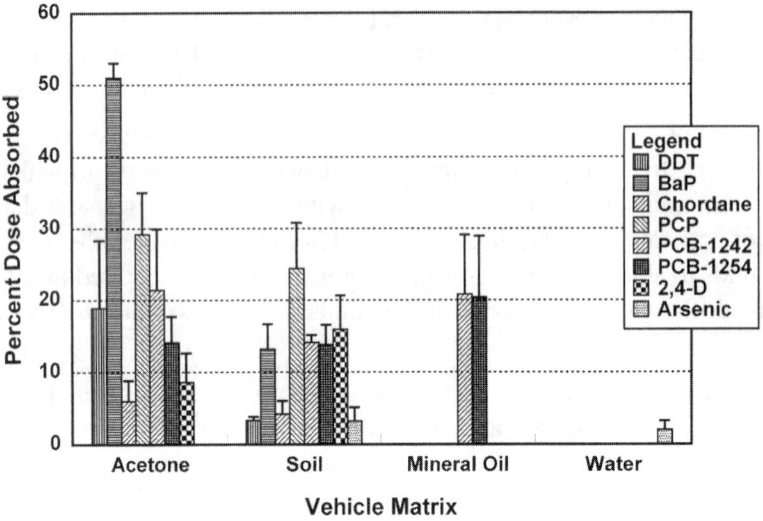

Figure 3-14. Percutaneous in vivo absorption from solvent vehicle or soil. Compounds were applied to the skin of rhesus monkeys at equal concentrations in soil or a liquid vehicle such as acetone, mineral oil or water and urine and feces collected for 24 hours. Adapted from Wester and Maibach, 1996.

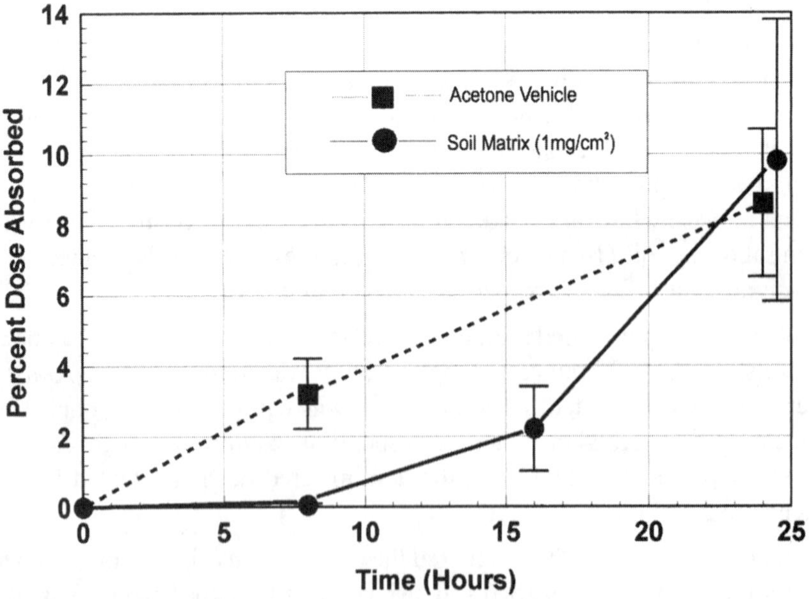

Figure 3-15. Kinetics of percutaneous absorption of 2,4-D in rhesus monkeys when applied in acetone (and left to evaporate) or in a soil matrix (adapted from Wester and Maibach, 1996).

that the percentage of the dose absorbed from the crude was 4 to 5 times greater than when an equivalent amount of B[a]P in crude oil was applied in soil.

Thus, different results concerning the effect of soil on percutaneous absorption have been found in these studies using various compounds and experimental protocols. Partitioning factors, duration of exposure, deposition rate and other unknown factors may be important for explaining these discrepancies. A theoretical and mathematical treatment of the physical chemistry involved in chemical-soil interactions with the skin has been presented in a series of articles (McKone, 1990, 1991; McKone and Howd, 1992). The ability to predict uptake accurately will likely improve as empirical data are collected in appropriately designed experiments.

Percutaneous absorption studies of volatile solvents in soil matrixes have been conducted using benzene, toluene and xylene (Skowronski et al., 1988, 1989, 1990). The solvents were either applied pure or mixed with soil and contained under a covered cap to prevent evaporation. When benzene was tested, clay soil decreased the dermal penetration, while the same soil had little effect on toluene absorption. In the xylene experiment, pure compound produced a slightly higher peak plasma concentration than for soil-absorbed xylene, but there was no delay in the time it took to reach maximum plasma concentration (1 to 2 hours). In the clay soil, the amount of xylene absorbed was equivalent to the pure xylene exposure.

A possible flaw with some of these studies is that they may have applied a soil loading upon the skin that is normally unlikely to adhere to people. Field studies involving a range of outdoor activities indicate that, at most, about 1 mg/cm^2 dry to slightly moist soil will adhere to skin (Kissel 1996a). Recent sampling of manual row crop harvesters found 1 to 1.6 mg/cm^2 as an average soil loading over both hands, although much of the contamination is on the palms (Boeniger, unpublished data). The loading does increase with moisture content, and other factors could affect this as well (Kissel et al., 1996).

The U.S. EPA and others have speculated that there is a finite thickness of soil upon the skin that a substance can diffuse through and result in absorption into the skin (U.S. EPA, 1992). This surface loading amount has been termed the "monolayer" thickness. Based on available data and expert judgment, the U.S. EPA estimated the monolayer thickness to be equivalent to about 5 mg/cm^2, which would have a 5 millimeters thickness if the soil had a density of one (U.S. EPA, 1992). More recent evidence indicates the monolayer occurs at a surface loading of only 2 mg/cm^2 for soil with a particle size less than 150 μm (Duff and Kissel, 1996). Experimental proof that soil loading affects cutaneous absorption was provided by Touraille et al., (1997). They found that the relative percent absorption of cyanophenol from contaminated soil at 7 hours was inversely proportional to soil loading. When the soil loadings were 5, 11, 40, or 140 mg/cm^2, the percent absorbed was 16, 10.9, 2.4, and 0.84, respectively. This indicates saturation of the process of chemical absorption from soil. The finding that a topical corticosteroid halcinonide cream that was applied at loadings above 5 mg/cm^2 did not appear to increase the rate of permeation corroborated this (Walker et al., 1991). Thus, the percutaneous absorption results from soil-bound chemical permeation studies performed in laboratories that applied large amounts of soil to the skin (super-monolayer concentrations) are questionable and probably not reflective of actual exposures.

WASHING

Ideally, contamination of the skin will be prevented by a "no contact" processing design and control, excellent housekeeping, and as a last resort, personal protective clothing. In reality, the skin will become contaminated to some degree, the extent to which is dependent on whether any of the above approaches have been implemented, and as a result, will require decontamination. Decontamination of the skin in general is necessary to prevent (1) percutaneous absorption and dermatitis from developing, and decontamination of the hands in particular is necessary to prevent (2) transfer of contaminants to the mouth leading to ingestion, and (3) transfer of toxics to other sensitive areas of the body. Chemical decontamination of the skin is a common and practical procedure that, in reality, has received insufficient critical and objective research. Some findings related to the above goals are reviewed below.

As with many procedures, there are ways and then there are better ways of doing things, even if it seems that little thought should be given to the task. For example, the temperature of wash water may play a significant role in the extent that irritants exert their effect on the skin. In one experiment, 10 volunteers were given a topical open application of 5% sodium lauryl sulfate in water to the volar forearm daily for four days, and the application site was left to dry (Berardesca et al., 1995). The temperature of the water that was applied was either 4°C, 20°C, or 40°C. Since open applications were performed, the temperature of the solution probably reached normal skin temperature rapidly. Nevertheless, even within this short time, statistically significant changes in the skin, as measured by objective test methods, were seen as a function of application temperature (see Figure 3-16). These results are supported by other studies (Ohlenschlaeger et al., 1996).

One occupationally relevant study involved the following protocol. Volunteers were asked to immerse one forearm into either hot water or cold water for 30 to 45 minutes, while the other arm was kept dry (control arm), after which the immersed arm was removed from the water and gently dried. Ethyl nicotinate, an irritating chemical, was applied to the skin of each arm in serial dilutions of 0.1 to 0.00125% to determine the minimal concentration needed to cause erythema. The result was that the concentration needed to cause an irritant response was on average 12 times lower for the hot water immersed arm and six times lower for the cold water immersed arm, compared to the dry arm (Cronin and Stoughton, 1962). There were considerable individual differences in the extent of effect of immersion on the response to this irritant, suggesting diverse innate susceptibility to hydration. Clearly, hydration, particularly with hot water, caused increased sensitivity to this model irritant. This experiment might simulate persons washing things in water, for instance, with exposure to irritants like soaps. To minimize irritation of the skin, wash water temperatures should probably never exceed 25° C, and frequent or prolonged contact with water should be avoided.

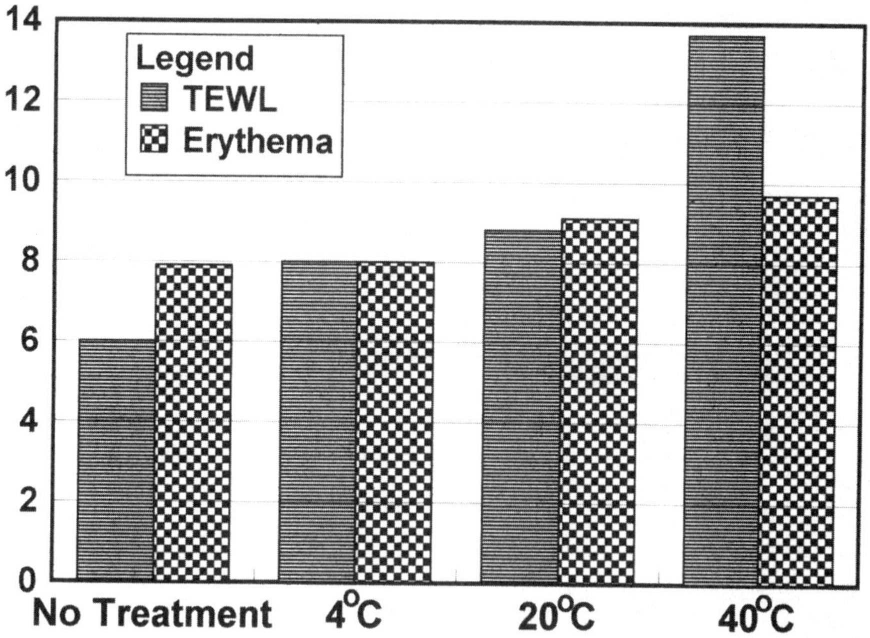

Figure 3-16. Effect of water temperature on transepidermal water loss (TEWL) and erythema when mild surfactant is applied to the skin. Ten volunteers allowed open application of 5% sodium lauryl sulfate in water to the volar forearm daily for four days and the application site was allowed to dry. The solution temperatures were 4°C, 20°C or 40°C. A dry area not exposed acted as the unexposed control. Adapted from Berardesca et al., 1995.

Washing in some experimental protocols appeared to enhance percutaneous uptake of the contaminant from skin. This phenomenon has been demonstrated often in *in vitro* and *in vivo* animal skin and has been termed the "wash in" effect. There has been some suggestion that human skin is more resistant to soaps and water from long-term conditioning (Bucks et al., 1991; Wester and Maibach, 1989), but some studies suggest the effect also occurs on human skin as well, at least for some compounds (Lang et al., 1981). Washing with water or other solvents might mobilize chemicals residing in the stratum corneum by chemically or mechanically disrupting the stratum corneum or acting as carriers (Barry, 1987; Barry, 1991). This wash-in effect is perhaps most dramatically depicted in Figure 3-17 in an *in vitro* experiment on the permeation of DDT through human skin (Moody et al., 1994). A four- to six-fold increase in permeation after washing was detected when using human skin. A 10- to 14-fold increase was seen using rat skin.

Selection of some solvents over others might improve decontamination efficiency, if the solubility of the contaminant in the solvent is expected to play a role. For example, the removal efficiency and percutaneous migration of TCDD into human skin *in vitro* was evaluated using

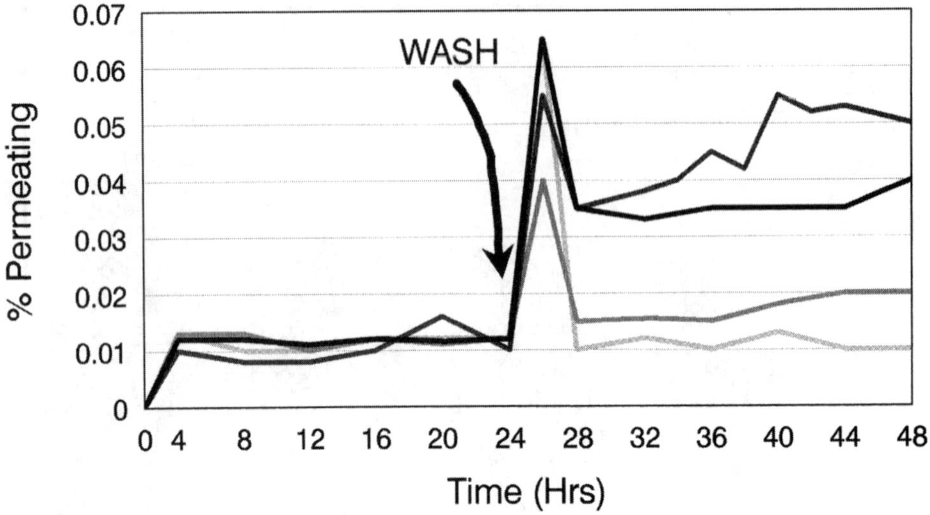

Figure 3-17. The flux of DDT into the receptor fluid in an in vitro experiment was increased when soap and water washing of the skin surface was performed. The figure shows the results of four separate replicates. Soap and water appear to mobilize DDT residing in the skin. Adapted from Moody et al., 1994.

four decontamination solvents, and the results compared to doing no cleaning of the skin. The duration of the experiment was 300 minutes. Each of the decontamination treatments were performed at 100 minutes post contamination. The decontamination treatments consisted of dry wipes with cotton balls (#1), mineral oil treatment for 10 minutes followed by dry wiping with cotton balls (#2), mineral oil treatment followed by acetone soaked cotton balls (#3), and soap and water wiping (#4). The results are shown in Table 3-14 (Weber et al., 1992). The least amount of TCDD was found in the stratum corneum when protocol 3 was used, but this did not result in the least amount in the epidermis and upper dermis. Rather, the mineral oil alone may have done just as well, since less TCDD was present in the lower layers. Soap and water wipes removed much of the TCDD from the stratum corneum, however, the amount found in the lower layers of the skin was only second to no decontamination at all. Since TCDD is a highly lipophilic compound, removal by a non-polar solvent like mineral oil, not water, would seem more rational, and is supported by these data. For other compounds, soap and water has been found to perform as well or better than some potent solvents and effectively reduces systemic absorption (Wester et al., 1990; Pelletier, 1990; Koizumi, 1991). Thus, one approach to decontaminating the skin should not be expected to work equally well for all kinds of chemical contaminants.

The effect of delaying washing following percutaneous absorption has been well studied. To generalize those findings, a large amount of chemical is absorbed immediately after applica-

Table 3-14
Penetration of TCDD into human skin after different decontamination protocols

	Percentage of ^{14}C-TCDD dose in skin layer and relative rank regarding lowest concentration							
Protocol	**Strautum corneum**	**rank**	**Epidermis**	**rank**	**Upper dermis**	**rank**	**Lower dermis**	**rank**
Control	43.7 ±4.5	4	4.06 ±1.11	5	2.14 ±0.49	5	0.51 ±0.12	5
1	44.5 ±5.7	5	1.84 ±1.04	3	0.8 ±0.18	3	0.27 ±0.08	2
2	29.2 ±1.9	3	1.08 ±0.18	1	0.51 ±0.07	1	0.38±0.05	3
3	16.6 ±2.5	1	1.35 ±0.22	2	0.58 ±0.08	2	0.43 ±0.20	4
4	22.3 ±6.5	2	2.17 ±0.48	4	0.91 ±0.21	4	0.23 ±0.05	1

Adapted from Weber et al., 1992

tion until about one hour later, when the chemical is applied neat in solvent. After about one hour of initial application, there is little difference in total absorption up to at least 24 hours later. However, if the initial skin concentration is great, increasing cumulative percutaneous absorption may be more likely to continue for an extended time (Hewitt et al., 1995; Wester et al., 1989).

Applying neat chemicals in solvents to skin may be relevant to chemical splashes or deposition of liquid aerosols on the skin. However, when the contaminant is present in a soil matrix, rapid absorption into the skin is not observed, and, in fact, there may be a considerable delay as discussed above for matrix effects (Wester and Maibach, 1996). For decontamination purposes, the soil matrix data suggest skin washing anytime up to 16 hours is effective in reducing absorption. Unfortunately, experimental evidence to confirm that late decontamination is effective in reducing absorption is lacking.

The subject of decontamination efficacy is too diverse and complex to fully address here, but the principles of solubility and partitioning should be remembered when choosing a wash solvent, and considering what effect that solvent might have on the skin, especially whether it can act as an irritant or accelerate absorption of contaminants.

Several excellent review articles have been written which should be referred to for further information about the variables related to percutaneous absorption (Schaefer et al., 1982; Wester and Maibach, 1983; Schwope, 1986; Wester and Maibach, 1987; Flynn, 1990; ECETOC, 1993).

DIAGNOSTIC AND EXPERIMENTAL TEST METHODS RELATED TO THE SKIN

The use of clinical diagnostic tests and experimental test methods related to skin permeation, and irritant and allergen screening, are not typically used directly by the occupational hygienist. Nevertheless, the hygienist can benefit from having some familiarity with these techniques, by knowing what techniques are available, and by knowing what the methods have to offer. This includes knowing what tests are available to appropriate experts who are consulted to provide assistance during a dermatitis outbreak, such as would be used by dermatologists to identify the causative agent, or what might be appropriate questions to ask a vendor related to performance or safety testing performed on a skin care product, or what toxicological test results might indicate about the risks of cutaneous exposure a new chemical might present if brought into the manufacturing process.

This section provides an overview of a range of methods for testing the effects of compounds on the skin, penetration through the skin, and of products protecting the skin. For a compendium of regulations, both national and international, regarding skin exposures and testing, please refer to Que Hee and Boeniger, 2000.

Allergens

Three types of tests are used to evaluate skin sensitization potential. *Predictive* tests are used to identify allergenic substances. The *diagnostic* test is used to determine what substance(s) may be causing dermatologic problems. Finally, *use* tests are intended for large-scale trials to provide assurance of the safety of chemicals in a particular combination before they enter the marketplace.

PREDICTIVE TESTING

Contact sensitizing chemicals can be identified before human exposure occurs by testing with one or more of several available animal assays. Historically, the most common of the animal tests included the Draize test (Draize, 1959), the Buehler test (Buehler 1965; Buehler and Griffith, 1975), and the Guinea Pig Maximization test (Magnusson and Kligman, 1969, 1970).

In the Draize guinea pig test, induction is performed by 10 intradermal injections, or 10 epicutaneous applications, given once every other day. Two weeks after the last injection is given, the animals are challenged by an intradermal test solution. This test is not used as commonly today because only potent allergens can be readily detected with it.

In the Buehler test, induction is attempted by occlusive topical application for a period of six hours, repeated three times at weekly intervals. Two weeks after the last application, the animals are challenged by a six hour occlusive patch.

The Guinea Pig Maximization (GPM) test attempts to induce sensitization in two stages. First, the test substance, with or without Freund's complete adjuvant, is injected into the shoulder region. Next, after one week, the test substance is applied to the same area under occlusion for 48 hours. Finally, two weeks after topical application, the animals are challenged by an occlusive patch for a period of 24 hours. Skin sites are examined 2, 24, and 48 hours after the test compound is removed. Red and swollen sites indicate sensitization. The GPM is the most sensitive of these three assays.

Both the Buehler test and GPM, as well as several alternatives, are currently accepted by regulatory authorities worldwide for screening the sensitization potential of chemicals. These tests have been reviewed in detail elsewhere (Botham et al., 1991; ECETOC, 1990; U.S. EPA, 1998). Several newer assays are gaining popularity and acceptance by regulatory authorities, including the mouse ear sensitivity test (MEST) and local lymph node assays (LLNA). The advantages of these newer tests is that they offer objective endpoints (Prevo et al., 1996; de Silva, 1996). The MEST test involves the measurement of challenged-induced increases in the ear thickness of previously sensitized mice. A response is regarded as positive if a 20% increase in mouse ear thickness occurs. Swelling can be easily quantified by measuring the ear thickness with calipers. Unlike the other tests that measure the elicitation phase response to the chemical, the LLNA evaluates the induction phase response. Animals are exposed to one of three concentrations of test chemical or vehicle on the dorsal surface of the ears for three consecutive days. Following two days of rest, the animals are injected intravenously with radiolabeled nucleotide (^{3}H-thymidine or ^{125}I-iododeoxyuridine). Five hours later, the lymph nodes draining the site are removed from the sacrificed animals and made into single cell suspensions. The amount of radioactivity incorporated into the test cell samples is an indication of induction potency, with at least a three-fold increase being considered positive. The LLNA was the first assay to be approved by the Interagency Coordinating Committee on the Validation of Alternative Test Methods (ICCVAM), a consortium of 14 federal agencies, and was recommended for acceptance by federal regulatory agencies (NIH, 1999). The EPA recognizes these screening tests as satisfactory for designating a compound as a positive sensitizer with respect to compliance with the Toxic Substances Control Act (TSCA), but if either test does not indicate sensitization, the test substance should be further tested using guinea pigs before designating it as a nonsensitizer (U.S. EPA, 1998).

These newer assays compared well to the GPM and Buehler tests, and the results of different assays can be complementary (Gad, 1988). Although these tests are generally predictive of whether a chemical can sensitize humans, they are not 100% accurate and human sensitization may still occur. It is also noteworthy that tests like the GPM assay do not always rank chemicals in the same order of potency as occur in the human assay (Basketter and Chamberlain, 1995).

USE TESTS

Pre-marketing tests on human volunteers is sometimes performed on products intended for large populations (Gerberick, 1994). The intent of human testing is to confirm safety at the administered test concentration, not sensitization potential. One such approach is to employ the human repeat insult patch test (HRIPT), which can provide an estimate of worst case response through an exaggerated or anticipated product misuse exposure scenario (Gerberick, 1994). The HRIPT should be designed to anticipate such usage factors as chemical dose, chemical bioavailability (e.g., matrix), concentration per surface area, duration of exposure, body location, presence of any skin penetration enhancer or vehicle, primary skin irritation potential, and the extent of occlusion that might occur.

DIAGNOSTIC TESTING

There are several test protocols for determining response associated with immediate hypersensitivity sensitization in humans. For patients with suggestive symptoms, the prick test is most often used for confirmation of the offending agent (Hannuksela, 1987). This consists of placing drops of allergen solutions on the skin of the forearm, upper arm, or upper back, and then piercing the sites of the drops using a special prick test lancet. After 15 to 20 minutes, the wheals are measured in two directions, perpendicular to each other. The result is compared to the positive response from a standard histamine solution. A positive reaction to the test allergen is recorded if the reaction is at least 3 mm and at least half the size of that produced by histamine. Other possible variations of the prick test procedure which are less standardized include the scratch test, and for special applications where the chemical is volatile, the scratch-chamber test, occlusive patch, and chamber tests might be used.

Delayed type allergic sensitization is commonly diagnosed by occlusive patch testing on the skin. This approach to testing for skin sensitization and appropriate test concentration dosages has evolved over a period of 100 years (Fischer and Maibach, 1987). For convenience of application, Finn Chambers® (Epitest, Helsinki, Finland), which are 12-mm in diameter and attached to a self-adhesive tape are often used. A filter paper disc is placed into each chamber and wetted with about 0.06 ml (1 drop) of test solution, or a 5-mm ribbon of petrolatum-based antigen is applied (no filter paper). The test materials are usually left in place for 24 or 48 hours, removed, and visually scored by the dermatologist according to degree of erythema, pustule formation, and swelling or oozing (Marks and Deleo, 1992). A newer application product consisting of 24 common allergens, referred to as the Pharmacia TRUE test®, offers advantages in that the allergens are ready to use in hydrophilic gels, coated on a water-impermeable sheet of polymer, and dried to a thin film. When the test patches are taped to the skin, the film is hydrated by perspiration and transformed into a gel from which the allergen migrates into the skin (Fischer and Maibach, 1989). The normal disappearance time for a positive patch test reaction is one to two weeks, but test reactions may remain visible for longer than one month.

Irritants

PREDICTIVE TESTING

For 50 years, the primary method for screening chemicals for skin irritancy worldwide has been some modification of the original Draize test (Draize et al., 1944; Draize, 1959). The method has established the albino rabbit as the animal of choice for predicting human dermal toxicity from topical applications (Prevo et al., 1996). The test results have long been accepted by all governmental bodies in the United States and European regulatory agencies, including the organization for Economic Cooperation and Development (OECD, 1981, 1992), as the primary indication of irritancy (Patrick and Maibach, 1994). Although widely adopted, the test is recognized to have limitations. Following its original protocol, it tends to over-predict irritation, and is generally unable to distinguish mild from moderate irritants. Rather, the rabbit model is best able to predict severe irritants.

The method entails placing a powder, liquid, gel, or film under an occlusive covering on the skin for 24 hours. In this test, a known amount (0.5 ml of a liquid or 0.5 gm of a solid or semisolid) of the test substance is introduced under a one square inch (6.45 cm^2) gauze patch. A total of 12 albino rabbits are tested, six with intact skin and six with abraded skin. After removing the covering, the sites are evaluated 0.5 and 48 hours later, grading the response according to the criteria in Table 3-15. The primary irritation index is calculated by summing the combined erythema and edema values for each site and dividing by the number of observations. This produces a score that is categorized according to the grouping shown in Table 3-16 (Draize, 1944). Other animal models can be used (but generally do not meet regulatory needs).

Table 3-15
Evaluation of skin reactions

Erythema and eschar formation		Edema formation	
Score	**Reaction**	**Score**	**Reaction**
0	No erythema	0	No edema
1	Very slight erythema	1	Very slight edema
2	Well-defined erythema	2	Well-defined edema
3	Moderate to severe erythema	3	Moderate to severe edema
4	Severe erythema to eschar formation (injuries in depth)	4	Severe edema (raised >1mm and extending beyond area of experiment)

Table 3-16
Irritation index

Index Score	Irritant Category
0.0 - 0.5	None or negligible
0.6 - 2.0	Mild
2.1 - 5.0	Moderate
5.1 - 8.0	Severe

The OECD modification of the Draize protocol entails applying 0.5 ml or 0.5 gram of the pure test substance on intact rabbit skin for not more than four hours. The substance is applied under an occlusive patch as before. As of 1995, the results of using this protocol for evaluating 176 substances had been collated (ECETOC, 1995).

A newly developed test, named the Corrositex®, can effectively assess the corrositivity of chemicals *in vitro* using a proteinaceous membrane and a color changing reagent that indicates breakthrough if dissolution of the membrane is caused by the test substance (Federal Register, 1999). The method was reviewed by the National Toxicology Program Interagency Coordinating Committee of the Validation of Alternative Methods, and is presently accepted by the U.S. Department of Transportation for the assessment of potentially corrosive compounds for labeling purposes (U.S. Dept. Transportation, 2000).

CLINICAL TESTING

Human testing can be performed following preliminary range-finding experiments with animals. Various types of occlusive coverings are available, and through comparison some differences in results have been found (York et al., 1995). Test guidelines for assessing skin irritation potential of cosmetic-finished products, which could be used to evaluate workplace skin care products as well, have been published (Walker et al., 1996). Several wash tests have been used to evaluate the irritation of soaps and synthetic detergents. Variations include immersion tests, repeated open tests, repeated occlusive tests, one-time occlusive tests (Tupker and Coenraads, 1996). These protocols typically entail visual observation of the skin condition, but objective endpoints can be measured. A brief review of alternative tests of the irritancy of compounds, including soaps and detergents, has been recently published (Simion, 1996). These include a 21-day cumulative irritation test for any compound and exaggerated washing tests for soaps and detergents.

More recently, the use of squamometry analysis has been reported to be sensitive to detecting surfactant-induced, sub-clinical skin alterations (Charbonnier et al., 1998). Squamometry is the combined use of several possible techniques for measuring the condition of corneocytes from the stratum corneum after adhesive tape strip samples are taken. After surfactants are used on the skin for a period of time, adhesive skin strip samples are taken and compared to unexposed skin surfaces. One treatment of the sample involves staining with toluidine blue-basic fuchisin solution. The disruption to corneocytes is measured objectively by reflectance colorimetry after staining, which selectively stains cells with non-intact membranes. The results of the staining are reported as a colorimetric index of mildness (CIM), which is related to exposure to a reference substance such as water (Goffin et al., 1997, 1997a). Another treatment of skin strip samples involves scoring intercorneocyte cohesion. The cells may come off the skin surface in conditions ranging from a large uniform sheet (best case) to large clumps with many isolated cells and occurrence of lysis (Charbonnier et al., 1998). Finally, the amount and distribution of dye found in cells is another parameter that can be scored.

Permeation

Skin permeation can be determined using *in vivo* or *in vitro* methods. There are advantages to using each. Because of potential differences between human and animal skin, and living versus dead skin, human *in vivo* testing would provide the most accurate measurement. However, because of the ethical concern of intentionally exposing humans to potentially toxic compounds, animal models and *in vitro* human skin techniques have been developed and are presently most often used.

When considering the choice of animal species for performing *in vivo* percutaneous permeation testing, a principle factor to consider is possible differences in skin permeability. These differences are inherent to the physical make-up of the stratum corneum (size and shape of corneocytes), number of hair follicles, lipid make-up, and thickness of the stratum corneum in each species.

Comparisons undertaken with a variety of compounds on the extent of percutaneous penetration in various animal species and humans have found that, generally, the species can be ordered as follows, from most permeable to least: mouse >rabbit >rat >guinea pig >weanling pig >human >monkey. However, this relative ranking is very dependent on the chemical tested. Nevertheless, the best animal models for human comparison with few exceptions are the weanling pig and rhesus monkey (Wester and Maibach, 1996).

Several approaches have been used for studying percutaneous absorption *in vivo*. These include recovery of radioactive labeled compounds in excreta or blood, surface recovery and surface disappearance measurement, or biological/pharmacological response studies. The use of labeled compounds entails comparing the appearance of the compound in excreta or blood after an intravenous versus a topical administration of the compound. Due to possible differences in the metabolism of the compound through each route, care must be given to label the compound at an intact site that is not affected by metabolism. Care must also be taken to determine if route of administration alters the urinary-fecal excretion ratio, as may occur with some compounds when topically administered (Carver and Riviere, 1989). Another approach is surface recovery studies, which determine the amount of topically applied compound that can be removed from the skin after a defined period of time. The assumption is that chemical not removed has been absorbed into the skin. This assumption has lead some researchers to erroneously high estimates of percutaneous absorption of volatile compounds due to evaporative loss. The compound can be kept in place under a patch or transdermal delivery device during the exposure period, or use of a sorbent to collect volatilized chemical, followed by wiping to remove the residue at the end of the exposure (Reifenrath and Spencer 1989; Reifenrath et al., 1989b). Surface disappearance studies may use radio-labeled compounds, FTIR, or photoacoustic methods to measure the change in concentration of the compound as it leaves the surface of the skin (Takamoto 1995; Higo et al., 1993; Touraille et al., 1997). Another technique is adhesive tape stripping, which has been employed to remove consecutive layers of the stratum corneum for chemical quantification. Rougier and co-workers have determined a high correlation between how much compound can be recovered within the skin after 30 minutes of exposure and how much is systemically absorbed and excreted in 96 hours (Rougier et al., 1983, 1985; Dupuis et al., 1986). Comparing *in vivo* tape stripping to *in vitro* tape stripping also correlates well (Trebilcock et al., 1994).

The rat *in vivo* animal model is favored by the U.S. EPA for percutaneous absorption studies (U.S. EPA, 1984). The guidelines call for using young male rats (these have twice the stratum corneum thickness of female rats and are more comparable to human skin), application of a minimum of three doses to non-occluded application sites on the back or shoulder, and a mass-balance approach where the total animal, excreta, and application site are included in the determination of recovery of the compound. The choice of rat was due to its convenient size, plentiful availability, low cost, and extensive data on toxic response and metabolism (Zendzian, 1989, 1991). Although rat skin is generally more permeable than human skin, it is not greatly different, and provides information on relative differences in permeation of chemicals through the skin.

A good deal of research has been reported on *in vitro* permeation methods. Generally, there is good agreement between *in vivo* and *in vitro* results if the studies have been performed carefully and all the many variables have been considered (Bronaugh and Collier, 1991). In the U.S., the Food and Drug Administration has had a major influence on providing guidelines for performing *in vitro* testing (Bronaugh and Collier, 1991a). Either animal or human skin can be surgically removed with a dermatome and used for *in vitro* percutaneous testing (Bronaugh et al., 1982). Human skin is obtained from surgical operations or cadavers. Some important considerations regarding the condition of *in vitro* skin include its original anatomical location, how it was treated prior to removal, its thickness, age of the donor, and length and conditions of storage prior to use. Techniques are available to test viability, using glucose utilization or enzymatic activity for instance (Kao et al., 1984; Macperson et al., 1991). Diffusion of tritiated water can be used to test barrier integrity (Franz and Lehman, 1990). Because inter-individual variation in skin permeability exists for all animal species, it is recommended to obtain samples from several individuals. During *in vitro* testing in one study, the overall mean inter-individual human skin variability was 66% ± 25, and the intra-individual variability was 43% ± 25 (Southweil et al., 1984). *In vivo* variability was appreciably less.

The test systems for *in vitro* testing basically consists of either a two-chamber device or only one chamber. The two chamber device separates the challenge solution from the receptor solution by the skin membrane. The one chamber system has a chamber beneath the skin, but is open to the environment above the skin. While the receptor chamber is stirred, samples are removed periodically to quantify the compound that may have penetrated the skin. Initially, only the static test cell was available to determine permeation through skin (Franz, 1975). Later, a flow-through cell system was introduced to automate sample collection from a one-chamber cell (Bronaugh and Stewart, 1985). Since the physiological receptor fluid is continually replaced, viability is maintained longer. Comparisons between the static and flow-through cells have been favorable (Clowes et al., 1994). When the challenge compound is highly lipophilic, it is important to choose a receptor fluid that is physiologically compatible and will not damage the skin. One solution was to add polyethylene glycol 20 oleyl ether to the receptor fluid (Bronough and Stewart, 1984). A special protocol for experimentally determining the effect of damaged skin on percutaneous absorption can supplement the data that is obtained from intact skin (Bronaugh and Stewart, 1985a). Details on setting up some of the various test apparatus are available elsewhere (Hawkins, 1990). A refinement that allows the collection of volatilized compounds has been developed to provide a total mass accounting of the compound after it has been applied to the skin (Hawkins and Reifenrath, 1984). At the present time, there are two *in vitro* test protocols in review that will, hopefully, harmonize testing among laboratories. These are being proposed by the Organization for Economic Cooperation and Development (OECD, 2000a, 2000b) and the U.S. Interagency Testing Committee (ITC) (Walker et al., 1996; Federal Register, 1999b).

Barrier Function

The disruption of the stratum corneum (SC) during various types of physical or chemical injury results in cytokine production and promotes an inflammatory skin reaction (Wood et al.,1992). Grubauer, et al., (1987) found that the water flux is responsible for signaling maintenance of homeostasis of the barrier by directing intercellular lipid synthesis and barrier recovery. Associated with disruption of the barrier function is a general increase in trans-epidermal water loss and swelling of the SC, accompanied by an increase in percutaneous penetration of exogenous compounds (Lamaud et al., 1984). The process of disruption may be multifaceted, that is, there are various types of damage that can occur and various ways the skin responds.

For instance, the extraction of skin lipids by solvents is not totally responsible for the disruption induced (Abrams et al., 1993).

Novel instrumental devices are commercially available to measure the changes in the surface skin that are related to the inflammatory process and barrier integrity. This is a relatively new field of dermatology, and the measurement of bio-physiological properties (commonly referred to as using "skin bioengineering" or "electronic biopsy" techniques) has undergone dramatic development during the past 10 years (Berardesca and Distante, 1995). These instrumental means of objectively measuring the skin can supplement the subjective evaluation of the skin by visual inspection (Edwards and Marks, 1995). However, proper interpretation of the results must be preceded by a good understanding of influencing variables and the limitations of each technique (Serup and Jemec, 1995). Interpretable results are very dependent upon a sound study protocol.

Transepidermal water loss (TEWL) and capacitance measurements are presently the most commonly used measures. TEWL, or evaporimetry, measures the water pressure gradient inside of a small hollow probe at two different fixed heights situated perpendicularly above the skin surface and within the zone of diffusion. The hydrosensors are paired with temperature sensors (thermistors). From the gradient, the evaporative TEWL value, in grams/m^2/hr, is digitally displayed in the main unit. The original instrument for TEWL was the Evaporimeter® (ServoMed, Kinna, Sweden), although newer instruments have become commercially available (Tewameter, Courage & Khazaka, Germany; DermaLab, Cortex Technology, Denmark). One of the most significant limitations of this approach is its sensitivity to variations in the microclimate that can affect perspiration. For consistent results, the ambient temperature and humidity must be controlled to about 20°C and 40%, respectively, and the subjects must be allowed to equilibrate under the controlled conditions for 20 to 30 minutes before testing. Even under controlled conditions, the intra-individual coefficient of variation (CV) of repeated studies in the same subject is about 9.1% (Blichmann and Serup, 1987). Inter-individual variation at the forearm was shown to be higher, with a CV of 31% to 57% (Blichmann, 1987). Regional differences in TEWL can be considerable, thus, it is important to test the same area for consistency (Cua et al.,1990). TEWL through the palm is much greater than other body regions. It is best to have individual "baseline" measurements of people before exposures, if possible. Guidelines for properly using this technique have been published along with performance results under a variety of conditions (Pinnagoda et al., 1990; Pinnagoda, 1994).

The TEWL method has been most often used to evaluate the response to acute exposure to irritants, and shows good correlation with regional skin permeability differences (Lotte et al., 1987). Preliminary evidence suggests that TEWL may be able to distinguish between allergic and irritant reactions after patch testing by measuring TEWL at 48 and 72 hours post application (Giorgini et al., 1996). TEWL does not appear sensitive to differentiating between skin involved in chronic dermatitis and healthy skin, possibly due to thickening of the chronically affected skin (Van der Valk and Tupker, 1994). Further research is needed to fully appreciate the relationships of TEWL and its relationship to alteration of SC and total skin thickness, or to differences in corneocyte size and SC patterns (Zienicke, 1992). An increase in TEWL does not necessarily imply that the SC is altered. Both the diffusion coefficient for water through the SC (possibly affected by modification of the SC) as well as the water activity gradient on either side of the SC (possibly affected by underlying inflammation and edema) could both affect TEWL rate.

Skin capacitance and conductance are electrical measurements related to internal and superficial water content, respectively. One commonly employed instrument used to measure skin capacitance is the Corneometer® (Courage & Khazaka, Cologne, Germany), which is reported to measure skin hydration down to a depth of 0.1 mm (Blichmann and Serup, 1988). A competing device is produced by Cortex Technology (Denmark). Conductance might be measured with the Skicon® (IBS Ltd, Tokyo, Japan), or the Nova Dermal Phase Meter (DPM) 9003 (Nova, Gloucester, MA) (Gabard and Treffel, 1994), which best measure very superficial moisture. Most studies indicate that the capacitance measurement tends to be less sensitive to moisture content in the surface stratum corneum, but to be more robust to variations in surface environmental conditions than conductance measurements (Tagami, 1994; Loden, 1994; Blichmann and Serup, 1988; Moss, 1996). However, there is a notable linear effect of external relative humidity on the capacitance hydration values that can be corrected for if the relative humidity is known. Otherwise, it is again recommended that measurements be obtained with uniform relative humidity during all readings, preferably at 40% to 50% (Barel and Clarys, 1995). Unlike TEWL, capacitance seems capable of measuring differences in later stage irritant dermatitis (Serup, 1995).

To date, using TEWL, capacitance or conductance measurements in the workplace as a screening tool has scarcely been reported, but might be most likely to be used in the future due to the portability and relatively low cost of the available devices. One approach might be to use these measures before and after starting work, or before and after an intervention to reduce irritant exposures. Other available methods include measurement of skin surface lipid and skin friction (Wolfram, 1989; Cua et al., 1995), skin hardness as measured by resistance to surface pressure (Falanga and Bucalo, 1993), and profilometry, which measures surface contour roughness or topography (Vieluf, 1992).

A final technology, which has been available for quite some time, is ultrasound echography (Altmeyer, 1992). Newer 20- and 30-MHz B-scanning (two-dimensional) provided with a computer software program for image analysis (Dermascan, Cortex Technology, Denmark) can be used to measure the epidermal hyper-reflective thickness and edema in the dermis as seen by hypo-echogenicity. B-scanning echography has been found to detect short-term occlusion effects for a longer duration after removing the covering than TEWL or capacitance measurements (Seidenari et al., 1996), and has been recommended for measuring late irritation responses (Serup, 1995).

SKIN PROTECTIVE (BARRIER) PRODUCT TEST PROTOCOLS

There are a variety of approaches that have been used to evaluate the effectiveness of topical skin protective product (SPP) lotions and creams. Manufacturers of these products may claim that theirs is effective in reducing chemical absorption, and irritation or allergic responses. Unfortunately, there are no standard approaches for evaluating effectiveness, but rather choice of protocol is a function of cost, laboratory ability, type of exposure, and type of outcome to be prevented. Some of the approaches are extensions of the *in vitro* and *in vivo* permeability assays described above, or tests for allergen or irritant response. Other approaches are rather novel.

One low-tech approach is to apply a colored dye upon skin previously treated with the SPP, as well as on skin not treated as a control. The efficacy of the SPP is determined by the degree to which the barrier prevents staining of the skin. There are a great number of dyes that might be used for this assay. Some of the water soluble dyes that have been used are eosin, methyl violet 2B, methyl blue, crystal violet, and oil red O (a lipophilic dye). The color intensity on the skin can be scored visually, or with the aid of skin reflectance measurement or colorimetry (Marks et al., 1989; Treffel et al., 1994; Olivarius et al., 1996; Zhai and Maibach, 1996). A variation of this approach is to apply a mineral oil that produces a brilliant violet fluorescence when examined in ultra-violet light. The oil can be applied over protective creams and, after subsequently washing the skin, the amount of oil adhering to the skin is determined (Cruickshank, 1948; Morris and Maloof, 1953).

In vivo testing of allergen response has been performed on human volunteers known to be sensitized to allergens. The basis of the test is to compare the allergic response in untreated skin to skin covered with the barrier product (Fischer and Rystedt, 1983; Blanken et al., 1987).

Percutaneous absorption, with and without treatment with a barrier product, can be determined *in vivo* with human volunteers by monitoring exhaled breath concentrations of the topically applied chemical. This test has been applied with toluene and m-xylene as the permeant, and is only applicable to volatile compounds (Guillemin et al., 1974; Lauwerys, 1978). Monitoring blood or urine concentrations as an indication of percutaneous absorption might also be conducted (Boman et al., 1982).

Some questions remain as to how bio-physiological measurements relate to all the possible effects on the skin. For instance, while erythema correlates to an increase in TEWL, and an increase in inflammation is generally associated with an increase in skin permeability, one cannot necessarily conclude that a reduction of inflammatory response and TEWL indicates that there is a corresponding reduction in penetration. In other words, there may be a mediation of irritant response that is not quantitatively related to penetration. Nevertheless, biophysical measures have been used to evaluate the efficacy of commercially available barrier creams *in vivo*, including corneometry, laser Doppler flow to measure subcutaneous blood flow, pH-metry, TEWL, colorimetry to provide objective measurement of erythema, and sebumetry (measure of sebum) (Grunewald et al., 1995a).

Another *in vivo* test measures the magnitude of irritant response when the skin is exposed to a standard amount of irritating chemical. Repeated topical dosing is performed for up to two weeks, and thus, the test is referred to as the repetitive irritancy test (RIT). Various inorganic acids, alkalies, and organic solvents have been tested in this way. The irritancy response is measured by visual scoring of erythema, and can be supplemented by objective bio-physiological techniques (Frosch et al., 1993, 1993a, 1993b; Frosch and Kurte, 1994; Schluter-Wigger and Elsner, 1996). Others have taken skin biopsies of the dosed areas and performed histopathological assessments of skin lesions (Mahmoud et al., 1985).

A fairly new *in vitro* technique uses a variation of corneosurfametry. Corneosurfametry is an assay that is initially performed by harvesting superficial corneocytes from human volunteers with a cyanoacrylate-coated slide. The resulting uniform layer of stratum corneum is exposed to offending chemicals which may disrupt the corneocytes for about two hours. Like squamometry, the disruption of corneocytes is measured objectively by reflectance colorimetry after staining with toluidine blue-basic fuchisin solution, which selectively stains non-intact cells. The resulting staining is reported as a colorimetric index of mildness (CIM), which is related to exposure to a reference substance such as water (Goffin et al., 1997, 1997a). To evaluate the effectiveness of skin protective products, the skin strips are first coated with the SPP, followed by an irritating chemical. After a suitable waiting period, the chemical and SPP is washed off and the stain is applied. The degree of skin staining, or CIM, is recorded and compared to staining of unprotected skin strips. Various SPPs can be evaluated quickly and inexpensively by the relative chemical protective (RCP) score (Goffin et al., 1998).

QUESTIONS

3.1. Why are occupational skin exposures tending to become more important over time?

3.2. What are some of the things that occupational health and safety professionals can do to identify the cause of a skin related problem?

3.3. If you noticed that an employee had signs of a rash on his skin, but does not make an issue of it by reporting the condition or seeking professional treatment, is this reason for any concern?

3.4. If both low molecular weight (LMW) and high molecular weight (HMW) compounds can effectively sensitize by the cutaneous route, why aren't more workers sensitized?

3.5. The company you work for is considering a new chemical product to be used in its widgets manufacturing line. It is a water-based formulation intended to reduce volatile organic compound (VOC) emissions. The old process is mostly automated, but you have observed that workers will occasionally touch contaminated work surfaces. Periodically, obviously wet parts have to be handled manually and the workers wear rubber gloves for this task. The MSDS for the new product includes typical precautions such as not to ingest, do not get it into your eyes, and avoid skin contact. There is no TLV® or other criteria for the primary ingredient, but the MSDS indicates that the product contains 20% by volume of this compound in water, which is also approximately the water saturation concentration for this chemical. You would like to better determine how much skin contact is tolerable and formulate your administrative, training, and personal protection program accordingly. You know that the MW is 225 and you obtain a P_{ow} of 2 from reference physical-chemical data sources. You also note that the no observable adverse effect level (NOAEL) is 0.05 mg/kg-d for the primary active ingredient. How would you calculate if there is a high or low likelihood that physical contact with this material could adversely affect workers' health?

3.6. How might the above estimate of risk be flawed when estimating what level of precautions are necessary?

3.7. You are out on the production floor one day and notice one of the employees has a rash on his hands. What course of action would you take? Discuss how you would go about identifying the problem, seeking a solution, and preventing others from developing similar conditions.

REFERENCES

Abrams, H.K.; Warr, P.: Occupational Diseases Transmitted via Contact with Animals and Animal Products. Indust. Med. Surgery 20:341-351 (1951).

Abrams, K.; Harvell, J.D.; Shriner, D.; et al.: Effect of Organic Solvents on in Vitro Human Skin Water Barrier Function. J. Invest. Dermatol. 101:609-613 (1993).

Adams, R.M.: Occupational Skin Disease, Grune and Straton, New York, pp. 27-57 (1990).

Altmeyer, P.: Ultrasound in Dermatology. Springer-Verlag, New York (1992).

American Conference of Governmental Industrial Hygienists (ACGIH®), TLVs® and BEIs®; Threshold Limit Values for Chemical Substances and Physical Agents and Biological Exposure Indices, Cincinnati, Ohio (1999).

Anonymous: North American Contact Dermatitis Group: Epidemiology of Contact Dermatitis in North America. Arch. Dermatol. 108:537-540 (1973).

Ancona, A.: Biological causes. In: Occupational Skin Diseases, 2nd Ed., R.M. Adams, Ed., pp. 89-112, W.B. Saunders Co., Philadelphia (1990).

Angelini, G.; Rigano, L.; Foti, C.; Vena, G.A.; Grandolofo, M.: Contact Allergy to Impurities in Surfactants: Amount, Chemical Structure and Carrier Effect in Reactions to 3-dimethylaminopropylamine. Contact Dermatitis 34:248-252 (1996).

Arakawa, H.; Lotvall, J.; Kawikova, I.; et al.: Airway Responses Following Intradermal Sensitization to Different Types of Allergens: Ovalbumin, Trimellitic Anhydride and Dermatophagoides farinae. Int. Arch. Allergy Immunology 108:274-280 (1995).

Artuc, M.; Reinhold, C.; Stuttgen, G.; Gazith, J.: A Rapid Measurement for Measuring Drug Enrichment in Epidermis. Arch. Dermatol. Res. 268:129-140 (1980).

Askin, D.P.; Volkmann, M.: Effect of Personal Hygiene on Blood Lead Levels of Workers at a Lead Processing Facility. Am. Ind. Hyg. Assoc. J. 58:752-753 (1997).

Barber, E.D.; Teetsel, N.M.; Kolberg, K.F.; Guest, D.: A Comparative Study of the Rates of in Vitro Percutaneous Absorption of Eight Chemicals Using Rat and Human Skin. Fundam. Appl. Toxicol. 19:493-497 (1992).

Barel, A.O.; Clarys, P.: Measurement of epidermal capacitance, In: Handbook of Non-Invasive Methods and the Skin, J. Serup and G.B.E. Jamec, Eds., pp.165-170, CRC Press, Boca Raton (1995).

Barratt, M.D.: Quantitative Structure-activity Relationships for Skin Permeability. Toxic. in Vitro, 9:27-37 (1995).

Barry, B.W.: Penetration Enhancers, Mode of Action in Human Skin. Pharmacol. Skin 1:121-137 (1987).

Barry, B.W.: Lipid-protein Partition Theory of Skin Penetration Enhancement. J. Controlled Release 15:237-248 (1991).

Basketter, D.A.; Chamberlain, M.: Validation of Skin Sensitization Assays. Fd. Chem. Toxic. 33:1057-1059 (1995).

Basketter, D.A.; Griffiths, H.A.; Wang, X.M.; et al.: Individual, Ethnic and Seasonal Variability in Irritant Susceptibility of Skin: The Implications for a Predictive Human Patch Test. Contact Dermatitis 35:208-213 (1996).

Basketter, D.A.; Cookman, G.; Gerberick, G.F.: Skin Sensitization Thresholds: Determination in Predictive Models. Food Chem. Toxicol. 35:417-425 (1997).

Baur, X.; Chen, Z.; Allmers H.; Raulf-Heimsoth, M.: Results of Wearing Test with Two Different Latex Gloves With and Without the Use of Skin-protective Cream. Allergy 53:441-444 (1998).

Beezhold, D.H.; Kostyal, D.A.; Wiseman. J.: The Transfer of Protein Allergens from Latex Gloves. AORNJ 59:605-613 (1994).

Behl, C.R.; Flynn, G.L.; Kurihara, T.; et al.: Hydration and Percutaneous Absorption I. Influence of Hydration on Alkanol Permeation Through Hairless Mouse Skin. J. Invest. Dermatol. 75:346-351(1980).

Behl, C.R., Linn, E.E.; Flynn, G.L.; et al.: Permeation of Skin and Escar by Antiseptics I: Baseline Studies with Phenol. J. Pharmaceut. Sci. 72:391 (1983).

Behrens, V.; Seligman, P.; Cameron, L.; et al.: The Prevalence of Back Pain, Hand Discomfort, and Dermatitis in the U.S. Working Population. Am. J. Publ. Health 84:1780-1785 (1994).

Berardesca, E.; Distante, F.: Bioengineering: Methods in the Irritant Contact Dermatitis Syndrome, pp. 313-316, P.G.M. van der Valk and H.I. Maibach, Eds., CRC Press, Boca Raton (1995).

Berardesca, E.; Vignoli, G.P.; Distante, F., et al.: Effects of Water Temperature on Surfact-induced Skin Irritation. Contact Dermatitis, 32:83 (1995).

Bickis, U.: Investigation of Dermally Induced Airway Hyper-reactivity to Toluene Diisocyanate in Guinea Pigs. Ph.D. Dissertation, Department of Pharmacology and Toxicology, Queens University, Kingston, Ontario, Canada (1994).

Biddle, J.; Roberts, K.; Rosenman, K.D.; Welch, E.M.: What Percentage of Workers with Work-related Illness Receive Workers' Compensation Benefits? JOEM, 40:325-331 (1998).

Bird, M.G.: Industrial Solvents: Some Factors Affecting Their Passage into and Through the Skin. Ann. Occup. Hyg. 24:235-244 (1981).

Birmingham, D.J.: Prolonged and Recurrent Occupational Dermatitis. Occupational Medicine: State of the Art Reviews, 1:349-355 (1986).

Bjornberg, A.: Irritant Dermatitis, In: Occupational and Industrial Dermatology, 2nd Ed., H.I. Maibach, Ed., Year Book Medical Publishers, Chicago (1987).

Blank, I.H.; Scheuplein, R.J.: Transport Into and Within the Skin. Br. J. Dermatol. 81 (suppl. 4), 4-10 (1969).

Blanken, R.; Nater, J.P.; Veenhoff, E.: Protection Effect of Barrier Creams and Spray Coating Against Epoxy Resins. Contact Dermatitis 16:79-83 (1987).

Blichmann, C.W.; Serup, J.: Reproducibility and Variability of Transepidermal Water Loss Measurement. Studies on the Servomed Evaporimeter. Acta. Derm. Venerol. (Stockh) 67:206-210 (1987).

Blichmann, C.W.; Serup, J.: Assessment of Skin Moisture: Measurement of Electrical Conductance, Capacitance and Transepidermal Water Loss. Acta Derm. Vernereol (Stockh) 68:284-290 (1988).

Boeniger, M.: Chemical Protective Clothing and the Skin: Practical Consideration of Possible Adverse Effects and Performance, In: Chemical Protective Clothing, 2nd Edition, Anna, D., Ed. AIHA Press, Fairfax, VA, in press (2002).

Boman, A.; Wahlberg, J.E.; Johanson, G.: A Method for the Study of the Effect of Barrier Creams and Protective Gloves on the Percutaneous Absorption of Solvents. Dermatologica 164:157-160 (1982).

Boman, A.; Hagelthorn, G.; Magnusson, K.: Percutaneous Absorption of Organic Solvents During Intermittent Exposure in Guinea Pigs. Acta Derm. Venereol. (Stockh) 75:114-119 (1995).

Bos, P.M.J.; Mik, G.; Bragt, P.C.: Critical Review of the Toxicity of Methyl N-butyl Ketone: Risk from Occupational Exposure. Am. J. Ind. Med. 20:175-194 (1991).

Botham, P.A.; Basketter, D.A.; Maurer, T.; et al.: Skin Sensitization - A Critical Review of Predictive Test Methods in Animals and Man. Fd. Chem. Toxic. 29:272-286 (1991).

Breit, R.; Turk, B.M.: The Medical and Social Fate of the Dichromate Allergic Patient. Br. J. Dermatol. 94:349-351 (1976).

Bress, W.C.; Bidanset, J.H.: Percutaneous in Vivo and in Vitro Absorption of Lead. Vet. Hum Toxicol. 33:212-214 (1991).

Bronaugh, R.L.; Stewart, R.F.; Congdon, E.R.: Methods for in Vitro Percutaneous Absorption Studies I. Animal Models for Human Skin. Toxicol. Appl. Pharmacol. 62:481-488 (1982).
Bronaugh, R.L.; Stewart, R.F.: Methods for in Vitro Percutaneous Absorption Studies II. Hydrophobic Compounds. J. Pharmaceut. Sci. 73:1255-1258 (1984).
Bronaugh, R.L.; Stewart, R.F.: Methods for in Vitro Percutaneous Absorption Studies, IV: The Flow-through Diffusion Cell. J. Pharmaceut. Sci. 74:64-67 (1985).
Bronaugh, R.L.; Stewart, R.F.: Methods for in Vitro Percutaneous Absorption Studies, V. Permeation Through Damaged Skin. J. Pharmaceut. Sci. 74:1062-1066 (1985a).
Bronaugh, R.L.; Collier, S.W.: In Vitro Percutaneous Absorption, In: Dermatotoxicology, 4th edition, F. Marzulli and H.I. Maibach, Eds., pp. 61-74, Taylor & Francis, Bristol, PA (1991).
Bronaugh, R.L.; Collier, S.W.: Protocol for in Vitro Percutaneous Absorption Studies, In: In Vitro Percutaneous Absorption: Principles, Fundamentals, and Applications, R. Bronaugh and H.I. Maibach, Eds., pp. 237-241, CRC Press, Boca Raton (1991a).
Bruze, M.; Almgren, G.: Occupational Dermatoses in Workers Exposed to Resins Based on Phenol and Formaldehyde. Contact Dermatitis 19:272-277 (1988).
Bucks, D.A.W., Hinz, R.S.; Sarason, R.; Guy, R.H.: In Vivo Percutaneous Absorption of Chemicals: A Multiple Dose Study in
Rhesus. Fd. Chem. Toxic. 28:129-132 (1990).
Bucks, D.A.W.; Maibach, H.I.; and Guy, R.H.: Vivo Percutaneous Absorption: Effect of Repeated Application Versus Single Dose. In: R. Bronough and H. Maibach, Eds. Percutaneous Absorption, Vol. 2, pp. 633-651, Marcel Dekker, New York, (1989).
Bucks, D.A.W., Wester, R.C., Mobayen, M.M.; et al.: In Vitro Percutaneous Absorption and Stratum Corneum Binding of
Alachlor: Effect of Formulation Dilution with Water. Toxicol. Appl. Pharmacol. 100:417-423 (1989b).
Bucks, D.A.W., Maibach, H.I., Guy, R.H.: Occlusion Does Not Uniformly Enhance Penetration in Vivo. In: In Vitro Percutaneous Absorption: Principles, Fundamentals, and Applications, R. Bronaugh and H.I. Maibach, Eds., pp. 77-93, CRC Press, Boca Raton (1991).
Bucks, D.A.W.: Predictive Approaches II: Mass-balance Procedure, In: Topical Drug Bioavailability, Bioequivalence, and Penetration. V.P. Shah and H.I. Maibach, Eds., pp. 183-195, Phenum Press (1993).
Buehler, E.V.: Delayed Contact Hypersensitivity in the Guinea Pig. Arch. Dermatol. 91:171-177 (1965).
Buehler, E.V.; Griffith, F.: Experimental Skin Sensitization in the Guinea Pig and Man, In: Animal Models in Dermatology, H.I. Maibach, Ed., pp. 56-66, Churchill Livingstone, Edinburgh (1975).
Bunge, A.L.; Cleek, R.L.: A New Method for Estimating Dermal Absorption from Chemical Exposure 2. Effect of Molecular Weight and Octanol-water Partitioning. Pharmaceut. Rev. 12:88-95 (1995).
Bunge, A.L.; Cleek, R.L.; Vecchia, B.E.: A New Method for Estimating Dermal Absorption from Chemical Exposure 3. Compared with Steady-state Methods for Predicting and Data Analysis. Pharmaceut. Rev. 12:972-982 (1995).
Burke, F.J.; Wilson, N.H.; Cheung, S.W.: Factors Associated with Skin Irritation of the Hands Experienced by General Dental Practitioners, Contact Dermatitis 32:35-8 (1995).
Burckhardt, W.: Experimentelle Sensibilisierung Des Meerschweinchens Gegen Terpentinol. Acta derm. venereol. 19:359-364 (1939).
Burnett, C.A.; Lushniak, B.D.; McCarthy, W.; Kaufman, J.: Occupational Dermatitis Causing Days Away from Work in U.S. Private Industry, 1993. Am. J. Indust. Med. 34:568-573 (1998).
Burrows, D.: Prognosis in Industrial Dermatitis. Br. J. Dermatol. 87:145-148 (1972).
Carver, M.P.; Riviere, J.E.: Percutaneous Absorption and Excretion of Xenobiotics after Topical and Intraveneous Administration to Pigs. Fund. Appl. Toxicol. 13:714-722 (1989).
Chang, S.K.; Reviere, J.E.: Percutaneous Absorption of Parathion in Vitro in Porcine Skin: Effects of Dose, Temperature, Humidity, and Perfusate Composition on Absorption Flux. Fund. Appl. Toxicol. 17:494-504 (1991).
Charbonnier, V.; Morrison, B.M.; Payre, M.; Maibach, H.I.: Open Application Assay in Investigation of Subclinical Irritant Dermatitis Induced by Sodium Lauryl Sulfate (SLS) in Man: Advantages of Squamometry. Skin Research and Technology 4:244-250 (1998).

Cleek, R.L.; Bunge, A.L.: A New Method for Estimating Dermal Absorption from Chemical Exposure. 1. General Approach. Pharmaceut. Res. 10:497-506 (1993).
Clowes, H.M.; Scott, R.C.; Heylings, J.R.: Skin Absorption: Flow-through or Static Diffusion Cells. Toxic. in Vitro 8:827-830 (1994).
Cohen, N.; Modai, D.; Khahil, A.; Golik, A.: Acute Resin Phenol-formaldehyde Intoxication. A Life Threatening Occupational Hazard. Human Toxicol. 8:247-250 (1989).
Cooley, J.E.; Nethercott, J.R.: Prognosis of Occupational Skin Disease. Occupational Medicine: State of the Art Reviews 9:19-24 (1994).
Corley, R.A.; Markham, D.A.; Banks, C.; et al.: Physiologically Based Pharmacokinetics and the Dermal Absorption of 2-butoxyethanol Vapor by Humans. Fund. Appl. Toxicol. 39:120-130 (1997).
Courtheoux, S.; Pechnenot, D.; Bucks, D.A.; et al.: Effect of Repeated Skin Administration on in Vivo Percutaneous Absorption of Drugs. Br. J. Dermatol. 115:49-52 (1986).
Cronin, E.; Stoughton, R.B.: Percutaneous Absorption: Regional Variations and the Effect of Hydration and Epidermal Stripping. Br. J. Dermatology. 74:265-272 (1962).
Cronin, E.: Clinical Patterns of Hand Eczema in Women. Contact Dermatitis 13:153-161 (1985).
Cronin, M.T.D.; Dearden, J.C.; Moss, G.P.; Murray-Dickson, G.: Investigation of the Mechanism of Flux Across Human Skin In Vitro by Quantitative Structure-permeability Relationships. Eur. J. Pharm. Sci. 7:325-30 (1999).
Cruickshank, C.N.D.: The Evaluation of Skin Cleansers and Protective Creams for Workmen Exposed to Mineral Oil. Brit. J. Industr. Med. 5:204-212 (1948).
CTFA: Proceedings of the Pharmacokinetics and Topical Applied Cosmetics Symposium, Cosmetics Toiletries and Fragrance Association, CTFA Monograph 2, Research Triangle Park (1983).
Cua, A.B.; Wilhelm, K.-P.; Maibach, H.I.: Cutaneous Sodium Lauryl Sulphate Irritation Potential: Age and Regional Variability. Br. J. Dermatol. 123:607-613 (1990).
Cua, A.B.; Wilhelm, K-P.; Maibach, H.I.: Skin Surface Lipid and Skin Friction: Relation to Age, Sex and Anatomical Region. Skin Pharmacol. 8:246-251 (1995).
Dary, C.C.; Blancato, J.N.; Castles, M.; et al.: Dermal Absorption and Disposition of Formulations of Malathion in Sprague-Dawley Rats and Humans, In: Biomarkers of Human Exposure to Pesticides, pp. 231-263, M.A. Saleh, J. Blancato and C.H. Nauman, Eds., ACS Symposium Series No. 542, American Chemical Society (1994).
Davies, R.H.; Parker, W.L.; Murdoch, D.P.: Aloe Vera as a Biologically Active Vehicle for Hydrocortisone Acetate. J. Am. Podiatric Med. Assoc. 81:1-4 (1991).
de Cock, J.; Heederik, D.; Hoek, F.; et al.: Urinary Excretion of Tetrahydrophtalimide in Fruit Growers with Dermal Exposure to Captan. Am. J. Indust. Med. 28:245-256 (1995).
deGroot, A.C.: Patch Testing: Test concentrations and vehicles for 3700 chemicals, 2nd edition, Elsevier, Amsterdam (1994).
Delepine, S.: Observations of the Effects of Exposure to Arsenic Trichloride upon Health. J. Ind. Hyg. 4:346-364 & 410-423 (1923).
deSilva, O.; Basketter, D.A.; Barratt; M.D.: Alternative Methods for Skin Sensitisation Testing. ATLA Abstracts 24:683-705 (1996).
Discher, D.P.; Kleinman, G.P.; Foster, F.J.: Pilot Study for Development of an Occupational Disease Surveillance Method. HEW Publ. No. (NIOSH)75-162 (1975).
DiVincenzo, G.; Hamilton, M.: Fate of n-Butanol in Rats after Oral Administration and Its Uptake by Dogs after Inhalation or Skin Application. Toxicology and Applied Pharmacology; 48:317-325 (1979).
Dow Chemical Company, Material Safety Data Sheet: Isonate (R) 125M Pure MDI, Midland, MI (1996).
Downie, A.; Khattab,T.M.; Malik, M.I.A.; Samara I.N.: A Case of Percutaneous Industrial Methanol Toxicity. Occup. Med. 42 :47-49 (1992).
Draize, T.H.; Woodland, G.; Calvery, H.O.: Methods for the Study of Irritation and Toxicity of Substances Applied Topically to the Skin and Mucous Membranes. J. Pharmacol. Exp. Ther. 82:377-390 (1944).
Draize, T.H.: Intracutaneous Sensitisation Test on Guinea Pigs, In: Appraisal of the Safety of Chemicals in Foods and Cosmetics, pp. 46-59. Association of Food and Drug Officials of the United States, Austin: Texas State Department of Health (1959).

Drexler, H.; Goen, T.H.; Angerer, J.: Carbon Disulphide. II. Investigations on the Uptake of CS_2 and the Excretion of Its Metabolite 2-Thiothiazolidine-4-Carboxylic Acid after Occupational Exposure. Internat. Arch. Occup. Environ. Health 67:5-10 (1995).
Driessen, L.H.H.M.: A Group of Eczema Patients: Five Years Later (in Dutch). Tijdschr. Soc. Geneesk 60:41-45 (1982).
Drill, S.; Konz, J.; Mahar, H.M.: The Environmental Lead Problem: An Assessment of Lead in Drinking Water from a Multi-media Perspective. Office of Drinking Water. EPA-570/9-79-003. U.S. Environmental Protection Agency, Washington, DC (1979).
Drinker, P.: Hydrocyanic Acid Gas Poisoning by Absorption Through the Skin. J. Ind. Hyg. 1:1-2 (1932).
Duff, R.M.; Kissel, J.C.: Effect of Soil Loading on Dermal Absorption Efficiency from Contaminated Soils. J. Toxicol. Environ. Health 48:93-106 (1996).
Duprat, P.; Gradiski, D.; Marignac, B.: Irritant and Allergic Action of Two Isocyanates: Toluene Diisocyanate (TDI) and Diphenyl Methane Diisocyanate (MDI). Eur. J. Tox. 9: 41-53 (1956)
Dupuis, D.; Rougier, A.; Roguet, R.; Lotte, C.: The Measurement of the Stratum Corneum Reservoir: A Simple Method to Predict the Influence of Vehicles on In Vivo Percutaneous Absorption. Br. J. Dermatol. 115:233-238 (1986).
Dutkiewicz, T.: The Absorption of Aniline Vapors Through the Skin, Medycyna Pracy 8: 25-26 (1957).
Edwards, C.; Marks, R.: Evaluation of Biomechanical Properties of Human Skin. Clin. Dermatol. 13:375-380 (1995).
Effendy, I.; Loeffler, H.; Maibach, H.I.: Baseline Transepidermal Water Loss in Patients with Acute and Healed Irritant Contact Dermatitis. Contact Dermatitis 33:371-374 (1995).
Elias, P.M.; Goerke, J.; Friend, D.S.: Mammalian Epidermal Barrier Layer Lipids: Composition and Influence on Structure. J. Invest. Dermatol. 69:535-546 (1977).
Elias, P.M.; Cooper, E.R.; Korc, A.; Brown, B.E.: Percutaneous Transport in Relation to Stratum Corneum Structure and Lipid Composition. J. Invest. Dermatol. 76:297-301 (1981).
Emmett, E.A.: Allergic Contact Dermatitis in Polyurethane Plastic Molders. J. Occ. Med. 18:802-804 (1976).
Enk, A.H.; Katz, S.: Early Events in the Induction Phase of Contact Sensitivity. Soc. Invest. Dermatol. 99:39S-41S (1992).
Erjefalt, I.; Persson, C.G.A.: Increased Sensitivity to Toluene Diisocyanate (TDI) in Airways Previously Exposed to Low Doses of TDI. Clin. Exp. Allergy 22:854-862 (1992).
Esswein, E.J.; Boeniger, M.F.; Hall, R.M.; Mead, K.: NIOSH Health Hazard Evaluation Report on Standard Industries, San Antonio, Texas, HETA 94-0268-2618, National Institute for Occupational Health, Cincinnati, Ohio (1996).
ECETOC, European Centre for Ecotoxicology and Toxicology of Chemicals: Skin Sensitization Testing, Monograph No. 14, Brussels (1990).
ECETOC, European Centre for Ecotoxicology and Toxicology of Chemicals: Percutaneous Absorption, Monograph No. 20, Brussels (1993).
ECETOC, European Centre for Ecotoxicology and Toxicology of Chemicals: Technical Report No. 66, Skin Irritation and Corrosion: Reference Chemicals Data Bank. Brussels, Belgium (1995).
Far, H.S.; Pin, N.T.; Kong, C.Y.; et al.: An Evaluation of the Significance of Mouth and Hand Contamination for Lead Absorption in Lead-acid Battery Workers. Int. Arch. Occup. Environ. Health 64:439-443 (1993).
Falanga, V.; Bucalo, B.: Use of Durometer to Assess Skin Hardness. J. Am. Acad. Dermatol. 29:47-51 (1993).
Federal Register, Vol. 64, Number 118, June 21, pp. 33109-33111 (1999).
Federal Register, Proposed Test Rule for In Vitro Dermal Absoption Rate Testing of Certain Chemicals of Interest to Occupational Safety and Health Administration, Vol. 64, Number 110, June 9, pp. 31074-31090 (1999b).
Federal Republic of Germany, Berufskankheitenrisiken, Bundesgestaltfur Arbeitsschutzund Arbeitsforschung (1990).
Feldman, R.J.; Maibach, H.I.: Penetration of ^{14}C Hydrocortisone Through Normal Skin. Arch. Dermatol. 91:661-666 (1965).
Feldman, R.J.; Maibach, H.I.: Percutaneous Penetration of Some Pesticides and Herbicides in Man. Toxicol. Appl. Pharmacol. 28:126-132 (1974).
Fenske, R.A.; Horstman, S.W.; Bentley, R.K.: Assessment of Dermal Exposure to Chlorophenols in Timber Mills. Appl. Ind. Hyg. 2:143-147 (1987).

Fischer, T.; Maibach, H.I.: Easier Patch Testing with True Test. J. Am. Acad. Dermatol. 20:447-453 (1989).

Fischer, T.; Rystedt, I.: Skin Protection Against Ionized Cobalt and Sodium Lauryl Sulphate with Barrier Creams. Contact Dermatitis 9:125-130 (1983).

Fischer, T.; Maibach, H.I.: Patch Testing in Allergic Contact Dermatitis, and Update, In: Occupational and Industrial Dermatology, 2nd Edition, H.I. Maibach, Ed., pp. 190-210, Year Book Medical Publishers, Chicago (1987).

Fiserova-Bergerova, V.; Pierce, J.T.; Droz, P.O.: Dermal Absorption Potential of Industrial Chemicals: Criteria for Skin Notation. Am. J. Indust. Med. 17:617-635 (1990).

Fitzgerald, D.A.; English, J.S.C.: The Long-term Prognosis in Irritant Contact Hand Dermatitis, In: Irritant Dermatitis, New Clinical and Experimental Aspects. Curr. Probl. Dermatol. Vol. 23, pp. 73-76, P. Elsner and H.I. Maibach, Eds., Basel, Karger (1995).

Florence, T.M.; Lilley, S.G.; Stauber, J.L.: Skin Absorption of Lead. Lancet 16:157-158 (1988).

Flynn, G.L.: Physiological Determinants of Skin Absorption, In: Principles of Route-to-Route Extrapolation for Risk Assessment., pp. 93-127, T.R. Gerrity and C.J. Henry, Eds., Elsevier Science Publ. (1990).

Flynn, G.L.; Stewart, B.: Percutaneous Drug Penetration: Choosing Candidates for Transdermal Development. Drug Develop. Res. 13:169-185 (1988).

Forslind, B.: A Domain Mosaic Model for the Skin Barrier. Acta. Derm. Venereol. (Stockh.) 74:1-6 (1994).

Franz, T.J.: On the Relevance of In Vitro Data. J. Invest. Dermatol. 64:190-195 (1975).

Franz, T.J.; Lehman, P.A.: The Use of Water Permeability as a Means of Validation for Skin Integrity in In Vitro Percutaneous Absorption Studies. J. Invest. Dermatol. 94:525-530 (1990).

Freeman, S.; Maibach, H.I.: Systemic Toxicity in Man Secondary to Percutaneous Absorption, In: The Environmental Threat to the Skin, pp. 249-263. R. Marks and G. Plewig, Eds., Martin Dunitz, London (1992).

Fregert, S.: Occupational Dermatitis in a 10 Year Material. Contact Dermatitis 1:96-107 (1975).

Fricker, C.; Hardy, J.K.: The Effect of an Alternate Environment as a Collection Medium on the Permeation Characteristics of Solid Organics Through Protective Glove Materials. Am. Ind. Hyg. Assoc. J. 55:738-742 (1994).

Frosch, P.J.; Schulze-Dirks, A.; Hoffman, M., et al.: Efficacy of Skin Barrier Creams (I). The Repetitive Irritation Test (RIT) in the Guinea Pig. Contact Dermatitis 28:94-100 (1993).

Frosch, P.J.; Schulze-Dirks, A.; Hoffman, M.; Axthelm, I.: Efficacy of Skin Barrier Creams (II). Ineffectiveness of a Popular “Skin Protector” Against Various Irritants in the Repetitive Irritation Test in the Guinea Pig. Contact Dermatitis 29:74-77 (1993a).

Frosch, P.J.; Kurte, A.; Pilz, B.: Efficacy of Skin Barrier Creams (III). The Repetitive Irritation Test (RIT) in Humans. Contact Dermatitis 29:113-118 (1993b).

Frosch, P.J.; Kurte, A.: Efficacy of Skin Barrier Creams (IV). The Repetitive Irritation Test (RIT) with a Set of 4 Standard Irritants. Contact Dermatitis 31:161-168 (1994).

Froslind, B.: A Domain Mosaic Model of the Skin Barrier. Acta Derm. Venereol (Stockh) 74:1-6 (1994).

Funckes, A.J.; Hayes, G.R.; Hartwell, W.V.: Urinary Excretion of Paranitrophenol by Volunteers Following Dermal Exposure to Parathion at Different Ambient Temperatures. J. Agricul. Food Chem. 11:455-457 (1963).

Gabard, B.; Treffel, P.: Hardware and Measuring Principle: The Nova™ DPM 9003, In: Bioengineering of the Skin: Water and Stratum Corneum, pp. 177-195, P. Elsner, E. Berardesca; H.I. Maibach, Eds., CRC Press, Boca Raton (1994).

Gad, S.C.: A Scheme for the Prediction and Ranking of Relative Potencies of Dermal Sensitizers Based on Data from Several Systems. J. Appl. Toxicol. 8:361-368 (1988).

Gallant, C.J.: A Long Term Follow-up Study of Patients with Hand Dermatitis Evaluated at St. Michael’s Occupational Health Clinic in 1981 and 1982. Doctoral thesis. The University of Toronto, Canada (1986).

Gantz, N.M.: Infectious Agents, In: Occupational Health. Recognizing and Preventing Work-related Diseases, 3rd Ed., pp. 355-379, B.S. Levy and D.H. Wegman, Eds. Little Brown and Co., Boston (1995).

Gerberick, G.F.: Risk Assessment of Sensitizing Agents. Arch. Toxicol. 16:95-101 (1994).

Ghosh, S.K.; Gohani, V.N.; Doctor, P.N.; Parikh, J.R.; Kashap, S.K.: Intervention Studies Against “Green Symptoms” among Indian Tobacco Harvesters. Arch. Environ. Health 6:316-317 (1991).

Giorgini, S.; Brusi, C.; Sertoli, A.: Evaporimetry in the Differentiation of Allergic, Irritant and Doubtful Patch Test Reactions. Skin Res. Technol. 2:49-51 (1996).

Glamme, J.; Sjostrom, B.: Industrial Thallium Poisoning. Svenska lakartidningen 52:136-1441 (1955).

Goffin, V.; Letawe, C.; Pierard, G.E.: Effect of Organic Solvents on Normal Human Stratum Corneum: Evaluation by the Corneoxenometry Bioassay. Dermatology 195:321-324 (1997).

Goffin V.; Pierard, G.E.; Henry F.; et al.: Sodium Hypochlorite, Bleaching Agents, and the Stratum Corneum. Ecotoxicology Environmental Safety, 37:199-202 (1997a).

Goffin, V.; Pierard-Franchimont, C.; Pierard, G.E.: Shielded Corneosurfametry and Corneoxenometry: Novel Bioassays for the Assessment of Skin Barrier Products. Dermatology 196:434-437 (1998).

Goh, C.L.; Chia, S.E.: Skin Irritability to Sodium Lauryl Sulfate - as Measured by Skin Water Vapour Loss - by Sex and Race. Clin. Exp. Dermatol. 13:16-19 (1988).

Goh, C.L.: Occupational Hand Dermatitis of Hairdressers in Tainan City. Occup. Environ. Med. 51:689-692 (1994).

Graves, C.J.; Edwards, C.; Marks, R.: The Occlusive Effects of Protective Gloves on the Barrier Properties of the Stratum Corneum, In: Irritant Dermatitis, New Clinical and Experimental Aspects. Curr. Probl. Dermatol. pp. 87-94, P. Elsner, H.I. Maibach, Eds., Basel, Karger (1995).

Grove, G.L.: Techniques for Assessing the Vulnerability and Repair Capacity of Human Skin In Vivo. Am. J. Indust. Med. 8:483-489 (1985).

Grubauer, G.; Feingold, K.R.; Elias, P.M.: Relationship of Epidermal Lipogenesis to Cutaneous Barrier Function. J. Lipid Res. 28:746-752 (1987).

Grunewald, A.M.; Gloor, M.; Gehring W.; Kleesz, P.: Damage to the Skin by Repetitive Washing. Contact Dermatitis 32:225-232 (1995).

Grunewald, A.M.; Gloor, M.; Gehring, W.; Kleesz, P.: Efficacy of Barrier Creams, In: Irritant Dermatitis, New Clinical and Experimental Aspects. Curr. Probl. Dermatol., P. Elsner and H.I. Maibach, Eds. 23:187-197, Basel, Karger (1995a).

Guillenmin, M.; Murset, J.C.; Lob, M.; Riquez, J.: Simple Method to Determine the Efficacy of a Cream Used for Skin Protection Against Solvents. Brit. J. Indust. Med. 31:310-316 (1974).

Halbert, A.; Gebauer, K.; Wall, L.: Prognosis of Occupational Chromate Dermatitis. Contact Dermatitis 27:214-220 (1992).

Halkier-Sorensen, L.; Menon, G.K.; Elias, P.M.: Cutaneous Barrier Function after Cold Exposure in Hairless Mice: A Model to Demonstrate How Cold Interferes with Barrier Homeostasis among Workers in the Fish-processing Industry. Br. J. Dermatol. 132:391-401 (1995).

Halkier-Sorensen, L.; Petersen, B.H.; Thestrup-Pedersen, K.: Epidemiology of Occupational Skin Diseases in Denmark: Notification, Recognition and Compensation, In: The Irritant Contact Dermatitis Syndrome, pp. 23-52, P.G.M. van der Valk and H.I. Maibach, Eds., CRC Press, Boca Raton (1996a).

Haklier-Sorensen, L.: Notified Occupational Skin Diseases in Denmark: Important Exposure Sources, Occupations and Trades. The Course from Notification to Compensation and Socio-economical Aspects, In: Occupational Skin Diseases, Contact Dermatitis, 35 (Suppl. 1):1-44 (1996b).

Haklier-Sorensen, L.: Preventive Activities. General Aspects and the Efficacy of Emollients and Moisturizers. Contact Dermatitis, Suppl. 1, 35:90-109 (1996c).

Hanninen, H.; Eskelinen, L.; Husman, K.; Nurminen, M.: Behavioral Effects of Long-term Exposure to a Mixture of Organic Solvents. Scand. J. Work Environ. Health 4:240-255 (1976).

Hannuksela, M.: Tests for Immediate Hypersensitivity, In: Occupational and Industrial Dermatology, 2nd Edition, pp. 168-178, H.I. Maibach, Ed., Year Book Medical Publishers, Chicago (1987).

Hansch, C.H.; Leo, A.: Substituent Constants for Correlation Analysis in Chemistry and Biology. Wiley, New York (1979).

Harvell, J.D.; Maibach, H.I.: Percutaneous Absorption and Inflammation in Aged Skin: a Review. J. Am. Acad. Dermatol. 31:1015-1021 (1994).

Harvey, D.T.; Hogan, D.J.: Common Environmental Dermatoses, In: Environmental Medicine, pp. 263-281, S.M. Brooks, Ed., Mosby, St. Louis (1995).

Hawkins, G.S.: Methodology for the Execution of In Vitro Skin Penetration Determination, In: Methods for Skin Absorption, pp.67-80, B.W. Kemppainen and W.G. Reifenrath, Eds., CRC Press, Boca Raton (1990).

Hawkins, G.S.; Reifenrath W.G.: Development of an In Vitro Model for Determining the Fate of Chemicals Applied to Skin. Fundam. Appl. Toxicol. 4:S133-S144 (1984).

Hayes, B.B.; Afshari, A.; Millecchia, L.; et al.: Evaluation of Percutaneous Penetration of Natural Rubber Latex Proteins. Toxicol. Sci. 56:262-70 (2000).

He, F.; Zhang, S.; Wang, H.; et al.: Neurological and Electroneuromyographic Assessment of the Adverse Effects of Acrylamide on Occupationally Exposed Workers. Scand. J. Work Environ. Health 15:125-129 (1989).

Health and Safety Executive, Self-reported work related illness. HSE Research Paper 33, pp. 37 (1993).

Health and Safety Commission, Health and Safety Statistics 1996/97. Government Statistical Service, HSE Books (1997).

Herman, M.; Thomas, F.; van-Hecke, W.: Medicolegal and Toxicologic Study of Two Cases of Fatal Intoxication in Agricultural Workers Due to Dinitroorthocresol (DNOC). Annales de Medecine Legale, 36:247-256 (1956).

Heron, R.J.L.: Worker Education in the Primary Prevention of Occupational Dermatoses. Occup. Med. 47:407-410 (1997).

Hewitt, P.G.; Hotchkiss, S.A.M.; Caldwell, J.: Decontamination Procedures after In Vitro Topical Exposure of Human and Rat Skin to 4,4'-methylenebis[2-chloroaniline] and 4,4'-methylenedianiline. Fund. Appl. Toxicol. 26:91-98 (1995).

Higo, N.; Naik, A.; Bommannan, D.B.: Validation of Reflectance Infrared Spectroscopy as a Quantitative Method to Measure Percutaneous Absorption In Vivo. Pharmaceut. Res. 10:1500-1506 (1993).

Hogan, D.J.; Dannaker, C.J.; Maibach, H.I.: The Prognosis of Contact Dermatitis. J. Am. Acad. Dermatol. 23:300-307 (1990).

Hogan, D.J.: The Prognosis of Occupational Contact Dermatitis. Occupational Medicine: State of the Art Reviews, 9:53-58 (1994).

Hogan, D.J.: Occupational Skin Disorders. Igaku-Shoin Medical Publishers, New York (1994a).

Hogberg, M.; Wahlberg, J.E.: Health Screening for Occupational Dermatitoses in House Painters. Contact Dermatitis 6:100-106 (1980).

Hogstedt, C.; Axelson, O.: Nitroglycerine-nitroglycol Exposure and the Mortality in Cardio-cerebrovascular Diseases among Dynamite Workers. J. Occup. Med. 19:675-678 (1977).

Hogstedt, C.; Stahl, R.: Skin Absorption and Protective Gloves in Dynamite Work. Linkoping University Medical Dissertations No. 8., 170 pp. (1980).

Hogstedt, C.: Has the Scandinavian Solvent Syndrome Controversy Been Solved? Scand. J. Work Environ. Health 20:59-64 (1994).

Holness, L.: Characteristic Features of Occupational Dermatitis: Epidemiologic Studies of Occupational Skin Disease Reported by Contact Dermatitis Clinics. Occup. Med: State of the Art Reviews, Vol. 9 (1):45-52, 1994.

Huang, J.; Wang, X.P.; Chen, B.M.: Immunological Effects of Toluene Diisocyanate Exposure on Painters. Arch. Environ. Contamin. Tox. 21:607-611 (1991).

Inaba, J.; Suzuki-Yasumoto, M.: A Kinetic Study of Radionuclide Absorption Through Damaged and Undamaged Skin of the Guinea Pig. Health Physics 37:592-595 (1979).

Ilyin, L.A.; Ivannikov, A.T.; Parfenov, Y.D.; Stolyarov, V.P.: Strontium Absorption Through Damaged and Undamaged Human Skin. Health Physics 29:75-80 (1975).

Jacobs, R.R.; Phanprasit, W.: An In Vitro Comparison of the Permeation of Chemicals in Vapor and Liquid Phase Through Pig Skin. Am. Ind. Hyg. Assoc. J. 54:569-575 (1993).

Jang, A.S.; Choi, I.S.; Koh, Y.I.; et al.: Increase in Airway Hyperresponsiveness among Workers Exposed to Methylene Diphenyldiisocyanate Compared to Workers Exposed to Toluene Diisocyanate at a Petrochemical Plant in Korea. Am. J. Ind. Med. 37:663-7 (2000).

Jarvholm, B.; Johansson, B.; Lavenius, B.; Torell, G.: Exposure to Triarylphosphate Andpolyneuropathy: A Case Report. Am. J. Ind. Med. 9:561-566 (1986).

Johanson, G.; Boman, A.: Percutaneous Absorption of 2-butoxyethanol Vapour in Human Subjects. Br. J. Industr. Med. 48:788-792 (1991).

Judge, M.R.; Griffiths, H.A.; Basketter, D.A.: Variation in Response of Human Skin to Irritant Challenge. Contact Dermatitis, 34:115-17 (1996).

Kanerva, L.; Jolanki, R.; Toikkanen, J.: Frequencies of Occupational Allergic Diseases and Gender Differences in Finland. Int. Arch. Occup. Environ. Health 66:111-116 (1994).

Kanerva, L.; Jolanki, R.; Toikkanen, J.K.: Statistics on Occupational Dermatoses in Finland, In: Irritant Dermatitis, New Clinical and Experimental Aspects. Curr. Probl. Dermatol., Vol. 23, pp. 28-40, P. Elsner and H.I. Maibach, Eds., Basel, Karger (1995).

Kanerva, L.; Toikkanen, J.; Jolanki, R.: Statistical Data on Occupational Contact Urticaria. Contact Derm. 35:229-233 (1996).

Kao, J.Y.; Hall, J.; Shugart, L.R.: An In Vitro Approach to Studying Cutaneous Metabolism and Disposition of Topically Applied Xenobiotics. Toxicol. Appl. Pharmacol. 75:289-298 (1984).

Karol, M.H.; Hauth, B.A.; Riley, E.J. ; Magreni, C.M.: Dermal Contact with Toluene Diisocyanate (TDI) Produces Respiratory Tract Hypersensitivity in Guinea Pigs. Tox. Appl. Pharm. 58:221-230 (1981).

Keczkes, K.; Bhate, S.M.; Wyatt, E.H: The Outcome of Primary Irritant Hand Dermatitis. Br. J. Dermatol.109:665-668 (1983).

Kimber, I.: The Role of the Skin in the Development of Chemical Respiratory Hypersensitivity. Tox. Letters 86:89-92 (1996).

Kintz, P.; Tracqui, A.; Mangin, P.: Accidental Death Caused by Absorption of 2, 4-dichlorophenol Through the Skin. Arch. Toxicol. 66:298-299 (1992).

Kissel, J.C.; Richter, K.Y.; Fenske, R.A.: Factors Affecting Soil Adherence to Skin in Hand-press Trials. Bull. Environ. Contam. Toxicol. 56:722-728 (1996).

Kissel, J.C.; Richter, K.Y.; Fenske, R.A.: Field Measurement of Dermal Soil Loading Attributable to Various Activities: Implications for Exposure Assessment. Risk Analysis 16:115-125 (1996a).

Kissel, J.C.; Shirai, J.H.; Richter, K.Y.; Fenske, R.A.: Empirical Investigation of Hand-to-mouth Transfer of Soil. Bull. Environ. Contam. Toxicol. 60:379-386 (1998).

Kligman, A.M.: The Identification of Contact Allergens by Human Assay. II. Factors Influencing the Induction and Measurement of Allergic Contact Dermatitis. J. Invest. Dermatol. 47:375-392 (1966).

Kligman, A.M.: Hydration Injury to Human Skin, In: The Irritant Contact Dermatitis Syndrome, P.G.M. van der Valk and H.I. Maibach, Eds., CRC Press, Boca Raton (1996).

Koh, D.; Lee, H.S.; Chia, H.P.; Phoon, W.H.: Skin Disorders among Hand Solderers in the Electronics Industry. Occup. Med. 44:24-28 (1994).

Koizumi, A.: Experimental Evidence for the Possible Exposure of Workers to Hexachlorobenzene by Skin Contamination. Br. J. Indust. Med. 48:622-628 (1991).

Kusama, T.; Itoh, S.; Yoshizawa, Y.: Absorption of Radionuclides Through Wounded Skin. Health Physics 51:138-141 (1986).

Kusters, E.: Biological Monitoring of MDA. Br. J. Indust. Med. 49: 72 (1992).

Lamaud, E.; Lambrey, E.; Schalla, W.; Schaefer, H.: Correlation Between Transepidermal Water Loss and Penetration of Drugs. J. Invest. Dermatol. 82:556 (1984).

Lammintausta, K.; Maibach, H.I.; Wilson, D.: Irritant Reactivity in Males and Females. Contact Dermatitis 17:276-280 (1987).

Lange, M.; Nitzsche, K.; Zesch, A.: Percutaneous Absorption of Lindane in Health Volunteers and Scabies Patients: Dependency of Penetration Kinetics in Serum upon Frequency of Application, Time and Mode of Washing. Arch. Dermatol. Res. 271:387-399 (1981).

Lantinga, H.; Nater, J.P.; Coenraads, P.J.: Prevalence, Incidence and Course of Eczema on the Hands and Forearms in a Sample of the General Population. Contact Dermatitis 10:135-139 (1984).

Larese, F.; Fiorito, A.; DeZotti, R.: The Possible Haematological Effects of Glycol Monomethyl Ether in a Frame Factory. Br. J. Ind. Med. 49:131-133 (1992).

Lauer, A.C.; Lieb, L.M.; Ramachandran, C.: Transfollicular Drug Delivery. Pharm. Res. 12:179-186 (1995).

Lehmann, P.; Holzle, E.; Melnik, B.; Plewig, G.: Human Skin Response to Irritants: The Effect of UVa and UVb Radiation on the Skin Barrier, In: The Environmental Threat to the Skin, pp. 203-209. R. Marks and G. Plewig, Eds., Martin Dunitz (1992).

Leung, D.Y.M.: Atopic Dermatitis: The Skin as a Window into the Pathogenesis of Chronic Allergic Diseases. J. Allergy Clin. Immunol. 96:302-318 (1995).

Leung, D.Y.M.; Diaz, L.A.; DeLeo, V.; Soter, N.A.: Allergic and Immunologic Skin Disorders. JAMA 278:1914-1923 (1997).

Lauwerys, R.R.; Dath, T.; Lachapelle, J.M.; et al.: The Influence of Two Barrier Creams on the Percutaneous Absorption of M-xylene in Man. J. Occup. Med. 20:17-20 (1978).

Leigh, J.P.; Miller, T.R.: Job-related Diseases and Occupations Within a Larger Worker's Compensation Data Set. Am. J. Indust. Med. 33:197-211 (1998).

Levin, O.L.; Behrman, H.T.: Cutaneous Hazards in the Fur Industry, Indust. Med. 7:673-678 (1938).

Lewis, R.: Researchers Deaths Inspire Actions to Improve Safety. Scientist. 11(21):1 ff. (1997) Oct 27.

Lien, E.; Gao, H.: QSAR Analysis of Skin Permeability of Various Drugs in Man as Compared to In Vivo and In Vitro Studies in Rodents. Pharmaceut. Res. 12:583-587 (1995).

Lilley, S.G.; Florence, T.M.; Stauber, J.L.: The Use of Sweat to Monitor Lead Absorption Through the Skin. Sci. Total Environ. 76:267-278 (1988).

Linden, C.: Allergic Contact Dermatitis from 4,4'-diisocyanato-diphenyl Methane (MDI) in a Molder. Contact Derm. 6:301-302 (1980).

Lockey, J.E.; Kelly, C.R.; Cannon, G.W.; et al.: Progressive Sytemic Sclerosis Associated with Exposure to Trichloroethylene. J. Occup. Med. 29:493-496 (1987).

Loden, M.; Lindberg, M.: Product Testing - Testing of Moisturizers, In: Bioengineering of the Skin: Water and the Stratum Corneum, pp. 275-289, P. Elsner, E. Berardesca and H.I. Maibach, Eds., CRC Press, Baton Raton (1994).

Loomis, D.; Browning, S.R.; Schenck, A.P.: Cancer Mortality among Electric Utility Workers Exposed to Polychlorinated Byphenyls. Occup. Environ. Med. 54:720-728 (1997).

Lotte, C.; Rougier, A.; Wilson, D.R.; Maibach, H.I.: In Vivo Relationship Between Transepidermal Water Loss and Percutaneous Penetration of Some Organic Compounds in Man: Effect of Anatomic Site. Arch. Dermatol. Res. 279:351-356 (1987).

Lovell, C.R.: Plants and the Skin. Blackwell Scientific Publications, London (1993).

Lumens, M.G.; Ulenbelt, P.; Herber, R.F.M.; Meyman, T.F.: The Impact of Hygienic Behavior and Working Methods on the Uptake of Lead and Chromium. Appl. Occup. Environ. Hyg. 9:53-56 (1994).

Lushniak, B.D.: The Epidemiology of Occupational Contact Dermatitis. Dermatologic Clinics, 13:671-680 (1995).

Macperson, S.E.; Scott,R.C.; Williams, F.M.: Percutaneous Absorption and Metabolism of Aldrin by Rat Skin in Diffusion Cells. Arch. Toxicol. 65:599-602 (1991).

Magnusson, B.; Kligman, A.M.: The Identification of Contact Allergens by Animal Assay. The Guinea Pig Maximization Test. J. Invest. Dermatol. 52:268-276 (1969).

Magnusson, B.; Kligman, A.M.: Allergic Contact Dermatitis in the Guinea Pig. Identification of Contact Allergens. Springfield, IL, Thomas (1970).

Magnusson, B.; Kligman, A.M.: Factors Influencing Allergic Contact Sensitization, In: Dermatotoxicology, 2nd Edition. F.N. Marzulli, Ed., Hemisphere Publ., Washington DC (1983).

Mahmoud, G.; Lachapelle, J.M.; Van Neste, D.: Histological Assessment of Skin Damage by Irritants: Its Possible Use in the Evaluation of a "Barrier Cream." Contact Dermatitis 11:179-185 (1985).

Maibach, H.I.; Feldmann, R.J.; Milby, T.H.; Serat, W.F.: Regional Variation in Percutaneous Penetration in Man: Pesticides. Arch. Environ. Health 23:208-211 (1971).

Maibach, H.I.; Feldman, R.: Systemic Absorption of Pesticides Through the Skin of Man, In: Occupational Exposure to Pesticides. Report to the Federal Working Group on Pest Management from the Task Group on Occupational Exposure to Pesticides (1974).

Maibach, H.I.: In Vivo Percutaneous Penetration of Corticoids in Man and Unresolved Problems in Their Efficacy. Dermatologica 152 (suppl. 1): 11-25 (1976).

Makki, S.: Percutaneous Absorption of Three Psoralens Commonly Used in Therapy: Effect of Skin Occlusion (In Vitro Study). Internat. J. Pharmaceutics 133:245-252 (1996).

MAK and BAT Values, DFG Maximum concentrations at the workplace and biological tolerance values for working materials. Deutsche Forschungsgemeinschaft. Commision for the Investigation of Health Hazard of Chemical Compounds in the Work Area, Weinheim, VCH (1997).

Malten, K.E.; Spruit, D.; Boemars, H.G.M.; Keizer, M.J.M.: Horny Layer Injury by Solvents. Berufsdermatosen 16:135-147 (1968).

Malten, K.E.: 4,4' Diisocyanato Dicyclohexl Methane (Hylene W): A Strong Contact Sensitizer. Contact Derm. 3:344-346 (1977).
Hannhold, R.; Dross, K.: Calculations for Molecular Lipophilicity: A Comparative Study. Quant. Struct.-Act. Relat. 15:403-409 (1996).
Marks, R.; Dykes, P.J.; Hamami, I.: Two Novel Techniques for the Evaluation of Barrier Creams. Br. J. Dermatol. 120:655-660 (1989).
Marks, J.G.; Deleo, V.A.: Contact and Occupational Dermatology, Mosby Year Book, St. Louis (1992).
Marzulli, F.N.; Maibach, H.I.: Allergic Contact Dermatitis, In: Dermatotoxicology, 5th Edition, F.N. Marzulli and H.I. Maibach, Eds. Taylor & Francis, Bristol, PA (1996).
Mathias, C.G.T.: The Cost of Occupational Skin Disease. Arch. Dermatol. 121:332-334 (1985).
Mathias, C.F.T.: Contact Dermatitis from Use or Misuse of Soaps, Detergents, and Cleansers in the Workplace. Stat. Art. Rev. Occup. Med. 1:205-218 (1986).
McAuliffe, D.J.; Blank, I.H.: Effects of UVa on the Barrier Characteristics of the Skin. J. Invest. Dermatol. 96:758-762 (1991).
McCulloch, W.F.; Braun, J.L.; Robinson, R.G.: Leptosphiral Meningitis. J. Iowa Med. Soc. 52:728-731 (1962).
McDougal, J.N.; Jepson, G.W.; Clewell, H.J.; et al.: Dermal Absorption of Organic Chemical Vapors in Rats and Humans. Fund. Appl. Toxicol. 14:299-308 (1990).
McFadden, J.P.; Wakelin, S.H.; Basketter, D.A.: Acute Irritation Thresholds in Subjects with Type I–Type VI Skin. Contact Dermatitis 38:147-149 (1998).
McGill, D.B.; Motto, J.D.: An Industrial Outbreak of Toxic Hepatitis Due to Methylenedianiline. New Eng. J. Med. 291:278-282 (1974).
McKone, T.E.: Dermal Uptake of Organic Chemicals from a Soil Matrix. Risk Analysis 10:407-419 (1990).
McKone, T.E.: The Precision of a Fugacity-based Model for Estimating Dermal Uptake of Chemicals from Soil, In: Hydrocarbon Contaminated Soils, Vol. 1. Remediation Techniques, Environmental Fate, Risk Assessment, Analytical Methodologies, Regulatory Considerations, pp. 555-574, E.J. Calabrese and P.T. Kostecki, Eds., Lewis Publ., Chelsea, MI (1991).
McKone, T.E.; Howd, R.A.: Estimating Dermal Uptake of Nonionic Organic Chemicals from Water and Soil I. Unified Fugacity-based Models for Risk Assessments. Risk Analysis 12:543-557 (1992).
Meding, B.; Swanbeck, G.: Epidemiology of Different Types of Hand Eczema in an Industrial City. Acta Derm. Venerol. (Stockh) 69:227-233 (1989).
Meding, B.; Swanbeck, G.: Consequences of Having Hand Eczema. Contact Dermatitis 23:6-14 (1990).
Meneghini, C.L.: Occupational Microbial Dermatoses, In: Occupational and Industrial Dermatology, pp. 95-107, H.I. Maibach and G.A. Gelin, Eds., Year Book Medical Publishers, Chicago (1982).
Meneghini, C.L.; Angelini, G.: Primary and Secondary Sites of Occupational Contact Dermatitis. Dermatosen. 32:205-207 (1984).
Menne, T.; Nilesen, N.H.: Epidemiology of Allergic Contact Dermatitis. In: Postgraduate Course in Allergological Aspects of Dermatology, pp. 37-48, A. Scheynius, Ed., Stockholm: EAACI-IAACI (1994).
Michaels, A.S.; Chandrasekaran, S.K.; Shaw, J.E.: Drug Permeation Through Human Skin. Theory and In Vitro Experimental Measurement. Am. Inst. Chem. Eng. 21:985-996 (1975).
Middleton, E.L.: Fatal Case of Poisoning by Ethylene Chlorhydrin. J. Ind. Hyg. 12:265 (1930).
Miselnicky, S.R.; Lichtin, J.L.; Sakr, A.: The Influence of Solubility, Protein Binding, and Percutaneous Absorption on Reservoir Formation in Skin. J. Soc. Cosmet. Chem. 39:169-177 (1988).
Moody, R.P.; Nadeau, B.; Chu, I.: In Vitro Dermal Absorption of Pesticides: VI. In Vivo and In Vitro Comparisons of the Organochlorine Insecticide DDT in Rat, Guinea Pig, Pig, Human and Tissue-cultured Skin. Toxic. in Vitro 8:1225-1232 (1994).
Morris, G.E.; Maloof, C.C.: Some Causes of Cutting Oil Dermatitis. Part II- A Study of Protective Creams. Industr. Med. Surg. 22:327-328 (1953).
Moss, J.: The Effect of 3 Moisturizers on Skin Surface Hydration: Electrical Conductance (Skicon-200), Capacitance (Corneometer Cm420), and Transepidermal Water Loss (TEWL). Skin Res. Technol. 2:32-36 (1996).
National Center for Health Statistics: Current Estimates from the National Health Interview Survey, 1988. Washington, DC: Series 10, No. 173, DHHS Publication No. (PHS) 89-1501 (1989).

NIH: The Murine Local Lymph Node Assay: A Test Method for Assessing the Allergenic Contact Dermatitis Potential of Chemicals/Compounds, NIH Publication No. 99-4494, National Toxicology Program Interagency Center for the Evaluation of Alternative Methods, Research Triangle Park, NC (1999).

NIOSH: Report on Occupational Safety and Health for Fiscal Year 1990. National Institute for Occupational Safety and Health (NIOSH), NTIS PB-93-215-184, pp. 30 (1993).

NIOSH: Health Hazard Evaluation Report, Distinctive Designs International, Inc., HETA 91-0386-2427, National Institute for Occupational Safety and Health (NIOSH), Cincinnati, Ohio (1994).

NIOSH, Health Hazard Evaluation and Technical Assistance Report, HETA 93-0436, Trus Joist MacMillian, Deerwood, Minnesta, National Institute for Occupational Safety and Health (NIOSH), Cincinnati, Ohio (1996).

NIOSH: State Employee Notification and Surveillance Occupational Registry (SENSOR) Program, National Institute for Occupational Safety and Health (NIOSH) Cincinnati, Ohio (1997).

Nethercott, J.R.: Fitness to Work with Skin Disease and the Americans with Disabilities Act of 1990. Occup. Med.: State of the Art Reviews, 9:11-19 (1994).

Nemert, B.; Lenaerts, L.: Exposure to Methylene Diphenyl Diisocyanate in Coal Mines. Lancet 341:318 (1993).

Newhouse, M.; McEvoy, D.; Rosenthal, D.: Percutaneous Paraquat Absorption: An Association with Cutaneous Lesion and Respiratory Failure. Arch. Dermatol. 114:1516-1519 (1978).

Nield, H.: Leptospirosis. Occup. Health 42:140-142 (1990).

Nielsen, J.: The Occurrence and Course of Skin Symptoms on the Hands among Female Cleaners. Contact Dermatitis 34:284-291 (1996).

Nixon, R., Frowen, K.: Allergic Contact Dermatitis Caused by Epoxy Resins. J. Occup. Health Safety: Australia & New Zealand 7:417-424 (1991).

O'Donaghue, J.L.: Alkanes, Alcohols, Ketones, and Ethylene Oxide, In: Neurotoxicity of Industrial and Commercial Chemicals, J.L. O'Donaghue, Ed., CRC Press, Boca Raton, 2:61-97 (1985).

OECD: OECD Test Guideline 404, Skin Irritation/Corrosion, Office of Economic Co-Operation and Development, Paris (1981).

OECD: OECD Test Guideline 404, Skin Irritation/Corrosion (1992 revision), Office of Economic Co-Operation and Development, Paris (1992).

OECD: OECD Guideline for Testing of Chemicals, Draft OECD Guideline, Dermal Delivery and Percutaneous Absorption: In Vitro Method, Office of Economic Co-Operation and Development, June (1996)

OECD: Office of Economic Cooperation and Development, Guideline for the Testing of Chemicals, Draft new guideline 427: Skin absorption: in vivo method. Draft Document, December (2000a).

OECD: Office of Economic Co-Operation and Development, OECD Environmental Health and Safety Publications, Series on Testing and Assessment, No. 28. Draft Guidance document for the conduct of skin absorption studies. December (2000b).

Ohi, G.; Wegman, D.H.: Transcutaneous Ethylene Glycol Monoethyl Ether Poisoning in the Work Setting. J. Occup. Med. 20:675-676 (1978).

Ohlenschlaeger, J.; Friberg, J.; Ramsing D.; Agner, T.: Temperature Dependency of Skin Susceptibility to Water and Detergents. Acta Derm. Venereol (Stockh) 76:274-276 (1996).

Olivarius, F.; Hansen, A.B.; Karlsmark, T.; Wulf, H.C.: Water Protective Effect of Barrier Creams and Moisturizing Creams: A New in Vivo Test Method. Contact Dermatitis 35:219-225 (1996).

Paludan, K.; Thestrup-Pedersen, K.: Use of the Polymerase Chain Reaction in Quantification of Interleukin 8 Mrna in Minute Epidermal Samples. Soc. Investigative Dermatol. 99:830-835 (1992).

Patil, S.M.; Singh, P.; Maibach, H.I.: Cumulative Irritancy in Man to Sodium Lauryl Sulfate: The Overlap Phenomenon. Internat. J. Pharmaceutics 110:147-154 (1994).

Patrick, E.; Maibach, H.I.: Dermatotoxicology, In: Principles and Methods of Toxicology, 3rd Edition, A.W. Hayes, Ed., Raven Press, New York (1994).

Pelletier, O.; Ritter, L.; Caron, J.: Effects of Skin Preapplication Treatments and Postapplication Cleansing Agents on Dermal Absorption of 2,4-dichloropheoxyacetic Acid Dimethylamine by Fischer 344 Rats. J. Toxicol. Environ. Health 31:247-260 (1990).

Petsonk, E.L.; Wang, M.L.; Lewis, D.M.; Siegel, P.D.; Husberg, B.J.: Asthma-like Symptoms in Wood Product Plant Workers Exposed to Methylene Diphenyl Diisocyanate. Chest 118:1183-93 (2000).

Pham, Q.T.; Cavelier, C.; Mur J.M.; Mereau, P.: Isocyanates at Levels Higher than Mac and Their Effect on Respiratory Function. Ann. Occ. Hyg. 21: 271-275 (1978).

Pinnagoda, J.; Tupker, R.A.; Agner, T.; Serup, J.: Guidelines for Transepidermal Water Loss (TEWL) Measurements: A Report from the Standardization Group of the European Society of Contact Dermatitis. Contact Dermatitis 22:164-178(1990).

Pinnagoda, J.: Standardization of Measurements in Bioengineering of the Skin: Water and the Stratum Corneum, pp. 59-63, P. Elsner, E. Berardesca, and H.I. Maibach, Eds. CRC Press, Baton Raton (1994).

Piotrowski, J.K.: Evaluation of Exposure to Phenol: Absorption of Phenol Vapour in the Lungs and Through the Skin and Excretion of Phenol in Urine. Brit. J. Industr. Med. 28:172-178 (1971).

Plotnick, H.: Analysis of 250 Consecutively Evaluated Cases of Workers' Disability Claims for Dermatitis. Arch. Dermatol. 126:782-786 (1990).

Potter, A.L.: The Successful Treatment of Two Recent Cases of Cyanide Poisoning. Brit. J. Indust. Med. 7:125-130 (1950).

Potts, R.O.; Francoeur, M.L.: The Influence of Stratum Corneum Morphology on Water Permeability. J. Invest. Dermatol. 96:495-499 (1991).

Potts, R.O.; Guy, R.H.: Predicting Skin Permeability. Pharmaceut. Res. 9:663-669 (1992).

Prevo, M.; Cormier, M.; Nichols, K.: Predictive Toxicology Methods for Transdermal Delivery Systems. Toxicology Methods, 6:83-98 (1996).

Qiao, G.L.; Chang, S.K.; Riviere, J.E.: Effects of Anatomical Site and Occlusion on the Percutaneous Absorption and Residue Pattern of 2,6 [Ring-14c] Parathion In Vivo in Pigs. Toxicol. Applied Pharmacol. 122:131-138 (1993).

Que Hee, S.; Boeniger, M.: Dermal Chemical Hazards. In: Essential Resources for Industrial Hygiene: A Compendium of Current Practice Standards and Guidelines, M.K. Harris, Ed., pp. 141-154, AIHA Press, Fairfax, VA (2000).

Quan, D.; Miabach, H.I.; Wester, R.C.: In Vitro Percutaneous Absorption of Glyphosate and Malathion Across Cotton Sheets into Human Skin. Toxicologist 14:107 (1994).

Rahman, M.S.; Hall, L.L.; Hughes, M.F.: In vitro percutaneous absorption of sodium arsenate in B6C3F Mice. Toxic. In Vitro 8:441-448 (1994).

Rajaniemi, R.: Clinical Evaluation of Occupational Toxicity of Methylmethacrylate Monomer to Dental Technicians. J. Soc. Occup. Med. 36:56-59 (1986).

Rajaniemi, R.; Pfaffli, P.; Savolainen, H.: Percutaneous Absorption of Methyl Methacrylate by Dental Technicians. Br. J. Ind. Med. 6:356-357 (1989).

Ramsing, D.W.; Agner, T.: Effect of Glove Occlusion on Human Skin (II). Long-Term Experimental Exposure. Contact Dermatitis 34:258-262 (1996).

Rattray, N.J.; Botham, P.A.; Hext, P.M.; Woodcock, D.R.; Fielding, I.; Dearman, R.J.; Kimber, I.: Induction of Respiratory Hypersensitivity to Diphenylmethane-4,4'-diisocyanate (MDI) in Guinea Pigs: Influence of Route of Exposure. Toxicology 88:15-30 (1994).

Rees, J.; Friedmann, P.S.; Matthews, J.N.: Contact Sensitivity to Dinitrochlorobenzene Is Impaired in Atopic Subjects. Arch. Dermatol. 126:1173-1175 (1990).

Reifenrath, W.G.; Spencer, T.S.: Evaporation and Penetration from Skin, In: Percutaneous Absorption: Mechanisms-Methodology-Drug Delivery, 2nd Edition, pp. 313-334, R.L. Bronaugh and H.I. Maibach. Eds., Marcel Dekker (1989).

Reifenrath, W.G.; Hawkins, G.S.; Kurtz, M.S.: Evaporation and Skin Penetration Characteristics of Mosquito Repellent Formulations. J. Am. Mosq. Control Assoc. 5:45-51 (1989b).

Rietschel, R.L.; Fowler, J.F.: Fisher's Contact Dermatitis, 4th Edition. Williams & Wilkins, Baltimore (1995).

Riihimaki, V.; Pfaffli, P.: Percutaneous Absorption of Solvent Vapors in Man. Scand. J. Work Environ. Health 4:73-85 (1978).

Riihimaki, V.: Percutaneous Absorption of m-Xylene from a Mixture of m-Xylene and Isobutyl Alcohol in Man. Scand. J. Work Environ. Health 5:143-150 (1979).

Roberts, M.S.; Anderson, R.A.; Swarbrick, J.: Permeability of Human Epidermis to Phenolic Compounds. J. Pharm. Pharmacol. 29:677-683 (1977).

Roberts, M.S.; Horlock, E.: Effect of Repeated Skin Application on Percutaneous Absorption of Salicyclic Acid. J. Pharm. Sci. 67:1685-1687 (1978).

Roberts, M.S.; Pugh, W.J.; Hadgraft, J.; Watkinson, A.C.: Epidermal Permeability-penetrant Structure Relationships. 1. An Analysis of Methods of Predicting Penetration of Monofunctional Solutes from Aqueous Solutions. Int. J. Pharm 126, 219-33 (1995).

Roels, H.; Buchet, J.P.; Truc, J.; Croquet, F.; Lauwerys, R.: The Possible Role of Direct Ingestion on the Overall Absorption of Cadmium or Arsenic in Workers Exposed to CdO or As2O3 Dust. Am. J. Ind. Med. 3:53-65 (1982).

Rosen, R.H.; Freeman, S.: Prognosis of Occupational Contact Dermatitis in New South Wales, Australia. Contact Dermatitis 29:88-93 (1993).

Rougier, A.; Dupuis, D.; Lotte, C.; et al.: In Vivo Correlation Between Stratum Corneum Reservoir Function and Percutaneous Absorption. J. Invest. Dermatol. 81:275-278 (1983).

Rougier, A.; Dupuis, D.; Lotte, C.; Roguet, R.: The Measurement of the Stratum Corneum Reservoir. A Predictive Method for In Vivo Percutaneous Absorption Studies: Influence of Application Time. J. Invest. Dermatol. 84:660 (1985).

Rustemeyer, T.; Frosch, P.J.: Occupational Skin Diseases in Dental Laboratory Technicians. Contact Dermatitis 34:125-133 (1996).

Rycroft, R.J.G.: Clinical Assessment in the Workplace: Dermatitis. Occup. Med. 46:364-366 (1996).

Samitz, M.H.; Pomerantz, H.: Studies of the Effects on the Skin of Nickel and Chromium Salts. A.M.A. Arch. Indust. Health, 18:473-479 (1958).

Sappalainen, A.M.; Rajaniemi, R.: Local Neurotoxicity of Methyl Methacrylate among Dental Technicians. Am. J. Ind. Med. 5:471-477 (1984).

Schaefer, H.; Zesch, A.; Stuttgen, G.: Skin Permeability. Springer-Verlag, Berlin (1982).

Scheuplein, R.J.; Blank, I.H.: Mechanism of Percutaneous Absorption. IV. Penetration of Nonelectrolytes (Alcohols) from Aqueous Solutions and from Pure Liquids. J. Invest. Dermatol. 60:286-296 (1973).

Scheuplein, R.J.; Bronaugh, R.L.: Percutaneous Absorption, In: Biochemistry and Physiology of the Skin, pp. 1255-95, L.A. Goldsmith, Ed., Oxford University Press (1983).

Schluter-Wigger, W.; Elsner, P.: Efficacy of 4 Commercially Available Protective Creams in the Repetitive Irritation Test (RIT). Contact Dermatitis 34:278-283 (1996).

Schwope, A.D.: Permeation of Chemicals through the Skin, In: Performance of Protective Clothing, ASTM STP 900, pp. 221-234, R.L. Barker and G.C. Coletta, Eds. Am. Soc. Testing & Materials, Phila. (1986).

Seidenari, S.; Belletti, B.; Pellacani, G.: Time Course of Skin Changes Induced by Short-term Occlusion with Water: Evaluation by TEWL, Capacitance and B-scanning Echography. Skin Res. Technol. 2:52-53 (1996).

Serup, J.: The Spectrum of Irritancy and Application of Bioengineering Techniques, In: Irritant Dermatitis, New Clinical and Experimental Aspects, Curr. Probl. Dermatol., 23:131-141, P. Elsner and H.I. Maibach, Eds., Basel, Karger (1995).

Serup, J.; Jemec, G.B.E.: Handbook of Non-invasive Methods and the Skin. CRC Press, Boca Raton (1995).

Shu, H.; Teitelbaum, P.; Webb, A.S.; et al.: Bioavailability of Soil-bound TCDD: Dermal Bioavailability in the Rat. Fundam. Appl. Toxicol. 10:335-343 (1988).

Shmunes, E.; Keil, J.E.: Occupational Dermatoses in South Carolina. A Descriptive Analysis of Cost Variables. J. Am. Acad. Dermatol. 9:861-866 (1983)

Siegers, C.P.; Sullivan, J.B.: Organometals and Reactive Metals in Hazardous Materials Toxicology, Clinical Principles of Environmental Health, pp. 928-936, J.B. Sullivan, Jr. and G.R. Krieger, Eds., Williams and Wilkins, Baltimore (1992).

Silver Platter Information, Chem Bank on CD, Norwood, Massachusetts (1997).

Simion, F.A.: In Vivo Models to Predict Skin Irritation, In: The Irritant Contact Dermatitis Syndrome, pp. 329-334, P.G.M. van der Valk and H.I. Maibach, Eds., CRC Press, Boca Raton (1996).

Sitkoff, D.; Sharp, K.A.; Honig, B.: Accurate Calculation of Hydration Free Energies Using Macroscopic Solvent Models. J. Phys. Chem. 98:1978-88 (1994).

Skowronski, G.; Turkall, R.M.; Abdel-Rahman, M.S.: Soil Absorption Alters Bioavailability of Benzene in Dermally Exposed Male Rats. Amer. Ind. Hyg. Assoc. J. 49:506-511 (1988).

Skowronski, G.; Turkall, R.M.; Abdel-Rahman, M.S.: Effects of Soil on Percutaneous Absorption of Toluene in Male Rats. J. Toxicol. Environ. Health 26:373-384 (1989).

Skowronski, G.; Turkall, R.M.; Kadry, A. R.; Abdel-Rahman, M.S.: Effects of Soil on the Dermal Bioavailability of M-xylene in Male Rats. Environ. Res. 51:182-193 (1990).

Smit, H.A.; Coenraads, P.J.: Epidemiology of contact dermatitis. Epidemiology of Clinical Allergy, M.L. Burr, Ed., Monogr. Allergy, 31:29-48, Basel, Karger (1993).

Smit, H.A.; Burdorf A.; Coenraads, P.J.: Prevalence of Hand Dermatitis in Different Occupations. Internat. J. Epidemiology 22:288-293 (1993a).

Smit, H.A.; van Rijssen, A.; Vandenbroucke, J.P.; Coenraads, P.J.: Susceptibility to and Incidence of Hand Dermatitis in a Cohort of Apprentice Hairdressers and Nurses. Scand. J. Work Environ. Health 20:113-121 (1994).

Smit, H.; Coenraads, P-J.; Emmett, E.: Dermatoses, In: Epidemiology of Work Related Diseases, pp. 143-164. C. McDonald, Ed., BMJ Publishing (1995).

Smith, J.G.: Paraquat Poisoning by Skin Absorption: A Review. Human Toxicology, 7:15-19 (1988).

Smith, A.B.; Bernstein, D.I.; London, M.A.; et al.: Evaluation of Occupational Asthma from Airborne Egg Protein Exposure in Multiple Settings. Chest 98:398-404 (1990).

Soares, E.R.; Tift, J.P.: Phenol Poisoning: Three Fatal Cases. J. Forensic Sci. 27:729-731 (1982).

Solomon, A.E.; Lowes, N.J.: Percutaneous Absorption in Experimental Epidermal Disease. Br. J. Dermatol. 100:717-722 (1979).

Southweil, D.; Barry, B.W.; Woodford, R.: Variations in Permeability of Human Skin Within and Between Specimens. Internat. J. Pharmaceutics 18:299-309 (1984).

Spencer, P.S.; Bischoff, M.C.: Skin as a Route of Entry for Neurotoxic Substances, In: Dermatotoxicology, 3rd Ed., pp. 625-640, F.N. Marzulli and H.I. Maibach, Eds., Hemisphere Publ., New York (1987).

Stadler, J.; Karol, M.H.: Use of Dose-response Data to Compare the Skin Sensitizing Abilities of Dicyclohexylmethane-4,4'-diisocyanate and Picryl Chloride in Two Animal Species. Tox. Appl. Pharm. 78:445-450 (1985).

Stanek, E.J.; Calabrese, E.J.; Barnes, R.; Pekow, P.: Soil Ingestion in Adults - Results of a Second Pilot Study. Ecotoxicol. Environ. Safety 36:249-257 (1997).

Starr, J.G.; Clifford, N.J.: Absorption of Pesticides in a Chronic Skin Disease. Arch. Environ. Health 22:396-400 (1971).

Stauber, J.L.; Florence,T.M.; Gulson, B.L.; Dale, L.S.: Percutaneous Absorption of Inorganic Lead Compounds. Sci. Total Environ. 145:55-70 (1994).

Stern, M.L.; Brown, T.A.; Brown, R.D.; Munson, A.E.: Contact Hypersensitivity Response to Isophorone Diisocyanate in Mice. Drug Chem. Tox. 12: 287-296 (1989).

Stewart, R.D.; Dodd, H.C.: Absorption of Carbon Tetrachloride, Trichloroethylene, Tetrachloroethylene, Methylene Chloride, and 1,1,1-trichloroethane Through the Human Skin. Ind. Hyg. J. 25:439-446 (1964).

Storres, F.J.: Technical and Ethical Problems Associated with Patch Testing. Clin. Rev. Allergy Immunol. 14:185-198 (1996).

Subcommittee on the Toxicology of Metals, Absorption and Excretion of Toxic Metals, Permanent Commission and International Association on Occupational Health, Nordisk Hygienisk Tidskrift, vol. 2, p. 70-104 (1971).

Surber, C.; Wilhelm, K-P.; Maibach, H.I.; et al.: Partitioning of Chemicals into Human Stratum Corneum: Implications for Risk Assessment Following Dermal Exposure. Fund. Appl. Toxicol. 15:99-107 (1990).

Susten, A.S.; Neimeier, R.W.; Simon, S.D.: In Vivo Percutaneous Absorption Studies on Volatile Organic Solvents in Hairless Mice II. Toluene, Ethylbenzene and Aniline. J. Appl. Toxicol. 10:217-225 (1990).

Swartzendruber, D.C.; Wertz, P.W.; Kitko, D.J.: Molecular Models of the Intercellular Lipid Lamellae in Mammalian Stratum Corneum. J. Invest. Dermatol. 92:251-257 (1989).

Tagami, H.: Hardware and Measuring Principle: Skin Conductance, In: Bioengineering of the Skin: Water and the Stratum Corneum, pp. 197-203, P. Elsner, E. Berardesca and H.I. Maibach, Eds., CRC Press, Baton Raton (1994).

Takamoto, R.; Yamamoto, S.; Namba, R., et al.: New Percutaneous Absorptiometry by a Laser Photoacoustic Method Using an Open-ended Cell. Analytica Chimica Acta 299:387-391 (1995).

Tanaka, K.I.; Takeoka, A.; Nishimura, F.; Hanada, S.: Contact Sensitivity Induced in Mice by Methylene Bisphenyl Diisocyanate. Contact Derm. 17: 199-204 (1987).

Taylor, J.S.: Protective Gloves for Occupational Use, pp. 255-265, CRC Press, Boca Raton (1994).
Tepperman, P.B.: Fatality Due to Acute Systemic Fluoride Poisoning Following a Hydrofluoric Acid Skin Burn. J. Occup. Med. 22:691-692 (1980).
Tola, S.; Hernberg, S.; Nikkanen, J.: Parameters Indicative of Absorption and Biological Effect in New Lead Exposure. A Prospective Study. Brit. J. Ind. Med. 30:134-141 (1973).
Touraille, G.D.; Marty, J.P.; Bunge, A.; Guy, R.H.: In Vitro Percutaneous Absorption of 4-cyanophenol from Contaminated Soil, Influence of Soil Loading and Exposure Time. Society of Toxicology Annual Meeting, Abstract 976, Cincinnati (1997).
Trebilcock, K.L.; Heylings, J.R.; Wilks, M.F.: In Vitro Tape Stripping as a Model for In Vivo Skin Stripping. Toxic. In Vitro 8:665-667 (1994).
Treffel, B.; Gabard, B.; Juch, R.: Evaluation of Barrier Creams: An In Vitro Technique on Human Skin. Acta Derm. Venereol. (Stockh) 74:7-11 (1994).
Treherne, J.E.: The Permeability of Skin to Some Electrolytes. J. Physiology (London) 133:171-180 (1965).
Tsuruta, H.: Percutaneous Absorption of Organic Solvents. III. On the Penetration Rates Of Hydrophobic Solvents Through the Excised Rat Skin. Industr. Health 20:335-345 (1982).
Tucker, S.B.; Key, M.M.: Occupational Skin Disease, In: Environmental and Occupational Medicine, 2nd Edition, W.N. Rom, Ed., Little, Brown & Co., Boston (1992).
Tupker, R.A.; Coenraads, P.J.: Wash Tests: Evaluation by Instrument Methodologies. Clinics Dermatol. 14:51-55 (1996).
Turjanmaa, K.; Alenius, H.; Makinen-Kiljunen, S.; et al.: Natural Rubber Latex Allergy. Allergy 51:593-602 (1996).
Turkall, R.M.; Skowronski, G.A.; Kadry, A.M.; Abdel-Rahman, M.S.: A Comparative Study of the Kinetics and Bioavailability of Pure and Soil-absorbed Naphthalene in Dermally Exposed Male Rats. Arch. Environ. Contam. Toxicol. 26:504-509 (1994).
Tyl, R.W.; York, R.G.; Schardein, J.L.: Reproductive and Developmental Toxicity Studies by Cutaneous Administration, In: Health Risk Assessment: Dermal and Inhalation Exposure and Absorption of Toxicants, R.G.M. Wang, J.B. Knaak, and H.I. Maibach, Eds., CRC Press, Boca Raton (1993).
Ulenbelt, P.; Lumens, M.E.; Geron, H.M.; et al.: Work Hygiene Behaviour as Modifier of the Lead Air-lead Blood Relation. Int. Arch. Occup. Environ. Health 62:203-207 (1990).
U.S. DOL: Occupational Injuries and Illnesses in the United States, 1991. U.S. Department of Labor, Bureau of Labor Statistics, Bulletin 2424 (1993).
U.S. Department of Transportation: Research and Special Programs Administration, Exemption DOT-E 10904 (Fourth Revision), Dec. 12 (2000).
U.S. EPA: Pesticide Assessment Guidelines, Subdivision K, Exposure and Reentry Protection. Springfield, Virginia: National Technical Information Service. NTIS Publication No. PB85-120962 (1984).
U.S. EPA: Health Effects Test Guidelines, OPPTS 870.2600, Skin Sensitization. EPA 712-C-98-197 (1998).
U.S. EPA: Dermal Exposure Assessment: Principles and Applications. Interim Report. EPA/600/8-91/011B.Washington, DC: Office of Research and Development (1992).
Upadhye, M.R.; Maibach, H.I.: Influence of Area of Application of Allergens on Sensitization in Contact Dermatitis. Contact Dermatitis 27:281-286 (1992).
Health 38:355-368 (1993).
Van der Valk, P.G.M.; Tupker, R.A.: Transepidermal Water Loss in Skin Disease with Special Reference to Irritant Contact Dermatitis, In: Bioengineering of the Skin: Water and the Stratum Corneum, P. Elsner, E. Berardesca, H.I. Maibach, Eds., CRC Press, Boca Raton (1994).
Vecchia, B.E.; Bunge, A.L.: Estimating Skin Permeability of Organic Chemicals from Aqueous Solutions, In: Transdermal Drug Delivery Systems, 2nd Ed., in press, J. Hadgraft and R.H. Guy, Eds. Marcel Dekker, New York (2001).
Veraldi, S.; Rizzitelli, G.; Schianchi-Veraldi, R.: Occupational Cutaneous Infections. Clinics in Dermatology, 10:225-230 (1992).
Vieluf, D.: Skin Roughness-measuring Methods and Dependence on Washing Procedure, In: Griesbach Conference Skin Cleansing with Synthetic Detergents: Chemical, Ecological, and Clinical Aspects, pp. 116-129, O. Braun-Falco and H.C. Korting, Eds., Springer-Verlag, Berlin (1992).

Wahlberg, J.E.: Health-screening for Occupational Skin Diseases in Building Workers. Berufodermatosen 17:184-198 (1969).

Wahlberg, J.E.: Prophylaxis of Contact Dermatitis. Seminars in Dermatology, 5:255-264 (1986).

Walker, A.P.; Basketter, D.A.; Baverel, M., et al.: Test Guideline for Assessment of Skin Compatibility of Cosmetic Finished Products in Man. Fd. Chem. Toxicol. 34:651-660 (1996).

Walker, J.D.; Whittaker, C.; McDougal, J.N.: Role of the Tsca Interagency Testing Committee in Meeting the U.S. Government Data Needs: Designating Chemicals for Percutaneous Absorption Rate Testing, In: Dermatotoxicology, 5th edition, pp. 371-381, F.N. Marzulli and H.I. Maibach, Eds., Taylor & Francis, Bristol, PA (1996).

Walker, M.; Chambers, L.A.; Hollingsbee, D.A.; Hadgraft, J.: Significance of Vehicle Thickness to Skin Penetration of Halcinonide. Internat. J. Pharmaceutics 70:167-172 (1991).

Wall, L.; Gerbauer, K.: A Follow-up Study of Occupational Skin Disease in Western Australia. Contact Dermatitis 524:241-243 (1991).

Wang, C.L.: The Problem of Skin Disease in Industry. Office of Occupational Safety and Health Statistics, U.S. Department of Labor (1978).

Wang, L-F.; Lin, J.Y.; Hsieh, K.H.; Lin, R.H.: Epicutaneous Exposure of Protein Antigen Induces a Predominant Th-2-like Response with High Ige Production in Mice. J. Immunology 156:4079-4082 (1996).

Ward, K.A.; Todd, D.; Thornton, C.; et al.: A Comparison of Expression of Surface-presenting Cells in Cutaneous Tissue Between Patients with Allergic, Irritant and Atopic Dermatitis. Contact Dermatitis 25:115-120 (1991).

Watt, A.D.: Hairdressers and Hepatitis B - A Risk of Inapparent Parenteral Infection. J. Soc. Occup. Med. 37:124-125 (1987).

Weber, L.W.D.; Zesch, K.; Rozman, K.: Decontamination of Human Skin Exposed to 2,3,7,8-Tetrachlorodibenzo-p-dioxin (TCDD) in Vitro. Arch. Environ. Health 47:302-308 (1992).

Weltfriend, S.; Bason, M.; Lammintausta, K.; Maibach, H.I.: Irritant Dermatitis (Irritation), In: Dermatotoxicology, 5th Edition. pp.87-118, F.N. Marzulli and H.I. Maibach, Eds., Taylor & Francis, Bristol, PA (1996).

Wester, R.C.; Noonan, P.K.; Maiback, H.I.: Frequency of Application on Percutaneous Absorption of Hydrocortisone. Arch. Dermatol. 113:620-622 (1977).

Wester, R.C.; Maibach, H.I.; Bucks, D.A.W.; Guy, R.H.: Malathion Percutaneous Absorption after Repeated Administration to Man. J. Pharm. Sci. 68:116-119 (1983).

Wester, R.C.; Maibach, H.I.: Cutaneous Pharmacokinetics: 10 Steps to Percutaneous Absorption. Drug Metab. Rev. 14:169-205 (1983).

Wester, R.C.; Maibach, H.I.: Clinical Considerations for Transdermal Delivery. Transdermal Delivery of Drugs, Vol. 1, pp. 72-78, R.C. Wester, Ed., CRC Press, Baton Rouge (1987).

Wester, R.C.; Maibach, H.I.: Dermal Decontamination and Percutaneous Absorption. In: Percutaneous Absorption, Mechanisms, Methodology, Drug Delivery. R.L. Bronaugh, H.I. Maibach, Eds., Marcel Dekker, New York (1989).

Wester, R.C.; Maibach, H.I.; Bucks, D.A.: Percutaneous Absorption and Skin Decontamination of Pcbs: In Vitro Studies with Human Skin and In Vivo Studies in the Rhesus Monkey. J. Toxicol. Environ. Health 31:235-246 (1990).

Wester, R.C.; Maibach, H.I.: Percutaneous Absorption of Hazardous Substances from Soil and Water, In: Dermatotoxicology, 5th Edition. pp.325-335, F.N. Marzulli and H.I. Maibach, Eds., Taylor & Francis, Bristol, PA (1996).

Wilhelm, K-P.; Surber, C.; Maibach, H.I.: Effect of Sodium Lauryl Sulfate-induced Skin Irritation on in Vivo Percutaneous Penetration of Four Drugs. J. Invest. Dermatol. 97:927-932 (1991).

Wilhelm, K.-P.; Vreitag, G.; Wolff, H.H.: Surfactant Induced Skin Irritation and Skin Repair: Evaluation of a Cumulative Human Irritation Model by Noninvasive Techniques. J. Am. Acad. Dermatol., 31:981-987 (1994).

Wilkinson, S.M.; Cartwright, P.H.; Armitage J.; English, J.S.: Allergic Contact Dermatitis from 1,6-diisocyanatohexane in an Anti-pill Finish. Contact Derm. 25:94-96 (1991).

Wilkinson, S.M.; English, J.S.C.: Hydrocortisone Sensitivity: Clinical Features of Fifty-nine Cases. J. Am. Acad. Dermatol. 27:683-687 (1992).

Williams, H.C.; Strachan, D.P., Eds. The Challenge of Dermato-Epidemiology, CRC Press, Boca Raton (1997).

Williamson, K.S.: A Prognostic Study of Occupational Dermatitis Cases in a Chemical Works. Br. J. Ind. Med. 24:103-113 (1967).

Wilschut, A.; ten Berge, W.F.; Robinson, P.J.; Mckone, T.E.: Estimating Skin Permeation. The Validation of Five Mathematical Skin Permeation Models. Chemospher. 30:1275-1296 (1995).

Wolfe, H.R.; Armstrong, J.F.; Staiff, D.C.; Comer, S.W.: Potential Exposure of Workers to Parathion Through Contamination of Cigarettes. Bull. Environ. Contam. Toxicol. 13:369-376 (1975).

Wolfram, L.J.: Frictional Properties of Skin, In: Cutaneous Investigations in Health and Disease: Noninvasive Methods and Instrumentation. J.L. Leveque, Ed., pp. 49-57, Marcel Dekker, New York (1989).

Wolfgang, U.; Pfahlberg, A.; Gefeller, O.; et al.: Prevalence and Incidence of Hand Dermatitis in Hairdressing Apprentices:
Results of the POSH Study. Int. Arch. Occup. Environ. Health 71:487-492 (1998).

Wood, L.C.; Jackson, S.M.; Elias, P.M.: Cutaneous Barrier Perturbation Stimulates Cytokine Production in the Epidermis of Mice. J. Clin. Invest. 90:482-487 (1992).

Woolhiser, M.R.; Munson, A.E.; Meade, B.J.: Immunological Responses of Mice Following Administration of Natural Rubber Latex Proteins by Different Routes of Exposure. Toxicol. Sci. 55:343-51 (2000).

World Health Organization Task Force on Environmental Health Criteria for Lead: Lead, WHO, Geneva (1977).

Wrbitzky, R.; Angerer, J.: N,N-dimethylformamide - Influence of Working Conditions and Skin Penetration on the Internal Exposure of Workers in Synthetic Textile Production. Int. Arch Occup. Environ. Health 71:309-316 (1998).

Wrangsjo, K.; Osterman, K.; van Hage-Hamsten, M.: Glove-related Skin Symptoms among Operating Theatre and Dental Care Unit Personnel (II). Clinical Examination, Tests and Laboratory Findings Indicating Latex Allergy, Contact Dermatitis, 30:139-43 (1994).

Yang, J.J.; Roy, T.A.; Krueger, A.J.; et al.: In Vitro and in Vivo Percutaneous Absorption of Benzo[a]pyrene from Petroleum Crude-fortified Soil in the Rat. Bull. Environ. Contam. Toxicol. 43:207-214 (1989).

York, M.; Basketter, D.A.; Cuthbert, J.A.; Neilson, L.: Skin Irritation Testing in Man for Hazardous Assessment - Evaluation of Four Patch Systems. Hum. Exper. Toxicol. 14:729-734 (1995).

Zendzian, R.P.: Skin Penetration Method Suggested for Environmental Protection Agency Requirements. J. Am. Coll. Toxicol. 8:829-835 (1989).

Zendzian, R.P.: Guidelines for Studying the Dermal Absorption of Pesticides: Background Document. Health Effects Division, Office of Pesticide Programs, United States Environmental Protection Agency, February (1991).

Zhai, H.; Maibach, H.I.: Effect of Barrier Creams: Human Skin In Vivo. Contact Dermatitis 35:92-96 (1996).

Zienicke, H.: Skin Hydration (Transepidermal Water Loss) - Measuring Methods and Dependence on Washing Procedure, In: Skin Cleaning with Synthetic Detergents, Chemical, Ecological, and Clinical Aspects. O. Braun-Falco, H.C. Korting, Eds., Springer-Verlag, Berlin (1992).

4

THE PRACTICE AND MANAGEMENT OF OCCUPATIONAL ERGONOMICS

Miriam Joffe, MS, PT and David C. Alexander, MSIE

This chapter was developed to introduce the history of ergonomics, present effective strategies to identify ergonomics-related problems, assess ergonomics-related risks factors, complete a comprehensive root cause analysis, prioritize cost-effective solutions, then communicate findings and potential solutions in the industrial setting. It serves as a reference for hygienists who are incorporating ergonomics into traditional safety and health programs.

The materials in this chapter have been used successfully by the authors over the past 10 years in a variety of manufacturing, assembly, transportation industries, and office settings. A variety of graphics were developed to illustrate key concepts and to facilitate the problem identification and solution development processes. Tables were developed to promote consistency of data gathering and assist the user with streamlining the problem solving and solution development processes.

A BRIEF HISTORY OF OCCUPATIONAL ERGONOMICS

Background

Occupational ergonomics is a multidisciplinary approach to studying how people work given their capabilities and, based on that knowledge, designing equipment, jobs, workstations, and end-products.

The priorities of occupational ergonomics are easily defined. The first priority is the prevention of occupational injuries to people at work. The second is the organization of work (both physical set-up and work methods) to improve worker efficiency and effectiveness while ensuring operator comfort. Every effort is made to design both an environment and systems that prevent injuries by ensuring that work tasks are within the capabilities of the worker. However, since ergonomics is a relatively new focus within industry, there are many areas where ergonomics-related problems are unfortunately an integral component of existing job tasks. These problems must be identified and corrected. The two-pronged approach to ergonomics is effective: work with engineers and designers to prevent future problems, while also working with manufacturing and operations personnel to identify current jobs with ergonomics-related risks so that existing problems will be corrected. The goal is to balance capabilities of people with job demands (see Figure 4-1). When demand exceeds capability, physical injury and process errors are more likely to occur.

While ergonomics is used primarily for injury and illness prevention, that is not its only benefit. Ergonomics job improvements frequently result in improvements to worker effectiveness with benefits in productivity, product quality, lower operating costs, increased uptime, and similar operational benefits. Based on the authors' experience, the economic value of ergonomics improvements can be as high as 10% of labor costs, although 3-6% is more typical and reasonable. Nonetheless, when a 6% increase in productivity is possible in addition to controlling occupational injuries, ergonomics becomes a powerful tool indeed.

Also important are the "operator comfort" aspects of ergonomics. These are often thought to be intangible benefits, but by making jobs fit within the capabilities of people, the overall

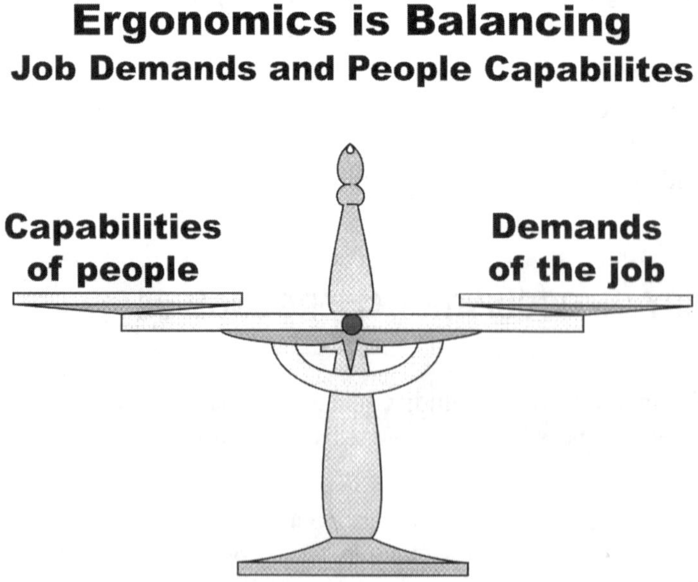

Figure 4-1. Balancing capabilities and demand.

stresses are reduced, and the jobs are less taxing on the individual. This makes the person less fatigued at the end of the work shift. As a result, the job is more attractive, there is higher job satisfaction, and ultimately problems with turnover and absenteeism diminish. While this economic impact is challenging to measure, it is likely to be significant where it is measured.

Ergonomics Versus Human Factors

Within the umbrella of occupational ergonomics, the terms *ergonomics* and *human factors* have been used interchangeably. Ergonomics traditionally refers to the physical limits of the individual, how the work and workplace design affect the person, the biomechanics of the work environment, and the ability to limit fatigue through optimal design. It includes the physiological response to physical workload, responses to environmental conditions (e.g., heat, noise, lighting, vibration), people's ability to utilize hand-eye coordination in complex psychomotor tasks, and visual monitoring of the work process (e.g., control panels). Human factors engineering, on the other hand, has its roots in experimental psychology with a focus on how people interact with equipment, job demands, environment, and workstations. The key focus is on system designs that minimize the potential for human processing errors.

With the advent of the computer, the gap in these two approaches is narrowing. The perception and interpretation of information on the screen is interrupted so that an input device can be used to affect the status on the screen. Work posture and data input involve biomechanics; font and icon size involve perception; and information layout on the screen involves cognitive psychology. By removing unnecessary effort and demand, improving information transfer between people and/or product (inspection), and improving how people accomplish tasks, both productivity and profit are likely to improve.

Along with productivity, concerns about human health/safety and job satisfaction have increased. This has led to increased interest in human factors and ergonomics as they apply to industry. Medical professionals are recognizing that direct treatment of symptoms is only one element in the rehabilitation of people returning to work. More important, identifying and modifying the source of the problem is the key to preventing recurrence of these symptoms.

History

Elements of ergonomics have been used as far back as the stone age. Close examinations of artifacts from this period demonstrate that people were designing hand tools to fit both the user and the specific task of the tool (Drillis, 1963). In the 1500s, a physician named Bernardino Ramazinni began to see a correlation between symptoms and work activities. In the 1860s, the term *ergonomics* was introduced into scientific literature by Wojciech Jastrzebowski, a Polish professor of natural sciences at the Agronomical Institute in Warsaw-Marymount. Ergonomics was derived from the Greek words ergo (work) and nomos (laws) and means "the laws (or study) of work" (Polish Ergonomics Society, 1979).

Then, in the early 1900s, Fredrick Taylor (Father of Industrial Engineering) introduced the "scientific" study of work. He broke the whole work process into individual tasks, assigned one person to perform each task, then studied the methods and amount of time needed to perform the task. His work is responsible for the first assembly lines such as the one Henry Ford used to start Ford Motor Company. Around this same time, Frank and Lillian Gilbreth developed time and motion studies and introduced the concept of dividing ordinary jobs into several small microelements called "therbligs" (Konz, 1990).

Despite these early advances, little attention was paid to this field until casualties and problems developed during World War II. The United States realized that sophisticated military equipment such as airplanes, sonar and radar stations, and tanks posed problems such as physical access to the cabins and/or cognitive problems leading to operator error. It is interesting to note that during the Korean War, more pilots were killed during training than during combat (Nichols, 1976).

In the 1950s, ergonomics became an independent discipline when the Ergonomics Research Society was formed in Great Britain and the Human Factor Society was formed in the United States. In 1961, Stockholm, Sweden hosted the first meeting of the International Ergonomics Association (Chapanis, 1990). Meanwhile in the United States, a joint meeting of the Aeromedical Engineering Association of Los Angeles and the San Diego Human Engineering Society was held in the mid-1950s. Finally, in 1957, the Human Factors Society was officially created in Tulsa, Oklahoma (Davis and Rodgers, 1983). This professional organization is now officially known as The Human Factors and Ergonomics Society.

Today, ergonomics and human factors engineering principles are consistently integrated into the design of equipment, layout, and work methods for U.S. military and aerospace industries. Additionally, other federal agencies also support research on non-military applications as listed below.

- The Federal Highway Administration investigates the design of highway and road signs.
- The National Highway Traffic Safety Administration uses this research for improved car design and crash tests as well as effects of drugs/alcohol on driving.
- NASA studies human capabilities and limitation in space. Since the early 1950s, every aerospace design and system has been required to have a concurrent human factors design. Simulators for space mission trainings have been developed. The Federal Aviation Administration incorporates these principles into aviation safety.
- The Department of the Interior is incorporating ergonomics principles in underground mining.
- The National Bureau of Standards monitors the safe design of consumer products.
- NIOSH studies ergonomics-related injuries at work, industrial safety, and work stress.

- The Nuclear Regulatory Commission incorporates ergonomics into the design requirements for nuclear power plants.

Other industries have been slow to understand the impact of ergonomics on productivity, health, safety, and job satisfaction. In the mid-1960s, Eastman-Kodak became the first manufacturing company to implement a substantial ergonomics program. It was followed by IBM (Big Blue), which began a formal ergonomics program in 1980. In the 21st Century, more companies are realizing the value of developing ergonomics as an integral part of their overall plans to keep experienced people on the job, reduce health care costs, improve productivity, and minimize product or equipment failure.

Integrating the Industrial Hygienist into Occupational Ergonomics

With recent downsizing and consolidation of responsibilities in industry, it is not uncommon to find that the company nurse, safety professional, or an interested employee is assigned to develop and manage the ergonomics program for the site. Sometimes the physical or occupational therapist already providing treatment for an injured employee is assigned to identify musculoskeletal problems and develop solutions for that employee's job. Other times the company calls on the on-site industrial hygienist to expand the assigned scope of work to include a comprehensive ergonomics program, even if it poses the need for continuing education or assistance from an ergonomics consultant.

Although some principles of ergonomics are included in core curricula for industrial hygiene, it will be important to recognize other skills that should be added for the competent practice of ergonomics in industry. Table 4-1 contains a basic ergonomics skills checklist. Ergonomists, professional organizations, and universities offer a variety of short courses, traditional courses, and self-study courses. Some of the more prominent professional organizations include the American Industrial Hygiene Association (*www.aiha.org*), the Institute of Industrial Engineers (*www.iienet.org*), the Human Factors and Ergonomics Society (*www.hfes.org*), the International Ergonomics Association (*www.iea.cc*), and the American Society of Safety Engineers (*www.asse.org*). Several organizations also offer advanced certification in ergonomics including the Board of Certified Professional Ergonomists (*www.bcpe.org*) and the American Society of Safety Engineers (*www.asse.org*).

UNDERSTANDING ERGONOMICS IN INDUSTRY

The Background Survey and the Structure of Industry (see also Volume 1, Chapter 12)

In many cases, product or processes may be new or unfamiliar to the industrial hygienist. This is particularly true when the hygienist is serving as a consultant to an organization, either internal or external. It may be helpful to refer to industry journals and other related engineering references. Review the company's (plant's) website (if available) to learn more about its prod-

Table 4-1
Problem solving skills checklist

Skills Needed for Ergonomics Problem Solving	check
Ability to identify neutral postures and the degree of deviation from neutral during job tasks.	√
Ability to identify job activities and motions that cause discomfort or musculoskeletal disorders (MSDs).	√
Ability to understand the anatomical and physiological basis for MSD development.	√
Ability to identify ergonomics risk factors associated with the job tasks.	√
Ability to identify root cause of risk.	√
Ability to redesign hand tools and user interfaces with optimal grip, weight, and force components.	√
Ability to develop optimal job rotation and job enlargement recommendations based on work schedules.	√
Ability to understand the influence of personal protective equipment (PPE) on posture and force.	√
Ability to use and interpret commonly used ergonomics and environmental analysis tools (e.g., NIOSH lifting guideline, psychophysical tables, 3-D motion analysis model, Ergo Job Analyzer™, light meter, WBGT, etc.).	√
Ability to understand how environmental factors affect human performance and productivity.	√
Ability to measure cycle time and its affect on productivity.	√
Ability to design communications methods to promote productivity and limit human error.	√
Ability to understand labeling, layout, and parts design to limit human error, cycle time, and bottlenecks.	√
Ability to document ergonomics risk factors, root causes and practical and cost-effective recommendations for company managers.	√
Ability to develop ergonomics educational programs that teach the principles of MSD symptom and ergonomics risk factor identification.	√

uct, structure, and commitment to safety. Determine if the company's business has a specialty, dying, or competitive market, and understand its product life. In preparing for the site visit, identify the hierarchical structure of the company, your key contacts, the person who will be your guide during the visit, and the person who will provide a tour of the facility.

THE CLIENT

Let the client[1] know that you appreciate the company's goals and needs. Identify whether your client is in the production, assembly, or service industry. Production and assembly industries are generally focused on machine performance with minimal staffing. In service industries, people are the "machines" and the companies are generally focused on the performance of people. In both cases, accomplishing the job quickly, without error, and without injury are goals of both the client and the ergonomist.

Understand the product and profit structure. A company that is rapidly changing, uses expensive state-of-the-art equipment, and runs high profit margins is more likely to spend money on improving next-generation machine and machine-interface designs (e.g., semiconductor and computer industries). A company whose technology and production do not change significantly may be more interested in improving the way employees interface with the machines. Even in this second scenario there will eventually be a need for machine upgrades or replacements when worn or damaged parts can no longer be replaced through the original equipment manufacturer (OEM) and custom-made parts become the last option. Therefore, ask the client if ideas for new tools and machines should be included in your assessment for future budgeting and decision-making. The client will appreciate recommendations that improve cycle time and quality, minimize rework, eliminate downtime, minimize staffing, keep experienced employees on the job, and improve business performance.

HIERARCHICAL STRUCTURE

The hierarchy generally starts with the Chief Operating Officer (COO) who oversees all departments of the facility. Reporting to the COO is the manufacturing manager who is responsible for all production or assembly activities. Next, the shift supervisors are assigned to manage personnel and production or assembly activities for each department. In many cases, there is a formal or informal senior (lead) for each shift who helps to train new employees and who problem solves when the shift supervisor is not available. In cases where the facility operates around the clock ("24/7" which means 24 hours/day, 7 days/week), a supervisor may only work the day shift, and the lead is responsible for activities on evening and night shifts. At the base of the hierachial structure is the line staff. In most cases, each employee is assigned to perform

---

[1] The client role may be traditional as in a consultant-client relationship or, for example, the hygienist's plant manager, the manager of a plant that a corporate hygienist visits, another professional at a plant within the hygienist's corporation, or many other possibilities.

one or a limited unit of specific duties on an on-going basis. In rare cases, employees are cross-trained to perform other tasks so they can routinely rotate between jobs, or so they can fill in elsewhere when another work area is particularly busy or when another employee is absent.

LABOR UNIONS

Many industries such as the steel, automotive, garment, and food industries have strong labor unions. On an annual basis, unions negotiate contracts with companies on issues dealing with work assignments, breaks, equipment needs, and seniority. It is helpful to identify key union leaders in the facility and work directly with them during the evaluation and solution development processes. Recommendations are more likely to be approved when employees and labor unions are involved throughout the process.

ECONOMICS

In most companies, funding for ergonomics is secured by the safety group or individual departments during annual budget requests. Expenditures are generally classified as capital or non-capital. Capital expenditures (usually more than $500) may require high-level management approval, whereas non-capital expenditures (usually less than $500) most often require only a manager's signature and may be available on a monthly basis. It is interesting to note that most simple engineering recommendations can be implemented for less than $500 without significantly impacting production. In contrast, most process and equipment changes require approval of multiple departments, may require machine or process "down time," and carry a price tag in excess of $500. If a company has multiple locations that use similar machines, processes, and layout, it would be wise to pilot changes and make improvements at one site before instituting these changes at other sites to minimize final costs. If process or equipment recommendations are cost-effective, they generally pay for themselves through productivity improvements. There are times, however, when changes are approved primarily for the benefit of employee health/safety and job satisfaction (e.g., sit/stand chairs, new floors, floor mats, or shoes with viscoelastic inserts to reduce standing fatigue) without primary regard to financial gains.

Understanding the Production and Assembly Process

PROCESS FLOW

Process flow (see Figure 4-2) is the linear progression of product or process from start to finish. Work should ideally be completed entirely at one work area, then continue to the next step without backtracking or rework. Backtracking and rework interrupt ideal process flow and may be damaging to cycle time, end-product, productivity, and profit.

Before recommending a change to product or process at any step in the process flow, evaluate potential impact to both upstream and downstream operations. Determine if the change will cause a bottleneck in production or if different equipment, hand tools, or workstations will be needed. (A bottleneck is a slow down in production that reduces the speed of product deliv-

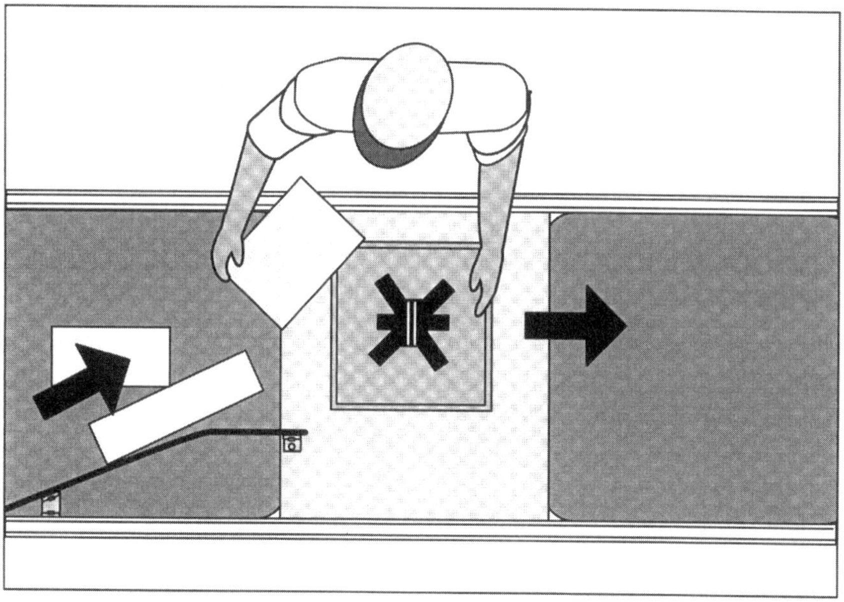

Figure 4-2. Process flow.

ery to the next step in the process.) If a bottleneck does occur, determine if the cause is human error, insufficient staffing, poor product quality, incorrect use or choice of tools, or a problem with delivery of parts.

Understand how a recommended change might impact the amount of time needed to complete the task or if additional training is required. For example, if you recommend a pallet jack for product transport to/from the work area, then determine if a dedicated line worker must leave production to transport carts, if additional staff is needed for this task, or if there is adequate lag time in the production process so that neither production nor staffing is impacted. Also, determine if special training is required to operate the pallet jack safely and if additional instruction is required for loading the product upstream and returning the pallet jack to the place of origin after use.

CYCLE TIME

Cycle time is the duration required to start, participate in, and complete the entire task one time. Cycle time can be influenced by accessibility and availability of parts, ease of using hand tools, ease of putting parts together, and eliminating unnecessary steps. Minimizing cycle time is always a desirable goal.

LAYOUT

Tool and machine placement can have a strong influence on posture, fatigue, and cycle time. Supplies that are frequently accessed should be placed within the primary reach zone (0 to 14 in. from the body), while infrequently accessed objects can be placed in the secondary reach zone (14 to 22.5 in. from the body, see later discussion) to minimize awkward postures and energy use. In some cases, placing supplies and machines at a distance that requires a short walk may be preferred as shown in Figure 4-3. The employee working at a table adjacent to a storage area would probably twist around to reach it. This configuration is designed to require the employee to turn completely around and walk to the storage table without twisting.

The other important issue with layout involves access to machine parts for routine maintenance. There must be adequate clearance from walls and other machines to permit reasonable access for both the employee and required tools or equipment (e.g., wrenches and vacuums).

TOOLS AND MACHINES

Optimally, access to machine controls and parts are designed with the user in mind. The smallest operator should always to able to reach the emergency cut-off switch and control panel without excessive reaching or use of a step stool. Machines should be designed so that unnecessary removal of parts is not required to reach the component that requires maintenance (e.g., stacking components requires the top section to be removed for access to lower sections). Tools

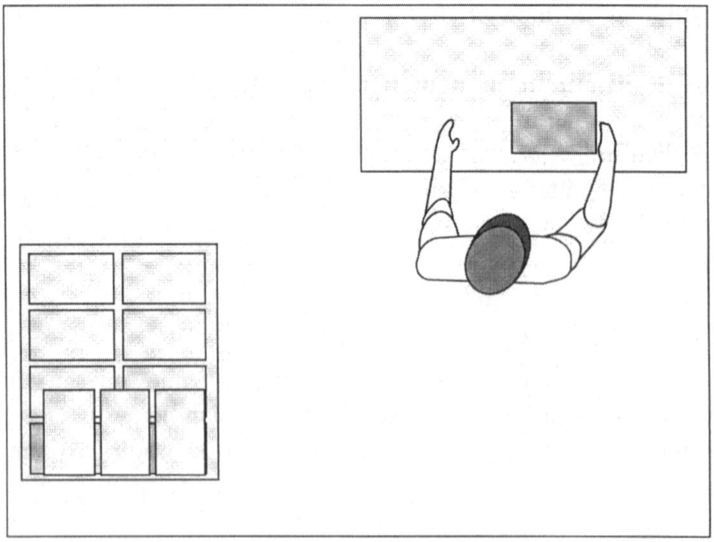

Figure 4-3. Layout that avoids twisting of the torso.

and machines should be designed to fit the needs of the job (e.g., size, weight, hand grip design, handle length, end detector, frequency of use, and durability), variations in operator size, and should be user friendly. This makes the job easier to do, requires less time, is less physically demanding, and minimizes mistakes.

Routine maintenance of tools and machines cannot be stressed enough. It is important that moving parts are properly cleaned and lubricated and that cutting instruments be kept sharp. This makes the tool easier to use, reduces cycle time, and provides a better end product. An unexpected machine breakdown abruptly stops production, leaving employees little to do during this downtime. It can also cause a product defect to occur and, most importantly, may cause injury. All are costly and undesirable and can be avoided with planning.

WORK SCHEDULES, STAFFING, BREAKS, AND OVERTIME

There are a variety of work schedules that a company may utilize to meet staffing needs. Work shifts are usually scheduled on an 8, 10, or 12-hour basis. In some cases, the employee always works the same shift. In other cases, employees work rotating shifts that may include weekends and evenings/nights. In union facilities, shift assignment is often dictated by seniority, leaving newer employees to work less desirable schedules. When there is suspicion that work schedule has influenced injury or performance, these data should be researched. It is important to note that most work performance research is based on the 8-hour work shift; therefore, effects of fatigue must be extrapolated for assessment of longer work shifts.

Look for staffing variations between shifts to understand if there is a correlation between injuries and staffing levels. Day shift is generally staffed with more employees to accommodate meetings, shipping schedules, and other special needs. Evening and night shifts may be staffed with fewer employees even though production rates remain unchanged. Routine or scheduled maintenance may be performed on any shift, however most maintenance crews work the day shift with a small back-up force for unscheduled or emergency needs on the off-shifts. Modifying the staffing schedule to accommodate routine maintenance activities will ensure optimal utilization of employees without rushing them or influencing them to take short cuts.

Productivity on each shift will be measured by some internal standard. Self-paced jobs that have overall quotas or productivity standards allow the employee to regulate rate of work and take micro-breaks as needed, whereas externally paced jobs (e.g., work arrives at timed intervals via conveyor) may not allow for needed breaks. Piece rate jobs or those with incentive for higher rate of production may influence employees to choose not to take breaks if there is an associated financial gain.

Work breaks will vary depending on job assignment, job autonomy, and level of responsibility. When the choice is left to the individual, there is a tendency to lump morning and afternoon breaks together with lunch for a longer break midway through the shift if the individual needs this time to take care of personal business, or if start-up/completion activities are too time consuming (e.g., donning/doffing PPE). Employees may also skip breaks altogether if production demands are too high. Studies show that employees who take short frequent breaks throughout the day are less fatigued, more productive, and more alert at the end of the shift than those who take only one long break.

Overtime required on a short-term basis may not be problematic, however, routine use of overtime often results in an increased rate of injuries. Overtime is often used when trained employees are off for vacation, medical leave, or when production needs peak. The company may prefer overtime to hiring temporary employees because of extensive training and skills required for these jobs. The option or requirement to work overtime may be mandatory, voluntary, or based on seniority. Since the incidence of musculoskeletal disorder (MSD) development (see later sections for a full discussion of MSD) is higher when there is a demand for excessive overtime, it may be useful to suggest that management proactively set policy limits on overtime to prevent overuse injuries.

ENVIRONMENTAL CONDITIONS

Environmental conditions such as lighting, noise, and temperature can either improve or negatively impact productivity. Adequate lighting is needed for optimal job performance and injury prevention. Ambient lighting must be adequate to eliminate shadows and glare, allowing employees to easily see and inspect equipment, tools, and accessories. Additional task lighting may be necessary to improve color contrast for accuracy during inspection tasks. When lighting is inadequate, the employee will likely work in a forward bent position to see better. The cost of lighting improvements is nominal compared to the cost of production defects and injury management.

Noise generated by machinery may interfere with auditory warning signals and verbal communication. In these environments, redundant warning signals (e.g., visual and auditory) may be required, and alternate or limited conversation may be necessary. In the United States, work areas where employees are exposed to noise at or above an 8-hour time weighted average (TWA) of 85 decibels (dBA) must institute OSHA's Hearing Conservation Program (29 CFR 1910.95). The program requires that these employees be tested and monitored for hearing loss, be provided with appropriate hearing protection, and be trained on the effects of noise on hearing, the purpose of hearing protection, and the purpose of audiometric testing.

Temperature may be difficult to regulate in the manufacturing environment, especially in older buildings. Heating and air conditioning may even be absent or insufficient. Temperatures may fluctuate to extremes with seasonal changes and affect whole body fatigue or fine motor movement of the fingers. For example, working outdoors or on the upper levels of a warehouse in the summer may expose the employee to extremely hot, humid, and fatiguing working conditions, while working in a meatpacking company exposes the fingers to chronic cold temperatures that can affect both sensation and manual dexterity. When temperatures are extreme, a regimented system of breaks may be required.

PSYCHOSOCIAL CONDITIONS

Multiple factors can influence the social atmosphere of the company and ultimately how well an employee likes the job. Where unions exist, they can be a strong and positive influence over the policy development and problem resolution processes in the company, giving the employee a feeling of empowerment, satisfaction, and safety. Unions can also be the source of conflict between employees and management for a variety of reasons. Coordinating ergonomics efforts between union leaders, employees, and management to develop acceptable outcomes can help to smooth out sub-optimal working conditions and improve overall job satisfaction. Employees who like both the job and fellow co-workers will tend to stay with the company longer and become more ambitious to find ways to improve production or service needs.

The long-term employees (typical in heavy manufacturing industries) can be both an asset and an impasse when it comes to injury prevention. Many believe that daily aches and pains are just part of the job and should not be reported. They have seen friends leave for medical reasons and not return as expected. They may incorrectly interpret this as a means for the company to flush out complainers without realizing that the injury may have progressed to a state so severe that returning to the same job would put the employee and/or process at risk for further damage. Strong management support for problem identification, early symptom reporting, education about symptom recognition, and workplace improvements will help those employees recognize that these injuries are preventable. Controlling MSDs will improve the social atmosphere of the workplace.

The On-Site Visit

Appropriate preparation for the site visit may be the key to a successful experience. An understanding of the information previously discussed and completion of critical paperwork should be a prerequisite to the site visit. Ensure that contracts and other required paperwork are reviewed for accuracy and signed as needed. Required paperwork may include a confidentiality agreement stating that information acquired about the company during the site visit can only be used for evaluation purposes and may not be shared with other individuals without the express permission of that company. Some facilities require safety training prior to admission. Safety training is usually site specific, usually on video or a computer, and may require a post-viewing test for comprehension. If photographic equipment is anticipated, verify that the company will permit pictures to be taken and whether photographic equipment can be brought in from the outside. Identify special protective clothing required by the site such as steel toe shoes, hardhat, gloves, safety glasses, and hearing protection. Hardhat, gloves, glasses, and hearing protection may be provided by the host since these items are generally on hand for special visitors. However, it is rare that shoes would be provided. Failure to have required PPE may result in a denial for admission to work areas of interest.

TOOLS AND SUPPLIES TO TAKE ON-SITE

It may be helpful to develop a checklist of needed supplies and equipment for the data collection process. First, secure personal protective equipment. If photography equipment is used, extra film and batteries are always a good idea. Identify the need for measuring devices such as a tape measurer, force gauge, or temperature/humidity gauge. For documentation, include a clipboard, blank forms, blank paper, and pen/pencil. The blank paper is useful for sketching ideas along side of experienced employees and supervisors. Even if photographs are taken, sketch the work area and document dimensions for the analysis and report writing processes. If the analysis is completed on site, reference documents should be readily available.

WHILE ON SITE

Meet key contacts, managers, and health services personnel to understand their needs and goals prior to initiating the investigation. Managers or other key personnel may identify additional needs beyond the initial contracted scope of work. Therefore, it is wise to review the scope of work before any other action is taken. Review the process to be used for the investigation along with how/when/to whom the report will be provided. A sample report may help clarify this process.

If possible, ask for a tour of the whole site to understand the entire process and working conditions. Once you have an understanding of the operations, review available data as described below.

A review of job descriptions should occur to understand the scope of the target jobs to be analyzed. Complete the data review process, which includes documents such as OSHA 200 Logs, workers' compensation data, summary of medical cases, first reports of injury, attendance records, maintenance records, and history of safety/ergonomics committee activities. Data review will be covered in more detail later in this chapter.

Once background data have been collected, observe the tasks as performed by employees. A friendly and non-intrusive introduction to the employees will help them feel comfortable with your presence. Ask for permission to speak with them and take pictures if needed. Observe several cycles and several employees performing the same job to identify variations. If possible, observe a new hire, experienced employee, tall employee, and small employee to identify variations in work method and anthropometry. Review job tasks with an employee or supervisor to ensure accuracy. If photographic equipment is used, film from the front, back, and both sides to get a full perspective of the work.

Prior to leaving, review collected data with the supervisor or experienced employee to ensure accuracy of task elements, task progression, and potential variations in tools or work methods. Review key ergonomics issues and potential solutions discussed during the evaluation process. Once company employees understand the intent of the evaluation, they are often able to provide additional detail that is helpful for solution development. Finally, identify a key contact person to answer questions after the site visit is complete if additional visits are not planned.

A CLASSIFICATION SYSTEM

Occupational ergonomics covers a broad range of characteristics of the human at work. It is so broad that examples and illustrations of typical applications (Figure 4-4) are necessary to understand its scope. Thus, one way to classify the differing aspects of ergonomics is in terms of the evolution of its typical applications. The most common ergonomics applications are injury prevention in production operations and maintenance and service tasks. As the ergonomics program matures, ergonomics applications transition to performance enhancement as measured by productivity and quality improvements. In addition, there are specialized areas of application that may be viewed as tangential applications such as office ergonomics, tool design, disability accommodation, and human factors design.

A second way to classify and understand ergonomics is to look at the human systems that traditionally contain the boundaries that limit human performance. There are six major human systems or functional areas of the human to evaluate when examining the human at work. The six areas are:

- The size of people
- The strength of people
- The control of work by people
- The endurance of people
- The environment in which people work
- The mental aspects of work

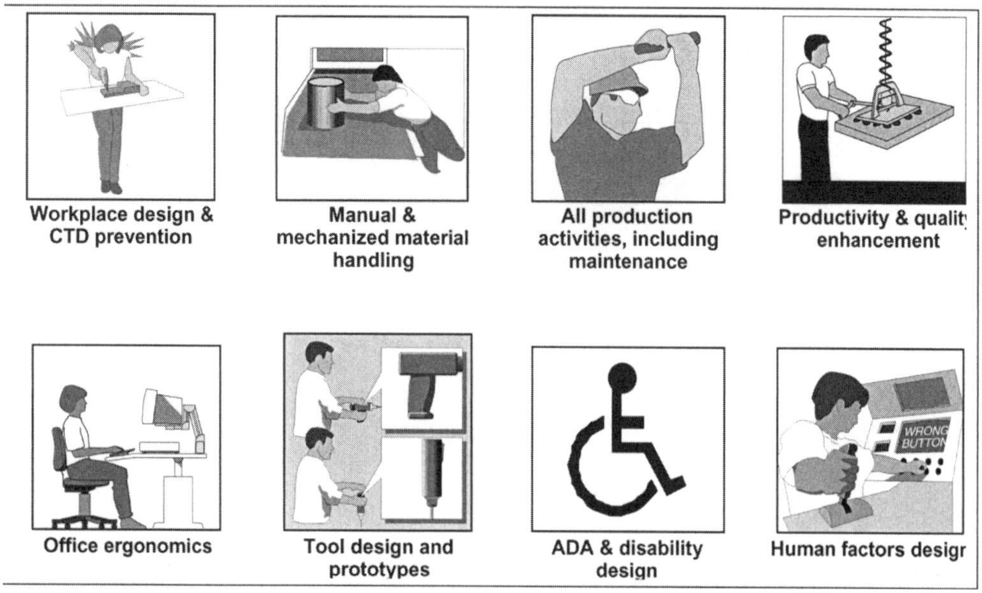

Figure 4-4. Examples of ergonomic applications.

In consideration of applying ergonomics to each of these areas, it is important to remember that ergonomics means fitting the job to the person, not the person to the job. While this may seem trite, it is so often violated in the workplace that it is worth repeating.

The Size of People (Anthropometry)

A brief look around a shopping mall or sporting event illustrates that people vary in size. The size of the person and "fit" at the workplace are major ergonomics issues. The person who is too tall, too wide, or too small relative to the workstation tools or workstation dimensions ends up adapting to its limitations. Quite often, the consequences are not good. Smaller people may not be able to reach items at the workplace (e.g., assembly parts, control knobs, and items placed on overhead shelves). Larger people, on the other hand, may have difficulty entering confined workplaces and may struggle for access through narrow passageways during maintenance activities. Figure 4-5 illustrates how height and reach differ between extremes of the population.

Anthropometrically, clearance and reach issues as shown in Figure 4-6 are among the first to be addressed. Larger people determine the clearance requirements. Clearance issues are important for hatches and vessel openings, as well as clearances between pieces of equipment, access for maintenance work, and clearance under overhead pipes and other obstructions. Smaller people determine reach requirements. Reach issues include accessing items on an overhead shelf, reaching down or forward into a parts bin, reaching for an overhead hoist, and reaching to activate a control button.

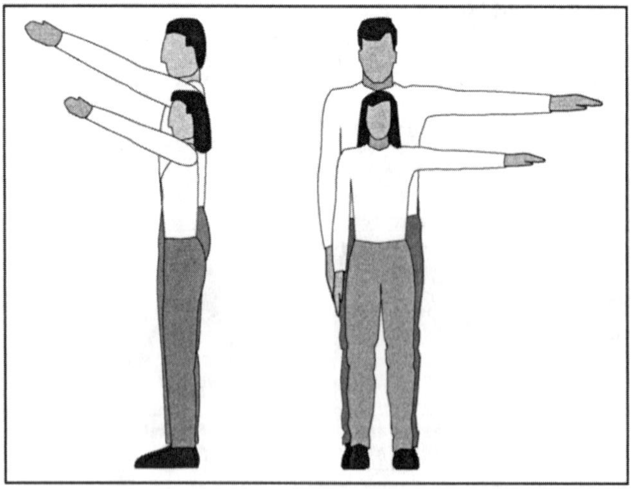

Figure 4-5. Height and reach variations.

In addition to the reach and clearance issues, people must be able to "fit" the workplaces that they use. Too many workplaces are the "one-size-fits-all" workplace where the person must adjust to the workplace by stooping or reaching. In Figure 4-7, the taller person on the left will probably get a backache from stooping over, while the shorter person on the right will have neck and shoulder aches from reaching up to do the work. The "one-size-fits-all" workplace is a common problem for industry, and can usually be resolved by providing adjustability at the workplace.

Think about driving a new car for the first time and the immediate need to adjust the seat and the mirrors before driving off. Why? These adjustments allow the driver to operate the vehicle both safely and comfortably. Even if for just a quick trip to the store, the driver will make adjustments to reach the controls, see through the windshield, and use the mirrors because

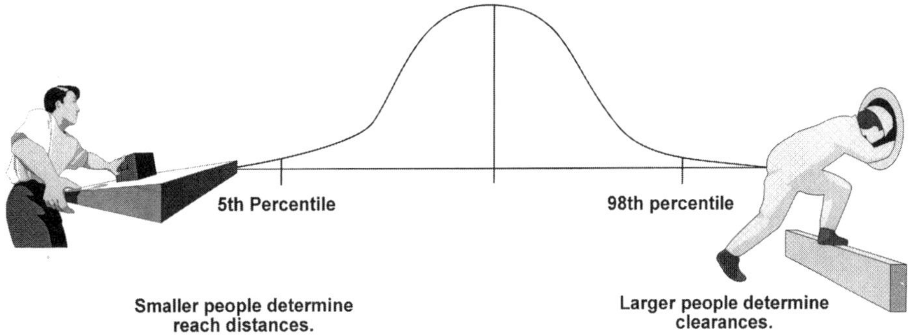

Figure 4-6. Design for reach and clearance.

Figure 4-7. Problems with "One-Size-Fits-All" workplaces.

the adjustments are both necessary and easy to make. Adjustability allows the workplace to fit a wide variety of workers. Figure 4-8, demonstrates that adjustability can be either intrinsic or extrinsic. The workstation on the left is intrinsically adjustable, raising or lowering the table to the correct height for the worker standing on the floor. The workstation on the right is extrinsically adjustable, raising the person up to the work height using a small standing platform. The platform is probably cheaper but may create other problems (e.g., a trip hazard). As Figure 4-8 illustrates, the workplace can be changed to fit the worker with some pretty simple techniques.

When the major elements of the workplace (usually the work surface and chair) are all adjustable, employees on all shifts are able to work in an upright and erect posture. The benefits of a well-designed and adjustable workplace are powerful. Parts and accessories are easy to reach, fatigue is reduced, muscle strain is minimized. This translates to improved productivity and quality of work.

The Strength of People (Musculoskeletal)

People have different levels of strengths and their capacities to perform lifting and other manual handling tasks are not equal. They lift, carry, push, pull, maneuver, handle, tug, tote, and move a variety of objects. The challenge from an ergonomics perspective is to ensure safe materials handling, yet not reduce the load so much that the job is no longer productive.

Lifting is the most common of the manual handling tasks. The weight of the object lifted, the frequency of the lift, and the worker's body position may all contribute to lifting problems.

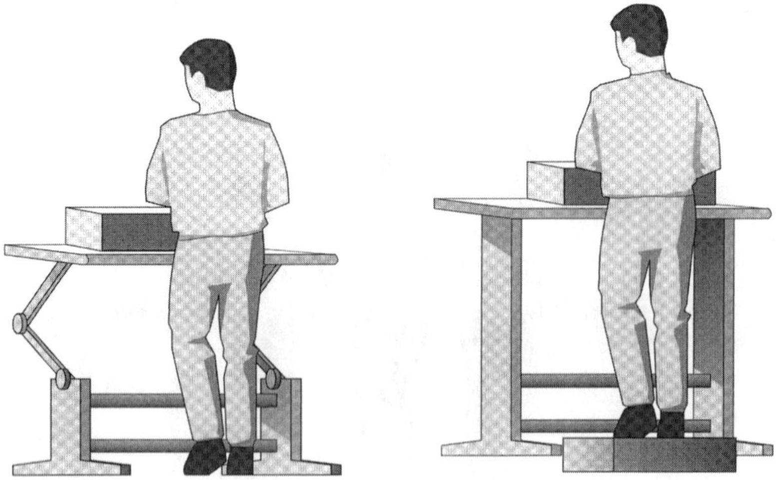

Figure 4-8. Adjustable workplaces.

A common problem is lifting an object stored low to the ground. Bending and stooping put the body at a mechanical disadvantage and increase the risk for injury. Figure 4-9 illustrates a person performing a lift from a low surface while at the same time reaching forward to the middle of a pallet. Several resulting problems are identified.

- The neck is forward and extended.
- The shoulders are protracted and unstable as they reach forward with a heavy load.
- The mid and lower back are fully flexed.
- The knees are fully flexed.
- The base of support is minimal and unstable.
- The weight of the package is multiplied by its distance from the body.

Simply observing good and bad postures will target the majority of problems that may occur when moving objects. However, when a more objective method of analysis is required, a variety of evaluation tools can be applied. A popular evaluation tool used for two-handed symmetrical lifting is the NIOSH Lifting Guideline that will be discussed in detail later in this chapter.

There are several practical solutions to minimize or eliminate lifting problems. For example, place an object on a stand or table to raise it off the floor. When possible, add adjustability

Figure 4-9. Problems with lifting objects stored close to the ground.

by using a spring-loaded pallet leveler, scissors lift, or other mechanical assistance. Slide or pull the object to the edge before lifting to avoid lifting from the center of the pallet.

One of the most important risk factors involving a lift is the horizontal reach or the distance from the hand (holding the object) to the center of the body. Reaching beyond arm's length may create a biomechanical disadvantage as the trunk bends forward and shoulder blades glide into a less stable position. Furthermore, the horizontal distance to the object multiplies the effect of the weight. A 10-pound object held at arm's length could be as stressful to the back as a 50-pound weight held in close to the body. The teeter-totter in Figure 4-10 shows the impact of the weight multiplied by its distance. The small girl offsets the weight of a large man when she sits far enough back on the teeter-totter.

Lifting is the most common type of manual materials handling problem, however, there are several other activities that are problematic as well. Carrying, pushing, pulling, box and drum handling, shoveling, and many other similar tasks are known to cause a variety of musculoskeletal symptoms. For example, pushcarts can be difficult to push or pull due to heavy weight if overloaded; casters that are broken or need maintenance do not move well over the ground; and cracked or uneven floors make it difficult to move the cart over an uneven surface. Problems associated with manual materials handling must be analyzed according to object weight and/or muscle force required to move the object, horizontal and vertical distance from the body, orientation to the body, fit between the hand and object, frequency, and duration of activities, regardless of the reason. The more critical problems must be identified, analyzed, and corrected to prevent musculoskeletal injuries.

Figure 4-10. Weight times distance is the true cost of lifting.

The Control of Work by People

Hands are primarily designed for fine motor activities and arms for carrying, whereas legs are designed for mobility. When the influence of posture, force, and repetition on the functional limits of the hands and arms are minimal, problems do not develop. However, when that limit is exceeded, symptoms develop. If too much force or endurance is required, if the demand on the hands for manual dexterity is too high, or if the task is highly repetitive, MSDs are likely to develop. For example, the hand is commonly used as a tool for performing tasks that range from heavy assembly operations to light keyboarding. High muscle force requirement coupled with awkward postures during these heavy operations can easily lead to MSD development. However, highly repetitive keyboarding coupled with awkward posture can be equally as disabling. The outcomes of these examples are often referred to as overuse injuries.

With overuse injuries, the primary risk factors are force, repetition, and posture. As the demand for hand and finger forces increase during gripping, pinching, or squeezing activities, so does the risk for symptoms development. As body segments move away from neutral posture, strength diminishes and the likelihood of symptom development increases. Couple either of these risk factors with repetition and the probability of symptom development increases more dramatically.

Although overuse injuries are quite common in the hands and wrists, the overuse pattern of MSDs on a cumulative basis can occur in any part of the body. Figure 4-11 illustrates how the overhead work of pipe fitters or drywall workers can result in shoulder pain, and how constant kneeling and squatting to reach low mounted valves can produce knee problems.

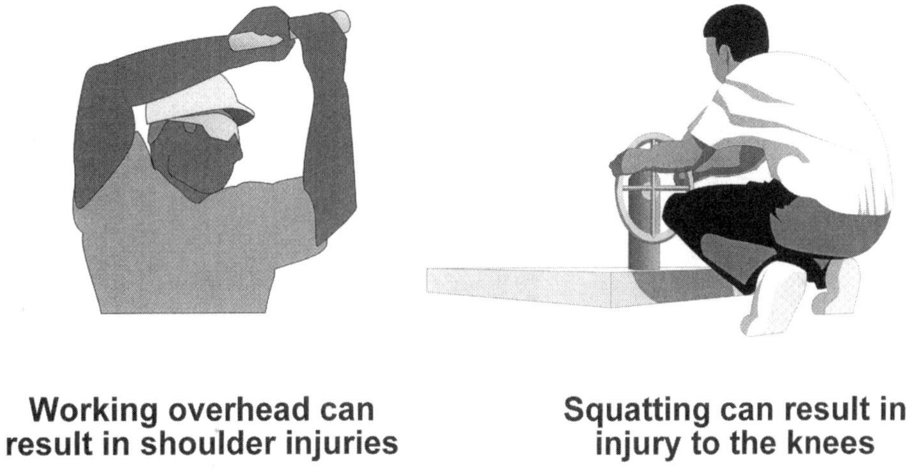

Working overhead can result in shoulder injuries
Squatting can result in injury to the knees

Figure 4-11. Awkward joint orientations can cause MSDs.

The cognitive issues of ergonomics are often referred to as "human factors." Some ergonomists discount these issues, but most are interested in ensuring that the cognitive demands of the job are within the capabilities of the people performing the job. There are tools and methodologies for designing the cognitive aspects of a job just as there are for designing the

A dependable option for preventing MSDs is matching the tool to the job. For example, Figure 4-12 illustrates that using a drill on a table that is too high causes shoulder elevation. Prolonged work in this posture will likely result in shoulder pain. Figure 4-13 demonstrates that a tool with a pistol grip is the tool of choice for drilling horizontally into a wall, while the in-line drill is better for drilling into a bench top. Therefore, the in-line drill is better suited for the work demonstrated in Figure 4-12. It allows for the proper fit between person and task, and permits use of tools with neutral body postures. It also improves hand control and the ability to manipulate products rapidly and reliably for manufacturing operations. It also improves the dexterity needed to perform precision operations such as with a microscope.

In addition to MSD issues, the control aspects of ergonomics include the ability to manipulate products rapidly and reliably for manufacturing operations, and the dexterity to perform precision operations such as with a microscope.

The Endurance of People (Physiology)

Endurance is often an overlooked ergonomics issue. Jobs that require endurance stress the overall physical capacity of the person. This is somewhat different from the musculoskeletal

Figure 4-12. Awkward posture while using a drill.

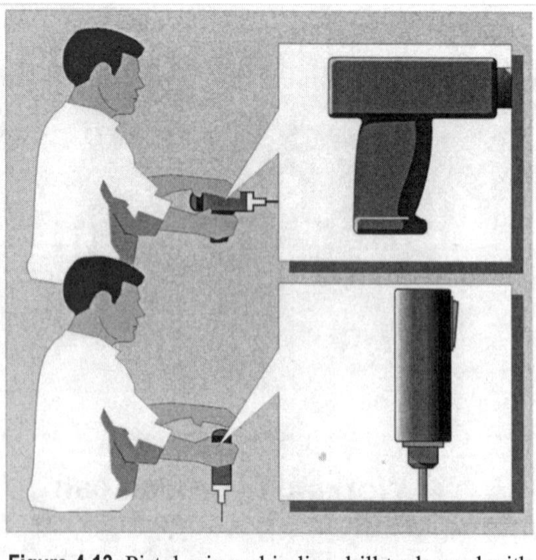

Figure 4-13. Pistol grip and in-line drill tools used with proper orientation and neutral body postures.

stress to a single body part that is experienced during lifting and repetitive handwork. Jobs that place a high demand on the cardiovascular system by requiring intensive physical activity or working in extreme temperatures are difficult to design because of variations in individual capacity for physically demanding work. One example of a physically demanding job is order picking. Order picking is common in the grocery industry where people may pick up, handle, and stock as many as 500 cases of food products per hour. The high rate of activity requires muscles to work overtime and may stress the heart to work harder, pumping oxygen to fatiguing muscles. In the case of high heat loading, prolonged exposure to high temperature and humidity can literally produce "killer" jobs. The body works hard and generates heat internally, raising the core body temperature. Couple this with the radiant heat from the sun (if outdoors) or from a hot production process and protective clothing, and eventually the cardiovascular system is overloaded. If the body cannot sufficiently dissipate the internal heat generation, then core body temperature may rise high enough to cause heat stress. Fortunately, most people experience early warning signs such as light-headedness, nausea, or fainting before serious problems occur.

The Comfort of People (Environment)

Environmental conditions have a strong influence over ergonomics issues. As with other variables, comfort and tolerance of temperature, lighting, noise, and vibration vary between people. These conditions must be assessed when balancing the work demand with the capacity of the worker. When the environment is hostile, performance may be diminished and the health and safety of the worker may be endangered. The environmental issues of concern to the ergonomist include hot and cold conditions, lighting, noise, and vibration.

The Memory of People (Cognition)

physical aspects of the job. Some of the more important issues include control panel design, signs and labeling, coding systems, and stereotypes (the expectations we have for the outcome of a control action). Errors here may impact safety, operations, and maintenance costs. A simple example of cognitive design issues involves the labeling of display instruments. Figure 4-14 shows a control on the left that measures PSI, however without further explanation, the operator may not necessarily know which parameter PSI actually measures, which machine it is measuring, or what PSI means. The improved design on the right identifies that pressure (Pounds per Square Inch) is being measured specifically for Pump No. 1. The addition of color coding and critical pressures marked on the borders helps the operator to simply glance at the gauge to identify if the pressure is too high. These simple and effective changes improve performance and minimize operator error. Further improvements might include color coding for the acceptable (green) and "warning" (yellow) ranges of pressure. The use of these colors supports a stereotype as established with the use of traffic signals. They improve the likelihood that the operator will respond accurately to changes in status.

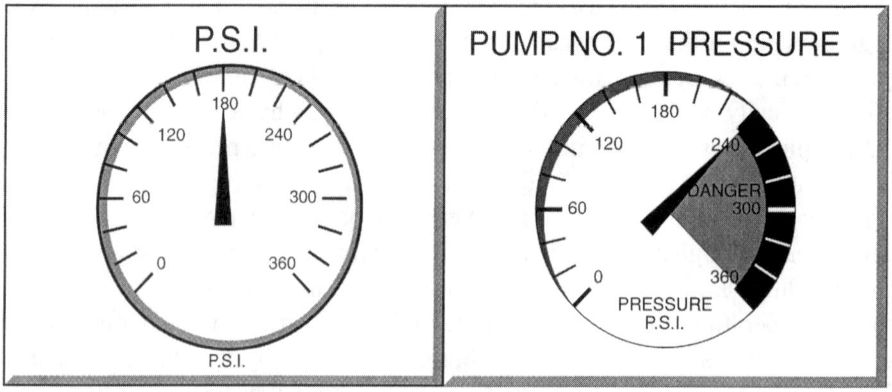

Bad Design Good Design

Figure 4-14. A display with poor labeling design and another with improved features.

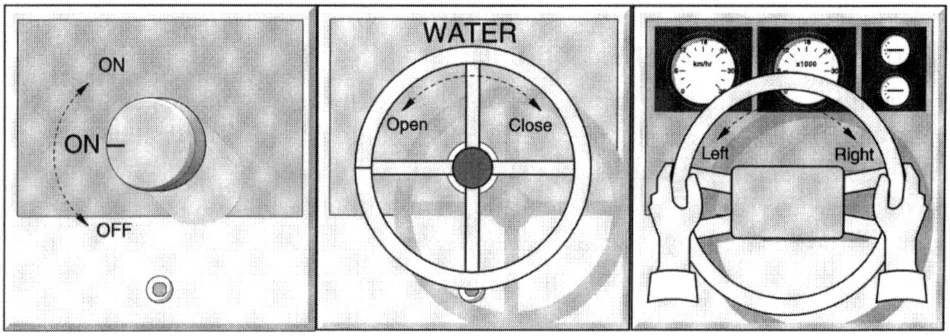

Figure 4-15. Some common stereotype controls.

Stereotypes are an important cognitive component that should be incorporated into ergonomic designs. There are three examples shown in Figure 4-15. The graphic on the left illustrates that electrical controls are expected to move clockwise to increase volume. The graphic in the center, however, illustrates that fluid controls increase flow when turning the handle counterclockwise. The stereotype of a steering wheel (graphic on the right) is so common that most people don't even recognize it as a design issue. This stereotype allows people to safely and consistently direct the path of motion for heavy equipment by turning the steering wheel in the desired direction of motion. Without this stereotype, the operator could easily turn to the wrong direction and cause an accident or injury. The most powerful, effective, and fully integrated stereotypes usually go unrecognized.

Stereotypes are often incorporated into (or perhaps originate from) national standards. For example, the U.S. national standard for a common wall electrical switch dictates that the up position indicates "on" and the down position indicates "off." However, Great Britain, Australia, and New Zealand reverse this stereotype using up as "off" and down as "on." With this in mind, when an international population uses the control, these differences may also contribute to operator error.

Cognitive errors resulting from common coding and labeling designs are not uncommon in operations such as shipping and receiving. The design of most product codes creates a haven for cognitive confusion. For example, when long strings of numbers are used, errors occur with reading, understanding, transmitting, and rewriting the numbers, particularly those digits in the middle of the string. The string of numbers "1234567" can easily be confused with "1234657" Where the 5 and 6 are transposed. Separating these numbers with spaces or dashes (e.g., telephone and credit card numbers) is a quick and inexpensive solution for this situation. The separation makes it quicker and easier for the eye to identify errors. Another common labeling problem occurs when different products have labels that look alike. For example, many product families have labels with similar looks to help the consumer identify the manufacturer (e.g., shampoo and conditioner bottles). This similarity may then create shipping errors at the warehouse if the employee picks the wrong product to ship.

WHAT ARE MUSCULOSKELETAL DISORDERS OR MSDs?

The terminology used to identify musculoskeletal disorders changes continually in the literature and may include: cumulative trauma disorders (CTDs), cumulative trauma injuries, repetitive motion injuries, repetitive strain injuries, repetitive stress disorders, overuse injuries, and musculoskeletal disorders (MSDs). Regardless of the name, the cause is the same: mechanical failure of a body part due to repeated microtrauma over time. Due to the slow onset and seemingly benign nature of the microtrauma, symptoms are most often ignored until they become chronic, painful, and often permanently damaging. Damage may occur to muscle, tendon, tendon sheath, ligament, spinal disc, nerve, joint capsule, and bone. Common diagnoses include, but are not limited to: tendonitis, tenosynovitis, bursitis, muscle strain, ligament sprain, peripheral compression neuropathy, spinal disc dysfunction, osteoarthroses, and degenerative joint disease.

Symptoms have been listed by a variety of names. Some are described in conjunction with the job performed, a particular sport associated with the symptoms, or the physician who first diagnosed the disorder. Examples of names associated with the job performed include: bricklayer's shoulder, pizza cutter's wrist, and gamekeeper's thumb (further description of these can be found in medical and ergonomics texts such as Putz-Anderson, 1987). Some of the sports-related names include: golfer's elbow (medial epicondylitis), tennis elbow (lateral epicondylitis), pitcher's shoulder (rotator cuff damage), and catcher's knee (strained knee ligaments). A well-known diagnosis named after its identifying physician is DeQuervain's Syndrome (stenosing tenosynovitis of the extensor and/or abductor tendons of the thumb).

MSDs are commonly associated with jobs requiring assembly, hand tool use, or extensive interface with equipment and machinery. The cause of these problems may be related to:

Workplace design

- overall fit of the person at the workplace
- extended reaches
- poor work postures
- poor layout resulting in repetitive work
- lack of accommodation to individual workers

Work organization

- low task variety
- inadequate recovery time

Methods and tools

- high forces
- rapid acceleration and velocity of dynamic motions
- repetitive motions
- vibration
- inadequate recovery times
- repeated or static parts manipulations with hand and wrist
- repeated or static postures of the arm, shoulder, and neck/back while stabilizing component parts

Lifting and material handling

- high forces
- poor postures
- repetitive actions
- poor load design

During the progression of symptom development, there are many opportunities to identify the injury, find its root cause, and develop corrective actions. Recognizing these early warning

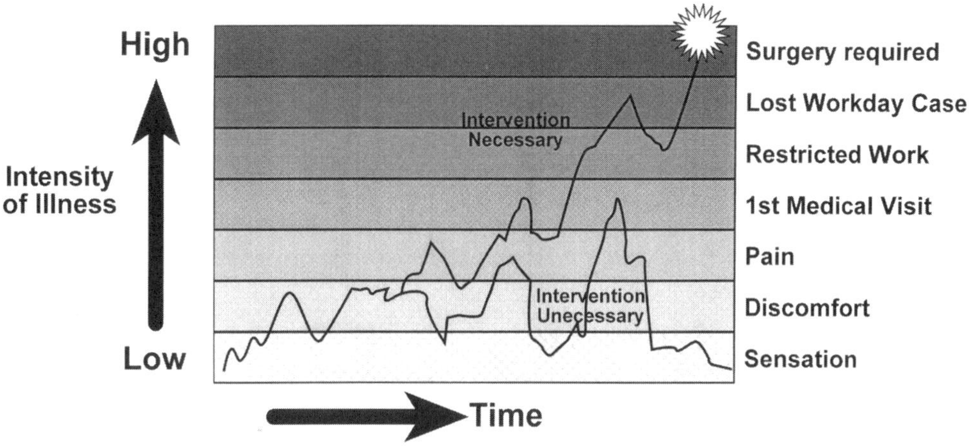

Figure 4-16. Progression of a musculoskeletal disorder.

symptoms, then changing how the person interfaces with the work or modifying the workplace allows these problems to be controllable and preventable. Figure 4-16 illustrates two paths for symptom progression.

It is clear that activities that occur both during and outside of work can influence an individual's tolerance to stressors and the potential for symptom development. However, when the body is allowed adequate time for rest and recovery, the soft tissue will regenerate and the cardiovascular system will return to normal function. Figure 4-17 uses a simple bucket to illustrate how stressors may exceed a person's capacity and how rest and recovery can prevent this overload.

PREVENTING WORKPLACE PROBLEMS BY DESIGN

Prevention is the application of good design practices before an injury occurs. Adjustability is an excellent prevention strategy, one of the best tools available, but total adjustability (with its high costs) is not always necessary.

Workplace problems can often be prevented when appropriate design criteria for the target population are incorporated. The first design principle requires that 95% of the population is accommodated, from the 2.5th percentile female to the 97.5th percentile male. In doing so, both a small female and a large male should be able to use the workplace, with the included adjustments, and without the need for additional accommodations or adaptive devices. In those rare occasions where an employee who is outside these percentiles still cannot work without stress, then special accommodations may be needed, but these are usually not designed into adjustable workplace features. Also note that in cases where safety is an issue, the design must meet the physical needs of 100% of the population. For example, the shortest employee must be

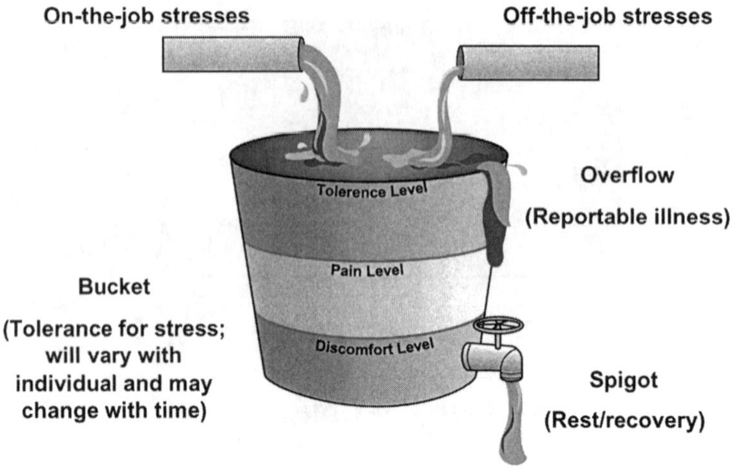

Figure 4-17. The bucket model of symptom development.

able to reach an emergency shut-off switch and the largest/widest employee must have adequate clearance through emergency passages.

The second design principle requires that reach and clearance needs are accommodated. Anthropometry tables are used to calculate design limits that meet the specific requirements for reach and clearance. These tables are easy to use, widely accepted, and do not add to the cost of design.

The third design principle requires that adjustability is incorporated to fit the necessary range of people working in a given environment. Although adjustable components add to the initial costs of outfitting a work area, these features offset the related injury costs of having multiple shifts and/or multiple operators use one piece of equipment.

If adjustability is not possible, then design for the average. Although inconvenient for people at either end of the spectrum, designing for the average size person meets the needs of more people than designing for either extreme.

When multiple workers use the same workstation throughout the course of a day or on different shifts, the importance of adjustable components becomes more critical. Some companies accommodate varying sizes of workers by providing a variety of different height workstations where the person can choose the one with the best fit. Still others assign each worker to a dedicated workstation that is modified to fit that individual's needs. Table 4-2 provides some design guidance for different situations when expensive adjustable components are not necessary.

Finally, the specific end user population must be identified when designing workstation and equipment layout. If the target user population resides in Latin America, then a design that

Table 4-2
Principles and practices for workplace design

Number of Workplaces for Each Worker	Examples	Design Principles and Practices
Multiple workplaces for each worker	Picking slots in warehouses Maintenance, electrical, and trades Service workers such as baggage handlers	Try to make stationary workplace (e.g., with carousel delivery) Use pareto analysis to determine more critical workplaces and design them ergonomically Let workers take modifications (e.g., ladders, step stools) with them to the job If retrofits are expensive, then concentrate on next generation of equipment
Several workplaces for each worker	Worker in manufacturing cell Worker servicing several workplaces	Determine more critical workplaces and set them up carefully Make inexpensive fixes for less frequently used workplaces
Dedicated workplace for each worker	Office workplace Machinist	Adjust workplace carefully and set up for each individual Be careful when using relief workers for vacations/time off
Several workers at a common workplace	Shift work in control rooms Two- and three-shift assembly workplaces	Easily (and quickly) adjustable workplaces "Memory seats" (like automobile seats) Identifying marks to help quickly and accurately position workplace elements
Many workers at a common workplace	Grocery checkouts (both checker and customer are users of workplaces) Airline check-in counters Hotel check-in (customer and clerk are both users of workplace)	Easily (and quickly) adjustable workplaces Design for middle of population Design tasks to require little adjustability (e.g., cashier checking task) Provide specific retrofit or accommodation (e.g., as in accommodation for someone with a disability) for special needs

fits large North American or European workers may not have the right fit. Due to the expense of retrofitting, the end user is usually forced to live with problems created by this error.

DESIGNING THE WORKPLACE

Using Anthropometry for Workplace Design

People vary in size, and these variations in body dimensions follow a normal distribution for each gender and ethnic background. The smallest people are referred to as 1st percentile, while the largest are in the 99th percentile. Dimensions are shown in Figures 4-18 and 4-19. As a general rule of thumb, the largest percentile values are roughly 33%-50% greater than the smallest percentile values. The stature (height) of larger males is 78.7 inches (200 cm), while the stature of smaller females is 55 inches (140 cm); the handbreadth of a large male is 5.9 inches (15 cm) and is 3.9 inches (10 cm) for a small female.

One note of caution is appropriate. There are many different sources of anthropometric data, and some studies are poorly performed or may contain only small sample populations. Before using anthropometric data with which you may not be familiar, be sure to learn the source and the quality of the data.

REACH AND CLEARANCE NEEDS IN WORKPLACE DESIGN

Reach is usually determined by the limits of a 5th percentile female worker. Optimally, vertical reach should occur no lower than knuckles nor higher than shoulders to prevent bending or overhead reaches. Forward reach will depend on the distance from the floor. This reach is at its maximum distance when the arm is straight out in front of the body, but diminishes above and below this point. This is known as the arch of motion about the shoulder joint and defines the safe reach limits.

Typically, a 95th percentile male determines appropriate clearance dimensions. The important clearance dimensions are stature, shoulder breadth, and body depth. For special situations, other anthropometric dimensions may be important (e.g., hand and arm clearance for a maintenance task). Both workplace clearances and workspace clearances are important. Workspace clearance refers to the workers' individual spatial needs. Workplace clearance is the unobstructed space required for entry into and egress from the workplace, for obtaining materials, and for performing maintenance at the workplace. Workplaces are less adjustable and flexible than workspaces. Workspaces are more personalized than workplaces.

Additional clearance may be required to accommodate clothing necessary for personal protection or cold weather. Some examples of these clearance dimensions include adding 3 inches when wearing a hard hat, 0.5 inches when using gloves (added to hand length and hand breadth), and 1 inch for shoes (added to stature, eye height, and knee height for clearance under a desk top while sitting).

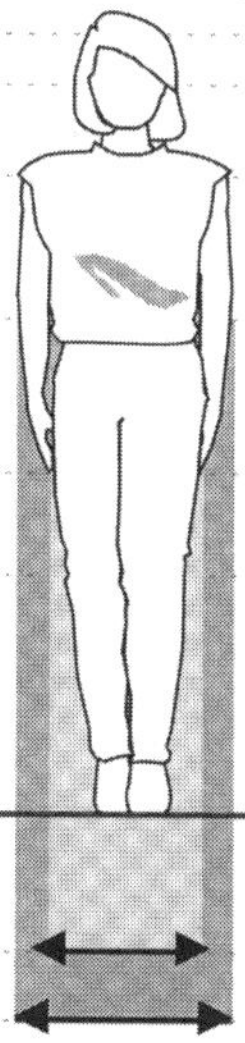

Body Feature	Percentile 5th	50th	95th	Standard Deviation
A. Height	58.9	63.2	67.5	2.6
B. Eye Height	54.4	58.6	62.7	2.4
C. Shoulder Height	47.7	51.6	55.9	2.4
D. Elbow Height	36.9	39.9	42.9	1.8
E. Knuckle Height	25.3	27.6	29.9	1.4
F. Knee Height	17.8	19.6	21.4	1.1

Floor

Body Feature	5th	50th	95th	Standard Deviation
G. Seat Breadth	12.3	14.3	17.2	1.5
H. Elbow—Elbow	12.4	15.1	19.3	2.1
I. Weight	101.9	134.7	198.2	5.4

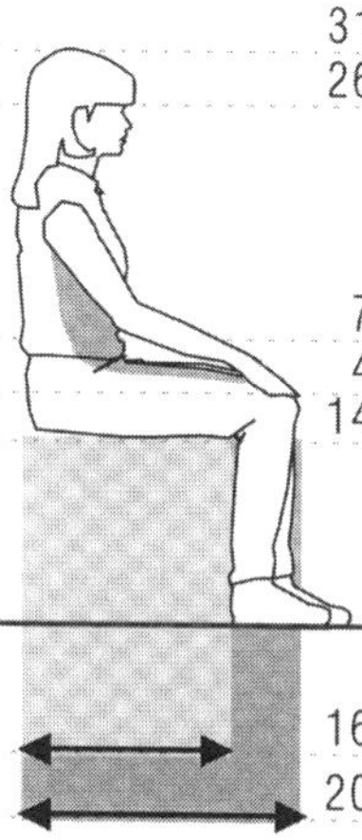

Body Feature	5th	50th	95th	Standard Deviation
a. Sitting Height*	31.0	33.5	35.7	1.4
b. Sittting Eye Height*	26.6	28.9	30.9	1.3
c. Elbow-rest Height*	7.1	9.2	11.1	1.1
d. Thigh Clearance*	4.2	5.4	6.9	0.7
e. Popliteal Height**	14.0	15.7	17.5	1.0
Floor				
f. Buttock—Popliteal	16.9	19.0	21.1	1.2
g. Buttock—Knee	20.4	22.4	24.6	1.2

* Measured from seat height.
** Measured from floor.

Figure 4-18. Selected anthropometric features of the North American female in inches. (NASA, 1978. Anthropometric Source Book. Vol. 1-III).

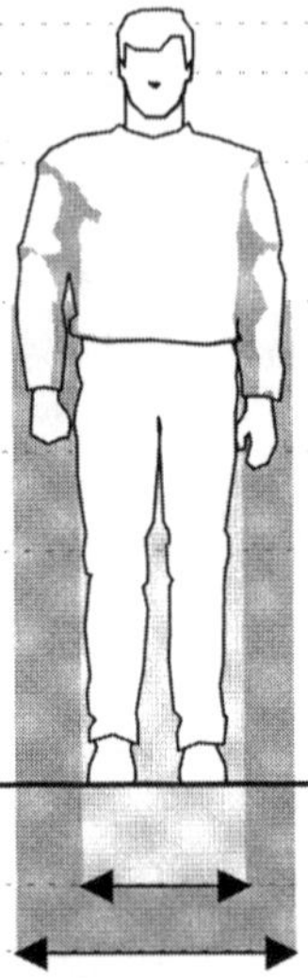

Body Feature	Percentile 5th	50th	95th	Standard Deviation
A. Height	63.7	68.4	72.7	2.7
B. Eye Height	59.5	63.9	68.0	2.7
C. Shoulder Height	52.1	56.2	60.0	2.4
D. Elbow Height	39.4	43.4	46.9	2.3
E. Knuckle Height	27.5	29.7	31.7	1.3
F. Knee Height	19.4	21.4	23.4	1.1
G. Seat Breadth	12.1	13.9	16.0	1.1
H. Elbow—Elbow	13.8	16.4	19.9	1.8
I. Weight	123.9	163.1	214.1	5.0

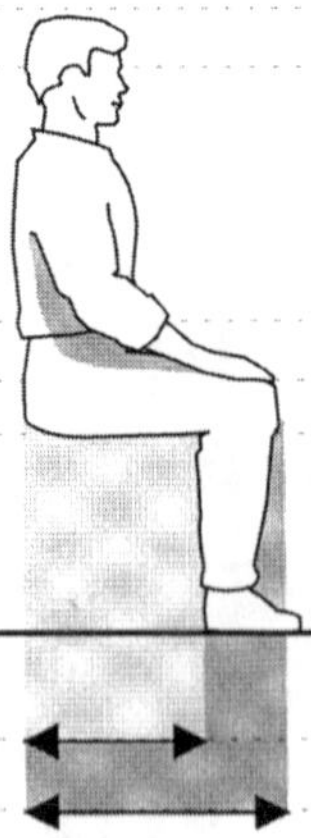

Body Feature	5th	50th	95th	Standard Deviation
a. Sitting Height*	33.2	35.7	38.1	1.5
b. Sittting Eye Height*	28.6	30.9	33.2	1.4
c. Elbow-rest Height*	7.5	9.6	11.6	1.2
d. Thigh Clearance*	4.5	5.7	7.0	0.7
e. Popliteal Height**	15.4	17.4	19.2	1.1
f. Buttock—Popliteal	17.4	19.5	21.6	1.2
g. Buttock—Knee	21.3	23.4	25.3	1.2

* Measured from seat height.
** Measured from floor.

Figure 4-19. Selected anthropometric features of the North American male in inches. (NASA, 1978. Anthropometric Source Book. Vol. I-III).

Many workplace design issues impact more than simple reach or clearance dimensions. For these situations, the workplace must accommodate both the larger and smaller people, as well as all others throughout the population. It is not possible to create a single workplace design that will fit such a wide range of people. Therefore, the workplace must have some type of adjustability inherent in its design.

POSTURE AND WORKPLACE DESIGN

Major body landmarks that include knuckles, elbows, and shoulders (see Figure 4-20) are used in design to assure good posture. Work performed below knuckle height requires the person to bend down. Work located above knuckle height ensures easy reach from an upright trunk posture.

Elbow height is important because it identifies the best range for performing most types of work. Approximately 90° elbow bend is optimal for light assembly-type activities. However, working with the hand well below the elbow height is preferred for very heavy work for added power and mechanical advantage to the muscles that straighten the elbow. Very fine or detailed work requires that the hands function above the elbow to aid sight.

The shoulder limits the suggested vertical height of work. Muscles and structures of the shoulder fatigue more readily and become less stable as the hand and elbow work above shoulder height. Soft tissue and nerve injuries such as tendonitis, bursitis, soft tissue entrapment or strain, and nerve compression in the neck/shoulder region are common when the limits of the shoulder are exceeded.

The working space for the hands is defined by two reach envelopes, formed by the rotation of arm segments around the elbow and shoulder joints, respectively. The primary reach envelope is determined by the movement at the elbow, with the forearm moving in all three planes and the elbow kept close to the body as shown in Figure 4-21. Work preformed within the primary reach envelope is recommended for most functional activities because it minimizes the effort.

The secondary reach envelope is also three dimensional and is defined by the swing of the entire arm around the shoulder joint as shown in Figure 4-21. Work in this reach zone should be limited to infrequent and non-load bearing activities because the structures of the shoulder are not as strong or stable with the arm extended. Reaches beyond the secondary reach envelope should be avoided for routine work tasks. Figures 4-22, 4-23, and 4-24 provide some practical anthropometry data on "reach envelopes" that can be used to supplement analysis and design.

The neutral posture is the anatomical position of the body at rest (Figure 4-25). The need to work beyond neutral posture is usually the result of a poorly designed workplace. When in neutral, muscles are in a resting length where agonist and antagonist muscles are balanced and relaxed. Neutral posture generally occurs at mid-range of joint motion. The key neutral postures used in ergonomics assessment are listed below.

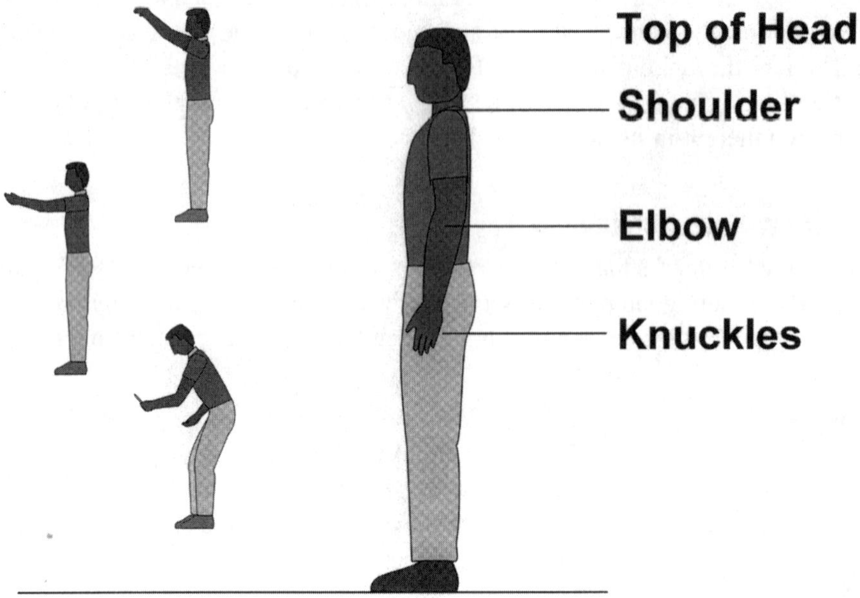

Figure 4-20. Natural body landmarks.

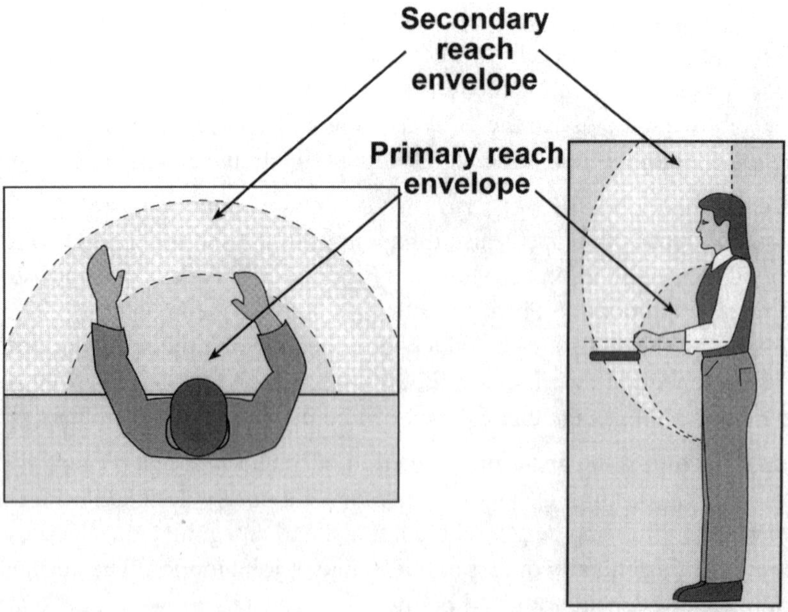

Figure 4-21. Primary and secondary reach envelopes.

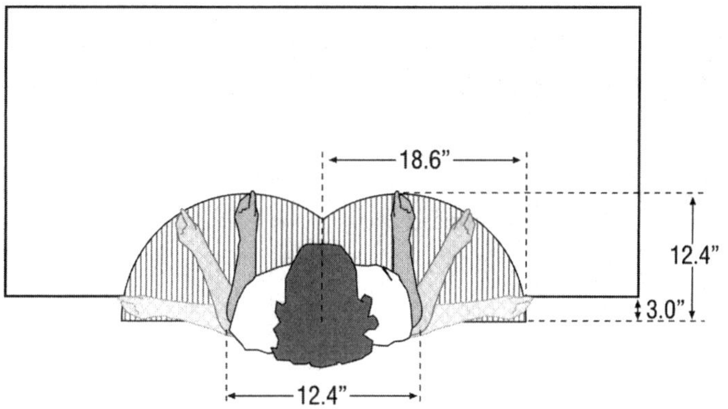

Figure 4-22. Primary forward reach envelope. (Adapted from NASA, 1978. Anthropometric Source Book, Vols. I-III.)

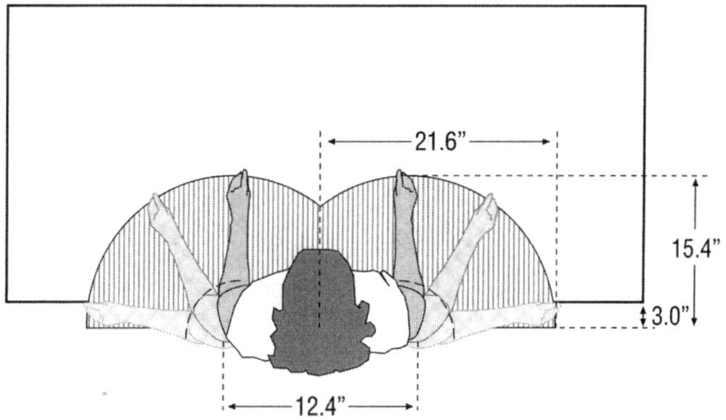

Figure 4-23. Practical primary forward reach envelopes, elbows 3" from the body, forearms supported. Distances in inches. (Adapted from NASA, 1978. Anthropometric Source Book, Vols. I-III.)

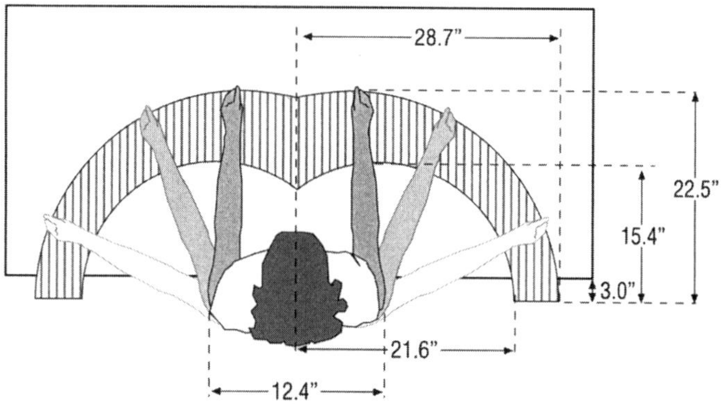

Figure 4-24. Secondary forward reach envelopes (arm fully extended). Distances in inches. (Adapted from NASA, 1978. Anthropometric Source Book, Vols. I-III.)

- Fingers – curled (flexed) as if loosely holding an 8-10 cm diameter cylinder or ball, and certainly not fully open (extended) or tightly closed (flexed)
- Thumb – slightly curled (flexed) at the first metacarpalphalangeal joint (knuckle) and lightly touching the index finger
- Wrist – the hand is straight on the forearm and is not flexed (palm toward the forearm), extended (back of the hand toward the forearm), bent sideways toward the little finger (ulnar deviation), or bent sideways toward the thumb (radial deviation); this is typically a handshake position
- Forearm – resting so the thumbs are pointed forward, and the palms are neither facing forward (supination) nor facing behind (pronation)
- Elbow – roughly bent at 90°, and neither extended (arm straight) nor tightly flexed (bent)
- Shoulder – upper arm hanging by the side of body, and not raised forward (flexed), away from the side of the body (abducted), or across the body (adducted)

Static work is a term that has no meaning from an engineering perspective. In the engineering world, work only occurs when there is motion, however static holding/work is of critical importance for ergonomics. Static work occurs when a body part is forced to support itself or hold an object still in space without motion (a prolonged isometric contraction). Static work occurs when performing activities in constrained workplaces or workplaces that simply don't fit the worker as shown in Figure 4-26. Instead of relying on the skeletal system to hold the body

Figure 4-25. Neutral postures.

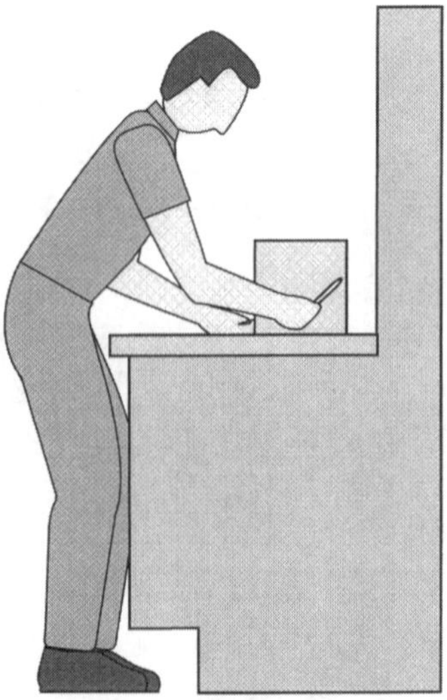

Figure 4-26. Static work.

upright, the person must hold a forward flexed posture at the trunk and a flexed posture of the shoulders to reach and manipulate the box. This posture rapidly increases energy demand and the need for blood flow to the muscles. During isometric contractions, blood flow is reduced to active muscle groups and they must, therefore, rely on a limited anaerobic system to generate needed energy. During anaerobic metabolism, muscles fatigue quicker and take longer to recover. As the person shows signs of fatigue, tremors and/or muscle pain may develop.

DESIGNING SITTING AND STANDING WORKSTATIONS

Design strategies are based on the type of work performed. Some activities are better suited for sitting workstations and others for standing workstations. Table 4-3 is provided to evaluate design alternatives for both standing and sitting workplaces.

A standing workplace is best for:

- tasks that require constant movement (e.g., stepping onto a platform to obtain parts during an assembly cycle or walking to another section of the work area to place finished products onto a pallet or conveyor such as case packing).
- tasks that require the application of force (e.g., lifting heavy boxes, pushing or pulling objects on a conveyor, or forcing objects together during assembly).
- tasks that require the handling of large work pieces (e.g., assembling large appliances such as washing machines, handling sheets and bedding in a textile operation, or handling large wooden products for furniture assembly).
- tasks that require long reaches (e.g., assembling large products such as automobiles and airplanes), operating an extensive number of controls in large analog control rooms, or reaching for parts in an assembly operation.

A sitting workplace is best for:

- tasks that require precision motions (e.g., fine assembly or work with high visual content such as inspection operations, especially with a microscope).
- tasks with long work duration and tasks that require working for more than one hour with no movement outside the work area or no need to reach for parts.
- situations where stability of the hands is required and the forearms can be stabilized on a tabletop or arm rests on the chair; these stabilizing devices promote greater precision and controlled movements of the hands and fingers.
- visual tasks so that the neck and trunk can be stabilized in neutral while the hands maneuver the part near the eye for work or inspection; referring to anthropometry tables, there is much less variability between the seat-to-eye dimension in sitting than the floor-to-eye dimension in standing, making it easier to accommodate a wide range of workers.

Table 4-3
Standing and sitting workplaces checklists

Standing Workplace Checklist	Sitting Workplace Checklist
• Can each worker stand up straight? • Is the work at an appropriate height (hands approximately 5 cm below elbows)? • Can the worker change positions periodically? • Are mats or floor coverings available? • Are stools and supports available? • Are footrests available?	• Can each worker sit up straight? • Is the work at an appropriate height (upper arms hanging, forearms supported)? • Is the work close enough to avoid long reaches? • Does the work avoid twisting to the side or back? • Can the worker change positions periodically? • Are there good chairs with adjustable seating? • Are footrests available? • Is there foot and leg clearance?

- workstations used by multiple workers becauase a wider variety of people can be more easily accommodated by a seated workplace; the overall dimensional variability is much less for seated workers than for standing workers, therefore less adjustability of the workplace is required from the equipment or workstation itself.

In some cases, a sit/stand workstation is preferred because it permits the operator to alternate between siting and standing. It also permits the operator to rest against a high stool-type support while legs remaining in mostly a standing position. This type of workstsation is recommended when frequent walking or position changes are required, however there are short periods of work performed directly at the workstation. It is also recommended when greater reach is required than can be achieved in the seated position. Once the type of workstation has been determined (sitting, standing, or sit/stand), then Figure 4-27 and Table 4-4 should be used to design workstation height. These dimensions are based on North American male and female anthropometry data for elbow height and on basic muscle anatomy and physiology.

Manual Materials Handling

INTRODUCTION

Workplace problems can be prevented by understanding and designing limits for how much a person should be allowed to carry. As previously mentioned, designing to the capacity of the worker population, creating optimal interfaces between the person and the tool, equipment, and product, and determnining when the load is too heavy to be carried safely are primary goals of reliable ergonomics designs. This general heading used to describe the transportation of items is "manual materials handling" (MMH), meaning that materials are moved by hand.

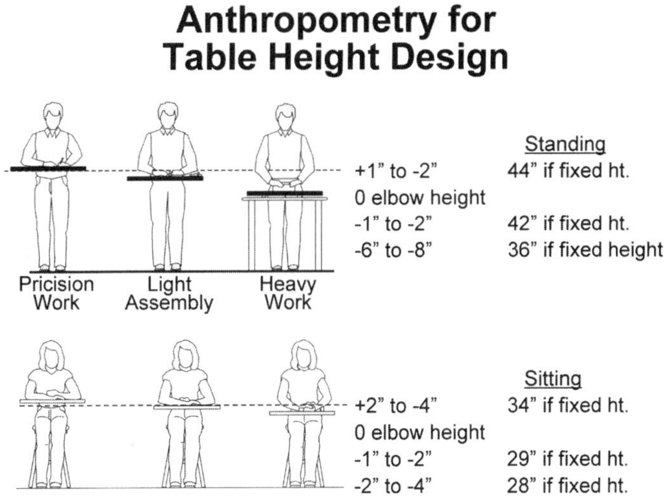

Figure 4-27. Standing and sitting work heights. Distances in inches. (Adapted from NASA, 1978. Anthropometric Source Book, Vols. I-III.)

Table 4-4
Recommended work heights (inches from floor)

Standing Work Criteria Work Height	Adjustable Standing	Fixed Standing Work Height	
		(Able to add platform to raise person)	(Platform cannot be added)
Precision work	39-50	50	44
Light assembly	36-47	47	42
Heavy work	30-42	42	36
Sitting Work Criteria	**If The Work Height Is Adjustable (with footrest)**	**If The Work Height Cannot Be Adjusted (and there is an adjustable chair and footrest)**	
Precision work	27-34	34	
Light assembly	23-29	29	
Heavy/manual work	21-28	28	

The task designer must follow guidelines on setting safe limits to protect both person and product. Although many of the neck, back, shoulder, and knee problems that occur during MMH activities appear to result from a single event (e.g., lifting one heavy box), the symptoms can usualy be traced back to a series of heavy lifts and awkward postures that take place over time.

By understanding the multiple factors that define an MMH system, it will be easier to identify where modifications can make the most impact in reducing the risk of injuries. Figure 4-28 illustrates the interactive nature of task characteristics, work practices, and material/container characteristics. The goal is to develop a balance between these factors and minimize the overall degree of hazard. This section will focus on evaluating, measuring, and setting limits for manual handling tasks.

DESIGN PRINCIPLES AND CONCEPTS FOR MANUAL MATERIALS HANDLING (MMH) TASKS

There are six basic principles used to design manual handling tasks that minimize the risk for MSDs. The first principle is that eliminating the need for manual handling is the first line of defense. Determine if materials can be placed or stored directly in the location of use so that secondary moving is not necessary. Delivering materials to a stationary or lift table can eliminate the need to lift from below knuckle height. Another common alternative is to purchase in bulk, thus allowing bulk handling systems to be used. Alternatively, make packages so heavy that people will not try to lift them manually.

When materials need to be moved, there are a variety of mechanical aids that can enhance human capabilities or completely eliminate the need for muscle power on the part of the operator. Mechanical aids can range from a sophisticated forklift truck to a simple pry bar. They

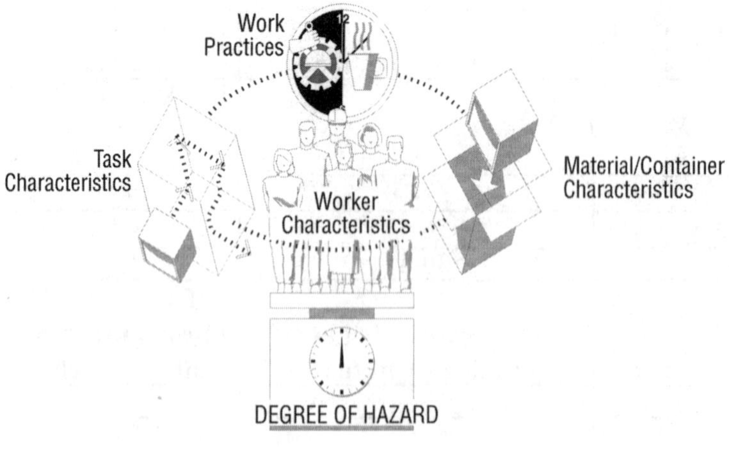

Figure 4-28. Interacting factors of a manual materials handling system.

should always be considered and may increase productivity based on the authors' experience. They should be prioritized for use if less than 75% of the industrial population can handle the load without overexertion. Examples include:

- Forklift trucks
- Scissors lifts; lift tables
- Hoists; overhead balancers
- Powered conveyors; dumb waiters
- Use layouts and equipment that eliminate manual handling
- Conveyors
- Gravity slides; chutes
- Tables; workbenches
- Cafeteria tray levelers

These lifting-assist devices may be very appropriate for use in the workplace; however, there may be some valid reasons why they are not properly utilized or accepted by the workforce.

- Awkward control of product
- Tasks take longer
- Counter balancer set improperly
- Poor handle design
- Unnatural "feel" of controls
- Only used by some of the operators
- Use requires stooping or extended reaches
- Gets in the way during other tasks
- High inertial forces make it difficult to stop the device

Assess the manual materials handling requirements before initiating a work task (see NIOSH Lifting Guide in next section). There are a variety of reliable measurement tools that can help the task designer understand if the task is safe for the user population. For example, when lifting, in addition to the direct weight of the object, consider the horizontal reach from the body and the height of the lift. Both of these factors can significantly reduce the maximum recommended weight to be lifted. For pushing tasks, the force required to initiate and sustain the push are more important than the weight of the object itself. Based on the requirements for the task, it may be necessary to either reduce the load handled, or make the load easier to handle. These options can be accomplished by:

- reducing the weight of each package;

- reducing the size of each package;
- purchasing in smaller quantities;
- putting less material in each container (use fill-to lines, add false bottoms, use smaller containers);
- reducing the weight of box, bag, or container;
- adding or relocating handles;
- keeping package size moderate;
- altering the shape of the package; and
- sliding rather than lifting.

Design the workplace for efficient and effeceive access to materials. For example, ensure that the operator can get close to the load. A rotating turntable can be used to eliminate an extended reach across a wooden pallet. Stands and tables can be used to keep material off the floor and place it between knuckle and elbow height. One goal of optimal workplace design for MMH is to handle the load in more optimal working positions. This can be done by:

- designing the workplace to keep body joints in neutral positions (back straight, shoulders down, elbows bent 90 degrees);
- designing the workplace to keep the load close to the body ("tummy touch");
- keeping the load bewtwen knuckle height and shoulder height;
- designing the workplace to avoid twisting (keep shoulders and hips aligned);
- stacking material on tables and work stands to keep it off the floor;
- providing space to walk around pallets rather than leaning over them; and
- using lazy susan turntables to bring materials to workers.

When possible, develop engineering options as a first choice for solutions. Although they may require more capital up front, they tend to be longer lasting, do not require repeated training and supervision for compliance, and often do not depend on the cooperation of the employee. When an engineering solution is too costly, is not technologically feasible or practical, cannot be instituted immediately, or does not fully resolve the problem, then administrative solutions can be considered.

Administrative solutions can be an effective method of reducing the risk of injury when appropriate training and supervision for compliance are used simultaneously. Some of the more effective solutions include the integration of a job rotation or job enlargement schedule. Best practice methods or techniques (e.g., handle one package at a time and keep the load in close to the body) are often developed to minimize force and optimize posture. The inclusion of task variation, short production breaks, and exercise may also be considered in this category. Other examples of good work practices and administrative controls include:

- handle items individually;

- avoid controlled positioning;
- enlarge the scope of the job so that less time is spent on manual handling;
- avoid redundant and unnecessary lifts; and
- never lower items and then re-lift them.

In addition to the above principles, there are some basic concepts that minimize the muscle force required to move an object:

- Holding the load close to the body
- Limiting vertical reach to between 20 inches (50 cm) and 50 inches (125 m) above the floor (knuckle to shoulder height)
- Limiting load dimensions to 16 inches (40 cm) in front of the body
- Working in neutral postures and avoiding twisting and bending
- Ensuring a balanced load
- Avoiding excessive weight and using mechanical assists for heavy loads
- Displaying package weights to prepare the worker before lifting
- Lifting comfortably, smoothly, and gradually and using proper body mechanics
- Using appropriate handle design for the task
- Staying in good physical condition and warming up before performing physical work
- Providing adequate recovery breaks based on heart rate and metabolic energy expenditure data

The degree of hazard during most manual handling activities will be influenced by a variety of factors, including the vertical location of the object, reach distance to the object, travel distance of the object from origin to destination, ease of fit between the hand and the object, frequency of lift, duration of lift, and degree of body twist required. Designing tasks to minimize the effect of these factors reduces the overall risk for symptom development. The NIOSH Lifting Guide can help accomplish this.

THE NIOSH LIFTING GUIDELINES FOR ASSESSING MMH REQUIREMENTS

One of the design principles detailed above identifies the need to measure risk factors resulting from manual handling tasks, especially lifting tasks. Although there is no fully reliable method to ensure that any lift is 100% safe, in 1981, the National Institute for Occupational Safety and Health (NIOSH, 1981) developed a widely recognized method of evaluating lifting tasks. This guide was revised to reflect new findings, and it became known as the Revised NIOSH Guidelines for Manual Lifting (aka NIOSH Lifting Guidelines) (Waters et al., 1993 and NIOSH, 1994).

It must be stressed that use of the Revised NIOSH Lifting Guidelines is limited to lifting activities that do not involve lifting or lowering under the following conditions:

- With one hand
- For more than eight hours
- While seated or kneeling
- In a restricted work space
- Unstable objects
- While carrying, pushing, or pulling
- With wheelbarrows or shovels
- With high speed motion (faster than about 30 inches (0.75 m)/second)
- On slippery floors
- In hot, cold, dry, or humid work environments

Also, the guide and calculations that follow assume that the effects of other manual handling activities (carrying, pushing, and pulling) are minimal. When not engaged in lifting activities, the employee is assumed to be at rest or performing non-lifting tasks. Finally, the workforce is assumed to be physically fit and accustomed to physical work.

When any of these restrictions are violated, another lifting analysis tool, such as the Job Severity Index (JSI) or psychophysical lifting data, should be used instead of the Revised NIOSH LIfting Guidelines. For more information on the Job Severity Index refer to Ayoub et al. (1983). For additional information on psychophysical data and its use, refer to Snook et al. (1991) and Rodgers et al. (1983).

The two outputs of the Revised NIOSH Lifting Guidelines are:

- the Recommended Weight Limit (RWL), and
- the Lifting Index (LI).

For a specific set of task conditions, the RWL is the weight nearly all workers (90%) can lift for a substantial period of time (8 hours) without an increased risk of developing lifting-related low back pain (LBP). The RWL is defined as:

$$RWL = 51 \text{ lbs} \times H_M \times V_M \times D_M \times A_M \times F_M \times C_M \qquad \text{[Equation 4-1]}$$

where 51 lbs (23 kg) is the load constant used in all calculations. It is the maximum weight that should be carried under ideal conditions and the other factors are all dimensionless multipliers determined from measured variables that are described later and used in the following equations. Thus, for a given set of conditions, the smaller the RWL, the greater the potential strain on the person. Note that all distances are in inches and 1 inch = 2.54 cm.

The Horizontal Multiplier (H_M) factors in the load distance from the spine. "H" is the horizontal distance from the body measured from the hands to the midpoint between the ankles.

The greater the distance, the greater the stress on the spine. Optimally, the load should be carried as close to 10 inches as possible and should not exceed 25 inches maximally.

$$H_M = 10 \div H \quad \text{[Equation 4-2]}$$

The Vertical Multiplier (V_M) factors in the vertical position of the lift measured from the floor. "V" is the vertical distance of the hands above the floor at the beginning of the lift. Lifts significantly below or above knuckle height (30 inches) increase stress on the spine.

$$V_M = 1 - 0.0075 \times |V\text{-}30| \qquad \text{[Equation 4-3]}$$

The Vertical Distance Travel Multiplier (D_M) factors in the vertical travel distance of the object between origin and destination. "D" is the absolute distance traveled from origin to destination. It must be a positive number.

$$D_M = (1.8 \div D) \div 0.80 \qquad \text{[Equation 4-4]}$$

The Asymmetry Multiplier (A_M) factors in the angle of asymmetry during the lift. "A" is the angle of asymmetry or angle of the degree of twist in degrees. Twist decreases the amount of force the spine can withstand without injury.

$$A_M = 1 - 0.0032 \times A \qquad \text{[Equation 4-5]}$$

Each of the above multipliers can also be determined by referencing the data in Figure 4-29 eliminating the need to calculate individual values using the equations provided above. The next two multipliers must be determined from Figure 4-29.

The Coupling Multiplier factors in the quality of handles or hand holds. Coupling (C) is the quality of interaction between the hand and the load. Good handles make the load easier to hold and support less risk of injury. Subjectively estimate if coupling is good, fair, or poor, and determine the C_M value based on the vertical height of the lift (V) relative to 30 inches (75 cm), using Figure 4-29.

The Frequency Multiplier factors in the frequency of lifts. "F" is the frequency of lifts per minute, based on the average of 15-minute samples. It is based on three variables that are reflected in the columns of the Frequency sub-table of Figure 4-29.

- Frequency (F) or the average number of lifts per minute during a 15-minute period.
- Duration of lifts (more details on classifying duration are given in Table 4-5).
- Vertical position of the lift relative to 30 inches (75 cm).

Look Up Tables

H (inches)	HM
<10	1.00
11	.91
12	.83
13	.77
14	.71
15	.67
16	.63
17	.59
18	.56

V (inches)	VM
0	.78
5	.81
10	.85
15	.89
20	.93
25	.96
30	1.00
35	.96
40	.93

D (inches)	DM
≤10	1.00
15	.94
20	.91
25	.89
30	.88
35	.87
40	.87

A (degrees)	AM
0	1.00
15	.95
30	.90
45	.86
60	.81
75	.76

Coupling Type	CM	
	V<30 inches	V≥30 inches
Good	1.00	1.00
Fair	.95	1.00
Poor	.90	.90

Freq. (F) lift/min	FM					
	Short		Moderate		Long	
	V<30	V≥30	V<30	V≥30	V<30	V≥30
≤0.2	1.00	1.00	.95	.95	.85	.85
0.5	.97	.97	.92	.92	.82	.81
1	.94	.94	.88	.88	.75	.75
2	.91	.91	.84	.84	.65	.65
3	.88	.88	.79	.79	.55	.55
4	.84	.84	.72	.72	.45	.45
5	.80	.80	.60	.60	.35	.35
6	.75	.75	.50	.50	.27	.27
7	.70	.70	.42	.42	.22	.22
8	.60	.60	.35	.35	.18	.18
9	.52	.52	.30	.30	.00	.15
10	.45	.45	.26	.26	.00	.13
11	.41	.41	.00	.23	.00	.00
12	.37	.37	.00	.21	.00	.00
13	.00	.34	.00	.00	.00	.00
14	.00	.31	.00	.00	.00	.00
15	.00	.28	.00	.00	.00	.00
>15	.00	.00	.00	.00	.00	.00

Figure 4-29. Look up tables for NIOSH Lifting Guidelines.

For example, if a lift is required for an average of 11 times per minute, with a duration of 1.5 hours and non-lifting time of 0.5 hours, and the vertical height of the lift is 34 inches (85 cm), then $F_M = 0.23$.

Once the RWL has been determined, the Lifting Index, or LI, is calculated. The LI is a term that provides a relative estimate of the level of physical stress defined by the relationship of the actual weight of the load lifted ("L", lbs.) and the recommended weight limit.

$$LI = L/RWL \qquad \text{[Equation 4-6]}$$

Using The NIOSH Lifting Guidelines

To assess the risk of injury or low back pain for a lifting task using the NIOSH Lifting Guidelines, record the required data (as listed above) while observing the lifting task, go to Figure 4-29 to find the appropriate multipliers, and then use equations 4.1 and 4.6 to determine the RWI and the LI, respectively. Figure 4-30 is useful for recording the data.

CALCULATIONS

Object Weight	Hand Location				Vertical Distance	Asymmetric Angle		Frequency Rate	Duration	Object Coupling
	Origin		Dest			Origin	Dest	lifts/min	Hrs	
L	H	V	H	V	D	A	A	F		C

RWL = 51 x HM x VM x DM x AM x FM x CM

ORIGIN RWL = [51] x [ ] x [ ] x [ ] x [ ] x [ ] x [ ] = [lbs.]

DEST. RWL = [51] x [ ] x [ ] x [ ] x [ ] x [ ] x [ ] = [lbs.]

DEST. LIFTING INDEX = Object Weight (L) / RWL = ________ = [ ]

DEST. LIFTING INDEX = Object Weight (L) / RWL = ________ = [ ]

Figure 4-30. Data collection form for NIOSH Lifting Guidelines.

Interpreting the Lifting Index (LI)

The design objective is to achieve a Lifting Index (LI) less than 1.0. An LI less than 1.0 does not necessarily require further intervention. However, for jobs with an LI bewteen 1.0 and 2.0, administrative controls (and possibly simply engineering controls) should be considered and judgment should be exercised, depending on the situation. If a group of healthy, young, and uninjured males work on a job with a calculated LI of 1.5, then the complex engineering controls may not be necessary. However, as the LI exceeds 2.0 and approaches 3.0, simple to complex engineering controls (and possibly supplemental administrative controls should be considered (see Figure 4-31).

Example of Using the NIOSH Lifting Guidelines

In this example, an employee is required to move a 31-lb. block from a table conveyor to an overhead conveyor (see Figure 4-32). Data for horizontal origin (location of the hand at the start of the lift) and destination (location of the hand at the end of the lift), vertical origin (location of the hand at the start of the lift) and destination (location of the hand at the end of the lift), travel distance (difference between vertical origin and destination), angle of asymmetry (amount of trunk twist), frequency and duration (exposure), and coupling (interface between the hand and the object) were determined from data supplied in Figure 4-33.

The results from these calculations demonstrate that both parameters for lifting from the starting position or origin (LI = 2.4) and the ending position or destination (LI = 3.3) exceed the design guideline for safe lifting. Since the lifting indices exceeded a value of 2 by a wide

Figure 4-31. Comparing the Lifting Index (LI) to the design guideline.

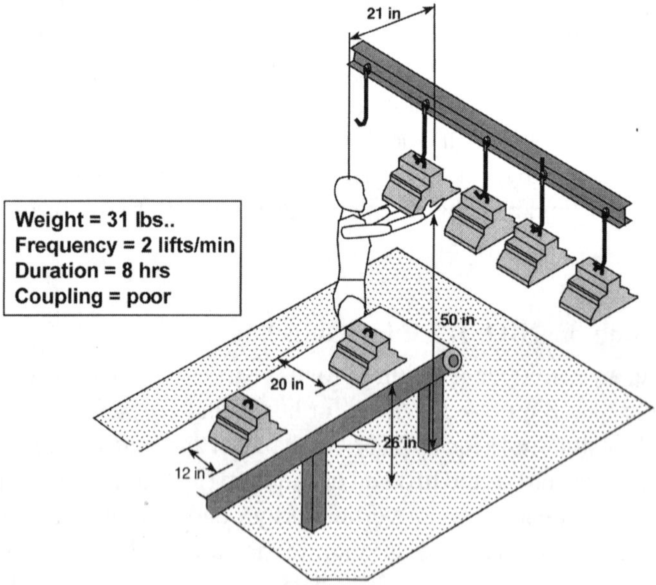

Figure 4-32. NIOSH Lifting Guidelines example, lifting blocks during an assembly production operation.

CALCULATIONS

Object Weight	Hand Location				Vertical Distance	Asymmetric Angle		Frequency Rate	Duration	Object Coupling
	Origin		Dest			Origin	Dest	lifts/min	Hrs	
L	H	V	H	V	D	A	A	F		C
31 lbs.	20"	26"	21"	50"	24"	0°	45°	2	8	poor

RWL = 51 x HM x VM x DM x AM x FM x CM

ORIGIN RWL = 51 x .50 x .97 x .89 x 1.0 x .65 x .90 = 12.9 lbs.

DEST. RWL = 51 x .48 x .85 x .89 x .86 x .65 x .90 = 8.7 lbs.

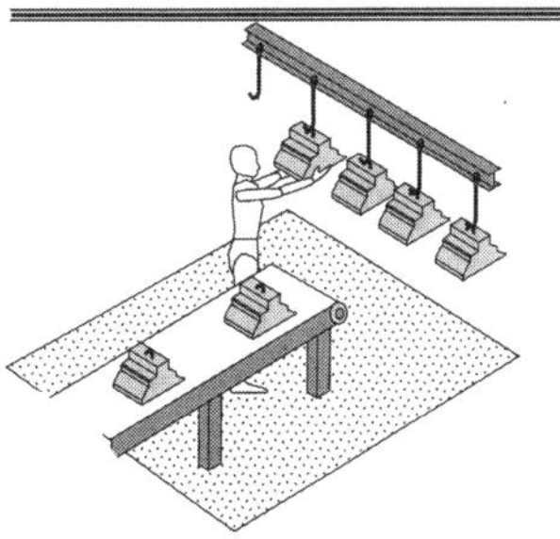

$$\text{Lifting Index (LI)} = \frac{\text{Object Weight (L)}}{\text{RWL}}$$

$$\text{LI (ORIGIN)} = \frac{31}{12.9} = 2.4$$

$$\text{LI (DEST)} = \frac{31}{8.7} = 3.6$$

Figure 4-33. NIOSH Lifting Guidelines example, calculations for lifting blocks during an assembly production operation.

margin, engineering controls are recommended at both locations. A close look at the multipliers identifies horizontal location and frequency of lift as the two most influential parameters affecting the safety of this lifting task.

Design Applications and the NIOSH Guidelines

The Revised NIOSH Lifting Equation has many applications other than evaluating single tasks or jobs including identifying specific design problems, comparing two job designs, prioritizing jobs, and quantifying design improvements. Specific design problems can be identified in conjunction with the resulting LI. Since 1.0 is the optimal value for each multiplier, then the multiplier having the lowest value indicates the biggest strain. Redesign efforts should focus on the smallest multiplier. For example, if V_M is the lowest multiplier, then the greatest improvement to the job would be made by raising the vertical position of the load close to knuckle height.

The LI has at least two other applications. It can be used as a criterion for selecting the best design from several potential job designs. The design having the lowest LI value is preferred. Also, for a group of problem jobs requiring some degree of lifting, a prioritized list of jobs with lifting problems can be developed. Once the LI for each job is calculated, rank the jobs in order of severity, then implement solutions for those jobs having the highest LI values.

Determining Push and Pull Limits in MMH

The push or pull force required to move an object is dependent on the friction between the object and the surface upon which it is moved and the quality of any interface between the two, e.g., pushing a heavy cart that has large casters in good condition over a polished linoleum floor versus pushing a cart of equal weight that has small, pitted casters over uneven asphalt. The push/pull force required to move an object can be measured using a standard force gauge. Once the actual force is measured, this value can be compared to design data to determine the level of risk or acceptability. Figure 4-34 and 4-35 are examples of such data that divide the level of risk into three zones.

Figures 4-34 and 4-35 can be used to identify whether the push or pull force measured for a task is acceptable relative to frequency. Push or pull forces that fall into Zone 3 indicate that engineering redesign is necessary. Forces falling into Zone 2 indicate that engineering redesign is recommended and should be considered prior to using any administrative controls because this activity is only safe for very strong men (25% of the male population). Forces that fall into Zone 1 indicate that the majority of people should be able to complete the task without an increase risk of injury. Obviously, any newly designed equipment should have forces within Zone 1. Note that the actual weight involved is not measured — only the measured push or pull forces go into the graph.

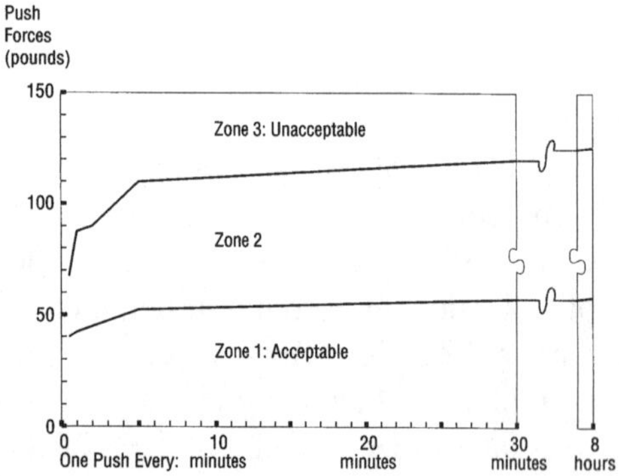

Pushing
• Elbow Height
• Initial Forces
Zone 3: Unacceptable
Zone 2: Defined by forces acceptable for 25% of male population (should have administrative or engineering controls).
Zone 1: Defined by forces acceptable for 75% of female population (acceptable for most work tasks).

Note: For other pushing tasks, see original material.

Figure 4-34. The acceptability of push forces in combination with frequency. (Snook, S.H., 1978, with permission).

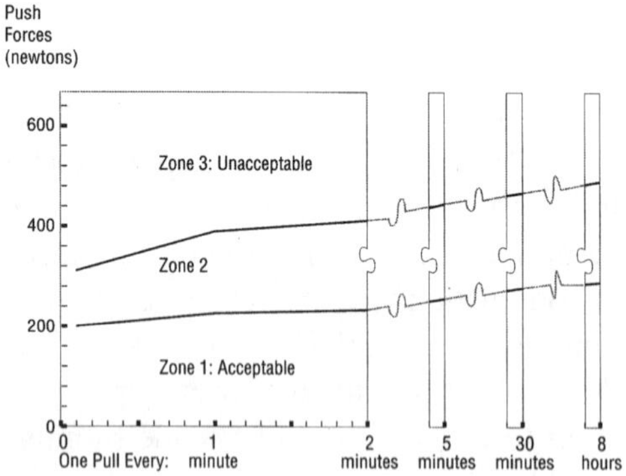

Pulling
• Elbow Height
• Initial Forces
Zone 3: Unacceptable
Zone 2: Defined by forces acceptable for 25% of male population (should have administrative or engineering controls).
Zone 1: Defined by forces acceptable for 75% of female population (acceptable for most work tasks).

Note: For other pulling tasks, see original material.

Figure 4-35. The acceptability of pull forces in combination with frequency. (Snook, S.H., 1978, with permission).

In addition, Snook (1978) also gives some recommended upper limits of horizontal force to start, maintain, and stop the motion of a handcart or truck as listed below. When these upper limits are exceeded, power equipment should be considered. Carts/trucks carrying over 1,100 pounds should have brakes.

- Start — 50 pounds
- Maintain for less than 10 feet — 40 pounds
- Maintain 1 minute continuously — 25 pounds
- Emergency stop within 3 feet — 80 pounds

DETECTING MUSCULOSKELETAL DISORDERS (MSDs)

There are three basic mechanisms used to identify the presence of musculoskeletal disorders in the workplace. These mechanisms provide information as to whether problems exist, and if so, the extent of those problems. They provide information about the overall injury, often pointing to particular injury site (wrist, elbow, etc.) and to the possible injury methodology (external trauma, overuse, joint posture, etc.). All of this information is valuable, and is certainly worth the cost in terms of time and effort.

1. Review of recent injury data such as OSHA or company injury and illness logs, workers' compensation claims, medical first aid visits, turnover, above average absenteeism, maintenance records.
2. Review of complaints of discomfort through personal interview and symptom surveys.
3. Direct observance of jobs and workplaces and review of job descriptions for identification of ergonomics risk factors.

Review of data using these mechanisms should be repeated periodically, probably more often at first, and then less frequently as injuries are controlled.

Data Review

A review of company records will help the investigator understand how and where significant injuries occur. Aside from the standard OSHA and workers' compensation data, evaluate first reports of injury, attendance records, maintenance records, and history of safety/ergonomics committee activities. Although OSHA records are kept for up to five years, it may only be helpful to review the two most recent years, especially given staffing and equipment changes. OSHA and insurance records provide information on cause and location of injury as well as lost or restricted work. They can identify the most expensive injuries. They can also identify trends related to work area, day of week, time of day, type of illness/injury, body parts, and total number of employees with similar incidences. Look for trends between repair and maintenance histories and the onset of injuries.

Workers may use personal time away from work rather than report a work-related injury because they fear repercussions related to job security, opportunity for promotions, or opportunity for overtime. A review of attendance records can help to identify important trends. Symptoms tend to be less bothersome after a weekend or scheduled time off, however they become more so as the work week progresses. An investigation of the pattern of absenteeism might show this trend.

Finally, safety/ergonomics committees will usually focus on company wide problems; therefore, a review of completed projects may also provide insight into known issues that may or may not have been resolved.

Symptom Identification

The authors have found that approximately one out of eight complaints of discomfort will convert to a recordable injury. With this is in mind, identifying complaints of discomfort, then relating discomfort of individual body parts to specific tasks will assist with accurate root cause identification.

Collecting employee input can be accomplished through one-to-one interviews, informal conversation between employee and supervisor, or formal reporting through symptom surveys. If there is a fear of reprisal, symptom surveys can be collected without the employee's name. It is important to gather the type and volume of information necessary to correlate discomfort with work activity.

Observing Jobs and Reviewing Job Descriptions

The analysis process continues with a review of the job description to understand the scope of work and direct observation of the job to categorize the essential tasks of the job.

Observation is best accomplished by watching both an experienced and inexperienced employee to fully understand the whole job and appreciate differences in perception and work method. The experienced employee will likely reveal "individually designed" modifications to the work or workplace because of problems that exist with the job. The inexperienced employee will likely reveal gaps in training, understanding of the job, and performance of the job without "homemade" modifications. After several initial observations, each task (the fundamental unit of analysis) can then be broken down into discrete task elements. These task elements help to identify the specific portion of the activity that exceeds acceptable limits (e.g., high, moderate, or low risk). Starting with high risk task elements, a correlation between risk and injury records will help the analyst determine how these tasks can best be modified through engineering and/or administrative changes.

As an example of task analysis, consider the job of food preparation.

Job Title: Food Services Specialist

Job Description: Prepare, assemble, and serve meals to school children; receive and store food and supplies; wash and store food preparation and serving items.

Target Essential Task: Assemble hamburgers.

- **Sub Tasks**: Retrieve meat from warmer and place on table, retrieve buns from storage and place on table, retrieve pan from storage and place on table, assemble hamburgers and place on pan, place pan into plastic bag, replace assembled burgers into warmer.

    Task Elements: Reach, grasp, and carry bun with right hand; place and hold bun bottom with left hand; pinch and carry bun top with right hand; pinch grip bun top and meat with right hand; move both hands together; grasp whole burger with left hand; reach, place, and release burger on tray.

When observing the worker performing the task, think in terms of the ergonomics "eye." In other words, look for the kinds of behaviors that you would predict if the workplace were not designed according to the anthropometrics data of the previous sections. It quickly points out postural problems and is especially useful when making a quick assessment of an existing work situation and the people performing the jobs. Here are some examples:

- When a worker changes something about the workplace to make it fit better, it is referred to as a "worker modified workplace." Often the change involves duct tape, foam padding, cardboard, or plywood, and is very easy to see. The reasons for the changes may be obvious, such as glare on a video screen or a tool handle with sharp edges, but interviewing the operators will help explain why the change was made.
- Unusual work positions such as stooping, kneeling, squatting, or bending should be noted during the initial observation. Unusual postures are a strong indication that something is wrong with the fit between the worker and the workplace.

Table 4-5
The duration characteristics of the frequency multiplier (to be used with Figure 4-29)

Short Duration	
	Lift time ≤ 1 hour and Non-lift time ≥ 1.2 × lift time
Moderate Duration	
	1 hour ≥ Lift time ≤ 2 hours and Non-lift time = .3 × lift time
Long Duration	
	Lift time = 2 to 8 hours and Non-lift time = standard breaks

- Physical contact between the workplace and the person is an indicator of poor fit. One example is an area of paint worn off the edge of a table because of repeated contact from the worker's body. This may indicate either the need for additional clearance or the need to move hard to reach parts closer.
- Padding at the workplace, just like body contact, may indicate a need to change the design of the workplace. However, it is important to identify whether padding solves the problem or forces the operator to stand further away from the work. Consider options to reduce the reach between operator and parts such as a cutout in the table-top or moving the work closer to the worker.

The identification of ergonomics-related problems at the workplace is very important but not difficult. Problems may develop when there is adjustability. Reach and height problems may also be created by design criteria that are based on the wrong target population (e.g., using North American anthropometry data for workstations used in China). The ergonomics eye should be used to identify unusual postures or poor fit while simply watching people work. Couple direct observations with the checklist of questions below to better understand how and why these workplace issues occur.

- Are there some "one-size-fits-all" work places?
- Where do reaching problems occur?
- Where do clearance problems occur?
- Where do people routinely work in non-neutral postures?
- Where do people work outside the normal reach envelopes?
- Where are static postures required?
- Do standing workplaces meet the checklist guidelines?

- Do sitting workplaces meet the checklist guidelines?
- Are there worker-modified work places?
- Is there excessive body contact between the worker and the work place?

CORRECTING WORKPLACE DESIGN PROBLEMS

Once workplace ergonomics issues are identified, modification of the workplace and/or work methods should be initiated. Solutions to workplace problems are easy to develop once key principles are reviewed. The main objective of a workplace change is to keep the trunk upright and make it easy for people to work in neutral posture.

First, take advantage of adjustable features that are already available at the workplace. Second, always specify adjustable components for new or replacement equipment. Adjustable features of the workstation allow the user to position equipment to meet individual dimensional needs. The user must be provided with enough information to understand the importance of adjusting the workplace both at the start and during the shift, especially if the workstation is used by others. Training for use of workplace equipment adjustment helps people "listen" to their own body signals and make small adjustments throughout the shift for both comfort and productivity. As a policy issue, some organizations deem it an unsafe act when workers fail to adjust the workplace at the start of the shift.

Worker modified workplaces provide insight into the source of the problem and ideas for solutions. These efforts to improve the "fit" should not be underestimated. Determine the reason for these modifications and interview the people who made the modifications to gain insight about the changes. Work with these individuals to further improve these modifications if appropriate. For example, a commercially available product may be available at the hardware store that is more durable than the towel and tape solution shown in Figure 4-36. Another example is enhancing a single height platform with several adjustable steps to accommodate people of various heights.

Legitimize improvements by securing peer and/or management approvals for the modification if necessary. Finally, always credit the people who initiated the improvements in the first place.

Take advantage of "ergonomic quick fixes," simple and easily implemented changes that usually cost less than $200 to $300 (U.S.). These changes generally require a simple work order approved by the first line supervisor. Figure 4-37 illustrates a simple but effective quick fix. A spacer or portable platform shown here can be used as a step to elevate a shorter worker, and it can also be used on top of the work bench to raise the work height for a taller worker. The spacer is inexpensive (under $10) and easy to use.

Some quick fixes examples are listed below:

- Alter worktables and work surfaces.
- Lower or raise the work or work surfaces.

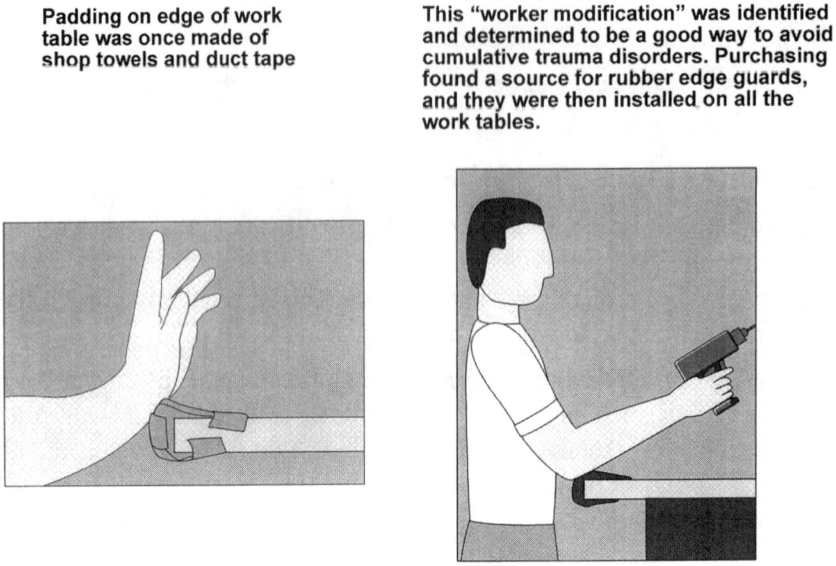

Padding on edge of work table was once made of shop towels and duct tape
This "worker modification" was identified and determined to be a good way to avoid cumulative trauma disorders. Purchasing found a source for rubber edge guards, and they were then installed on all the work tables.

Figure 4-36. Workplace improvement using worker modified workplace.

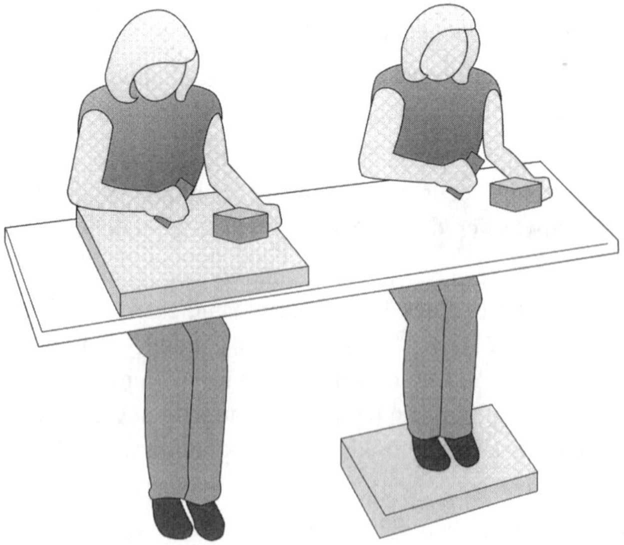

Figure 4-37. Example of an ergonomic quick fix.

- Use spacers to raise work surfaces.
- Add platforms to raise shorter workers.
- Pad equipment that has sharp or pointed edges; however, if padding becomes an obstacle as indicated by frequent body contact, then add more clearance not more padding.
- Provide knee and toe clearance.
- Use special tools for reaching.

Jigs and fixtures are often listed in the quick fix category, especially when they are user friendly and inexpensive. A simple fixture can be used to hold and stabilize a component part. This eliminates local muscle fatigue that otherwise occurs with static holding. It also frees both hands to complete the required task on a simultaneous or alternating basis. The fixture can be used to reorient the component part with direct access, allowing the operator to work in neutral arm and trunk postures. The first example on the left in Figure 4-38 illustrates how a fixture can free the hand from static holding (excessive muscle force), and the middle example illustrates how a triangular block can reposition the part to a better orientation and posture. The third example shows a mechanized or automated process that may eliminate static holding, eliminate awkward postures, and improve precision of the end product. Sometimes this provides the best long-term solution.

Adequate access to parts, supplies, and between equipment is necessary to make the workplace usable and avoid awkward postures, especially twisting. Figures 4-39 and 4-40 illustrate

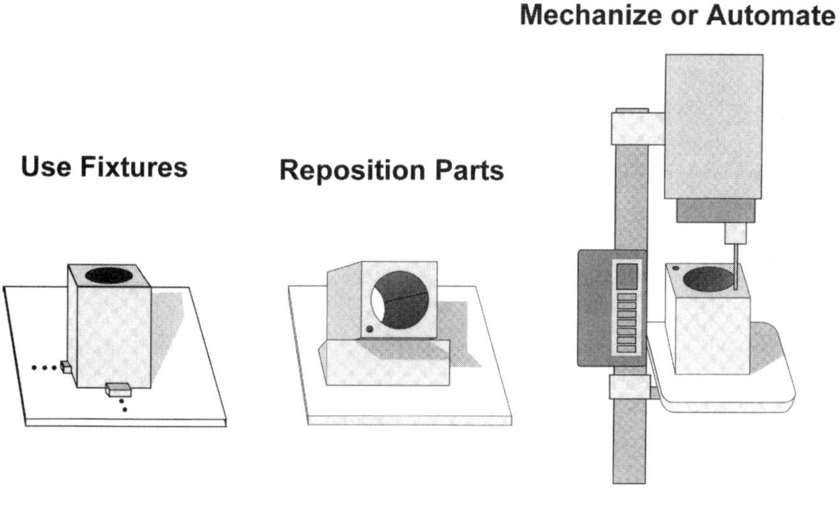

Figure 4-38. Job aids to minimize force and awkward postures.

Figure 4-39. Provide accessible reach for materials.

Figure 4-40. Ensure that visual access is available.

the value of storing objects within reach and ensuring visual access and direct line of sight. Ideas to enhance access are listed below:

- Place objects close enough for the smallest person to reach comfortably and without bending forward.
- Relocate tools and materials to the primary zone if frequently used, and to the secondary reach zone if used less frequently.
- Tilt work surfaces forward to reduce long reaches.
- Remove obstructions to avoid reaching around them.

Clearance needs are often overlooked during initial design and installation. As a result, it often takes more time and effort to complete service or production activities and extra time to perform even routine maintenance assignments. An operations design with long or obstructed reach to control panels or knobs and one that provides little clearance to move quickly and easily between equipment is inefficient and may cause symptom development. These inefficiencies may cause employees to work in awkward postures and will likely increase the cycle time and subsequent cost of production. Based on the authors' experience, in maintenance tasks, as much as 10-30% of work time can be spent on getting to the part that needs repair. For both operations and maintenance tasks, clearance between equipment, walls, poles, and other obstacles must be sufficient for the largest employee. The same need holds for leg, knee, and foot clearance underneath tables and fixtures.

Clearance and reach requirements are often resolved with a sit/stand workplace design as shown in Figure 4-41. It offers the best compromise between sitting and standing needs. This allows the worker to sit for short periods of time, then stand to reach for accessories or move

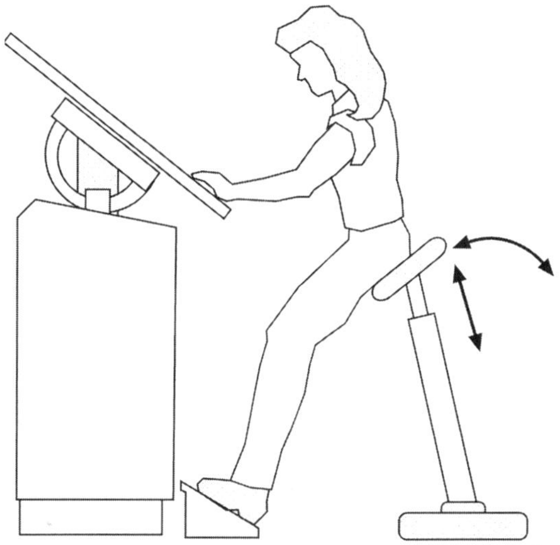

Figure 4-41. Example of a sit/stand workplace.

about the workplace, and return to the workstation for short periods of rest for the low back, legs, and feet.

If adjustability cannot be incorporated into the initial design of the work area, then adding an adjustable feature later may improve the usability of some equipment. For example, placing a machine on top of a scissors lift to raise or lower the machine improves workstation design. This allows the operator to stand upright and adjust the machine to optimal work height. If machine height cannot be adjustable, then platforms to raise the operator or raise the work surface height may be reasonable options. A "lean-to" chair (see Figure 4-41) can be used to lower the hand height of the operator. The worst case scenario is never incorporating adjustability into the design, installation, or operations phase of work.

Administrative Controls

Administrative controls should always be considered when engineering controls cannot be used or when they do not adequately resolve the problem. There are so many manual material handling tasks (especially lifting tasks) occurring in industry that every lifting task may not be considered and/or engineered properly. At the same time, the Revised NIOSH Lifting Guideline does not protect the entire working population (the RWL is designed for 90% of healthy workers). In all of these cases, administrative controls add a margin of safety that is often inexpensive and further reduces the risk of injury.

Whenever administrative controls are instituted, initial and on-going training will be required. (Note that methods training is never a solution for a poor layout.) Training the worker to keep the load close is futile when the workplace requires excessive reaching to pick up the material.

Training for employees should include three key points. First, employees should understand the specific risk of injury that may occur if objects are handled in a careless or unskilled fashion. Second, employees must understand the value of lifting methods and devices to reduce unnecessary muscle fatigue. Third, the employees should understand their own physical capacities and know when to ask for assistance. Training on lifting technique and/or best practice methods should include the widely accepted and basic principles previously discussed. In particular, it is important to emphasize:

- Keep the load close to the body to reduce the muscle force required by the arms and upper body.
- Avoid twisting and maintain neutral postures as much as possible to reduce muscle strain and mechanical stress to the hands.
- Use secure handholds to reduce stress on hands and fingers.
- Lift one object at a time to prevent an overload situation. Let momentum (ballistic movements) help with motion initiation.
- Know when to ask for assistance.

- Plan ahead to ensure and maintain stable footing to prevent inadvertent slips, trips, or falls that could damage both person and product.

Skills training provides an additional alternative. This allows workers to get the "feel" for the product with less risk. For example, handling empty or partially filled boxes allows a person to gain experience with the most effective grasp. Skills training is particularly effective for ballistic movements and drum handling.

Some administrative solutions focus on matching the size and capabilities of the worker to the required tasks. This is also known as worker selection. Matching people to jobs appears to be simple and easy, but it is not. Effective selection of people for jobs requires extensive testing and validation. It is not recommended as a means to overcome excessive physical demands of a job.

Finally, administrative controls such as job rotation (moving to a different work task that does not require lifting) and job enlargement (adding non-lifting tasks to the primary job responsibilities) may be initiated to reduce the exposure to difficult or frequent lifting tasks. Other administrative controls may incorporate the use of specific work methods (e.g., lifting with both hands), the use of special or specialty tools (e.g., pry bars), slowing and monitoring the pace of the work to minimize overexertion, and the mandated use of work/rest cycles to limit exposure.

Again, although administrative controls are very important in the overall scheme, it is best to rely on engineering solutions before spending too much time developing administrative solutions.

THE ERGONOMICS PROBLEM SOLVING PROCESS

Introduction

The purpose of this problem solving guide is to present an organized ergonomic problem solving process. There are six steps in this process:

1. **Identification** or picking the job to fix.
2. **Investigation** or finding the problem.
3. **Development of Alternate Solutions** or creating ways to fix it.
4. **Selection of the Most Appropriate Solution(s)** or picking the best solution(s).
5. **Implementation** or resolving the problem
6. **Following Up** or verifying the solution.

Step 1: Identification

The purpose of this step is review all jobs to identify ergonomics-related issues, then to select one of the most critical jobs (based on injuries and complaints) for further investigation. The problem jobs, or those with the highest injury/complaint severity, are then targeted for investigation.

FINDING THE PROBLEM JOBS

Review the injury data

The most obvious way to find problem jobs is to determine where injuries have already occurred. Examine two years of data from the OSHA 300 log, the medical visits log, accident reports, and/or workers' compensation summaries for quantity, severity, and patterns. It is advisable to review more than one data source to gain better reliability.

Identifying the specific job where the injury occurred is the first step. Unfortunately, the information documented in the OSHA 300 log may be too vague to be of much value. For example, the job title may be listed as "Welding Line 4," but there are 12 different jobs on Welding Line 4. It may not be easy to identify which one of the 12 jobs requires further investigation. The Accident Report, or **First Report of Injury (OSHA 101)**, can often help to clarify this. Once the data review is completed, group the jobs with higher than average injury or illness rates.

Talk to people who do the job

Listen to people describe what they do and how they do it. Which jobs are most difficult? Which jobs have the highest turnover? Which jobs are the most physically demanding? These jobs will most likely be where many of the injuries are occurring, confirming the data from the OSHA logs. Sometimes an operator will identify a problematic job where injuries have not yet occurred, but the likelihood seems high. These may be new jobs that have existed only a few weeks or months, or where employees have developed a "work around" to minimize or avoid the high-risk parts of the job.

Use your "ergonomics eye"

Walk through the work areas and look for awkward postures, unusual working positions, heavy tools, or equipment that is difficult to operate or maintain. These may be indicators of jobs where risks are present but injuries have not yet occurred. While there is the need to objectively measure and document problems in a workplace, it is also possible to subjectively look at a worker and the workplace and derive useful information (see Figure 4-42). This technique can save time and money during job evaluations, and is an effective and simple tool. Obviously, it depends on experience.

The "worker-modified" workplace should always be identified. These modifications should also be improved wherever possible. Examples of worker-modified workplaces include:

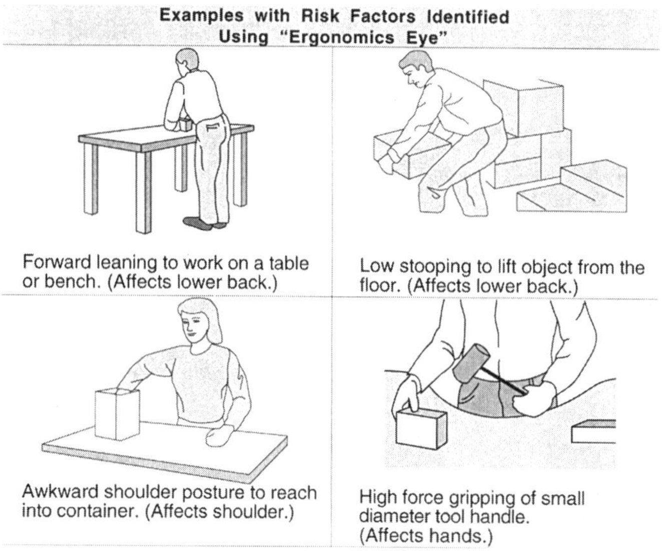

Figure 4-42. Examples of the "ergonomics eye" at work, identifying posture and force issues.

- foam rubber taped to the back or seat of a chair to increase the support or cushion
- tape placed around a tool handle to increase the diameter
- pallets stacked under a container in order to raise the height
- cardboard placed around computer monitor to avoid glare

PRIORITIZE THE PROBLEM JOBS

Once the problem jobs have been identified, prioritize the list according to risk for an ergonomics-related injury. A good qualitative approach is shown in Table 4-6. It provides objective boundaries.

Once each job has been categorized, the prioritization for each separate job should be adjusted based on exposure. For this purpose, the amount of exposure can be thought of as a combination of the number of employees who perform the job or task and the amount of time required for each person to perform that job or task. As the exposure level increases (increasing number of employees and/or amount of hours), the priority level should also increase. Figure 4-43 illustrates that as the number of employees exposed to the job stressors increases and as the number of exposure hours increases, so should the individual priority of the job or task.

Table 4-6
One approach to prioritizing ergonomics risks

Risk Level	Characteristics of Risk Level
High	• Some injuries, many complaints, and/or high turnover • Forceful gripping (pinch > 12.5 lbs., power > 50 lbs.) • Extreme joint and body postures • Lifting or carrying more than 50 lbs. • Pushing more than 60 lbs. • Over 5,000 repetitions at single joint during single work shift
Moderate	• Few injuries and some complaints • Stressful joint and body postures during the work • Significant gripping (pinch > 7.5 lbs., power > 25 lbs.) • Liftibg or carrying more than 35 lbs. • Pushing more than 45 lbs. • Over 3,000 repetitions at single joint during single work shift
Low	• Few complaints and few, if any, injuries • Body and joint postures are mild rather than stressful

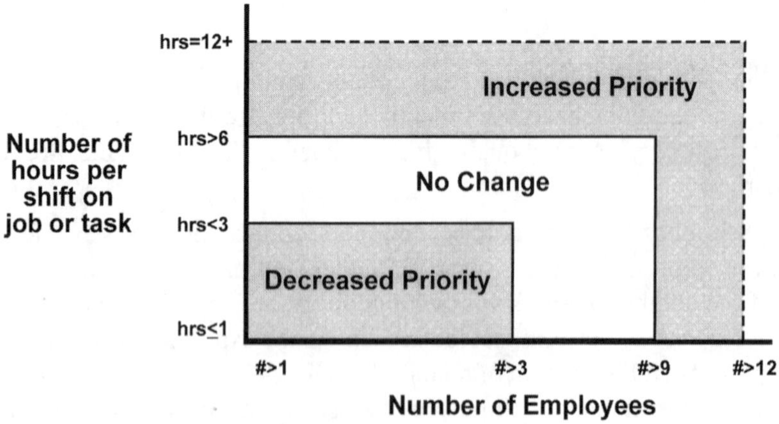

Figure 4-43. Categorizing ergonomics risk levels.

SELECTING A PRIORITY JOB OR TASK

Once the list of problem jobs in the facility or department has been narrowed down and prioritized, recognize that dependent on experience, the number one ranked job may be to difficult to solve. Select a solvable problem that has potential for significant injury or cost prevention in order to gain the trust of management and workers. Identify jobs that have the biggest payoff in terms of injury prevention, jobs that affect the greatest number of people, and those that will have the greatest affect on operating performance. Select a manageable project, one that does not appear to be overwhelming, and one within your area of expertise. Finally, if the project is to be a team effort, ensure that each team member agrees on the target job.

Step 2: Investigation

The investigation step will highlight the aspects of the job that are causing injuries or complaints and, therefore, require some type of modification. This is accomplished through direct observation, data review, and root cause analysis.

GET JOB OVERVIEW

The investigation begins with a thorough review of the entire job. Details of this review are documented for future reference and analysis. Understanding the process upstream and downstream of the job in question will be most valuable when solutions are developed. Changes to the work area or work methods should not negatively impact upstream or downstream operations. An example of a job overview is presented in Table 4-7.

REVIEW INJURY DATA

Again, review the injury data sources (e.g., OSHA 200 Log and accident report) for at least the last two years, this time focusing only on injuries that occur on the job under investigation. The OSHA 200 log will identify total injuries, however, the accident reports have detailed information about the work area, cause of injury or discomfort, and body parts involved.

At this point, the "ergonomics eye" will begin to focus on specific injured body part(s). For example, if the injury data show that there have been four right wrist injuries in the last year on the particular job, then attention to activities and motions of the right wrist becomes a focal point of observation.

Table 4-8 is provided as a template for recording the body part injured, status of OSHA recordability, number of lost (LWD) or restricted (RWD) work days, a short description of the injury, and the suspected cause of injury. Further definitions follow:

Body part injured. Focus only on ergonomics-related injuries. This excludes slips, trips, and falls. Note whether the records specify the left, right, or both sides. This will be valuable information when you are observing the job later.

Table 4-7
Example of job overview

Job Name: Packing Associate – Line #3	Description of Job: After cart unrolls the fabric, associate uses pole to remove air and wrinkles
Date: December 1996	

Table 4-8
Example of an injury data summary format

Classify the Case (check most serious only)			No. of days injured or ill worker was:		About the Case			
Days away from work	Job transfer or restriction	Other record-able case	Restricted Work Days	Lost Work Days	What was employee doing just before the incident occurred?	What happened?	What was the injury or illness?	What object or substance directly harmed the employee?
1				1	Pushing carts 500' for 4 hours	Tingling at right wrist	Carpal tunnel	Laundry cart

OSHA recordable injuries. Injuries can only be listed as recordable if they are documented in the OSHA 300 Log. Injuries that appear only on an accident report and not on the OSHA 300 Log are not considered recordable cases.

Number of lost or restricted workdays cases. A lost work day case indicates a more significant or severe injury than a restricted day case. Although the injured employee may see a physician in either case, (direct and indirect) costs for a lost work day case are much higher and, therefore, command a higher priority.

Description of injury. Details in the injury log may be listed in specific terms, such as "tendonitis in shoulder," or they may be presented in general terms, e.g., "sore shoulder." Further inquiries of the medical department or area supervisor are advisable if further detail is necessary.

Specific cause of injury. Injury logs will often list the specific part of the job that caused the injury. For example, the job may be "shipping laborer" but the specific activity may be "lifting part #TS-20 to second shelf." This is valuable information for when you observe the job being performed.

REVIEW COMPLAINTS

Employees often complain about discomfort in specific body parts even if actual injuries are not reported or do not occur. These complaints help the investigator focus on work activities that involve those body parts. For example, if there are several complaints of pain in the lower back, then special attention should be paid to evidence of stooping, forward leaning, twisting, or lifting tasks that could cause this discomfort. In most cases, the only difference between a complaint and an injury is time.

There are several reliable methods commonly used to gather worker complaints. If a company has a formal structure for workers to submit job complaints, this can often provide good information about ergonomics-related problems. Enter these data in a format similar to Table 4-8.

Speaking directly with the employees and supervisors is an excellent way to gather complaints about the job. Look for links between areas of pain or discomfort and work activities. Pay attention to frequency and duration of these tasks as well as recent changes or failures with equipment. Table 4-9 can be used as a template for documenting employee complaint information. The first column lists the body part where discomfort exists. The second column lists the employee's perceived pain level. Ask the employee to describe the level of pain on a scale, such as 1 to 10. The third column lists the tasks or activities that the employee states have caused the pain.

OBSERVE AND ANALYZE JOB

Careful observation of the employee performing the job is the most important part of the investigation. While videotaping can be helpful for reviewing specific aspects or body parts stressed by the job, it should never be a substitute for live observation because there are many things in the periphery that the camera does not catch. The observation process can be divided into four primary components discussed below.

Table 4-9
A complaints recording form

Body Part	Pain Level (1-10)	Tasks that usually cause the pain?
Lower back	4	Reaching and leaning to align fabric
Left wrist	5	Gripping pole (operator left handed)

Break down job tasks

Each job can be broken down into its component tasks. Most jobs consist of one to three different tasks. Of these three tasks, usually one will be the primary task while the others are secondary. Table 4-10 is provided as a template for documenting these tasks and the percentage of time required for each task. The percentage of time will help the investigator understand how much the employee is exposed to the risk factors of that task and where to focus attention during observation. It may also be helpful to sketch the workplace or to record any workplace measurements, such as table height or reach distances. This information will be useful during the solution development step.

Identify the risk factors

Once the job is broken into its major tasks, the risk factors should be assessed for each task in a format similar to Table 4-11. The example in Table 4-11 identifies risk factors only for task #1 or the "remove wrinkles" task.

For most jobs, the primary risk factors will include one or more of the following:

- awkward posture (e.g., long or high reaches, stooping, and twisting)
- force (e.g., gripping, pressing, and pushing)
- repetitions (e.g., high frequency and inadequate recovery time)

When observing an employee who is performing the task in question, list the body part affected and its associated risk factors. For example, at a standing workstation, you may notice that someone works in a forward leaning or stooping posture. In the "Risk Factors" column write "static forward leaning" or "static stooping." The abbreviated terminology used in Table 4-11 should be consistent. Continue with the observation until all risk factors of concern have been listed on the worksheet.

Table 4-10
Example of job tasks description and workplace layout form

Job Tasks	Percentage of time/shift	Sketch of Workplace
1. Remove air and wrinkles from fabric.	85%	
2. Enter production data and fill out report.	10%	
3. Clean up.	5%	

Table 4-11
Example of risk factors listed by body part

Body part	Key Risk Factors (in abbreviated terminology)	
Hand & wrist	• High force gripping • Pinch gripping • Awkward wrist posture (e.g., backwards, side, forward)	• Pressing or hammering with palm • Tool pressing into palm
Elbow	• Forceful rotation of hand	• Forceful extension of elbow
Shoulder	• Long forward reach • Reaching above shoulder • Static reaching	• Reaching behind body • Long reach to the side • Holding material away from body
Neck	• Repetitive twisting • Static extension i.e., looking up	• Static flexion i.e., looking down
Back	• Static forward leaning • Stooping	• Twisting at wrist • Extreme bending at waist
Legs	• Prolonged standing • No foot rest - seated • Frequent squatting • Using foot pedal while standing	• Standing on unpadded surface • Frequent kneeling on unpadded surface
Feet	• Prolonged standing	• Frequent squatting

After all of the risk factors are identified for the task, circle or highlight the most critical or severe risk factor(s). If injuries have occurred on the job, the critical risk factors will usually be those that have caused or contributed to the injury. If injuries or complaints have not occurred, the critical factors will be those that create the highest degree of physical stress.

Determine root cause for each risk factor

In general, the presence of risk factors leads to ergonomics-related injuries. Table 4-12 illustrates the link between an injury, its risk factors, and associated root cause. Understanding how and why the risk factor exists is known as a root cause analysis. The root cause for each risk factor must be determined. Identify what it is about the job that causes or creates the risk factor. For example, if a person must reach forward into a small bin to get a part, a risk factor for the shoulder may be "fully extending arm." For this risk factor, the root cause might be "location of parts bin." The location of the bin determines the reach distance, requiring the worker to extend the arm.

List initial solution ideas

For recording field data in Table 4-11, it may be useful to add columns for Root Causes and Initial Solution Ideas so that all the investigation information can flow through the problem solving steps easily. An initial list of solutions generated during the walkthrough/evaluation need not be completed while observing the worker. Other solutions will develop while brainstorming.

Solutions seek to modify the root cause. For example, if the location of the parts bin (root cause) is causing a long reach (risk factor), then one potential solution is to locate the bin closer to the worker. Fixing the root cause eliminates or reduces the risk factor and prevents or significantly reduces the likelihood of future injury.

Make a note of solution ideas that may develop during the observation. Discuss these ideas with the employee and supervisor for several reasons. The people who perform the job

Table 4-12
Three examples of injury, risk factor, and root cause links

Injury	Risk Factor	Root Cause
Low back pain	Static stooping	Bench too low
Shoulder pain	Overhead & extended reach	Location of parts bin
Wrist pain	High force torque	Hand tool design

understand the process very well, know the history of maintenance problems and modifications, know of problems that may not be identified otherwise, can provide valuable input as to the effectiveness of the solution, and may be able to add to the initial ideas. Including them in the process can have a profound impact on their acceptance of proposed solutions.

Step 3: Development of Alternate Solutions

TYPES OF SOLUTIONS

It is good practice to develop a list of solutions in the event that the preferred solution is not feasible. Solutions include modification of equipment, tools, workstations, or work methods that contribute to excessive work demands. Solutions can also be listed in two main categories: engineering solutions and administrative solutions. Both types of solutions should be considered.

Engineering Solutions

Engineering solutions typically involve physical and long lasting changes to the workplace. Changing a tool handle, adding a scissors lift device, and raising a workstation height are three examples of engineering solutions. Engineering solutions are preferred where possible, because they physically change how the work is performed without relying on the cooperation, memory, or skill level of the operator. Figure 4-44 provides three examples of ergonomics-related problems and several potential engineering solutions for each. Notice that each of the example engineering solutions physically changes the workplace design or how the task is performed.

Administrative Solutions

Administrative solutions typically involve only the workers and not the workplace. Although engineering solutions are preferred, administrative solutions can be effective control measures for the short term. Administrative controls do not resolve the root cause of the problem, however they can reduce a person's exposure to the stressor. In cases where an engineering control is not technologically feasible, takes too long to build or install, is too expensive, or does not completely resolve the issues, administrative solutions are recommended. While it may take several weeks to design and implement an engineering solution, an administrative control (e.g., job rotation, job enlargement, or best practice method) can be quick and easy to implement. Table 4-13 contains three ergonomics-related problems with several potential administrative solutions for each.

BRAINSTORMING NEW SOLUTIONS

Brainstorming with a group of people is a great way to generate a large number of creative and effective solutions. The ideas should be evaluated after the list is made. Table 4-14 provides some guidance for brainstorming solutions.

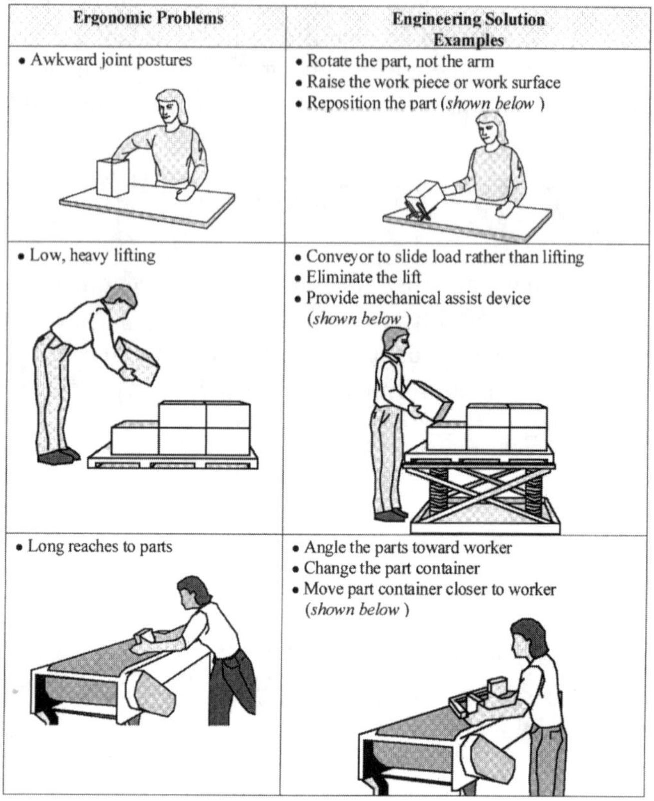

Figure 4-44. Examples of engineering solutions.

Table 4-13
Examples of administrative solutions

Ergonomic Problem	Administrative Solution Examples
• Awkward joint postures	• Use best or preferred methods
• Heavy lifting	• Allow frequent rest pauses • Training in proper techniques • Require two people to lift certain objects
• High repetition tasks	• Frequent worker rotation • Allow self-pacing

Table 4-14
Tips for brainstorming new solutions

- ***Visualize the neutral postures***
  If the risk factor is an awkward posture, visualize the neutral posture that is your design objective. Think of alternatives that will improve the operator's current posture toward a more neutral one.

- ***Suggest both engineering and administrative controls***
  Although administrative controls are not suitable as long-term solutions, they can improve the job until a permanent, long-term fix is implemented.

- ***Involve people who do the job***
  People who do the job often think of the best solutions or provide valuable input. Do not overlook these people. They may also be more accepting of new ideas if they are part of the development process.

- ***Check to see if other people are working on the problem***
  Many times a team or an individual will begin to work on a problem that is already being processed by engineering or maintenance. Check with engineers, maintenance, or technicians who are assigned to this area.

- ***Be creative***
  Think outside the box. Let your imagination go beyond the current workstation or methods.

- ***Critique ideas later***
  Without judging cost, merit, or feasibility, get all ideas out on the table and written down. This list will be whittled down later.

- ***Review equipment catalogues for ideas***
  Many times just seeing the available tools or equipment can spark a solution idea.

- ***Identify "worker modified" workstations***
  The "homemade" solutions are often very effective, e.g., foam padding added to table edges or chairs, platforms made of empty pallets, cardboard positioned over computer screens to deflect glare.

Step 4: Selection of the Most Appropriate Solution(s)

At this point in the process, a well-defined problem and a substantial list of solutions that eliminate or reduce the problem should exist. There are many strategies for selecting the most appropriate solution: the most effective, the safest, the cheapest, and the quickest. Roughly plotting each solution's cost versus effectiveness is a good way of visually comparing several alternatives. Effectiveness is the degree to which the solution fixes the root cause and minimizes the risk factors. The preferred strategy is to select the solution having the lowest cost and highest effectiveness. Effectiveness can be evaluated on a qualitative scale and ranked as shown in Table 4-15.

Once each solution has been given an effectiveness rating, the cost of each solution should be estimated. The cost of each solution should at least include estimates of:

- Purchased equipment (e.g., tools, chairs, and components)
- Design time (internal department or outside contractor?)
- Fabrication time
- Installation time

Once the effectiveness and cost have been assigned to each solution alternative, these values should be plotted as shown in Figure 4-45. Solutions in the upper left corner are often the most powerful and always the easiest to sell to management.

Figure 4-45 shows nine alternative solutions to a given problem. Solution #1 has a moderate cost and an effectiveness rating of 4, whereas Solution #8 has a low cost and low effectiveness rating. Using this comparison method, the best choices would be #1, #4, and #7. From this point, several other potential decision factors should be considered to narrow the gap and identify the top two or three solutions. These factors are listed in Table 4-16.

Step 5: Implementation

It is advisable to take the best two or three solutions to the "Implementation" step. If the preferred solution is rejected due to limited money, time, or resources, other effective solutions should be readily available. The process, time, money, and resources needed to implement the chosen solution(s) must be carefully planned. It is usually at this point that the costs and effectiveness of implementing the solutions are presented for management approval.

OBTAIN SOLUTION APPROVAL FOR CHOSEN SOLUTION

Approval for solution development comes from two levels. First, the people doing the work must agree to support new ideas because they have to live with the outcome of the changes. Second, managers must agree to finance the changes.

Table 4-15
Example of a ranking system for effectiveness of solutions

Ranking	Description
5	Job stressors fully eliminated
4	Job stressors eliminated for normal working conditions
3	Job stressors significantly reduced
2	Some job stressors reduced
1	Few or no job stressors reduced

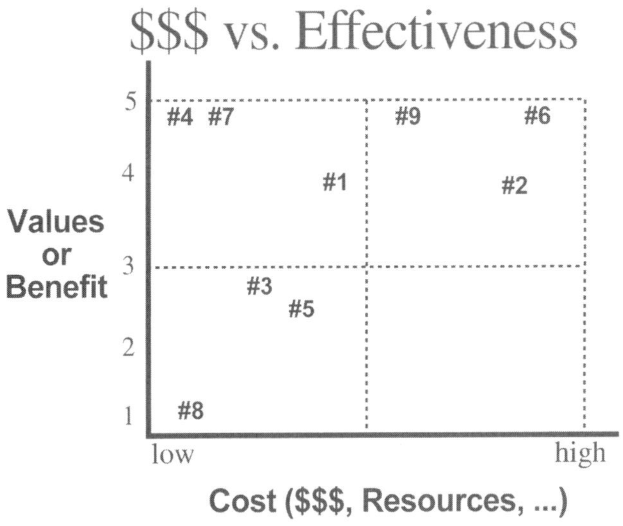

Figure 4-45. Example of an effectiveness/cost matrix.

Table 4-16
Categorizing decision fctors

Decision factor	Comments
1. Productivity & Quality	By implementing some solutions, not only are the ergonomic risks resolved, but productivity and/or quality is also enhanced. This decision factor is a good one to use when deciding between two seemingly equal "low cost-high value" solutions.
2. Timeliness of Implementation	Those solutions that could be more quickly implemented would be more attractive. When using this factor to evaluate solutions, solution "effectiveness" should also be used. The quickest solution may not always be the best one. Some solutions that completely resolve the ergonomic concerns may require design, fabrication, and installation.
3. Usability	A solution having low usability is one that the cost oc using it (e.g., time, effort) outweighs the benefits gained (e.g., injury avoided). An example of a solution with low usability is a lift assist device that is difficult to use and takes longer to use than lifting the object by hand.
4. Safety	Does the solution create any new potential safety hazards? For example, the use of a step or riser to resolve a work height issue could be considered a safety hazard if the workstation is in a high traffic area. This is also a good factor to use as a "tie-breaker" between two seemingly equal solutions.

Operators

The information that operators will want to know will include:

- why the job/workstation is going to change (i.e., the problem)
- how will the job/workstation change (i.e., the solution)

If the people doing the job are involved in the development process, then these questions should already be answered. If not, convey this information through a formal or informal group discussion meeting. Both the workers and the development team/manager should remain "open" to suggestions that improve the solutions.

Managers

The managers will want a definition of the problem, solution benefits, and solution costs. First define the problem by identifying the number of ergonomics-related injuries and complaints that are associated with the job, then identify the primary ergonomics risk factors. Next, define the benefits of the proposed solutions. While related to effectiveness, this needs to be more detailed than the effectiveness assessment performed in the last step. Table 4-17 provides a list of benefits and examples that may be useful for securing management support.

Table 4-17
Categorizing benefits

Potential Benefits	Examples
1. Avoided current expenses	• Annual cost of injuries on the specific job
2. Elimination or reduction of risk factors	• Reduced reach from 17 in. to 6 in. • Eliminated twisting to get parts • Reduced carrying distance from 40 ft. to 5 ft. • Allowed operator to work either seated or standing • Eliminated need for worker to support weight of the tool
3. Enhanced productivity	• Decreased worker fatigue • Decreased machine or process downtime • Improved machine "changeover" - less time required • Decreased wasted time walking to get material
4. Improved quality	• Using the correct tools decreased damage to product • Lowered fatigue, thus greater attentiveness • Reduced shipping errors
5. Enhanced work life	• Lowered employee turnover • Reduced overtime work due to absent workers • Reduced cost of hiring and training replacement workers

Estimate the cost of proposed solutions for comparison with benefits. When compiling estimates of costs for modifications that will be made internally, be sure to consult the department performing the work and ask for a time and cost estimate. For example, if the company machine shop will be asked to make a tool, get an estimate of the cost and delivery time. Other possible cost considerations are shown in Table 4-18.

HOW TO ORGANIZE THE INFORMATION FOR MANAGEMENT

Whether the project is approved through a formal or informal presentation, information must be organized in a simple format. It should follow the outline below:

The Problem

The problem can be illustrated through the use of sketches, digital images, or videotape. Sketches and digital images should include annotation of workplace dimensions and the risk factor in action (if possible). Narration of videotaping is recommended and should include a description of the job basics and key risk factors.

Review only the primary risk factors, root causes, and solution ideas with the audience. Provide summary sheets of the proposed project if a large number of risk factors have been identified.

The Solution(s)

A short description of the solution(s) should be provided, listing the following information as needed:

Table 4-18
Potential costs of workplace modifications

Costs Categories	Examples
Equipment costs	• Items purchased "off the shelf" (e.g., hand tools, lift tables, carts) • Material cost for items to be fabricated "in house"
Design costs	• Labor to design equipment or tool by "in-house" personnel (e.g., engineering, design, maintenance) • Design fee by "outside" source
Fabrication costs	• Labor for "in-house" personnel to build something (e.g., maintenance, technicians, machine shop) • Fee for "outside source" to create or build equipment
Installation costs	• Moving existing equipment, placing new equipment • Running electrical or air lines • Assembly of purchased equipment (e.g., conveyor)

- Proposed modifications to existing equipment (e.g., shorten conveyor or modify tool handle)
- Purchase of new equipment (e.g., carts, tool balancers or chairs, lift tables)
- Proposed changes to work methods (e.g., 1 hour rotation, heavier material on second shelf, or standing job rather than sitting)
- Proposed changes to workplace (e.g., raise table by 6 inches or redesign of the workstation)

A sketch of the solution will help to illustrate the basic concept. Again, two or three alternate solutions are recommended in the event that the first choice is rejected.

The Benefits and Costs

Benefits should be presented with as much detail as possible. Use Tables 4-15 through 4-18 and Figure 4-45 to organize this presentation. Solutions that potentially prevent future injury should be presented with a dollar estimate of compensation or medical costs. Solutions that reduce or eliminate ergonomics risk factors should point out the body parts affected and detail how the risk factors are affected.

Group and present cost information by gross categories of "materials" and "labor." These should be further reduced to items such as equipment, design, fabrication, and installation.

DEVELOP THE IMPLEMENTATION PLAN

Once a solution is approved, formulate a detailed action plan to implement the solution. This plan should designate levels of responsibility and target deadlines. The individual responsibilities for completing each component, action items needed, and persons assigned should be clearly documented. Responsibilities should be assigned to individuals rather than general departments. The level of detail for the actions does not have to be precise, but it should be sufficient to carry out the implementation.

Accurate and realistic deadlines for each part of the implementation plan should be identified after working with the individuals who agree to work on the project. This gains buy-in to the project and holds everyone accountable for assigned actions.

Step 6: Following Up

A timely follow-up to the implementation ensures that the solutions successfully resolve the problem and do not create new problems. Use the "ergonomics eye" to observe employees working in the modified area. Look for the presence of ergonomics risk factors and ask these employees for their opinions about the changes. Ask employees if risk factors have been eliminated or reduced to an acceptable level, if they believe further modification is necessary, if new or different problems have been created, and if they have additional ideas for improving the solution. For example, a long reach problem may be resolved by relocating the material, but perhaps the new location requires the person to twist and reach behind.

Occasionally, one or more employees on the particular job will feel that the solution is not working as intended. When this happens, several actions can be taken. First, identify if other employees share this opinion or if special circumstances exist for the one employee. Second, modify the components of the implemented solution that were problematic. Third, if the solution is not acceptable even after a second or third modification, re-evaluate the other potential solutions. A fourth option is to request "expert help," perhaps an ergonomist outside the company.

THE CERTIFICATION PROCESS FOR PROFESSIONAL ERGONOMISTS

Introduction

For the past few centuries, credentialing has been used to certify the technical competence of individuals to perform professional work. According to NOCA (1996), certification is a voluntary process by a non-governmental agency that recognizes individuals for advanced knowledge and skill. In 1990, the Board of Certification in Professional Ergonomics (BCPE) was incorporated. Its mission is to develop a certification process to ensure a defined level of competence. Two years later, the BCPE began the certification process used today.

Certification in ergonomics has several key benefits. It aids the employer in the hiring process because it identifies full-time and contract employees with appropriate skills in ergonomics. It is a first step to help an employer immediately recognize the basic qualifications of potential employees. It helps companies to distinguish between qualified ergonomics consultants and individuals who simply sell products with ergonomic features but who do not possess an acceptable level of technical competency in ergonomics, and develops a "pool" of practitioners who have met baseline standards. It also helps to reinforce personal accomplishment in the field of ergonomics and in some cases supports company criteria for advancement and pay increases.

Certification Levels

Several levels of certification are offered by the BCPE. The information listed below contains only a cursory review of qualifications for each level of certification. Contact the BCPE (*www.bcpe.org*) directly for complete information on the policies, practices, and procedures for certification.

Industrial hygienists who are interested in the certification process, who engage in ergonomics on a part-time or adjunct basis, and who have not completed Master's level coursework in ergonomics, are most often interested in qualifying for the Certified Ergonomics Associate level of certification. This certification was developed to provide a technical level of certification to meet the growing need for certified ergonomists who use commonly accepted tools and techniques for analysis and enhancement of human performance in existing systems, but are not required to solve complex and unique problems or develop advanced measurement tools. In addition, the more advanced Associate Ergonomics Professional/Associate Human Factors Professional (AEP/AHFP) and Certified Professional Ergonomist/Certified Human Factors Professional (CPE/CHFP) certifications have also been developed. It should be noted that the BCPE regards ergonomics as synonymous with human factors and human factors engineering.

QUESTIONS

4.1. What are the two priorities of ergonomics?

4.2. Why is it important to balance capacity and demand in ergonomics design?

4.3. Why is ergonomics a strategic business advantage?

4.4. How do the principles of ergonomics and human factors engineering complement each other?

4.5. Ergonomics is derived from which two Greek words.

4.6. What are the key benefits of understanding a client's product line when providing ergonomics consulting services?

4.7. What are the client's main ergonomics and business needs?

4.8. What are two main reasons for considering the economic impact of ergonomics improvements?

4.9. How do cycle time and process flow improvements affect productivity?

4.10. How would the location of a machine influence the cost of routine maintenance?

4.11. How does the work/rest cycle impact productivity and fatigue?

4.12. How are anthropometry measurements used to design workstation dimensions and location of supplies?

4.13. Which has a greater impact on the muscle force required to lift a load, horizontal or vertical location?

4.14. Why would it be better to squeeze the trigger of a hand tool with the wrist in a straight or neutral posture rather than bent toward the palm?

4.15. What are the benefits of using stereotypes when designing instructions, signage, and displays?

4.16. What are four main causes of musculoskeletal disorders?

4.17. Should the industrial hygienist be concerned with physical activities or hobbies performed off the job?

4.18. When is it more appropriate to use the primary rather than secondary reach zone when designing a workplace?

4.19. What is the difference between static and dynamic work and which is preferred for long duration work activities?

4.20. If asked to design a workstation for a large control panel, would the optimal design have the operator sit, stand, or be in another position?

4.21. What is the best workstation design for the following situations: a) boxing and packing boxes in a warehouse, b) repairing watches, and c) telemarketing?

4.22. Many ergonomics improvements or solutions cost under $500. What are some examples of quick fixes to improve workstation problems?

4.23. What are the five main principles for preventing ergonomics-related workplace problems?

4.24. The first step in problem solving is identifying the problem. If there are no reported ergonomics related recordable injuries, would it be reasonable to assume that ergonomics risk factors are not sufficient to warrant further investigation?

4.25. What are the three main principles of injury investigation?

4.26. What is the benefit of developing a long list of solutions if only one will be used?

4.27. What is a good strategy for planning the implementation process?

4.28. A worker must lift boxes from a 15 inch high conveyor and place them on a shelf. The boxes stop 24 inches from the end of the conveyor. Would raising the conveyor or bringing the box closer to the edge of the conveyor have more influence on reducing the NIOSH Lifting Index?

4.29. Does a Lifting Index of 1.5 indicate the need for a major engineering change?

4.30. Is it better to push or pull carts or heavy loads?

4.31. What are the major benefits of the certification process for ergonomics?

4.32. Which certification level is most appropriate for the full time industrial hygienist who has primary responsibility for ergonomics at a given company?

FURTHER READINGS

There are many excellent texts and publications in the field of occupational ergonomics. The interested reader may wish to review some of the following works.

Alexander, D. C.: The Practice and Management of Industrial Ergonomics. Englewood Cliffs, N.J.: Prentice-Hall. 385 pages (1986).

Chaffin, D. B.; Anderson, G.B.J.: Occupational Biomechanics, New York: John Wiley & Sons. 454 pages (1984).

Grandjean, E.: Fitting the Task to the Man: An Ergonomic Approach. 4th ed. London: Taylor & Francis. 363 pages (1988).

Konz, S.: Work Design: Industrial Ergonomics. Worthington, Ohio: Publishing Horizons. 543 pages (1990).
Kroemer, K.H.E.; Kroemer, H.J.; Kroemer-Elbert, K.E.: Engineering Physiology. Bases of Human Factors; Ergonomics. 2nd ed. Amsterdam: Elsevier. 247 pages (1986).
Panero, J.; Zelnik, M.: Human Dimensions and Interior Space. New York: Watson-Guptil Publications. 320 pages (1979).
Rodgers, S.H., Ed.: Ergonomic Design for People at Work, Vol. I, Van Nostrand Reinhold (now Wiley and Sons), New York (1986).
Sanders, M.S.; McCromick, E.J.: Human Factors in Engineering and Design, McGraw-Hill. 790 pages (1993).

REFERENCES

Auburn Engineers, Inc.: Ergonomics Design Guidelines. Auburn, AL: Auburn Engineers, Inc. 374 pages (1997).
Ayoub, M.M.; Selan, J.L.; Jiang, B.C.: A Mini-Guide for Lifting, Texas Tech University Lubbock, TX, USA (1983).
Chapanis, A.: The International Ergonomics Association: Its First 30 Years, Ergonomics, Vol. 33, pp. 275–282 (1990).
Chapanis, A.: The Research for Relevance in Applied Research, in Singleton, W.T., Fox, J.G. and Witfield, D. (Eds.), Measurement of Man at Work, pp. 1–14, London: Taylor & Francis (1971).
Davis, H.; Rodgers, S.: Ergonomic Design for People at Work, Vol. 1: pp. 3-9. Van Nostrand Reinhold Company (now Wiley and Sons), New York.
Drillis, R.J.: Folk Norms and Biomechanics, Human Factors, Vol. 5, pp. 427–441 (1963).
Konz, S.: Vision of the Workplace: Part II, International Journal of Industrial Ergonomics, Vol. 10, pp. 139–160 (1992b).
NASA: Anthropometric Source Book. Vol. I-III (Reference Publication 1024). Edited by the staff of the Anthropology Research Project, Webb Associates, 1167 pages. Yellow Springs, Ohio: NSA Scientific and Technical Information Office, (1978).
Nichols, D.L.: Mishap Analysis in Ferry, T.S. and Weaver, D.A. (Eds.), Directions in Safety, Springfield, Illinois: Charles C. Thomas (1976).
NIOSH: Work Practices Guide for Manual Lifting, NIOSH Technical Report, No. 81-122, National Institute for Occupational Safety and Health, Cincinnati, OH (1981).
NIOSH: Applications Manual for the Revised NIOSH Lifting Equation. DHHS (NIOSH) Publication No. 94-110, 119 pages. National Institute for Occupational Safety and Health, Cincinnati, OH, USA (1994).
Polish Ergonomics Society: *Ergonomia* 2 (1). 7th International Ergonomic Association Congress (1979).
Putz-Anderson, V.; ed.: Cumulative Trauma Disorder Manual for the Upper Extremity, 128 pages. NIOSH Publications. Cincinnati, Ohio (1987).
Rodgers, S. H.; Ed.: Ergonomic Design for People at Work, Vol. II, Van Nostrand Reinhold (now Wiley and Sons) New York (1986).
Snook, S.H.: The Design of Manual Material Handling Tasks. Ergonomics, 21 (No. 2): 975-978 (1978) http://www.tandf.com.
Snook, S.H.; Ciriello, V.M.: The Deisgn of Manual Handling Tasks: Revised Tables of Maximum Acceptable Weights and Forces, Ergonomics, (34)9 (1991).
Waters, T.R.; Putz-Anderson, V.; Garg, A.; Fine, L.J.: Revised NIOSH Equation for the Design and Evaluation of Manual Lifting Tasks, Ergonomics 36:749-776 (1993).

5

BIOLOGICAL AGENTS – RECOGNITION

Jeroen Douwes, PhD, Peter S. Thorne, PhD, Neil Pearce, PhD, and Dick Heederik, PhD

DEFINITION OF BIOLOGICAL AGENTS

Biological agents are usually defined as agents from plant or animal matter or from microorganisms. Airborne and settled particulate material of biologic origin is often referred to collectively as organic dust, and sometimes the term bioaerosol is also used.

Airborne contagion by infectious microorganisms was recognized as an important route of transmission for some biological agent associated diseases more than a century ago. Later, it was recognized that microorganisms might not only be infectious, but may also elicit responses through toxins produced as part of their metabolism or toxins that are part of the structure of the organism itself. It is now well understood that the effects of exposure to biological agents can include contagious infectious diseases, acute toxic effects, allergies, and cancer.

Sometimes, a particular niche is created in an industrial process through specific interactions of the materials used, the temperature range, and the acidity of the environment, that results in exposure to one particular group of microorganisms. An example is the composting industry, where regular exposure occurs to the thermophilic fungus *Aspergillus fumigatus*, which can lead to allergic and toxic responses in exposed composting workers. Other examples of specific niches created as part of an industrial process, are oil emulsions used in the metal industries to cool objects during drilling, honing, etc. The composition of the emulsion, the acidity, and temperature can preferentially support the growth of specific microorganisms.

In other situations, exposure occurs to complex mixtures of toxins and allergens, and a wide range of potential health effects have to be considered. A classic example is the grain processing industry where exposure occurs to grain dust, containing toxins and allergens from the cereal grains, a wide range of microorganisms and their toxins, insects, etc. This exposure has been mainly associated with non-allergic asthma and organic dust toxic syndrome, but specific allergic respiratory symptoms have also been described.

An important development in terms of new work related exposures to biological agents is the use of microorganisms in biotechnological processes. Traditionally, microorganisms have been used in food processing industries to modify structure, texture, or taste of food products. The use of yeast (Saccharomyce serevisiae) in bread processing is one of the well-known examples. Isolation of enzymes from microorganisms that have beneficial effects has been the next step. Today, there is a clear trend to increase production of these enzymes by genetic modification of the organisms. This leads to highly purified enzyme preparations for industrial use. However, many of those enzymes are potent allergens that may cause allergic asthma and rhinitis in workers handling the enzymes and/or the products that are prepared from these enzyme preparations.

Biological agents comprise some classic pathogens, as well as well-known organic dusts from agricultural activities, but also include more recently created agents from biotechnological processes.

INFECTIOUS DISEASES

As an integral part of the ecosystem, we cannot be isolated from its smallest creatures, while at work or away. The workplace, broadly defined, has always been an important source of infections and infectious diseases. Infections can arise from hazards that are specific to that worksite. For some occupations, the hazards are readily recognized, while for others they may be obscure. Veterinary employees, health care workers, and biomedical researchers studying infectious agents have exposures that are largely known, and infection control strategies can limit their risks. For others, such as farmers, foresters, or soldiers, specific risks may be less obvious. Sex workers (prostitutes) face considerable risks but with minimal infection controls.

The world population has grown from 350 million at the time of the Black Death in 1350 to 6 billion today. As population densities have increased, so has the potential for epidemics transmitted by person to person contact. Occupational infections may arise as a result of work settings bringing people into close proximity with other people or animals, thus facilitating the transmission of microorganisms. Occupational infectious diseases attributable to the clustering of people affect workers in daycare centers, schools, nursing homes, hospital wards, correctional facilities (prisons), dormitories, military barracks, shelters for the homeless, cruise ships, and refugee camps (MMWR 1990b, 1997b, 1997c, Frost 1998). In recent years, many office settings have placed large numbers of workers in a "great room," partially enclosed in small cubicles. Such dense occupancy may increase transmission of diseases. This is especially true for diseases with annual outbreaks, such as influenza and "winter stomach flu." The latter, although carrying the moniker "flu" is unrelated to the influenza virus, and is usually associated with the genus Norwalk-like viruses (family Caliciviridae) (LeBaron et al., 1990).

People who engage in foreign travel as part of their work face greater risks of developing a variety of common local communicable diseases. This is particularly true of those traveling to less developed nations and remote areas. Soldiers, aid workers, engineers, journalists, diplomats, and researchers are included in this gap. The risks stem primarily from contact with carriers and exposures through consumption of contaminated food and water. In order to reduce the risk of infection, these groups of workers are often immunized against an array of infectious agents prior to international travel (CDC Travelers' Website).

Many workers are at risk of infectious diseases that rarely affect the general population. Examples include zoonotic diseases such as Q-fever, rabies, leptospirosis, and brucellosis faced by abattoir (slaughterhouse) workers, zookeepers, animal handlers, and veterinarians (MMWR 1997a, 1998a, Regnery 1995). Workers such as foresters, field biologists, game wardens, hunters and trappers who spend time in wooded areas, experience tick-borne and mosquito-borne illnesses with increased frequency over the general population. These illnesses include the arboviral encephalitides, Rocky Mountain Spotted Fever, Lyme Disease, and ehrlichiosis (Azad and Beard 1998, CDC Arboviral Website, MMWR 1990a).

Agents

An infection involves more than simple colonization. Infection means that an organism is present in the host and is replicating, with the result that the host develops subclinical or clinical disease. Pathogenicity refers to the ability of an agent to cause disease, and virulence refers to the communicability, morbidity, and mortality of an agent in populations, or the clinical severity in individuals. Occupational infectious diseases arise from viruses, bacteria (including rickettsia, chlamydia, and more complex bacteria), fungi, protozoa and helminths, listed in order of increasing size and complexity.

Viruses are 25 to 250 nm microorganisms incapable of growth or reproduction outside of living cells. They are not considered living, and are not classified within the five traditional kingdom taxonomy (Monera, Protista, Fungi, Plantae, Animalia) or the modern molecular sequence-based three domain taxonomy (Archaea, Eubacteria, Eukaryota). They reproduce by utilizing the host cell to replicate their genetic material (NCBI Taxonomy Website).

Bacteria are prokaryotic unicellular microorganisms forming the domain Eubacteria. Rickettsia (order: Rickettsiales) and chlamydia (order: Chlamydiales) are obligate intracellular parasitic bacteria less complex than most other bacteria. Most bacteria live in the environment as autotrophic (using inorganic materials as nutrients), saprophytic (obtaining nutrients from dead organic matter), or parasitic organisms, and range in size from 150 nm to 250 μm (Holt et al., 1994).

Fungi are a large group of microorganisms in the domain Eukaryota. Fungi resemble plants in that they extract energy from their environment, but they lack chlorophyll. They exist as saprophytes or parasites and include molds, yeasts, rusts, and smuts. Fungi consist of a thallus, or vegetative body that grows by extending filaments, referred to as hyphae. Collectively, the hyphal mass comprises the mycelium. At further stages of development, the mycelium can form other structures such as mushrooms or rhizomorphs. In soil, the hyphal mass can cover huge expanses and weigh hundreds of kilograms. Reproduction can be asexual or sexual. Asexual reproduction occurs via budding or sporulation. Spores can be released into the air in great numbers and serve to establish colonies at distant sites.

Also, in the domain Eukaryota are the parasites. The term "parasite" is a convenience designation and includes primarily protozoa and helminths. The parasitic protozoa are a diverse grouping of usually unicellular organisms parasitic to humans and animals. Protozoa often have very complex life cycles that include sexual and asexual reproduction and varying morphologies in the reservoir (e.g., pigs), vector (e.g., mosquitoes), and host (e.g., humans).

The helminths are macroscopic multicellular worms and flukes that infect higher organisms. Included are nematodes, trematodes, and cestodes.

Transmission

The process of infection represents the transmission of an infectious agent from a reservoir to a susceptible host, directly or indirectly. Between the reservoir and the host may be a source (e.g., feces), an intermediate host (e.g., animal), or a vector (e.g., insect). Transmission occurs through direct contact, a common vehicle, airborne transmission, or vector-borne transmission.

- Transmission by contact occurs between a source and the ultimate host through direct contact, as person to person or animal to person or through indirect contact such as with transfer from individual to individual via fomites or inanimate objects (e.g., a contaminated syringe).
- Transmission through a common vehicle refers to transmission from an individual to many hosts via a vehicle such as food or water.
- Airborne transmission occurs locally from large aerosols (> 10 μm) that do not remain airborne for long. More significant are the smaller aerosols and droplet nuclei less than about 5 μm that remain airborne for longer periods and may travel considerable distances. These can facilitate infection from a single individual to a large number of hosts.
- Vector-borne transmission is most commonly associated with mosquitoes, flies, fleas, ticks, and lice. Vectors carry the agent from reservoir to host or from reservoir to a source that may then infect a host. This route of transmission is exemplified by mosquitoes taking a blood meal from an infected host (e.g., birds) and passing the agent to a naive host (human). Flies also carry fecal microorganisms to a food item subsequently consumed by humans.

The route of entry of the agent into the host may be via the skin (surface or percutaneous injection), through mucous membranes, by inhalation into the upper or lower respiratory tract, via the urinary tract or via the gastrointestinal tract. Environmental factors can play a role in the transmission of an agent and include temperature, moisture levels, available nutrients, air movement, and effectiveness of environmental controls. In some cases, environmental factors may interact with host susceptibility factors to increase the communicability of an infectious agent.

Workers at Risk

Infectious diseases are a concern for a wide variety of workers (Garibaldi and Janis 1992, Gerberding and Holmes 1994, NIOSH Website). Table 5-1 lists the infectious diseases that affect those workers outside of health care, child and eldercare, biomedical research and commercial sex workers. Included with the diseases is a listing of the workforces at greatest risk of the disease, the organisms and their classification, and the principal routes of exposure that may lead to infection. A cross indicates federally notifiable diseases in the United States after the disease name. Extensive information is available elsewhere for each disease and each microorganism listed in Tables 5-1 and 5-2. For further details regarding disease vectors, intermediate hosts, their life cycles, disease signs, symptoms, diagnosis, and treatment the reader is referred to texts in infectious disease, internal medicine, and clinical microbiology (e.g., Faucci et al., 1998, Mandell et al., 2000; Murray, 1999), and the excellent web sites maintained by the U.S. Center for Disease Control and Prevention.

Some workers are at risk because their job brings them into contact with many others, increasing the opportunities for transmission of infections (see Table 5-1). Other workers are exposed to animals or animal carcasses in the course of their work, and are at risk for zoonotic diseases (diseases communicated from animals to humans) and those diseases for which animals may harbor the vector. These would include veterinarians, zookeepers, animal handlers, farmers, ranchers, livestock workers, trappers, hunters, abattoir workers, rendering plant workers, tanners, and pet store clerks. Those workers who spend a significant portion of their time outdoors, especially in wooded or prairie areas, are at heightened risk for developing tick-borne and mosquito-borne infectious diseases. These diseases include Lyme Disease, Rocky Mountain Spotted Fever, the arboviral encephalitides (Moore et al., 1993), and ehrlichiosis (Mueller-Anneling et al., 2000). Several of the diseases listed in Table 1 arise from inhalation of fungal spores in the course of handling decaying matter, feces, compost, or soil. These diseases include aspergillosis, histoplasmosis, blastomycosis, coccidioidomycosis, and adiaspiromycosis (MMWR 1993, 1999, NIOSH 1997). Finally, it is important to note that one can contract these diseases in the pursuit of leisure activities or in the course of every day life apart from the work environment. Examples include hiking in wooded areas or raising animals as a hobby. The diseases listed here are those for which various jobs significantly elevate the risks.

Infectious diseases experienced by workers in health care, nursing homes, and daycare represent the diseases of the populations they serve. Table 5-2 lists the most important diseases in these work forces organized by the class of agent, with a listing of the specific causal organism, its classification, and the principal route of exposure. Increasingly, these labor forces are immunized and educated in disease prevention measures, with the effect that the most serious diseases in this population are on the decline. It is now common for health care workers in developed nations to receive vaccinations for Hepatitis B, measles, mumps, rubella, pertussis,

Table 5-1
Occupational infectious diseases exclusive of those posing risks primarily to health care workers, day care workers, and sex workers

Disease	Workforces at Greatest Risk	Organism(s)	Disease type	Route of Exposure
Acinetobacter pneumonia	foundry workers, machinists	*Acinetobacter calcoaceticus*	bacterial	inhalation
Adiaspiromycosis	grain handlers, warehouse workers, farmers	*Chrysosporium parvum*	fungal	inhalation
Amebiasis (amebic dysentery)	travelers, aid workers, soldiers	*Entamoeba histolytica*	protozoal	fecal-oral, food/waterborne, fomites
Anthrax†	farmers, abattoir workers,	*Bacillus anthracis*	bacterial	dermal, inhalation
Arboviral Encephalitides: Eastern equine encephalitis†, EEE Western equine encephalitis†, WEE LaCrosse encephalitis†, LAC St. Louis encephalitis†, SLE Powassan encephalitis, POW Tick-borne encephalitis, TBE	foresters, field biologists, farmers, veterinarians, animal handlers	Alphavirus (EEE, WEE) Bunyavirus (LAC) Flavivirus (SLE, POW, TBE)	viral	mosquito bites (EEE, WEE. SLE, LAC), tick bites, (POW, TBE)
Ascariasis	sewage workers, farmers, abattoir workers	*Ascaris lumbricoides*	helminthic	fecal-oral, food/waterborne
Aspergillosis	compost workers, farmers, grain handlers	*Aspergiollus niger, A. fumigatus, A. flavus*	fungal	inhalation
Blastomycosis	hunters, trappers, field biologists	*Blastocystis dermatitidus*	fungal	inhalation, innoculation, dog bite
Brucellosis†	ranchers, abbatoir workers, rendering plant workers, veterinarians, farmers, dairy farmers	*Brucella melitensis, B. abortus, B. suis, B. canis*	bacterial	dermal
Cat-scratch disease	veterinarians, animal handlers, pet store clerks	*Bartonella henselae*	bacterial	cat scratch or bite
Coccidioidomycosis†	archeologists, construction workers	*Coccidioides immitis*	fungal	inhalation
Colorado tick fever	hunters, field biologists	*Orbivirus*	viral	tick bite
Cryptococcosis	pigeon breeders, animal handlers, zookeepers	*Cryptococcus neoformans*	fungal	inhalation of bat and bird droppings

Table 5-1 (Cont'd)
Occupational infectious diseases exclusive of those posing risks primarily to health care workers, day care workers, and sex workers

Disease	Workforces at Greatest Risk	Organism(s)	Disease type	Route of Exposure
Cryptosporidiosis†	sewage workers, ranchers, animal handlers	*Cryptosporidium parvum*	protozoal	fecal-oral
Dengue fever	soldiers, aid workers, travelers	*Flavivirus*	viral	mosquito bite
Echinococcosis	farmers, veterinarians, pet store clerks, animal handlers, sheep farmers	*Echinococcus granulosis, E. multilocularis*	helminthic	fecal-oral
Equine morbillivirus	veterinarians, animal handlers, farmers, stable hands	*Morbillivirus*	viral	inhalation
Fish handler's disease, Seal finger	fish, abattoir workers, poultry handlers, veterinarians, rendering plant workers	*Erysipelothrix rhusiopathiae*	bacterial	dermal
Fish tank granuloma	fishermen, pet store clerks, commercial divers	*Mycobacterium marianum*	bacterial	inoculation
Hantavirus pulmonary syndrome†	warehouse workers, farmers, animal handlers, field biologistss	*Hantavirus*	viral	inhalation
Helicobacter	sewage workers	*Helicobacter pylori*	bacterial	ingestion
Hepatitis	sewage workers	*Hepatitis A, B, C, E, G virus*	viral	inoculation, fecal-oral
Histoplasmosis	poultry handlers, pigeon	*Histoplasma capsulatum*	fungal	inhalation
Human monocytic ehrlichiosis	foresters, field biologists, farmers,	*Ehrlichia chaffeensis (HME),*	bacterial	tick bite
Human granulocytic ehrlichiosis	veterinarians	*Ehrlichia Sp. (HGE)*		
Legionellosis† (Legionnaire's disease, Pontiac	ventilation engineers, machinists, grocers, office workers	*Legionella pneumophilia, Legionnella* Sp.	bacterial	inhalation
Leishmaniasis, cutaneous	soldiers, aid workers, travelers	*Leishmania donovani, L. major, L. aethiopica, L. mexicana, L. braziliensis, L. donovani*	protozoal	sandfly bite
Leptospirosis	veterinarians, farmers, sewage workers, trappers, abattoir workers, miners	*Leptospira interrogans*	bacterial	inoculation, bite
Lyme disease†	foresters, hunters, trappers, ranchers	*Borrelia burgdorferi*	bacterial	tick bite
Malaria	soldiers, aid workers, travelers	*Plasmodium falciparuum, P. malaria, P. ovale, P. vivax*	protozoal	mosquito bite
Melioidosis	rice farmers, soldiers	*Pseudomonas pseudomallei*	bacterial	inoculation

Table 5-1 (Cont'd)
Occupational infectious diseases exclusive of those posing risks primarily to health care workers, day care workers, and sex workers

Disease	Workforces at Greatest Risk	Organism(s)	Disease type	Route of Exposure
Monkey B virus	zookeepers, animal handlers, veterinarians	*Herpes B virus*	viral	monkey bites or scratches
Nonophyetiasis	fish handlers	*Nanophyetus salmincola*	helminthic	ingestion
Nocardiosis	field biologists, foresters	*Nocardia asteroides, N. braziliensis*	bacterial	inoculation, inhalation
Orf	sheep farmers, veterinarians, abattoir workers	*Orf virus*	viral	dermal, inoculation
Pasteurellosis	veterinarians, pet store clerks, animal handlers, zookeepers	*Pasteurella multocida, P.* Sp.	bacterial	animal bite, inoculation
Plague[†]	farmers, ranchers, animal handlers, field biologists, zookeepers	*Yersinia pestis*	bacterial	flea bite
Psittacosis[†] (Ornithosis)	pet store clerks, pigeon breeders, zookeepers, veterinarians, poultry handlers	*Chlamydia psittaci*	bacterial	inhalation of bird feces
Q-fever	farmers, ranchers, veterinarians, abattoir workers, rendering plant workers, dairy farmers	*Coxiella burnetti*	bacterial	inhalation, ingestion of contaminated milk
Rabies[†]	veterinarians, ranchers, trappers, farmers, animal handlers, aid workers	*Rabies virus*	viral	animal bite
Rat-bite fever	animal handlers, poultry handlers	*Streptobacillus moniliformis, Spirillum minus*	bacterial	animal bite, inoculation
Relapsing fever	field biologists, veterinarians, trappers	Borrelia hermsii, B. turicatae	bacterial	tick bite
Rocky Mountain spotted fever[†]	field biologists, hunters, foresters, ranchers, trappers, farmers	Rickettsia rickettsii	bacterial	inoculation
Salmonellosis	veterinarians, cooks, pet store clerks, poultry handlers	Salmonella *Sp.*	bacterial	fecal-oral
Silicotuberculosis	miners, sand blasters, construction workers, foundry workers	*Mycobacterium tuberculosis* with silica	bacterial	inhalation
Sporotrichosis	lumber sorters, foresters, ranchers, horticulturists	*Sporothrix schenkii*	fungal	inhalation
Streptococcal flu, hearing loss	swine farmers	*Streptococcus suis*	bacterial	dermal, inhalation
Streptococcal pneumonia[†§]	dairy farmers	*Streptococcus zooepidemicus*	bacterial	ingestion of contaminated milk
Swine influenza	swine farmers	Swine influenza virus	viral	inhalation

Table 5-1 (Cont'd)
Occupational infectious diseases exclusive of those posing risks primarily to health care workers, day care workers, and sex workers

Disease	Workforces at Greatest Risk	Organism(s)	Disease type	Route of Exposure
Tetanus†	farm workers	*Clostridium tetani*	bacterial	inoculation
Toxoplasmosis	abattoir workers, animal handlers, veterinarians, pet store clerks	*Toxoplasma gondii*	protozoal	fecal-oral, inoculation
Tuberculosis†	farm workers, law enforcement officers	*Mycobacterium tuberculosis homine*	bacterial	inhalation
Tuberculosis†	farmers, veterinarians, abattoir workers, zookeepers	*Mycobacterium bovis*	bacterial	inhalation, ingestion of contaminated milk
Tularemia	hunters, trappers, ranchers, abattoir workers, farmers, soldiers, veterinarians, cooks	*Francisella tularensis*	bacterial	arthropod bite, inhalation
Typhus, epidemic	soldiers, aid workers, travelers	*Rickettsia prowazekii*	bacterial	louse bite
Typhus, murine	grain handlers, warehouse workers, field biologists, foresters, animal handlers	*R. typhi, R. felis*	bacterial	flea bite
Yellow fever†	soldiers, aid workers, travelers	Flavivirus	viral	mosquito bite
Yersiniosis	animal handlers, pet store clerks, zookeepers	*Yersinia enterocolitica, Y. pseudotuberculosis*	bacterial	inoculation, ingestion

†Notifiable diseases at the national level in the United States (MMWR 1997c)

§Invasive, systemic, or drug-resistant cases only.

Table 5- 2
Occupational infectious diseases affecting workers in health care, nursing homes, day cares, dentistry, and biomedical research

Disease	Organism	Family	Vector, route of exposure
Viral diseases			
Acute viral gastroenteritis (stomach flu)	Norwalk-like viruses	Caliciviridae	Fecal-oral, food/waterborne outbreaks, fomites, inhalation of aerosols
Adult T-cell leukemia	Human T-cell lmyphotrophic virus (HTLV-I, HTLV-II)	Retroviridae	Contact with body fluids
AIDS[†] (acquired immunodeficiency syndrome)	Human immunodeficiency virus (HIV-I)	Retroviridae	Contact with body fluids
CMV mononucleosis	Cyomegalovirus	Herpesviridae	Contact with body fluids
Fifth disease	Parvovirus B19	Parvoviridae	Contact with respiratory secretions
Hepatitis A[†]	Hepatitis A virus	Picornaviridae	Fecal-oral, food/waterborne outbreaks
Hepatitis B[†]	Hepatitis B virus	Hepadnaviridae	Contact with blood products
Hepatitis C[†]	Hepatitis C virus	Flaviviridae	Contact with blood products
Hepatitis E	Hepatitis E virus	Caliciviridae	Fecal-oral, food/waterborne outbreaks
Hepatitis G	Hepatitis G virus	Flaviviridae	Contact with blood products
Infectious mononucleosis (mono)	Epstein Barr virus	Herpesviridae	Transfer of saliva, contact with secretions
Influenza (flu, grippe)	Influenza virus	Orthomyxoviridae	Inhalation of small aerosols
Measles[†] (Rubeola)	Measles virus	Paramyxoviridae	Inhalation of small aerosols
Mumps[†]	Mumps virus	Paramyxoviridae	Contact with saliva, inhalation of small aerosols
RSV pneumonia	Respiratory synctial virus	Paramyxoviridae	Inhalation of large aerosols
Rubella[†] (German measles)	Rubella virus	Togaviridae	Inhalation of small aerosols
Varicella (chickenpox)	Varicella-zoster virus	Herpesviridae	Inhalation of small aerosols
Bacterial diseases			
Haemophilus influenza[†§] (HIB)	*Haemophilus influenza type* B	Pasteurellaaceae	Inhalation of aerosols, contact with respiratory secretions
Impetigo, toxic shock syndrome	*Staphylococcus aureus,* S. sp.	Staphylococcaceae	Dermal invasion
Meningococcal Disease[†]	*Neisseria meningitidis*	Neisseriaceae	Inhalation of large aerosols, contact with respiratory secretions
Pertussis[†] (whooping cough)	Bordetella pertussis	Alcaligenaceae	Inhalation of aerosols
Pseudomembranous colitis	Clostridium difficile	Clostridiaceae	Fecal-oral

Table 5- 2 (Cont'd)
Occupational infectious diseases affecting workers in health care, nursing homes, day cares, dentistry, and biomedical research

Disease	Organism	Family	Vector, route of exposure
Bacterial diseases (Cont'd)			
Streptococcal pharyngitis, impetigo, scarlet fever†§	*Streptococcus* Sp. (group A or B)	Streptococcaceae	Formites-oral
Tuberculosis†	*Mycobacterium tuberculosis*	Mycobacteriaceae	Inhalation of droplet nuclei
Parasitic diseases			
Cryptosporidiosis†	*Cryptosporidium parvum*	Cryptosporidiidae	Fecal-oral, food/waterborne
Giardiasis	*Giardia lamblia*	Hexamitidae	Fecal-oral, food/waterborne

†Notifiable diseases at the national level in the United States (MMWR 1997c)
§Invasive, systemic, or drug-resistant cases only.

tetanus, and annually for influenza (Bolyard et al., 1998). Infectious diseases such as acute viral gastroenteritis, influenza, and streptococcal pharyngitis annually effect many health-care workers but are rarely fatal among immunocompetent individuals.

In the U.S., the occupational infectious diseases of greatest concern are hepatitis B, hepatitis C, multidrug-resistant tuberculosis, and AIDS. This is largely due to their severity and prevalence in the patient population. According to the U.S. Centers for Disease Control and Prevention, 56 health care workers in the U.S. have been documented as having seroconverted to HIV following occupational exposures (CDC Hepatitis Website). In addition, there is great concern regarding the risk of contracting multidrug-resistant tuberculosis for the 6 million U.S. health care workers and for those working in correctional facilities. There are an estimated 9,000 deaths/yr. in the U.S. due to chronic liver disease caused by hepatitis C virus and 5,000 deaths/yr. from hepatitis B. The CDC estimates that there are 36,000 new hepatitis C infections in the U.S. per year and 140,000 to 320,000 new hepatitis B infections. The proportions of infections that are symptomatic are 25% to 30% for hepatitis C and 50% for hepatitis B (CDC 1996). The prevalence of infection with Hepatitis C virus among health care workers is 1% to 2%, about 10-fold lower than the prevalence for HBV infection (MMWR, 1998b). The availability and widespread use of a vaccine against HBV has reduced incident cases of hepatitis B.

Diseases affecting sex workers are many of those listed in Table 5-2, as well as a wide variety of sexually-transmitted diseases and diseases associated with drug abuse and poverty. These include vulvovaginitis associated with *Candida albicans*, dermal and systemic diseases caused by Herpes Simplex Virus, syphilis from *Treponema pallidum,* urethritis from *Neisseria gonorrhoeae* and *Chlamydia trachomatis*, venereally-transmitted cytomegalovirus (CMV)-mononucleosis, Human Papiloma Virus-associated cervical cancer and AIDS, coupled with its associated opportunistic infections. For descriptions of sexually transmitted infectious diseases experienced by sex workers, the reader is referred to infectious disease textbooks (Mandell et al., 2000; Murray, 1999).

Workers who study agents of biological warfare and bioterrorism have peculiar risks specific to the agents they study. Included are *Bacillus anthracis, Yersinia pestis, Brucella* Sp., Alpha viruses, Filoviruses (Marburg, Ebola), Arenaviruses, and microbial and plant toxins. These workers are vaccinated against 20 or more agents, work under extreme biosafety conditions (BL-4, see chapter 7) with a myriad of administrative and engineering controls, and have a high degree of medical surveillance.

Foreign travelers (journalists, diplomats, engineers, tour guides, etc.), aid workers, and soldiers are at risk for diseases that are often rare in their country of origin. Included are amebiasis, dengue fever, leishmaniasis, malaria, tuberculosis, typhus, yellow fever, and many other diseases specific to certain geographic regions. There are country-specific travel recommendations and vaccinations available at most academic medical centers and through the U.S. CDC (CDC Travelers' Website).

Incidence

For certain diseases that carry a high incidence, such as acute viral gastroenteritis from Norwalk-like viruses or influenza (influenza virus), cases may arise from both community outbreaks and occupational exposures. Three of the most prevalent life threatening infectious diseases associated with occupational exposures in the U.S. are tuberculosis, viral hepatitis from Hepatitis B Virus, and Lyme disease (CDC Vectors Website). Of course, these diseases also occur outside of the workplace. Some infectious diseases with an occupational association have exhibited a steady incidence over the past two decades (e.g., Legionellosis). Others occur episodically depending upon environmental conditions (e.g., Hantavirus pulmonary syndrome), epizootic outbreaks (epidemics among animals), or human error (e.g., foodborne illnesses).

There is little data available from which to determine incidence or prevalence for most occupational infectious diseases. In the United States, cases of federally notifiable diseases must be reported to the CDC. Table 5-3 lists data for 1997 from the CDC for those infectious diseases that pose a significant occupational risk. These cases are reported from the entire U.S. population (a denominator of 267 million in 1997), but are important risks for the civilian workforce of 138 million. The degree to which each of these case numbers represents occupational exposure varies by disease and is largely unknown. For several specific diseases and job titles, there are incidence data that were collected to demonstrate the efficacy of preventive

Table 5-3
Reported cases in the United States (1997 data) for infectious diseases with a significant occupational component. Data specific for occupational cases are not available.‡

Disease	Reported U.S. Cases
Arboviral encephalitides	47
Brucellosis	98
Cryptosporidiosis	2,566
Hepatitis B	10,416
Legionellosis	1,163
Lyme disease	12,801
Plague	4
Psittacosis	33
Rabies	2
Rocky Mountain spotted fever	409
Tetanus	50
Tuberculosis	19,851

‡The denominator for calculating population-wide incidence was 267,637,000.

measures. For instance, for hepatitis B in health care workers, there were about 8,700 cases per year (145 cases per 100,000) in the United States prior to widespread vaccination and implementation of educational and administrative programs. The CDC has documented that by 1994, this had declined to under 1,000 annual cases (16 cases per 100,000).

RESPIRATORY DISEASES

Respiratory symptoms due to organic dust exposure can range from acute mild conditions, which (at least initially) hardly affect daily life, to severe chronic respiratory diseases that require specialist care. Small acute changes in lung function and mild respiratory symptoms are predictive of chronically affected respiratory health, and are, therefore, also of potentially major concern. However, since these early mild respiratory symptoms do not usually lead to absence of work, they often go unnoticed or are ignored by hygienists and physicians. One such example is moderate exposure to endotoxin, which leads to relatively mild work related symptoms and small, acute (reversible) cross-shift changes in lung function, but not to sick leave. After long term exposure, however, these symptoms will develop into a more serious condition characterized by chronic airway obstruction and possibly bronchitis. These more serious conditions can be prevented when minor symptoms are taken seriously, and consequently preventative action occurs.

There are two major divisions of respiratory disease (see chapter 1 for more details). **Chronic obstructive respiratory disease** is characterized by airway obstruction, subsequent expiratory airflow limitation, and the presence of respiratory symptoms. **Chronic restrictive airway diseases** (e.g., fibrosis, hypersensitivity pneumonitis) can occur as well (either in combination, or separately from airway obstruction), but are less common in *organic* dust exposed workers. The more serious obstructive conditions can be grouped into two main categories, i.e., Chronic Obstructive Pulmonary Diseases (COPD; including chronic bronchitis and emphysema) and asthma. In COPD, airway obstruction is mainly irreversible, whereas in asthma, airway obstruction is mainly reversible.

Generally, occupationally related respiratory symptoms result from airway inflammation caused by specific exposures to toxins, pro-inflammatory agents, or allergens. Based on the underlying inflammatory mechanisms and subsequent symptoms, a distinction between allergic and non-allergic respiratory diseases can be made. Non-allergic respiratory symptoms reflect a non-immune-specific airway inflammation, whereas allergic respiratory symptoms reflect an immune-specific inflammation in which various antibodies (IgE, IgG) can play a major role in the inflammatory response. For instance, in the literature, the terms 'asthma' and 'asthma-like disorders' are often used referring to variable airflow limitations with allergic (IgE mediated) and non-allergic underlying mechanisms, respectively (both conditions will be further described later in this chapter). In practice, it is not always easy to make a distinction between these disease mechanisms because symptoms are often very similar (shortness of breath, wheezing, acute cross shift decrease in lung function, etc.), and exact pathology is usually unknown or, at best, only poorly characterized. In fact, asthma is commonly defined as 'variable airflow obstruction' and the different mechanisms for variable airflow obstruction (allergic and non-allergic) are then considered as subtypes of the general condition of 'asthma'. For example, it is known that exposure to plicatic acid in wood dust from western red cedar causes asthma (variable airflow obstruction) in saw mill workers, but underlying inflammatory mechanisms are currently poorly understood, and both allergic and non-allergic mechanisms have been suggested. The use of skin prick tests and serological determinations of specific antibodies can be used to establish whether allergic responses may be involved. However, specific tests for occupational allergens may not always be commercially available. Techniques such as nasal lavages or sputum induction may be used to assess the nature of the inflammatory responses in the airways. These latter techniques, however, are currently only used in research settings, and not available to the hygienist or physician.

Apart from the classifications described above, various diseases typical for organic dust exposure have been described. Workers exposed to organic dust may develop various respiratory symptoms that can, based on the specific symptom patterns and underlying pathophysiology, be divided into hypersensitivity pneumonitis, organic dust toxic syndrome, and mucous membrane irritations. These conditions may result in, or be part of, the previously mentioned categories of pulmonary diseases (COPD and asthma), and underlying patho-physiology can be distinctively different from non-organic dust induced respiratory symptoms such as the pneumoconioses caused by silica and asbestos. Depending on the nature of the exposure (which usually is a very heterogeneous mix of toxic, pro-inflammatory, and allergenic components), the exposure levels, and host conditions, individual workers may develop different symptoms that may occur on the same occasion or at different times. Thus, in real-life, the hygienist or physician will encounter a mixture of these symptoms and diseases (both allergic and non-allergic) in workers from one industry, and sometimes even within individual workers. It is important to acknowledge that pre-existing respiratory conditions or other host factors (e.g., atopy [see later], smoking) may modify the risk of developing work related respiratory symptoms. For instance, an asthmatic worker that has developed asthma due to house dust mite exposure in his or her home may experience work related exacerbation of his or her asthma due to organic dust exposure at levels that do not normally induce any symptoms in other 'healthy' workers.

As previously mentioned, recognition of these various respiratory conditions may not always be easy and specific knowledge of both exposure and the various respiratory health effects are thus necessary. In the next paragraphs, a more detailed description of organic dust-induced occupational respiratory diseases will be given, and where possible, simple diagnostic tools will be suggested that can be applied in hygiene surveys to detect early signs of occupational respiratory diseases. In addition, where known, causative agents will be identified and examples will be given.

Background Immunology

Since immunological and inflammatory reactions underlie most (if not all) airway diseases caused by biological agents, a brief and simplified immunology review is given, but only for those pathways of the immune system that are of relevance to understanding the following sections of this chapter. For a more extensive and complete background on the immunology of the airways, we refer to recent textbooks on immunology and allergology (e.g., Roitt, 1997).

Immune responses are essential to protect the host against foreign agents, in particular exposures to a large variety of infectious microbial agents in our environment, including viruses, bacteria, fungi, and parasites. Usually these responses go unnoticed, not causing any symptoms, but sometimes immune responses can occur in an exaggerated or inappropriate form causing tissue damage that may, in the case of the airways, cause various respiratory diseases such as those described in this chapter. This may occur either due to unusually high exposures to certain offending agents found in certain occupations (e.g., bacterial endotoxin exposure in farmers), or may occur because a worker is hypersensitive to certain agents (e.g., flour allergens in bakeries). For the purpose of this chapter, we make a distinction between two types of immune responses: 1) immune responses in which hypersensitivity plays a major role, involving prior sensitization and antibody mediated inflammation; and 2) non-specific immune responses in which prior sensitization is not required and inflammation occurs as a result of exposure to pro-inflammatory or toxic agents. The latter type will be discussed further in the section on non-allergic respiratory diseases. In each case, a variety of cells and soluble factors are involved in the subsequent inflammatory reactions. Many of the soluble factors (or cytokines) are produced by activated cells in order to activate, attract or inhibit other cell types.

In the context of the diseases discussed in this chapter, the most important cells are T lymphocytes, B lymphocytes, mast cells, neutrophils, eosinophils, and macrophages. T cells have a very important function in controlling immune responses; the most important functions include: 1) killing infected cells (e.g., virally infected cells); 2) activating various other cells involved in the immune response such as macrophages; and 3) helping with B cell production of antibodies. B cells produce various kinds of antibodies (IgA, IgG, IgE, IgM) upon stimulation by antigens and T cells (T helper cells). Mast cells are cells that both synthesize and store histamine, proteoglycans, and proteases (important mediators of inflammation) in granules and have surface receptors that bind IgE. Neutrophils, eosinophils, and macrophages have an important role in destroying antigens and pathogens by phagocytosis. (However, eosinophils are only weakly phagocytic, but they are important in damaging large pathogens such as intestinal worms.) Eosinophils are usually present in elevated concentrations in type I mediated airway diseases (allergic asthma and rhinitis), whereas in non-allergic airway diseases these cells are usually not present. Activated cells (e.g., eosinophils, mast cells, basophils; all cells that play an important role in allergic asthma and rhinitis) can degranulate and release mediators of inflammation leading to some of the clinical effects observed in airway diseases (such as bronchoconstriction).

HYPERSENSITIVITY

As noted above, hypersensitivity requires prior sensitization, and only susceptible workers will develop the disease, whereas others will not be affected. In most forms of hypersensitivity (type I, II, III), specific antibodies are involved that recognize the offending agent and mediate a further immune response, which, in the lungs, may result in airway inflammation and subsequent respiratory symptoms. Antibodies are bi-functional molecules with one part of the molecule being responsible for antigen recognition and the other for interaction with various cells of the immune system. They circulate in the blood and can mediate local immune responses in the area where the offending agent is entering the host (e.g., the lungs). Hypersensitivity can be divided into four types, but only type I, III, and IV will be discussed here since only these types are involved in the diseases described in this chapter.

Type I

Type I hypersensitivity is often used synonymously with the term allergy. However, in this chapter, we have defined allergy as any hypersensitivity response involving the production of specific antibodies (i.e., IgE and IgG). Type I hypersensitivity involves repeated exposure to allergens in susceptible or atopic workers (see definition of atopy later in this chapter) resulting in IgE production by B cells. This process is called atopic sensitization. Repeated exposures may result in acute symptoms in some (but not all) sensitized workers, which is due to the interaction of allergen with IgE sensitized mast cells and basophils. An acute reaction may be followed by a delayed response involving eosinophil recruitment. Typical examples of type I allergy include hay fever (allergic rhinitis), allergic asthma, eczema, and anaphylactic reactions to bee venom, for example.

Type III

Type III hypersensitivity is caused by IgG antibody immune complexes. Unlike IgE in type I allergy, everyone has high concentrations of IgG antibodies against a large variety of antigens. However, only in susceptible subjects and with repeated exposures of type III inducing antigens (e.g., certain microbial agents) this may lead to IgG immune complexes that are not eliminated, and that can be deposited at different sites in the body where they can induce an inflammatory reaction causing symptoms. Type III reactions (in combination with type IV hypersensitivity, see below) are suspected in cases of hypersensitivity pneumonitis (HP) or farmers lung.

Type IV

Type IV (or delayed) hypersensitivity involves an antigen – T cell interaction, which in turn recruits various other cells of the immune system to the site where the antigen exposure took place. The reaction is more delayed than type I or III hypersensitivity. Formation of granulomas in the lung parenchyma, as observed in HP or farmers lung, is one of the characteristic features of type IV hypersensitivity. Increased T cell levels (in the lung or skin) is another important characteristic of type IV immunity. As with type I and III responses, Type IV immune reactions also require some form of prior sensitization. However, the mechanisms underlying this sensitization process are not very well understood.

Allergic Respiratory Diseases

Several allergic airways diseases can be distinguished including allergic asthma, allergic rhinitis, and hypersensitivity pneumonitis (HP, or extrinsic allergic alveolitis (EAA) as it is often called). Asthma and rhinitis are common allergic airways diseases both in the general and occupational population. HP is a less common condition that is usually not observed in the general population, but may incidentally occur in occupational environments after massive exposures to microorganisms (and fungal spores and actinomycetes in particular). Although hypersensitivity mechanisms are suspected, the exact underlying pathophysiological mechanisms for HP are not clear. Underlying inflammatory mechanisms for allergic asthma and rhinitis are much better understood and involve a process of specific allergic (IgE) sensitization and local allergic inflammation in the upper (rhinitis) or lower (asthma) airways causing subsequent symptoms.

ALLERGIC RHINITIS

Allergic rhinitis is caused by an allergic inflammation in the upper respiratory tract mucosae. Usually the conjunctival mucosa is also affected. Typical symptoms of allergic rhinitis include sneezing, rhinorrhoea, and/or nasal congestion often accompanied by tears, itchy eyes, and red and swollen eyelids (conjunctivitis). Episodic symptoms are the hallmark of allergic rhinitis. Washing swabs or biopsy of the mucosa will usually reveal eosinophils, but some neutrophils are usually also present (Schenker et al., 1998). Rhinitis is associated with asthma, although they are separate conditions and the overlap between the two conditions is not very great, at least not in the general population (ISAAC, 1998). Although it is assumed that many people are affected by rhinitis, it usually does not lead to severe medical conditions and/or hospital admissions. The main focus in this section is asthma.

ALLERGIC ASTHMA

The word "asthma" comes from a Greek word meaning "panting" (Keeney, 1964), but reference to asthma can also be found in ancient Egyptian, Hebrew, and Indian medical writings (Ellul-Micallef, 1976; Unger and Harris, 1974). Asthma has puzzled and confused physicians and patients from the time of Hippocrates to the present day. There were clear observations of patients experiencing attacks of asthma in the 2nd century, and evidence of disordered anatomy in the lung as far back as the 17th century (Willis, 1678). Ramazzini was probably one of the first (early 18th century) to describe occupational asthma in his book '*De Morbis Artificum Diatriba*' where he, for example, speaks of those who are exposed to flour dust, which makes them liable to 'coughs, short of breath, hoarse and finally asthmatic,' or "bakers asthma" as we know it today.

The definition of asthma has become more complex as our understanding of its pathophysiology has increased. However, despite this increased complexity, the basic characteristic features of reversible airflow obstruction by which one recognizes or diagnoses the disease, has changed little over recent years. The definition initially proposed at the Ciba Foundation conference in 1959 (CIBA, 1959) and endorsed by the American Thoracic Society in 1962 (ATS, 1962) is that "asthma is a disease characterized by wide variation over short periods of time in resistance to flow in the airways of the lung." Although these features receive less prominence in some current definitions, as the importance of airways inflammation is appropriately recognized, they still form the basis of the recent World Health Organization/National Heart, Lung and Blood Institute definition (Gina, 1994) of asthma as:

> "... a chronic inflammatory disorder of the airways in which many cells play a role, in particular mast cells, eosinophils and T lymphocytes. In susceptible individuals, this inflammation causes recurrent episodes of wheezing, breathlessness, chest tightness and cough, particularly at night and/or in the early morning. These symptoms are usually associated with widespread but variable airflow limitation and are at least partly reversible either spontaneously or with treatment. The inflammation also causes an associated increase in airway responsiveness to a variety of stimuli."

These three components: chronic airways inflammation, reversible airflow obstruction, and enhanced bronchial reactivity, form the basis of current definitions of allergic asthma. They also represent the major pathophysiological events leading to the symptoms of wheezing, breathlessness, chest tightness, cough, and sputum production by which physicians clinically diagnose this disorder. Although the mechanisms by which these pathophysiological events occur are still not completely understood, a theoretical paradigm has evolved in which it is believed that the fundamental etiological mechanism is that allergen exposure produces atopic (or IgE mediated) sensitization (as defined later), and that continued exposure results in asthma through the development of airways inflammation, which leads to bronchial hyperresponsiveness (BHR; as defined later) and reversible airflow obstruction. However, it is acknowledged that not all cases of asthma fit this paradigm, e.g., many occupational causes of asthma do not appear to involve atopy as will be discussed later. Therefore, a distinction is made between allergic and non-allergic asthma.

Occupational asthma

Occupational asthma can be defined as a "disease characterized by variable airflow limitation and/or airway hyperresponsiveness due to causes and conditions attributable to a particular occupational environment and not to stimuli encountered outside the workplace" (Bernstein et al., 1993). Occupational asthma can lead to temporary sick leave, and in severe cases, even to permanent disability.

Allergy, Allergic inflammation, Atopy, and Bronchial hyperresponsiveness

Since allergy, allergic inflammation, atopy and bronchial hyperresponsiveness are main features of both allergic asthma and rhinitis, these issues are defined and discussed separately in this paragraph. The focus is on type I or IgE mediated allergy only, since IgE mediated inflammation is the main cause of allergic asthma and rhinitis.

Allergy

Allergy can be defined as acute or chronic symptoms and illness due to exposure to external agents (allergens) to which the exposed subject is hypersensitive, because of a preceding specific immunologic sensitization. In case of allergic asthma and rhinitis, immunological sensitization is characterized by the production of specific IgE antibodies directed against a particular allergen. However, not all IgE sensitized subjects experience symptoms after exposure. It is, therefore, important to distinguish on the one hand 'allergy' as the occurrence of allergic symptoms and/or well-diagnosed allergic disease, and on the other hand 'allergic sensitization' as a specific feature of someone's immune status, assessed by in vivo or in vitro diagnostic tests (see below). IgE sensitization occurs only in part of the allergen exposed population (mainly atopics; see below), and it can take weeks to years between first encounter with an allergen and the development of specific sensitization and the subsequent development of an allergy.

Occupational allergy can be defined as allergic disease in which both sensitization to the allergen *and* the induction of symptoms are caused by exposure to an allergen at the workplace. The cause-effect relation is usually rather obvious, since exposure to the 'occupational allergen' practically only occurs in the work environment. However, in some cases, allergen exposure levels may only slightly differ between the work and the home or outdoor environment, e.g., for pollens in greenhouses, or for cat allergens in a pet shop or a veterinarian practice. It may then occur that non-occupational exposure in the past has caused sensitization without serious health effects at 'regular' exposure levels in the home environment, whereas relatively high occupational exposure levels lead to work-related allergic symptoms. In principle, it would also be possible that sensitization has occurred at the workplace, but without obvious work-related symptoms, whereas symptomatic disease develops upon exposure at home or in the outdoor environment (and not specifically at work). However, the latter situation is unlikely since the characterization and identification of 'occupational' allergens in fact implies that exposure is higher, and usually much higher, at the workplace than elsewhere.

Allergic inflammation

Exposure to aero-allergens can lead to respiratory symptoms in allergic individuals via complex inflammatory mechanisms often occurring only minutes after being exposed. This is known as the early phase allergic reaction and symptoms develop as a result of mast cell degranulation and release of inflammatory mediators (caused by allergen-IgE antibody complexes at the surface of the mast cells), thus leading to contraction of smooth muscle tissue in the airways that subsequently leads to decreased lung function and symptoms of wheeze, shortness of breath, chest tightness, and coughing. During the late phase of the allergic reaction (4-8 hours after exposure), eosinophil-related inflammatory reactions are particularly important. This reaction is most likely induced by mediators released during the early phase of the allergic reaction, and is characterized by (but does not always involve) the development of a non-specific bronchial hyperreactivity (an exaggerated bronchoconstrictor response to a wide variety of non-specific and allergic stimuli; see below) that can continue to exist for several days. Repeated exposures can result in a more permanent bronchial hyperreactivity. Many asthmatics only experience an early phase allergic reaction, which is not followed by a late phase reaction. It is not clear why some asthmatics develop late phase allergic reaction and others don't. Upper airway obstruction (rhinitis) is not caused by smooth muscle contraction (the nose does not have smooth muscle tissue), but is due to the swelling of the nasal mucosa as a result of local inflammation and subsequent edema. More detailed information on the allergic inflammatory processes involved in allergic asthma and rhinitis can be found in many recent textbooks on immunology and allergology (e.g., Roitt, 1997)

Atopy

'Atopy' is a common term for type I (IgE-mediated) sensitization and/or type I allergic reactions. In population studies, the term 'atopy' is used to indicate the genetic predisposition of individuals to produce increased levels of specific or total IgE after exposure to '*common allergens*' such as house dust mite and pet and various food allergens. Atopy is often assessed by using skin prick tests or specific and/or total serum IgE against common allergens. In population studies, atopics are sometimes defined based on a family history of asthma, eczema, or other allergic diseases, but this definition is now less used and the term 'atopy' is increasingly restricted to that given above. Depending on the definition, 20% to 40% of the people living in affluent countries are atopic. Atopy is strongly associated with an increased risk to become sensitized against occupational allergens, and the subsequent development of allergic occupational airway symptoms such as allergic asthma and rhinitis. However, it is also common in occupational environments that subjects develop specific sensitization and allergies without being atopic.

Bronchial hyperresponsiveness

Bronchial hyperresponsiveness is an exaggerated bronchoconstrictor response to a wide variety of non-specific and allergic stimuli. As previously mentioned, asthmatics may develop a non-specific bronchial hyperresponsiveness (BHR), and it has been suggested that asthma should be defined as symptomatic BHR (BHR can be assessed by specific histamine or metacholine challenges). However, although BHR is clearly related to asthma, and may be involved in many of the pathways by which variable airflow obstruction may occur, variable airflow obstruction may occur independently of BHR, and vice versa (Pearce et al., 1998a). Thus, they remain separate phenomena both of which usually involve inflammation of the airways (Chung, 1986). In fact, available evidence from random population surveys indicates that symptom questionnaires are a better predictor of asthma than BHR (or the combination of BHR and symptoms) when compared with a careful clinical diagnosis of asthma (Pekkanen and Pearce, 1999).

RISK FACTORS FOR ALLERGIC RHINITIS AND ASTHMA

Atopic predisposition is an important risk factor for occupational sensitization and the subsequent development of atopic respiratory diseases such as rhinitis and allergic asthma. The atopic predisposition is probably primarily genetically determined, but in addition, modified by events during the further development and maturation of the immune system early in life. In addition to allergen exposure, exposure to 'adjuvant factors,' which actually function as effect modifiers in the relation between allergen exposure and sensitization may be a risk factor, which could explain why not all 'atopic' individuals develop occupational sensitization at the same exposure levels. Evidence for the existence of such adjuvant factors comes mainly from experimental animal studies, while for human populations very few clear data are available. These adjuvant factors can include diet, pre-existing or concomitant disease, and smoking behavior. Particularly with regard to smoking there is no agreement: some studies have found smoking to be a significant risk factor for occupational sensitization, while in others no significant effects were noted, or even negative associations with the prevalence of IgE to common or occupational allergens. Of special relevance for occupational medicine and hygiene are findings suggesting that work-related exposure to fumes, gases, or specific chemicals, for example, may enhance the risk of allergic sensitization to inhaled occupational allergens (Kapsenberg, 1996; Løvik et al., 1996). This may be true for widely varying agents like diesel exhaust particles, formaldehyde, and disinfectants. Table 5-4 gives an overview of potential risk factors for IgE sensitization to occupational allergens.

The most important risk factors for work-related allergic *symptoms* are type I sensitization to an allergen and exposure levels at the workplace (see Table 5-5). Note that, for induction of symptoms, the instantaneous exposure and short-term duration are much more important than for the induction of sensitization. For sensitization, an often prolonged exposure period is required (for sometimes up to several years). The risk of allergic respiratory symptoms further depends on several other parameters like bronchial or nasal hyperreactivity and a number of 'enhancing' factors that are very similar to the factors mentioned as possible 'adjuvants' in Table 5-4. While the presence of bronchial or nasal hyperreactivity enhances the risk of allergen-specific allergic reactions, it should be noted that it can also enhance the risk of (non-allergic) reactions towards non-allergic exposures (e.g., endotoxins) at the workplace. Similarly, 'common atopy' (IgE sensitization against commonly available allergens) is often recognized as a risk factor for symptoms, particularly in the absence of demonstrable IgE sensitization to the specific allergen, but this relationship to atopy is mainly due to its strong association with airway reactivity.

Table 5-4
Risk factors for IgE sensitization to occupational allergens

- *atopic predisposition*: 'atopy' defined by
    - presonal or family history of atopic disease
    - positive skin prick test(s) to common allergen(s)
    - specific IgE to common allergen(s)

- *allergen exposure*
    - high airborne concentrations at workplace
    - high frequency and short-term duration: e.g., job tasks (hrs/weeks)
    - long-term: job duration (yrs)

- *'adjuvant' factors*
    - life style factors: diet, smoking
    - history of non-atopic airway disease
    - other occupational exposures:
      e.g., diesel exhaust particles, gases, disinfectants

NB: *For most factors, only circumstantial or partially conclusive evidence is available.*

OCCURRENCE OF ALLERGIC ASTHMA AND RHINITIS

In several countries, occupational respiratory disease registries have been established that can be used to assess the incidence of occupational asthma. However, definition of occupational asthma can be quite different between registries, and may, in some cases, strongly depend on whether employees are compensated for a confirmed diagnosis of occupational asthma. In addition, generally no distinction between allergic and non-allergic asthma is made. Well known registers are: the British Surveillance of Work-related and Occupational Respiratory Disease (SWORD) project and a more intensive scheme for occupational asthma in the West Midlands region known as SHIELD; a Finnish registry maintained by the Finnish Institute of Occupational Health; the U.S. Sentinel Event Notification System for Occupational Risks (SENSOR); as well as several others in Canada and Germany (Meredith and Nordman, 1996). These registries report incidences of occupational asthma ranging from two to 15 cases per 100,000 persons per year (see Table 5-6). However, these incidence rates are likely to be underestimated by at least a factor 2-3 (Meredith, 1993). Interestingly, occupations with exposures to biological agents (animal allergens, enzymes, flour and grain, wood dust, molds, other plants) make up a considerable proportion of all registered occupational asthma cases in various countries (UK, Finland and Canada; reviewed by Meredith and Nordman, 1996). No registries for rhinitis are available.

Table 5-5
Risk factors for work-related allergic symptoms

- *occupational sensitization*:
  - allergen-specific IgE
  - positive skin prick test (SPT)

- *allergen exposure:*
  - high airborne concentrations at workplace
  - high frequency and duration: job tasks (hrs/weeks)

- *non-specific bronchial (or nasal) hyperreactivity*
  - positive BHR test: histamine or metacholine provocation
  - personal history of allergic airway symptoms
  - reported hypersensitivity to exercise, cold air, dust, smoke, etc.

- *atopic predisposition*: 'atopy' defined by
  - personal or family history of atopic disease
  - positive skin prick test(s) to common allergen(s)
  - specific IgE to common allergen(s)

- '*enhancing factors*'
  - concomittant work-related exposures (gases, fumes, disinfectants)
  - smoking, diet, psychosocial stress

NB: *For most factors, only circumstantial or partially conclusive evidence is available.*

Occupational *allergic* asthma and rhinitis can be found in a large variety of occupational settings such as compost facilities, agricultural and related industries, food industry, detergent industry, medical and public health sector, laboratory animal facilities, bio-pesticide industry, etc. Table 5-7 gives an overview of the most important occupational environments (and most relevant allergens) with an increased risk for occupational allergic asthma and rhinitis. However, many more occupations and industrial environments can potentially contribute to the development of allergic airway diseases, and with the introduction of new industries producing or using products of modern biotechnology (that can potentially act as potent allergens) the number of occupational environments with an increased risk for allergic asthma and rhinitis may even grow. Some of the most widely published examples of occupational allergic asthma include bakers asthma (Houba et al., 1998), latex asthma in health care workers due to the widespread use of latex gloves in health care facilities (Poley and Slater, 2000), and asthma in animal care workers in research institutes (Bush et al., 1998). As noted before, asthma inci-

Table 5-6
Incidence of occupational asthma in various countries, and in different years

Country	Incidence (per 100,000)	Reference
UK	2.0	Meredith, 1993
USA (Michigan)	2.9	Rosenmann et al., 1997
UK (West Midlands)	3.0	Gannon et al., 1991
Finland	3.6†	Keskinen et al., 1978
UK	3.7	Meredith et al., 1996
Germany	4.2†	Baur et al., 1998a
UK (West Midlands)	4.3	Gannon et al., 1993
Canada (Quebec)	6.3	Provencher et al., 1997
Sweden	8.1†	Toren, 1996
Finland	8.1†	Vaarannen et al., 1985
Canada (British Columbia)	9.2	Contreras et al., 1994
Finland	15.0†	Kanerva et al., 1994
Finland	15.2†	Nordman, 1994

†Incidence rates are calculated based on registries for the purpose of compensation for occupational diseases.

Table 5-7
Occupational Type I allergens and occupational environments with increased risk for their workers to develop Type I sensitization and respiratory allergy and asthma

Allergen (source)	Occupational environments with increased risk
Molds	Compost facilities, agriculture and related industries
Microbial enzymes	Biotechnology industry and primary enzyme producers
	Food and feed industry, e.g., bakeries
	Detergent industry
Plant proteins:	
- Pollens of 'new' flowers and vegetables	Agri- and horticulture
- Wheat	Bakery industry
- Soy, corn, etc.	Animal feed industry
- Latex proteins	Medical and public health sector, and other occupations where workers regularly use latex gloves
Mammalian proteins	Animal farming and veterinary occupations
	Pet shops
	Laboratory animal facilities
Invertebrate proteins	Agriculture
	Biopesticides industry (moths, spiders, etc.)

dence and prevalence data are available (Table 5-6), however, it is not clear how many of these asthma cases are attributable to allergic responses.

RECOGNITION AND DIAGNOSIS OF OCCUPATIONAL *ALLERGIC* ASTHMA AND RHINITIS

Before a diagnosis of *allergic* asthma or rhinitis can be made it is necessary to establish whether the worker indeed suffers from asthma or rhinitis (regardless of whether it is caused by allergic or non-allergic mechanisms). A second approach is needed to establish whether allergic or non-allergic mechanisms are involved (non-allergic respiratory diseases will be discussed later in this chapter).

How to recognize and diagnose asthma and rhinitis

There are several simple diagnostic tools available that may help to establish an accurate diagnosis of asthma or rhinitis. When studying a group of workers, a questionnaire may be the first choice. Written questionnaires have been the principal instrument for measuring asthma symptom prevalence in community or occupational surveys, and in homogenous populations these have been standardized, validated, and shown to be reproducible (Burney et al., 1989). A number of symptoms, including wheezing, chest tightness, breathlessness, and coughing (with or without sputum), are recognized by physicians as indicative of asthma. Symptoms such as stuffy, runny, irritative nose, sneezing, and itching, burning, watering eyes are indicative of rhinitis. Table 5-8 gives an overview of relevant questions to diagnose respiratory diseases including asthma (based on modified version of the European Community Respiratory Health Survey (ECHRS) questionnaire (Burney et al., 1994)), rhinitis, bronchitis, and organic dust toxic syndrome (ODTS) (bronchitis and ODTS will be discussed later in the section on non-allergic respiratory diseases). Since respiratory diseases such as asthma and rhinitis involve symptoms that occur from time to time rather than the presence or absence of symptoms on a particular day, most questionnaires define 'current symptoms' as symptoms at any time in the previous 12 months (or in case the worker has worked less than a year in his or her current occupation, during the time the worker has been employed in his or her current occupation). In addition to symptom reports occurring at any time in the previous 12 months, it is of interest to ask how often these symptoms occur (e.g., daily/almost daily; 1 to 2 times per week; 1 to 2 times per month, never/seldom).

Most of these questions are not specific for work induced asthma, and thus may also detect asthma that is not work-related or work-related exacerbations of pre-existing asthma. Healthy individuals will reply with 'never/seldom' on all questions (although many of them may have had at least some asthma symptoms during the previous year), whereas workers with occupational asthma and rhinitis will answer with at least 1 to 2 times per month for most questions. Depending on exposure levels and severity of the disease, symptoms may occur more frequently. When symptoms disappear or lessen during the weekends and holidays, it is a good indication that symptoms are indeed work-related (workers with non-work related asthma or rhinitis will also express similar symptoms, but symptoms usually do not lessen or disappear during weekends and holidays).

Table 5-8
Suggestions for questions to diagnose or recognise occupational respiratory diseases for use in occupational health surveys

Asthma (including both allergic and non-allergic asthma)*
(Based on modified phase I screening questionnaire for the European Community Respiratory Health Survey (ECRHS); Burney et al., 1994)

1. Have you had a wheezing or whistling in your chest at any time in the last 12 months?
   IF 'NO,' GO TO QUESTION 2; IF 'YES,'

1.1 Have you been at all breathless when the wheeze noise was present?

1.2 Have you had this wheezing or whistling when you did not have a cold?

2. Have you had a feeling of tightness in your chest at any time in the last 12 months?
3. Have you had an attack of shortness of breath at any time in the last 12 months?
4. Have you had an attack of coughing at any time in the last 12 months?
5. Have you had an attack of asthma in the last 12 months?
6. Are you currently taking any medicine (including inhalers, aerosols, or tablets) for asthma?

Allergic rhinitis and mucous membrane irritations*

1. Do you have any nasal allergies, including hay fever (applies only to allergic rhinitis)?
2. Have you had one or more of the following symptoms in the last 12 months: stuffy, runny, irritative nose, or sneezing?
3. Have you had itching, burning, or watering eyes in the last 12 months?
4. Have you had dry cough in combination with nose and eye irritations (applies only to MMI)?

Organic dust toxic syndrome (ODTS)*

1. Have you, during the past 12 months, had sudden episodes of flu-like symptoms such as fever, chills, malaise, muscle or joint pains, <u>and felt completely well within 1 to 2 days?</u>

Bronchitis

1. Do you cough up phlegm almost daily for at least part of the year?

IF 'YES,'

1.1 How many months a year do you have this cough?

1.2 How many consecutive years have you had this cough?

* *To assess the severity of the disease (asthma, rhinitis, MMI, or ODTS), the worker could be asked how often these symptoms appear (or how often they use medication), e.g., Daily/almost daily; 1 to 2 times per week; 1 to 2 times per month; never/seldom.*

* *To assess whether symptoms are work-related, the workers may be asked whether symptoms disappear during weekends and holidays.*

Questionnaires are useful for epidemiological surveys, but are a crude measure in clinical practice. Often a more objective assessment, such as lung function testing using a spirometer, is required to diagnose asthma. Cross-shift lung function tests repeated through the workweek are preferred (e.g., Mondays after a weekend or holiday, Wednesday, and Friday). Cross-shift decrease in forced expiratory flow in one second (FEV_1, a good lung function parameter to assess airway obstruction; see chapter 1 for more details) on workdays is a good indication of asthma. Acute airway obstruction can (instead of measuring FEV_1 using spirometry) also be monitored by measuring peak flow (PEF) using portable peak flow meters. Workers can self-monitor their peak flow during the day for one or more weeks (including the weekends, allowing the assessment of work related patterns). In order to detect occupational asthma, a PEF reading should be performed at least every third hour and during (or shortly after) attacks of wheeze and cough, starting in the morning before work, during the workday, and after work until bedtime. It is imperative that there is a thorough instruction, and that people are informed that missing readings are expected in order to avoid false recordings (Sigsgaard et al., 1994). Although portable peak flow meters are not as accurate as a spirometer, they are an excellent tool in the recognition of occupational asthma since many measures of one individual can be obtained over a long period. In addition, they are cheap and easy to use, making them suitable instruments for surveys among large groups of workers. Peak flow meters have successfully been used in a large number of studies to show work related airway obstruction (both allergic and non-allergic; Hollander et al., 1998; Zock et al., 1998). Repeated nasal peak flow measurements performed by the subjects on themselves can be used to assess an increase in nasal resistance indicating work related rhinitis.

How to diagnose whether symptoms of asthma and rhinitis are type I allergic symptoms

Once it is established that the worker is suffering from asthma or rhinitis, it is important to assess whether the symptoms are induced by allergic (IgE mediated) or non-allergic responses. Recognition of work-related type I or IgE mediated allergic respiratory symptoms such as *allergic* asthma and rhinitis can be based on their typical features as discussed in previous paragraphs. A systematic approach may comprise the following steps:

1. Evaluation of the presence of allergens at the workplace. Are substances handled that have known or suspected allergenic properties? In addition to the well-known bio-allergens (see Table 5-7), attention should be given to the use of 'new' proteins, e.g., recently introduced enzymes or other products of modern biotechnology.

2. Evaluation of the risk of exposure. Is airborne dispersion of allergens, as dust particles or mists, likely, and if so, during which job tasks or work conditions?

3. Are symptoms compatible with typical type I allergy as described for allergic asthma and rhinitis? Do they occur in direct association with possible allergen exposure, i.e., during or shortly after the tasks or other activities identified in step 2?

4. Are workers with symptoms specifically sensitized to the suspected allergen, and is sensitization rare among workers without symptoms? When no *in vivo* or *in vitro* test is available, do workers with symptoms show a typical 'risk profile', including atopy, BHR, and/or a history of allergic respiratory disease?

Diagnostic tools for recognizing occupational type I allergy are summarized in Table 5-9.

For individual patients, the physician's diagnosis of 'occupational type I allergy' is usually based on the combination of symptoms and a positive skin prick test (SPT) or serum IgE test using the occupational allergen. As indicated earlier, this can be very difficult if the allergen has not been specifically identified or isolated for testing. However, for the most common occu-

Table 5-9
Diagnostic tools to assess occupational Type I (IgE-mediated) allergy

- *symptoms*
  - typical symptoms of upper or lower respiratory allergy
  - symptoms during or shortly after work with suspected allergen
- *demonstration of occupational sensitization*
  - serologic IgE test
  - skin prick test (SPT)
- *systematic evaluation of symptoms in time*
  - diaries recording monitoring of symptoms and job tasks
  - PEF recordings
- *allergen-specific provocation*
  - nasal provocation: symptoms, acoustic rhinometry
  - bronchial provocation: lung function, symptoms

pational allergens, test kits for specific IgE in serum or SPT are available commercially. Although these tests can give a good indication, they do not prove or disprove an etiologic role for occupational allergen exposure and sensitization.

If strictly required (e.g., because of workers' compensation), such proof can be provided by a rigorously controlled allergen provocation test, which may be considered as the 'gold standard.' In such a test, respiratory and other symptoms, lung function, etc. are monitored during and after provocation in an exposure chamber with graded doses of aerosolized allergen. This is a procedure that should be performed under strict medical supervision because of the risk of severe anaphylactic reactions, or of life-threatening broncho-obstruction. Alternatively, as noted before, systematic evaluation of symptoms and/or lung function may be performed with the use of diaries and peak flow meters or other devices for self-monitoring of lung function. This approach can be very useful to demonstrate the work- and task-related effects of certain allergenic exposure on the airways.

A third possibility to confirm the diagnosis of allergic asthma or rhinitis is the demonstration of clinical improvement when the worker is away from work. However, if a patient with suspected occupational allergy shows non-specific airway hyperreactivity, and the workplace is also characterized by exposure to dust, fumes, or other irritants, the beneficial effect of absence from work is not necessarily due to reduced allergen exposure but could be related to non-allergens. Thus, clinical improvement after implementation of more specific measures to reduce allergen exposure (e.g., replacement of certain enzyme-containing preparations, technically improved application systems using solutions and no dry powders, closed systems, or improved ventilation at sites of allergen handling) is much stronger evidence, and also allows the worker to continue his or her present job at the same, but improved, occupational environment.

Often, one diagnostic tool will not give a satisfying diagnosis, and more options have to be explored.

HYPERSENSITIVITY PNEUMONITIS

Hypersensitivity pneumonitis (HP) is a generic term used to describe an acute, subacute, or chronic pulmonary condition with delayed febrile systemic symptoms, manifested by an influx of inflammatory cells and the formation of granulomas in the lung parenchyma (Curtis and Schuyler, 1994). HP is also known as Extrinsic Allergic Alveolitis (EAA), and, depending on the specific work environment where the disease has been observed, various other names have been introduced to describe the disease (e.g., farmer's lung, pigeon breeder's lung, mushroom grower's lung, maple bark stripper's disease, etc.). Symptoms characteristic for HP are very similar to those described for non-allergenic organic dust toxic syndrome (ODTS, described later in this chapter). The most important differences between both diseases have been summarized in Table 5-10, and will be discussed in the section on ODTS.

Table 5-10
Most important differences between ODTS and HP

ODTS	HP
Incidence is relatively high, particularly amongst farmers and other workers exposed to high organic dust levels.	Incidence is low.
Every worker can develop symptoms with sufficiently high exposures (dry cough, chills, fever, malaise, dyspnea, myalgia, chest tightness, headache, mild respiratory impairment).	Only a portion of the workers will develop symptoms (fever, chills, malaise, dry cough, dyspnea, and in a later stage, fatigue and weight loss, and severe respiratory impairment).
Prior sensitization is not required.	Prior sensitization is required.
Symptoms usually last less than 24 hours, and are not as severe as HP, and the worker will usually not seek medical treatment.	Symptoms usually last 12 to 36 hours, but can continue for days, or even weeks or months, in the more chronic form of HP. Symptoms are severe, and the worker will normally seek medical treatment.
Clusters of cases are common.	Clusters of cases are uncommon.
A full recovery from the disease is normal, and the worker can continue working in the same workforce. Symptoms may reappear after massive exposures.	A full recovery is not always possible, and often requires the worker to leave the occupation. Once sensitized, an attack of HP may be triggered only by very limited exposures.
Causal agents: heavy exposure to organic dust containing microbial agents, of which bacterial endotoxin may be most important.	Causal agents: repeated exposures to microbial agents, particularly from fungi and actinomycetes, but also certain animal proteins (e.g., pigeon serum).
Non-immune inflammatory mechanisms play a role in which neutrophil levels in the airways and the blood are increased.	Both immune (involving antibodies) and non-immune inflammatory mechanisms may play a role. Both neutrophil and lymphocyte levels in the airways are elevated.
Serum precipitin tests against a panel of antigens that may cause HP are often negative.	Serum precipitin tests against a panel of antigens that may cause HP are often positive, but in many cases, a negative test may be found as well (these tests have rather low specificity and sensitivity).
Chest X-rays are usually normal, or show minimal interstitial infiltration. infiltrates.	Chest X-rays often show interstitial

Typical symptoms during the *acute form* of HP include acute systemic symptoms such as fever and chills, as well as pulmonary manifestations such as chest tightness, dyspnea and cough. This acute form is the most frequent presenting form and the easiest to characterize, although symptoms of acute HP closely resemble those of ODTS. Symptoms develop 4 to 6 hours after inhalation of the offending substance and usually last for one to several days. With recurrences, weight loss can occur, and dyspnea becomes progressively more continuous (Von Essen and Donham 1997, Salvaggio 1997, Schenker et al., 1998). In the *subacute form* of HP, dyspnea and cough are manifested gradually over several weeks or months, and it can progress to severe respiratory impairment (both obstructive and restrictive, which can be measured by drop in FEV_1 and FVC, respectively). Fatigue and weight loss are often prominent symptoms in this form of the disorder (Richerson et al., 1989, Von Essen and Donham, 1997). The *chronic form* of HP is probably the long-term sequela of one or both of the other forms (Schenker et al., 1998), and is often afebrile and associated with dyspnea, malaise, weakness, weight loss, and cough. In this stage, pulmonary function abnormalities range from diffusion defects with restrictive dysfunction to varying degrees of obstructive dysfunction and very severe disease can result in fibrosis (Salvaggio, 1997). Thus HP is a serious disease that can lead to temporary sick leave in the acute form, and permanent lung damage and disability in the sub-chronic and chronic forms of the disease (Pickering and Newman Taylor, 1994).

The underlying immunological mechanisms of HP are complex and only partly understood. Like other allergic diseases, prior sensitization is required in order to develop HP. HP has long been considered an IgG mediated allergic disease (type III allergy) involving IgG antigen antibody-complexes, complement activation, and infiltration and activation of neutrophilic granulocytes. However, to date it is acknowledged that several types of immunologically induced tissue injury (including type III and IV induced injury) as well as non-immunological factors are operative in disease pathogenesis (including T cell and macrophage activation), either simultaneously or at different stages of the disease (Salvaggio, 1997). The initial response observed following exposure is an increase in neutrophils in the airways, subsequently followed by a rapid increase in T-lymphocytes. Various cytokines including TNF-α, IL-1, IL-6, and GM-CSF are increased but their precise role remains unclear (Schenker et al., 1998). For a more extensive review on disease mechanisms, the readers may wish to consult the review by Salvaggio (1997).

Susceptible subgroups

Although workers vary in their susceptibility to develop HP, it is not yet well understood why some workers develop the disease and others don't. Genetic factors may be important. Atopic individuals do not seem to be more susceptible than non-atopics. HP is less likely to occur in smokers than in non-smokers (Warren, 1977).

Occurrence

HP is found most frequently among farmers, but workers in other occupations with high exposures to organic dusts containing high levels of fungi can also develop HP. These include occupations in grain harvesting and processing, wood and timber industry, mushroom packing and harvesting, raising pigeons or other birds, waste processing and composting, office work (when exposed to contaminated air conditioning systems), cheese processing, and many others. No good data are available on incidence or prevalence of the disease mainly because diagnoses of HP are often not reliable (no firm diagnostic criteria are established yet; see below), and it often happens that symptoms of ODTS are confused with HP. However, ODTS is much more common than HP, and is essentially a different disease for which prior sensitization is not required to develop the disease (unlike HP). Generally, HP is much less common than any of the other respiratory diseases discussed in this chapter.

Exposures

The most important etiologic agents include antigens present in the spores of thermophilic actinomycetes fungi (e.g., *Saccharopolyspora rectivirgula* [formerly *Micropolyspora faeni*], *Thermoactinomyces candidus, Thermoactinomyces vulgaris,* etc.), and fungal antigens of many common fungal genera such as *Aspergillus* (particularly *A. fumigatus* and *A. umbrosus*), *Penicillium, Cladosporium*, and many others (Von Essen and Donham, 1997). Exposure to these agents often occurs during the handling of microbial contaminated products such as grain, hay, wood chips, and other vegetable matter.

Diagnostic tools

Since the underlying pathology is not entirely clear, the diagnosis of 'extrinsic allergic lung disease' or 'hypersensitivity pneumonitis' is now mainly based on a combination of typical symptoms, lung function test, chest radiograph, and (IgG) precipitin test, and a specific occurrence at a workplace where high levels of putative allergens like mold spores have been demonstrated. Note that several of the tests used are not specific, and can only be used as an indication. For example, strong IgG responses (as measured with serum IgG precipitin tests) to inhaled occupational allergens appear to be rather common in exposed workers, including those that do not have the disease. At present, diagnostic criteria for HP have not been firmly established, although various guidelines for the diagnosis of HP have been published (Richerson et al., 1989; Terho, 1986; Schenker, 1998). Most important diagnostic criteria include: 1) appropriate exposure (see above); 2) dyspnea on exertion; 3) inspiratory crackles; and 4) elevated lymphocytes in the lung. Supportive findings of HP include: 1) recurrent febrile episodes; 2) infiltrates on chest x-rays; 3) decreased lung diffusion capacity; 4) precipitating antibodies to HP inducing antigens; 5) granulomas on lung biopsy; and 6) improvement with contact avoidance (Schenker et al., 1998).

Non-allergic Respiratory Diseases

The most important organic dust induced respiratory conditions lacking a specific underlying immunological mechanism (thus not inducing an antibody mediated response) such as described for the allergic respiratory diseases are: non-allergic asthma, organic dust toxic syndrome (ODTS), mucous membrane irritation (MMI), chronic bronchitis, and chronic airflow obstruction. In addition, organic dust exposure may non-specifically exacerbate pre-existing respiratory conditions such as allergic asthma.

In general, with these diseases, prior sensitization to a specific agent is not required to mediate symptoms, and thus tests for antibodies to specific agents are negative and subjects will develop symptoms without a latency period (in contrast to allergic diseases where a latency period is common). Although there may be differences in susceptibility between individuals, it is assumed that when the exposure is sufficient, symptoms may develop in most, if not all, workers, unlike allergic respiratory diseases in which only a proportion of all workers will develop symptoms. Moreover, regularly exposed workers may develop a short-lived tolerance for acute effects which does not, however, prevent the worker from developing chronic effects.

The underlying inflammatory responses are often directed against constituents of bacteria and fungi, but can also be directed against certain plant products (described later). These agents usually interact with specific receptors on inflammatory and other cells. Macrophages and neutrophils (in contrast to eosinophils in type I allergic respiratory diseases) are the key inflammatory cells involved. In addition, the complement system (one of the mediators of inflammatory reactions that can be activated by the presence of certain microbial surfaces) is believed to play a role as well. Both mechanisms are part of the basal immune response to react to foreign agents (particularly micro-organisms) in order to protect the host from infections. With relatively low daily life exposures, inflammatory responses are very mild and the individual will not experience any symptoms. However, in case of high exposures, the normal defense system may 'over-react' resulting in a harmful host response mediated by various endogenous cytokines and metabolites (IL-1, IL-6, IL-8, TNF-α, reduced oxygen species, arachadonic acid, PAF, etc.), causing acute respiratory symptoms. Chronic inflammation of the airways, which does occur in chronically exposed workers, may subsequently result in chronic respiratory symptoms and chronic loss of lung function.

In the next paragraphs, a more detailed description of the various non-allergic respiratory diseases is given. Although described here as separate conditions, one should bear in mind that, more often than not, many of these diseases or syndromes may occur at the same time in the same work environment, and often even within the same individual workers. Moreover, most of these conditions are at least partially overlapping in terms of underlying inflammatory mechanisms and subsequent symptomatology.

NON-ALLERGIC ASTHMA

The term non-allergic asthma or asthma-like syndrome is used to describe an acute non-allergic airway response characterized by reversible airway obstruction and symptoms such as cough, chest tightness, wheeze, and/or dyspnea but without persistent hyperresponsiveness and eosinophilia. Although many researchers prefer the term 'asthma like syndrome' to make the distinction between the traditional allergic asthma symptoms and asthma symptoms that are caused by non-allergenic agents and non-specific immune inflammatory mechanisms, the symptoms, as well as the variable airflow obstruction associated with this condition, meet all criteria of asthma.

Affected workers often compare the feeling of chest tightness to that of a chest cold (Merchant and Bernstein, 1993). A transient increase in non-specific airway responsiveness may be apparent (Schenker et al., 1998). The underlying intense airway inflammation is one in which neutrophils and pro-inflammatory cytokines dominate (Rylander, 1986). Symptoms may be very similar to those reported by workers that suffer from allergic asthma, and cross-shift lung function tests will show a similar acute decrease in Forced Expiratory Volume in one second (FEV_1) as observed in allergic asthmatics. Symptoms will develop over the course of the workday, and will be most pronounced after four to eight hours of exposure. A maximum drop in FEV_1 is usually observed six to eight hours after exposure (Rylander, 1997a).

In contrast to workers with work related allergic asthma, previously unexposed workers will develop symptoms without any prior sensitization or latency period. At the early stage of the disease, the acute respiratory symptoms, as well as cross-shift fall in FEV_1, is worst in the beginning of the week after a two-day exposure-free period during the weekend (Schenker et al., 1998). Although the exposure remains the same, the worker's condition progressively improves over the week with no symptoms and cross-shift lung function changes at the end of the week. This characteristic also clearly differentiates it from the classic allergic asthma. At this stage of the condition, symptoms do not usually lead to sick leave and workers will usually not consult a physician for treatment, which makes it difficult for the hygienist and/or physician to recognize the condition. However, several epidemiological studies have shown that small cross-shift decreases in FEV_1 (as is observed with asthma-like syndrome) may predict later development of chronic airflow limitation (Ebi-Kryston, 1989, Heederik et al., 1992), demonstrating the importance of recognizing the early stage of the disease to allow subsequent appropriate action to be taken.

At a later stage of the disease, workers will experience respiratory symptoms every day of the workweek and an acute decline in lung function will now be observed every workday as well. Symptoms may include productive cough at this stage, and it is more likely that the worker will consult a physician for treatment. Also, the disease may now lead to a decrease in working capacity, and because the clinical picture has shifted more to one resembling an allergic occupational asthma, the correct diagnosis of non-allergic asthma is harder to make. Non-allergic asthma may be accompanied by other syndromes such as organic dust toxic syndrome, and mucous membrane irritations. The main differences between allergic asthma and non-allergic asthma are summarized in Table 5-11.

Table 5-11
Most important differences between occupational allergic asthma and non-allergic asthma

Non-allergic Asthma	Allergic (Type I or IgE-mediated) Asthma
Every worker can develop symptoms. Prior sensitization is not required.	Only a proportion of the workers (particularly atopics) will develop symptoms. Prior sensitization is required.
Symptoms and airflow obstruction (cross-shift decrease in FEV_1 and PEF) most pronounced on Mondays (after weekends or holidays), at least in early stage of the illness.	No distinct pattern in symptoms or airflow obstruction during the workweek. Condition does, however, usually improve during the weekends.
Likely to be accompanied with other disease conditions, such as ODTS and MMI, that are most often caused by the same exposure agents.	Less likely to be accompanied by ODTS and MMI.
Symptoms and airway obstruction develop 4 to 8 hours after exposure.	Symptoms and airway obstruction develop almost immediately after exposure. However, an immediate response may be followed by a delayed response several hours later.
Non-immune inflammatory mechanism in which neutrophils are predominant.	Immune IgE-mediated inflammatory mechanism in which eosinophils are predominant.
Causal agents: pro-inflammatory agents, of which bacterial endotoxin may be most important.	Causal agents: occupational allergens.

Susceptible subgroups

Although both atopic and non-atopic individuals can develop non-allergic asthma with sufficient exposure, there is some evidence that atopic individuals are more susceptible to develop the disease (Sepulveda et al., 1984). Also, smokers may have an increased risk to develop work-related non-allergic asthma, and symptoms in smokers are usually more severe and more frequent than in equally high exposed non-smokers (Merchant et al., 1974; Molyneux and Berry, 1968).

Occurrence

One of the earliest recognized forms of non-allergic asthma (or asthma-like syndrome) is byssinosis, an airway disease commonly observed among workers exposed to textile fibers such as cotton, flax, hemp, jute, sisal, and silk. It has long been thought that this disease was a specific characteristic of working with textile fibers, and particularly cotton. However, in the past few decades, a large number of studies have shown a similar symptom pattern with associated lung function changes resembling byssinosis, in workers of a number of other industries (mainly agricultural or associated industries). These industries were all characterized by high exposures to organic dust, usually containing high concentrations of microbial contaminants. Examples of industries where non-allergic asthma due to biological agents is prevalent include the cotton, grain, swine confinement, potato processing, sugar beet processing, waste, poultry, and animal feed industry, as well as many others listed in Table 5-12. The prevalence of symptoms may be as high as 50% among exposed populations (Schenker et al., 1998). Thus worldwide, millions of workers may experience acute work-related symptoms of non-allergic asthma of which a portion will develop chronic airway obstruction (and possibly bronchitis). Currently, it is not known exactly what portion may develop more serious chronic airway obstruction, but it is expected to be quite substantial.

Exposures

In various studies performed in the cotton, grain, animal feed, swine confinement, and fiberglass industries, dose-response relationships have been observed between symptoms, cross-shift decline in lung function (FEV_1 and PEF), and concentrations of inhalable or respirable dust or specific components therein (Kennedy et al., 1987; Donham et al., 1989; Smid et al., 1994; Sigsgaard et al., 1992; Milton 1996). Several components specific for the industrial environment where asthma was prevalent have been suggested such as, for example, components from cotton bracts and grains. However, bacterial endotoxin seems common to all these environments, and it is, therefore, often suggested that this agent may be the most important causal factor responsible for the development of non-allergic asthma. In a number of studies in which both endotoxin and dust was measured, the strongest association was observed with airborne endotoxin exposure (Vogelzang et al., 1998; Milton et al., 1996). In fact, in some studies, no relationship was shown for dust exposure, whereas strong effects were found with endotoxin exposure (Smid et al., 1994). Moreover, it has been shown in various experimental exposure

Table 5-12
Environments with potentially elevated exposures to organic dust in which pro-inflammatory and/or toxic components such as endotoxins, β(1→3)-glucans, mycotoxins, etc., may play an important role in the development of non-allergic respiratory symptoms

Agricultural environments
Livestock farming (pigs, cows, horses, chicken)
Mushroom cultivation
Grain, field crops, and hay
Transportation and storage of livestock and agricultural produce
Manure handling
Industrial environments
Animal feed industry
Vegetable and animal-fibre processing (cotton, flax, hemp, wool, etc.)
Food industry (potato, sugarbeet processing, etc.)
Wood industry
Paper production
Fermentation industry
Slaughterhouses
Metal cutting industries (contaminated cutting and machining fluids)
Waste processing
Garbage collection
Composting
Recycling industry
Sewage treatment
Manure treatment
Buildings
Contaminated ventilation systems
Contaminated humidifiers

Partially adopted from Jacobs, 1997

studies that endotoxin on its own can induce acute airway obstruction and asthma-like symptoms (Michel et al., 1989, 1997). It is likely, however, that endotoxin is not the only factor responsible, and many other components, of which several are described later in this chapter (e.g., mold ß(1,3)-glucans, bacterial petidoglycans, plant tannins etc.), may also play a role.

Diagnostic tools

As with allergic occupational asthma, it is important to first establish whether the worker is suffering from asthma and whether it is indeed work-related. Simple diagnostic tools (questionnaire and lung function testing) to establish this are described in the section dealing with allergic asthma. Questions that can be used to assess asthma prevalence in a working population are summarized in Table 5-8.

Once it is established that the worker is indeed suffering from asthma, it is important to assess whether the symptoms are induced by allergic (IgE mediated) or non-allergic responses (not involving antibodies). Recognition of non-allergic asthma is usually based on: 1) a negative diagnosis of allergic asthma; and 2) the typical features of non-allergic asthma as discussed above (and summarized in Table 5-11). For example, a cross-shift decrease in forced expiratory flow in one second (FEV_1), measured by using spirometry or peak flow (PEF) measured with personal peak flow meters (see section on allergic asthma), on Mondays becoming less notable during the rest of the week is a good indication of non-allergic occupational asthma. However, cross-shift decreases monitored for each day in the week without a notable difference between the beginning and the end of the workweek may indicate allergic asthma or a more advanced stage of asthma-like syndrome.

In addition, the nature of the underlying airway inflammation (allergic versus non-allergic) can be studied by using techniques such as bronchial alveolar lavage (BAL), nasal lavage (NAL), and sputum induction (Hunter et al., 1999; Douwes et al., 2000a). Small volumes of saline solution are used to rinse the upper or lower airways. The retained fluid cells can then be counted and differentiated, and inflammatory markers determined. The presence of elevated concentrations of neutrophils and certain cytokines such as IL-1, IL-6, IL-8, TNF-α and MPO are indicative of non-allergic asthma, whereas increased levels of eosinophils and IL4 and IL5 may be indicative for allergic asthma. However, these techniques are difficult to perform, and analyses can only be done in specialized laboratories, and are, therefore, not suitable for routine use in industrial hygiene.

ORGANIC DUST TOXIC SYNDROME

Organic dust toxic syndrome (ODTS) is an acute febrile illness, characterized by an increase in body temperature (fever), shivering, dry cough, chest tightness, dyspnea, headache, muscle and joint pains, fatigue, nausea and general malaise (Donham and Rylander, 1986; Von Essen et al., 1990a, Rylander 1997a). The symptoms resemble those of influenza, but symptoms usually disappear the following day. The disease is common among workers heavily exposed to organic dust. The term 'organic dust toxic syndrome' was chosen in 1985 at a consensus conference and is now generally accepted (Donham and Rylander, 1986). However, the term ODTS is somewhat misleading since other non-organic exposures such as zinc fume (as encountered by welders or moulders), isocyanates (as encountered by hard plastic workers), or Teflon are also able to induce these symptoms. Therefore, several other names have been suggested including 'toxic alveolitis' (Malmberg et al., 1988) and 'inhalation fever' (Rask-Andersen, 1992). In the past, this syndrome of symptoms was named after the particular environment in which these symptoms were observed. These diseases such as 'metal fume fever,' 'mill fever,' 'grain fever,' 'humidifier fever,' 'silo unloaders syndrome,' etc., are all based on an acute intense inflammation with similar resulting pathology (Rylander, 1994a). In the older literature, terms such as 'precipitin negative farmer's lung' or 'pulmonary mycotoxicosis' were used to describe the symptoms of ODTS.

The condition appears four to eight hours after exposure and is usually associated with asthma-like symptoms. Symptoms are very similar to those described for hypersensitivity pneumonitis (HP), but the underlying inflammatory mechanisms are different (Von Essen et al., 1990a). It is often very difficult to distinguish between both diseases because their symptoms are so similar. In ODTS, increased numbers of neutrophils and cytokines are found in the airways but no lymphocytes (unlike HP where lymphocytes are increased). An increase in leukocytes (mainly neutrophils) and inflammatory mediators has also been observed in the blood after exposure, explaining some of the systemic effects (fever, joint pains, fatigue, etc.). Symptoms usually last from less than 24 hours to 2 to 3 days and are usually not as severe as in HP. Some workers have multiple episodes of ODTS (Von Essen and Donham, 1997). ODTS occurs in subjects without evidence of hypersensitivity, and with high enough exposure concentrations, all exposed workers may develop the disease. In fact, often several persons working together in the same environment become ill. Repeated exposures can induce adaptation, and ODTS may not reappear until there is an unusually large exposure or there has been a period of absenteeism, such as during holidays (Rylander, 1997a). ODTS usually leads to sick leave but many workers do not seek medical care for this condition.

Although no chronic irreversible effects are known to be associated with ODTS, it may be that ODTS is the first step towards HP. Some believe that these disorders represent parts of a spectrum of responses to organic dust exposure rather than completely distinct clinical entities (Von Essen and Donham, 1997). However, the main difference with HP is that the recovery is complete and that future episodes can be prevented by avoiding massive exposures to organic dusts. Thus, after their illness the affected workers can work in the same environment being exposed to organic dust in lower concentrations without a recurrence of symptoms, whereas HP has important long-term consequences with the possible development of permanent lung damage and disability if exposure does not cease (Pickering and Newman Taylor, 1994). The most important differences between ODTS and HP are summarized in Table 5-10.

Susceptible subgroups

Differences between subgroups (e.g., smokers, atopics) have not been studied in depth. It may be, however, that atopic subjects are more sensitive than non-atopic subjects (Schenker 1998).

Occurrence

The incidence of ODTS is much higher than for HP. In most environments where workers develop asthma-like symptoms, workers may also develop ODTS symptoms. However, exposures to induce ODTS are generally higher than those needed to induce asthma-like symptoms. Also, ODTS may occur in the same environments as where workers may develop HP.

ODTS is a common illness particularly among farmers. It is estimated that about 30% of all farmers experience symptoms of ODTS at least once in their lives (Von Essen and Donham, 1997). A Scandinavian study estimated the yearly incidence of ODTS in farmers to be 1/100 (Malmberg et al., 1988). Another industry that is well known for the occurrence of ODTS among its workers is the textile fiber processing industry. Estimates of the prevalence of ODTS among new workers in the cotton, flax, hemp, and kapok industry range from 10% to 80% (Newman Taylor and Pickering, 1994).

ODTS may occur in any work place where workers get exposed to organic dust at levels that are clearly in excess of normal exposures, particularly exposures to dusts that have been highly contaminated by micro-organisms. Thus, ODTS may be prevalent in a large variety of industries such as the swine confinement, dairy farming, grain processing, potato processing, sewage treatment, wood milling, mushroom cultivation, composting, textile fiber, and many more industries as listed in Table 5-12.

Other environments where ODTS may occur are environments that use humidifiers or air-conditioning systems that are heavily contaminated with micro-organisms. ODTS (or 'humidifier fever' or 'air conditioner lung disease' or 'printers' fever') has been associated with air-conditioning systems in homes and office environments, as well as in industries that require carefully controlled humidity, such as printing, stationary, and manufacture of textiles (Newman Taylor and Pickering, 1994). Also, saunas and steam baths wherein the water is heavily contaminated with micro-organisms, may induce symptoms of ODTS (or 'bath water fever'; Muittari et al., 1980).

Exposures

ODTS is clearly associated with high exposures to micro-organisms and/or their components and with the handling of materials that are heavily contaminated by fungi or bacteria such as straw, silage, grain, and wood chips. Agents causing ODTS have not been fully identified but endotoxin is considered to be important. Endotoxin challenges have been shown to induce symptoms similar to those of ODTS (Michel, 1997). However, as described for non-allergic asthma, endotoxin is not expected to be the only agent. Mycotoxins and other fungal products have often been suggested to play a role, but to date, there is little or no evidence to support this hypothesis. The involvement of certain plant products (from grains and textile fibers) can also not be excluded. Clearly, more research is needed to identify all other agents that also play a role in the pathogenesis of ODTS.

Diagnostic tools

Although the individual symptoms of ODTS are generally not very specific, the combination of these symptoms – together with the duration of the illness – is usually well recognizable. It is important to acknowledge that the disease is short-lived and the exposures are usually quite high. If symptoms persist for a longer time (weeks) and progressively deteriorate, then improve during weekends and/or holidays, it is more likely that the worker suffers from HP. One should bear in mind, however, that HP is much less common and symptoms and subsequent disease are more severe. In contrast to HP, the majority of subjects with ODTS do not have abnormal lung function tests, precipitin tests are not different from those of control subjects, and chest x-rays generally look normal, or are reported as showing minimal interstitial infiltrates (Table 5-10).

A question to assess the prevalence of ODTS in a working population is suggested in Table 5-8. The question, based on the typical symptoms associated with ODTS, has not been widely validated, and its sensitivity and specificity is not known.

MUCOUS MEMBRANE IRRITATIONS

The term mucous membrane irritations (MMI) is used to describe a syndrome that is characterized by dry cough and irritations in the eyes, nose, and throat. A non-immune inflammation (dominated by neutrophils) of the nasal mucosa accompanied by an increase in nasal resistance (irritant rhinitis) is usually observed as well. Thus, MMI shares many features with allergic rhinitis making it difficult to correctly diagnose this condition. Although the population prevalence is not known, mucous membrane irritations are among the most common observed symptoms in subjects exposed to organic dusts, particularly in farmers and workers in related industries. Similar symptoms have also been reported for office workers where it might be part of the 'sick building syndrome.' Endotoxins are assumed to play a role in the pathogenesis of MMI, but many other agents including fungi and non-organic dusts, gases, fumes, and vapors may also play a role. MMI is often seen in subjects with non-allergic asthma.

Table 5-8 gives an example of questions that could be asked to diagnose MMI. Again, symptoms are fairly similar with the symptoms observed in allergic rhinitis, and misclassification is likely to occur based only on this question. Healthy subjects will reply with 'never/seldom,' whereas workers with MMI will answer with 'at least 1 to 2 times per month.' Depending on exposure levels and severity of the mucous membrane irritations, symptoms may occur more frequently. Determination of cell types and cytokines in nasal lavages or nasal swabs can be used to make a distinction between allergic rhinitis and MMI or irritant rhinitis. Also, symptoms typical for allergic rhinitis often occur almost immediately after exposure, whereas symptoms of MMI or irritant rhinitis caused by biological agents will usually develop hours after exposure.

CHRONIC BRONCHITIS

Chronic bronchitis is defined as cough with phlegm (productive cough) for at least three months per year for two years or more. Chronic bronchitis may be present with or without airway obstruction. The underlying mechanism is an increase in mucus secreting glands and alterations in the characteristics of the mucus itself (Parkes, 1994). Airway inflammation is present and an increased number of neutrophils can be observed (Von Essen et al., 1990b). Subjects with chronic bronchitis may also have symptoms such as dyspnea, chest tightness, and wheezing, and there is a considerable overlap in symptoms between allergic asthma, non-allergic asthma, and bronchitis (Schenker, 1998). Smoking is one of the most important causes of chronic bronchitis. However, occupational exposures to both organic and non-organic dust for prolonged periods of time may also cause chronic bronchitis. The risk of developing chronic bronchitis increases with age and years of exposure. Symptoms of subjects with chronic bronchitis are often exacerbated by workplace exposures and may be experienced as a long lasting 'chest cold' (Von Essen and Donham, 1997). Chronic bronchitis that is accompanied with airway obstruction can lead to a decrease in working capacity and absenteeism.

Susceptible subgroups

Subjects with a previous history of HP and asthma-like syndrome may be at increased risk to develop work-related chronic bronchitis (Dalphin et al., 1993; Lalancette et al., 1993).

Occurrence

Work-related chronic bronchitis has been demonstrated in many farmers, particularly those that are involved in animal confinement farming (e.g., swine and poultry confinement environments) with prevalence rates of approximately 25% to 30% (Schenker, 1998). Other environments where chronic bronchitis is prevalent include the grain and animal feed industry, the tea processing industry, wood milling industry, and cotton processing industry (Chan-Yeung et al., 1992; Castellan et al., 1981; Mandryk et al., 2000; Kennedy et al., 1987).

Exposures

Development of chronic bronchitis is associated with cumulative dust exposures of both organic and non-organic nature. Endotoxin is recognized as a very important agent associated with the risk for bronchitis from organic dust exposure (Rylander 1997a, Kennedy 1987). Various studies, particularly in grain and swine confinement workers, have shown that endotoxin exposure is more strongly associated with chronic bronchitis than dust exposures (Vogelzang et al., 1998, Schwartz et al., 1995). It is assumed, however, that endotoxin alone may not be the only factor, and that exposure to other agents may be required for the disease to develop (Von Essen, 1988; Rylander, 1997a). The involvement of certain plant products (e.g., grains) cannot be excluded. More research is needed to identify all agents that play a role in the pathogenesis of organic dust induced chronic bronchitis.

Diagnostic tools

A diagnosis of bronchitis can be made based on its operational definition (productive cough for three months per year for at least two years) by asking the bronchitis questions suggested in Table 5-8.

Subjects whose symptoms do not exactly meet this operational definition of bronchitis may, however, also suffer from the same condition, therefore, the definition should not be used too rigidly. A positive reply to the first part of the question (Table 5-8) can be considered a good indication for chronic bronchitis even if the reply to the second part of the question does not (entirely) meet the formal definition (Rylander, 1997a). Obviously, as noted before, smoking is a main contributor to bronchitis and should be considered before concluding that bronchitis is work-related.

CHRONIC AIRFLOW OBSTRUCTION

Several studies have shown associations between chronic airflow obstruction as measured by spirometry (particularly FEV_1) and organic dust exposures (Kennedy et al., 1987; Heederik et al., 1991; Rylander and Bergstrom, 1993; Smid et al., 1992; Zejda et al., 1994; Schwartz et al., 1995). Some of the most widely published examples include studies in pig farming, grain handling, animal feed industry, and cotton industry. One recent study in the swine confinement industry showed an excess lung function decline in 171 pig farmers over a three-year period with exposure to bacterial endotoxin. The mean decline in FEV_1 was 73 ml/yr, and a doubling of the endotoxin exposure was associated with an additional decline in FEV_1 of 19 ml/yr (Vogelzang et al., 1998).

Chronic airflow limitations are essentially the same as observed in mineral dust and fume exposed workers, and result from both airway obstruction and loss of elastic recoil in the parenchyma (Schenker et al., 1998). This condition is preceded by chronic non-allergic inflammation of the airways due to chronic exposures of organic dusts and specific microbial agents. However, it should be noted that it is now appropriately recognized that allergic inflammation as observed in allergic asthma may also lead to chronic airflow obstruction.

In most of the studies referenced above, the association between FEV_1 reduction and exposure (cumulative lifetime exposure or current exposure) was stronger for endotoxin than it was for dust, suggesting that endotoxin may be causally related. However, other agents may be involved as well.

Chronic impaired lung function can be monitored by repeated lung function testing (using a spirometer) over the course of several years. Alternatively, chronic loss of lung function can also be detected by comparing baseline lung function test results with reference values of healthy controls, adjusted for age and height. Since smoking is another important cause of chronic lung function decline, data should always be adjusted for smoking. Potentially, a healthy worker effect may bias these comparisons, making it less likely to detect chronic respiratory health effects. Therefore, longitudinal studies are preferred.

AGENTS CAUSING ALLERGIC AND NON-ALLERGIC RESPIRATORY DISEASES

Allergenic Agents

All agents that can induce specific immune responses (resulting in the production of specific antibodies) are also potential allergens. The term *'allergen'* can refer to a single molecule, a mixture of molecules, or a particle from which allergen molecules can be eluted. The latter may be dead material like animal skin scales or mite fecal particles, or viable or living (but non-pathogenic) propagules such as pollen grains, bacteria, or mold spores. Pathogenic infectious agents are usually not called allergens, although many infectious agents can induce severe illness just through activation of non-immune or immune-specific inflammatory reactions. However, these reactions are not included in the usual definition of 'allergic diseases,' which, by definition, can be said to be 'non-infectious' disorders. The only exception may be the respiratory pathology caused by spores of various molds, especially *Aspergillus fumigatus*, which can cause both a lung mycosis (where fungal growth takes place in the lung) and simultaneously induce severe IgE mediated allergic reactions as well as hypersensitivity pneumonitis. Thus, allergens can comprise a large variety of macromolecular structures ranging from low molecular weight sensitizers (mainly chemicals such as diisocyanates) to high molecular weight sensitizers (such as polymeric carbohydrates and proteins). Occupational respiratory allergy may, therefore, occur in each workplace with airborne exposure to such molecules. Table 5-7 gives an overview of occupational Type I allergens (limited to biological agents only) and occupational environments having increased risk for workers to develop type I sensitization and respiratory allergy, asthma, and rhinitis.

FUNGI AND BACTERIA

Fungi are well-known sources of type III (or IgG inducing) allergens. The species involved include many common genera such as *Penicillium* and *Aspergillus*, which can be found in most indoor home and work environments, but do occur in some work environments at markedly higher levels. Moldy hay for example, is the source of allergens causing farmer's lung. Similar disorders have been noted among mushroom growers, and incidentally, among compost workers, who are also exposed to very high mold spore levels. At other workplaces more specific mold exposure may occur, like exposure to *Rhizopus* or *Faenia spp*. in wood and sawmill workers. A specific exposure with high risk of occupational disease is that to *Aspergillus fumigatus,* a fungus that preferentially grows at 37°C, and not only induces allergic sensitization and symptomatic allergic lung disease, but also invades the respiratory tract and causes an infectious mycosis, known as allergic broncho-pulmonary aspergillosis, especially in immunocompromised subjects, such as patients with chronic inflammatory disease or HIV infection, or in patients that have undergone treatment with cytostatic drugs.

Most bacteria or bacterial agents are not very potent allergens with the exception of the 'mold-like' spore forming actinomycetes which can be found in high concentrations in occupational environments where moldy hay, compost, grain, etc. is handled and/or processed.

Many fungal species have also been described as producers of type I allergens, and IgE sensitization to the most common outdoor and indoor fungal species like *Alternaria*, *Penicillium*, *Aspergillus*, and *Cladosporium spp*. is strongly associated with allergic respiratory disease, especially asthma. However, there is very little evidence that supports an important role for type I allergy to fungi in occupational respiratory disease, even in workplaces with high mold exposure levels. One reason may be that the most potent allergenic proteins that have been identified as 'major allergens' in *in vitro*-produced mold extracts are not the same as those that workers are actually exposed to. This may potentially explain the non-detectable allergen levels in many airborne or settled dust samples and the lack of positive skin prick test or IgE results in symptomatic workers. Thus, the symptoms may still be caused by IgE inducing allergens, but this cannot be proven because the test system to detect an IgE response (serum IgE test or skin prick test) makes use of the wrong allergen.

ISOLATED MICROBIAL PRODUCTS: ENZYMES

Enzymes derived from fungi and bacteria and produced by biotechnological companies for use in washing powders and both the human and animal food industry are well known allergens inducing strong IgE responses. Sensitization to such products is known since the pioneering work of Flindt et al. (1969), who described occupational asthma in the detergent industry. Possibly most well known is alcalase ('subtilisin') from *Bacillus subtilis*, but also various other proteases and lipases used in washing powders have been described as potent type I allergens (Johnsen et al., 1997; Vanhanen et al., 2000; Cullinan et al., 2000). In the 1980s and 1990s, the use of enzymes has markedly increased and extended to other areas, particularly the food industry. Populations at risk are not only workers in the enzyme producing industries, but particularly the workers in industries where enzyme preparations are applied, such as bakeries. For example, sensitization to α-amylase from *Aspergillus oryzae*, an important compound of so-called 'bread improver' mixtures in dough making, has been recognized as an important cause of baker's asthma and rhinitis (in addition to the allergenic proteins of wheat flour) (Houba et al., 1998). Other enzymes used in bakeries are cellulases, hemicellulases, and xylanases, and all have been described as incidental or frequent causes of occupational IgE sensitization. Other industries with potential enzyme exposure may be breweries and producers of juices, and other food industries, although inhalatory exposure in most of these industries may be much lower than in the relatively dusty bakeries and flourmills. Another example is the animal feed industry, which is also characterized by relatively high dust levels and the use of fungal enzymes. Recent studies have shown that the use of enzymes in animal feed production (particularly the phosphatase 'phytase' from a recombinant strain of *Aspergillus niger*) may be associated with a remarkably high risk of specific IgE sensitization and probable ensuing work-related respiratory illness (Doekes et al., 1999).

PLANTS

Best known common allergens of plant origin are pollens of grasses, trees, and various weed species. Although a few studies have reported respiratory symptoms due to grass pollen allergy in lawn cutters, for example (Gautrin et al., 1994), these agents are not generally known as occupational allergens. However, in modern agriculture, introduction of new plants, especially when grown in large numbers in greenhouses, may be the cause of sensitization and type I allergy, which in some cases may lead to very severe respiratory disease (van der Zee et al., 1999).

Most other plant-derived allergens are primarily known as food allergens, like potato, wheat, soy, and lupine allergens. Wheat allergens are well known as airborne allergens causing occupational allergy in bakery workers (bakers' asthma). Allergy to airborne soy allergens has been described as the major cause of the asthma epidemics among the general population in Barcelona, which was induced by clouds of soy dust blown from unloading ships and from silos in the harbour to the inner city (Anto et al., 1993). However, no evidence is available on the prevalence of soy sensitization and allergy among workers in the animal feed industry, for example, although exposure levels in soy factories and the animal feed industry in general, are remarkably higher. Also, workers in the potato processing industry are exposed to high levels of airborne potato proteins, but type I anti-potato sensitization is very rare in the potato processing industry. Similarly, animal and grain farmers, although exposed to high dust and plant allergen levels, and thus theoretically at high risk, show remarkably low prevalences of specific IgE sensitization to these agents.

Proteins in the rubber tree juice (*Hevea brasiliensis*) have received extensive attention during the last decade, as 'Natural Rubber Latex' allergens, to which very high numbers of health and hospital workers are sensitized, due to the use of latex gloves (Turjanmaa et al., 1996). Since the use of latex gloves has increased and extended also in other professions and work environments, the impact of latex allergy may be large. Although sensitization may often occur through skin contact — especially since occlusion may increase the permeability of the skin for glove-associated proteins — latex proteins attached to small powder particles on many types of gloves increases the risk of airborne exposure (Swanson et al., 1994), and many examples of respiratory illness due to latex allergy have recently been described (Baur et al., 1998b).

ANIMALS

The most potent common allergens are mainly animal proteins, such as the house dust mite fecal proteins *Der p1*, *Der f1*, etc., and the cat skin dander protein *Fel d1*. Occupational contact with animals is, therefore, nearly always a risk for IgE sensitization to the specific animal that one is working with, including horses, cows, and pigs in animal farming, or cat and dog in veterinary occupations. Pet handlers may also be at increased risk, and the considerably increased risk of occupational type I allergy to mice and/or rats among laboratory animal workers is very well established (Kruize et al., 1997). Also, in these small animals, the major allergenic proteins may be skin derived, although according to various studies, the urinary proteins may be strong allergens as well.

In addition to mammalian allergen exposure from their livestock, farmers are also often highly exposed to allergens from storage mites and, according to various studies, this could be a major cause of occupational allergy and asthma (Iversen et al., 1990). Other examples of type I sensitization to invertebrates have been reported in small workforces of companies producing insects, spiders, mites, etc. for use as 'bio-pesticides' (Lugo et al., 1994; Seward 1999). To date, no epidemiologic studies have been published, but the high frequency of case reports suggests that the risk of developing occupational type I allergy in these companies may be very high.

Pro-inflammatory and Toxic Agents

Although various substances in organic dust have been identified as possible causal agents in non-immune airway inflammation and related respiratory health effects, there are only a few that are well characterized, of which almost all are of microbial origin. Most information is available on bacterial endotoxin, and it is believed that bacterial endotoxin is one of the major causative agents for non-allergic organic dust induced diseases (Douwes and Heederik, 1997). In addition, there are several other microbial components that are suspected to be involved in non-allergic respiratory symptoms and diseases, of which fungal ß(1→3)-glucans and mycotoxins may be the most important (Rylander et al., 1992; Douwes et al., 2000b; Di Paolo et al., 1993). Bacterial peptidoglycans and exotoxins have also been suggested (Verhoef and Kalter, 1985). In addition to microbial agents, some plant components have been suggested to be relevant, such as plicatic acid and abeitic acid in western red cedar and pine trees, respectively (Chan-Yeung, 1994; Ayars et al., 1989), and tannins in cotton and wood (Merchant and Bernstein, 1993). However, very limited information is available for this last group of agents on health effects.

Thus, micro-organisms are one of the most important sources of pro-inflammatory and toxic agents in organic dust, and because micro-organisms are ubiquitous in nature, these agents are commonly present in various environments. For occupational diseases generally only airborne levels of these agents are relevant. Pro-inflammatory and toxic components such as endotoxins, ß(1→3)-glucans, and mycotoxins become airborne during manufacturing or handling of organic materials. Microbial contaminated plant material as well as animal faeces contribute largely in organic dust related exposure to those agents. Elevated exposures are, therefore, prevalent mainly in agricultural and related industries. Air humidifiers in buildings and recycled industrial process waters may, however, also be an important source of exposure. Exposure to pathogenic plant agents mainly occurs in industries where these plants are being processed. Table 5-12 gives a summary of environments in which elevated exposures to endotoxins, ß(1→3)-glucans, mycotoxins, other microbial toxins, and certain pathogenic plant components can be expected. Whereas, exposure information for particular environments in general is sparse for most individual components, exposure data is abundant for endotoxin. In Table 5-13, an overview is given of endotoxin exposures in various occupational environments. It should be noted that allergen exposures may also be high in many of the occupational environments listed in Tables 5-12 and 5-13.

In the following paragraphs, a more detailed description is given for some of the agents that are believed to be most relevant in terms of non-allergic organic dust induced respiratory health effects.

BACTERIAL ENDOTOXINS

Identity

Endotoxins are integral components of the outer membrane of gram-negative bacteria and are composed of proteins, lipids, and lipopolysaccharides. The term 'endotoxin' refers to the toxin as present on the bacterial cell wall, which is often liberated as a result of cell lysis. In the environment, airborne endotoxins are usually associated with dust particles or aqueous aerosol with a broad size distribution. However, data have shown that the smaller size aerosol fractions contain greater amounts of endotoxins per weight of aerosol.

Lipopolysaccharide (LPS) of gram negative bacteria refers to a class of pure lipid carbohydrate molecules (free of protein and other cell wall components) that are responsible for most of the biologic properties characteristic of bacterial endotoxins. LPS are stable water-soluble molecules composed of lipid and polysaccharide. This group of macro molecules has been detected in such taxonomically remote groups of gram negative bacteria as *Enterobacteriaceae*, *Pseudomonadaceae*, and *Rhodospirillaceae*, but has not been found in cell walls of gram positive bacteria, mycobacteria or fungi (Morrison, 1979; Rietschel et al., 1985).

The lipid moiety of LPS is termed 'lipid A' and is responsible for the toxic properties of LPS. The composition of Lipid A deviates significantly from other lipids in biological membranes, and among various bacterial species the composition is remarkably constant. The hydrophobic polysaccharide moiety is composed of O-specific side chains and core sugars. The composition of O-specific chains varies considerably between bacterial species. The composition of the core is relatively constant and is identical or closely related for large groups of bacteria (Morrison, 1979; Rietschel et al., 1985). Although most of the biologic effects can be reproduced by purified LPS, it is not correct to assume that this term is always preferable to the term endotoxin in defining biologic responses to this bacterial product. The terms 'endotoxins'

Table 5-13
Endotoxin exposures in various occupational environments

Type of Industry	Dust fraction	n	Mean endotoxin concentration (EU/m^3)*
Grain elevator and animal feed industry			
DeLucca et al., 1987	Respirable	69	0-7.4
Smid et al., 1992	Inhalable (P)	530	12-285
	Inhalable (A)	79	19
Pig farmers			
Clark et al., 1983	Total	18	400-2,800
Atwood et al., 1987	Total	166	1,200-1,280
	D_{50}<8.5μm	166	1,050-1,150
Donham et al., 1989	Total	57	2,400
	Respirable	57	2,300
Preller et al., 1995	Inhalable	350	920
Dairy farmers			
Kullman et al., 1998	Inhalable	194	647
Chicken farmers			
Clark et al., 1983	Total	7	1,200-5,000
Thelin et al., 1984	?	25	1,300-10,900
Jones et al., 1984	Total	7	240-590
	Respirable	7	38-98
Poultry slaughter houses			
Lenhart et al., 1990	Inhalable	17	2,500
	Respirable	19	130
Hagmar et al., 1990	Total	24	400-7,800
Cotton industry			
Rylander and Morey, 1982	Respirable	36	200-3,700
Kennedy et al., 1987	PM < 15μm	62	20-5,300
Potato processing industry			
Zock et al., 1995	Inhalable (P)	195	9-102
	Inhalable (A)	68	1-4,000
Sugar beet processing industry			
Forster et al., 1989	?	?	2.5-32

*EU = Endotoxin Units (10 EU is approximately 1 ng)

P = personal sampling; A - Area sampling

and 'lipopolysaccharides' are, therefore, often used interchangeably in the scientific literature (Morrison, 1979).

Health effects

Thomas' (1974) suggestions that endotoxins are "read by our tissues as the very worst of bad news" and that in response to these molecules "we are likely to turn on every defence at our disposal," elaborate beautifully the toxic potential of these macromolecules (Morrison, 1979). Pernis et al. (1961) and Cavagna et al. (1969) were probably the first to recognize the role of endotoxins in the so called 'Monday morning malaise' among cotton workers (early stage of byssinosis). Many reports have suggested a causal role for endotoxins and gram negative bacteria in the etiology of byssinosis, ODTS and non-allergic asthma (Cincotai, 1977; Castellan, 1987; Haglind, 1984; Rylander, 1985a). However, although endotoxin is regarded as a very important component in the etiology of these diseases, it is not believed to be the only factor.

Acute effects

Clinical symptoms after intravenous administration of endotoxin in human volunteers consist of joint aches and fever, shivering, and other influenza-like symptoms. In addition, leukocytosis can be observed (Merchant et al., 1975). Symptoms disappear the next day and repeated exposure results in similar, but less intense, symptoms. Obviously, multiple exposures may lead to some kind of tolerance. Greisman & Hornick (1973) described an endotoxin tolerance that lasted five weeks after a long-term exposure. Experiments showed that 90 to 120 minutes after intravenous administrations of endotoxin in volunteers, fever reactions occurred which lasted three to four hours. A maximum elevation in body temperature of 1.9° C was demonstrated using 2 ng endotoxin derived from *Salmonella abortus equi* per kg body weight. Others (Elin, 1981) were able to demonstrate fever reactions after intravenous administration of endotoxin concentrations as low as 0.1-0.5 ng/kg body weight using *E. coli* endotoxin. In addition, it was shown that granulocytosis occurred at even lower endotoxin concentrations than needed for the fever reactions (Wolff, 1973). This demonstrates that endotoxin is a very toxic component that is able to provoke its toxic effects at very low levels.

Data on inhalation experiments on human volunteers are also available. Pernis et al. (1961) themselves inhaled endotoxin aerosols derived from *E. coli* in doses of 5, 10, and 20 µg. Acute effects were dry cough and shortness of breath accompanied with a decrease in FEV_1. Inhalation of endotoxin from other microbial genera resulted in similar symptoms accompanied with light fever reactions and malaise. Cavagna et al. (1969) demonstrated a more than 10% decrease in FEV_1 in two of eight volunteers that were exposed to a dose of 80 µg *E. coli* endotoxin, while Jamison and Lowry (1986) demonstrated an 11% decrease in lung diffusion capacity in volunteers (n=6) exposed to a dose of 12 µg *Enterobacter agglomerans* endotoxin. This was later confirmed at an even lower dose (0.8-4 µg endotoxin) in four volunteers. Several studies have demonstrated stronger airway reactions in asthmatic bronchial hyperreactive persons after experimental exposure (Muitari, 1980; Zwan et al., 1982; Michel et al., 1992a). LPS induced bronchial obstruction demonstrated in these bronchial hyperreactive subjects was shown to be associated with a non-specific immune inflammatory process (Michel et al., 1992b) rather than with an IgE mediated inflammation (Michel et al., 1992a).

There are some larger studies in which workers and naive healthy subjects were experimentally exposed to endotoxin-containing cotton dust, in addition to these smaller studies where subjects were exposed to pure endotoxin (Rylander et al., 1985b; Castellan et al., 1987). These studies showed similar health effects including fever, malaise, chest tightness, and breathing difficulties that were clearly associated with the level of endotoxin exposure, but not with the level of dust exposure. The largest of the two studies (Castellan et al., 1987) provides a good quantitative basis for the assessment of a no effect level. A no effect level for endotoxin exposure based on a 0% FEV_1 decrease was calculated to be 9 ng/m^3. An 'effective concentration' required for 50% of the exposed population to achieve a 5% or larger decrease in FEV_1 (EC_{50}) was calculated to be approximately 100 ng/m^3. An EC_{50} for a 10% decrease in FEV_1 was suggested to occur at a concenration of 1μg/m^3.

Endotoxin-related acute lung function changes and symptoms, as reported in the above summarized experimental studies, have been confirmed in a large number of field studies conducted in various environments such as swine confinement workers, animal feed and grain workers, fiberglass workers, cotton mill workers, potato processing workers, and poultry slaughterhouses (Donham et al., 1989; Kennedy et al., 1987; Milton et al., 1995, 1996; Rylander and Morey, 1982; Smid et al., 1994, Thelin et al., 1989; Zock et al., 1998). In most of these studies, endotoxin exposure was more strongly associated with the studied health effects than dust exposure. Most of the experimental and population studies thus strongly support the hypothesis that inhaled endotoxin has a causative role in the acute pulmonary response in a dose dependent way. However, it cannot be ruled out that other constituents in organic dust may also be of importance in the development of acute pulmonary effects.

Chronic effects

Apart from the acute effects of airborne endotoxin exposure, some large epidemiological studies in the cotton industry (Kennedy et al., 1987; Rylander et al., 1993), the grain processing and animal feed industry (Smid et al., 1992), and the swine confinement industry (Vogelzang et al., 1998; Heederik et al., 1991) give clear indications for endotoxin dose-related chronic respiratory disorders indicated by decreased lung function (FEV_1 and FVC), and increased prevalence of respiratory symptoms and bronchial hyper responsiveness. A finding of major importance in the field studies is that chronic lung function effects were more clearly related to endotoxin exposure than to dust exposure, suggesting an important role for endotoxins in the etiology of occupation related chronic lung diseases. Interestingly, these studies also show that chronic lung function impairment is related with endotoxin exposure at similar levels as described for acute effects in experimental studies. The exposure limits suggested in some of these studies

(Kennedy et al., 1987; Smid et al., 1992) are less than 20 ng/m^3, which are in remarkable agreement with the experimental results described earlier (9 ng/m^3; Castellan et al., 1987).

Mechanisms

Endotoxins are capable of inducing a wide range of inflammatory reactions in vitro and in vivo in man. Inflammatory mechanisms have been extensively reviewed by Ulmer (1997). Briefly, the harmful host responses to endotoxins are mediated by endogenous cytokines and metabolites (IL-1, IL-6, IL-8, TNF-α, reduced oxygen species, arachidonic acid, PAF, etc.), which are released by various cells in the airways, i.e., macrophages, vascular cells, epithelial cells, and polymorphonuclear cells. Macrophages/monocytes carry specific LPS binding receptors (CD14) that play a crucial role in the activation of these cells and the subsequent inflammatory reactions. These endotoxin induced inflammatory responses in the airways can cause acute (and chronic) airway obstruction resulting in obstructive respiratory symptoms and decreased lung function.

PEPTIDOGLYCANS

Peptidoglycans are cell wall components of all bacteria but are most prevalent in the cell wall of gram-positive bacteria. They have been postulated as a possible causative agent for respiratory health effects because of their pro-inflammatory properties. These properties are suggested to resemble many of the pro-inflammatory properties described for endotoxins (Verhoef and Kalter, 1985). One study showed that peptidoglycans can be found in hospital and home air conditioner filters (Fox and Rosario, 1994). However, relationships between occupational peptidoglycan exposure and respiratory health has, to date, not been systematically studied and, therefore, its role in organic dust induced respiratory health effects is not clear.

FUNGAL ß(1→3)-GLUCANS

Identity

ß(1→3)-glucans are glucose polymers with variable molecular weight and degree of branching that may appear in various conformations, i.e., triple helix, single helix, or random coil structures of which the triple helix appears to be the predominant form (Williams, 1997). ß(1→3)-glucans originate from a large variety of sources, including most fungi, some bacteria, most higher plants and many lower plants (Stone and Clarke, 1992). They are water insoluble structural cell wall components of these organisms, but may also be found in extracellular secretions of microbial origin. Glucans may account for up to 60% of the dry weight of the cell wall of fungi, of which the major part is ß(1→3)-glucan (Klis, 1994). The ß(1→3)-glucan content of fungal cell walls has been reported to be relatively independent of growth conditions (Rylander, 1997b).

Health effects

Recently it has been suggested that ß(1→3)-glucans play a role in bio-aerosol induced inflammatory responses and resulting respiratory symptoms (Rylander et al., 1992; Williams, 1994; Fogelmark et al., 1994). Only a few human exposure studies are available. One small population study (n=39) suggested a relationship between airborne ß(1→3)-glucan levels measured in various environments (a daycare center, post office and two school buildings) and questionnaire assessed symptoms such as dry cough, throat and eye irritation, and itching skin (Rylander et al., 1992). The reported associations were not adjusted for other potential risk factors. A subsequent study showed a small increase in the severity of symptoms of nose and throat irritations, in 26 subjects experimentally exposed for four hours to aerosolized ß(1→3)-glucan in a concentration of approximately 210 ng/m^3 (Rylander, 1996). Symptoms were assessed by questionnaire directly after exposure. No effects on FEV_1 or airway responsiveness were found for the whole group, nor for subgroups that were defined based on atopic status (n=12). In the same study, it was shown that in 16 non-atopic and non-symptomatic individuals, exposed to particulate ß(1→3)-glucan (appr. 210 ng/m^3 for 4 hrs), a statistically significant but very small decrease of FEV_1 was found directly and 3 days after exposure, whereas no significant association was found with airway responsiveness (Rylander, 1996).

Several other small studies have been performed in the home environment, a daycare center and among household waste collectors, suggesting a relation with respiratory symptoms and airway inflammation in exposed individuals (Rylander et al., 1997c, 1997d, 1994b; Thorn et al., 1998a). Another study in 129 individuals suggested a relation between indoor ß(1→3)-glucan levels and increased serum MPO levels (myeloperoxidase; an inflammatory mediator produced by neutrophils), and increased prevalence of atopy and decreased lung function (FEV_1) (Thorn et al., 1998b). One larger epidemiological study showed an association between peak flow variability and ß(1→3)-glucan levels in house dust among young children with respiratory symptoms (Douwes et al., 2000b). However, although some evidence has been presented suggesting a role for ß(1→3)-glucan in occupational lung diseases, more studies are needed to confirm this.

Mechanisms

ß(1→3)-glucans can initiate a variety of biological responses in vertebrates such as stimulation of the reticulo-endothelial system, activation of neutrophils, macrophages and complement, and possibly activation of eosinophils, resulting in enhancement of host-mediated induced resistance to infections and antitumor activity (Stone and Clarke, 1992). It has also been demonstrated that ß(1→3)-glucans can induce T lymphocyte activation and proliferation in experimental animals.

At present, few studies have been published on ß(1→3)-glucan related lung pathophysiology, and, therefore, biological mechanisms involved in presumed respiratory health effects remain unclear. Two studies in which guinea pigs were experimentally exposed to an aerosol of ß(1→3)-glucan indicated no neutrophilia in the airways after a single exposure, as is normally observed after endotoxin exposure (Fogelmark et al., 1992, 1994). A combined aerosol exposure to ß(1→3)-glucan and endotoxin showed that the inflammatory response to the endotoxin exposure was reduced by ß(1→3)-glucans in a dose-related fashion. Repeated exposures for a period of five weeks to an aerosol of ß(1→3)-glucans, endotoxin, or a combination of both resulted in an increase in inflammatory cells in the airways of exposed animals. The combined exposure resulted in an even larger increase in inflammatory cells than for the two compounds separately. Thus, it may be, that in order to mediate symptomatology, glucans need to act in combination with other etiologic agents such as endotoxin (Williams, 1997).

MYCOTOXINS

Mycotoxins or fungal toxins are low molecular weight biomolecules produced by fungi that are toxic to both animals and humans. They appear to give fungi a competitive advantage over other organisms. Mycotoxins are known to interfere with RNA synthesis and may cause DNA damage. Individual fungal species may produce various mycotoxins depending on the substrate. In case, of for instance, *Penicillium*, one such compound is penicillin, a strong antibiotic. Other well known mycotoxins are the carcinogenic mycotoxins of which aflatoxin from *Aspergillus* is probably the best known example. Numerous other mycotoxins have been classified (Krough, 1984; Nelson et al., 1983) possessing distinct chemical structures and reactive functional groups, including primary and secondary amines, hydroxyl or phenolic groups, lactams, carboxylic acids, and amides.

Many of the mycotoxins have health effects in animals when contaminated feed is ingested. However, when animals inhale the mycotoxins, health effects usually do not seem to occur (Von Essen and Donham, 1997). Very little is known about occupational airborne exposures of mycotoxins and respiratory health effects. Mycotoxins of *Fusarium, Aspergillus,* and *Penicillium* genera are known to be present in the inhalable fraction of airborne corn dust (Sorenson et al., 1990), cotton dust (Salvaggio et al., 1986) and grain dust (Lacey et al., 1994). It is not clear, however, whether these components in the dust contribute to the frequently reported respiratory symptoms by workers in the cotton or grain industry. In experimental animals, acute inhalation toxicity was demonstrated for T-2 mycotoxin (Creasia et al., 1990), and some mycotoxins are toxic to alveolar macrophages (Sorenson et al., 1986). One study reported acute respiratory distress (as well as acute renal failure) in humans after inhaling ochratoxin from *Aspergillus ochraeceus* in a granary (Di Paolo et al., 1993). Others have suggested a role for trychothecenes, highly toxic mycotoxins produced by fungal species of the *Stachybotrys genus,* in the development of acute pulmonary hemorrhage and hemosiderosis in infants (life threatening toxic reaction in which lung bleeds occur) (Montana et al., 1997). Whether mycotoxins play a major role in organic dust induced occupational respiratory diseases is, however, not clear at present.

OTHER MICROBIAL COMPONENTS

Several other microbial components are known toxins that may be involved in occupational respiratory health effects including exotoxins and phytotoxins. Exotoxins (other than the before-mentioned mycotoxins) are bioactive molecules, usually proteins, which can be shed into the environment during growth or lysis of bacteria. They are, unlike endotoxins, not part of the cell wall structure of microorganisms. They are generally associated with infectious diseases such as botulism, cholera, and tetanus, but they also include the group of phytotoxins (plant toxins produced by various bacteria). However, there are currently no indications that these components may in fact contribute to respiratory health problems in the workplace.

PLANT COMPONENTS

There are a large number of occupational environments in which workers are exposed to vegetable dust including the agricultural, grain, cotton, hemp, flax, and wood industry among many others. The prevalence of non-allergic respiratory health symptoms among the workers in those industries is usually high and the nature of the symptoms is often very similar between various industries (mainly symptoms of ODTS, non-allergic asthma [byssininosis] and chronic bronchitis). In most cases, the workers do not only get exposed to specific components derived from the plants that are being processed, but also get exposed to the microorganisms living on these plants.

Many of the non-allergic symptoms expressed by those workers can be explained by high exposure levels to microbial components such as bacterial endotoxin and possibly fungal ß(1,3)-glucans. However, a role for pathogenic plant agents in the development of non-allergic (as well as allergic) symptoms cannot be excluded. It has been demonstrated that exposure to aerosols of cotton dust extracts in which endotoxin was removed, induced airway obstruction (Buck et al., 1986). Various other studies have shown that water-soluble as well as non-water-soluble extracts of cotton bract can cause airway inflammation and obstruction. Several substances from the cotton bract have been identified to act as pro-inflammatory agents, including tannins, terpenoid aldehydes, and others (Merchant and Bernstein, 1993). Besides microbial components, various specific plant components, many of which have not yet been identified, may contribute to the reported respiratory symptoms as well. Most plants also contain large amounts of ß(1→3)(1→4)-glucans (in contrast to ß(1→3)(1→6)-glucans in fungi), particularly grains. Some studies have shown that plant glucans are bioactive molecules (Stone and Clarke 1992), but it is not clear to date whether these molecules can indeed induce non-allergic respiratory effects.

Another example of an industry in which workers develop respiratory symptoms due to vegetable exposures is the wood industry (saw milling, furniture making, etc.). Several respiratory diseases arise as a result of exposure to wood dust, including asthma, bronchitis, chronic airflow obstruction as well as ODTS and HP, of which the latter two are assumed to be caused by microbial components present in the dust. Airborne wood dust exposure can also cause other health effects such as mechanical irritation to eyes and nose, dermatitis, rhinitis, and nasal cancer (Whitehead et al., 1981).

Respiratory health effects have also been demonstrated for wood dusts of many other tree species including both soft and hard woods (e.g., western red cedar, pine, oak, mahogany, ebony, California redwood, maple, etc.; Chan-Yeung, 1993). Western red cedar and pine are, however, studied most in relation to occupational respiratory health effects. The prevalence of occupational asthma caused by wood dust exposure of western red cedar ranges from 1.1%–13.5%. (Chan-Yeung et al., 1984; Brooks et al., 1981). Plicatic acid, a low molecular weight component, is the chemical component of western (and eastern) red cedar and Japanese cedar that is assumed to be the causative agent for asthma, while in pine, the primary irritant is assumed to be abietic acid. Pathophysiological mechanisms by which these components induce respiratory symptoms are not entirely clear, and both IgE mediated and non-allergic mechanisms have been suggested (Chan-Yeung, 1994).

CANCER

It is only in recent years that the role of biological agents in occupational carcinogenesis has received attention. Studies of occupational cancer are complex because there are few, if any, complete carcinogens, i.e., occupational exposures increase the risk in developing cancer, but the actual development of cancer is not certain, and it may occur 20 to 30 years after exposure (Pearce et al., 1998b). It is commonly assumed that occupational exposures, as with other cancer risk factors, can increase the risk of developing cancer either by causing mutations in DNA or by various "epi-genetic" mechanisms of cancer promotion usually involving increased proliferation of partially transformed cells. These two general features of cancer causation (DNA mutation, and increased cell proliferation) can be caused by a variety of factors including viruses and other biological agents. For example, in laboratory studies it has been found that retroviruses can cause similar mutations in DNA to those caused by certain chemicals. However, in most instances the mechanisms by which biological agents can cause occupational cancer are unknown.

Table 5-14 summarizes current evidence on possible biological causes of occupational cancer. In many instances, it is known that a particular industry or occupation is associated with an increased risk of a specific cancer type, and the etiologically relevant exposure is unknown, but is likely to be of biological origin. The table does not include possible carcinogens of biological origin in which the exposure involves viral, bacterial, or parasitic diseases transferred between humans. These include cancers caused by viruses such as HIV, hepatitis B, HTLV-I and HTLV-II. Exposure to these viruses can occur in workers in health care, nursing homes, daycare, dentistry, and biomedical research.

Table 5-14
Industrial processes, occupations, and specific exposures associated with cancer in humans

Industry (Iarc Group*)	Exposure (Iarc Group*)	Cancer Type
Specific exposures		
Industries in which mold-contaminated materials are handled (e.g., peanut processing, livestock feed processing)	Aflatoxins(1)	Liver
	Ochratoxin A (2B)	Urothelial urinary tract
	Toxins derived from *Fusarium*	Esophagus
Farming-related occupations		
Farmers	Unknown	Hematologic cancers, lip, stomach, prostate, connective tissue, brain
Meat workers, meat inspectors, veterinarians	Unknown (probably viral)	Hematologic cancers
Wood-based industsries		
Furniture and cabinet making (1), carpentry and joinery (2B)	Wood dust	Sinonasal
Wood industry, sawmills	Wood dust, solvents, preservatives	Sinonasal, lip, lung, liver, Hodgkin's disease
Pulp and paper industry	Paper dust	
Carpentry and joinery	Wood dust	Nasal cavity
Other manufacturing industries using biological materials		
Boot and shoe manufacture and repair (1)	Leather dust, benzene	Sinonasal, leukemia
Leather industry (1)	Leather dust, benzene	Sinonasal, leukemia
Textile manufacturing (2B)	Dyes, dusts from fibres and yarns	Nasal cavity, bladder
Rubber industry (1)	Natural rubber components, solvents, aromatic amines	Bladder, leukemia

*IARC groups are: (1) carcinogenic in humans; (2A) probably carcinogenic in humans (2B) possibly carcinogenic in humans; (3) not classifiable; (4) probably not carcinogenic in humans.

Mycotoxins

The only clearly established biological occupational carcinogens are the mycotoxins. These occur in industries in which mold contaminated materials are handled (Schenker et al., 1998). Perhaps the best known carcinogenic mycotoxin exposure is aflatoxin from Aspergillus flavus, which is an established human carcinogen, particularly with regard to liver cancer (Hayes et al., 1984; Bray et al., 1991; Sorenson et al., 1984). Ochratoxin A is also considered a possible human carcinogen (National Toxicology Program, 1991).

The most relevant route of exposure to aflatoxin and ochratoxin is by ingestion, but exposure can also occur by inhalation in industries such as peanut processing or live stock feed processing, and in industries in which grain dust exposure occurs (Selim et al., 1997). Workers in live stock feed processing have been found to have an increased risk of liver cancer as well as cancers of the biliary tract, salivary gland, and multiple myeloma (Olsen et al., 1988).

Another important mold species capable of producing potentially carcinogenic mycotoxins is *Fusarium moniliforme* (Bacon et al., 1989; Schenker et al., 1998), which grows well on corn, wheat, sorghum, barley, bananas and rice, but in the field produces fumonisins only when grown on corn (Schenker et al., 1998). Some evidence suggests a link to human esophageal cancer (Chang et al., 1992).

Farming-related Occupations

There is consistent evidence from a number of countries that farmers are at increased risk for certain specific cancers, including hematologic cancers, lip, stomach, prostate, connective tissue and brain cancer (Blair et al., 1992; Khuder et al., 1998). The most commonly hypothesized explanations involve exposure to pesticides, or exposure to oncogenic viruses or other biological agents carried by farm animals.

Although epidemiologic evidence for the role of oncogenic animal viruses in human cancer is currently weak (Blair et al., 1985), a number of potentially zoonotic viruses exist in the agricultural environment. These include the herpes virus that causes Marek's disease in poultry (Witter, 1972), the avian leukosis virus (Burmeister and Purchase, 1970), and papilloma viruses in cattle (Jarrett et al., 1978; Jarrett, 1980; Smith and Campo, 1985).

Most interest has centered on the bovine leukemia virus (BLV), an exogenous C-type retrovirus closely related to HTLV-I (Jakobs, 1983) that has been established as the etiologic agent of the adult form of bovine lymphosarcoma (BLS) (Miller et al., 1969). BLV has been found in herds in most countries, and has also been found in meat and unpasteurized milk (Ferrer et al., 1981). BLV can also infect sheep and goats, and can cause cancer in sheep (Jakobs, 1983). It is capable of infecting human cells in tissue culture (Diglio and Ferrer, 1976), but human studies have not shown evidence of infection (Donham et al., 1977). A Swedish case-control study found that leukemia patients had BLV in their cattle herds more frequently than did neighbourhood controls (Kvarnfors et al., 1975). However, a subsequent United States case-control study (Donham et al., 1987) found that the prevalence of antibodies to BLV in dairy

Table 5-15
Relative risk estimates for studies showing excess risks of hematologic cancers in abattoir workers*

Study	Site Code	ICD Code	Exposed Cases	Relative Risk	95% Confidence Interval
Smith et al. (1984)	STS	171	19	2.8	1.1-7.3
Smith and Pearce (1986)	STS	171	11	1.6	0.8-3.5
Pearce et al. (1987)	NHL	200	24	1.8	0.9-3.5
Pearce et al. (1987)	NHL	202	19	1.7	0.9-3.4
Johnson et al. (1986a)	NHL	200	1	0.4	0.0-2.2
	NHL,MM	202,203,208	5	2.2	0.7-5.1
Johnson et al. (1986b)	HD	201	4	2.4	0.8-7.5
Pearce et al. (1986)	AML	205.0	9	2.5	1.1-5.7

*STS, soft tissue sarcoma; NHL, non-Hodgkin's lymphoma; MM, multiple myeloma; HD, Hodgkin's disease; AML, acute myeloid leukemia.

herds exposed to cases of acute lymphatic leukemia (ALL) was actually lower than that found in herds with which the controls had had contact.

Thus, most studies of farm animal viruses have not found an association with human cancer, although this may be due to the difficulties of assessing exposure to these viruses in farmers. Possible leads have been suggested by recent studies of abattoir workers, a group that may have high exposure to animal viruses (Pearce et al., 1988). Studies in New Zealand (Pearce et al., 1986; Smith et al., 1984; Pearce et al., 1986, 1987; Smith and Pearce, 1986) and Baltimore, MD (Johnson et al., 1986a) generally show an increased risk of hematologic cancers in slaughterhouse workers (Table 5-15). Johnson et al. (1986b) also found a three-fold risk of mortality from myeloid leukemia and non-Hodgkin's lymphoma among female workers in the meat department of retail food stores. Elevated lung cancer rates in butchers and meatcutters have also been observed in several countries (Reif et al., 1989), and have also been attributed to exposures in the wrapping process (Johnson et al., 1986a, 1986b, 1991, 1994). However, Johnson et al. (1991) note that the increased risks of lung cancer in their studies were confined to black males working in the stock yard and killing/dressing departments where exposure to zoonotic viruses is highest (Johnson et al., 1995). In addition, a cohort study in the United Kingdom (Coggon et al., 1989) found that the lung cancer risk was highest in workers exposed to warm (freshly slaughtered) meat. A recent study of lung cancers in butchers found no evidence that human papilloma virus (HPV) infection was a significant causal factor (Al-Ghamdi et al., 1995).

Veterinarians are another occupational group with exposure to farm animals. There is inconsistent evidence that veterinarians may be at increased risk for cancer of hematologic malignancies. Matanowski and Lilienfeld (1976) reported an SMR (see chapter 10) of 1.48 for Hodgkin's disease in a mortality study of 20,000 veterinarians. Blair and Hayes (1982) studied 5,016 deaths in white male veterinarians and found elevated risks for Hodgkin's disease (PMR= 1.87, 95 confidence interval 1.11-2.96) and cancer of other lymphatic tissue (PMR = 1.92, 95% confidence interval 1.26-2.82). Interestingly, the PMR for all hematologic malignancies was elevated in meat inspectors (PMR = 3.36, 95% confidence interval 1.44-6.57), a finding consistent with the increased risks in abattoir workers.

Wood-based Industries

A number of studies have found associations between exposure to wood dust and various specific cancers. In particular, several studies reviewed by Demers and Boffetta (1998) have found an excess risk of sinonasal cancer, particularly adenocarcinoma, associated with occupational exposure to wood dust, in furniture, cabinet making, carpentry and joinery, and in other wood related jobs including sawmills. Although risk increases with duration of exposure, even men employed for less than five years appear to be at increased risk (Demers and Boffetta, 1998).

Currently, it is unclear whether the excess cancer risks associated with wood dust exposure are due to the wood dust itself, the various chemicals applied to the wood, or other carcinogenic agents involved in working with wood. The risk of sinonasal cancer, particularly adenocarcinoma, appears to be very high amongst workers who have been exposed to dust from hardwoods, while the risks associated with softwood dusts are less certain (IARC, 1995). This may explain why relative risks in North American studies are generally lower than in European ones (Brinton et al., 1984; Finkelstein, 1989).

Several other cancer sites that have been associated with wood dust exposure include nasopharygeal, other pharyngeal, laryngeal, lung, stomach, and colorectal cancers as well as hematologic malignancies (Demers and Boffetta, 1998). However, unlike the situation with sinonasal cancer, a consistent pattern has not emerged with any of these other cancer sites.

Other Manufacturing Industries Using Biological Materials

Nasal adenocarcinoma has been associated with employment in boot and shoe manufacture and repair (IARC, 1981, 1987), and there is also evidence that an increased risk exists for other types of nasal cancer. The highest risk is for people who worked in the dustiest occupations, particularly those with heavy exposure to leather dust. For example, a mortality study of employees in the boot and shoe manufacturing industry in three towns in the United Kingdom (Acheson et al., 1982) showed a large excess of deaths from nasal cancer which was almost totally confined to employees in the preparation and finishing rooms, where most of the dusty occupations occurred. Although the only consistent evidence is for nasal adenocarcinoma, some studies have also suggested associations between exposure to leather and mucinous adenocarcinoma in the nose and ethmoidal cancer, bladder cancer, kidney cancer, leukemia, and cancers of the lung, oral cavity and pharynx and stomach (IARC, 1987). A few cases of leukemia have also been reported during the manufacture of leather goods other than boots and shoes, but these are believed to be due to exposure to benzene. There is little information specifically on workers in tanneries. IARC has therefore classified boot and shoe manufacture and repair as being an industry in which there is sufficient evidence of carcinogenicity in humans; this evidence particularly relates to nasal adenocarcinoma, and leather dust appears to be the main exposure. There is inadequate evidence for carcinogenicity in leather goods manufacture, and in leather tanning and processing (IARC, 1987).

Textile manufacturing involves exposure to organic dust (cotton, flax, hemp) and to various dyes and their components, as well as to bacteria and endotoxins, solvents and other chemicals. An IARC review (IARC, 1990) concluded that there was limited evidence that working in the textile industry involves a carcinogenic risk. This evaluation was based mainly on findings of bladder cancer amongst dyers (possibly due to exposure to dyes) and amongst weavers (possibly due to exposure to dust from fibres and yarns) and of cancer of the nasal cavity among weavers (possibly due to exposure to dust from fibres and yarns) and among other textile workers. There is some evidence that the prevalence of respiratory cancers is lower among cotton workers, and this has been linked to beneficial effects of LPS (Henderson and Enterline, 1973; Merchant and Ortmeyer, 1981; Rylander, 1990).

A large number of studies have been conducted in rubber industries in Canada, China, Finland, Norway, Sweden, Switzerland, the United Kingdom, and the USA (IARC, 1982, 1987).

It has been well established that workers employed in this industry before 1950 have a high risk of bladder cancer, probably from exposure to aromatic amines. Leukemias have been associated with exposure to solvents and with employment in back processing, tyre curing, synthetic rubber production and vulcanisation (IARC, 1987). Excesses of lymphomas have been noted among workers exposed to solvents, and other cancers, including lung, renal, stomach, pancreas, oesophagus, liver, skin, colon, larynx and brain have been reported as occurring in excess in workers in various product areas and departments. Thus there is consistent evidence of an excess of bladder cancer in rubber workers and suggestive evidence for excess risk for a number of other cancer sites. However, it is currently unknown whether these excess risks occur from exposures to biological agents (including agents derived from natural rubber), or due to various chemicals used in rubber manufacture.

ACKNOWLEDGMENTS

Jeroen Douwes is supported by a research fellowship from the Netherlands Organization for Scientific Research (NWO). Neil Pearce is supported by the New Zealand Health Research Council.

QUESTIONS

5.1. Define 'biological agents' and list three types of major health effects that are associated with work related exposure to these agents.

5.2. Name at least three groups of workers who have an increased risk of develop ing an infectious disease and give an example of the disease they may develop and name the most important route of exposure.

5.3. Respiratory diseases caused by biological agents can be divided into allergic and non-allergic conditions. What are the main differences between both groups of respiratory conditions and give at least two examples for each group?

5.4. Describe which diseases (caused by biological agents) can occur in sawmill workers and which exposures may cause them.

5.5. Describe which diseases (caused by biological agents) can occur in workers working in laboratory animal facilities.

5.6. Which agent is believed to be one of the most important causative agents for non-allergic organic dust induced diseases, and in which environments can this agent be found in elevated concentrations?

5.7. Which other microbial agents have been suggested to play a role in non-allergic organic dust induced diseases?

5.8. Name at least three industrial environments with an increased risk of cancer due to exposures of biological agents.

REFERENCES

Acheson, E.D.; Pippard, E.C.; Winter, P.D.: Nasal cancer in the Northamptonshire boot and shoe industry: Is it declining? Br J Cancer 46: 940-6 (1982).

Al-Ghamdi, A.A.; Sanders, C.M.; Keefe, M.; Coggon, D.; Maitland, D.J.: Human papilloma virus DNA and TP53 mutations in lung cancers from butchers. Br J Cancer 72: 293-7 (1995).

American Thoracic Society (ATS) Committee on Diagnostic Standards. Definitions and classification of chronic bronchitis, asthma and pulmonary emphysema. Am Rev Respir Dis 85: 762-768 (1962).

Anto, J.M.; Sunyer, J.; Reed, C.E.; Sabria, J.; Martinez, F.; Morell, R.; Codina, R.; Rodriguez-Roisin, R.; Rodrigo, M.J.; Roca, J.; Saez, M.: Preventing asthma epidemics due to soybeans by dust-control measures. N Engl J Med 329:1760-1763 (1993).

Attwood, P.; Brouwer, R.; Ruigewaard, P.; Versloot, P.; de Wit, R.; Heederik, D.; Boleij, J.S.: A study of the relationship between airborne contaminants and environmental factors in dutch swine confinement buildings. Am Ind Hyg Assoc J 48:745-51 (1987).

Ayars, G.H.; Altman, L.C.; Frazier, C.E.; Chi, E.Y.: The toxicity of constituents of cedar and pine woods to pulmonary epithelium. J Allergy Clin Immunol 83:610-618 (1989).

Azad, A.F.; Beard, C.B.: Rickettsial pathogens and their arthropod vectors. Emerg Infect Dis 4(2):179-186 (1998).

Bacon, C.W.; Marijanovic, D.R.; Norred, W.P.; Hinton, D.M.: Production of fusarin C on cereal and soybean by *Fusarium moniliforme*. Appl. Environ. Microbiol 55:2745-2748 (1989).

Baur, X.; Degens, P.; Wever, K.: Occupational obstructive airway diseases in Germany. Am J Ind Med 33:454-462 (1998a).

Baur, X.; Chen, Z.; Allmers, H.: Can a threshold limit value for natural rubber latex airborne allergens be defined? J Allergy Clin Immunol 101:24-27 (1998b).

Bernstein, I.L.; Chan-Yeung, M.;, Malo, J-L.; Bernstein, D.I.: Definition and classification of asthma. In: Bernstein IL, Chan-Yeung M, Malo J-L, Bernstein DI (eds). Asthma in the workplace, 1st Ed, Marcel Dekker, New York, (1993).

Blair, A.; Malker, H.; Cantor, K.P.; Burmeister, L.; Wiklund, K.: Cancer among farmers: A review. Scand J Work Environ Health 11:397-407 (1985).

Blair, A.; Hoar Zahm, S.; Pearce, N.E.; Heineman, E.F.; Fraumeni, J.F.: Clues to cancer etiology from studies of farmers. Scand J Work Environ Health 18:209-15 (1992).

Blair, A.; Hayes, Jr., H.M.: Mortality patterns among US veterinarians; 1947-1977: An expanded study. Int J Epidemiol 11:391-397 (1982).

Bolyard, E.A.; Tablam, O.C.; Williams, W.W.; Pearson, M.L.; Shapiro, C.N.; Deitchman, S.D.: Guideline for infection control in health care personnel. Am J Infect Control, 26:289-354 (1998).

Bray, G.A.; Ryan, D.H. (eds).: Mycotoxins, cancer, and health. Louisiana State University Press. Baton Rouge. LA. (1991).

Brinton, L.A.; Blot, W.J.; Becker, J.A.; Winn, D.M.; Browder, J.P.; Farmer, J.C.; Fraumeni, J.F.: A case-control study of cancer of the nasal cavity and paranasal sinuses. Am J Epidemiol 119: 896-906 (1984).

Brooks, S.M.; Edwards, J.J.; Apol, A.; Edwards, F.H.: An epidemiologic study of workers exposed to western red cedar and other wood dust. Chest 80:30-32 (suppl) (1981).

Buck, M.J.; Wall, J.H.; Schachter, E.N.: Airway constrictor response to cotton bract in the absence of endotoxin. Br J Ind Med 43:220-226 (1986).

Burmeister, B.R.; Purchase, H.G.: Occurrence, transmission and oncogenci spectrum of the avian leukosis viruses. Bibl Haematol 36:83-95 (1970).

Burney. P.G.J.; Laitinen. L.A.; Perdrizet, S. et al.: Validity and repeatability of the IUATLD (1984) bronchial symptoms questionnaire: an international comparison. Eur Resp J 2:940-945 (1989).

Burney, P.G.I.; Luczynska, C.; Chinn, S.; Jarvin, D.: The European Community Respiratory Health Survey. Eur Respir J 7:954-960 (1994).

Bush, R.K.; Wood, R.A.; Eggleston, P.A.: Laboratory animal allergy. J Allergy Clin Immunol 102:99-112 (1998).

Castellan, R.M.; Boehlecke, B.A.; Peterson, M.R.; Thedell, T.D.; Merchant, J.A.: Pulmonary function and symptoms in herbal tea workers. Chest 79:81s-85s (1981).

Castellan, R.M.; Olenchock, S.A.; Kinsley, K.B.; Hankinson, J.L.: Inhaled endotoxin and decreased spirometric values. N Eng J Med 317:605-9 (1987).

Cavagna. G.; Foa. V.; Vigliani. E.C.: Effects in man and rabbits of inhalation of cotton dust or extracts and purified endotox-ins. Br J Ind Med 26:314-21 (1969).

CDC Arboviral Website. Centers for Disease Control and Prevention. Division of Vector Borne Infectious Diseases, Information on Arboviral Encephalitides web site: www.cdc.gov/ncidod/dvbid/arbor/arbdet.htm

CDC Hepatitis Website. Centers for Disease Control and Prevention. Hepatitis Branch, Hepatitis web site: www.cdc.gov/ncidod/diseases/hepatitis

CDC Travelers' Website. Centers for Disease Control and Prevention. Travelers' Health web site: www.cdc.gov/travel/index.htm

CDC Vectors Website. Centers for Disease Control and Prevention. Division of Vector Borne Infectious Diseases, Lyme Disease web site: www.cdc.gov/ncidod/dvbid/lymeinfo.htm

CDC. Hepatitis surveillance report No. 56. Atlanta: U.S. Department of Health and Human Services, Public Health Service, Centers for Disease Control and Prevention. 1-33 (1996).

Chang, F.; Syrjänen, S.; Wang, L.; Syrjänen, K.: Infectious agents in the etiology of esophageal cancer. Gastroeneterology 103:1336-1348 (1992).

Chan-Yeung, M.; Vedal. S.; Kus, J.; MacLean, L.; Enarson, D.; Tse, K.S.: Symptoms, pulmonary function, and bronchial hyperreactivity in Western red cedar workers compared with those in office workers. Am Rev Respir Dis 130:1038-1041 (1984).

Chan-Yeung, M.; Dimmich-Ward, H.; Enarson, D.A.; Kennedy, S.M.: Five cross-sectional studies of grain elevator workers. Am J Epidemiol 136:1269-1279 (1992).

Chan-Yeung, M.: Western Red cedar and other wood dusts. In: eds Bernstein IL, Chan-Yeung M, Malo JL, Bernstein DI. Asthma in the workplace. Marcel Dekker, New York, (1993).

Chan-Yeung, M.: Mechanisms of occupational asthma due to western red cedar (*thuja plicata*). Am J Ind Med 25:13-18 (1994).

Chung, K.F.: Role of inflammation in the hyperreactivity of the airways in asthma. Thorax 41: 657-62 (1986).

Ciba Foundation Guest Symposium. Terminology definitions, classification of chronic pulmonary emphysema and related conditions. Thorax 14, 286-99 (1959).

Cincotai, F.F.; Lockwood, M.G.; Rylander, R.: Airborne micro-organisms and prevalence of byssinotic symptoms in cotton mills. Am Ind Hyg Assoc J 38:554-59 (1977).

Clark, S.; Rylander, R.; Larsson, L.: Airborne bacteria, endotoxin and fungi in dust in poultry and swine confinement buildings. Am Ind Hyg Assoc J 44:537-41 (1983).

Coggon, D.; Pannett. B.; Pipard, E.C.; Winter, P.D.: Lung cancer in the meat industry. Br J Ind Med 46: 188-91 (1989).

Contreras, G.R.; Rousseau, R.; Chan-Yeung, M.: Occupational respiratory diseases in British Clumbia, Canada in 1991. Occup Env Med 51:710-712 (1994).

Creasia, D.A.; Thurman, J.D.; Wannemacher, R.W.; Bunner, D.L.: Acute inhalation toxicity of T-2 mycotoxin in the rat and guinea pig. Fundam Appl Toxicol 14:54-59 (1990).

Cullinan, P.; Harris, J.M.; Newman Taylor, A.J.; Hole, A.M.; Jones, M.; Barnes, F.; Jolliffe, G.: An outbreak of asthma in a modern detergent factory. Lancet 2:356:1899-900 (2000).

Curtis, J.L.; Schuyler, M.: Immunologically mediated lung disease. Textbook of pulmonary diseases, 1994, Chapter 28, pp 689-744.

Dalphin, J.C.H.; Pernet, D.; Dubiez, A.; Debieuvre, D.; Allemand, H.L.; Depierre, A.: Etiologic factors of chronic bronchitis in dairy farmers. Case control study in the Doubs region of France. Chest 103:417-421(1993).

DeLucca, A.J. II[A]; Godshall, M.A.; Palmgren, M.S.: Gram-negative bacterial endotoxins in grain elevator dusts. Am Ind Hyg Assoc J 45:336-39 (1984).

Demers, P.A.; Boffetta, P.: Cancer risk from occupational exposure to wood dust. IARC Technical Report No.32. Lyon: IARC, (1998).

Diglio, C.A.; Ferrer, J.F.: Induction of syncytia by the bovine C type leukemia virus. Cancer Res 36:1056-1067 (1976).

DiPaolo, N.; Guarnieri, A.; Loi, F.; Sacchi, G.; Mangiarotti AM, M Dipaolo. Acute renal failure from inhalation of mycotoxins. Nephron 64:621-625 (1993).

Doekes, G.; Kamminga, N.; Helwegen, L.; Heederik, D.: Occupational IgE sensitization to phytase, a phosphatase derived from *Aspergillus niger*. Occup Environ Med 56:454-459 (1999).

Donham, K.J.; VanDer Maaten, M.J.; Miller, J.M.; Kruse, B.C.; Rubino, M.J.: Sereopidemiologic studies on the possible relationships of human and bovine leukemia: Brief communication. J Natl Cancer Inst 59:851-853 (1977).

Donham, K.J.; Rylander, R.: Epilogue: Health effects of organic dusts in the farm environment. Am J Ind Med 10:339-340 (1986).

Donham, K.J.; Burmeister, L.F.; Van Lier, S.F.; Greiner, T.C.: Relationships of bovine leukemia virus prevalence in dairy herds and density of dairy cattle to human lymphocytic leukemia. Am J Vet Res 48:235-238 (1987).

Donham, K.; Haglind, P.; Peterson, Y.; Rylander, R.; Belin, L.: Environmental and health studies of farm workers in Swedish swine confinement buildings. Br J Ind Med 46:31-37 (1989).

Douwes, J.; Heederik, D.: Epidemiologic investigations of endotoxins. *Int J Occ Env Health* 3:S26-S31 [supplement] (1997).

Douwes, J.; Dubbeld, H.; van Zwieten, L.; Wouters, I.; Doekes, G.; Heederik, D.; Steerenberg, P.: Upper airway inflammation assessed by nasal lavage in compost workers: A relation with bio-aerosol exposure. Am J Ind Med 37:459-469 (2000a).

Douwes, J.; Zuidhof, A.; Doekes, G.; van der Zee, S.; Wouters, I.; Boezen, H.M.; Brunek-reef, B.: (1→3)-ß-D-gluc-an and endotoxin in house dust and peak flow variability in children. Am J Resp Crit Care Med 162:1348-1354 (2000b).

Ebi-Kryston, K.L.: Predicting 15 year chronic bronchitis mortality in the whitehall study. J Epidemiol Commun Health 43:168-72 (1989).

Elin, R.J.; Wolff, S.M.; McAdam, K.P.; e.a.: Properties of reference Escherichia coli endotoxin and its phtalylated derivative in humans. J Infect Dis 144:329-36 (1981).

Ellul-Micallef, R.: Asthma: a look at the past. Br J Dis Chest 70: 112-6 (1976).

Essen, S.G. von; Robbins, R.A.; Thompson, A.B.; Ertl, R.F.; Linder, J.; Rennard, S.: Mechanisms of neutrophil recruitment to the lung by grain dust exposure. Am Rev Respir Dis 138:921-927 (1988).

Essen, S.G. von; Robbins, R.A.; Thompson, A.B.; Rennard, S.I.: Organic dust toxic syndrome: An acute febrile reaction to organic dust exposure distinct from hypersensitivity pneumonitis. Clin Toxic 389-420 (1990a).

Essen, S.G. von; Thompson, A.B.; Robbins, R.A.; Jones, K.K.; Dobry, C.A.; Rennard, S.I.: Lower respiratory tract inflammation in grain farmers. Am J Ind Med 17:75-76 (1990b).

Essen, S.G. von; Donham, K.J.: Respiratory diseases related to work in agriculture. In: Langley, R.L., McLymore, R.L., Meggs, W.J., Roberson, G.T. (editors). Safety and health in agriculture, forestry, and fisheries. Government Institutes, Rockville, MD (1997).

Faucci, A.S.; Braunwald, E.; Isselbacher, K.J.; Wilson, J.D.; Martin, J.B.; Kasper, D.L.; Hauser, S.L.; Longo, D.L.: Harrison's Principles of Internal Medicine, 14th Edition, McGraw-Hill, New York, (1998).

Ferrer, J.F.; Kenyon, S.J.; Gupta, P.: Milk of dairy cows frequently contains a leukemogenic virus. Science 213:1014-1016 (1981).

Finkelstein, M.M.: Nasal cancer among North American woodworkers: another look. J Occup Med 31: 899-901 (1989).

Flindt, M.L.H.: Pulmonary disease due to inhalation of derivatives of bacillus subtilis containing proteolytic enzyme, Lancet I:1177-1181 (1969).

Forster, H.W.; Crook, B.; Platts, B.W.; Lacey, J.: Topping MD. Investigation of organic aerosols generated during sugar beet slicing. Am Ind Hyg Assoc J 50:44-50 (1989).

Fogelmark, B.; Goto, H.; Yuasa, K.; Marchat, B.; Rylander, R.: Acute pulmonary toxicity of inhaled (1→3)-ß-*D*-glucan and endotoxin. Agents Action 35:50-55 (1992).

Fogelmark, B.; Sjöstrand, M.; Rylander, R.: Pulmonary inflammation induced by repeated inhalations of ß(1→3)-D-glucan and endotoxin. Int J Exp Path 75:85-90 (1994).

Fox, A.; Rosario, R.M.T.: Quantitation of muramic acid, a marker for bacterial peptidoglycan, in dust collected from hospital and home air-conditioninf filters using gas chromatography-mass spectrometry. Indoor Air 4:239-247 (1994).

Frost, F.; Craun, G.F.; Calderon, R.L.: Increasing hospitalization and death possibly due to *Clostridium difficile* diarrheal disease. Emerg Infect Dis 4(4):619-625 (1998).

Gannon, P.F.G.; Burge, P.S.: A preliminary report of a surveillance scheme of occupational asthma in the West Midlands. Brit J Ind Med 48:579-582 (1991).

Gannon, P.F.G.; Burge, P.S.: The SHIELD scheme in the West Midlands region, United Kingdom. Brit J Ind Med 5:791-796 (1993).

Garibaldi, R.; Janis, B.: Occupational infections. In: Environmental and Occupational Medicine, Rom WN, Editior, Little, Brown and Co., Boston, pp. 607-617 (1992).

Gautrin, D.; Vandenplas, O.; De Witte, J.D.; L'Archeveque, J.; Leblanc, C.; Trudeau, C.; Pulin, C.; Arnoud, D.; Morand, S.; Comtois, P.; et al.: Allergenic exposure, IgE-mediated sensitisation, and related symptoms in lawn cutters. J Allergy Clin Immunol 93:437-45 (1994).

Gerberding, J.L.; Holmes, K.: Microbial agents and infectious diseases. In: Textbook of Clinical Occupational and Environmental Medicine. Rosenstock L and Cullen MR, Editors, WB Saunders, Philadelphia, pp. 699-716 (1994).

GINA. Global Strategy for Asthma Management and Prevention. NHLBI/WHO Workshop Report. Global Initiatives for Asthma. NHLBI, Washington, DC, (1994).

Haglind, P.; Rylander, R.: Exposure to cotton dust in an experimental cardroom. Br J Ind Med 41:340-43 (1984).

Hagmar, L.; Schütz, A.; Hallberg. T.; Sjöholm, A.: Health effects of exposure to endotoxins and organic dust in poultry slaughter-house workers. Int Arch Occup Environ Health 62:159-64 (1990).

Hayes, R.B.; Van Nieuwenhuize, J.P.; Raatgever, J.W.; Ten, A.N.D.; Kate, F.J.W.: Aflatoxin exposures in the industrial setting: an epidemiological study of mortality. Food Chem. Toxicol 22:39-43 (1984).

Heederik, D.; Brouwer, R.; Biersteker, K.; Boleij, J.S.M.: Relationship of airborne endotoxin and bacteria levels in pig farms with the lung function and respiratory symptoms of farmers. 62:595-601 (1991).

Heederik, D.; Kromhout, H.; Kromhout, D.; Burema, J.; Biersteker, K.: Relations between occupation, smoking, lung function, and incidence and mortality of chronic non-specific lung disease: the Zutphen Study. Brit J Ind Med 49:299-308 (1992).

Henderson, V.; Enterline, P.E.: An unusual mortality experience in cotton textile workers. J Occup Med 15:717-719 (1973).

Hollander, A.; Heederik, D.; Brunekreef, B.: Work-related changes in peak expiratory flow among laboratory animal workers. Eur Respir J 11:929-936 (1998).

Holt, J.G.; Krieg, N.R.; Sneath, P.H.A.; Staley, J.T.; Williams, S.T.: Bergey's Manual of Determinative Bacteriology, 9th Edition. Williams and Wilkins, Baltimore, (1994).

Houba, R.; Doekes, G.; Heederik, D.: Occupational respiratory allergy in bakery workers: a review of the literature. Am J Ind Med 34:529-546 (1998).

Hunter, C.J.; Ward, R.; Woltmann, G.; Wardlaw, A.J.; Pavord, I.D.: The safety and success rate of sputum induction using a low output ultrasonic nebuliser. Respir Med 93:345-348 (1999).

IARC Working Group. Wood, leather and associated industries. IARC Monographs on the evaluation of the carcinogenic risk of chemicals to humans. Vol 25. Lyon: IARC (1981).

IARC Working Group. Overall evaluations of carcinogenicity: an updating of the IARC Monographs Volumes 1 to 42. IARC Monographs on the evaluation of the carcinogenic risk of chemicals to humans. Supplement 7. Lyon: IARC, (1987).

IARC Working Group. Flame Retardants and Textile Chemicals and Exposures in the Textile Industry. IARC Monographs on the evaluation of the carcinogenic risk of chemicals to humans. Vol 48. Lyon: IARC, (1990).

IARC Working Group. Wood dust and formaldehyde. IARC Monographs on the evaluation of the carcinogenic risk of chemicals to humans. Vol 62. Lyon: IARC, (1995).

IARC Working Group. The Rubber Industry. IARC Monographs on the evaluation of the carcinogenic risk of chemicals to humans. Vol. 28. Lyon: IARC, (1982).

IARC Working Group. The Rubber Industry. IARC Monographs on the evaluation of the carcinogenic risk of chemicals to humans. Suppl. 7. Lyon: IARC, (1987).

ISAAC Steering Committee (Writing Committee: Beasley, R., Keil, U., Von Mutius, E., Pearce, N.). Worldwide variation in prevalence of symptoms of asthma, allergic rhinoconjunctivitis and atopic eczema: ISAAC. Lancet 351:1225-32 (1998).

Iversen, M.J.; Korsgaard, J.; Hallas, T.E.; Dahl, R.: Mite allergy and exposue to storage mites and house dust mites in farmers. Cin Exp Allergy 20:211-219 (1990).

Jacobs, R.: Endotoxins in the environment. Int J Occup Environ Health 3:S3-S5 [Supplement] (1997).

Jakobs, R.M.: Bovine lymphoma. In: Olson RG, Krakowska S, Blakeslee JR Jr (eds). Comparative pathobiology of viral diseases. Boca Raton, FL: CRC Press, (1983).

Jarrett, W.F.H.; McNeil, P.E.; Grimshaw, W.T.R.; et al.: High incidence area of cattle cancer with a possible interaction between an environmental carcinogen and a papilloma virus. Nature 274:215-7 (1978).

Jarrett, W.F.H.: Bracken fern and papilloma virus in bovine alimentary cancer. Brit Med Bull 36:79-81 (1980).

Johnsen, C.R.; Sorenson, T.B.; Ingemann, L.A.; Bertelsen Secher, A.; Andreasen, E.; Kofoed, G.S.; Fredslund Nielsen, L.; Gyntelberg, F.: Allergy risk in an enzyme producing plant: a retrospective follow up study. Occup Environ Med 54:671-5 (1997).

Johnson, E.S.: Nested case-control study of lung cancer in the meat industry. J Natl Cancer Inst 83:1337-9 (1991).

Johnson, E.S.: Cancer mortality among workers in the meat department of supermarkets. Occup Environ Med 51:541-7 (1994).

Johnson, E.S.; Dalmas, D.; Noss, J.; Matanowski, G.M.: Cancer mortality among workers in abattoirs and meatpacking plants: an update. Am J Ind Med 27: 389-403 (1995).

Johnson, E.S.; Fischman, H.R.; Matanowski, G.M.; Diamond, E.: Cancer mortality among white males in the meat industry. J Occ Med 28:23-32 (1986a).

Johnson, E.S.; Fischman, H.R.; Matanowski, G.M.; Diamond, E.: Occurrence of cancer in women in the meat industry. Brit J Ind Med 43:597-604 (1986b).

Jones, W.; Morring, K.; Olenchock, S.A.; Williams, T.; Hickey, T.: Environmental study of poultry confinement buildings. Am Ind Hyg Assoc J 45:760-66 (1984).

Kanerva, L.; Jolanki, R.; Toikkanen, J.: Frequencies of occupational allergic diseases and gender differences in Finland. Int Arch Occup Environ Health 66:111-116 (1994).

Kapsenberg, M.L.: Chemicals and proteins as allergens and adjuvants. Toxicology Letters 86:79-83 (1996).

Keeney, E.L.: The history of asthma from Hippocrates to Meltzer. J Allergy 35: 215-226 (1964).

Kennedy, S.M.; Christiani, D.C.; Eisen, E.A.; Wegman, D.H.; Greaves, I.A.; Olenchock, S.A.; Ye, T.T.; Lu, P.L.: Cotton dust and endotoxin exposure-response relationships in cotton textile workers. Am Rev Respir Dis 135:194-200 (1987).

Keskinen, H.; Alanko, K.; Saarinen, L.: Occupational asthma in Finland. Clin Allergy 8:569-579 (1978).

Khuder, S.A.; Mutgi, A.B.; Schaub, E.A.: Meta-analyses of brain cancer and farming. Am J Ind Med 34: 252-60 (1998).

Klis, F.M.: Review: Cell wall assembly in Yeast. Yeast 10:851-869 (1994).

Krough, P., editor: Mycotoxins in food. Academic press. San Diego, (1984).

Kruize, H.; Post, W.; Heedrik, D.; Martens, B.; Hollander, A.; Beek van der, E.: Respiratory allergy in laboratory animal workers: a retrospective cohort using pre-employment screening data. Occup Env Med 54:830-835 (1997).

Kullman, G.J.; Thorne, P.S.; Waldron, P.F.; Marx, J.J.; Ault, B.; Lewis, D.M.; Siegel, P.D.; Olenchock, S.A.; Merchant, J.A.: Organic dust exposures from work in dairy barns. Am Ind Hyg Assoc J 59:403-414 (1998).

Kvarnfors, E.; Henricson, B.; Hugoson, G.: A statistical study on farm and village level on the possible relations between human leukemia and bovine leucosis. Acta Vet Scand 16:163-9 (1975).

Lacey, J.; Auger, P.; Eduard, W.; Norn, S.; Rohrbach, M.S.; Thorne, P.S.: Tannins and mycotoxins. Am J Ind Med 25:141-144 (1994).

Lalancette, M.; Carrier, G.; Laviolette, M.; Ferland, S.; Rodrique, J.; Begin, R.; Cantin, A.; Cormier, Y.: Farmer's lung: long-term outcome and lack of predictive value of bronchoalvelar lavage fibrosing factors. Am Rev Respir Dis 148:216-221 (1993).

LeBaron, C.W.; Furutan, N.P.; Lew, J.F.; Allen, J.R.; Gouvea, V.; Moe, C.; Monroe, S.S.: Viral agents of gastroenteritis: public health importance and outbreak management. MMWR 39(RR-5);1-24 (1990).

Lenhart, S.W.; Morris, P.D.; Akin. R.E.; Olenchock, S.A.; Service, W.S.; Boone, W.P.: Organic dust, endotoxin, and ammonia exposures in the North Carolina poultry processing industry. Appl Occup Eviron Hyg. 5:611-8 (1990).

Løvik, M.; Dybing, E.; Smith, E.: Environmental chemicals and respiratory hypersensitization: A synopsis. Toxicology Letters 86:211-222 (1996).

Lugo, G.; Cipolla, C.; Bonfiglioli, R.; Sassi, C.; Maini, S.; Cancellieri, M.P.; Raffi, G.B.; Pisi, E.: A new risk of occupational disease: allergic asthma and rhinoconjunctivitis in persons working with beneficial arthropods. Preliminary data. Int Arch Occup Health 65:291-294 (1994).

Malmberg, P.; Rak-Andersen, A.; Hoglund, S.; Kolmodin-Hedman, B.; Guernsey, J.R.: Incidence of organic dust toxic syndrome and allergic alveolitis in Swedish farmers. Int Arch Allergy Appl Immunol 87:47-54 (1988).

Mandell, G.L.; Bennett, J.E.; Dolin, R.: Mandell, Douglas and Bennett's Principles and practice of infectious diseases, 5th Edition. Philadelphia, Churchill Livingston, (2000).

Mandryk, J.; Alwis, K.U.; Hocking, A.D.: Effects of personal exposures on pulmonary function and work-related symptoms among sawmill workers. Ann Occup Hyg 44:281-9 (2000).

Matanowski, G.M.; Lilienfeld, A.: How veterinarians die. Presented at the 104th meeting of the American Public Health Association. Miami Beach, FL: APHA, (1976).

Merchant, J.A.; Lumsden, J.D.; Kilburn, K.H.; O'Fallon, W.M.; Germino, F.H.; McKenzie, W.N.; Baucom, D.; Currin, P.; Stilman, J.: Intervention studies of cotton steaming to reduce biological effects of cotton dust. Br J Ind Med 31:261-274 (1974).

Merchant, J.A.; Halprin, G.M.; Hudson, A.R.; et al.: Responses to cotton dust. Arch Environ Health 30:322-229 (1975).

Merchant, J.A.; Ortmeyer, C.: Mortality of employees of two cotton mills in North Carolina. Chest 79:6s-11s (1981).

Merchant, J.A.; Bernstein, I.L.: Cotton and other textile dusts. In: eds Bernstein IL, Chan-Yeung M, Malo JL, Bernstein I. Asthma in the workplae. Marcel Dekker, New York, (1993).

Meredith, S.: Reported incidence of occupational asthma in the United Kingdom, 1989-90, J Epidem Community Health 47:459-463 (1993).

Meredith, S.; Nordman, H.: Occupational asthma: measures of frequencies from four countries, Thorax 51:435-440 (1996).

Michel, O.; Duchateau, J.; Sergysels, R.: Effect of inhaled endotoxin on bronchial reactivity in asthmatic and normal subjects. Am Phys Soc 1059-64 (1989).

Michel, O.; Ginanni, R.; Sergysels. R.: Relation between the bronchial obstructive response to inhaled lipopolysaccharide and bronchial responsiveness to histamine. Thorax 47:288-91 (1992a).

Michel, O.; Ginanni, R.; Le Bon, B.; Content, J.; Duchateau, J.; Sergysels, R.: Inflammatory response to acute inhalation of endotoxin in asthmatic patients. Am Rev Respir Dis 146:352:57 (1992b).

Michel, O.: Human challenge studies with endotoxins. Int J Occup Env Health 3:s18-s25 [supplement] (1997).

Miller, J.M.; Miller, L.D.; Olsen, C.; Gilette, K.G.: Virus like particles in phytohemaglutinin simulated cultures with reference to bovine lymphosarcoma. J Natl Cancer Inst 43:1297-1305 (1969).

Milton, D.K.; Amsel, J.; Reed, C.E.; Enright, P.L.; Brown, L.R.; Aughenbaugh, G.L.; Morey, P.R.: Cross-sectional follow-up of a flu-like respiratory illness among fiberglass manufacturing employees: Endotoxin exposure associated with two distinct sequelae. Am J Ind Med 28;469-488 (1995).

Milton, D.K.; Wypij, D.; Kriebel, D.; Walters, M.D.; Hammond, S.K.; Evans, J.S.: Endotoxin exposure-response in a fiberglass manufacturing facility. Am J Ind Med 29:3-13 (1996).

MMWR. Common source outbreak of relapsing fever - California. MMWR 39(34):579-586 (1990a).

MMWR. Group A beta-hemolytic streptococcal pharyngitis among U.S. Air Force trainees – Texas, 1988-89. MMWR 39(1):11-13 (1990b).

MMWR. Coccidioidomycosis – United States, 1991-1992. MMWR 42(2):21-24 (1993).

MMWR. Compendium of psittacosis (chlamydiosis) control, 1997. MMWR 46(RR-13):1-13 (1997a).

MMWR. Outbreak of invasive group A Streptococcus associated with varicella in a childcare center – Boston, Massachusetts, 1997. MMWR 46(40):944-948 (1997b).

MMWR. Summary of notifiable diseases, United States, 1997. MMWR 46(54):1-87 (1997c).

MMWR. Rat-bite fever – New Mexico, 1996. MMWR 47(05):89-91 (1998a).

MMWR. Recommendations for prevention and control of Hepatitis C virus (HCV) infection and HCV-related chronic disease. 47(RR-19):1-7 (1998b).

MMWR. Blastomycosis acquired occupationally during prairie dog relocation – Colorado, 1998. MMWR 48(5):98-100 (1999).

Molyneux, M.B.K.; Berry, G.: The correlation of cotton dust exposure with the prevalence of respiratory symptoms. In: Proceedings of the international conference on respiratory disease in textile workers, 1968, Alicante Spain, pp. 177-183.

Montana, E.; Etzel, R.A.; Allan, T.; Horgan, T.E.; Dearborne, D.G.: Environmental risk factors associated with pediatric idiopathic pulmonary hemorrhage and hemosiderosis in a Cleveland community. Pediatrics 99:E51-E58 [supplement] (1997).

Moore, C.G.; McLean, R.G.; Mitchell, C.J.; Nasci, R.S.; Tsai, T.F.; Calisher, C.H.; Marfin, A.A.; Moore, P.S.; Gubler, D.J.: Guidelines for arbovirus surveillance programs in the United States. DHHS, Centers for Disease Control and Prevention. 1-81 (1993).

Morrison, D.C.; Ryan, J.L.: Bacterial endotoxins and host immune responses. Adv in Immun 28:293-434 (1979).

Muittari, A.R.; Rylander, R.; Salkinoja-Salonen, M.: Endotoxin and bathwater fever. Lancet 89 (1980).

Mueller-Anneling, L.; Gilchrist, M.J.; Thorne, P.S.: Ehrlichia chaffeensis antibodies in white-tailed deer, Iowa 1994 & 1996. Emerging Infectious Diseases 6(4):397-400 (July-August 2000).

Murray, P.R.; Baron, E.J.; Pfaller, M.A.; Tenover, F.C.; Yolken, R.H.: Manual of Clinical Microbiology, 7th Edition, ASM Press, Washington, DC, (1999).

National Toxicology Program (1991). sixth Annual Report on Carcinogens. Summary. U.S. Government Printing Office. Washington, D.C. 17-19:309-311.

NCBI Taxonomy Website. National Center for Biotechnology Information. The NCBI Taxonomy Homepage: *www.ncbi.nlm.nih.gov/Taxonomy/taxonomyhome.html*

Nelson PE, Toussoun TA, Marassas WFO. Fusarium species: an illustrated manual for identification. Penn State University Press. University Park. PA, (1983).

Newman Taylor AJ, CAC Pickering. Occupational asthma and byssinosis. In: ed WR Parkes. Occupational lung disorders. Butterworth-Heinemann. Oxford, UK, (1994).

NIOSH. Histoplasmosis: Protecting workers at risk. DHHS (NIOSH) Publ. No. 97-146, (1997).

NIOSH. NORA/National Occupational Research Agenda: Infectious Diseases. *www.cdc.gov/niosh/nrinfo.html*

Nordman, H.: Occupational asthma – time for prevention. Scan J Work Environ Health 20:108-115 (1994).

Olsen, J.H.; Dragsted, L.; Autrup, H.: Cancer risk and occupational exposure to aflatoxins in Denmark. Br J Cancer 58:392-396 (1988).

Parkes, W.R.: Chronic bronchitis, airflow obstruction and emphysema. In: ed WR Parkes. Occupational lung disorders. Butterworth-Heinemann. Oxford, UK, (1994).

Pearce, N.E.; Sheppard, R.A.; Howard, J.K.; Fraser, J.; Lilley, B.M.: Leukemia among New Zealand agricultural workers: a Cancer Registry based study. Am J Epidemiol 124:402-9 (1986).

Pearce, N.E.; Sheppard, R.A.; Smith, A.H.; Teague, C.A.: Non-Hodgkin's lymphoma and farming: an expanded case-control study. Int J Cancer 39:155-61 (1987).

Pearce, N.E.; Smith, A.H.; Reif, J.S.: Increased risks of soft tissue sarcoma, malignant lymphoma and acute myeloid leukemia in abattoir workers. Am J Ind Med 14:63-72 (1988).

Pearce, N.; Beasley, R.; Burgess, C.; Crane, J.: Asthma epidemiology: principles and methods. Oxford University Press, New York (1998a).

Pearce, N.; Boffetta, P.; Kogevinas, M.: (Introduction to occupational cancer. In: Stellman JM (ed). ILO Encyclopaedia of Occupational Safety and Health. 4th ed. Geneva: ILO, Vol. 1, pp 2.1-2.18 (1998b).

Pekkanen, J.; Pearce, N.: Defining asthma in epidemiological studies. Eur Respir J, 14:951-9 (1999).

Pernis, B.; Vigliani, E.C.; Cavagna, C. e.a.: The role of bacterial endotoxins in occupational disease caused by inhaling vegetable dusts. Brit. J. Industr. Med 18:120-129 (1961).

Pickering, C.A.C.; Newman Taylor, A.J.: Extrinsic allergic alveolitis (hypersensitivity pneumonia). In: ed WR Parkes. Occupational lung disorders. Butterworth-Heinemann. Oxford, UK, (1994).

Poley, Jr., G.E.; Slater, J.E.: Latex allergy. J allergy Clin Immunol 105:1054-62 (2000).

Preller, L.; Heederik, D.; Kromhout, H.; Boleij, J.S.; Tielen, M.J.: Determinants of dust and endotoxin exposure of pig farmers: development of a control strategy using empirical modelling. Ann Occup Hyg 39:545-57 (1995).

Provencher, S.; Labrèche, F.P.; De Guire, L.: Physician based surveillance system for occupational respiratory diseases: the experience of PROPULSE, Québec, Canada. Occup Env Med 54:272-276 (1997).

Ramazzini, B.: Disease of workers (translated by Wright WC, DeMorbis artificium diatriba). New York: Hafner, (1994).

Rask-Andersen , A.: Inhalation fever: A proposed unifying term for febrile reactions to inhalation of noxious substances. Brit J Ind Med 49:40 (1992).

Regnery, R.; Tappero, J.: Unraveling mysteries associated with cat-scratch disease, bacillary angiomatosis, and related syndromes. Emerg Infect Dis 1(1):16-21 (1995).

Reif, J.S.; Pearce, N.E.; Fraser, J.: Cancer risks among New Zealand meat workers. Scand J Work Environ Health 15:24-9 (1989).

Rietschel, E.T.; Brade, H.; Kaca, W. e.a.: Newer aspects of the chemical structure and biological activity of bacterial endotoxins. In: Bacterial Endotoxins; Structure, Biomedical Significance, and Detection With the Limulus Amebo-cyte Lysate Test. Alan Liss, New York, 189:31-50 (1985).

Richerson, H.B.; Bernstein, I.L.; Fink, J.N.; Hunninghake, G.W.; Novey, H.S.; Reed, C.E.; Salvaggio, J.E.; Schuyler, M.R.; Schwartz, H.J.; Stechschulte, D.J.: Guidelines for the clinical evaluation of hypersensitivity pnemonitis. J Allergy Clin Immunol 84:839-844 (1989).

Roitt, I.: Essential immunology. 9th edition, Blackwell Science Ltd, UK, (1997).

Rosenman, K.D.; Riley, M.J.; Kalinowski, D.J.: A state-based Surveillance system for work-related asthma. J Occup Env Medicine 39:415-425 (1997).

Rylander, R.; Morey, P.: Airborne endotoxin in industries processing vegetable fibers. Am Ind Hyg Assoc 43:811-12 (1982).

Rylander, R.: Organic dust and lung reactions - Exposure characteristics and mechanisms for disease. Scan J Work Environ Healh 11:199-206 (1985a).

Rylander, R.; Haglind, P.; Lundholm, M.: Endotoxin in cotton dust and respiratory function decrement among cotton workers in an experimental cardroom. Am Rev Respir Dis 131:209-213 (1985b).

Rylander, R.: Lung diseases caused by organic dusts in the farm environment. Am J Ind Med 10:221-227 (1986).

Rylander, R.: Environmental exposures with decreased risks for lung cancer. Int J Epid 3 suppl 1:S67-S72 (1990).

Rylander, R.; Persson, K.; Goto, H.; Yuasa, K.; Tanaka, S.: Airborne Beta-1,3-Glucan may be related to symptoms in sick buildings. Indoor Environ 1:263-267 (1992).

Rylander, R.; Bergstrom, R.: Bronchial reactivity among cotton workers in relation to dust and endotoxin exposure. Ann Occup Hyg 37:57-63 (1993).

Rylander, R.: Organic dusts - from knowledge to prevention. Scan J Work Environ Health. 20:116-22 (1994a).

Rylander, R.; Hsieh, V.; Courteheuse, C.: The first case of sick building syndrome in Switzerland. *Indoor Environment* 3:159-162 (1994b).

Rylander, R.: 1996. Airway- responsiveness and chest symptoms after inhalation of endotoxin or (1→3)-ß-*D*-Glucan. Indoor Built Environ 5:106-11 (1996).

Rylander, R.: Evaluation of the risks of endotoxin exposures. Int J Occup Environ Health 3:s32-s36 (1997a).

Rylander, R.: (1→3)-ß-*D*-glucan in some indoor air fungi. Indoor Built Environ 1997b;6:-291-294.

Rylander, R.: Airborne (1→3)-ß-*D*-glucan and airway disease in a Day-care center before and after renovation. Arch Env Health 52:281-285 (1997c).

Rylander, R.: Investigations of the relationship between disease and airborne (1→3)-ß-*D*-glucan in building. Mediators of inflammation 6:257-277 (1997d).

Salvaggio, J.E.; Neil, C.E.O.; Butcher, B.T.: Immunologic responses to inhaled cotton dust. Environ Health Perspect 66:17-23 (1986).

Salvaggio, J.E.: Extrinsic allergic alevolitis (hypersensitivity pneumonitis): past, present and future. Clin Exp Allergy 27 [suppl 1]:18-25 (1997).

Schenker, M.B.; Christiani, D.; Cormier, Y.; Dimich-Ward, H.; Doekes, G.; Dosman, J.; Douwes, J.; et al.: Respiratory Health hazards in agriculture. Am J Respir Crit Care Med 158: S1-S76 [Supplement] (1998).
Schwartz, D.A.; Thorne, P.S.; Yagla, S.J.; Burmeister, L.F.; Olencjock, S.A.; Watt, J.L.; Quinn, T.J.: The role of endotoxin in grain dust-induced lung disease. Am J Respir Crit Care Med 152:603-608 (1995).
Selim, M.I.; Juchems, A.M.; Popendorf, W.: Potential predictors of aurborne concentrations of aflatoxin B1. J Agromedicine 4: 91-8 (1997).
Sepulveda, M-J.; Castellan, R.M.; Hankinson, J.L.; Cocke, J.B.: Acute lung function response to cotton dust in atopic and non-atopic individuals. Br J Ind Med 41:487-491 (1984).
Seward, J.P.: Occupational allergy to animals. Occup Med 14:285-304 (1999).
Sigsgaard, T.; Pedersen, O.F.; Juul, S.; Gravesen, S.: Respiratory disorders and atopy in cotton, wool and other textile mill workers in Denmark. Am J Ind Med 22:163-184 (1992).
Sigsgaard, T.; Abell, A.; Jensen, L.D.: Work related symptoms and lung function measurements in paper mill workers exposed to recycled water. Occ Hyg 1:177-189 (1994).
Smid, T.; Heedrik, D.; Houba, R.; e.a.: Dust and endotoxin related respiratory effects in the animal feed industry. Am Rev Respir Dis 146:1474-79 (1992).
Smid, T.; Heederik, D.; Houba, R.; Quanjer, P.H.: Dust- and endotoxin-related acute lung function changes and work-related symptoms in workers in the animal feed industry. Am J Ind Med 25:877-88 (1994).
Smith, A.H.; Pearce, N.E.; Fisher, D.O.; Giles, H.J.; Teague, C.A.; Howard, J.K.: Soft tissue sarcoma and exposure to phenoxyherbicides and chlorophenols in New Zealand. J Natl Cancer Inst 73: 1111-7 (1984).
Smith, A.H.; Pearce, N.E.: Update on soft tissue sarcoma and phenoxyherbicides in New Zealand. Chemosphere 15:795-8 (1986).
Smith, K.T.; Campo, M.S.: The biology of papillomaviruses and their role in oncogenesis. Anticancer Res 5: 31-48 (1985).
Sorenson, W.G.; Jones, W.; Simpson, J.; Davidson, J.I.: Aflatoxin in respirable airborne peanut dust. J Toxicol Environ Health 14: 525-33 (1984).
Sorenson, W.G.; Gerberick, G.F.; Lewis, D.M.; Castranova, V.: Toxicity of mycotoxins in the rat pulmonary macrophage in vitro. Environ Health Perspect 66:45-53 (1986).
Sorenson, W.G.: Mycotoxins as potential occupational hazards. Develop Ind Micr 31:205-211 (1990).
Stone, B.A.; Clarke, A.E.: Chemistry and biology of (1→3)-ß-glucans. La Trobe University Press, Victoria Australia, 1992.
Swanson, M.C.; Bubak, M.E.; Hunt, L.W.; Yunginger, J.W.; Warner, M.A.; Reed, C.E.: Quantification of occupational latexaeroallergens in a medical center. J Allergy Clin Immunol 94:445-451 (1994).
Terho, E.O.: Diagnostic criteria for farmer's lung disease. Am J Ind Med 10:329 (1986).
Thelin, A.; Tegler, Ö.; Rylander, R.: Lung reactions during poultry handling related to dust and bacterial endotoxin levels. Eur J Respir Dis 65:266-71 (1989).
Thomas, L.: The lives of a cell. Notes of a biology watcher. The Viking Press, New York 78 (1974).
Thorn, J.; Rylander, R.: Airway inflammation and glucan in damp rowhouses. Am J Resp Crit Care Med 157:1798-1803 (1998a).
Thorn, J.; Rylander, R.: Airways inflammation and glucan exposure among household waste collectors. Am J Ind Med 33:463-470 (1998b).
Torén, K.: Self reported rate of occupational asthma in Sweden 1990-2. Occup Environ Med 53:757-761 (1996).
Turjanmaa, K.; Alenius, H.; Mäkinen-Kiljunen, S.; Reunala, T.; Palosuo, T.: Natural rubber latex allergy. Allergy 51:593-602 (1996).
Ulmer, A.J.: Biochemistry and cell biology of endotoxins. Int J Occup Env Health 3:-s8-s17 (1997).
Unger, L.; Harris, M.C.: Stepping stones in allergy. Annals Allergy 32: 214-30 (1974).
Vaaranen, V.; Vasama, M.; Alho, J.: Occupational diseases in Finland 1984, Institute of Occupational Health, Helsinki (1985).
Vanhanen, M.; Tuomi, T.; Tiikainen, U.; Tupasela, O.; Voutilainen, R.; Nordman, H.: Risk of enzyme allergy in the detergent industry. Occup Environ Med 57:121-5 (2000).

Verhoef, J.; Kalter, E.: Endotoxic effects of peptidoglycan. In: ten Cate JW, Büler HR, sturk A (eds). Bacterial endotoxins: structure, biomedical significance, and detection with the Limulus Amebocyte Lysate test. New York, Alan R. Liss Inc, 101-112 (1985).

Vogelzang, P.F.; van der Gulden, J.; Folgering, H.; Kolk, J.J.; Heederik, D.; Peller, L.; Tielen, M.J.; van Schayk, C.P.: Endotoxin exposure as a major determinant of lung function decline in pig farmers. Am J Resp Crit Care Med 157:15-8 (1998).

Warren, C.P.W.: Extrinsic allergic alveolitis: A disease commoner in non-smokers. Thorax 32:567-69 (1977).

Whitehead, L.W.; Ashikaga, T.; Vacek, P.: Pulmonary function status of workers exposed to hardwood or pine dust. Am Ind Hyg Assoc J 42:178-186 (1981).

Williams, D.L.: (1→3)-ß-D-glucans. In: Organic dusts: exposure, effects, and prevention. Rylander R, Jacobs RR (Eds). Lewis Publishers, pp 83-85 (1994).

Williams, D.L.: Overview of (1→3)-ß-*D*-glucan immunobiology. Mediators of inflammation 6:247-250 (1997).

Willis, T.: Practice of Physick, Pharmaceutice Rationalis or the Operations of Medicine in Humane Bodies. London, 1678.

Witter, R.L.: Epidemiology of Marek's disease - a review. In: Biggs PM, de The G, Payne LN (eds). Oncogenesis and herpesviruses. Lyon: IARC, (1972).

Wolff, S.M.: Biological effects of bacterial endotoxins in man. In Bacterial Lipopolysaccharides, ed. Kass EH and Wolff SM. Univer-sity of Chicago Press, 251-256 (1973).

Zee van der, J.S.; Kager de K.S.; Kuipers, B.F.; Stapels, S.O.: Outbreak of occupational allergic asthma in a stephanotis floribunda Nursery. J Allergy Clin Immunol 103:950-952 (1999).

Zejda, J.E.; Barber, E.; Dosman, J.A.; et al.: Respiratory health status in swine producers relates to endotoxin exposure in the presence of low dust levels. J Occup Med 36:49-56 (1994).

Zwan van der, J.C.; Orie, N.G.; Kauffmann, H.F.; Wiers, P.W.; de Vries, K.: Bronchial obstructive reactions after inhalation with endotoxin and precipitinogens of Haemophilis influenzae in patients with chronic non-specific lung disease. Clin allergy 12:547-59 (1982).

Zock, J.P.; Heederik, D.; Kromhout, H.: Exposure to dust, endotoxin and microorganisms in the Dutch potato processing industry. Ann Occup Hyg 39:841-854 (1995).

Zock, J.P.; Hollander, A.; Heederik, D.; Douwes, J.: Acute lung function changes and low endotoxin exposures in the potato processing industry. Am J Ind Med 33:384-391 (1998).

6

BIOLOGICAL AGENTS – MONITORING AND EVALUATION OF BIOAEROSOL EXPOSURE

Dick Heederik, PhD, Peter S. Thorne, PhD, and Jeroen Douwes, PhD

INTRODUCTION

Bioaerosols comprise microorganisms as well as toxins or allergens that originate from higher plants or animals (see Chapter 5 for a definition). The assessment of exposures to bioaerosols offers distinct challenges from those for inorganic aerosols and chemical agents. Pathogenic microorganisms may be hazardous at extremely low levels, while other organisms may only become important health hazards at orders of magnitude higher concentrations. Some organisms and spores are extremely resilient, while others may be easily degraded in the sampling process. Certain fungal spores are easily identified and counted, while many bacteria are difficult to distinguish. Sensitive and specific methods are available for the quantitation of some biological agents, but there are no good methods for others.

Several methods for sampling and analysis of bioaerosols, especially sampling of viable microorganisms, have been developed since the early 1950s. Early research was carried out largely because of military concerns regarding the use of biological warfare. Sampling equipment first described in those days is still being used, despite some of the technical disadvantages, e.g., wall losses. Another issue is that most bioaerosol samplers do not comply with current size-selective criteria for respirable or inhalable dust sampling. For microorganisms, decisions regarding sampling duration are strongly influenced by the viability of the microorganisms. Because microorganisms are exposed to considerable stress during the sampling and analytical process, viable sampling is often based on short sampling times usually not exceeding 20 minutes, usually considerably shorter, to reduce viability loss. This may introduce a considerable measurement error that has to be dealt with in the overall exposure assessment strategy to avoid invalidating the representativenss of the sampling program. In addition, since one can usually not predict which microorganisms will be present in the environment to be sampled, one cannot reliably provide appropriate media and culture conditions *a priori.*

An important issue is that sampling and subsequent analysis may strongly influence the quality of the bioaerosol sample. This interaction is dependent on the specific sampler used and analytical procedures applied, and is especially of concern when microorganisms are being sampled. For instance, choice for a particular sampling volume and time determines the viability of microorganisms in the sample. Selection of specific media determine the growth conditions and, depending on the medium chosen, the growth conditions are optimal or sub-optimal for specific species. This might affect the generalizability of the measurement outcomes, but at least influences the results. In addition, storage and extraction procedures are often much more critical in the case of bioaerosol measurements, particularly when focusing on specific labile biological agents such as viruses, proteins, endotoxin, etc. Repeated freezing and thawing of samples for example may cause damage to the agents of study resulting in substantial underestimation of the true exposure. Storage conditions (including temperature, humidity, duration, etc.) of samples for viable microbial determinations are very critical since these conditions can seriously affect the viability of microbes. Therefore, interpretation of bio-aerosol measurement results is impossible without detailed information about the sampling and analytical procedures, and is, therefore, less straightforward than for many other agents. Selection of the measurement method, and for many hygienists in the field, the collaborating laboratory, is crucial. This issue may have little bearing on exposure data gathered evaluation of control measures where absolute values are not needed, but greatly affects health hazard evaluations.

This chapter will give an overview of the available methods for sampling and analysis of bio-aerosols. The methods include viable and non-viable sampling of microorganisms. Sampling and analysis of aeroallergens or toxins in bioaerosol receives considerable attention. In the last paragraphs, some issues regarding interpretation of data will be discussed in the context of health hazard evaluations.

ASSESSMENT METHODS FOR MICROORGANISMS

Measurement of microorganisms relies upon collection of a sample into or onto solid, liquid, or agar media with subsequent microscopic, microbiologic, biochemical, immunochemical or molecular biological analysis (Eduard and Heederik, 1998). Two distinctly different approaches are being distinguished for evaluation of microbial exposure: "culture-based methods," and "non-culture methods." Culture-based methods rely on the sampling of culturable microorganisms. Non-culture based methods, sometimes referred to as total bioaerosol methods, attempt to enumerate organisms without regard to viability.

Culture-based Methods

A variety of devices for microbial bioaerosol sampling have been developed over the years (see Table 6-1 and Figure 6-1). These devices and their use have been described previously (Buttner and Stetzenbach, 1991; Buttner and Stetzenbach, 1993; Chang et al., 1994; Jensen et al., 1992; Juozaitis et al., 1994; Nevalainen et al., 1992; Thorne et al., 1992; Eduard and

Table 6-1
Microbial bioaerosol sampling devices

Viable Sampling Methods	Examples of Devices Available
Stationary Jet-to-Agar Impactors	Andersen Microbial Sampler Burkard Portable Air Sampler for Agar
Plates	MicroBio Air Sampler
Rotating Media Slit-to-Agar Impactors	Slit-to-Agar Biological Air Sampler Mattson/Garvin Air Sampler Casella Airborne Bacteria Sampler
Centrifugal Agar Impaction Samplers	RCS PLUS Centrifugal Air Sampler
Liquid Impingers	Ace All-Glass Impinger SKC BioSampler Burkard Multi-Stage Liquid Impinger
Air Filtration Methods	Nuclepore Filtration and Elution Method Inverted Filter Method
Non-Viable Sampling Methods	
Stationary Slit-to-Slide Spore Samplers	Burkard Personal Volumetric Air Sampler Zefon Air-O-Cell Sampling Cassette
Moving Slit-to-Slide/Tape Spore Samplers	Burkard Recording Air Sampler Burkard Recording Volumetric Spore Trap Allergenco Air Sampler
Rotating Impaction Surface Sampler	Rotarod Spore Sampler
Virtual Impactor Spore Sampler	Burkard High Throughput Jet Spore Sampler
Analysis Methods for Liquid Impinger Solutions or Air Filter Samples	
Immunoassay Methods for Microbial Surface Antigens	
Non-Specific Staining Methods with Enumeration by Microscopy or Flow Cytometry	
Fluorescence *In Situ* Hybridization with Enumeration by Microscopy or Flow Cytometry	
Polymerase Chain Reaction Analysis	

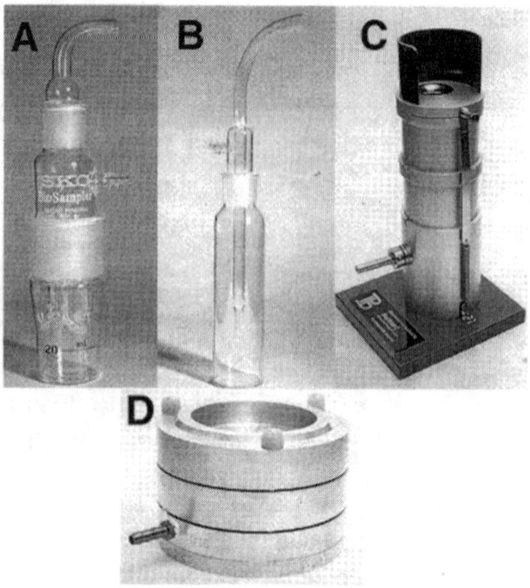

Figure 6-1. Some popular bioaerosol samplers. A) BioSampler; B) AGI-30; C) May 3-stage liquid impinger; D) Two-stage Anderson microbial sampler.

Heederik, 1998; Thorne and Heederik, 1999). There are three standard approaches to active sampling of culturable bio-aerosols:

IMPACTOR METHODS[1]

With impactor sampling, bioaerosols moving in the airstream pass through a round jet or a slit and impact onto agar in a Petri dish or on some other type of nutrient surface where they can grow, multiply and form colonies that can subsequently be enumerated. Correction of the counts is generally required with this method of enumeration to adjust for the probability of more than one viable organism passing through a single jet, and then upon growth, merging and being counted as a single colony. The probability of occurrence of this event increases as the concentration of culturable microorganisms increases. A measurement device-specific adjustment of colony counts has to be made. This is usually referred to as 'positive hole correction.' Correction tables for this correction have been published in the literature, but are usually delivered by the suppliers of sampling equipment upon purchase of a sampler (Andersen, 1958; Leopold, 1988; Macher, 1989). Multistage devices typically have 100 to 400 jets per stage and

---

[1] For a general review of aerosol physics, see Chapter 9 of Volume I. Other information on general aerosol sampling instruments, including impactors, can be found in Chapter 26 of Volume I.

allow size discrimination by sequentially increasing the velocity through the jet and decreasing the jet to plate spacing. Larger organisms deposit on the higher stages and spores as small as 0.6 μm are typically impacted on the final stage. Some jet-to-agar and slit-to-agar impaction samplers slowly rotate the collection media (held in a Petri dish) during sampling or advance a nutrient-coated strip to allow longer sampling times.

LIQUID IMPINGER METHODS[2]

Liquid impingers, such as the Ace all-glass impinger (AGI) and the SKC BioSampler, collect microorganisms by impinging the airstream onto the agitated surface of the liquid collection fluid. Most microorganisms and spores will be retained in the collection fluid such that when this fluid is plated onto a suitable agar in Petri dishes, colonies will emerge. Serial dilutions of the collection fluid into sterile isotonic solutions are generally required prior to plating to allow the numbers of colonies on the plates to be in a quantifiable range (30 to 300 per plate is desirable). Many different solutions have been used and several investigators have tested the efficacy of one fluid over another (Marthi and Lighthart, 1990; Thorne et al., 1994). The most commonly used sterile fluids are water, peptone water (1% peptone in distilled water with 0.01% Tween 80 and 0.005% antifoam A), betaine water (5 mM betaine with Tween 80 and antifoam A), phosphate-buffered saline with Tween 80, or diluted tryptic soy broth with antifoam A. Some investigators use water during sampling and will add nutrients after sampling. (Duchaine et al., 2000).

The BioSampler has a single inlet that separates into three nozzles that are oriented with the curved side of the glass reservoir. Thus, when air is drawn through the sampler, airborne particles pass through the nozzles and are entrained in the swirling impinger fluid via impaction and centrifugation (Willeke et al., 1998). Laboratory studies have compared the BioSampler performance to the AGI and suggest better removal efficiency and improved preservation of culturability for fragile organisms with the BioSampler (Lin et al., 1999, 2000). The multistage liquid impinger (MSLI) works in an analogous fashion to the AGI and the BioSampler but allows fractionation of the bioaerosol into three size fractions: less than 4 μm, 4 to 10 μm, and greater that 10 μm (Lange et al., 1997a; May and Harper, 1957).

AIR FILTRATION METHODS

Several sampling methods in common use rely upon air filtration to remove bioaerosols from a sampled air volume. Generally, the aerosols are removed by a combination of impaction on the filter surface as well as interception, electrostatic attraction, and diffusion deposition. After sampling, filters are agitated or sonicated in a solution to re-entrain the microbes. This solution is then serially diluted and plated onto appropriate culture media.

[2] General information on aerosol sampling using filters, impactors, etc. can be found in Chapter 22 of Volume I.

Growth Media

With all of the methods described above, polluted air is drawn into the sampling device that strips bioacrosols from the air stream. After sample collection, organism colonies of bacteria and fungi are grown on culture media held at a defined temperature in an incubator. Colonies are counted manually or with the aid of image analysis techniques as they emerge usually over a 3 to 7 day period. Laboratory and field blanks are included to control for contamination from the culture media or sample handling.

For all of the viable bioaerosol assessment methods described above, the culture medium is selected to test for broad-spectrum bacteria or fungi or to select for specific groups, genera, or species. The most commonly used media and incubation temperatures are: tryptase soy agar (TSA) or R2A (a low nutrient medium) with cycloheximide for broad spectrum mesophilic bacteria at 22-35°C, blood agar for mesophilic bacteria of human origin at 37°C, TSA for thermophilic bacteria at 50-55°C, eosin methylene blue agar (EMB) or MacConkey's medium (MAC) for Gm- bacteria at 22-35°C, and malt extract agar (MEA), rose bengal streptomycin (RBS) agar, or DG-18 for fungi at 22-30°C. Recipes for these media appear in standard texts (Difco, 1984; Gerhardt et al., 1994) and dry mixes are commercially available. If one is sampling to isolate a particular genus then media, antimicrobials, and growth conditions must be tailored to support the growth of that genus and to exclude competing organisms. This usually requires input of experts in environmental microbiology.

In many situations there are some *a priori* indications of what to sample for. Cooling towers have been associated with increased contamination by *Legionella pneumophila* serotypes, and sampling in cooling towers implies that because of this particular exposure risk at least part of the sampling effort will have to be focused on identifying the presence of this microorganism. However, *Legionella pneumophila* grow on selective plates under very specific conditions in order to reduce competition by other microorganisms, and this requires specialized microbiological input in order to guarantee that the species one is interested in will actually be detected when present.

After counting, results from viable sampling measurements are usually expressed as Colony Forming Units per cubic meter of air sampled (CFU/m^3). Results can only be interpreted if ancillary information is available with regard to sampling location, sample duration, sampler type, culture medium, culturing time, and incubation temperature.

Non-culture Based Methods

The various techniques that have been described in the literature can be divided into (1) microscopy and flow cytometry for counting individual spores or cells, and (2) chemical/bio-chemical methods for the measurement of specific microbial components. Flow cytometry is a process of measuring the characteristics of cells such as size and fluorescence as they pass single file in a moving fluid stream through a laser-illuminated detection field. For enumeration of microorganisms, cells are first labeled with a fluorescent probe that facilitates the detection.

SPORE COUNT METHODS

Many fungal spores have distinctive morphology at the genus level that allows enumeration with some specificity. Samples are collected using spore samplers such as the Burkard Personal Volumetric Air Sampler, the Burkard High Throughput Jet Spore Sampler, or the Zefon Air-O-Cell Spore Sampling (see Table 6-1). The former device is a portable, battery operated, slit-to-glass slide impactor that samples at 10 L/min. The High Throughput Jet Spore Sampler collects particles by virtual impaction with sedimentation into a sealed chamber. It operates at a very high flow rate (850 L/min). The Air-O-Cell is a particle-sampling cassette designed for collection and analysis of bioaerosols and particles for microscopic analysis. Particles are collected from the 15 L/min inlet air stream by impaction onto a glass cover slip coated with a sticky material. Particulate matter that has been collected on coated glass slides is then generally stained using lactol phenol cotton blue and counted under light microscopy by a trained analyst (Solomon, 1976). Spores are identified with the aid of reference samples and spore atlases (Smith, 1986; Samson et al., 1995) and quantified. Typically, at least the following designations are provided by the analyst: *Penicillium/Aspergillus*, *Alternaria*, *Cladosporium*, *Fusarium*, *Botrytis*, ascomycetes, basidiomycetes, zygomycetes, pollens, and unclassified. The Pan American Aerobiology Association has a more detailed target list of spore classifications (see website in Appendix). Methods for calculating airborne concentrations in cfu/m^3 have been described previously. Comparison of these methods to Anderson microbial sampler culture samples has demonstrated significant correlation for total fungi ($r=0.80$, $p<0.001$) and for Cladosporium ($r=0.65$, $p<0.001$) in indoor environments (DeKoster and Thorne, 1995).

DIRECT COUNT METHODS

These methods have also been described as non-specific, non-culture methods and are based on the observation that particular fluorescent labels can attach to or intercalate into resident nucleic acids (in DNA or RNA), thereby labeling the microorganisms in a sample. An important advantage of most direct count methods is that dust sampling equipment can be used, allowing respirable and inhalable personal exposure monitoring strategies. Indoor air samples are collected by air filtration or liquid impingement followed by fixation and staining of the organisms. Fixation with formaldehyde or glutaraldehyde serves to stabilize the structure of the microorganisms while preventing any further growth. Samples prepared for counting by flow cytometry should first be reviewed by microscopy to ensure that the sampled organisms are not agglomerated and that there is minimal non-specific staining of debris.

Counting is performed by microscopy under epifluorescence illumination or by flow cytometry to assess total counts of the microbes in the air as described below. Several recently developed techniques have improved the sensitivity of this bioaerosol direct count method. Laser scanning confocal microscopy (LSCM) is a recent advance that provides higher resolution and enhanced sensitivity over traditional methods. LSCM uses illumination from an expanded laser beam that is focused by the optics to a tiny spot beam. This spot beam is then used to scan the object via an x-y scanning mechanism. Incident laser light is passed through excitation filters and a dichroic mirror to direct the light onto a specimen labeled with fluorescent probes (e.g., fluorescent oligonucleotide probes, fluorescent antibodies, or fluorescent covalently binding dyes). The incident spot beam stimulates fluorescent emission from the specimen that is of a longer wavelength and is deflected by the dichroic mirror to the photomultiplier detection system. Reflected light from the specimen is not directed to the detector. A confocal aperture in front of the detector rejects the emitted fluorescent that came from points on the specimen out of the focal plane. New fluorochromes with higher degrees of emission intensity have been developed that improve the enumeration process since fewer probes need to bind to the cells in order for detection to occur. Monomeric and dimeric cyanines can be successfully used to stain the bacteria or bacterial cell walls with limited interference from background fluorescence from foreign materials, bacterial degradation debris, and other autofluorescing dust particles collected during the bioaerosol sampling.

Acridine orange (AO) has been routinely used as the staining agent for quantifying bacteria by counting individual organisms under fluorescence microscopy using the direct count method (Eduard and Aalen, 1988; Eduard et al., 1990; Kullman et al., 1998; Lange et al., 1997a; Lange et al., 1997b; Palmgren et al., 1986; Thorne et al., 1992; Thorne et al., 1994). This dye is widely used because it is inexpensive and has a characteristic narrow excitation band and relatively low background illumination. However, when attempting to enumerate small bacteria by fluorescence microscopy, differentiating cells from fluorescing dust particles may be subjective and tedious. In addition, AO binds to DNA in a non-specific manner and autofluorescence of organic debris can lead to over-estimation of bacterial counts. Acriflavine dye staining methods have been suggested to yield a more uniform and reproducible count of bacteria (Bergstrom et al., 1986). Several dyes have been developed for use in nucleotide chemistry but also have utility for bioaerosol enumeration. They bind to nucleic acids through intercalation, and thus label cells through DNA and RNA fluorescence. These poly-functional intercalating dyes have been used effectively at much lower concentration levels than monomeric dyes for intercalation with DNA (Ogura et al., 1994). It has been demonstrated recently that labeling with 4'-6-diamidino-2-phenylindole (DAPI) and counting organisms by flow cytometry can improve the sensitivity of this method by several orders of magnitude (Lange et al., 1997a). This makes direct counting a practical method for use in the evaluation of bioaerosols in indoor air when information on organism taxa is not needed.

Special Issues in the Evaluation of Microbial Exposure

There are a number of significant problems with measuring bioaerosol exposures and these are summarized in Table 6-2.

None of the sampling devices available for viable sampling is ideal for its purpose. Most viable sampling devices are relatively unchanged over time compared to gravimetric dust sampling devices. The choice for a specific sampler is, therefore, usually not taken only on the basis of the information about the performance or the specific particle size fraction sampled.

Table 6-2
Recognized problems with assessment of bioaerosol concentrations*

General problems:
Seasonal, temporal and spatial variability are large
Few methods are available to assess total bacterial concentration and taxa
Most available methods are labor intensive and tedious
Viable bioaerosol sampling methods:
Some methods require short sampling times
Some samplers have excessive loss of organisms through physicochemical trauma, re-entrainment, or wall losses
Airflow calibration is difficult with some sampling devices
Viability loss may occur through bombardment or solubilization of toxic vapors
Antibiosis and colony masking may lower colony yield
Selection of media and culture conditions affects colony growth
Viability may be irrelevant to the toxicity
Non-culture bioaerosol sampling methods:
Debris may be falsely enumerated as microorganisms
Large variability for methods employing microscopic enumeration
Agglomeration of organisms interferes with counting accuracy
Quantitative identification of specific organisms is difficult
Viability may be required for the disease process
Molecular biology assessment methods:
Dependent upon the access of the oligonucleotide probe to the organismal nucleic acids
FISH methods dependent upon sufficient rRNA in the bioaerosol
PCR methods are difficult to perform quantitatively
PCR for common organisms can be easily contaminated
Can not determine viability of the organisms
Immunoassay aeroallergen assessment methods:
Dependent upon the specificity of the antibody probe for the antigen
Standard antigens are not available for many aeroallergens
Non-specific binding of the antibody with sample debris

*Modified from Thorne and Heederik, 1999

Over the years, certain practices developed, and in many hygiene studies the Andersen microbial sampler was used. This is probably the most widely used sampler for viable measurement despite some of the obvious disadvantages such as high wall losses and low efficiency. Another problem with viable sampling is that reliable methods for personal sampling are not common. All glass impingers have been used extensively as an alternative to jet-to-agar samples and instead of direct plating methods, because small impingers are available for personal sampling.The disadvantage is the possibility of artificially inflated spore counts after culturing. This occurs because larger particles, which contain more than one viable spore (propagules), break up and form individual colonies after plating in a dilution series.

It is traumatic for microorganisms to undergo the process of sampling in any active sampling device. Trauma can reduce culturability for viable sampling and impede identification for non-viable methods. Most sampling devices require impaction on a solid or liquid surface at a high enough velocity to cause the aerosol to exit airflow streamlines and deposit. This velocity for two common devices, the AGI and the AMS in the N6 configuration is 26,500 cm/sec and 2400 cm/sec, respectively (Nevalainen et al., 1993). When particles strike the surface of the Petri dish or enter the collection fluid, their microenvironment changes in terms of temperature, hydration, osmolarity, nutrients, and oxygen tension. There is differential resilience to these shocks across bacterial and fungal species. In addition, some fungal spores are hydrophobic, while others can trap air on their surface making them buoyant. Trapping of these spores in a liquid impinger may be difficult because of re-entrainment into the air stream and loss from the sampler. Thus, whether one is using a slit-to-agar sampler, a jet-to-agar sampler, or a single or multistage liquid impinger, only a portion of the viable organisms present in the sampling environment will be detected.

Some commonly used sampling methods allow only short-term sampling due to their relatively low, upper limits of detection. For instance, agar plates can overgrow easily in plate impactor samplers such as the widely used Andersen sampler. Suppose an impactor sampler with 250 impaction holes is used to draw a sample for only two minutes at a flow rate of 40 L/min, and that the plate appears to be overgrown after three days of incubation at 24°C. This means that more than 250 colonies have been counted at a sampled volume of 80 liter. This is equivalent to a CFU count of 3125 per cubic meter. To raise this upper limit of detection, shorter sampling times are needed.

When bioaerosols are collected in a sampler, competitive organisms may inhibit each other through production of toxins. For viable bioaerosol sampling, this means that the range of airborne organisms present may not be observed in the sample or the numbers of viable bioaerosols will be underestimated. In a similar fashion, the choice of media on which to culture the organisms can also affect the emergence of colonies. Some organisms such as obligate plant pathogens, some anaerobes, and many mushroom spores will not grow under standard culture conditions. For those organisms that can be cultured, the media may not yield colonies of the organisms in the same proportion by genus or species as they existed in the environment that was sampled. The collection medium used in liquid impingers may differentially favor or disfavor the culturability of bioaerosols in the sampling environment. No standard solutions exist for many of these methodological issues; they remain to be resolved.

Vapors in the sampled environment may concentrate in the collection media and differentially affect the viability of organisms. For instance, AGI air sampling in a poultry or swine barn containing 35 ppm ammonia (24 mg/m^3) using de-ionized water as the collection medium may lead to a substantial change in the pH of the fluid and loss of sensitive organisms. With 20 minutes of sampling at 12.5 L/min in an environment with 35 ppm ammonia and the collection fluid at 20°C, the impinger solution would be expected to increase to 0.01 N or more and the pH to exceed 10.6. In addition, sampling in dry environments may lead to unacceptable loss of collection fluid and concentration of organisms (Lin et al., 1999). For samplers that impact organisms onto agar with fixed agar plates, an organism collected early in the sampling period may continue to be bombarded for the rest of the sampling period. For liquid impingers, compounds highly soluble in water may accumulate in the impinger solution. With air filtration, sampling in dry environments leads to desiccation of the microbes reducing their culturability.

New Molecular Biological Approaches for Analysis of Microorganisms

Analysis of markers of exposure or direct measurement of toxins and aero-allergens, or application of molecular biological approaches receive more and more attention. The use of advanced methods, such as PCR based technologies, FISH (see section on Fluorescent In Situ Hybridization Techniques (FISH)), and immunoassays (see section on ASSESSMENT METHODS FOR BIO-ALLERGENS (High Molecular Weight Sensitizers)) have opened new possibilities for detection and speciation regardless of whether the organisms are culturable (Lange et al., 1997a; Eduard and Heederik, 1999).

POLYMERASE CHAIN REACTION TECHNIQUES (PCR)

Recent developments involve polymerase chain reaction (PCR) techniques that allow one to amplify small quantities of target DNA typically by 10^6 to 10^{10} times, and determine in a qualitative manner the presence of specific microorganisms. PCR analysis has been automated and can provide turn around times of two to three hours rather than the three to five days required for culture techniques (Atlas, 1991). PCR has been shown to be useful for analysis of water samples for microbial contamination (Bej et al., 1990) and clinical specimens (Eisenach et al., 1991; Kolk et al., 1992). Viruses that cause human diseases, such as influenza, are spread through the air via droplet nuclei. These are 5 to 10 μm aerosols containing large numbers of viral particles that are produced by coughing or sneezing. Droplet nuclei containing RNA or DNA viruses can be collected by air filtration or impinger methods and assayed primarily by PCR and related molecular methods.

The application of qualitative PCR to sampling the airborne environment has been demonstrated (Alvarez et al., 1994; Khan and Cerniglia, 1994). Quantitation of PCR for bioaerosol samples can be accomplished using most probable number (MPN) techniques with PCR or through the real-time PCR using sequence detection systems. In the classical MPN method for bacteria, replicate cultures of serial dilutions of a sample in an appropriate medium are incubated for several days and then graded as + or – for colony growth. The fraction of culture tubes or Petri dishes at each dilution that exhibit microbial growth is recorded. The number of cultures of bacterial cells that are positive for growth will follow a Poisson distribution. Thus, one may estimate the number of cells that were present in the original sample.

It is possible to apply this technique in combination with PCR to quantify microorganisms. In this method lysed bacteria contribute DNA strands for amplification and the PCR products appear as a band on an agarose gel. Theoretically, a single organism can result in a detectable band although in practice the detection limit is closer to 10 organisms. Previous researchers have studied saprophytic soil bacteria (Degrange and Bardin, 1995; Picard et al., 1992), genetically engineered microorganisms (Recorbet et al., 1993), and pathogens in the intestinal environment (Miwa et al., 1997). In these studies, DNA was extracted and purified from the sample and MPN-PCR was performed on purified DNA. These purification steps lead to important losses of DNA and impair the ability to attain good quantitation by MPN-PCR.

A recent development is the application of sequence detection systems such as the TaqMan system. With this approach, one detects and monitors target DNA sequences in real time. During PCR, an oligonucleotide probe containing both a reporter fluorophore and a quencher dye, anneals between the forward and reverse primers. When the probe is cleaved at the 5' end by the DNA polymerase, the reporter fluorophore is separated from the quencher dye and a sequence-specific signal is generated. As the PCR amplification proceeds, more of the reporter fluorophore is released, and the fluorescence intensity increases. If, for instance, primers are selected that are complementary to, and bind specifically to *Legionella pneumophilia*, then the fluorescence increase during PCR will be dependent upon the number of *L. pneumophilia* cells present in the sample. Application of quantitative PCR for analysis of air samples containing microorganisms is still under development by several specialized research laboratories, but is expected to find applications in situations where specific infectious microorganisms may be present.

FLUORESCENT IN SITU HYBRIDIZATION TECHNIQUES (FISH)

Advances in molecular biology have made *in situ* hybridization of microorganism-derived ribosomal ribonucleic acid (rRNA) labeled, oligonucleotide probes, a potentially useful method for characterization of organisms in bioaerosols at the family, genus, or species level, regardless of their culturability. Oligonucleotide probes are short, single-stranded, synthetic deoxynucleotide (DNA) base sequences. Oligonucleotide probes can now be made quickly and economically, there are publicly accessible databases of rRNA sequences, and a variety of non-radioactive fluorescent labels can be used avoiding the inherent handling problems associated with autoradiography. Potential advantages of FISH techniques include identification of non-culturable organisms, highly sensitive and species-specific enumeration, and automated analysis.

The identification of single microorganisms by FISH uses probes that are complementary to unique sequences of 16s rRNA contained within every ribosome. These strands of nucleotides will bind to complementary sequences on rRNA. The nucleotide bases pair as follows: adenine pairs with uracil and guanine pairs with cytosine. rRNA consists of phylogenetically conserved and variable nucleotide regions allowing determination of taxon at any desired level: group, genus, or species. Strands with as few as 16 bases in length can be made that will bind to one, and only one, microorganism species. Each individual active bacterial cell contains approximately 10^4 to 10^5 ribosomes with the same number of identical copies of rRNA. The number of rRNA copies within a bacterium is proportional to the rate of growth. Thus, the rRNA represents a natural amplified target site for probe hybridization. Detection of hybridization has been accomplished with epifluorescence microscopy (Delong et al., 1989), scanning confocal microscopy for higher sensitivity (Bauman et al., 1990), and flow cytometry for automated enumeration (Amann et al., 1990). These techniques were developed using pure cultures at high concentrations at the optimal growth stage.

Recently, quantification of bioaerosols using FISH coupled with flow cytometry has been demonstrated for laboratory-generated bioaerosols and air samples collected in the field using both the AGI and the MSLI (Lange et al., 1997a). Specific quantitation was accomplished using three oligonucleotide probes with different fluorophores: a general probe complementary to all eubacteria, a specific probe complementary to all species within the genus *Pseudomonas*, and a nonsense probe that controlled for non-specific binding. This produced differential labeling of pseudomonads from other eubacteria and facilitated specific enumeration by flow cytometry. The application of FISH to environmental bioaerosol samples is presently limited due to the lower numbers of rRNA targets in these cells compared to cultured organisms captured in the logarithmic growth phase. Work is in progress to develop methods to amplify the DNA *in situ* using PCR and then to introduce labeled probes specific for the DNA. New techniques for specific quantitation of bioaerosols regardless of viability using molecular biology methods are likely to produce major advances in exposure assessment to bioaerosols.

ASSESSMENT METHODS FOR CONSTITUENTS OF MICROORGANISMS

In addition to counting culturable and non-culturable microbial propagules in air or settled dust, molecular constituents or metabolites of microorganisms can be measured as an estimate of microbial exposure. Toxic and/or allergenic components can be measured, but also non-toxic molecules may serve as markers of either large groups of microorganisms or of specific microbial genera or species. Some markers for assessment of mold concentrations include ergosterol (Seitz et al., 1977; Miller and Young 1997), chitin (Donald and Morcha, 1977), and for bacteria muramic acid or bacterial peptidoglycans are sometimes measured as markers for bacterial biomass present (Fox et al., 1994; Mielniczuk et al., 1995). Genus- or species-specific markers include mold extracellular polysaccharides measured with specific enzyme immunoassays (Douwes et al., 1999), allowing the partial identification of the mold genera and assay of specific mold allergens using the same technique. Other agents such as β(1→3)-glucan (Aketagawa et al., 1993; Douwes et al., 1996) and bacterial endotoxin are being measured because of their toxic potency.

Most of the non-culture-based techniques described above are in an experimental phase and have not been routinely applied and/or are not commercially available, with the exception of endotoxin. Some of these methods may have disadvantages such as being too unspecific (chitin, $\beta(1\rightarrow3)$-glucan), or are too laborious for use in occupational health surveys or large field studies (scanning electron microscopy, ergosterol). Important advantages include (1) the stability of most of the measured components, allowing longer sampling times for airborne measurements, and frozen storage of samples prior to analysis, (2) the use of laboratory standards, and (3) the enhanced possibility to assess reproducibility. A few of the more frequently used markers will be reviewed in greater detail in the following paragraphs.

Endotoxins

ANALYSIS OF ENDOTOXIN

Analytical chemistry methods for quantification of lipopolysaccharides (LPS) have been developed employing gas chromatography-mass spectrometry (GC-MS) (Sonesson et al., 1988, 1990). These methods require special LPS extraction procedures and have not been widely used. Immunoassays have also been developed but have not been widely adopted either. Antibodies produced against LPS recognize the polysaccharide part (Rietschel et al., 1996) and not the toxic lipid A moiety. More importantly, since the biological activity of endo-toxins and LPS from various organisms varies depending upon the chemical structure of the particular endotoxin molecule, absolute chemical and immunochemical methods may not re-present the potency as well as a functional bioassay.

The discovery by Bang in 1956 that Gm- bacteria could induce gelation of lysate prepared from blood cells of the horseshoe crab, *Limulus polyphemus* (Bang, 1956), led to the development of the so-called *Limulus amebocyte* lysate (LAL) gel clot assay for endotoxin (Levin and Bang, 1968). The LAL contains a mixture of proenzymes that are activated by endotoxin in a cascading set of reactions (Iwanaga, 1993). Modern versions of this assay are based on the activation of the clotting enzyme present in mixed lysates of a large number of *Limulus polyphemus*.

Several LAL-based assays have been developed including turbidimetric and chromogenic methods (Iwanaga, 1979; Teller et al., 1979). The turbidimetric method measures an increase in light scattering in proportion to endotoxin concentration, whereas the chromogenic methods measure an increase in chromophore release from synthetic substrates with increasing endotoxin concentrations. The turbidimetric techniques are most frequently used in the kinetic mode, where the rate of increase in turbidity is measured over time (Remmilard et al., 1987). The chromogenic methods traditionally used acid-quenched endpoint measures, where the absorbance at a given wavelength following a specific incubation period is compared to standard endotoxin concentrations treated in the same manner (Dunér, 1993). However, to date the endpoint version of the chromogenic test is considered inferior compared to the kinetic version in which the increase in optical density over time is measured.

There are now well-developed commercial LAL-based endotoxin assays using kinetic chromogenic types of measurements in microtiter plates (Cohen and McConnell, 1984; Milton et al., 1992). These have been evaluated as to their robustness (Douwes et al., 1995; Hollander et al., 1993; Milton et al., 1997; Thorne et al., 1997). A new biochemical method different from the LAL assay is currently under development by several companies. This is a direct fluorescence method that uses a labeled, recombinant DNA-prepared, LPS binding protein reactive with endotoxin.

Endotoxin activity in the LAL assay is highly correlated to the production of pyrogenicity in the rabbit (Baseler et al., 1983; Ray et al., 1991; Weary et al., 1980) and to pulmonary effects in the guinea pig, mouse and human (Deetz et al., 1997; Gordon et al., 1994; Michel et al., 1997; Rylander, 1994; Schwartz et al., 1994; Schwartz et al., 1995; Thorne et al., 1996). Thus, because the test results correlate strongly with physiologic responses, the LAL assay is considered a highly sensitive bioassay with physiologic relevance. With the LAL test, predominantly cell wall-dissociated endotoxins are detectable (Sonesson et al., 1990). In contrast, animal inhalation experiments have shown that cell-bound endotoxins have similar or even increased toxicity compared to free endotoxin (Burrell and Shu-Hua, 1990; Duncan et al., 1986). Different Gm-bacterial species or even strains have shown differing degrees of inducing the LAL coagulation reaction.

The LAL assay is at present the most widely accepted assay for endotoxin measurements that has been adopted as the standard assay for endotoxin detection in pharmaceutical products by the U.S. Food and Drug Administration (FDA, 1980). The European Committee for Standardization (CEN, 2001) has published a first draft protocol for endotoxin measurement in the air. This protocol proposes the use of the more recent kinetic chromogenic versions of the LAL-assay. Versions of this assay are very sensitive and have a broad measurement range (0.01 - 100 Endotoxin Units (EU)/ml » 1 pg/ml - 10 ng/ml). The detection limit for airborne endotoxin is approximately 0.05 EU/m^3 (5 pg/m^3).

Since different test batches may give different results, an internal standard must be used. The U.S. FDA has established a Reference Standard Endotoxin (currently RSE: *E. coli*-6 {EC-6}) as part of their standardization procedures. The activity of all endotoxin analyses must, therefore, be referenced to RSE:EC-6 to be considered valid.

Large differences in both the hydrophilic, and to a lesser extent, the lipid A moiety between endotoxins of different bacterial species or strains make a comparison on weight basis almost meaningless (Cooper, 1985). Results are, therefore, expressed in Endotoxin Units (EU). The RSE:EC-6 is based on purified LPS from *E. coli* and is expressed in Endotoxin Units (EU) which is a measure for LAL activity. Since the RSE is expensive and in limited supply most laboratories use a Control Standard Endotoxin (CSE) that is standardized to the RSE. The CSE is normally included in commercial LAL assay kits and allows results to be expressed in EU.

SAMPLING AND ANALYTICAL PROCEDURES

Environmental monitoring of endotoxins is usually performed by sampling airborne dust on filters with a subsequent aqueous extraction. Several types of filters are commonly used: glass fiber, polycarbonate, cellulose, polyvinylchloride, polytetrafluoroethylene, or polyvinylidene difluoride. For filter extraction, most laboratories use pyrogen-free water or buffers like Tris or phosphate triethylamine, with or without detergents such as Tween-20, Tween-80, Triton-X-100 and saponin. The use of dispersing techniques may be beneficial in the LAL-assay. The most common method for extraction is rocking or sonication of filters in extraction media, or a combination of both.

Although studies have been published on optimization of filter choice, filter extraction methods, extraction buffers and choice of glassware (Douwes et al., 1995; Gordon et al., 1992; Olenchock et al., 1989; Milton et al., 1990; Novitsky et al., 1986; Thorne et al., 1997), no generally accepted protocol exists. The European Committee for Standardization is currently developing a standard protocol and the American Society for Testing and Materials (ASTM) has established a provisional standard method for metal working environments.

Another approach is to analyze endotoxins in collection fluid from liquid impingers. This allows for analysis of airborne microorganisms and endotoxins from the same sample. No standard method exists for sample storage and extraction. All glassware used during extraction, storage, and analyses should be rendered pyrogen free by heating at 190°C for four hours.

ASSAY INTERFERENCE

A number of studies have investigated interferences with the LAL assay and have attempted to minimize those (Douwes et al., 1995; Hollander et al., 1993; Thorne et al., 1997). These studies have demonstrated that results may vary depending upon the sample matrix, the extraction method, and the assay method. Other constituents present in the sample may interfere with the LAL assay and cause inhibition or enhancement of the test, or aggregation and adsorption of endotoxins, resulting in under- or over-estimation of the concentration.

In addition to LPS, β(1-3)-glucan, which is present in the cell walls of most fungi and some plants and bacteria, is also capable of triggering the LAL coagulation system (via an alternative activation pathway of the enzyme complex, the so-called factor G) (Iwanaga, 1993; Rylander et al., 1994) unless the glucan clotting pathway is deliberately blocked. Cell wall peptidoglycans of Gm+ bacteria can also activate the LAL system via factor G, though orders of magnitude higher concentrations are required (Brunson et al., 1976; Mikami et al., 1982; Morita et al., 1981; Tanaka et al., 1991). Interference by LAL-reactive glucans and peptidoglycans may differ for assay kits from various suppliers, though this issue has not been dealt with in detail.

Techniques such as spiking with known quantities of purified endotoxin and analysis of dilution series of the same sample have been described to deal with these interferences (Hollander et al., 1993; Milton et al., 1990, 1992; Whitakker, 1988). Studies in the laboratories of the authors have demonstrated within-laboratory coefficients of variation (see Chapter 7, Volume 1) between 15 and 20% in routine assay work. Several inter-laboratory comparison studies have been performed and demonstrate much greater variability.

One in-depth comparison of two laboratories experienced in the LAL assay compared results of assays performed on sets of 12 simultaneous samples under varying conditions (Thorne et al., 1997). Regression analyses demonstrated variability attributable to the assay method (kinetic vs. endpoint chromogenic LAL assay), air sampling filter type (glass fiber vs. polycarbonate) and interactions between the assay method and filter type. Intra-laboratory correlation coefficients were 0.92 across extraction methods and 0.98 across filter type. Reliability coefficients within methods for analysis of 12 paired filters were all greater than 0.94.

Two multi-laboratory round robin studies have been performed. The first included six laboratories (Reynolds et al., 2002). Excluding one laboratory that was an outlier, the remaining laboratories agreed to within ½ log unit. The correlation coefficients for pairs of these five labs were all in excess of 0.85. However, between and within lab standard deviations are the more common parameters reported (see standard practice E691 of ASTM on interlaboratory data analysis).

A second round robin test was performed with 8 to 10 laboratories analyzing endotoxin in air samples from generated cotton dust (Chun et al., 1998). When laboratories used their typical extraction and analysis methods, the results were within 0.83 log unit (again excluding one outlier). When the extraction method was harmonized in a subsequent round of testing, the results still spanned 0.77 log unit. In spite of the above-mentioned difficulties, it is generally accepted that a harmonized LAL-based endotoxin assay method will allow adequate assessments of exposure for standard setting.

Glucans

There are currently three principal methods in use for the assay of β(1→3)-glucans. Two are based upon the bioactivity of this molecule in the factor G-mediated *Limulus* coagulation pathway (Iwanaga, 1993; Morita et al., 1981). The third method is an immunoassay (see section on Assessment Methods for Bio-Allergens, high molecular weight sensitizers; Douwes et al., 1996). The amebocyte lysate from *Limulus polyphemus* or its Asian relatives *Tachypleus tridentatus, T. giga,* or *T rotundicauda*, coagulates upon contact with LPS via factor C and β(1→3)-glucans via factor G. Thus, an LAL assay for glucans can be designed by either disabling the factor C pathway (Roslansky and Novitsky, 1991) or by extraction and purification of components of the factor G pathway from the lysate (Nagi et al., 1993; Obayashi et al., 1995). Commercial kits using this approach are now available. An additional approach is under study that utilizes a monoclonal antibody against the LPS-factor C complex in the LAL assay to inactivate the factor C pathway allowing the glucans to stimulate the cascade via factor G. Recently, a protein that specifically binds β(1→3)-glucans was isolated from the amebocyte lysate of *T. tridentatus* (Tamura et al., 1997). This *Tachypleus*-glucan binding protein (T-GBP) was used in an immunoassay for detecting glucans.

The third method is an immunoassay (see section on Assessment Methods for Bio-Allergens, high molecular weight sensitizers; Douwes et al., 1996). In this assay, β(1→3)-glucans in the sample inhibit the binding of affinity-purified rabbit anti-glucan antibodies to the glucans coating a microtiter plate. Quantitation is achieved by labeling the rabbit antibodies with an enzyme-linked anti-rabbit antibody. There should continue to be advancements in methods for the analysis of β(1→3)-glucans since they may serve as important markers for indoor fungal contamination.

Mycotoxins

Mycotoxins possess distinct chemical structures and reactive functional groups that include primary and secondary amines, hydroxyl or phenolic groups, lactams, carboxylic acids and amides. Standard analytical methods for mycotoxin analysis from grains, feeds, and food products have been developed, thoroughly reviewed by professional associations and international agencies, and results from inter-laboratory studies have been compared (FAO, 1990; Van Egmond, 1989). A combination of analytical methodologies must be used to evaluate the wide spectrum of mycotoxins. These include thin layer chromatography (TLC), high-performance liquid chromatography (HPLC), solid phase extraction (SPE), gas chromatography (GC), UV spectrophotometry, nuclear magnetic resonance spectroscopy (NMR), GC-mass spectrometry (MS), and LC-MS. Accepted methods for the analysis of mycotoxins are documented in the official methods book of the Association of Official Analytical Chemists (Scott, 1990) and the International Agency for Research on Cancer (Van Egmond, 1991). Simultaneous determination methods for aflatoxins, ochratoxin A, sterigmatocystin, and zearalenone have been evaluated and detection limits for multianalytes have been determined (Soares and Rodriguez-Amaya, 1989). Several analytical methods such as TLC, HPLC, GC-MS, radio immunoassay (RIA), and enzyme linked immunosorbant assay (ELISA) have been reviewed for the analysis of mycotoxins such as aflatoxins, deoxynivalenol, fumonisins, zearalenone and their metabolites (Bauer and Gareis, 1989; Richard et al., 1993). These immunologic techniques (RIA, ELISA) may be particularly useful for screening large numbers of samples for an array of mycotoxins. There have been only a small number of methods developed specifically for sampling mycotoxins in indoor air environments (Jarvis, 1990).

Extracellular Polysaccharides (EPS)

Nearly all soil biomass is produced by microorganismal degradation of higher organic structures into humus. These organic structures include chitins, glucans, and mannans, heat stable glycoproteins often referred to as Extra Cellular Polysaccharides (EPS). Both bacteria and fungi are important producers, but fungi contribute the greatest proportion (Schlegel, 1993). As fungi enzymatically degrade material such as cellulose, starch, and pectin they produce polysaccharides. Some of the material consumed is incorporated into the cell wall of the fungi. Chemical methods for the analysis of fungal polysaccharides have been developed for the food and animal feed industry, but these have not found wide application in indoor air studies. Although EPS can induce an IgG-response both in man and experimental animals, there is, at present no evidence for a pathogenic role of EPS in mold induced allergic or inflammatory reactions. Thus, EPS may be primarily regarded as a potentially good marker for fungal biomass in environmental samples.

Two very important molds that appear indoors and can induce allergy are *Aspergillus* and *Penicillium*. These genera are particularly important in buildings where water damage is evident. These organisms are ascomycetes that have a characteristic conidial stage that gives rise to large numbers of inhalable allergenic spores. Their cell walls consist primarily of α(1-3)-glucan and β(1-5)-galactan (Leal et al., 1992). Since these organisms are also food storage molds, there has been research into methods for monitoring food for their presence. Antibodies have been developed against certain extracellular polysaccharides from these organisms (β-D-galacofuranosyl residues from mycelial galactomannans) (Kamphuis et al., 1992; Notermans and Soentoro, 1986) that were then used to develop an inhibition ELISA to detect EPS-*Aspergillus/Penicillium* in house dust samples (Douwes et al., 1998). Epidemiologic studies are underway in which the EPS-*Aspergillus/Penicillium* assay is being used as an exposure variable (see Wouters et al., 2000).

Ergosterol

Ergosterol, a primary cell-membrane sterol of most fungi, may be another suitable marker of fungal mass in air samples (Miller and Young, 1997; Newell, 1992, Dillon et al., 1999). The quantity of ergosterol in a fungal spore is a function of surface area and growth conditions (Newell, 1992). The ratio of ergosterol to spore mass for 11 species of fungi commonly found in indoor air has, however, been demonstrated to be reasonably constant at 1 ± 0.25 mg/g (Miller and Young, 1997). In most environments, ergosterol will be a specific measure of fungal mass because it is not present in vascular plants, although it is found in algae and protozoa (Newell, 1992). Ergosterol is determined by gas chromatograpy with mass spectrometric detection (GC/MS). There is experience in measuring ergosterol in house dust and air (Miller and Young, 1997; Newell, 1992; Saraf et al., 1997; Dillon et al., 1999).

Fungal Volatile Components

Volatile organic compounds produced by fungi may be suitable markers of fungal growth (Dillon et al., 1996). It has been suggested that these compounds may also cause respiratory symptoms, but no supporting evidence has been offered (Norback et al., 1994). They are collected on a solid sorbent (Anasorb 747), extracted with methylene chloride and determined by GC/MS. The method is sensitive, allowing the determination of a concentration of about 10 ng/m3 in 25-L air samples. 3-Methyl furan has been used as a measure of active fungal growth; 1-octene-3-ol as a measure of inactive growth; and geosmin as an indicator of either active or inactive growth (Dillon et al., 1999).

ASSESSMENT METHODS FOR BIO-ALLERGENS (HIGH MOLECULAR WEIGHT SENSITIZERS)

Immunoassays for Measurement of Allergens

Antibody-based immunoassays, particularly enzyme-linked immunosorbent assays (ELISA), are widely used for the measurement of aeroallergens and allergens in settled dust in buildings. These assays use antibodies, with specificity for the target aeroallergen, to bind the allergen to a micro-titer plate. Detection is provided via reaction of a substrate with an enzyme-linked second antibody that immunologically recognizes and binds to the anti-aeroallergen primary antibody. The antibody-bound enzyme reacts with the substrate to produce a spectral shift or color change that can be measured in a microplate reader.

Specific antibodies for these assays may be collected from a seropositive patient or a group of patients (a patient pool) or may be raised in rabbits immunized repeatedly with crude or purified antigen accompanied by an adjuvant. Cloned cells derived from mouse lymphocytes producing a single antibody can produce monoclonal antibodies *in vitro*. If non-human antibodies are being used, validation studies are needed to ensure that the antibody is directed to the allergen of interest, for instance by immunoblotting techniques (see Figure 6-2). Monoclonal antibodies offer enhanced specificity over polyclonal antibodies from animals or humans.

Samples are generally assayed at multiple dilutions and a titer is determined based on the highest dilution where the optical density exceeds a defined threshold. Standard allergens are

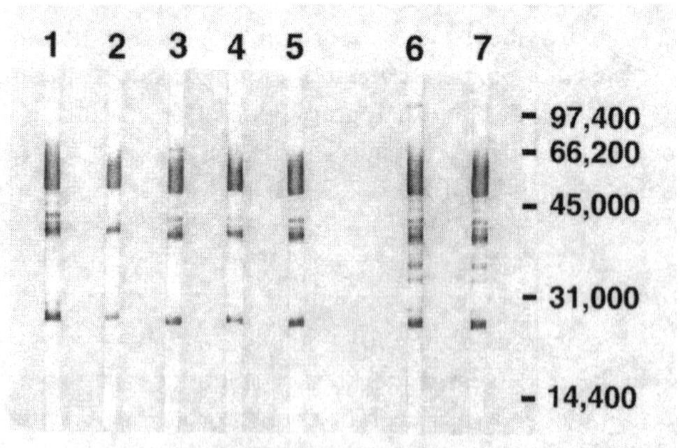

Figure 6-2. Immunoblotting of a commercially available α-amylase preparation. Reaction with human IgE from sensitized bakers (1-5) and affinity purified rabbit polyclonal antibodies (6-7). Both antibody sources react with similar allergens, illustrating that an assay to measure amylase in the air using one of the two antibody sources will measure the same allergens in the air (from Houba et al., 1997).

used as a basis for assigning units to the assay. ELISA's can also be run as inhibition assays where standard allergen is added to the sample at defined concentrations and the loss of antibody binding is assessed. In radio-immunoassays (RIA), radio-labeling is used for detection rather than enzymatic reaction with a substrate.

The principals of these methods are described in standard clinical immunology texts and their use in aeroallergen detection has been reviewed (Pope et al., 1993). To date, the house dust mite allergens, *Der p* I, *Der f* I, and *Der p/f* II have been most widely investigated and the methods have been well described (Platts-Mills and Chapman, 1987; Platts-Mills and de Weck, 1989; Price et al., 1990). Methods for assessment of exposure to allergens from animals (Eggleston et al., 1989; Hollander et al., 1997; Schou et al., 1991; Swanson et al., 1985; Virtanen et al., 1986; Wood et al., 1988), cockroaches (Pollart et al., 1991), storage mites (Iverson et al., 1990; van Hage-Hamsten, 1992), and Latex rubber (Miguel et al., 1996) have also been published. Assays are also available for the measurement of bio-technologically produced allergens such as fungal α-amylase derived from *Aspergillus oryzae* (Houba et al., 1997).

Before using assays under field conditions, the sensitivity for other potential allergens from the same environment needs to be established in order to avoid false positive results due to these other allergens. This is usually done by inhibition tests with extracts from materials present in the environment of interest (see Figure 6-3).

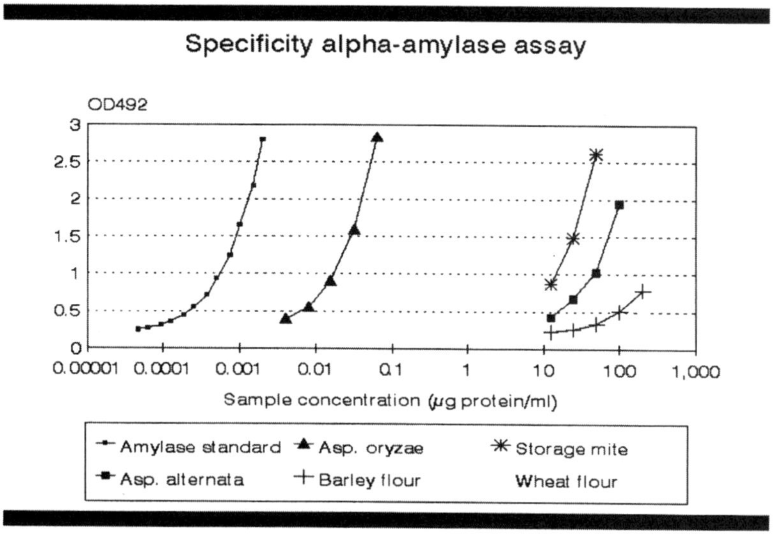

Figure 6-3. Evaluation of the specificity of an immunoassay for measurement of α-amylase as described by Houba et al. (1997). Different inhibition curves have been produced for a range of agents also present in the baking environment (OD492, measured optical density at 492 nm vs. sample concentration). An extract from Aspergillus oryzae, the organism which is being used for production of α-amylase shows a relatively good inhibition. Other agents inhibit only at very high concentrations, which makes interference of the assay by these agents under normal conditions unlikely.

Little is known about the validity and comparability of immunoassays that have been used to measure allergen levels in the air. Two studies clearly indicate that differences in extraction procedures and the use of extraction buffers contribute to differences between assay results up to at least a factor 10 (Zock et al., 1996; Hollander et al., 1999). The most important determinants that contribute to differences between assays seem to be the antibody source (monoclonals vs. polyclonals), type of immunoassay (inhibition vs. sandwich), and the standard used (purified allergen vs. crude extract). Although few comparative studies exist, the available evidence suggests that while the correlation of allergen levels obtained from different assays is very good, large systematic differences seem to occur (Hollander et al., 1998; Renström et al., 1999). The available literature is limited to a few allergens, and generalizations cannot be made. Validation and comparison of different assays by inter-laboratory comparisons are important issues that have not received the attention needed. These experiences are applicable to any allergen, including those from the house dust mite. This means that comparison of assay results with *ad hoc* exposure standards or even other published work is only valid when measurement and analytic conditions are comparable.

EVALUATION OF BIOAEROSOL DATA FOR HEALTH HAZARD EVALUATIONS

How to Sample

Usually, an exposure assessment strategy gives answers to the questions who should be sampled, where, and when (see Volume 1, Chapter 14). The answers are usually similar, although in some cases differences exist in compraison with strategies for chemical agents. Generally, there is a strong preference for air sampling during normal operations. However, in some indoor air studies, collection of short-term air samples has been performed by aggressive re-aerosolization of settled dust (Pope et al., 1993; Rylander et al., 1992). For studies of some aeroallergens such as house dust mites, vacuum collection of dust from floors, carpets, upholstery, or pillows may be more predictive of occupant health status since these aeroallergens are typically associated with larger particles that settle rapidly (Platt-Mills et al., 1992). Thus, one may find indoor aeroallergen levels expressed per cubic meter of air, per square meter of surface, or per milligram of collected dust.

Sampling Duration and Sampling Frequency

Viable sampling is usually short term, because longer sampling times will affect the viability of the microorganism. Thus, while one has a strong preference to sample for a period of a full work-shift this is not possible. There are large temporal, seasonal, and spatial variations in bioaerosol composition and level that relate to the specific activities that disperse the bioaerosols as well as those conditions that lead to sporulation of organisms. The short duration of viable sampling reflects this extreme variability.

One of the few papers that deals with this issue is a study on variability in viable microorganism levels in houses (Verhoef et al., 1992). The authors of this paper took repeated samples over a five week interval in 25 houses. The sampling time was two minutes with the N6 modification of an Andersen sampler. The total CFU was measured as well as the number of CFU for specific mold species. The inter-house and intra-house variability in exposure was calculated using analysis of variance for repeated measurements (see Table 6-3).

The table shows that the intra-house variance (or the variability in exposure from time to time) is high. For instance, the intra-house variance for the total CFU count is 0.7 and expressed as a Geometric Standard Deviation (GSD), is 2.3 (exp $0.7^{0.5}$). For *Aspergillus sp.*, the complete variance in exposure is explained by intra house variability and had a GSD of 4.3, which is extremely large. The information on the intra-house variance (or day to day variance in exposure) can be used to calculate confidence intervals for single measurements, or used in conventional power calculations to calculate the number of samples necessary to distinguish a predefined difference between two houses. Variance ratios of 3-4 imply that the variability over time is large compared to the real differences between houses. To a large extent, this is the result of very short sampling times. Measurements over a longer period would have shown a smaller within house variance, but sampling over a longer time using variable sampling methods is usually not feasible. As a result, many samples have to be taken to obtain a reliable estimate of the average exposure per house and to be able to distinguish differences in exposure between houses. One can calculate an appropriate sample size using information about the variance components (variability over time and between houses). For example, consider studies in which relationships between determinants of exposure, e.g., humidity, occupancy, or health effects are regressed on the measured exposure. For such a study, given the above variabilities, 27-36 samples per house would be required to estimate the average exposure sufficiently reliably so as to obtain less than 10% bias in the estimated slope versus the true slope. Put another way, assume that the true intra-house variability is 1.0, approxiamtely the value in Table 6-3. Note

Table 6-3
Intra- and inter-house variance of log values of total number of Colony Forming Units and counts of specific mold species

	Intra-house variance	Inter-house variance	Variance ratio
Total CFU	0.7	0.2	4
Aspergillus	2.1	0	∞
Cladosporium	1.3	0.3	4
Penicillium	0.5	0.4	1
Wallemia	2.0	0.1	19

Source: Verhoeff et al., 1992

that this is a log value and so the geometric standard deviation is 2.7. If one wanted to estimate the mean for a single house with an accuracy such that the upper confidence limit on the estimate of the mean did not exceed 2 times the mean, then one would need approximately 10 measurements (see Chapter 15, Volume 1). (Heederik and Attfield, 2000).

These results show how variable CFU count measurements are, and that the aspect of measurement effort needs consideration. A practical conclusion from this information is that one should be careful in concluding that exposure levels are different between different environments, if the evaluation has been based on a small number of measurements. The same is true for comparisons between species specters obtained from different environments. The natural fluctuation within and between environments is large, and often a considerable number of repeated measurements per environment is needed to be able to come to a reliable comparison. In addition, small differences between different environments are more likely to be the effect of random fluctuations than large differences, in terms of orders of magnitude.

Exposure Standards

Generally speaking, health based exposure standards do not exist for comparison with bioaerosol exposure data. In the past, a few exposure standards have been proposed, for instance, for the carcinogen aflatoxin and the sensitizer subtilisin, an enzyme derived from *Bacillus subtilus* (ACGIH®, 1980). The latter proposal has been criticized mainly because sensitization has been observed in some studies below the level of the TLV® (for a review see Brisman, 1994). Recently, health based standards have been proposed for bacterial endotoxin, and it is to be expected that standards will be developed for some high molecular weight sensitizers such as wheat allergens, and possibly fungal α-amylase and latex allergens (Heederik et al., 1999). It is not likely that large series of health based exposure standards will be developed, apart from some of the aforementioned examples. The examples involve closely related molecules (endotoxins) with similar health effects, or individual protein molecules (fungal α-amylase) for which clear-cut exposure-response relationships have been described. The risk associated with an exposure to a heterogeneous population of microorganisms cannot be described easily and, because of the heterogeneity, cannot be described satisfactorily in simple quantitative terms. Moreover, for infectious agents individual susceptibility plays a key role. Even if standards can be developed, especially in the near future, using exposure data gathered using PCR and FISH techniques, generalizibility will be limited. Therefore, exposure data have to be interpreted dependent on the context of the exposure.

Sensitization risk has been shown to increase with increasing aeroallergen exposure in cross-sectional studies in several industries (see Figure 6-4). Only a few epidemiological studies allow such analyses because of the absence of good exposure data. A tentative analysis of studies with allergen exposure data suggests that there is a wide range in sensitization potency between various allergens. The allergen exposure in these studies has been evaluated by one laboratory, and results have been corrected for the protein content of the allergen standard (Heederik et al., 1999). Sensitization against rat urinary proteins occured in the pg/m^3 range

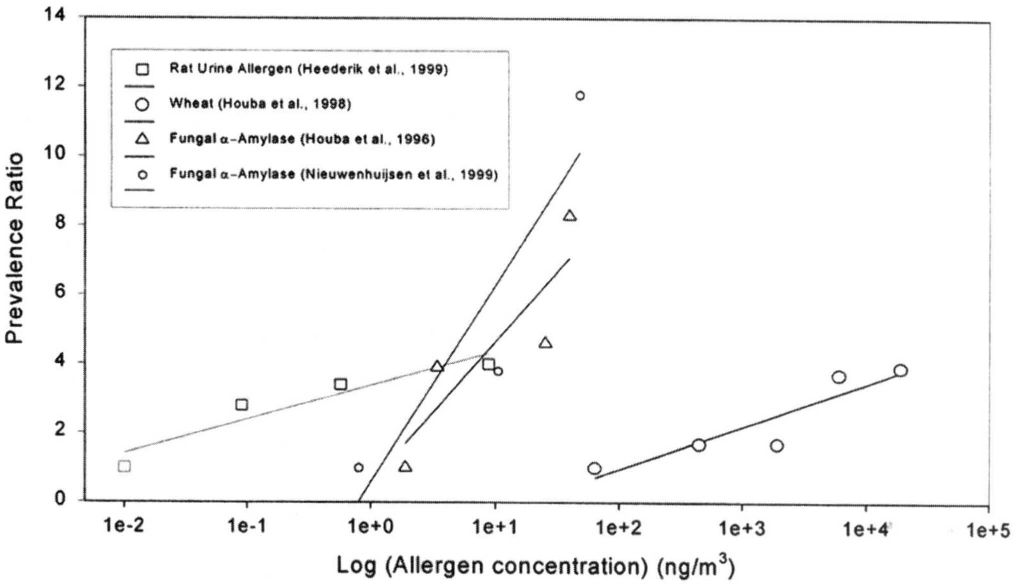

Figure 6-4. Exposure response relationships for sensitization against rat urinary allergens (sensitization defined as IgE > 0.7 kU/L) (average exposures of 0, 0.09, 0.57, 8.8 ng/m^3, respectively), wheat allergen exposure (sensitization defined as IgE > 0.35 kU/L) and fungal α-amylase (sensitization defined as IgE > 0.35 kU/L and Skin Prick Test wheal > 3mm) (from Heederik et al., 1999).

(Heederik et al., 1999). Fungal a-amylase sensitizes in the low ng/m^3 range as shown in two cross-sectional studies among bakery workers (Houba et al., 1996, Nieuwenhuijsen et al., 1999), whereas sensitization against wheat flour occurs in the mg/m^3 range as observed in a cross-sectional study among bakery workers (Houba et al., 1998). Sensitization against animal proteins from pigs and cows suggest exposure levels also in the mg/m^3 range. Results of these epidemiological studies illustrate the potential to relate dose and response when valid exposure data are available in combination with appropriate health data. Standard setting seems possible for those allergens where such scientific evidence of exposure response relationships exists. Until such standards become available, hygienists have to rely on information about exposure determinants and apply *ad hoc* trial and error exposure control strategies.

On the other hand, there is clear consensus about the health effects that result from an (occupational) endotoxin exposure. Despite the variability between laboratories, there is strong evidence in the literature for an exposure-response relationship. This has led to a series of documents with proposals for 'ad hoc' or even legally-binding exposure standards. The Dutch Expert Committee on Occupational Standards (DECOS) proposed a health based standard (MAC value) based on acute respiratory effects of 50 Endotoxin Units/m^3 averaged over 8 hours (DECOS, 1998). After advice from unions and employers, a higher value of 200 Endotoxin

Units/m^3 has been officially adopted, and became the legally binding standard on July 1, 2001. The endotoxin concentration has to be measured using the (draft) European CEN measurement protocol.

Practical Evaluation of Bioaerosol Levels

Typical strategies to evaluate exposures are to compare bioaerosol exposure data with background levels, or compare areas with symptomatic and asymptomatic individuals. The first approach is simple and straightforward but may be misleading when measurement techniques and strategies are not comparable between the exposure site and the reference site. Evaluation of exposure differences between exposure or symptom groups is a more powerful tool, but is more time consuming and complex and also requires evaluation of symptoms using questionnaires or other health parameters. Comparisons using self-reported symptoms can lead to biased results and the reader is referred to epidemiological textbooks on this topic.

When comparisons are made with background levels, for instance outdoor levels, preferably some background level measurements have to be included. Reference material from the literature can seldom be used because of differences in climatic and meteorological conditions during the measurements, and differences with regard to the measurement protocol (viable versus non-viable sampling, sampler type, analysis, etc.). Exposure data can be evaluated quantitatively (*"is the level different compared to background levels?"*) or qualitatively (*"is the spectrum of microorganisms different compared to the background samples?"*).

Interpretations, regardless of the comparison group, should be based on multiple samples, because the analytical variability, especially of viable samples, is usually in the range of 20-100%, and the space-time variability in the environment is high as well. Unfortunately, power calculations are seldom applied but extremely useful to give an impression of how large a measurement series is needed to detect a significant difference between contaminated and background areas. Since variability exists in both the exposure values and the standard (in this case the controls), sample sizes are necessarily larger than in the case of sampling for chemicals and subsequent comparison to a fixed standard (see Volume 1, Chapter 15).

One should be especially aware of upper and lower detection limits for viable sampling. Calculated CFU levels are, in some cases, based on only one or two colonies on the impaction medium, especially at low sampling volumes or sampling times. The resulting confidence intervals will be extremely wide, and this problem should be avoided by increasing the sampling time. This problem generally occurs in environments with low exposure levels such as office and other indoor environments.

An explicit evaluation of the quality of the gathered data in terms of comparability with other available data, reproducibility, and representativeness is warranted for each survey. However, even if the quality of the data is considered appropriate, and when sufficient exposure data are available, interpretation remains complex and less straightforward than for chemical agents.

Basically, statistical approaches for bioaerosol exposure are similar to those available for chemical exposures. However, the large variability in space and time has to be considered and will have its impact on statistical inferences. Bioaerosol exposures are ubiquitous and everyone is exposed throughout life. This issue of background exposure risk and resulting health risk complicates a risk assessment process and makes interpretation of the findings less straightforward than for chemicals.

SOURCES FOR BIOAEROSOL SAMPLERS AND OTHER INFORMATION

Andersen Instuments Inc.
http://www.graseby.com/products.htm

Burkard Sampler
http://www.burkard.co.uk/

Spiral Biotech
http://www.spiralbiotech.com

Rotorod sampling equipment
http://www.multidata.com/

For assorted filter holders:
http://www.bgiusa.com

The International Association for Aero-biology
http://www.isao.bo.cnr.it/aerobio/iaa/index.html

The Pan American Aero-biology Association
http://www.paaa.org

Information from the European Committee for Standardization on Bioaerosol Exposure Standards:
http://www.cenorm.be

Maruha Corp., Tokyo, Japan; Seikagaku Corp., Tokyo, Japan

SUPPLIERS FOR LAL BASED ENDOTOXIN OR GLUCAN ASSAYS

Associates of Cape Cod and Seikagaku America

http://www.acciusa.com

Cambrex Corporation (Cambrex Biosciences is comprised of BioWhittaker, Inc., BioWhittaker Molecular Applications, Inc. and LumiTech, Ltd.)

http://www.cambrex.com

BioWhittaker Europe

http://www.biowhittaker.be

Charles River Laboratories

http://www.criver.com/products/invitro/endotoxin/index.html

QUESTIONS

6.1. Describe the limitations of viable sampling of microorganisms versus non-viable sampling.

6.2. Which factors have an effect on the observed viable count when an exposure measurement has been made with a plate impactor type of sampler?

6.3. In what situation can a PCR-based assay be best used to identify microorganisms?

6.4. What problems exist that limit the comparability of measurements made by immunoassays to evaluate exposure to high molecular weight sensitizers?

6.5. No exposure thresholds have been observed for the risk of sensitization due to exposure to high molecular weight sensitizers. As a result, what pragmatic approach needs to be used?

REFERENCES

American Conference of Governmental Industrial Hygienists: In Documentation of the Threshold Limit Values. ACGIH®, Cincinnati, Ohio, USA, pp. 374-375 (1980).

Aketagawa, J.; Tanaka, S.; Tamura, H.; Shibata, Y.; Sait, H.: "Activation of limulus coagulation factor G by several (1→3)-β-D-glucans: Comparison of the potency of glucans with identical degree of polymerization but different conformations. J Biochem 11-3-:683-686 (1993).

Alvarez, A.J.; Buttner, M.P.; Toranzos, G.A.; Dvorsky, E.A.; Toro, A.; Heikes, T.B.; Mertikas-Pifer, L.E.; Stetzenbach, L.D.: Use of solid-phase PCR for enhanced detection of airborne micro-organisms. Appl. Environ. Microbiol., 60, 374-376 (1994).

Amann, R.I.; Binder, B.J.; Olson, R.J.; Chisholm, S.W.; Devereux, R.; Stahl, D.A.: Combination of 16S rRNA-targeted oligonucleotide probes with flow cytometry for analyzing mixed microbial populations. Appl. Environ. Microbiol., 56, 1919-1925 (1990).

Andersen, A.A.: New Sampler for the collection, sizing, and enumeration of viable airborne particles. J. Bacteriol., 76, 471-484 (1958).

Associates of Cape Cod, Endo-Fluor Assay, Woods Hole, MA, USA.

Atlas, R.M.: Environmental applications of the polymerase chain reaction. ASM News, 57, 630-632 (1991).

Bang, F.B.: A bacterial disease of Limulus polyphemus. Bull. John Hopkins Hosp., 98, 325-350 (1956).

Baseler, M.W.; Fogelmark, B.; Burrell, R.: Differential toxicity of inhaled gram-negative bacteria. Infect. Immun., 40, 133-138 (1983).

Bauer, J.; Gareis, M.: Analytical methods for mycotoxins. DTW - Deutsche Tierarztliche Wochenschrift., 96, 346-350 (1989).

Bauman, J.G.J.; Bayer, J.A.; Van Dekken, H.: Fluorescent in-situ hybridization to detect cellular RNA by flow cytometry and confocal microscopy. J. Microscopy, 157, 73-81 (1990).

Bej, A.K.; Steffan, R.J.; DiCesare, J.; Haff, L.; Atlas, R.M.: Detection of coliform bacteria in water by polymerase chain reaction and gene probes. Appl. Environ. Microbiol., 56, 307-314 (1990).

Bergstrom, I.; Heinnanen, A.; Salonen, K.: Comparision of acridine orange, acriflavine, and bisbenzimide stains for enumeration of bacteria in clear and humid waters. Appl. Environ. Microbiol., 51, 664-667 (1986).

Brisman, J.: Arbetsmiljö Institutet. The Nordic Expert Group for Criteria Documentation of Health Risks from Chemicals. 111. Industrial Enzymes. Arbete Och Hälsa 28 (1994).

Brunson, K.W.; Watson, D.W.: Limulus amebocyte lysate reaction with streptococcal pyrogenic exotoxin. Infect. Immun., 14, 1256-1258 (1976).

Burrell, R.; Shu-Hua, Y.: Toxic risks from inhalation of bacterial endotoxin. Brit. J. Ind. Med., 47, 688-91 (1990).

Buttner, M.P.; Stetzenbach, L.D.: Evaluation of four aerobiological sampling methods for the retrieval of aerosolized Pseudomonas syringae. Appl. Environ. Microbiol., 57, 1268-1270 (1991).

Buttner, M.P.; Stetzenbach, L.D.: Monitoring airborne fungal spores in an experimental indoor environment to evaluate sampling methods and the effects of human activity on air sampling. Appl. Environ. Microbiol., 59, 219-226 (1993).

Chang, C.-W.; Hwang, Y.-H.; Grinshpun, S.A.; Macher, J.M.; Willeke, K.: Evaluation of counting error due to colony masking in bioaerosol sampling. Appl. Environ. Microbiol., 60, 3732-3738 (1994).

Chun, D.T.; Chew, V.; Bartlett, K.; Gordon, T.; Jacobs, R.R.; Larsson., B.-M.; Larsson, L.; Lewis, D.M.; Liesivuori, J.; Michel, O.; Milton, D.K.; Rylander, R.; Thorne, P.S.; White, E.M.: Report on a round robin endotoxin assay study using cotton dust. In press (1998).

Cohen, J.; McConnell, J.S.: Observations on the measurement and evaluation of endotoxemia by a Limulus lysate microassay. J. Infect. Dis., 150, 916-924 (1984).

Cooper, J.F.: Ideal properties of an LAL reagent for pharmaceutical testing. In: Bacterial endotoxins; structure, biomedical significance, and detection with the Limulus amebocyte lysate test. Alan R. Liss, Inc., New York, 241-249 (1985).

Dutch Expert Committee on Occupational Standards (DECOS). Endotoxins: Health based recommended exposure limit. A report of the Health Council of the Netherlands. Rijswijk: Health Council of the Netherlands, publication no. 1998/03WGD (see also http://www.gr.nl) (1998).

Deetz, D.C.; Jagielo, P.J.; Thorne, P.S.; Bleuer, S.A.; Quinn, T.J.; Schwartz, D.A.: The kinetics of grain dust-induced inflammation of the lower respiratory tract. Am. J. Respir. Crit. Care Med., 155, 254-259 (1997).

Degrange, V.; Bardin, R.: Detection of Nitrobacter populations in soil by PCR. Appl. Env. Microbiol. 61:2093-2098 (1995).

DeKoster, J.A.; Thorne, P.S.: Bioaerosol concentrations in non-complaint, complaint, and intervention homes in the Midwest. Am. Ind. Hyg. Assoc. J. 1995; 56:573-580 (1995).

DeLong, E.F.; Wickman, G.S.; Pace, N.R.: Phylogenetic Stains: Ribosomal RNA-based probes for the identification of single cells. Sci., 243, 1360-1363 (1989).

Dillon, H.K.; Miller, J.D.; Sorenson, W.G.; Douwes, J.; Jacobs, R.R.: A review of methods applicable to the assessment of mold exposure to children. Environmental Health Perspectives 107:473-480 [supplement] (1999).

Dillon, H.K.; Heinsohn, P.A.; Miller, J.D.; eds.: Field Guide for the Determination of Biological Contaminants in Environmental Samples. American Industrial Hygiene Association, Fairfax, VA (1996).

Difco Manual, 10th Ed.: Difco Laboratories, Ann Arbor, MI (1984).

Donald, W.W.; Mrocha, C.J.: Chitin as a measure of fungal growth in stored corn and soybean seed. Cereal Chem 54:466-474 (1977).

Douwes, J.; Doekes, G.; Montijn, R.; Heederik, D.; Brunekreef, B.: Measurement of β(1→3)-glucans in occupational and home environments with an inhibition enzyme immunoassay. Appl. Environ. Microbiol., 62, 3176-3182 (1996).

Douwes, J.; Versloot, P.; Hollander, A.; Heederik, D.; Doekes, G.: Influence of various dust sampling and extraction methods on the measurement of airborne endotoxin. Appl. Environ. Microbiol., 61, 1763-1769 (1995).

Douwes, J.; van der Sluis, B.; Doekes, G.; van Leusden, F.; Wijnands, L.; van Strien, R.; Verhoeff, A.; Brunekreef, B.: Mold extracellular polysaccharides in house dust as a marker for mold exposure: Relations with culturable molds, reported home dampness and respiratory symptoms. In press (1998).

Douwes, J.; McLean, D.; van der Maarl, E.; Heederik, D.; Pearce, N.: Worker exposures to airborne dust, endotoxin and beta(1,3)-glucan in two New Zealand sawmills. Am. J. Ind. Med. 38:426-430 (2000).

Douwes, J.; van der Sluis, B.; Doekes, G.; van Leusden, F.; Wijnands, L.; van Strien, R.; Verhoeff, A.; Brunekreef, B.: Fungal extracellular polysaccharides in house dust as a marker for exposure to fungi: Relations with culturable fungi, reported home dampness, and respiratory symptoms. J. Allergy Clin. Immunol. 103(3 Pt. 1):494-500 (1999).

Douwes, J.; Zuidhof, A.; Doekes, G.; van der Zee, S.C.; Wouters, I.; Boezen, M.H.; Brunekreef, B.: (1→3)-β–D-glucan and endotoxin in house dust and peak flow variability in children. Am. J. Respir. Crit. Care Med. 162(4 Pt. 1):1384-54 (2000).

Duchaine, C.; Meriaux, A.; Thorne, P.S.; Cormier, Y.: Assessment of particulates and bioaerosols in Eastern Canadian sawmills. Am. Ind. Hyg. Assoc. J., 61, 727-732 (2000).

Duncan, R.L., Jr.; Hoffman, J.; Tesh, V.L.; Morrison, D.C.: Immunologic activity of lipopolysaccharides released from macrophages after the uptake of intact E coli in vitro. J. Immunol., 136, 2924-29 (1986).

Dunér, K.I.: A new kinetic single stage LAL method for the detection of endotoxin in water and plasma. J. Biochem. Biophys. Meth., 26, 131-142 (1993).

Eduard, W.; Aalen, O.: The effect of aggregation on the counting precision of mould spores on filters. Ann. Occup. Hyg., 32, 471-479 (1988).

Eduard, W.; Lacey, J.; Karlsson, K.; Palmgren, U.; Ström, G.; Blomquist, G.: Evaluation of methods for enumerating microorganisms in filter samples from highly contaminated occupational environments. Am. Ind. Hyg. Assoc. J., 51, 427-436 (1990).

Eduard, W.; Heederik, D.: Methods for quantitative assessment of airborne levels of non-infectious microorganisms in highly contaminated work environments. Am. Ind. Hyg. Assoc. J. 59, 113-127 (1988).

Eggleston, P.A.; Newill, C.A.; Ansari, A.A.; Pustelnik, A.; Lou, S.R.; Evans, III, R.I.; Marsh, D.G.; Longbottom, J.L.; Corn, M.: Task related variation in airborne concentrations of laboratory animal allergens: Studies with Rat n I. J. Allergy Clin. Immunol., 84, 347-352 (1989).

Eisenach, K.B.; Sifford, M.D.; Cave, M.D.; Bates, J.H.; Crawford, J.T.: Detection of mycobacterium tuberculosis in sputum samples using a polymerase chain reaction. Am. Rev. Respir. Dis., 144, 1160-1163 (1991).

FAO. FAO Food and Nutrition Paper: Manuals of food quality control. 10. Training in mycotoxins analysis, 14, 1-113 (1990).

Fox, A.; Wright, L.; Fox, K.: Gas chromatography-tandem mass spectrometry for trace detection of muramic acid, a peptidoglycan chemical marker in organic dust. J. Microb. Meth. 22:-11-26 (1994).

Gerhardt, P.; Murray, R.G.E.; Wood, W.A.; Krieg, N.; Eds.: Methods for general molecular bacteriology. American Society for Microbiology Press, Washington, DC (1994).

Gordon, T.: Role of the complement system in the acute respiratory effects of inhaled endotoxin and cotton dust. Inhal. Tox., 6, 253-66 (1994).

Gordon, T.; Galdanes, K.; Brosseau, L.: Comparison of sampling media for endotoxin contaminated aerosols. Appl. Occup. Environ. Hyg., 7, 427-436 (1992).

Heederik, D.; Doekes, G.; Nieuwenhuijsen, M.: Exposure assessment of high molecular weight sensitizers: Contribution to occupational epidemiology and disease prevention. Occup. Environ. Med., 1999 in press (1999).

Heederik, D.; Attfield, M.: Characterization of dust exposure for the study of chronic occupational lung disease – A comparison of different exposure assessment strategies. Am. J. Epid.; 151: 982–990 (2000).

Heide van der, S.; Kauffman, H.F.; de Vries, K.: Cultivation of fungi in synthetic and semi-synthetic liquid medium, I. Growth characteristics of the fungi and biochemical properties of the isolated antigenic material. Allergy 40:586-91 (1985).

Hollander, A.; Gordon, S.; Renström, A.; Thissen, J.; Doekes, G.; Larson, P.H.; Malmberg, P.; Venables, K.M.; Heederik, D.: Comparison of methods to assess airborne rat and mouse allergen levels. I. Analysis of air samples. Allergy 54(2):142-9 (1999).

Hollander, A.; Heederik, D.; Doekes, G.: Respiratory allergy to rats: Exposure-response relationships in laboratory animal workers. Am. J. Crit. Care Med., 155, 562-567 (1997).

Hollander, A.; Heederik, D.; Versloot, P.; Douwes J.: Inhibition and Enhancement in the Analysis of Airborne Endotoxin Levels in Various Occupational Environments. Am. Ind. Hyg. Assoc. J., 54, 647-653 (1993).

Hollander, A.; Renström, A.; Gordon, S.; Thissen, J.; Doekes, G.; Larsson, P.; Venables, K.; Malmberg, P.; Heederik, D.: Comparison of three methods to assess airborne rat and mouse urinary allergen levels. Submitted to Allergy, first review (1998).

Houba, R.; Heederik, D.; Doekes, G.: Wheat sensitization and work-related symptoms in the baking industry are preventable. An epidemiologic study. Am. J. Respir. Crit. Care Med. 158(5 Pt. 1):499-503 (1998).

Houba, R.; Heederik, D.I.; Doekes, G.; van Run, P.E.: Exposure-sensitization relationship for alpha-amylase allergens in the baking industry. Am. J. Respir. Crit. Care Med. 154(1):130-6 (1996).

Houba, R.; van Run, P.; Doekes, G.; Heederik, D.; Spithoven, J.: Airborne α-amylase allergens in bakeries. J. Allergy Clin. Immunology 99: 286-92 (1997).

Iversen, M.; Korsgaard, J.; Hallas, T.E.; Dahl, R.: Mite allergy and exposure to storage mites and house dust mites in farmers. Clin. Exp. Allergy, 20, 211-219 (1990).

Iwanaga, S.: Chromogenic substrates for horseshoe crab clotting enzyme, its application for the assay of bacterial endotoxins. Haemostasis, 12, 183-188 (1979).

Iwanaga, S.: The Limulus clotting reaction. Curr. Opinion Immunol., 5, 74-82 (1993).

Jarvis, B.B.: Mycotoxins and indoor air quality. In: Biological contaminants in indoor environments, ASTM STP 1071, Morey P.R., Feeley J.C. and Otten J.A., Eds, ASTM, Philadelphia, PA. 201-214 (1990).

Jensen, P.A.; Todd, W.F.; Davis, G.N.; Scarpino, P.V.: Evaluation of eight bioaerosol samplers challenged with aerosols of free bacteri. Am. Ind. Hyg. Assoc. J., 53, 660-667 (1992).

Juozaitis, A.; Willeke, K.; Grishpun, S.A.; Donnelly, J.: Impaction onto a glass slide or agar versus impingement into a liquid for the collection and recovery of airborne microorganisms. Appl. Environ. Microbiol., 60, 861-870 (1994).

Kamphuis, H.J.; Ruiter de, G.A.; Veeneman, G.H.; Boom van, J.H.; Rombouts, F.M.; Notermans, S.H.W.: Detection of Aspergillus and Penicillium extracellular polysaccharides (EPS) by ELISA: Using antibodies raised against acid hydrolyzed EPS. Antonie van Leeuwenhoek, 61, 323-332 (1992).

Khan, A.A.; Cerniglia, C.E.: Detection of Pseudomonas aeruginosa from clinical and environmental samples by amplification of the exotoxin A gene using PCR. Appl. Environ. Microbiol., 60, 3739-3745 (1994).

Kolk, A.H.J.; Schuitema, A.R.J.; Kuijper, S.; van Leeuwen, J.; Hermans, P.W.M.; van Embden, J.D.A.; Hartskeerl, R.A.: Detection of mycobacterium tuberculosis in clinical samples by using polymerase chain reaction and a nonradioactive detection system. J. Clin. Microbiol., 30, 2567-2575 (1992).

Kullman, G.J.; Thorne, P.S.; Waldron, P.F.; Marx, J.J.; Ault, B.; Lewis, D.M.; Siegel, P.D.; Olenchock, S.A.; Merchant, J.A.: Organic dust exposures from work in dairy barns. Am. Ind. Hyg. Assoc. J. 59:403-413 (1998).

Lange, J.L.; Thorne, P.S.; Lynch, N.L.: Application of flow cytometry and fluorescent in situ hybridization for assessment of exposures to airborne bacteria. Appl. Environ. Microbiol., 63, 1557-1563 (1997a).

Lange, J.L.; Thorne, P.S.; Kullman, G.J.: Determinants of viable bioaerosol concentrations in dairy barns. Ann. Agric. Environ. Med. 4:187-194 (1997b).

Leal, J.A.; Guerrero, C.; Gómez-Miranda, B.; Prieto, A.; Bernabé, M.: Chemical and structural similarities in wall polysaccharides of some Penicillium, Eupeniccilium and Aspergillus species. FEMS Microbiol. Lett., 90, 165-168 (1992).

Leopold S.S.: "Positive hole" statistical adjustment for a two stage, 200-hole-per-stage Andersen air sampler. Am. Ind. Hyg. Assoc. J., 49, 88-89 (1988).

Levin, J.; Bang, F.B.: Clottable protein in Limulus: Its localization and kinetics. Thromb. Diath. Haemorrh., 19, 186-197 (1968).

Lin, X.; Reponen, T.A.; Willeke, K.; Grinshpun, S.A.; Foarde, K.; Ensor, D.: Long-term sampling of airborne bacteria and fungi into a non-evaporating liquid. Atmos. Environ. 33, 4291-4298 (1999).

Lin, X.; Reponen, T.A.; Willeke, K.; Wang, Z.; Grinshpun, S.A.; Trunov, M.: Survival of airborne microorganisms during swirling aerosol collection. Aerosol Sci. Technol. 32:184-196 (2000).

Macher, J.M.: Positive-hole correction of multiple-jet impactors for collecting viable microorganisms. Am. Ind. Hyg. Assoc. J., 50, 561-568 (1989).

Marthi, B.; Lighthart, B.: Effects of betaine on enumeration of airborne bacteria. Appl. Environ. Microbiol., 56, 1286-1289 (1990).

May, K.R.; Harper, G.J.: The efficiency of various liquid impinger samplers in bacterial aerosols. Br. J. Ind. Med., 14, 287-297 (1957).

Michel, O.; Nagy, A.M.; Schroeven, M.; Duchateau, J.; Neve, J.; Fondu, P.; Sergysels, R.: Dose-response relationship to inhaled endotoxin in normal subjects. Am. J. Respir. Crit. Care Med., 156, 1157-64 (1997).

Mielniczuk, Z.; Mielniczuk, E.; Larsson, L.: Determination of muramic acid in organic dust by gas chromatography-mass spectrometry. J. Chromatogr. B. Biomed. Appl. 670:167-172 (1995).

Miguel, A.G.; Cass, G.R.; Weiss, J.; Glovsky, M.M.: Latex allergens in tire dust and airborne particles. Environ. Health Perspec., 104, 1180-1186 (1996).

Mikami, T.; Nagase, T.; Matsumoto, S.; Suzuki, S.; Suzuki, M.: Gelation of Limulus amoebocyte lysate by simple polysaccharides. Microbiol. Immunol., 26, 403-409 (1982).

Miller, J.D.; Young, J.C.: The use of ergosterol to measure exposure to fungal propagules in indoor air. Am. Ind. Hyg. Assoc. J. 58:39-43 (1997).

Miller, J.D.; Young, J.C.: The use of ergosterol to measure exposure to fungal propagules. Amer. Ind. Hyg. Assoc. J. 58: 39-43 (1997).

Milton, D.K.; Johnson, D.K.; Park, J.H.: Environmental endotoxin measurement: Interference and sources of variation in the limulus assay of house dust. Am. Ind. Hyg. Assoc. J. 58, 861-867 (1997).

Milton, D.K.; Feldman, H.A.; Neuberg, D.S.; Bruckner, R.J.; Greaves, I.A.: Environmental Endotoxin Measurement: the Kinetic Limulus Assay with Resistant-Parallel-Line Estimation. Envir. Res., 57, 212-230 (1992).

Milton, D.K.; Gere, R.J.; Feldman, H.A.; Greaves, I.A.: Endotoxin Measurement: Aerosol Sampling and Application of a New Limulus Method. Am. Ind. Hyg. Assoc. J., 51, 331-227 (1990).

Miwa, N.; Nishina, T.; Kubo, S.; Atsumi, M.: Most probable number method combined with nested polymerase chain reaction for detection and enumeration of enterotoxigenic *Clostridium perfringens* in intestinal contents of cattle, pig and chicken. J. Vet. Med. Sci. 59:89-92 (1997).

Morita, T.S.; Tanaka, S.; Nakamura, T.; Iwanaga, S.: A new (1-3)-ß-Glucan-mediated coagulation pathway found in Limulus amebocytes. FEBS Lett., 129, 318-21 (1981).

Nagi, N.; Ohno, N.; Adachi, Y.; Aketagawa, J.; Tamura, H.; Shibata, Y.; Tanaka, S.; Yadomae, T.: Application of Limulus test (G pathway) for the detection of different conformers of (1→3)-ß-D-glucans. Biol. Pharm. Bull., 16, 822-828 (1993).

Nevalainen, A.; Pastuszka, J.; Liebhaber, F.; Willeke, K.: Performance of bioaerosol samplers: Collection characteristics and sampler design considerations. Atmos. Environ., 26A, 531-540 (1992).

Nevalainen, A.; Willeke, K.; Liebhaber, F.; Pastuszka, J.; Burge, H.; Henningson, E.: Bioaerosol sampling. In: Aerosol measurement – principles, techniques, and applications. Willeke K. and Baron P., Eds., Van Nostrand Reinhold, New York, 471-492 (1993).

Newell, S.Y.: Estimating fungal biomass and productivity in decomposing litter. In: The Fungal Community (Carroll, G.C., Wicklow, D.T., eds.). New York: Marcel Dekker, Inc., 535-539 (1992).

Nieuwenhuijsen, M.J.; Heederik, D.; Doekes, G.; Venebles, K.M.; Newman Taylor, A.J.: Exposure-response relations of alpha-amylase sensitisation in British bakeries and flour mills. Occup. Environ. Med. 56(3):197-201 (1999).

Notermans, S.; Soentoro, P.S.S.: Immunological relationship of extracellular polysaccharide antigens produced by different mould species. Antonie van Leeuwenhoek, 52, 393-401 (1986).

Norback, D.; Edling, C.; Wieslander, G.: Asthma symptoms and the sick building syndrome (SBS)—The significance of microorganisms in the indoor environment. In: Health Implications of Fungi in Indoor Environments (Samson, R.A., Flannigan, B., Flannigan, M.E., Verhoeff, A.P., Adan, O.C.G., Hoekstra, E.S., eds). Amsterdam:Elsevier, 229-239 (1994).

Novitsky, T.J.; Schmidt-Gegenbach, J.; Remillard, J.F.: Factors affecting recovery of endotoxin absorbed to container surfaces. J. Parenteral Sci. Tech., 40 284-286 (1986).

Obayashi, T.; Yoshida, M.; Takeshi, M.; Goto, H.; Yasuoka, A.; Iwasaki, H.; Teshima, H.; Kohno, S.; Horiuchi, A.; Ito, A.; Yamaguchi, H.; Shimada, K.; Kawai, T.: Plasma (1→3)-ß-D-glucan measurement in diagnosis of invasive deep mycosis and fungal febrile episodes. Lancet, 345, 17-20 (1995).

Ogura, M.; Koo, K.; Mitsuhashi, M.: Use of the fluorescent dye YOYO-1 to quantify oligonucleotides immobilized on plastic plates. Biotechniques, 16, 1032-1033 (1994).

Olenchock, S.A.; Lewis, D.M.; Mull, J.C.: Effects of different extraction protocols on endotoxin analysis of airborne grain dusts. Scand. J. Work Environ., 15, 430-435 (1989).

Platt-Mills, T.A.; Thomas, W.R.; Aalberse, R.C.; Vervloet, D.; Chapman, M.D.: Dust mite allergens and asthma: Report of a second international workshop. J. Allergy Clin. Immunol., 89, 1046-1060 (1992).

Palmgren, U.; Strom, G.; Blomquist, G.; Malmberg, P.: Collection of airborne microorganisms on nuclepore filters, estimation and analysis -CAMNEA method. J. Appl. Bacteriol., 61, 410-406 (1986).

Picard, C.; Ponsonnet, C.; Paget, E.; Nesme, X.; Simonet, P.: Detection and enumeration of bacteria in soil by direct DNA extraction and polymerase chain reaction. Appl. Env. Microbiol. 58:2717-2722 (1992).

Platts-Mills, T.A.E.; Chapman, M.D.: Dust mites: immunology, allergic disease, and environmental control. J. Allergy Clin. Immunol., 80, 755-775 (1987).

Platts-Mills, T.A.E.; de Weck, A.L.: Dust mite allergens and asthma – A worldwide problem. J. Allergy Clin. Immunol., 83, 416-427 (1989).

Pollart, S.M.; Smith, T.F.; Morris, E.C.; Gelber, L.E.; Platts-Mills, T.A.E.; Chapman, M.D.: Environmental exposure to cockroach allergens: Analysis with monoclonal antibody-based enzyme immunoassays. J. Allergy Clin. Immunol., 87, 505-510 (1991).

Pope, A.M.; Patterson, R.; Burge, H.; Eds.: Indoor allergens: Assessing and controlling adverse health effects, National Academy Press, Washington, DC (1993).

Price, J.A.; Pollock, I.; Little, S.A.; Longbottom, J.L.; Warner, J.O.: Measurement of airborne mite antigen in homes of asthmatic children. Lancet, 336, 895-897 (1990).

Ray, A.; Redhead, K.; Seikirk, S.; Poole, S.: Variability in LPS Composition, Antigenicity and Reactogenicity of Phase Variants in Bordetella pertussis. FEMS Microbiol. Lett., 63, 211-217 (1991).

Recorbet, G.; Picard, C.; Normand, P.; Simonet, P.: Kinetics of the persistence of chromosomal DNA from genetically engineered Escherichia coli introduced in soil. Appl. Env. Microbiol. 59: 4289-4294 (1993).

Remillard, J.F.; Case Gould, M.; Roslansky, P.F.; Novitsky, T.J.: Quantitation of endotoxin in products using the LAL kinetic turbidimetric assay. In: Detection of bacterial endotoxins with the Limulus amebocyte lysate test. Alan R. Liss, Inc., New York, 197-210 (1987).

Renström, A.; Gordon, S.; Hollander, A.; Spithoven, J.; Larsson, P.H.; Venables, K.M.; Heederik, D.; Malmberg, P.: Comparison of methods to assess airborne rat and mouse allergen levels. II. Factors influencing antigen detection. Allergy 54(2):150-7 (1999).

Renström, A.; Hollander, A.; Gordon, S.; Thissen, J.; Doekes, G.; Larsson, P.; Venables, K.; Malmberg, P.; Heederik, D.: Comparison of methods to assess airborne rat or mouse allergen levels: II Factors influencing antigen detection. Submitted to Allergy, first review (1998).

Reynolds, S.J.; Thorne, P.S.; Donham, K.J.; Croteau, E.A.; Kelly, K.M.; Lewis, D.; Whitmer, M.; Heederik, D.; Douwes, J.; Connaughton, I.; Koch, S.; Malmberge, P.; Larson, B-M.; Milton, D.K.: Interlaboratory Comparison of Endotoxin Assays Using Agricultural Dusts. Am. Ind. Hyg. Assoc. J. (2002).

Reynolds, S.J.; Thorne, P.S.; Donham, K.J.; Croteau, E.; Milton, D.K.; Lewis, D.M.; Heederik, D.; Connaughton, I.; Malmberg, P.; Larsson, L.; DeKoster, J.; Kelley, K.; Woolson, R.: International interlaboratory comparison of endotoxin assays using agricultural dusts. In press (1999).

Richard, J.L.; Bennett, G.A.; Ross, P.F.; Nelson, P.E.: Analysis of naturally occurring mycotoxins in feedstuffs and food. J. Animal Sci., 71, 2563-2574 (1993).

Rietschel, E.T.; Brade, H.; Holst, O.; Brade, L.; Muller-Loennies, S.; Mamat, U.; Zahringer, U.; Beckmann, F.; Seydel, U.; Brandenburg, K.; Ulmer, A.J.; Mattern, T.; Heine, H.; Schletter, J.; Loppnow, H.; Schonbeck, U.; Flad, H.D.; Hauschildt, S.; Schade, U.F.; Di Padova, F.; Kusumoto, S.; Schumann, R.R.: Bacterial endotoxin: Chemical constitution, biological recognition, host response, and immunological detoxification. Curr. Topics Microbiol. Immunol., 216, 39-81 (1996).

Roslansky, P.F.; Novitsky, T.J.: Sensitivity of Limulus amebocyte lysate (LAL) to LAL-reactive glucans. J. Clin. Microbiol., 29, 2477-2483 (1991).

Rylander, R.; Persson, K.; Goto, H.; Yuasa, K.; Tanaka, S.: Airborne β,1-3-glucan may be related to symptoms in sick buildings. Indoor Environ., 1, 263-267 (1992).

Rylander, R.: Endotoxins. In: Rylander R. and Jacobs R.R., Eds. Organic dusts: Exposure, effects and prevention. Lewis Publishers, Boca Raton, FL (1994).

Rylander, R.; Goto, H.; Yuasa, K.; Fogelmark, B.; Polla, B.: Bird droppings contain endotoxin and (1-3)-β-D glucans. Internat. Arch. Allergy Immunol., 103, 102-104 (1994).

Samson, R.A.; et al., eds.: Introduction to Foodborne Fungi, 4th Edition. Baarn, The Netherlands, **Centraalbureau** voor Schimmelcultures (1995).

Saraf, A.; Larsson, L.; Burge, H.; Milton, D.: Quantification of ergosterol and 3-hydroxy fatty acids in settled house dust by gas chromatography-mass spectrometry: Comparison with fungal culture and determination of endotoxin by a Limulus amebocyte lysate assay. Appl. Environ. Microbiol. 63:2554-2559 (1997).

Schlegel, H.G.: General microbiology, 7th edition, Translated by M. Kogut, Cambridge Univ. Press, Cambridge UK, 447 (1993).

Schou, C.; Svendsen, U.G.; Lowenstein, H.: Purification and characterization of the major dog allergen, Can f I. Clin. Exper. Immunol., 21, 321-328 (1991).

Schwartz, D.A.; Thorne, P.S.; Jagielo, P.J.; White, G.E.; Bleuer, S.A.; Frees, K.L.: Endotoxin responsiveness and grain dust-induced inflammation in the lower respiratory tract. J. Appl. Physiol. 267 (Lung Cell. Mol. Physiol. 11), L609-L617 (1994).

Schwartz, D.A.; Thorne, P.S.; Yagla, S.J.; Burmeister, L.F.; Olenchock, S.A.; Watt, J.L.; Quinn, T.J.: The role of endotoxin in grain dust-induced lung disease. Am. J. Respir. Crit. Care Med., 152, 603-608 (1995).

Scott, P.M.: Natural poisons. In: AOAC Official Methods of Analysis. 1990, Ch. 49, 1185 (1990).

Seitz, L.M.; Mohr, H.E.; Burroughs, R.; Sauer, D.B.: Ergosterol as an indicator of fungal invasion in grains. Cereal Chem. 54:1207-1217 (1977).

Smith, E.G.: Sampling and Identifying Allergenic Pollens and Molds. Blewstone Press, San Antonio, TX (1984).

Smith, E.G.: Sampling and Identifying Allergenic Pollens and Molds - Volume II. Blewstone Press, San Antonio, TX (1986).

Soares, L.M.; Rodriguez-Amaya, D.B.: Survey of aflatoxins, ochratoxin A, zearalenone, and sterigmatocystin in some Brazilian foods by using multi-toxin thin-layer chromatographic method. Journal of AOAC Intl., 72, 22-26 (1989).

Solomon, W.R.: Assessing fungal prevalence in domestic interiors. J. Allergy Clin. Immunol. 56:235-242 (1975).

Sonesson, A.; Larsson, L.; Fox, A.; Westerdahl, G.; Odham, G.: Determination of environmental levels of peptidoglycan and lipopolysaccharide using gas chromatography-mass spectrometry utilizing bacterial amino acids and hydroxy fatty acids as biomarkers. J. Chromatogr. Biomed. Appl., 431, 1-15 (1988).

Sonesson, A.; Larsson, L.; Schütz, A.; Hagmar, L.; Hallberg, T.: Comparison of the Limulus Amebocyte Lysate Test and gas chromatography-mass spectrometry for measuring lipopolysaccharides (endotoxins) in airborne dust from poultry processing industries. Appl. Environ. Microbiol., 56, 1271-78 (1990).

Sporik, R.B.; Arruda, L.K.; Woodfolk, J.; Chapman, M.D.; Platts-Mills, T.A.E.: Environmental exposure to Aspergillus fumigatus allergen (Asp f I). Clin. Exp. Allergy 32:326-31 (1993).

Swanson, M.C.; Agarwal, M.K.; Reed, C.E.: An immunochemical approach to indoor aeroallergen quantitation with a new volumetric air sampler: Studies with mice, roach, cat, mouse, and guinea pig antigens. J. Allergy Clin. Immunol., 76, 724-729 (1985).

Tamura, H.; Tanaka, S.; Ikedo, T.; Obayashi, T.; Hashimoto, Y.: Plasma (1→3)-ß-D-glucan assay and immunohistochemical staining of (1→3)-ß-D-glucan in the fungal cell walls using a novel horseshoe crab protein (T-GBP) that specifically binds to (1→3)-ß-D-glucan. J. Clin. Lab. Anal., 11, 104-109 (1997).

Tanaka, S.; Aketagawa, J.; Takahashi, S.; Shibata, Y.; Tsumuraya, Y.; Hashimoto, Y.: Activation of a limulus coagulation factor G by (1-3)-ß-D-glucans. Carbohyd. Res., 218, 167-174 (1991).

Teller, J.D.; Key, K.M.: A turbidimetric LAL assay for the quantitative determination of Gram negative bacterial endotoxins. In: Biomedical applications of the horseshoe crab. Alan R. Liss, Inc., New York, 423-433 (1979).

Thorne, P.S.; DeKoster, J.A.; Subramanian, P.: Pulmonary effects of machining fluids in guinea pigs and mice. Am. Ind. Hyg. Assoc. J., 57, 1168-72 (1996).

Thorne, P.S.; Kiekhaefer, M.S.; Whitten, P.; Donham, K.J.: Comparison of bioaerosol sampling methods in barns housing swine. Appl. Environ. Microbiol., 58, 2543-2551 (1992).

Thorne, P.S.; Lange, J.L.; Bloebaum, P.D.; Kullman, G.J.: Bioaerosol sampling in field studies: Can samples be express mailed? Am. Ind. Hyg. Assoc. J., 55, 1072-1079 (1994).

Thorne, P.S.; Reynolds, S.J.; Milton, D.K.; Bloebaum, P.D.; Zhang, X.; Whitten, P.; Burmeister, L.F.: Field evaluation of endotoxin air sampling assay methods. Am. Ind. Hyg. Assoc. J., 58, 792-799 (1997).

Thorne, P.S.; Heederik, D.: Indoor bioaerosols - sources and characteristics. In: Salthammer T, editor. Organic Indoor Air Pollutants - Occurrence, Measurement, Evaluation. Weinheim, Germany: Wiley/VCH, 275-288 (1999).

Thorne, P.S.; Heederik, D.: Assessment methods for bioaerosols. In: Salthammer T, editor. Organic Indoor Air Pollutants - Occurrence, Measurement, Evaluation. Weinheim, Germany: Wiley/VCH, 85-103 (1999).

Van Egmond, H.P.: Current situation on regulations for mycotoxins. Overview of tolerances and status of standard methods of sampling and analysis. Food Add. Contam., 6, 139-188 (1989).

Van Egmond, H.P.: Methods for determining ochratoxin A and other nephrotoxic mycotoxins. IARC Scientif. Publ., 115, 57-70 (1991).

van Hage-Hamsten, M.: Allergens of storage mites. Clin. Exp. Allergy, 22, 429-431 (1992).

Verhoeff, A.; Wijnen, J.H.; van Brunekreef, B.; Fischer, P.; Reenen-Hoekstra, E.S.; van Samson, R.A.: Presence of viable mould propagules in indoor air in relation to house damp and outdoor air. Allergy, 47: 83-91 (1992).

Virtanen, T.; Louhelainen, K.; Montyjarvi, R.: Enzyme-linked immunosorbent assay (ELISA) inhibition method to estimate the level of airborne bovine epidermal antigen in cowsheds. Int. Arch. Allergy Appl. Immunol., 81, 253-257 (1986).

Weary, M.E.; Donohue, G.; Pearson, F.C.; Story, K.: Relative potencies of four reference endotoxin standards as measured by the Limulus amoebocyte lysate and USP rabbit pyrogen tests. Appl. Environ. Microbiol., 40, 1148-1151 (1980).

Whittaker Bioproducts, Inc.: LAL review. Walkersville, MD., 4 (1988).

Willeke, K.; Lin, X.; Grinshpun, S.A.: Improved aerosol collection by combined impaction and centrifugal motion. Aerosol Sci. Technol. 28, 439-456 (1998).

Wood, R.A.; Eggleston, P.A.; Lind, P.; Ingemann, L.; Schwartz, B.; Graveson, S.; Terry, D.; Wheeler, B.; Adkinson, Jr., N.F.: Antigenic analysis of household dust samples. Am. Rev. Respir. Dis., 137, 358-363 (1988).

Zock, J.-P.; Hollander, A.; Doekes, G.; Heederik, D.: The influence of different filter elution methods on the measurement of airborne antigens. Am. Ind. Hyg. Assoc. J., 57, 567-570 (1996).

7

BIOLOGICAL AGENTS – CONTROL IN THE OCCUPATIONAL ENVIRONMENT

Kenneth F. Martinez, MSEE, CIH and Peter S. Thorne, PhD

INTRODUCTION

Microorganisms such as molds and bacteria are normal inhabitants of the environment; they exist in the soil we walk on, in the air we breathe, the water we drink, and even the food we eat. The saprophytic varieties (those utilizing the organic products of other organisms, living or dead, as a food source) generally predominate. Under the appropriate conditions (optimum temperature, pH, sufficient moisture, and available nutrients), saprophytic populations of bacteria and molds can increase in number, or amplify. In the air, they can exist as discrete particles (e.g., as individual fungal spores and bacteria, or their aggregates). These organisms can also aggregate, attach to other particles, form aerosols, deposit on surfaces or growth substrates. Many microbial species have been documented to tolerate various environmental extremes of temperature and pressure allowing them to acclimate even to the most harsh conditions. Certain bacteria, called extremophiles, have been identified in the lower depths of the ocean growing amidst intense volcanic activity. Mycobacteria have been found in mummified tissues. It is clear that microorganisms (fungi and bacteria) are ubiquitous occupants throughout all of the diverse environments of this world and are continual companions on the roads that we travel.

Many species of microorganisms are beneficial to, if not a necessary part of, the environment and to our very existence. Saprophytic molds and bacteria are composters of organic material and are essential actors in the carbon and nitrogen cycles. Many essential nutrients are recycled through microbial actions. Microbial flora in our gastrointestinal tract aid metabolism and help protect us from pathogens.

Humans have made use of microbes over time. Yeasts have been used for millennia to produce alcoholic beverages and to leaven bread. More recently, the metabolic systems of specific microbial populations have been used to produce pharmacologically important products (e.g., penicillin and citric acid), food products (e.g., amylase), and a number of other chemical agents. Unique uses of the microbial community include the application of bacterial species in the decontamination of hazardous substances (i.e., oil spills in the ocean) and as an insecticide (i.e., *Bacillus thuringiensis*) to control cabbage worm, cotton boll worm, and chicken louse. Our everyday encounters with most members of the saprophytic community generally do not pose a significant health risk to the immunocompetent worker population (healthy adults). However, select occupational environments present unique exposure concerns due to the nature of the microorganisms encountered, the microbial concentrations observed, and the susceptibility of the exposed population. Within the healthcare industry, attention has focused on human infections or infectious agents of which the obligate parasites and facultative saprophytes (including the primary pathogens and opportunistic pathogens) are the primary concern (Burge, 1989). In recent years, bioaerosols (the term given to microorganisms and/or their products entrained in/on airborne particles) have become prominent safety and health issues in agriculture, biotechnology, industrial settings, and more recently, the non-industrial indoor environments. Much of the concern for these types of exposures has focused on the ability of certain microbiological species to elicit allergic or inflammatory responses in susceptible individuals.

THE BASICS

The practice of industrial hygiene is based on the ability of an investigator to *anticipate, recognize*, *evaluate*, and *control* exposures to hazardous agents (chemical, biological, and/or physical) in the occupational environment. Hazard recognition may include the identification of significant agents from past exposures and the anticipation of exposures to agents which may be infectious, toxigenic, or allergenic. The investigator must have a fundamental understanding of the physical, chemical, and toxicological properties of the agent in order to accurately predict the route and extent of the exposure, and resulting health effects. Additionally, a comprehensive knowledge of the suspect etiologic agent will lead to the selection of appropriate evaluation tools and recommendations for control strategies that could prevent future exposures. For example, investigators from the Centers for Disease Control and Prevention (CDC) were requested in 1996 by a district health department for assistance in evaluating an unusual increased incidence of unexplained pneumonia within the community (CDC, 1996). Thorough epidemiologic study of the community and positive diagnosis of legionellosis among three individuals in the population focused investigation efforts on a large home-improvement center. Environmental sampling resulted in a positive culture for *Legionella pneumophila* collected from a whirlpool spa filter that was a positive match for two of the three legionellosis cases. Control recommendations included regular inspection, maintenance with biocides, and routine filter changes or decontamination. A similar episode occurred at a large floral show in the Netherlands in February 1999 resulting in 242 cases and 28 fatalities (Boshuizen et al., 2000). In this instance, molecular biology based assessment methods allowed identification of the source within two days, thus reducing the severity of the epidemic.

In some occupational settings, these characteristic properties of agents causing illness may not be clearly defined. In the indoor environment (e.g., office buildings), exposures to reservoirs of toxigenic fungi have been associated with the development of occupant disorders including respiratory and central nervous system effects (Johanning et al., 1996; Hodgson et al., 1998). However, a specific organism or toxigenic product could not be exclusively linked with the identified health effects. Without the recognition of a specific etiologic agent, decisions regarding the remediation and subsequent control of affected areas in these situations is complicated. The potentially toxic nature of the suspect agents suggests a conservative approach that includes the complete removal and decontamination of all fungal reservoirs using rigorous isolation strategies and personal protective equipment.

The evaluation phase of an investigation provides qualitative and/or quantitative data to support or refute the hypothesis of occupational exposures to the suspected agent. A series of hypotheses, postulating the presence of an agent (or agents) of putative illness, a pathway for potential exposure, and evidence for exposure must be framed to support a sampling strategy (Macher et al., 1999). The data collected as a result of the investigative plan can range from documentation of safety and health deficiencies observed during a walk-through investigation to the results of environmental (bulk, surface, and/or air) samples analyzed for microbial content. The collected data are then evaluated to appropriately estimate the risk to the worker population and to suggest possible control methodologies. There may exist opportunities when the collection of environmental data may be applicable to verify the efficacy of implemented control measures, e.g., the collection of air samples for fungal spores after remediation of a fungally-contaminated building. However, the limitations of microbiological sampling methodologies and the paucity of exposure criteria for microbiological agents precludes their use as the sole determinants of the success of the control measures. Success should be objectively determined based on the efficacy of containment (as measured by parameters including removal of dust and debris, pressure differentials, air flow, surrogate indicators of airborne concentrations, and/or other appropriate "tools") and the hazard of the microbiological agent when known.

The control of occupational exposures to chemical, biological, and physical agents is accomplished by the application of engineering measures, work practices, and personal protective equipment (PPE). These measures, practices, and/or equipment are applied at the source of the contaminant generation, to the general workplace environment, or at the significant exposure point of an individual. The application of engineering measures at the source provides the most effective control of both, occupational and environmental contaminants. Source control is amenable to situations where the generation of the etiologic agent is localized. For example, a patient diagnosed with active tuberculosis may undergo aerosolized drug-treatment within the confines of an isolation tent filtered by a high efficiency particulate air (HEPA) filter. Similarly, small local exhaust ventilation units have been suggested for use during the application of lasers and electro-surgical devices (U.S. DHHS, 1996). However, large-scale fungal contamination of the interior surfaces of peripheral walls in an office building are not conducive to engineering control measures applied at the source of generation. In these situations, the control strategies used are a hybrid of local exhaust and general dilution ventilation.

Substitution with a less hazardous material is the preferred approach to providing a safe work environment. Research activities in microbiology laboratories can be conducted on attenuated or avirulent strains that reduce or eliminate the risk of worker infection. As with many other situations involving microbiological agents, this option may not be available since the agents are usually not an intentional process material. However, the nutrients upon which the organisms feed may be amenable to substitution with substrates that are less conducive to growth. When resolving microbiological contamination of building materials in an indoor environment, materials that contain a high proportion of cellulose may be replaced with other composites. In some instances, work practices can be modified to minimize the potential for contaminant generation and subsequent exposure. Appropriate work practices are a major component of the biosafety program used in microbiology laboratories to reduce the risk of occupational exposures. The application of mechanical pipettors has significantly reduced the possibility of inadvertent infection compared to the technique of employing "mouth pipetting." In 1991, the Occupational Safety and Health Administration (OSHA) published the final rule on occupational exposures to blood borne pathogens based, in large part, on the concept of "universal precautions" (OSHA, 1991). Universal precautions are work practice modifications that treat all human blood and certain other human body fluids as if known to be infected with human immunodeficiency virus (HIV), hepatitis B virus (HBV), and other bloodborne pathogens.

Under those circumstances where source control is not a practical solution, modifications to the general work environment can provide the next level of control. The techniques employed include dilution ventilation, aerosol (e.g., dust) suppression, and improved housekeeping activities. The concept of dilution is one control strategy that has been applied to isolation rooms occupied by patients that have been diagnosed with infectious diseases transmissible through inhalation (e.g., tuberculosis). Riley has suggested that the probability of infection within a ventilated room can be predicted by the application of the Reed-Frost equation (Riley et al., 1989):

$$C = S(1 - e^{-Iqpt/Q}) \qquad \text{[Equation 7-1]}$$

where *C* is the number of susceptible individuals (*S*) who become infected, *I* is the number of sources (actively infected cases), *p* is the pulmonary ventilation rate (volume per unit time), *t* is the exposure time in minutes, *Q* is the removal rate through dilution with fresh air (cubic feet per minute), and *q* is the quanta (number of doses of airborne infection) added to the air per unit time by an actively infectious case. The risk of infection is reduced by insuring that the airborne concentration is below a known airborne infectious dose.

The last level of control attempts to separate the exposed worker from the chemical, biological, or physical agent. Separation can be attained by the application of isolation environments (e.g., remote control rooms, isolation booths, and supplied-air cabs). In the previous example of a fungally-contaminated building, large sections of contaminated wall are encapsulated with polyethylene barrier walls that simulate a large biological safety cabinet with contaminated air being exhausted through HEPA filters. The generation source is contained within the structure to protect the environment outside; while the concentrations of airborne fungal spores within the structure are continuously reduced to some equilibrium level based on the calculated volumetric introduction of treated (filtered; possibly dehumidified) outdoor air. Separation can also be achieved by employing PPE including chemically impervious clothing and respirators approved by the National Institute for Occupational Safety and Health (NIOSH). However, PPE should always be considered the "last line of defense" to control occupational exposures.

The selection of an appropriate NIOSH-approved air-purifying respirator is determined by knowledge of the suspected air contaminants (NIOSH, 1996). The size of most bacteria range from 0.3 to 30 micrometers (μm), most disease-causing bacteria range from 0.2 to 1.2 μm in diameter and 0.4 to 14 μm in length; fungal conidia may range from 1 to 50 μm in diameter (Wistreich et al., 1988; Levetin, 1995; Thorne et al., 1999). Viruses and rickettsias are in the sub-micrometer size range, 0.003 to 0.2 μm. Through various mechanisms, these organisms can be disseminated as individual cells, spores, or fragments (or agglomerates there of) or in association with soil/dust, water, or some other particle (as is generally the case with bacteria, viruses, and rickettsias). Particulate respirators certified under Code of Federal Regulations (CFR) Part 84 should offer adequate protection to the wearer against these agents in the majority of instances. Adequate protection is afforded by a proper face seal as determined through a quantitative or qualitative fit test. In selecting an appropriate particulate respirator for use during a specific task, and with known microbiological agents, the NIOSH Respirator Decision Logic Sequence should be applied (NIOSH, 1987). Respirators must be used in accordance with a complete respiratory protection program as specified in OSHA Standard 29, CFR 1910.134 (OSHA, 1992). OSHA requires that respiratory protection programs include written standard operating procedures; respirator selection on the basis of hazard; fit testing; user instruction and training; respirator cleaning, disinfection, storage, and inspection; surveillance of work area conditions; evaluation of the respirator protection program; medical review; and use of certified respirators.

The design of effective exposure controls for microbiological agents depends on an appropriate understanding of the differences between the clinical endpoints (i.e., infection versus immunologic response) and the mechanisms of microbial growth and dissemination. To elicit a response (infectious or immunologic) in a susceptible individual, the microorganism must be present (reservoir), must be capable of propagation to concentrations necessary to induce a response (amplification), and must be dispersed to the susceptible individual (dissemination). Knowledge that the suspected agent is infectious (resulting from a definitive diagnosis of the clinical symptoms) generally carries an awareness of the etiologic agent, including the exposure pathway. For example, engineering the patient environment to reduce the ambient microorganism concentration has been shown to reduce the incidence of nosocomial infections (Sheretz et al., 1987). The identification of the specific etiologic agent can result in focused control efforts using well-known engineering and administrative control methodologies. However, in many indoor environments, the specific microbiological agents that stimulate the immune response in susceptible individuals to affect a disease state have not been definitively documented. To complicate our understanding of the indoor environment further, the clinical endpoints resulting from exposures to many microbiological agents have not been well-defined. This limited understanding of the etiologic agents and the clinical endpoints results in the application of generalized controls that affect the environment to (presumably) minimize the proliferation of all microorganisms. Generalized control strategies are not as effective as source focused control efforts, such as those applied in the health care industry.

CLINICAL ENDPOINTS (HEALTH EFFECTS)

Infectious Disease

Infectious disease has been defined as an interference with the normal functioning of a host's physio-chemical process resulting from the activities of a microorganism living within the host's tissues or on its surface (Wistreich et al., 1988). Infection specifically refers to the microorganism's presence in the host, the ability to replicate, and the result of the host developing subclinical or clinical disease. It is possible for an individual to be infected with a pathogen and show no (asymptomatic) clinical response. This phenomenon, defined as a carrier state, proves to be the most difficult to control resulting from the inability to definitely identify the source. Infection is characterized by a microorganism's pathogenicity and virulence, the pathogen dose, and the susceptibility or resistance state of the host (Equation 7-2).

$$\text{Infection} \propto \frac{\text{Virulence} \times \text{Dose}}{\text{Host Resistance}} \qquad \text{[Equation 7-2]}$$

In healthcare environments, the risk of infection demands special consideration. Patients or workers with suppressed immune systems (due to certain diseases, targeted immune suppressants, anti-cancer drug, or irradiation treatments) are more susceptible to infection by microbiological agents. This predisposition not only increases the likelihood of infection, but can also result in an increase in the severity of the disease. Additionally, healthcare environments present other distinctive factors that increase the probability of transmission, including a common source of food, water, and air and, unique to specific environments/departments, concentrations of infectious individuals, special environmental infection sources, and the presence of drug resistant microorganism strains (Parker, 1973).

Microorganisms that have been implicated in hospital infections include various viruses, bacteria, and fungi. Based on data from the CDC National Nosocomial Infections Surveillance System (1986-1990), from cultures obtained in 89% of reported nosocomial infections, 85% involved aerobic bacteria; 4% anaerobic bacteria; 9% fungi; and the remaining 2% included a miscellaneous group of viruses, protozoa, and parasites (Martone et al., 1992). Documented cases of viral nosocomial infections (through direct contact or inhalation of contaminated airborne particles) include varicella-zoster virus, respiratory syncytial virus, adenovirus, HIV, hepatitis B, hepatitis C, and herpes (Gustafson et al., 1982; Chancock et al., 1961; Brummitt et al., 1988; Snydman et al., 1976; Perl et al., 1992). Certain bacterial species, predominantly Gram-negative organisms including *Escherichia coli*, *Enterobacter* spp., *Pseudomonas* spp., and *Proteus* spp., have been associated with bacteremia and death in hospital environments. Gram-negative bacteria are primarily transmitted through contact, with contaminated fluids, moist objects, or ingestion. Many gram-positive bacteria also pose significant nosocomial concerns. *Staphylococcus aureus* is reportedly responsible for approximately 10% of all nosocomial infections (CDC, 1982). Recently, tuberculosis has received increased attention as an important nosocomial disease due to the emergence of multidrug-resistant strains, and the report of several outbreaks in healthcare settings (CDC, 1990).

Fungal agents pose unique nosocomial concerns. Their ubiquitous and hardy existence have weakened the ability to recognize, and subsequently control, indoor exposures. *Aspergillus* spp. are the primary environmental fungal nosocomial hazards; they are opportunistic pathogens in individuals with suppressed immune systems. Lentino et al. reported 10 cases of acquired aspergillosis among immunosuppressed patients resulting from treatment of hematologic malignancy, advanced age, or drugs taken to prevent rejection of renal allografts (Lentino et al., 1982). It was suggested that road construction activities adjacent to the hospital may have disturbed spore-containing soil that could not be appropriately filtered by window air-conditioning units. *Aspergillus* spp. were recovered from the selected air-conditioning units.

Immunologic Response

Immunologic responses are activated by an individual's reaction to particular antigenic constituents of a given microbial species. These responses and the subsequent expression of allergic disease are based on the type and extent of the exposures and, in part, on a genetic predisposition (Pickering, 1992). Allergic respiratory diseases [including allergic rhinitis, allergic asthma, allergic bronchopulmonary aspergillosis (ABPA), and hypersensitivity pneumonitis (HP) resulting from exposures to microbiological agents have been documented in agricultural, biotechnology, office, machining, and home environments (Vinken et al., 1984; Malmberg et al., 1985; Topping et al., 1985; Edwards, 1980; Weiss et al., 1971; Hodgson et al., 1985; Fink et al., 1971; Banazak et al., 1974; Campbell et al., 1979).

Symptoms vary with the type of allergic disease: (1) Allergic rhinitis is characterized by paroxysms of sneezing; itching of the nose, eyes, palate, or pharynx; nasal stuffiness with partial or total airflow obstruction; and rhinorrhea with postnasal drainage. (2) Allergic asthma is characterized by episodic or prolonged wheezing and shortness of breath due to bronchial narrowing and mucus hypersecretion. (3) ABPA is characterized by symptoms of cough, lassitude, low grade fever, wheezing, and occasional expectoration of mucous (Burge, 1988; Kaliner et al., 1987). Exposures to large concentrations of airborne microorganisms may result in an acute form of hypersensitivity pneumonitis (HP, also known as extrinsic allergic alveolitis), which is characterized by chills, fever, malaise, cough, and dyspnea (shortness of breath) generally appearing four to eight hours after exposure. Chronic HP is thought to be induced by a continuous low-level exposure, and onset occurs without chills, fever, or malaise but is characterized by progressive shortness of breath with weight loss (Jordan, et al., 1987). Chronic HP of sufficient duration results in progressive loss of lung function. The disease is not reversible and patients must be removed from exposure. This can be accomplished either by removing the patient from the presence of the sensitizing agent, or removing the sensitizing agent from the patient, to halt progression of the disease.

Acceptable levels of airborne microorganisms have not been established for a) total culturable or countable bioaerosols; b) specific culturable or countable bioaerosols other than infectious agents; c) infectious agents; or d) assayable biological contaminants for the reasons identified below (ACGIH®, 2000). Total culturable or countable bioaerosols exist as a complex mixture of different microbial, plant, and animal particles that produce widely varying human responses, depending on the susceptibilities of the individual. This fact, and the lack of a single sampling method that can adequately characterize all bioaerosol components, limits the ability to produce sufficient information to describe the exposure-response relationships that can predict human health effects. The scientific data that are available consist of case reports and qualitative exposure assessments, which are generally insufficient as well. The absence of good epidemiologic data on exposure-response relationships related to specific culturable or countable bioaerosols has resulted from the derivation of data from indicator measurements as opposed to actual effector agents. Also, the variations of bioaerosol components and concentrations among differing occupational and environmental settings, combined with the limitations of sampling methodologies, have resulted in the inaccurate representation of workplace exposures. Human dose-response relationships are available for some infectious bioaerosols. However, facilities associated with increased risks for airborne disease transmission rely on the application of engineering and administrative controls to minimize the potential for workplace exposure. In the future, some assayable, biologically derived contaminants (e.g., endotoxin) may lend themselves to the development of numeric criteria due to advances in assay methods.

Relationships between health effects and environmental microorganisms must be determined through combined medical, epidemiologic, and environmental evaluations. The current strategy used by NIOSH for on-site evaluation of indoor environments where microorganisms are the suspected etiologic agent involves a comprehensive inspection of the building to identify sources of microbial contamination and routes of dissemination. In those locations where contamination is visibly evident or suspected, bulk samples may be collected to identify the predominant species (fungi, bacteria, and thermoactinomycetes). In limited situations, additional environmental samples (e.g., air samples) for microorganisms may be collected to test a specific hypothesis, such as to document the presence and dissemination of a suspected microbial contaminant or to determine if microorganisms from reservoirs or amplifiers are entering indoor air. Elevated airborne concentrations of the contaminant in the high-symptom area, compared to outdoor and low-symptom areas, and anomalous ranking among the microbial taxa may suggest that the contaminant is responsible for the health effects.

Disease Incidence

For certain diseases that carry a high incidence, such as acute viral gastroenteritis from Norwalk-like viruses or influenza (influenza virus), cases may arise from both community outbreaks and occupational exposures. Three of the most prevalent life-threatening infectious diseases associated with occupational exposures in the U.S. are tuberculosis, viral hepatitis from hepatitis B virus, and Lyme disease. Of course, these diseases also occur outside of the workplace. Some infectious diseases with an occupational association have exhibited a steady incidence over the past two decades (e.g., Legionellosis). Others occur episodically depending upon environmental conditions (e.g., Hantavirus pulmonary syndrome), epizootic outbreaks (epidemics among animals), or human error (e.g., foodborne illnesses).

There is little data available from which to determine incidence or prevalence for most occupational infectious diseases. In the United States, cases of federally notifiable diseases must be reported to the CDC. Table 7-1 lists data for 1997 from the CDC for those infectious diseases that pose a significant occupational risk. These cases are reported from the entire U.S. population (a denominator of 267 million), but are important risks for the civilian workforce of 138 million. The degree to which each of these case numbers represents occupational exposure varies by disease and is largely unknown. For several specific diseases and job titles, there are incidence data that were collected to demonstrate the efficacy of preventive measures. For instance, for hepatitis B in health care workers, there were about 8,700 cases per year (145 cases per 100,000) in the United States prior to widespread vaccination and implementation of educational and administrative programs. The CDC has documented that by 1994, this figure had declined to under 1,000 annual cases (16 cases per 100,000).

Prevention

There are many approaches to the prevention of occupational infections and these have been implemented with varying degrees of success. Preventive measures tend to have the greatest efficacy in settings where workers are concentrated, where the workforce is well structured, where the hazards are recognized, and where there are sufficient resources to provide physical controls, training programs, and surveillance. To a large extent the health care system meets these conditions. In hospitals, health clinics, and dental clinics, implementation and enforcement of universal precautions, vaccinations, use of disposable products, proper waste handling procedures, and detailed administrative controls have increased the safety of the workplace. Still there are an estimated 600,000 needle stick injuries annually among health care workers in the United States. One-third of these occur during disposal of the syringe or sharp object. The risk of infection after a needle stick is 0.3% for HIV, 6 to 30% for hepatitis B virus, and 5 to 10% for hepatitis C virus. In contrast to health care settings, many occupations have little structure

Table 7-1
Reported cases in the United States (1997 data) for infectious diseases with a significant occupational component. Data specific for occupational cases are not available.‡

Disease	Reported U.S. Cases
Arboviral encephalitides	47
Brucellosis	98
Cryptosporidiosis	2,566
Hepatitis B	10,416
Legionellosis	1,163
Lyme disease	12,801
Plague	4
Psittacosis	33
Rabies	2
Rocky Mountain spotted fever	409
Tetanus	50
Tuberculosis	19,851

‡The denominator for calculating population-wide incidence was 267,637,000.

and workers are disseminated, making preventive measures difficult to develop and implement. Examples include workers in agriculture, fishing, forestry, composting and waste handling, and animal handlers.

Table 7-2 lists six components necessary for an effective infection control program. Control strategies generally encompass engineering controls such as high performance ventilation systems, biosafety cabinets, special equipment (e.g., retracting needles), specially-designed waste containers, and apparatus to isolate the worker from the hazard. Personal measures such as

Table 7-2
Components of effective workplace infection control programs

- Identify occupational infection risks and develop control measures
- Educate workers about infection control and personal responsibility
- Coordinate infection control personnel to effectively monitor cases and control outbreaks
- Institute a program of immunization for agents that have an approved vaccine
- Institute an administrative structure to monitor compliance and efficacy and ensure confidentiality
- Provide health care and counseling for workers who sustain exposures to develop occupational infectious diseases

hand washing, immunization, and use of personal protective equipment (gloves, safety glasses, respirators, etc.) are also important. Administrative controls include worker screening, criteria for returning to work following illness or exposure, emergency response for prevention of exposures or outbreaks, controlled access to hazardous areas, and biosafety enforcement. An effective system for surveillance should also be implemented. For diseases transmitted from pets, livestock, and zoo animals, immunization of the animals can help protect those who handle them.

Where infectious diseases are transmitted by an arthropod vector, protective clothing, insect and tick repellants containing n,n-diethyl meta-toluamide (DEET) or pyrethrins, and careful inspection for ticks are important for disease risk reduction. Avoiding work at dawn or dusk can significantly reduce mosquito bites. Chemoprophylactic therapies and vaccinations are important for soldiers, travelers, and those who work outside. However, some of these drugs are not recommended for immunocompromised workers, pregnant workers, or individuals with certain enzyme deficiencies. It has been shown that taking chloroquine as a prophylactic treatment for malaria can interfere with the development of protective antibodies following vaccination against rabies. There are also concerns regarding heavy use of dermally-applied insecticides concurrent with chemoprophylactic therapies and vaccinations.

Community or worksite-based vector control programs can be helpful. For mosquito control, draining swamps, disposing of discarded tires, covering water reservoirs, and use of insecticides are helpful for primary prevention. Many countries have instituted surveillance programs using sentinel animals for tracking mosquito borne diseases. As an example, the United States monitors small groups of captive chickens disseminated throughout regions where Arboviruses appear. A high correlation between seroconversion of the chickens for the agents that cause Western equine encephalitis and disease in horses and humans in those same regions has been observed. Rodent control through elimination of food sources; rat-proofing of buildings;

eradication programs using traps, fumigants, and poisons; habitat elimination; and proper rubbish containment and collection are highly effective control measures. Another important avenue toward lowering rodent populations is reducing poverty and improving the social structure within a community.

Occupational infectious diseases cause significant morbidity and mortality and result in personal suffering and lost productivity. While they can affect any worker, certain occupations carry greater risks. Recognition of the hazards and implementation of appropriate preventive measures can reduce the incidence and severity of these occupational diseases.

MICROBIAL AGENTS IN HEALTH CARE FACILITIES

The occupational hazards encountered by health care professionals are varied. These hazards include exposures to biological, chemical, and physical agents, as well as ergonomic challenges posed by the manipulation and transport of patients. In a 1988 document (NIOSH, 1988) that focused on protecting the safety and health of health care workers, NIOSH determined that, compared to the civilian workforce, hospital workers have a greater percentage of workers' compensation claims for sprains and strains, infectious (e.g., hepatitis) and parasitic diseases, dermatitis, mental disorders, eye diseases, influenza, and toxic hepatitis. Probably, the most occupationally unique and high risk hazards are those posed by exposure to infectious agents. In some instances, the risk of occupationally acquired infections is of such concern that comprehensive, agent focused guidelines have been published by the CDC. Specifically, these guidelines have addressed occupational exposures to *Mycobacterium tuberculosis*, HIV, and HBV (CDC, 1994; CDC, 1989). In 1991, OSHA published their final rule on occupational exposures to blood borne pathogens (OSHA, 1991). Additionally, in 1997, OSHA published a draft standard for enforcement procedures for occupational exposures to tuberculosis (OSHA, 1996).

The effective management and control of bioaerosols in hospital environments requires a comprehensive understanding of the possible antigenic or infectious nature of microorganisms, the susceptibility of the host, and the ability of the investigator to predict dissemination pathways based on the available information. Designing the patient environment to reduce the ambient microorganism concentration has been shown to reduce the incidence of nosocomial *Aspergillus* infections (Sheretz et al., 1987). Invasive medical procedures significantly enhance the ability of an organism to invade and colonize a host. Based on current estimates, surgical wound infections account for one-fourth of all nosocomial infections. Studies conducted by the CDC have estimated the national rate of nosocomial infections (among all patients admitted to hospitals for reasons other than infections) to be approximately 5% (Eickhoff et al., 1969; Haley et al., 1985).

The diversity of health effects (immunogenic or infectious) of microbiological agents in the hospital environment dictates a multi-disciplinary control approach that includes infection control personnel, biosafety officers, and industrial hygienists. Infection control personnel focus primarily on nosocomially acquired infections through the activities of an infection control program. An organized program should include (Fleming, 1989):

- Collection and analysis of infection control data (surveillance)
- Planning, implementation, and evaluation of infection prevention and control measures
- Education of individuals about infection risk, prevention, and control
- Development and revision of infection control policies and procedures
- Investigation of suspected outbreaks of infection
- Provision of consultation on infection risk assessment, prevention, and control strategies

Issues related to chemical or biological hazards other than nosocomial control are coordinated through the biosafety officer and/or industrial hygienist. Responsibilities of the various disciplines may overlap in certain situations. A typical example includes the recognition, evaluation, and control of microbiological contamination in a hospital ventilation system. In this example, an epidemiologic assessment may indicate a cluster of aspergillosis among patients. Environmental investigation confirms the presence, amplification, and dissemination of microbiological reservoirs. Subsequent interaction with the engineering staff can then foster solutions that (1) protect the workers and patients during remediation activities and (2) prevent the recurrence of similar events.

Infectious diseases contracted through the airborne route are the most amenable to the industrial hygiene precepts of recognition, evaluation, and control. The rapid clinical diagnosis of a disease in an individual results in the *recognition* of not only the etiologic agent, but may suggest the identification of the source as well. Knowledge of the etiologic agent generally means information on the route of transmission. Infectious diseases can be contracted through five primary pathways (i.e., oral, contact, penetration, respiratory, and via vectors such as insect bite). The first four pathways are generally the predominant focus in health care settings.

Airborne Pathogens

The strategies used to control occupational infectious disease exposures transmitted through the airborne pathway are best exemplified by the 1994 CDC tuberculosis guidelines. The increasing worldwide incidence of tuberculosis has resulted in a global recognition of the resurgence of this disease (see Figure 7-1) (WHO, 1996). The 1994 CDC tuberculosis guidelines were designed to address the prevention of tuberculosis transmission in health care facilities using a hierarchical control structure including administrative, engineering, and PPE control measures. Clearly, the early identification, isolation, and treatment of individuals having active tuberculosis is the basis of an effective control program. These concepts are addressed in the development of the administrative control measures. However, engineering control measures can be effective supplements during the time between case identification and disease resolution

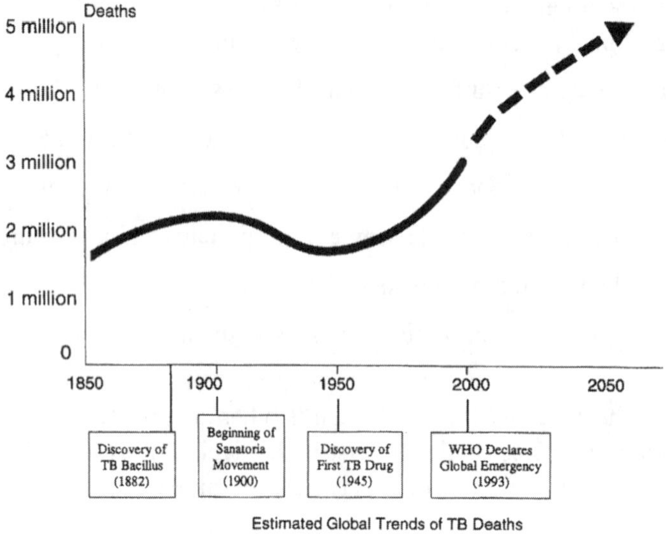

Figure 7-1. Global trends of tuberculosis deaths (adapted from WHO, 1996).

subsequent to treatment. Engineering controls can establish an environment that confines the airborne infectious particles in an isolated area. The engineering control measures employed include general and local exhaust ventilation, the application of ultraviolet germicidal irradiation (UVGI) or HEPA filtration or both, and respiratory protection.

Local exhaust ventilation (LEV) is applicable to those situations where control at the source can significantly reduce the concentration of the generated aerosol. Procedures that are amenable to LEV include sputum induction, bronchoscopy, intubation, irrigation, and aerosolized drug therapy. These activities induce the patient to cough (or otherwise release aerosols) that potentially contain significant concentrations of tubercle bacilli. For example, sputum induction can be conducted inside a booth designed to filter the air surrounding the patient before exhausting it back into the room (see Figure 7-2). Enclosures (i.e., booths and tents) create an encapsulated (enclosed) environment around the patient. The containment barrier is defined by the enclosure structure and by negative pressure (with respect to the atmosphere outside of the enclosure) induced by exhausting air. The air can be exhausted directly to the outdoors or recirculated into the room. However, it is critical that the exhausted airstream (especially for recirculated air) be appropriately filtered to ensure removal of airborne tubercle bacilli (typically in the size range between 1 to 5 μm). Air exhausted outdoors should be filtered if there is a possibility of contaminated air to re-enter the building or if it is exhausted into an outdoor area that could potentially expose other individuals. For infectious agents that pose a significant airborne risk (e.g., *Mycobacterium tuberculosis*), filtration should be accomplished

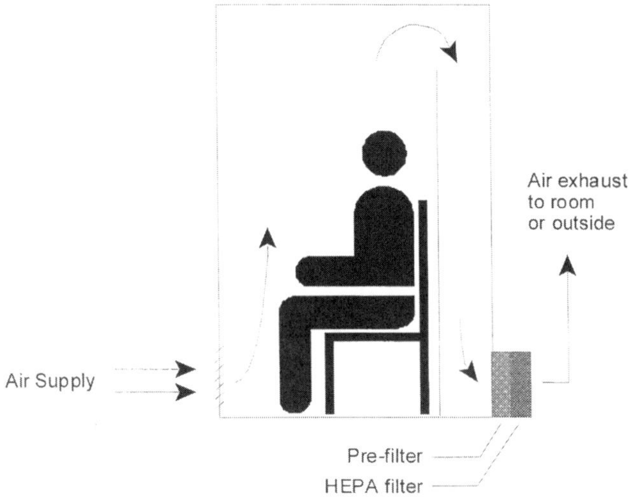

Figure 7-2. Isolation booth for sputum induction.

with HEPA filters which are 99.97% effective in removing particles greater than or equal to 0.3 µm in diameter.

General ventilation introduces a clean airstream into the space occupied by the infectious individual to dilute the contaminant concentration to a level that reduces the risk of exposure to airborne tubercle bacilli. The introduced air may originate from a conditioned outdoor source or from a highly filtered, recirculated source. Effective dilution assumes a homogeneous mixture in the occupied space, i.e., the provision of airflow patterns that promote the mixing of the air and contaminants. Contaminant dilution is a function of the exhausted volumetric air flow, the volume of the room, and time. Disseminated tubercle bacilli have been assumed to move in air similar to gases. (Note, the tracer gas, sulfur hexafluoride, has been suggested to be an acceptable surrogate to track the movement of *M. tuberculosis* in air (Decker, 1995).) This assumption allows the estimation of the removal efficiencies over varying values of air changes per hour (ACH). The time (units of minutes) to achieve varying removal efficiencies as a function of ventilation rate (ACH computed from the exhaust flow rate divided by the room volume) may be determined by the following purge equation:

$$\Delta t = \left(-\frac{1}{ACH}\, ln \left(\frac{C_2}{C_1} \right) \right) \times 60 \qquad \text{[Equation 7-3]}$$

where C_1 is the initial concentration and C_2 is the concentration after a known time interval (Δt). Table 7-3 provides calculations of the minutes required for varying removal efficiencies (de-

Table 7-3
ACH required for removal efficiencies [$100(1-C_2/C_1)$] of airborne contaminants

	Minutes		
ACH	90%	99%	99.9%
1	138	276	414
2	69	138	207
3	46	92	138
4	35	69	104
5	28	55	83
6	23	46	69
7	20	39	59
8	17	35	52
9	15	31	46
10	14	28	41
11	13	25	38
12	12	23	35
13	11	21	32
14	10	20	30
15	9	18	28
16	9	17	26
17	8	16	24
18	8	15	23
19	7	15	22
20	7	14	21
25	6	11	17
30	5	9	14
35	4	8	12
40	3	7	10
45	3	6	9
50	3	6	8

Source: Adapted from CDC, 1994

fined as $(1 - C_2/C_1) \times 100$ and ACH computed from Equation 7-3. CDC has recommended a volumetric air change rate in tuberculosis isolation rooms of 12 ACH for new construction and 6 ACH for renovated rooms. These values were selected based on studies that indicated that rates greater than 6 ACH were likely to produce incrementally greater reduction of bacterial concentrations compared with lower ventilation rates.

Air should be introduced and, subsequently exhausted, in a manner that moves the uncontaminated airstream from the health care worker's breathing zone through the patient's breathing zone (see Figure 7-3). Therefore, supply/exhaust diffuser placement becomes critical to the mixing and general movement of the air in the room. Poor mixing may result in short-circuiting (between supply and exhaust diffusers) and/or stagnation zones that do not allow contaminants to be efficiently removed. The values calculated in Table 7-3 assume perfect mixing of the air within the space (i.e., mixing factor = 1). However, perfect mixing does not occur; for very poor air distribution the mixing factor can range to 10, increasing the values in Table 7-3 accordingly.

Isolation rooms, by definition, are designed to contain generated contaminants within the confines of the room. This containment is achieved by the physical barriers created by the room structures (i.e., walls, ceilings, doors, and windows) and by the creation of negative pressure differentials with respect to adjacent areas. By exhausting air at a volumetric flow rate that is greater than what is supplied to the room, negative pressure differentials are created. These differentials are designed to create a directional flow of air from clean to contaminated areas. Balancing the ventilation system so that the exhaust flow rate is 10% or 50 cubic feet per minute (cfm, 1416 liters per minute [lpm]) greater than the supply air flow rate should achieve a pres-

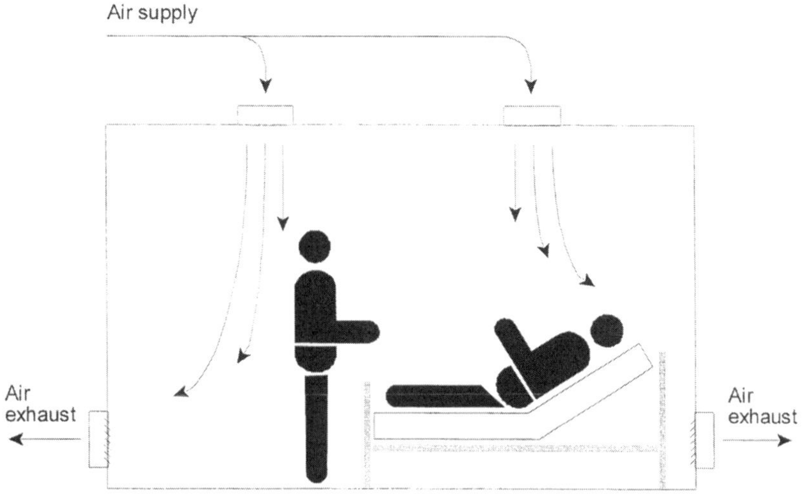

Figure 7-3. General ventilation for isolation rooms.

sure differential of at least 0.001 inches water gauge (0.25 Pa). The aggregate flow restrictions resulting from leakage areas within the enveloping structure of the room determine the attainable pressure differential. If the minimum pressure level cannot be achieved within the limitations of the ventilation system, the degree of room leakage should be assessed and subsequently corrected. Negative pressure isolation rooms are designed to properly operate with the entry/exit door closed. Doors that remain open will significantly impact the ability of the ventilation system to maintain the proper pressure differential. However, it has been experimentally demonstrated that the opening and closing of the entry/exit door and the movement of personnel into and out of the isolation room have little effect on the migration of air into or out of the room (Hayden et al., 1998).

Preventive maintenance is critical to the performance of the containment systems within isolation rooms. Verification of the room pressure status should be conducted on a routine basis (daily at a minimum) to ensure that containment is maintained. Pressure status can be monitored qualitatively or quantitatively. Qualitative assessment is conducted by observing the direction of chemically generated smoke at leakage points around the entry/exit door. The movement of smoke into the isolation room from the adjacent hall or room indicates negative pressure. Quantitative information can be attained with the aid of differential pressure-sensing devices. These devices can be used periodically (monitoring by an individual moving from room to room) or continuously (pressure sensors permanently installed across the hall/isolation room connecting wall).

Administrative measures are the basis of an effective infectious disease control program and engineering measures provide supplemental control. However, general ventilation does little to reduce the exposure encountered at close range to the active tuberculosis patient. Exposures at close range can result from inhaling a concentrated aerosol bolus released by the patient during coughing and sneezing. In those situations when administrative and/or engineering controls are not capable of adequately reducing the potential for occupational exposures to airborne infectious agents, respirators should be used. The selection of the appropriate respirator is based on the expected size range of the microbiological agent and the hazard associated with that agent. The CDC recommends that the minimum respirator for encounters (i.e., in isolation rooms, during vehicle transport, or during cough-inducing procedures) with individuals who have or are suspected of having tuberculosis should meet the following criteria:

- ability to filter particles 1 μm in size in the unloaded state with a filter efficiency of greater than or equal to 95%
- the ability to be qualitatively or quantitatively fit tested in a reliable way to obtain a face–seal leakage of less than or equal to 10%
- the ability to fit the different facial sizes and characteristics of wearers
- the ability to be checked for facepiece fit upon donning, in accordance with OSHA standards and good industrial hygiene practice

Although the filter media used in many current surgical masks may meet the first criterion, they may not meet the latter three. The flat designs often seen in these masks (with elastic straps that secure the mask around the head or ears) cannot provide the quality seal of molded (including "duck bill" designs) respirators. In experimental studies challenging surgical masks with artificially generated aerosols of *Bacillus subtilis* subsp *niger*, inadequate face seal accounted for approximately 85% of the penetration (Johnson et al., 1994). Additionally, these masks are not easily amenable to donning checks for facepiece fit. As a result, surgical masks worn by health care workers are not appropriate respiratory protection against airborne *Mycobacterium tuberculosis* or other infectious microbiological agents. A study of the filter efficiency of respirators and surgical masks challenged with mycobacterial aerosols showed a geometric mean penetration rate of 22 percent for non-approved (i.e., devices that have not been certified under 30 CFR Part 11) surgical masks (Brosseau et al., 1997). Another study demonstrated that aerosol penetration for sub-micrometer size particles may range from 5 to 100 percent for the eight evaluated surgical masks (Weber et al., 1993). Such masks are designed to reduce the potential for contamination of the patient operating field by capturing large bacterial laden aerosols emitted from the mouth and nose of the medical staff. However, these masks can be worn by tuberculosis infected patients to minimize their expulsion of tubercle bacilli.

NIOSH–approved respirators with an N95 designation (as defined by the current NIOSH certification procedures 42 CFR 84 effective July 10, 1998) would meet or exceed the CDC standard performance criteria. The filter material of the N95 particulate respirator has been certified to be at least 95% efficient at removing particles at the most penetrating size (approximately 0.3 μm). The CDC guidelines state that a facility's risk assessment may identify settings where the estimated risk for transmission of *Mycobacterium tuberculosis* may be at such a high level that the standard selection criteria may not be stringent enough. Careful risk assessment techniques should be employed to determine if a higher level of protection is required.

Bloodborne Pathogens

The hazards posed by occupational exposures to infectious agents in blood and bodily fluids have become globally recognized issues. On a historical time-scale, this increased awareness is a recent event. Since the beginning of the 1920s, published reports of laboratory-associated infections have grown steadily, with an increasing percentage caused by exposures to hepatitis (Pike, 1979). The CDC has estimated that, in 1987, the total number of HBV infections in the United States was 300,000 per year, with approximately 25% of those infected developing acute hepatitis. Additionally, the CDC has estimated that 12,000 health care workers (whose job duties put them at risk for exposures to blood) will become infected with HBV each year resulting in the yearly hospitalization of 500-600 workers and approximately 250 deaths (CDC, 1989). However, as a result of vaccine use and adherence to other preventive measures, there has been a 90% reduction in the number of HBV-infected health care personnel from 1985 to 1994 (Shapiro, 1995). Within the health care industry, certain situations (i.e., autopsies) are known to present the greatest risk of exposure to both bloodborne pathogens and other infectious agents. During autopsies, the increased risk to the pathologist and pathology technicians may be due to the presence of unknown etiologic agents in the decedent and exposure to large amounts of infectious material resulting in the release of these agents during invasive procedures. The infectious diseases of particular concern during autopsies have included tuberculosis, hepatitis B (HBV) and C virus, HIV, Creutzfeldt-Jacob disease, group A streptococcal infection, gastrointestinal infections, and possibly meningitis and septicaemia (especially meningococcal) (Healing et al., 1995; Young and Healing, 1995).

In 1991, OSHA published a final rule on occupational exposures to bloodborne pathogens (OSHA, 1991). The bloodborne pathogens rule is in part based on the concept of "universal precautions." Universal precautions are defined as the treatment of all human blood and certain other human body fluids as if known to be infected with HIV, HBV, and other bloodborne pathogens. Under the OSHA regulation, blood is defined as human blood, human blood components, and products made from human blood. Bloodborne pathogens include any pathogenic agents that are present in human blood and that can cause disease in humans. The definition of other potentially infectious materials are inclusive of human body fluids; any unfixed tissue or organ from a human; HIV-containing cell or tissue cultures, organ cultures, and HIV- or HBV-containing culture medium or other solutions; and blood, organs, or other tissues from experimental animals infected with HIV and HBV.

If employees are potentially (i.e., "reasonably" anticipated) exposed to human blood or bodily fluids in the performance of their normal duties, then an exposure control plan should be developed. Components of the plan should include exposure determinations, schedules and methods of implementing compliance methods, HBV vaccination and post-exposure evaluation and follow-up, hazard communication to employees, recordkeeping, and written procedures for circumstances surrounding exposure incidents. The determination of exposure must list all job classifications in which all employees have occupational exposures, all job classifications in which some employees have occupational exposures, and all tasks and procedures or groups of closely related tasks and procedures in which occupational exposure occurs.

Exposure control methods should always be based on the precept of "universal precautions." There are three primary categories of control: engineering and work practice controls, personal protective equipment, and housekeeping. Engineering controls include the provision of hand-washing facilities; appropriate containers for needles and sharps disposal and for specimens of blood or potentially infectious materials; and the use of laboratory equipment that minimizes splashing, spraying, spattering, generation of aerosol droplets (this can also be achieved through work practices), and potential direct contact exposures (e.g., the use of mechanical pipetting devices instead of mouth pipetting). Work practice controls rely on the vigilant awareness and continuous cooperation of the worker to identify potential exposure situations and subsequently apply established protective practices. These practices should include hand washing after removal of PPE or following contact with blood or potentially infectious materials; the prohibition of eating, drinking, and smoking, except in designated areas; and the exclusion of food and drink from storage areas reserved for blood and potentially infectious materials. Additional work practice controls can be developed to fit each unique exposure situation as long as "universal precautions" are maintained.

Various levels of PPE may be employed depending on the "reasonably" anticipated exposure risk. The equipment includes gloves, gowns, laboratory coats, face shields or masks and eye protection, and mouthpieces, resuscitation bags, pocket masks, or other ventilation devices. Masks, eye protection, and face shields should be worn whenever splashes, spray, splatter, or aerosol droplets of blood or other potentially infectious materials are generated or anticipated. It is the employer's responsibility to make these items accessible to employees; ensure their use; clean, launder, or dispose of used items; and repair or replace them as necessary. If a garment has been penetrated by blood or other potentially infectious materials, the item should be removed immediately (or as soon as feasible). All PPE should be removed and placed in an appropriate container prior to leaving the immediate work area. Disposable (single-use) gloves may be used but should be replaced as soon as possible when contaminated or if their barrier properties are compromised. These damaged items should *never* be reused. Utility gloves can be decontaminated for reuse if the integrity of the glove is not compromised.

The last control strategy involves housekeeping practices that maintain the work site in a clean and sanitary condition. All equipment and working surfaces should be cleaned and decontaminated immediately after contact with blood or other potentially infectious material. Additionally, work surfaces should be decontaminated upon completion of procedures, immediately after overt contamination or spill of blood or other potentially infectious materials, and at the end of the work shift if contamination occurred since the last cleaning. Work surface protective coverings should be removed after overt contamination or at the end of the work shift. Receptacles intended for reuse that have a reasonable likelihood of becoming contaminated should be visually inspected and decontaminated on a regularly scheduled basis, and immediately if visibly contaminated by blood or potentially infectious materials. Broken glassware should be extracted by mechanical devices, not with the hands if suspected of contamination. All contaminated sharps and other regulated waste should be placed in appropriately designated containers. Laundry that is contaminated with blood or other potentially infectious materials should be handled as little as possible, with an effort to minimize agitation, and should be placed in appropriate bags or containers.

Monitoring Environmental Contaminants in the Healthcare Setting

In specific instances, environmental microbiological sampling can serve as an essential adjunct to the infection control program. Using various environmental sampling methods (i.e., bulk, surface, and air), Lentino et al. established a correlation between aspergillosis case clusters among renal transplant patients, adjacent road construction, and the recovery of *Aspergillus fumigatus* and *A. flavus* from window air conditioners. Sawyer et al. collected air samples of varicella-zoster virus (VZV) to document the dissemination of viral particles outside of negative-pressure isolation rooms occupied by VZV patients (Sawyer et al., 1994). However, the *evaluation* of the health care environment (characterized by the collection of exposure data) is not routinely conducted. There is no evidence that routine environmental microbiological sampling is necessary for the maintenance of good practices in hospital environments, nor has it been shown that routine sampling has significantly influenced the incidence of nosocomial infections. In 1970, the CDC and the American Hospital Association (AHA) changed their original recommendations for microbiological sampling of the environment to advocating the discontinuation of most routine environmental sampling in hospitals (CDC, 1970; AHA, 1974). Environmental sampling has only been indicated for (1) documenting the effectiveness of various sterilization processes, (2) monitoring infant formulas prepared in the hospital, (3) checking for high levels of contamination on certain patient-care supplies, and (4) as a supplemental tool used in the investigation of an infection outbreak or other specific problem (Mallison and Haley, 1981).

MICROBIOLOGY LABORATORIES AND BIOSAFETY LEVEL

Laboratories engaged in microbiological research and/or analytical support services present unique occupational exposure risks (see Figure 7-4). The variety, pathogenicity, and concentrations of the microbial agents can be significantly greater than in other occupational settings. For example, *Mycobacterium tuberculosis* (Mtb) has been identified in recent years as posing a significant risk to laboratory personnel. Studies have shown that the incidence of Mtb infection in those who work with Mtb in the laboratory is 3 to 5 times higher than the incidence among laboratory personnel who do not work with the bacterium (Reid, 1957; Capewell et al., 1988; Harrington and Shannon, 1976).

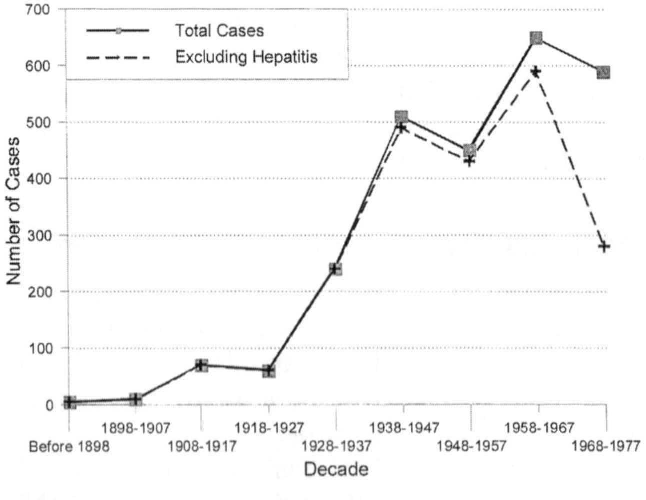

Figure 7-4. Cases of laboratory associated infections (adapted from Pike, 1979).

One of the more common routes of infection of laboratory-acquired illnesses has been attributed to the inhalation of aerosols (Sewell, 1995). Some aerosol-generating procedures that have been shown to produce droplet nuclei in the respirable size range include: pouring of cultures and supernatant fluids, using fixed volume automatic pipettors, mixing a fluid culture with a pipette, dropping tubes or flasks of cultures, spilling suspensions from pipettes, and breaking tubes during centrifugation (Kenney and Sabel, 1968; Stem et al., 1974; McKinney et al., 1991). Additional concerns for microbiologists processing clinical samples include: (1) the increasing numbers of multiple drug resistant organisms, and (2) the increasing numbers of individuals who are co–infected with HIV.

In 1974, the CDC published the first in a series of booklets outlining categorization levels for facilities and practices to be used for activities with infectious agents (CDC, 1974). These guidelines have evolved into the CDC/National Institutes of Health publication, *Biosafety in Microbiological and Biomedical Laboratories* (CDC/NIH, 1993). This booklet discusses the concepts of primary and secondary containment and their application to four levels of biosafety for microbial agents based on hazard and the specific function or activity of the laboratory. Combinations of microbiological practices, laboratory facilities, and safety equipment (including biological safety cabinets [BSC]) are described within each of the four categories (see Table 7-4). Within each biosafety level, the management of infectious agents is facilitated by the precepts of isolation and containment. As with the control of exposures to chemical agents, containment should be designed to protect the laboratory workers, other individuals not involved with laboratory activities, and the outside environment. The concept of containment can be further sub-divided into primary and secondary levels. Primary containment should protect laboratory personnel and the immediate working environment through the application of appropriate work practices and safety equipment. Secondary containment ensures that infectious agents do not contaminate the environment external to the laboratory via the implementation of appropriate facility design and operational procedures.

Biosafety Level (BSL) 1 provides the least protective environment for microbiological activities as it does not include specifications for primary or secondary containment barriers (defined below). BSL 1 laboratories are encountered in such settings as academia, where the work is with microbial agents not known to cause disease in healthy humans. Subsequent increases in the hazard of the microbial agents will result in the need for more protective biosafety levels. BSL 4 facilities are intended for activities with dangerous and exotic microbial agents that pose significant risk of aerosol-transmitted and life-threatening infectious disease. Examples of microorganisms that are included in BSL 4 are the Marburg, Ebola, Lassa, and Machupo viruses. These facilities include provisions for limited access, specialized training for the handling of high-risk microbial agents, facility isolation through the use of physical barriers and negative pressure relationships with adjacent areas, and all activities confined to Class III biological safety cabinets (BSC – defined below) or Class II BSCs in conjunction with personnel equipped with one-piece positive pressure suits ventilated by a life support system. Varying combinations of work practices, facility design, and safety equipment are encountered with BSL 3 and BSL 4 (see Table 7-4).

Table 7-4
Summary of recommended biosafety levels for infectious agents

Biosafety Level	Agents	Practices	Safety Equipment (Primary Barriers)	Facilities (Secondary Barriers)
1	Not known to cause disease in healthy adults.	Standard Microbiological Practices.	None required.	Open bench top sink required.
2	Associated with human disease, exposure via autoinoculation, ingestion, mucous membrane exposure.	BSL-1 practice plus: Limited access; Biohazard warning signs; "Sharps" precautions; Biosafety manual defining any needed waste decontamination or medical surveillance policies.	Primary barriers = Class I or II BSCs or other physical containment devices used for all manipulations of agents that cause splashes or aerosols of infectious materials; PPE; laboratory coats; gloves; face protection as needed.	BSL-1 plus: Autoclave available.
3	Indigenous or exotic agents with potential for aerosol transmission; disease may have serious or lethal consequences.	BSL-2 practice plus: Controlled access; Decontamination of all waste; Decontamination of lab clothing before laundering; Baseline serum.	Primary barriers - Class I or II BSCs or other physical containment devices used for all manipulations of agents; PPE; protective lab clothing; gloves; respiratory protection as needed.	BSL-2 plus: Physical separation from access corridors; Self-closing, double-door access; Exhausted air not recirculated; Negative airflow into laboratory.
4	Dangerous/exotic agents which pose high risk of life-threatening disease, aerosol-transmitted lab infections; or related agents with unknown risk of transmission.	BSL-3 practices plus: Clothing change before entering; Shower on exit; All material decontaminated on exit from facility.	Primary barriers = All procedures above in combination with full-body, air-supplied, positive pressure personnel suit.	BSL-3 plus: Separate building or isolated zone; Dedicated supply/exhaust, vacuum, and decontamination systems; Other requirements outlined in the text.

Source: from CDC 1993

Primary Barriers

Primary barriers (safety equipment) are designed to ensure the isolation and containment of infectious material and include BSC, enclosed containers, and other engineering controls (e.g., sharps disposal containers and mechanical pipettors). BSCs should be reserved for laboratory activities with potentially infectious agents or for activities that generate from uncontrolled aerosols generated by laboratory activities. These cabinets are specially designed local exhaust ventilation hoods that include high-efficiency particulate (HEPA) air filters. Air from the cabinet work space is HEPA-filtered before it is recirculated back to the interior of the cabinet and/or exhausted to the laboratory or ducted to a dedicated exhaust system. BSCs are categorized into three distinct classes based on differences in specific performance characteristics: Class I, Class II, and Class III. For each class, the type of protection afforded (personal protection and/or product protection) can vary.

Class I BSCs provide a negative pressure, semi-enclosed environment that exhausts air through a HEPA filter back into the laboratory or outside to a dedicated exhaust system (see Figure 7-5). Room air enters the cabinet through a fixed front opening. The magnitude of the air velocity entering the cabinet is designed to prevent microbial aerosols released within the cabinet from escaping into the room. At a minimum, the air flow for Class I BSCs should be adequate to provide a face velocity at the opening of the cabinet of 75 feet per minute (fpm, 0.4 meters per second). These cabinets are not designed to provide purified air to the cabinet work area. It is possible that microbially contaminated air from the laboratory or cross-contamination of cultures within the cabinet work area may result. Therefore, these cabinets should not be

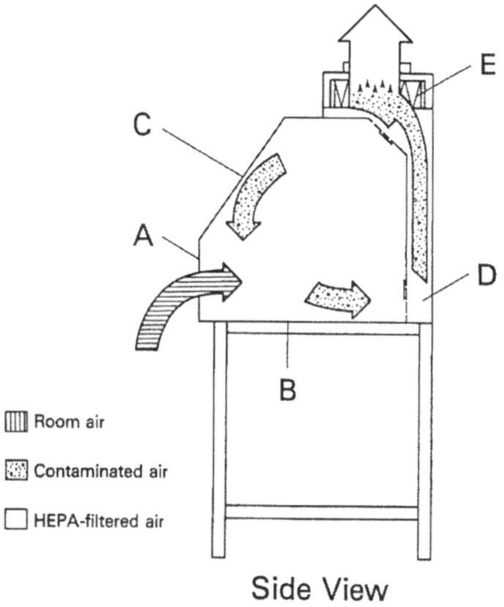

Figure 7-5. *Class I Biological Safety Cabinet* A. front opening; B. work surface; C. window; D. exhaust plenum; E. HEPA filter (adapted from CDC/NIH, 1993).

used for laboratory activities where susceptibility to contamination by other microbial agents exists.

Class II BSCs are most often recommended as a primary containment device. They are designed to protect the inner working environment of the cabinet (i.e., provide a clean airstream for contamination sensitive laboratory activities) *and* significantly minimize the escape of microbial contaminants into the surrounding laboratory. These cabinets are classified into Type A or B as defined by the cabinet construction, airflow velocities and patterns, and exhaust system design. Both types utilize a downward laminar flow of HEPA-filtered air that separates at the working surface to slots in the back and the front of the cabinet (see Figure 7-6).

Type A cabinets use an internal system of ducts, HEPA filters, and a fan to recirculate a substantial portion (up to 70%) of HEPA filtered air back to the cabinet workspace and to exhaust the remaining 30% back into the laboratory. Most Class II, Type A cabinets have dampers to modulate this 30/70 division of airflow. This recirculation makes these cabinets suitable for microbiological activities but not for work with volatile or toxic chemicals and radionuclides. The face velocities of these systems should be maintained at a minimum of 75 fpm (0.4 m/s). It is possible to duct the exhaust from a Type A cabinet out of the building. However, it must be done in a manner that does not alter the balance of the exhaust system, which could result in a disturbance of the internal cabinet airflow or other cabinets attached to the system. The usual method is to use a "thimble," or canopy hood, which maintains a small opening around the cabinet exhaust filter housing (CDC/NIH, 1995).

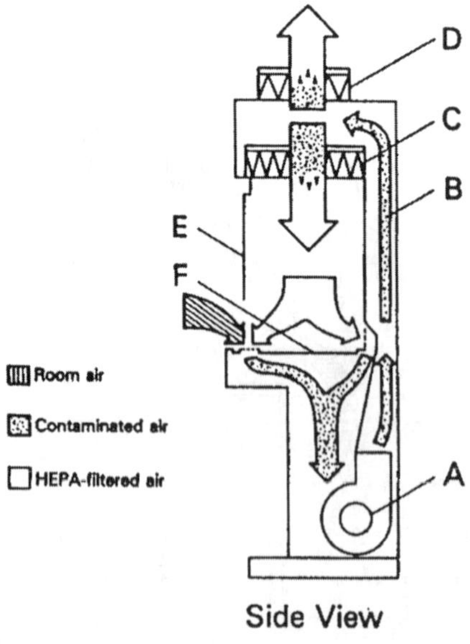

Figure 7-6. *Class III, Type A Biological Safety Cabinet*
A. blower; B. rear plenum; C. supply HEPA filter; D. exhaust; E. sash; F. work surface (adapted from CDC/NIH, 1993).

Type B cabinets are similar in design to Type A, however, this cabinet allows for separate exhaust of up to 100% of HEPA-filtered air to the building dedicated exhaust system. Additionally, the plenum is maintained under negative pressure. These modifications ensure that contaminated materials (including toxic chemicals and radionuclides) are contained in the cabinet assembly until exiting the exhaust system. For all Type B cabinets, face velocities should be maintained at a minimum of 100 fpm (0.5 m/s).

To reflect the degree of exhaust to a dedicated, external system, Type B cabinets are further divided into three sub-categories. Type B1 systems recirculate 30% of the air through a HEPA filter back into the cabinet work space with the remaining 70% being exhausted via a separate fan and duct work (see Figure 7-7). Type B2 is a total exhaust cabinet, that is, no air is recirculated (see Figure 7-8). All air from the cabinet work space is ducted to a dedicated exhaust fan. This type of cabinet is expensive to operate since it exhausts large quantities of conditioned room air. If the exhaust fails, this type of cabinet will be pressurized via the supply fan, resulting in a flow of air from the cabinet back into the laboratory. Cabinets built since the early 1980s incorporate an interlock system to prevent the supply blower from operating whenever the exhaust flow is insufficient. Presence of such an interlock system should be verified; existing systems can be retrofitted if necessary. Exhaust air movement should be monitored by a pressure-independent device. Type B3 cabinets are identical to a Class II, Type A cabinet with the exception that the 30% component is ducted to a dedicated exhaust system. All positive pressure contaminated plenums within the cabinet are surrounded by a negative air pressure

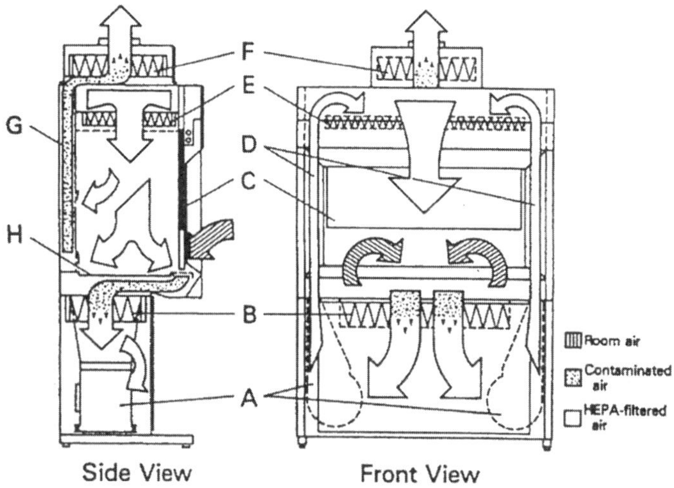

Figure 7-7. *Class II, Type B1 Biological Safety Cabinet* A. blowers; B. supply HEPA filters; C. sliding sash; D. positive pressure plenums; E. additional supply HEPA filter or back-pressure plate; F. exhaust HEPA filter; G. negative pressure exhaust plenum; H. work surface (adapted from CDC/NIH, 1993).

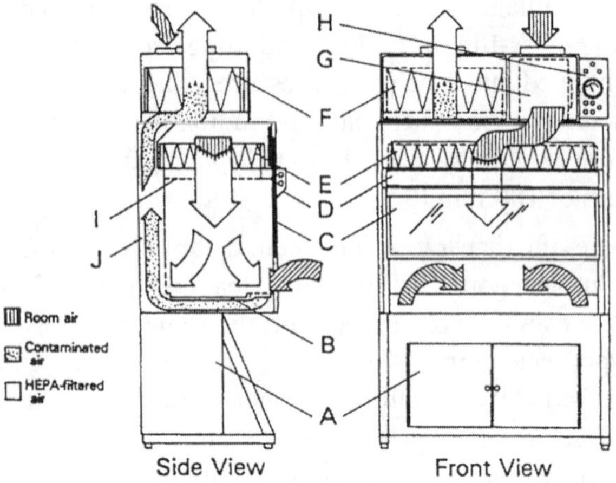

Figure 7-8. *Class II, Type B2 Biological Safety Cabinet* A. storage cabinet; B. work surface; C. sliding sash; D. lights; E. supply HEPA filter; F. exhaust HEPA filter; G. supply blower; H. control panel; I. filter screen; J. negative pressure plenum (adapted from CDC/NIH, 1993).

plenum. Thus, leakage in a contaminated plenum will be into the cabinet and not into the environment. (see Figure 7-9).

It is critical that work practices within the Class II cabinets ensure an unobstructed flow of air to the exhaust slots. Cabinets should never be used to store laboratory materials. Work activities should not be conducted too near the front opening as this could disrupt the clean air sheath that protects the laboratory worker. Additionally, activities around the BSC that could affect airflow into the front slot (thereby breaking the protective containment of the system) should be closely monitored. Disruptive activities may include the repeated insertion and withdrawal of worker arms, supply air diffusers, or cooling fans with focused airstreams in the vicinity of the cabinet, the opening and closing of entry doors, and personnel movements in front of the BSC. In Type B cabinets, since the air that flows to the rear grille is discharged directly into the exhaust system, activities that may generate hazardous chemical vapors or particulates should be conducted towards the rear of the cabinet.

Blowers on exhaust systems should be located at the terminal end of the duct work. A failure in the building exhaust system may not be apparent to the user, as the supply blowers in the cabinet will continue to operate. A pressure-independent monitor should be installed to sound an alarm and shut off the BSC supply fan, should failure in exhaust air flow occur. Since this feature is not supplied by all cabinet manufacturers, it is prudent to install a sensor in the exhaust system as necessary. To maintain critical operations, laboratories using Type B BSCs should connect the exhaust blower to the emergency power supply.

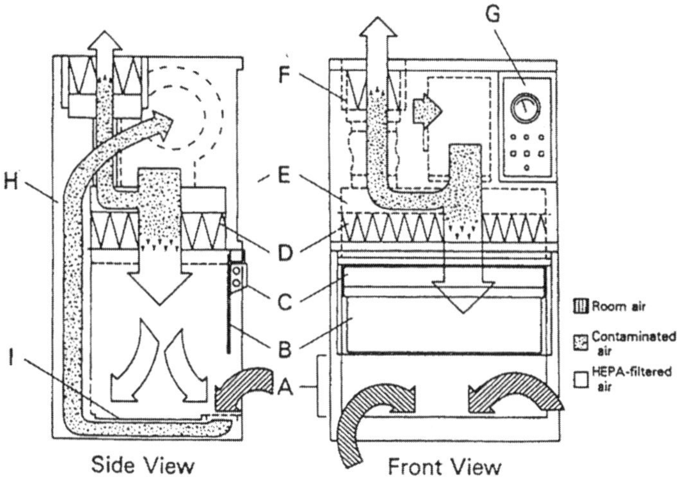

Figure 7-9. *Class II, Type B3 Biological Safety Cabinet, table-top model* A. front opening; B. sliding sash; C. light; D. supply HEPA filter; E. positive pressure plenum; F. exhaust HEPA filter; G. control panel; H. negative pressure plenum; I. work surface (adapted from CDC/NIH, 1993).

Class III BSCs are totally enclosed, ventilated systems that operate under negative pressure. Because these systems are designed to be used with extremely hazardous microbial agents, the seals and their subsequent ability to maintain pressure is integral to the containment performance. Worker access to the cabinet interior is facilitated through rubber gloves attached to ports located on the front (see Figure 7-10). Air entering the cabinet first passes through a HEPA filter. Contaminated air that exhausts the cabinet must pass through two HEPA filters placed in series or a single HEPA filter backed up by an incinerator before being exhausted to the outdoors. Equipment that is used to perform functions within the cabinet must be an integral part of the system due to the containment requirements for working with these extremely hazardous agents (i.e., a gastight enclosure). The entry for equipment into the cabinet is facilitated through sealed air locks; equipment removal must be through a double-doored autoclave (or other decontaminating, air lock systems) or through a "dunk tank" containing a liquid disinfectant.

The critical nature of BSCs in the microbiological laboratory requires that they be constructed according to design specifications and that they meet minimum standards of performance. The National Sanitation Foundation (NSF International) Standard No. 49 for Class II (Laminar Flow) Biohazard Cabinetry provided the first independent standard for design, manufacture and testing of Class II BSCs. Specifically, NSF Standard 49 outlines design, construction, and performance criteria (NSF, 1992). This Standard for biological safety cabinets establishes performance criteria for biological safety cabinets and provides the minimum require-

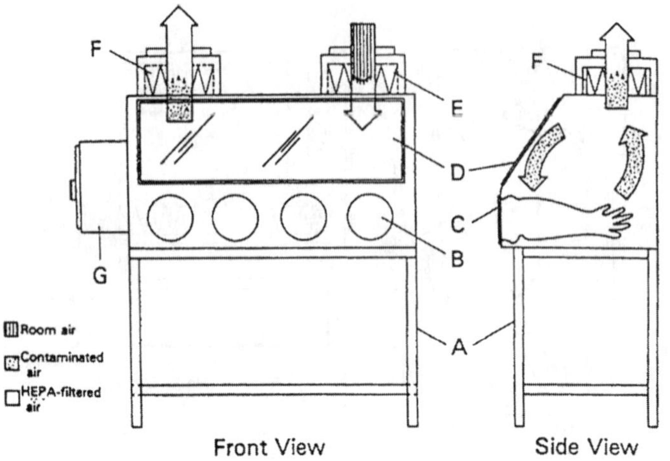

Figure 7-10. *Class III Biological Safety Cabinet* A. stand; B. glove ports; C. O-ring for attaching arm-length gloves to cabinet; D. slopped glass viewing window; E. supply HEPA filter; F. exhaust HEPA filter (second HEPA exhaust is not shown); G. double-ended autoclave (adapted from CDC/NIC, 1993).

ments that are accepted in the United States. Cabinets that meet the standard and were certified by the NSF bear an NSF 49 Seal.

When one buys a Class II BSC, an NSF certification seal should be affixed. Once the cabinet is installed, it becomes the responsibility of the user to maintain the BSC in conformance with the applicable performance criteria. The Class II BSC should undergo a field certification upon installation and at scheduled intervals thereafter. Field re-certification of Class II BSCs should be undertaken on an annual basis, if the cabinet is relocated, or when the HEPA filters are changed. While the NSF standard does not cover field testing of BSCs, it is common for many of its test methods and parameters to be applied in the field, and these are included in Annex "F" of the document. On-site testing following the recommendations for field testing in NSF Standard 49 must be performed by qualified personnel. The NSF maintains a list of companies that manufacture Class II Biosafety Cabinets as well as a list of accredited Field Certifiers (NSF, 1998). Cabinets should contain on the front of their surfaces, explicit statements of certification, limitations regarding cabinet usage, and the next date for re-certification. A list of agents authorized to service the interior of the cabinet should also be kept and prominently posted on the cabinet structure.

All users of cabinets should receive basic training with subsequent periodic refresher courses regarding overall cabinet usage, biological agent manipulation techniques, proper interpretation of cabinet limitation statements (and any subsequent revisions), and certification statements. Periodic review by laboratory management to ascertain adherence to proper BSC usage, laboratory techniques, and training effectiveness should also be performed. Records regarding cabinet purchase, intended use, agents authorized for manipulation within the cabinet, training documentation, and certification/operational status, should be readily accessible to users and to laboratory management.

Secondary Barriers

Secondary barriers (facility design) provide the next level of control beyond the isolation of the contaminant at the source. These barriers are designed to protect personnel inside and outside the laboratory environment from inadvertent exposures (those uncontrolled by the primary barriers). Additionally, the implementation of more complicated engineering control measures (i.e., ventilation) can minimize the escape of agents into the environment outside of the facility.

The degree of hazard of the infectious agent will dictate the types of controls used. For agents classified for work at BSL 1, specific recommendations for secondary barriers include the availability of handwashing stations; non-porous (i.e., impervious to water) bench-tops or work surfaces that are resistant to acids, alkalis, organic solvents, and moderate heat; and laboratory equipment and furniture (including placement) that facilitate cleaning. For BSL 2 agents, additional equipment suggestions include the availability of autoclaves for the proper decontamination of materials and equipment used in the laboratory.

Higher risk agents and activities require the application of physical barriers to help prevent the migration of contaminants from laboratory work areas to other areas of the facility. However, physical barriers alone cannot guarantee containment. The use of general ventilation to create pressure differentials (i.e., with respect to adjacent rooms) between activity zones will minimize opportunities for contaminant escape when doors are opened. BSL 3 recommendations include the physical separation of the laboratory work environment from access corridors. Additionally, access points should be designed with two sets of self-closing doors to act as a "buffer" zone between contaminated and clean areas. Dedicated, ducted exhaust and supply ventilation systems are used to create the negative pressure differentials necessary to move air from clean to contaminated areas. The exhausted air should be discharged outside away from points where re-entrainment of contaminated air could occur (i.e., outdoor air intakes, entry doors, and windows) or away from locations that are occupied. At no time should the exhaust air be recirculated to other areas of the building. Filtration or other means of decontamination may be applied before exhausting the air.

Due to the extremely hazardous nature of work with BSL 4 agents, the requirements for laboratory facilities include very stringent design criteria regarding isolation and containment. Laboratory work areas should be physically-separated from other activities by means of a dedicated building or isolated zones that have been clearly identified. Laboratory personnel accessing or exiting the laboratory must do so through changing rooms separated by a shower facility. All access doors should be self-closing and lockable. All penetration points in the structural envelope of the facility must be tightly controlled. The construction of walls, floors, and ceilings should be done in a manner that forms a sealed shell that facilitates cleaning and decontamination (i.e., fumigation) and eliminates the possibility of animal or insect intrusion. Windows should be resistant to breakage. Other penetration points for consideration include water drainage lines and ventilation lines (e.g., sewer vents). Water drainage line traps should be filled with demonstrably effective chemical disinfectants as a means to seal potential escape routes. Similarly, ventilation exhaust lines should be equipped with HEPA filters.

Materials, supplies, and equipment that are not transported into or out of the laboratory via the changing rooms must pass through double-door autoclaves, fumigation chambers, or ventilated air-locks. Similarly, all materials must be decontaminated before removal from the laboratory environment either through double-door autoclaves or, for materials that cannot be autoclaved, pass-through dunk tanks, fumigation chambers, or other decontamination systems. Liquid effluents (excluding those from the showers and toilets) must be decontaminated through physically and biologically validated heat treatment systems. After decontamination, effluents may be discharged to the sanitary sewer.

Ventilation systems used to supply conditioned air to the laboratory work environment, as well as exhaust air (general removal systems or those ducted from the BSCs) should be dedicated exclusively to the BSL 4 laboratory. Air that is exhausted from the laboratory should never be recirculated back into the work space or any other part of the facility. Additionally, the placement of exhaust exit points should be in locations that guard against re-entrainment of air into the building, e.g., through ventilation system air intakes outdoors. Although ventilation lines are equipped with HEPA filters, system integrity failures can occur (e.g., filters may have slipped off of the seal or there could be breaks in the filter material). These could cause contaminated air to enter the building. Ventilation systems should be balanced in a manner that moves air from clean to less clean or contaminated areas. This includes negative pressure differentials created between the laboratory and designated, uncontaminated areas of the building, and from designated low risk areas to areas of greater potential risk. It is critical that personnel involved in the maintenance and operation of the ventilation system recognize that any change in a component of the system can have a dramatic effect on the pressure relationships between adjacent locations. A component may include the exhaust fans, central HVAC units, dampers, and/or doors and windows. For this reason, access to the ventilation systems (i.e., exhaust fans and HVAC units) should be given to designated personnel only. Monitoring systems should be installed that include alarms that indicate malfunctions of the system. Interlocks between the supply and exhaust systems can help ensure that negative pressure (or minimally a neutral pressure differential) is maintained. Regular preventive maintenance and monitoring is critical to the continued containment capabilities of the system.

In those circumstances when the laboratory work activities limit the practicality of Class III BSCs, a specially designed room or environment can be used along with the use of specially designed PPE as a component of the primary barrier system. One-piece positive pressure suits are worn by personnel entering the containment area. These suits encapsulate the worker and are ventilated with fresh air from a life support system. The supplied air should be clean, low humidity, conditioned (i.e., cool), and free from oil. Backup systems should be incorporated into the life support system in case of equipment failure with alarms to indicate to the wearer that failure has occurred. Access to the containment area is gained through an airlock fitted with airtight doors. The availability of a chemical shower allows decontamination of the environmental suit exterior prior to exit. The shell of the containment area is designed as a large BSC Class III cabinet, i.e., exhausted air must pass through two HEPA filters placed in series or must be incinerated. All materials leaving the containment area must be properly decontaminated (e.g., through double-doored autoclaves, fumigation chambers, or chemical dunk tanks). The room must be operated under negative pressure, and interlocks must be installed between the supply and exhaust ventilation systems to ensure that negative (or neutral) pressure is maintained. Work activities within the containment area should be conducted in Class I or II BSCs, even with the protection afforded by the encapsulating environmental suit.

More detailed discussions may be found in the CDC *Biosafety in Microbiological and Biomedical Laboratories*, the American Society for Microbiology *Laboratory Safety: Principles and Practice*, and the American Industrial Hygiene Association *Biosafety Reference Manual* (Fleming et al., 1995; Heinsohn et al., 1995).

NON-INDUSTRIAL INDOOR ENVIRONMENTS

Unlike the health care industry where the disease and, subsequently, the etiologic agent is identified, the nature and extent of health outcomes resulting from exposures to microbiological agents in indoor environments have not been well defined. This makes the assessment and control of microbiological contamination a difficult process. Difficulties arise from the diverse nature of the causative agents, the variability of the airborne concentrations of the organisms, the sensitivity and reliability of the evaluation tools, and the lack of a clear dose-response relationship between exposures and health effects, and the variation in susceptibility among those exposed. However, the collection of appropriate and sufficient information by various members of an investigation team (including clinicians, epidemiologists, and environmental consultants, i.e., industrial hygienists) will suggest plausible hypotheses regarding the existence of microbiological reservoirs and their potential effect on the population. These hypotheses can then be tested through the collection of environmental data, health questionnaires, and/or clinical tests. Information must be collected with consideration of the determinants of bioaerosol health effects and their interactions. The likelihood and severity of adverse health effects from bioaerosol exposure depend on (a) the biological materials present; (b) the concentrations in air of such materials, preferably at the time of the exposure that caused the effect; (c) the disease outcomes associated with exposure to the bioaerosols; and (d) the general health and immunologic status of the exposed individuals.

The findings from the initial investigation may result in a re-evaluation of the original hypotheses, with the possible collection of additional data. The collected information can reveal environmental conditions suitable for microbiological growth, the location of microbiological reservoirs, and possible dissemination mechanisms. During the initial investigation, the search should focus only on those agents that can be associated with the reported health effects.

For example, in identified clusters of legionellosis, the investigator would not conduct an investigation in search of fungi, dust mites, or other aeroallergens. All of the collected information is subsequently collated and critically analyzed towards the development of recommendations including the remediation of current microbiological reservoirs and the control of future contamination.

The necessary elements of microbiological proliferation are the presence of the biocontaminant; the susceptibility of a material to contamination by microorganisms, nutrients, and moisture; and the appropriate environmental conditions (i.e., temperature and humidity) for the growth of the specific microorganism. Many biocontaminants are ubiquitous inhabitants of the human environment, and therefore, their presence must always be assumed. Material susceptibility to microbial colonization is a function of its porosity. Increasing the number and/or size of pores increases the available surface area for the deposition of particles containing organic debris and hinders cleaning of the surface. Additionally, porous materials more easily absorb moisture. A suitable localized environment (i.e., having adequate nutrients, available moisture, and an appropriate temperature) when combined with highly porous materials, provides optimum conditions for microorganisms to grow. For example, interior duct lining downstream of the ventilation system cooling coils is highly porous. Low efficiency filters upstream of the coils allow the passage of particles that are deposited in the duct lining. In addition, water-saturated air downstream of the coils (caused by the reduction of the air temperature down to, or below, the dew point) condenses onto metal ducting and is absorbed into duct lining. The result is often the growth of fungal colonies in the interior of the ventilation system.

Nutrient sources for bacterial or fungal growth are common in the indoor environment. Examples include accumulated dirt on the insides of ventilation ducts; wood and paper products such as books, cardboard containers, wallpaper, and gypsum wallboard; and other plant and animal materials such as cotton fabrics, wicker baskets, jute carpet backing, sloughed skin cells, and leather. Housekeeping activities and improved ventilation system filtration can help to reduce the deposition and accumulation of particles on surfaces. As stated above, microbial growth and survival also depend on the water activity of the substrate. Water activity is defined by the vapor pressure exerted by the moisture in the material, expressed as a percentage of the saturation vapor pressure of pure water at the same temperature (Flannigan, 1992). Fungi are grouped by their preference for high or low amounts of available water, described, respectively, as hydrophilic, mesophilic, xerotolerant, and xerophilic (Burge, 1995). The documentation of water incursion (either chronic, intermittent, or as an isolated event) can direct a search for microbial reservoirs to damp or water-damaged areas of a building. Microorganisms are also grouped according to their temperature preferences or tolerances, e.g., bacteria are described as psychrophilic, mesophilic, thermotolerant, or thermophilic, respectively, from low to high temperature preference.

Information regarding the moisture content of building materials and/or the relative humidity and temperature within a building may help investigators anticipate what kinds of microorganisms may be present. Su et al. found an association between certain groups of fungi and environmental factors (e.g., soil fungi in homes with dirt-floor crawlspaces; hydrophilic fungi in homes with collected water; and above-ground decay fungi in homes using gas stoves, which may add moisture to the indoor air and encourage people to ventilate their homes with outdoor air) (Su et al., 1992). Similar research may direct attention to particular environments in the search for certain microorganisms and, conversely, may clue investigators to look for certain bioaerosols and related health effects when they notice specific environmental conditions.

Control

The control of moisture incursion, nutritional substrates, and/or temperatures to appropriate levels in the indoor environment will decrease the ability of microorganisms to proliferate. However, microbes are known to survive (even thrive) extreme temperature variants. For example, *Aspergillus fumigatus* has been shown to have a temperature growth range of 12-52°C (Cooney and Emerson, 1964). Therefore, it appears unlikely that the indoor environmental temperature in the occupied spaces can be controlled so as to significantly affect the proliferation of most microorganisms. Given the abundance of nutrient rich media in indoor environments, the exclusion of which also appears unlikely, moisture incursion remains the only practically affectable control factor. To be effective, the investigator must be conscious of water incursion in two forms, liquid and vapor, and how these forms penetrate the building envelope.

Moisture can pass through the building envelope by one or more of four transport mechanisms: 1) liquid flow, 2) capillary action, 3) air movement, and 4) vapor diffusion (Lstiburek and Carmody, 1994). Each transport mechanism can produce unique communities of microbial contamination in specific building locations depending on the frequency of the water intrusion, the quantity of water, and the micro-climate into which it is introduced. For example, ground water intrusion through basement foundation cracks can introduce large "flooding" quantities of water resulting from hydrostatic pressures on the building exterior. These hydrostatic pressures are generally associated with high ground water tables (e.g., resulting from heavy rain). This type of intrusion is predominantly intermittent. However, if the problem is not promptly remediated, the quantity of moisture remaining can promote the growth of hydrophilic fungi (e.g., *Fusarium* sp. and *Stachybotrys chartarum*), yeasts, and Gram negative bacteria. On the other hand, water vapor (i.e., high relative humidity) introduced through a ventilation system's outdoor air intakes can moisten porous interior duct liners providing suitable environments for fungal growth. Martinez et al., in a study of a large office building with a microbiologically contaminated ventilation system, identified inherent system design flaws (improper cooling coil temperature, low-efficiency filters, and porous duct lining) that permitted the development, amplification, and dissemination of microorganisms (predominantly *Penicillium* species) (Martinez et al., 1995). The moistened material and cool temperatures induced by the cooling coils in the ventilation systems selected for psychrotolerant and mesophilic (moisture) and/or xerotolerant species (e.g., *Penicillium* and *Cladosporium*). Carpets installed onto concrete slab flooring provide a niche for summer mold growth when water vapor condenses on the concrete and carpet backing.

The structural design of the building envelope will have the greatest effect on the amounts of moisture incursion. Significant intrusion points can occur at foundation cracks, through openings in exterior walls (i.e., around doors and windows and through the vapor barrier), and at breaks in the roof. Liquid flow and capillary action can occur at all penetration points of the building envelope including foundations, exterior walls, and roofs, and is generally the result of groundwater, rain, or snow melt. Moisture intrusion through air movement is predominantly governed by air pressure relationships induced by the building ventilation system. Moisture-laden air enters through the outdoor air intakes or through cracks in the building envelope via ventilation-induced negative pressures. Air pressure differentials in a building can also be induced by the stack effect — air movement caused by vertical thermal convections in a building — and by pressures induced by wind on outside building surfaces. Vapor diffusion into a building occurs as a result of the inappropriate installation of vapor barriers in exterior walls or around the foundation.

The application of specific moisture controls varies according to the transport mechanism. Liquid flow and capillary action are minimized by the application of deterrents or barriers that prevent water migration through breaks in the building envelope structure. For example, ground water intrusion can be controlled with the use of drainage systems that reduce the hydrostatic pressure around building foundations coupled with exteriorly applied moisture resistant coatings. Rain water intrusion is reduced by the appropriate application of drainage screens, flashing, gutters, and downspouts. Capillary action is a consequence of the surface tension of a liquid between two closely spaced adjacent materials, e.g., two panels of wood siding. By controlling the available capillary moisture, sealing of the capillary pores or making them larger, or providing a receptor of the capillary moisture, the amounts of moisture intrusion by this mechanism can be minimized. These types of water incursion are generally not limited by the climatic conditions of different geographic areas. The design concepts should control ground and rain water intrusion without regard to temperate zones (i.e., heating, cooling, or mixed climates). However, the mechanisms of air movement and vapor diffusion are controlled or influenced by the local climatic conditions.

Buildings are designed to protect the occupants from the elements outdoors. However, geographical regions throughout the United States exhibit distinct climatic conditions and building designs should be adjusted accordingly. A simplified categorization system that can be applied to moisture control is based on the geographically predominant environmental control system shown in Figure 7-11. Architectural design concepts (regarding vapor diffusion) that are used in the northern climates (ventilation defined heating zones) should not be used for building designs in the southern regions (ventilation defined cooling zones). The climate zone has been defined according to the number of heating degree days or hours of wet-bulb temperature. A recommendation for heating zones is defined by 4000 heating degree days (base of 65°F [18°C]) or more. Cooling zones (warm, humid climates) are defined by 1) 3000 or more hours during

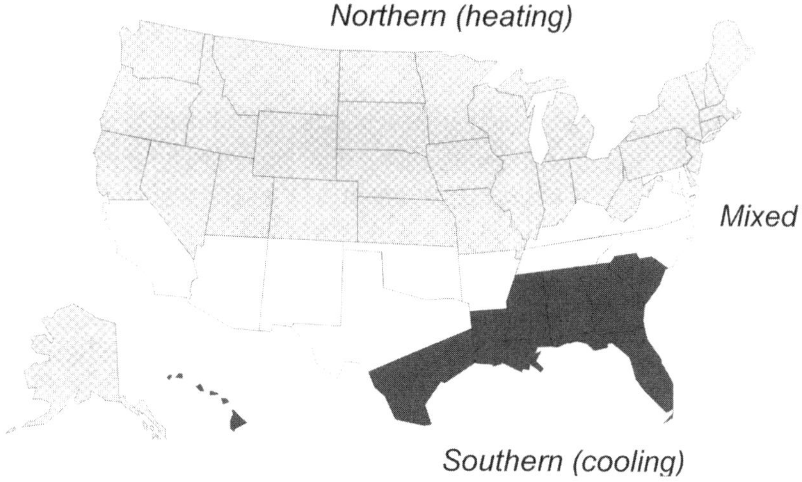

Figure 7-11. Climate zones used for moisture recommendations (adpated from Lstiburek and Carmody, 1994).

the warmest six consecutive months of the year of 67°F (19°C) or higher wet-bulb temperature and/or 2) 1500 or more hours during the warmest six months of the year of 73°F (23°C) or higher wet-bulb temperature. Mixed zones fall within those not defined as heating or cooling.

Vapor-retarding systems are usually installed at or near the surface that is exposed to higher water vapor pressure (Pope et al., 1993). In the northern climates, these systems are normally applied to the interior surface of the envelope wall space in front of the insulation (higher vapor pressure generally occurs in the occupied spaces) (ASHRAE, 1997). This practice controls the ex-filtration of moisture laden interior air out through the exterior wall. Applying the same technique in a southern climate can result in moisture condensation at this surface because water vapor is migrating from outdoor (humid) locations into the interior spaces. The resultant moisture availability, combined with the building materials (i.e., paper of the gypsum board), provides an environment conducive to the growth of microorganisms. Therefore, in the southern regions in air-conditioned buildings, vapor-retarding systems should be applied to the exterior surface of the envelope (higher water vapor pressure exists outdoors). These differences in design also apply to air movement. Positively pressurizing a building in a southern climate will ensure that moisture laden air will not infiltrate the interior of a building. In the northern climates, buildings should operate under a neutral or slightly negative pressure, however, this may enhance the infiltration of contaminants that are detrimental to the indoor environment (e.g., products from combustion appliances, radon gas emitted from soil, outdoor fungal spores, etc.).

Remediation

Remediation of microbially-contaminated building surfaces will result in the disruption of microbiological reservoirs. The airborne dissemination of these bioaerosols can pose a significant exposure concern for the remediation workers. Additionally, these aerosols can be spread to uncontaminated areas of a building, increasing the hazard for the remaining occupants and adding to the difficulty of clean-up. Therefore, it is important that all remediation activities be conducted with an awareness of the potential bioaerosol exposures and with minimal disturbance of contaminated materials. Specifically, controls must be instituted that protect both the *worker* and the *adjacent environment.*

Given the level of disruption that may occur during microbiological remediation work, engineering controls applied at the source should be the primary control measure. Remediation activities should be conducted in a manner that minimizes the disturbance of microbiological reservoirs. However, as the extent of the microbial contamination becomes larger, reservoir dissemination becomes unavoidable due to the removal of building materials. Under these conditions, isolation barriers are required to contain airborne spores and other biological matter. Barriers alone disrupt the pathways between remediation zones and adjacent environments, but disseminated aerosols almost invariably find breaks in any barrier system. Therefore, negative pressure relative to adjacent areas is induced in the remediation zone to ensure containment. It is critical that the exhausted airstreams be appropriately filtered (i.e., HEPA filters) to guard against the re-entry of microbially-contaminated air back into the zone of remediation and/or to other areas that are considered uncontaminated.

Specific control guidelines have been recommended by the New York City Department of Health (NYCDH) for the remediation of fungi from contaminated building materials (see Table 7-5) (NYC Dept. of Health, 2000). The recent revision of the guidelines was intended to be inclusive of all fungi as opposed to the 1993 version, which only focused on *Stachybotrys chartarum*. The expansion to other fungi was based on the following: (1) *Stachybotrys chartarum* cannot be considered as uniquely toxic in indoor environments since many fungi produce mycotoxins, some of which are identical compounds; (2) individuals engaged in remediation activities of widespread fungal contamination may be at risk of developing organic dust toxic syndrome or hypersensitivity pneumonitis; and (3) fungi can cause allergic reactions. These guidelines define five levels of abatement characterized by the scale of the contaminated surface area. All remediation efforts should be conducted during unoccupied periods. Remediation workers should be properly trained in the potential health hazards, appropriate work practices, and correct use of engineering measures. Appropriate PPE should be worn by those individuals involved in all aspects of clean-up activities. Workers should be medically cleared.

Under the NYCDH guidelines, Level 1 abatement requires minimal engineering control measures (other than the use of dust suppression techniques), and relies predominantly on worker practices of appropriately trained building maintenance staff to minimize the disturbance and subsequent airborne dissemination of fungal spores. Contaminated materials should be sealed in plastic bags and disposed of as sanitary waste (Morey, 1993). Level 2 abatement employs

Table 7-5
Suggested guidelines for the remediation of fungi

Abatement Level	Extent of Microbial Contamination	Remediation/Isolation Level
1	< 10 ft^2	Work performed by trained maintenance staff in unoccupied area. Use of N-95 respirator, gloves, and eye protection. Containment unnecessary; dust suppression methods recommended. Damp wiping of surfaces. Contaminated materials removed in sealed plastic bags.
2	10 to 30 ft^2	Work performed by trained maintenance staff in unoccupied area. Use of N-95 respirator, gloves, and eye protection. Work area covered in plastic and sealed with tape. Dust suppression methods recommended. Work and egress areas HEPA vacuumed and cleaned. Contaminated materials removed in sealed plastic bags.
3	30 to 100 ft^2	Health and safety professionals consulted. Remediation personnel properly trained in the handling of hazardous materials. Use of N-95 respirator, gloves, and eye protection. Work area covered in plastic and sealed with tape. Seal ventilation ducts and grills. Dust suppression methods recommended. Work and egress areas HEPA vacuumed and cleaned. Contaminated materials removed in sealed plastic bags. For dust generation move to Level 4.
4	> 100 ft^2	Health and safety professional consulted. Remediation personnel properly trained in the handling of hazardous materials. Use of full-face, HEPA filtered respirators; full-body covering; and gloves. Complete isolation of work area with critical barriers, negative pressure, airlocks, and decontamination room. Contaminated materials removed in sealed plastic bags (outsides must be HEPA vacuumed or damp wiped). Work and egress areas HEPA vacuumed and damp wiped. Air monitoring should be conducted.
5 (HVAC)	< 10 ft^2	Work performed by trained maintenance staff in unoccupied area. Use of N-95 respirator, gloves, and eye protection. Work area coveredin plastic and sealed with tape. HVAC system shut down. Dust suppression methods recommended. Contaminated growth supporting materials removed in sealed plastic bags. Workand surrounding areas HEPA vacuumed and cleaned.
	> 10 ft^2	Health and safety professional consulted. Remediation personnel properly trained in the handling of hazardous materials. Use of N-95 respriator, gloves, and eye protection; for > 30 ft^2, use of full-face, HEPA filtered respirators; full-body covering; and gloves. Complete isolation of work area with critical barriers and negative pressure; airlocks and decontamination room for > 30 ft^2. Contaminated materials removed in sealed plastic bags (outsides must be HEPA vacuumed or damp wiped). Work and egress areas HEPA vacuumed and damp wiped. Air monitoring should be conducted.

Adapted from New York City Department of Health, 2000

engineering measures (i.e., polyethylene enclosures) to contain spore dissemination. Additionally, the work and egress areas require cleaning with HEPA-filtered vacuum equipment prior to the application of detergent solutions. Level 3 requires that remediation be conducted by specialized individuals who have been trained in the handling of hazardous materials. All work areas and those directly adjacent should be covered with polyethylene sheeting and taped (i.e., sealed) including ventilation supply and exhaust grills/diffusers. If the abatement procedures are expected to generate considerable dust or visible fungal contamination is heavy, then Level 4 procedures should be instituted; a health and safety specialist experienced with microbiological agents should be consulted in addition to the use of remediation workers trained in the handling of hazardous materials equipped with appropriate PPE. Appropriate PPE includes full-face respirators with HEPA cartridges, full-body disposable clothes covering, and gloves. The design of containment structures should be comprehensive including total enclosure, exhaust ventilation through HEPA filters to create negative pressure, airlocks, and a decontamination room. Air monitoring is recommended prior to re-occupancy.

Level 5 remediation protocols apply to HVAC systems. The protocols for contaminated areas less than 10 ft^2 (0.9 m^2) are similar in concept to Level 2 remediation activities. Additionally, it is recommended that the ventilation system be shut down prior to work and that contaminated porous materials be removed. For HVAC systems with contamination greater than 10 ft^2 (0.9 m^2), abatement includes combinations of engineering measures, worker practices, and administrative controls similar to Level 4 remediation protocols to ensure fungal spore containment during remediation activities. Isolation barriers are constructed of polyethylene walls; containment is maintained with HEPA-filtered fans that exhaust to the outside. The handling of contaminated materials is conducted in a manner consistent with the handling of hazardous chemicals to ensure the protection of the remediation workers and adjacent environment.

Remediation workers should use PPE appropriate for the hazards to which they may be exposed. Such decisions require *a priori* awareness of potentially hazardous agents, significant exposure routes (e.g., inhalation, dermal contact, or ingestion), and possible concentrations of the biological materials. For example, disturbance of obvious fungal growth and large accumulations of organic matter (bird, bat, or rodent droppings) can be a significant exposure risk for remediation workers. Even the inspection and/or collection of water samples from operating cooling towers in legionellosis investigations can place investigators at risk of exposure. The first step in risk assessment is a visual evaluation of the possible type and extent of contamination, subsequently leading to a determination of the level of protection needed. For example, remediation work on small, localized patches of mold growth on ceilings or walls should be conducted with appropriate respirators, eye protection, and gloves. In contrast, investigators entering an attic with large accumulations of bird or bat droppings may need full-face, powered air-purifying respirators, disposable protective clothing with hoods, gloves, and disposable shoe coverings (see Figure 7-12) (Lenhart, 1994).

In many circumstances, a disposable N-95 NIOSH-approved respirator should offer adequate protection provided that the facepiece fits tightly, ensuring that contaminants do not

Figure 7-12. PPE for remediation workers removing bird or bat droppings.

enter through leaks between the respirator and a wearer's face. (The N-95 designation indicates that the filter material has been shown to remove 95% of particles greater than 0.3 μm.) The size of airborne fungal spores generally ranges from 1 to 50 μm. Other bioaerosols generally fall within a similar size range. A relatively intense exposure is usually necessary to affect non-sensitized individuals. However, some environments may require a higher level of PPE due to the concentrations of the microbial agents and their disease potential. Lenhart et al. has developed a work practice and PPE selection guideline for remediation workers involved in the removal of material potentially contaminated with *Histoplasma capsulatum* (Lenhart et al., 1997).

Various levels of "clearance" sampling may provide information on the quality of the remediation efforts. Results from air sampling can be evaluated by comparing those from the remediated areas to those from adjacent areas and outdoor locations. Sampling can include the collection of bulk, surface, and/or air samples for the suspected etiologic agent(s). However, due to the lack of health-based exposure criteria and the limitations of microbiological sampling protocols, the interpretation of the sampling results is complicated. Establishing acceptable residual concentrations after remediation activities ideally requires risk assessment procedures that include knowledge of the possible clinical endpoints (i.e., infectious, allergic, or toxigenic) and their disease severity, as well as the contribution from outdoor sources and viability of the suspected agent. However, if workers have been immunologically sensitized they may not be able to return to a workplace that has been characterized as sufficiently "clean" for the non-sensitized worker. Bulk and surface samples are collected to identify the predominant taxa of bacteria and fungal contaminants on source materials. In certain situations, air samples can be collected to document the continued airborne presence of a suspect microbial contaminant.

GENERAL INDUSTRY

Agriculture

The contribution of microorganisms to adverse occupational health outcomes is well-recognized, especially in agriculture. "Farmer's lung" has been associated with exposures to various grain saprophytes, including thermoactinomycetes and thermotolerant fungi (e.g., *Aspergillus* sp.), and also mites (Schenker et al., 1998; Kotimaa et al., 1984; Taylor, 1987). These microorganisms are ubiquitous inhabitants of the environment and thrive in the presence of abundant organic substrates found during the harvest season (and during the storage of grains) and suitable amounts of moisture. The magnitude and specific nature of the agricultural activities (i.e., the use of heavy farm machinery such as harvesters that can generate large concentrations of airborne dust) limit the application of source control measures. In these instances, isolation strategies can significantly reduce the occupational exposures to the machine operator. Tractor cabs equipped with HEPA-filtered ventilation systems have been shown to reduce the outdoor air concentrations of particulates by a factor of greater than 100 for particles larger than 1.4 μm in diameter (NIOSH, 1997a, 1997b). However, the efficacy of cab filtration may be hindered by work practices such as frequent door opening and carrying of dust into the cab on work clothes. Where isolation is not a practical alternative, particulate respirators equipped with a minimum of NIOSH certified N95 filters can help to reduce the inhalation of antigenic materials by the worker. Higher levels of respiratory protection may be appropriate for environments that pose increased risks to the occupational population.

Post-harvest, drying of crops before storage can help to reduce the ability of microorganisms to develop a growth niche. The moisture content of storage materials will impact the diversity of the microbial community, i.e., moist grains may contain a greater proportion of thermotolerant and thermophilic microorganisms including *Aspergillus* sp., *Micropolyspora* sp., and thermoactinomycetes (NIOSH, 1986). Additionally, properly ventilated storage bins can limit the available moisture. It has been suggested that the addition of 1% propionic acid can inhibit microbial colonization and limit increased temperatures that select for thermoactinomycetes.

Automated agricultural processes may be more amenable to the application of source control measures. The focused pathways of the process materials during transport and the physical barriers created by the process equipment limit the aerosol generation points. During an evaluation of a sugar beet manufacturer, the installation of LEV over the beet slicers and conveyor belt system was found to significantly reduce the airborne concentrations of total dust and bacteria (Forster et al., 1989). A subsequent decrease in the specific IgG antibody titers of exposed workers was also observed after the installation of the ventilation system. In a separate study of a sugar beet refinery, high exposures to species of *Aspergillus*, *Penicillium*, and *Cladosporium* were identified in pellet loaders and pellet silo workers (Jensen et al., 1993). Subsequent control recommendations included a redesign of the silo to enhance product turnover, modification of the pellet conveyor to prevent spillage, the incorporation of ventilation to control dust from the conveyor, and the use of ventilated spouts during truck and railroad car filling operations.

Animal confinement buildings have been shown to pose significant exposure risks to a diversity of antigenic materials including microorganisms (Cormier et al., 1990; Donham et al., 1989; Jones et al., 1984; Clark et al., 1983; Kullman et al., 1998). The microorganisms observed in confinement buildings are predominately composed of bacteria in contrast to grain handling processes, which have a higher proportion of fungal spores. As with grain handling, source control strategies are limited. Dust control measures may include improved management priorities (e.g., lower stock densities), contaminant dilution through the application of ventilation, and the use of respiratory protection (Donham, 1993). Respirator selection is dependent on the anticipated exposures. Full-face piece respirators with high efficiency filters have been recommended for poultry workers during poultry-catching activities due to levels of bacterial endotoxin encountered (Lenhart, 1998). These respirators have the advantage of affording protection to the eyes in addition to purifying the inhaled air. In those instances when the poultry workers elect to use half-face respirators, eyecup goggles should be provided. Combinations of particulate and ammonia filter cartridges may be appropriate for environments that have demonstrated high levels of ammonia.

The type of feed delivery systems used in animal confinement buildings can have a significant effect on the contribution of dusts from the feed materials. Pelleted feeds have been shown to produce lower dust concentrations than dry meals or wet slurries (Crook et al., 1991). Floor feeding, high moisture feed corn, and indoor feed grinding have resulted in high total and respirable dust concentrations in pig confinement buildings (Holness et al., 1987). Automatic feed delivery systems have been suggested to contain dusts (Myers, 1998). Limited success has been reported with dust suppression systems. The application of a water spray has been shown to reduce the aerosol concentrations in turkey rearing confinements (Cravens et al., 1981). In another study, the addition of a small amount (up to one quart) of water to hay bales prior to their being chopped into material for dairy cow bedding reduced dust and specific dust component concentrations five-fold (Jones et al., 1995). Oil misting systems have been shown to reduce the dust levels in swine confinement buildings by 50%. These systems typically use a 0.5% oil-in-water emulsion automatically misted 12 times per day for 12 seconds at a rate of 7 gallons of oil per pig per day (Nonnenmann et al., 1999). However, such water additive treatments may be ineffectual during freezing weather. Additionally, the introduction of water into an organic-rich environment will increase the potential for the growth of many microbial species (i.e., Gram negative bacteria and hydrophilic fungi).

Biotechnology

The etiologic agents of disease may be contaminants that have found a suitable environment to proliferate in (as in agricultural environments) or may consist of an active component of the manufacturing process. Topping et al. documented sensitization among workers in a biotechnology plant producing citric acid using *Aspergillus niger)* (Topping et al., 1985). The results of this study indicated that the risk was not only from exposures to fungal spores, but also to proteinaceous products found in the culture fluid. In the pharmaceutical industry, Lagier et al. identified a case of occupational asthma in a worker exposed to penicillamine (Lagier et al., 1989). The primary occupational biological hazard in the biotechnology industry is the potential for process microorganisms and their metabolic products to produce an immunologic response in susceptible individuals. Intermediate processing chemicals used during the manufacture of specific products of biotechnology can also pose occupational exposure hazards (e.g., amyl acetate used to extract penicillin).

The effective containment of the potential hazards in the biotechnology industry is dependent on the equipment designs employed in existing chemical process technology. The equipment designs must provide for containment of the microorganisms (viable and non-viable forms), biologically active products or intermediates, and processing chemicals such as extraction solvents. The level of containment is determined by the anticipated risks associated with these agents. Specific factors affecting containment include the selection of appropriate fermentor and associated equipment, suitable exhaust gas treatment, design considerations for vessel overpressurization and relief, suitable inoculation and sampling systems, and the collection and inactivation of condensate that may contain viable microorganisms (Van Houten, 1990).

Anticipating this reliance on chemical process technology, a 1988 study characterized the engineering controls used in conventional enzyme fermentation processes (Martinez et al., 1988). Sample locations selected to reflect worker exposures to process microorganisms included the laboratory (where culture transfers were conducted), inoculum and fermentor tanks (sample ports and agitator shafts), filtering operations, and background locations. The study results indicated that controls are most needed around high-energy operations, including separation equipment (i.e., filters and centrifuges), fermentor agitator shafts, and manual sampling ports (see Figure 7-13). These operations may not be amenable to complete sealing, enclosure, or isolation. Significantly lower bacterial concentrations were observed at a rotary vacuum drum filter compared with concentrations at the filter press. These differences appeared to have resulted from the application of local exhaust ventilation, the inherently better containment characteristics of drum filters, and operator work practices (dislodging filter cake from the filter press plates at the end of each cycle). Rotary vacuum drum filters have been reported to be the most widely used filter in the fermentation industry (Belter, 1979). In contrast, a centrifuge would be expected to produce large concentrations of microbial aerosols. However, as reported in the enzyme manufacturing study, effective process enclosure and the application of local exhaust ventilation at the biomass discharge point resulted in bacterial concentrations significantly below those of the filter process (although above those of the rotary vacuum drum filter). Alternative methods of solids removal include precipitation, coagulation and floculation, chro-

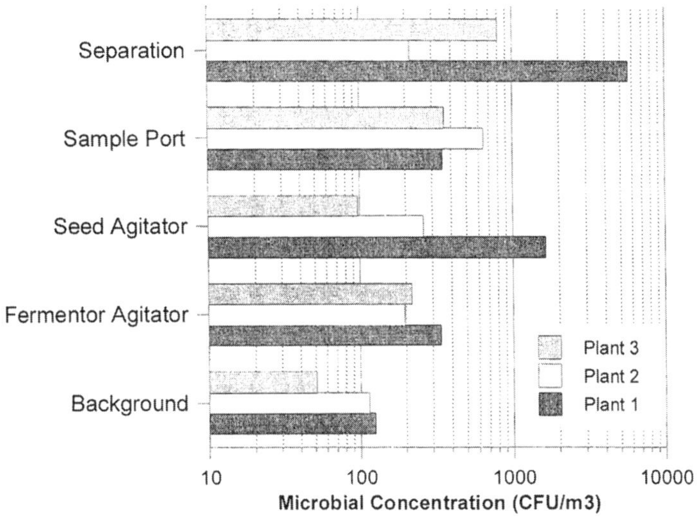

Figure 7-13. Airborne bacterial concentrations at select unit processes in enzyme manufacturing operations.

matography, electrophoresis, and ultracentrifugation. These methods should be applied only after consideration of the agent risk and the inherent containment abilities of the technology. Rapid advancements in the biotechnology industry and use of microorganisms that pose significantly increased risks have resulted in large-scale equipment that offers improved containment capabilities. A study of biohazards associated with using a large-scale zonal centrifuge on oncogenic viruses indicated minimal risks to laboratory personnel during optimal operation (Baldwin et al., 1975). However, faulty seals did result in the detection of high concentrations of phage in the turbine air exhaust and the seal coolant system.

Exhaust gases from the fermentor tanks can be another major emission source of the production microorganisms. The aeration of fermentor broths produces a foam on the surface of the liquid that results in the continuous bursting of bubbles. It has been demonstrated that the droplets of dilute solutions formed by the bursting of bubbles can enrich the concentration of microorganisms by factors of 10 to 1000 times (Blanchard and Syzdek, 1982; Wangwongwatanta et al., 1990). The control of vented bioaerosols within the fermentor tanks produced by this and other mechanical aerosolization processes is achieved primarily by the application of sterilizing (i.e., high efficiency) filtration systems proceeded by "roughing" systems (e.g., cyclones, scrubbers, and/or condensers) (Sayer et al., 1994). Data from the enzyme manufacturing study revealed that scrubbing systems alone may not be effective in controlling vented bioaerosols from the fermentors. The seals around the agitator shafts may be another emission source for the process microorganisms. Double mechanical steam seals appear to provide inherently better containment than packed seals.

The work practices of the operators can also be a determining factor in the potential for occupational exposure. Duiring the collection of fermentor tank sample volumes in the enzyme manufacturing study, operators were observed purging the sample port with a "blast" of pressurized steam prior to the collection of a broth sample. The steam served to clean and decontaminate the interior surfaces of the pipes to ensure a pure sample for subsequent analysis. However, the contact time between the steam and residual microbial populations in the sample line were not adequate to kill the microorganisms and, therefore, resulted in a dissemination of viable aerosols into the surrounding environment. Work practices are most reliable when used in combination with effective engineering measures such as isolation of the machinery from the operator or automation of the process. For example, microbial exposures during the separation of solids can be reduced by limiting operator interaction with those processes or, if this is not possible, the observance of proper and safe work practices.

Large-scale spills of fermentor tanks can be effectively controlled by concrete dikes constructed around the periphery of the tank. Spilled material is directed to a sump and subsequently pumped to a holding tank for inactivation of the microorganisms. In-line sterilizers may also prove effective prior to the disposal of the biological material. Bulk samples of the inactivated material should be microbiologically examined to validate the efficacy of the sterilization techniques. Spill responders should be equipped with PPE including impervious clothing, gloves, autoclavable boots, and appropriate respirators (e.g., a self-contained breathing apparatus).

SUMMARY

This chapter described controls instituted by the health care industry to combat infectious diseases of concern among patients and HCWs. The primary control strategy has been the implementation of administrative measures that promote aseptic work practices and rapid and effective drug treatments for affected individuals. However, engineering strategies have been shown to be effective supplements in the control of infectious agents. These engineering control measures include providing sharps containers for contaminated needles, and the use of specialized ventilation systems designed for isolation rooms that provide negative pressure to contain patient-generated aerosols. Biotechnology relies on the application of engineering controls similar to those used in the chemical industry. Because many of the unit processes are similar to those employed in chemical industry, it is reasonable to assume that the same engineering control technology is generally applicable, albeit in some cases with modifications. Non-industrial indoor environments present a greater challenge due to the limited understanding of the etiologic agent(s) and the clinical endpoints. However, the recommended approach focuses on controlling environmental conditions (e.g., available moisture) at levels unfavorable for microbiological growth.

Regardless of the industry-specific nature of the microbial exposures and health outcomes, the basic precepts of administrative and engineering control strategies can limit exposures. The complicating issue is that the clinical endpoint can be infectious, toxic, or allergic disease or some combination of these. Knowledge of the causative agent and an understanding of the exposure pathway can focus the choice of the control strategies employed. It is imperative that the control practitioner , whether an industrial hygienist, biosafety specialist, infection control practitioner, or engineer, effectively communicate with other involved professionals (i.e., physicians and epidemiologists) to ensure that all pertinent information is used to design or develop appropriate control efforts.

QUESTIONS

7.1. As a control measure, how can substitution be applied to microbiological hazards to provide for a safer working environment?

7.2. Understanding the differences between the clinical endpoints of disease and the mechanisms of microbial growth and dissemination are critical to the design of effective exposure controls. For a susceptible individual to elicit a response, what three things must occur?

7.3. Describe the relationship between infection, virulence, dose, and host resistance.

7.4. Numeric criteria do not currently exist for the interpretation of environmental measurements of biological agents. Explain the reasons why, covering total culturable or countable bioaerosols, specific culturable or countable bioaerosols other than infectious agents, and infectious agents or assayable biological contaminants.

7.5. What are the three most prevalent life-threatening infectious diseases associated with occupational exposures in the United States?

7.6. In what settings do preventive measures tend to have the greatest efficacy?

7.7. The control of infectious aerosols in the healthcare industry has evolved to include the use of general dilution ventilation to minimize the risk of occupational exposure. Describe the desired movement of airstreams within an isolation environment. Include a discussion of a mixing factor.

7.8. Define bloodborne pathogens and the concept of universal precautions.

7.9. Describe and define Class II BSCs.

7.10. What are the elements required for the proliferation of microorganisms within the indoor environment?

7.11. What are the transport mechanisms by which moisture can enter a building?

7.12. What are the primary goals for remediation?

7.13. Farmer's Lung has been attributed to what microorganisms?

7.14. How do operator work practices effect occupational exposures in the biotechnology industry?

REFERENCES

American Conference of Governmental Industrial Hygienists: 2000 Threshold Limit Values for Chemical Substances and Physical Agents and Biological Exposure Indices. ACGIH®, Cincinnati, OH (2000).

American Hospital Association: Statement on microbiologic sampling in the hospital. Hospitals 48:125-126 (1974).

American Society for Heating, Refrigerating, and Air-Conditioning Engineers: Thermal and moisture control in insulated assemblies - applications. In: The ASHRAE Handbook - Fundamentals. American Society of Heating, Refrigerating and Air-Conditioning Engineers, Atlanta, GA (1997).

Baldwin, C.J.; Lemp, J.F.; BarBeito, M.S.: Biohazards assessment in large-scale zonal centrifugation. App Microbiol 29(4):484-490 (1975).

Banazak, E.F.; Barboriak, J.; Fink, J.; et al.: Epidemiologic studies relating thermophilic fungi and hypersensitivity lung syndrome. Am Rev Respir Dis 110:585-591.

Belter, P.A.: General procedures for isolation of fermentation products. In: H.J. Peppler, D. Perlman, Eds. Microbial Technology: Fermentation Technology. 2nd ed. Vol. 2, page 403. Academic Press, New York, NY (1979).

Blanchard, D.C.; Syzdek, L.D.: Water-to-air transfer and enrichment of bacteria in drops from bursting bubbles. Appl Environ Microbiol 43:1001-1005 (1982).

Boshuizen, H.C.; Neppelenbroek, S.E.; Vliet, J.A.; et al.: Serological findings and health complaints in exhibitors working on the 1999 Westfriese Flora in Bovenkarspel. National Institute of Public Health and the Environment of The Netherlands, RIVM Report #213690006 (2000).

Brosseau, L.M.; McCullough, N.V.; Vesley, D.: Mycobacterial aerosol collection efficiency of respirator and surgical mask filters under varying conditions of flow and humidity. Appl Occup Environ Hyg 12(6):435-445 (1997).

Brummitt, C.F.; Cherrington, J.M.; Katzenstein, D.A.; et al.: Nosocomial adenovirus infections: Molecular epidemiology of an outbreak due to adenovirus 3a. J Infect Dis 158(2):423-432 (1988).

Burge, H.A.; Ed.: Bioaerosols, p. 1-23. Lewis Publishers, Boca Raton, FL (1995).

Burge, H.A.: Environmental allergy: definition, causes, control. Engineering Solutions to Indoor Air Problems, pp. 3-9. American Society of Heating, Refrigeration and Air-Conditioning Engineers, Atlanta, GA (1988).

Burge, H.A.: Indoor air and infectious disease. Occupational Medicine: State of the Art Reviews 4(4):713-721 (1989).

CDC (Centers for Disease Control): Microbial environmental surveillance in the hospital, National Nosocomial Infections Study Report. Atlanta, GA: U.S. Department of Health and Human Services, Public Health Service, Centers for Disease Control and Prevention (1970).

CDC (Centers for Disease Control) : Classification of etiologic agents on the basis of hazard, 4th Edition. U.S. Department of Health, Education, and Welfare, Public Health Service (1974).

CDC (Centers for Disease Control): National Nosocomial Infections Study Report, annual summary 1979. Centers for Disease Control and Prevention, Atlanta, GA (1982).

CDC (Centers for Disease Control and Prevention): Guidelines for prevention of transmission of human immunodeficiency virus and hepatitis B virus to health-care and public-safety workers. MMWR 38(No. S-6) (1989).

CDC (Centers for Disease Control and Prevention): Guidelines for preventing the transmission of tuberculosis in health-care settings, with special focus on HIV-related issues. MMWR 39(RR-17):1-29 (1990).

CDC/NIH (Centers for Disease Control and Prevention/National Institutes of Health): Biosafety in microbiological and biomedical laboratories. U.S. Department of Health and Human Services, Public Health Service (1993).

CDC (Centers for Disease Control and Prevention): Guidelines for preventing the transmission of *Mycobacterium tuberculosis* in health care facilities. MMWR 43(No. RR-13) (1994).

CDC/NIH (Centers for Disease Control and Prevention/National Institutes of Health): Primary containment for biohazards: selection, installation and use of biological safety cabinets. U.S. Department of Health and Human Services, Public Health Service (1995).

CDC (Centers for Disease Control and Prevention): Legionnaires disease associated with a whirlpool spa display – Virginia, September-October, 1996. MMWR 46(4):83-86 (1997).

Campbell, I.A.; Cockcroft, A.E.; Edwards, J.H.; et al.: Humidifier fever in an operating theater. Br Med J 27:1036-1037 (1979).

Capewell, S; Leaker, A.R.; Leitch, A.G.: Pulmonary tuberculosis in health service staff – Is it still a problem? Tubercle 69(2): 113–118 (1988).

Chancock, R.W.; Kim, H.W.; Vargosko, A.J.: Respiratory syncytial virus 1. Virus recovery and other observations during 1960. Outbreak of bronchiolitis, pneumonia and other minor respiratory illness in children. JAMA 176:647-653 (1961).

Clark, S.; Rylander, R.; Larsson, L.: Airborne bacteria, endotoxin and fungi in dust in poultry and swine confinement buildings. Am Ind Hyg Assoc J 44(7):537-541 (1983).

Code of Federal Regulations [1992]. OSHA respiratory protection. 29 CFR 1910.134. Washington, D.C.: U.S. Government Printing Office, Federal Register.

Cooney, D.G.; Emerson, R.: Thermophilic fungi, p. 188. Freeman, San Francisco, CA (1964).

Cormier, Y.; Tremblay, G.; Meriaux, A.; et al.: Airborne microbial contents in two types of swine confinement buildings in Quebec. Am Ind Hyg Assoc J 51(6):304-309 (1990).

Cravens, R.L.; Beaulieu, H.J.; Buchan, R.M.: Characterization of the aerosol in turkey rearing confinements. Am Ind Hyg Assoc J 42:315-318 (1981).

Crook, B.; Robertson, J.F.; Travers Glass, S.A.; et al.: Airborne dust, ammonia, microorganisms, and antigens in pig confinement houses and the respiratory health of exposed farm workers. Am Ind Hyg Assoc J 52(7):271-279 (1991).

Decker, J.: Evaluation of isolation rooms in health care settings using tracer gas analysis. Appl Occup Environ Hyg 10(11):887-891 (1995).

Donham, K.; Haglind, P.; Peterson, Y.; et al.: Environmental and health studies of farm workers in Swedish swine confinement buildings. Br J Ind Med 46:31-37 (1989).

Donham, K.J.: Respiratory disease hazards to workers in livestock and poultry confinement structures. Seminars in Respiratory Medicine 14(1):49-59 (1993).

Edwards, J.H.: Microbial and immunological investigations and remedial action after an outbreak of humidifier fever. Br J Ind Med 37:55-62 (1980).

Eickhoff, T.C.; Brachman, P.S.; Bennett, J.V.; et al.: Surveillance of nosocomial infections in community hospitals: I. Surveillance methods, effectiveness, and initial results. J Infect Dis 120:305 (1969).

56 Fed. Reg. 235 [1991]. Occupational Safety and Health Administration: occupational exposure to bloodborne pathogens; final rule. 29 CFR 1910.1030.

Fink, J.N.; Banaszak, E.F.; Thiede, W.H.; et al.: Interstitial pneumonitis due to hypersensitivity to an organism contaminating a heating system. Ann Intern Med 74:80-83 (1971).

Flannigan, B.: Approaches to assessment of the microbial flora of buildings. In: IAQ 92, Environments for people. American Society of Heating, Refrigerating, and Air-conditioning Engineers, Inc. pp. 139-145 (1992).

Fleming, D.O.: Hospital epidemiology and infection control: The management of biohazards in health care facilities. In: Liberman, D.F.; Gordon, J.G.; eds. Biohazards management handbook. New York, NY: Marcel Dekker, Inc., pp. 171-182 (1989).

Fleming, D.O.; Richardson, J.H.; Tulis, J.J.; et al., Eds.: Laboratory safety: principles and practice. 2nd Edition. American Society for Microbiology, Washington, D.C. (1995).

Forster, H.W.; Crook, B.; Platts, B.W.; et al.: Investigation of organic aerosols generated during sugar beet slicing. Am Ind Hyg Assoc J 50(1):44-50 (1989).

Gustafson, T.L.; Lavely, G.B.; Brawner, E.R.; et al.: An outbreak of airborne nosocomial varicella. Pediatr 70:550-556 (1982).

Haley, R.W.; Culver, D.H.; White, J.W.; et al.: The nation-wide nosocomial infection rate: A new need for vital statistics. Am J Epidem 121:159 (1985).

Harrington, J.M.; Shannon, H.S.: Incidence of tuberculosis, hepatitis, brucellosis, and shigellosis in British medical laboratory workers. Br Med J 1: 759–762 (1976).

Healing, T.D.; Hoffman, P.N.; Young, S.E.J.: The infection hazards of human cadavers. Commun Dis Rep 5(*5*):R61-R68 (1995).

Heinsohn, P.A.; Jacobs, R.R.; Concoby, B.A.; Eds.: Biosafety reference manual. 2nd ed. American Industrial Hygiene Association, Fairfax, VA (1995).

Holness, D.L.; O'Blenis, E.L.; Sass-Kortsak, A.; et al.: Respiratory effects and dust exposures in hog confinement farming. Am J Ind Med 11:571-580 (1987).

Hayden, C.D.; Johnston, O.E.; Hughes, R.T.; et al.: Air volume migration from negative pressure isolation rooms during entry/exit. Appl Occup Environ Hyg 13(7):518-527 (1998).

Hodgson, M.J.; Morey, P.R.; Attfield, M.; et al.: Pulmonary disease associated with cafeteria flooding. Arch Environ Health 40(2):96-101 (1985).

Hodgson, M.J.; Morey, P.; Leung, W.; et al.: Building-associated pulmonary disease from exposure to *Stachybotrys chartarum* and *Aspergillus versicolor*. JOEM 40(3):241-249 (1998).

Jensen, P.A.; Todd, W.F.; Hart, M.E.; et al.: Evaluation and control or worker exposure to fungi in a beet sugar refinery. Am Ind Hyg Assoc J 54(12):742-748 (1993).

Johanning, E.; Biagini, R.; Hull, D.; et al.: Health and immunology study following exposure to toxigenic fungi (*Stachybotrys chartarum*) in a water-damaged building. Int Arch Occup Environ Health 68:207-218 (1996).

Johnson, B.; Martin, D.D.; Resnick, I.G.: Efficacy of selected respiratory protective equipment challenged with *Bacillus subtilis* subsp. *niger*. Appl Environ Microbiol 60(6):2184-2186 (1994).

Jones, W.; Morring, K.; Olenchock, S.A.; et al.: Environmental study of poultry confinement buildings. Am Ind Hyg Assoc J 45(11):760-766 (1984).

Jones, W.G.; Dennis, J.W.; May, J.J.; et al.: Dust control during bedding chopping. Appl Occup Environ Hyg 10(5)467-475 (1995).

Jordan, F.N.; deShazo, R.: Immunologic aspects of granulomatous and interstitial lung diseases. JAMA 258(20):2938-2944 (1987).

Kaliner, M.; Eggleston, P.A.; Mathews, K.P.: Rhinitis and asthma. JAMA 258(20):2851-2873 (1987).

Kenny, M.T.; Sabel, F.L.: Particle size distribution of Serratia marcescens aerosols created during common laboratory procedures and simulated laboratory accidents. Appl Microbiol 16: 1146–1150 (1968).

Kotimaa, M.H.; Husman, K.H.; Terho, E.O.; et al.: Airborne molds and actinomycetes in the work environment of farmer's lung patients in Finland. Scand J Work Environ Health 10:115-119 (1984).

Kullman, G.J.; Thorne, P.S.; Waldron, P.F.; et al.: Organic dust exposures from work in dairy barns. Am Ind Hyg Assoc J 59:403-413 (1998).

Lagier, F.; Cartier, A.; Dolovich, J.; et al.: Occupational asthma in a pharmaceutical worker exposed to penicillamine. Thorax 44:157-158 (1989).

Lenhart, S.W.: Recommendations for protecting workers from *Histoplasma capsulatum* exposure during bat guano removal from a church's attic. Appl Occup Environ Hyg 9:230-236 (1994).

Lenhart, S.W.; Schafer, M.P.; Singal, M.; et al.: Histoplasmosis: Protecting workers at risk. Cincinnati, OH: U.S. Department of Health and Human Services, Public Health Service, Centers for Disease Control, National Institute for Occupational Safety and Health, DHHS (NIOSH) Publication No. 97-146 (1997).

Lenhart, S.W.: Livestock rearing: Poultry and egg production. In: Stellman J.M., Ed. Encyclopedia occupational health and safety. 4th ed. Vol. 3. Geneva, Switzerland: International Labor Office, page 70.23 (1998).

Lentino, J.R.; Rosenkranz, M.A.; Michaels, J.A.; et al.: Nosocomial aspergillosis: A retrospective review of airborne disease secondary to road construction and contaminated air conditioners. Am J Epidemiol 116(3)430-437 (1982).

Levetin, E.: Fungi. In: Burge, H.; Ed. Bioaerosols, p. 87. Lewis Publishers, Boca Raton, LA (1995).

Lstiburek, J.; Carmody, J.: Moisture control handbook: principles and practices for residential and small commercial buildings. Van Nostrand Reinhold, New York, NY (1994).

Macher, J.; Amman, H.A.; Burge, H.A.; et al.; Eds.: Bioaerosols: Assessment and Control. American Conference of Governmental Industrial Hygienists, Cincinnati, OH (1999).

Malmberg, P.; Rask-Andersen, A.; Palmgren, U.; et al.: Exposure to microorganisms, febrile and airway-obstructive symptoms, immune status and lung function of Swedish farmers. Scand J Work Environ Health 11:287-293 (1985).

Mallison, G.F.; Haley, R.W.: Microbiologic sampling of the inanimate environment in U.S. hospitals, 1976-1977. Am J Med 70:941-946 (1981).

Martinez, K.F.; Sheehy, J.W.; Jones, J.H.; et al.: Microbial containment in conventional fermentation processes. Appl Ind Hyg 3(6):177-181 (1988).

Martinez, K.; Seitz, T.; Lonon, M.; et al.: Application of culturable sampling methods for the assessment of workplace concentrations of bioaerosols. Inhal Toxicol 7:947-959 (1995).

Martone, W.J.; Jarvis, W.R.; Culver, D.H.; et al.: Incidence and nature of endemic and epidemic nosocomial infections. In: Bennett J.V., Brachman P.S., Ed. Hospital Infections, pp. 577-596. Little, Brown, and Company, Boston, Massachusetts (1992).

McKinney, R.W.; Barkley, W.E.; Wedurn, A.G.: The hazard of infectious agents in microbiologic laboratories. In: Disinfection, Sterilization, and Preservation, Fourth Edition, Chapter 43. pp. 749-756 (1991).

Morey, P.R.: Microbiological contamination in buildings: precautions during remediation activities. In: ASHRAE indoor air quality '93: environments for people. American Society of Heating, Refrigerating and Air-Conditioning Engineers, Atlanta, GA (1993).

Myers, M.L.: Livestock rearing: pigs. In: Stellman J.M., Ed. Encyclopedia occupational health and safety. 4th ed. Vol. 3, page 70.22. International Labor Office, Geneva, Switzerland (1998).

Nonnenmann, M.W.; Rautiainen, R.H.; Donham, K.J.; et al.: Vegetable oil sprinkling as a dust reduction method in a swine confinement. Proceedings of the International Symposium on Dust Control in Animal Production Facilities, Aarhus Denmark, 271-278 (1999).

NIOSH: Microbial flora and fauna of respirable grain dust from grain elevators. Cincinnati, OH: U.S. Department of Health and Human Services, Public Health Service, Centers for Disease Control and Prevention, National Institute for Occupational Safety and Health, DHHS (NIOSH) Publication No. 86-104 (1986).

NIOSH: NIOSH guide to industrial respiratory protection. Cincinnati, OH: U.S. Department of Health and Human Services, Public Health Service, Centers for Disease Control and Prevention, National Institute for Occupational Safety and Health, DHHS (NIOSH) Publication No. 87-116 (1987).

NIOSH: Guidelines for protecting the safety and health of health care workers. Cincinnati, OH: U.S. Department of Health and Human Services, Public Health Service, Centers for Disease Control and Prevention, National Institute for Occupational Safety and Health, DHHS (NIOSH) Publication No. 88-119 (1988).

NIOSH: NIOSH guide to the selection and use of particulate respirators certified under 42 CFR 84. Cincinnati, OH: U.S. Department of Health and Human Services, Public Health Service, Centers for Disease Control and Prevention, National Institute for Occupational Safety and Health, DHHS (NIOSH) Publication No. 96–101 (1996).

NIOSH: In-depth survey report: control technology for agricultural environmental enclosures at John Deere Manufacturing Company, Inc. Cincinnati, OH: U.S. Department of Health and Human Services, Public Health Service, Centers for Disease Control and Prevention, National Institute for Occupational Safety and Health, NIOSH Report No. ECTB 223-13a (1997).

NIOSH: In-depth survey report: control technology for agricultural environmental enclosures at Nelson Manufacturing Company, Inc. Cincinnati, OH: U.S. Department of Health and Human Services, Public Health Service, Centers for Disease Control and Prevention, National Institute for Occupational Safety and Health, NIOSH Report No. ECTB 223-11b (1997).

NSF (National Sanitation Foundation): Standard 49 Class II (laminar flow) biohazard cabinetry. Ann Arbor, MI: National Sanitation Foundation (1992).

NSF: NSF listings: biohazard cabinetry: class II cabinet certification/field certifier accreditation. Ann Arbor, MI: National Sanitation Foundation (1998).

New York City Department of Health: Guidelines on Assessment and Remediation of Fungi in Indoor Environments. New York, NY: New York City Department of Health, Bureau of Environmental & Occupational Disease Epidemiology (2000).

OSHA: Enforcement procedures and scheduling for occupational exposures to tuberculosis. CPL 2.106. Washington, D.C.: U.S. Department of Labor, Occupational Safety and Health Administration, Office of Health Compliance Assistance (1996).

Parker, M.T.: Transmission in hospitals. In: Hers JF, Winkler KC, ed. Airborne transmission and airborne infection: Concepts and methods presented at the VIth International Symposium on Aerobiology, pp. 438-444. John Wiley and Sons, New York, NY (1973).

Pickering, C.A.; Immune respiratory disease associated with the inadequate control of indoor air quality. Indoor Environ 1:157-161 (1992).

Pike, R.M.: Laboratory-associated infections: incidence, fatalities, causes, and prevention. Ann Rev Microbiol 33:41-66 (1979).

Perl, T.M.; Haugen, T.H.; Pfaller, M.A.; et al.: Transmission of herpes simplex virus type 1 infection in an intensive care unit. Ann Intern Med 117(7):584-586 (1992).

Pope, A.M.; Patterson, R.; Burge, H.; Eds.: Engineering control strategies. In: Indoor allergens: assessing and controlling adverse health effects, p. 206. National Academy Press, Washington, D.C. (1993).

Reid, D.P.: Incidence of tuberculosis among workers in medical laboratories. Brit Med J 2: 10–14 (1957).

Riley, R.L.; Nardell, E.A.: Cleaning the air: the theory and application of ultraviolet air disinfection. Am Rev Respir Dis 139:1286-1294 (1989).

Sayer, P.; Burckle, J.; Macek, G.; et al.: Regulatory issues for bioaerosols. In: Lighthart B, Mohr AJ, eds. Atmospheric microbial aerosols: theory and application, page 331. Chapman & Hall, New York, NY (1994).

Sawyer, M.H.; Chamberlin, C.J.; Yuchiao, N.W.; et al.: Detection of varicella-zoster virus DNA in air samples from hospital rooms. J Infect Dis 169:91-94 (1994).

Schenker, M.B.; Christiani, D.; Cormier, Y.; et al.: Respiratory Health Hazards in Agriculture. Am J Resp Crit Care Med Suppl 158 (Part2):S1-S76 (1998).

Sewell, D.L.: Laboratory-associated infections and biosafety. Clin Microbiol Rev 8(3):389-405 (1995).

Shapiro, C.N.: Occupational risk of infection with hepatitis B and hepatitis C virus. Surg Clin North Am 75:1047-1056 (1995).

Sherertz, R.J.; Benlani, A.; Kramer, B.S.; et al.: Impact of air filtration on nosocomial aspergillus infections: Unique risk of bone marrow transplant recipients. Am J Med 83:709-718 (1987).

Snydman, D.R.; Hindman, S.H.; Wineland, M.D.; et al.: Nosocomial viral hepatitis B: A cluster among staff with subsequent transmission to patients. Ann Intern Med 85:573-577 (1976).

Stem, E.L.; Johnson, J.W.; Vesley, D.; et al.: Aerosol production associated with clinical laboratory procedures. Amer J Clin Path 62:591–600 (1974).

Su, H.J.; Rotnitzky, A.; Burge, H.A.; et al.: Examination of fungi in domestic interiors by using factor analysis: Correlations and associations with home factors. Appl Environ Microbiol 58:181-186 (1992).

Taylor, A.J.N.: Respiratory allergy in farmers. The Practitioner 231:1146-1150 (1987).

Thorne, P.S.; Heederik, D.: Indoor bioaerosols - sources and characteristics. In: Organic Indoor Air Pollutants - Occurrence, Measurement, Evaluation, pp. 275-288. Salthammer, T.; Ed. Wiley/VCH, Weinheim, Germany (1999).

Topping, M.D.; Scarsbrick, D.A.; Luczynska, C.M.; et al.: Clinical and immunological reactions to *Aspergillus niger* among workers at a biotechnology plant. Br J Ind Med 42:312-318 (1985).

U.S. Department of Health and Human Services: Control of smoke from laser/electric surgical procedures. Washington, D.C. (1996).

Van Houten, J.: Safe and effective spill control within biotechnology plants. In: Hyer W.C., Ed. ASTM STP 1051 Bioprocessing safety: worker and community safety and health considerations, page 91. American Society for Testing and Materials, Philadelphia, PA (1990).

Vinken, W.; Roels, P.: Hypersensitivity pneumonitis to *Aspergillus fumigatus* in compost. Thorax 39:74-74 (1984).

Wangwongwatanta, S.; Scarpino, P.V.; Willeke, K.; et al.: System for characterizing aerosols from bubbling liquids. Aero Sci Tech 13:297-307 (1990).

Weber, A.; Willeke, K.; Marchioni, R.; et al.: Aerosol penetration and leakage characteristics of masks used in the health care industry. Am J Inf Cont 21(4):167-173 (1993).

Weiss, N.S.; Soleymani, Y.: Hypersensitivity lung disease caused by contamination of an air-conditioning system. Ann Allergy 29:154-156 (1971).

Wistreich, G.A.; Lechtman, M.D.: The procaryotes: their structure and organization. In: Microbiology, 5th ed., p. 186. MacMillan Publishing Company, New York, NY (1988).

Wistreich G.A.; Lechtman, M.D.: Microbial virulence. In: Microbiology. 5th ed., p. 647 Macmillan Publishing Company, New York, NY (1988).

WHO (World Health Organization): Groups at risk: WHO report on tuberculosis epidemic 1996. Geneva, Switzerland: World Health Organization, Global Tuberculosis Programme (1996).

Young, S.E.J.; Healing, T.D.: Infection in the deceased: a survey of management. Commun Dis Rep 5(5):R69-73 (1995).

8

BIOLOGICAL MONITORING IN OCCUPATIONAL HEALTH

Glenn Talaska, PhD

INTRODUCTION

Workers in contaminated environments absorb materials via the respiratory and dermal routes, and by ingestion. Biological monitoring uses the individual as the dosimeter to estimate total exposure. Biological monitoring involves measurement in blood, urine, or breath of *determinants* resulting from exposure; the levels are compared with consensus or legal recommendations to determine if worker health is being protected. Biological monitoring can be useful in several ways that air sampling cannot: exposures from all routes can be estimated, as can differences in individual response to the same dose. Another unique potential of biological monitoring is that past exposures can be reconstructed as during a spill and unplanned cleanup, when no air samples were taken. For these reasons, industrial hygienists in many parts of the world are increasingly looking to biological monitoring in exposure assessment.

Despite potential advantages, biological monitoring has not been widely employed to estimate exposure by industrial hygienists in the United States. A biological monitoring program requires an understanding of the physiology (toxicokinetics) of exposure, but toxicokinetics have not been part of traditional hygiene training. Fortunately, the level of training needed is not beyond anyone with scientific training. This chapter begins with a review of the strengths and weaknesses of the current exposure assessment paradigm, and then discusses the strengths and weaknesses of biological monitoring and what is needed to conduct a biological monitoring program, so that it will be used to increase the precision of exposure estimates.

AIR SAMPLING: STRENGTHS AND WEAKNESSES

Volume 1 focuses on sampling the ambient environment to predict exposure. A significant fraction of industrial hygienists regard air sampling as the major, if not the sole, exposure assessment tool. Historically, there are several important reasons why dusts, mists, fumes, vapors and other airborne materials have been a focus of industrial hygienists. These materials were in the air, often at very high concentrations; air sample collection and analysis was, and is, easy, relative to collecting body fluids; entry into the body via the lungs is facilitated by the anatomy and physiology of these organs; and, health effects did occur and could be related to the workers' long term exposure. There are exposures that affect the respiratory tract directly, like silica, asbestos, acids, and other irritants, for which there is no alternative to air sampling because a relationship between the airborne exposure levels and effect can be directly demonstrated. For example, by the time the buffering capacity of the blood is overcome by an exposure to an acid mist, the lungs will be so severely damaged that the person is fatally injured. For these agents, ambient levels directly predict effects.

The air sampling tradition has brought with it many advantages. A tremendous research effort has been brought to bear to deal with such problems as size and placement of sampling equipment (breathing zone or area samples, for example), improvement of pumps, battery packs and collection devices in the sampling train, in an effort to obtain a representative sample (Volume 1). Today, pumps can be expected to operate for a full shift, and workers will generally wear a sampling train with minimal protest if they anticipate that the purpose of the sampling is to improve or control exposure. Air sampling has become easier for the practicing hygienist to perform. The advisory and regulatory standards for allowable air concentrations of contaminant that have developed over the years give hygienists a place to stand relative to good practice, and are an integral part of the air sampling tradition. Unfortunately, these advantages have encouraged the use of air sampling to the point of exclusion of any other means of evaluating the work environment.

Is there a need to have alternative exposure evaluation strategies? Clearly there is. The guiding precept of hygiene remains "to protect the health of the worker" (see Code of Ethics, Volume 1). While collecting an air sample is a relatively easy task, interpretation of the results becomes less straightforward if there is a potential for: significant exposure by other routes; significant variability in individual response to the same exposure; chronic, cumulative effects related to an internally produced metabolite of the agent.

Interactions Between the Person, the Agent, and the Environment

The response of any individual to an exposure is the outcome of interaction between factors of the person, the agent, and the environment in which they interact (see Figure 8-1). A model used in medical microbiology to understand why some people will become ill and others will not following an exposure to an infectious agent can be adapted usefully to the occupational setting (Jawetz et al., 1978). Changes in factors related to these three components will alter the interaction and increase or decrease response accordingly. Genetics may play a significant role in some cases, but often dose and exposure route are the main effectors. Dose is a key factor in response. The central question to be addressed is, "What do we mean by dose and exposure?" Can the amount of material in the environment represent the dose? Or is the key

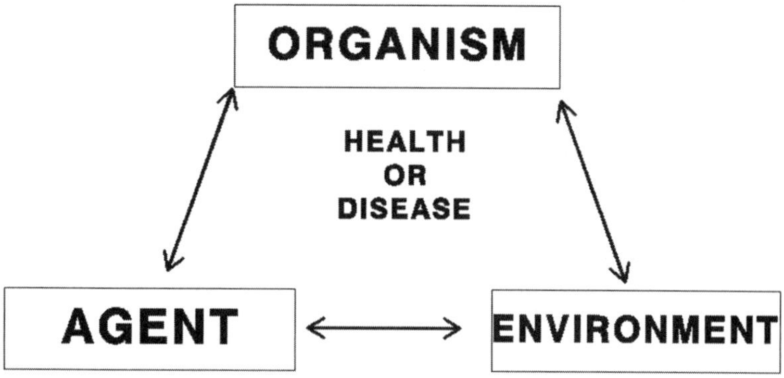

Figure 8-1. Model of interaction of host, agent and environmental factors that can provide health or disease in an individual.

factor of dose the amount of material that actually enters the body, or, is it the fraction of the latter that actually interacts with the target of toxicity?

PERSON/ORGANISM

Workers are not a homogeneous group even though they have more commonalities than the population as a whole. Persons who enter an industrial environment are changed by it. To a greater or lesser degree they accept the ethic of the workplace and the discipline of the tools they operate. If exposed, they also absorb chemical agents and are affected by them. A person who absorbs even a low level of solvent will show the subtle, reversible effects that the solvent has on the fluidity of the nerve cell membrane (Snyder and Andrews, 1996). Fortunately, most often these changes do not dramatically affect the ability to make judgments. Workers also bring to the workplace the influence of their "outside life." Factors such as age, body size, personal habits, hormonal and general health status, percent body fat, and breathing rate affect absorption of compounds, while nutrition, general health, genetics of metabolism, and damage repair can alter the disposition of the absorbed dose. Personal health status can have a considerable impact; persons who are sleep-deprived or stressed may also respond differently than those who are not.

Uptake of a compound by inhalation from an ambient air level of a compound is generally assumed to be uniform for all workers. This is not the case. Opdam and Smolders (1986) investigated the difference in uptake in inhaled dose for perchloroethylene and found that there was a two-fold variability between individuals in the six persons tested. The amount of body fat is generally considered to play a key role in the uptake of compounds, but the level of lipids in the blood may also prove to be very important. A person who eats a very fatty meal for breakfast or lunch will have a correspondingly large level of lipids in their blood for several hours as the dietary fat is processed in the liver and transported to storage sites. If an individual is exposed to a lipophilic compound during this time, absorption may be markedly increased. Riihimäki and co-workers (1979) have shown that blood xylene concentrations are increased following a meal. Pang et al. (1980) saw that blood concentration of dichloromethane was correlated with blood triglyceride concentrations.

Individuals also differ significantly in their absorption through the skin. Van Rooij et al. (1993a and 1993b) reported a four-fold range of absorption of a coal tar preparation through the skin, although they reported that the anatomical site of application contributed more to the total variability seen. Drexler et al. (1995) published that the slope of the line relating external exposure to internal dose was distinctly higher in workers with damaged skin than those with undamaged skin. There is also significant variability in the metabolism of absorbed agents. Autrup (1990) and Butler et al. (1989a, 1989b) reported a 10-fold variability in the ability of human livers to metabolize PAH and aromatic amines, respectively, in vitro. These data indicate that there are a variety of personal, physiological and genetic factors that impact on absorption, distribution, metabolism and elimination and the health status of the exposed person.

AGENT

Toxicity and dose are the two most important considerations when assessing the effects of an agent, but these are affected by other factors. Physical form, including whether the compound is a solid, liquid or gas, particle size, and vapor pressure, are key factors that determine whether a compound will be absorbed. The chemical form of a compound, its water and fat solubility and its ability to cross biological membranes are also important factors in toxicity and effect. Most materials are more toxic if they are in a form that can be readily absorbed; work environments that favor the formation of vapors or fine particulates are especially hazardous. It is widely recognized that some materials are more toxic in one chemical form relative to another; there are vastly different TLVs® for chromium III and VI compounds, for example, reflecting the fact that the former is less well absorbed into target cells.

ENVIRONMENT

Workplace and environmental factors can have a substantive effect on both the person and the agent. Some of these factors include route of exposure, temperature, humidity, work rate, and concurrent exposures. Piotrowski (1957) saw that the internal dose of aniline increased exponentially as skin temperature increased from 30°C to 37°C. The physical demands a task places on a worker may have an impact on the uptake and disposition of chemical agents. Increasing the work rate from 0 to 150 watts increases lung ventilation seven-fold and cardiac output three-fold while the blood flow to the brain remains constant and that to the liver decreases slightly (Droz et al., 1991). Consider what might happen in the case of exposure to a water-soluble, airborne solvent like acetone or ethanol. These compounds are active CNS depressants. Ethanol is detoxified by hepatic metabolism. Because of water solubility, increased breathing rate would also increase the mass absorbed. Since the blood flow to the liver is decreased, metabolism would also be decreased while blood flow to the brain would remain unchanged. Effects of less water-soluble compounds would be less dramatic because the absorption of these compounds from the air is perfusion dependent. Thus, they are limited by the amount of blood flow to the lungs that increases three-fold or less with increasing work rate. Baelum et al. (1987) studied the effects of exercise on the absorption of toluene and the elimination of its major metabolite, hippuric acid. With exercise performed intermittently during the exposure at an average rate of 75 watts, the peak level of urinary toluene metabolites was nearly twice as when there was no exercise. This would indicate that more compound has been absorbed under these conditions. Similarly, Lapare et al. (1993) noted that the amount of toluene and xylene volunteers absorbed increased by 30 and 27.8 %, respectively, when the volunteers exercised for 10 minutes at 50 watts under conditions of controlled exposure to both compounds. Drexler et al. (1995) reported that overweight workers exposed to carbon disulphide excreted significantly less metabolite into the urine in post shift samples collected at the work site than would be expected. It appeared that the remainder of CS_2 partitioned into the fat during the shift and the metabolism and excretion was delayed until the worker reached home after work. Monster et al. (1979) reported that minute volume (ventilation) was a much better predictor of uptake in volunteers exposed to 1,1,1-trichloroethane than was total body weight or body fat and explained greater than 90% of the variability in pulmonary uptake.

Exposure by All Routes

Dermal exposure is common in industry. Dermatoses account for about 70% of all occupational diseases (Suskind, 1997). While most of these (80%) are due to the effects of primary irritants such as acids, alkalis, and solvents on the skin, their frequency indicates how common dermal exposure is in industry. Currently, at least 150 compounds with ACGIH® TLVs® carry the skin notation, i.e., indicating that the skin can be a significant source of exposure. This number is an underestimate of the compounds that can be readily absorbed through the skin. Tsuruta (1990) predicted that three compounds (trichloromethane, 1,2-dichloroethane, and dichloromethane) of a series of six chlorinated hydrocarbons without skin notations would be at least as well absorbed through the skin as was carbon tetrachloride, which carries a skin notation. It is hard to imagine that the active principals from coal tar, asphalt and compounds like benzo(b)fluoranthrene and benzo(a)pyrene are not well absorbed through the skin (they have no skin notation), particularly since recent data indicates in several cases that this is the major route by which the carcinogenic principals gain entry into the body (Herbert et al., 1990, Jongeneelen et al., 1988a and 1988b). The problem of dermal absorption is particularly important for chemical carcinogens. Greater than 50% of the A1 or A2 carcinogens in the 1996 TLVs® and BEIs® book carry the skin notation (ACGIH®, 1996). The rate of absorption through thc skin is influenced by the agent itself (Tsuruta, 1990), personal factors like dermal injury (Drexler et al., 1995) and environmental factors like temperature and humidity (Piotrowski, 1957). The rate of dermal absorption varies greatly between individuals and by site. These factors indicate that dermal exposure is important in industry and can contribute greatly to total exposure, while not at all being accounted for by air sampling.

Chronic Effects and Self Selection in the Workplace

When acute effects are the major exposure concern, some self-selection of sensitive workers can be expected. For example, individuals less adaptable to heat stress will learn quickly that they are not suitable for bakery or foundry work; nor will persons with existing allergies to formaldehyde become embalmers. The same is not true when chronic effects are the major concern because neither workers nor management may realize that individuals are more likely to exhibit effects that require 20 to 30 years of cumulative exposure and damage to become manifest and, therefore, not allow self-selection out of the workforce. The imputation of chronic health effects from a limited set of air samples on an individual is suspect. The damage that accumulates from long-term exposure to certain compounds is only slowly reversible, if it is at all. For example, DNA adducts of carcinogens often persist in the target organs for the lifetime of the cell, which is often over 100 days (Clayson and Lawson, 1987). The damage caused by compounds like cadmium which has a biological half-life of at least 10 years (Jarup et al., 1983) or n-hexane, which attacks peripheral neurons, cells thought to be non-dividing, is essentially irreversible. Saltzman and others have noted that in order to accurately assess this type of exposure, sampling duration must be at least 1/4 the biological half-life (Saltzman, 1996). In the case of DNA adducts, this would mean a sampling time of at least 200 hours, depending on the cell type. For irreversible damage, it is clear that air sampling would be inappropriate to predict health effect.

These problems beg several other questions regarding individual response predictions from aggregate group exposure data. How can effects seen in a subset of individuals from a larger group apparently exposed to the same air levels of a compound be explained by air sampling alone? Are these workers "more sensitive," or more exposed because of other exposure routes? What can be done when the air levels appear to be within exposure guidelines, yet some workers are exhibiting symptoms? Again, sensitive individuals, or more exposure? While most would agree that there are persons who appear to be exposed below the current recommended standards yet still exhibit symptoms, and others who appear to be highly exposed who do not develop disorders, whether the difference is due to genetics, physiology, or true internal exposure is difficult to ascertain without data. Air sampling results by themselves cannot deal with these problems adequately. The hygienist might be tempted to say that when the air sampling results are within legal, company, or advisory guidelines, the person sampled is safe (see Figure 8-2). This is definitely not in the spirit or letter of the guidelines (below).

The TLVs® are fluid entities, changing as information regarding exposure and effects become known. For example, the TLV® for lead has decreased from 150 μg/m³ to 50 μg/m³

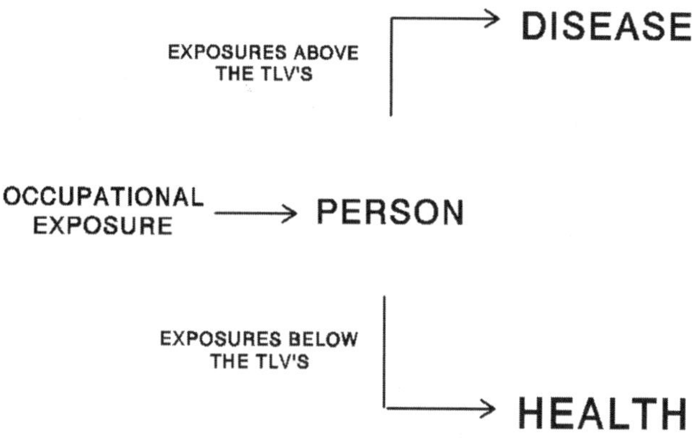

Figure 8-2. Simplified (and inappropriate) model for health effects relative to the TLVs®.

between 1946 and 1996 and the TLV® for benzene has decreased from 100 ppm to the proposed 0.5 ppm during the same time. While animal studies have been used to establish many TLVs®, the responses of workers are a key component in their re-evaluation. Limits have been lowered because at least some workers have responded to levels earlier thought to be safe.

To summarize, the advantages of using air sampling to control occupational exposure include: 1) A long tradition of use; 2) Good worker acceptance; 3) Standards with the force of law; 4) Availability of good equipment; 5) availability of trained sampling personnel from educational programs; and, 6) Relatively easy analysis. Air sampling is particularly useful for primary irritants and those agents that affect the site of absorption, acutely or chronically, but do not affect more distal tissues. These advantages are offset by several very important disadvantages, which include inability to: 1) account for all exposure routes; 2) account for individual differences in absorption, or metabolism, i.e., individual sensitivity; 3) account for the impact of personal protective equipment; and 4) deal with concomitant exposures. These weaknesses of air sampling will limit its use in 21st century hygiene practice.

BIOLOGICAL MONITORING AND BIOMARKERS

Biological monitoring is the measurement of agents, metabolites, or the effects of agents or metabolites in the breath, urine, or blood of the exposed worker. Other secretions and samples have been proposed, but have not found general use. Biomarkers are the measured determinants. Biological monitoring is a means of evaluating the effect of the workplace on the worker, using the exposed person as the dose-integrator. Biological monitoring has been used for many years in hygiene. Robert Kehoe at the Kettering Laboratory in the University of Cincinnati was among the pioneers of biological monitoring, using blood lead levels to determine the routes of exposure, and the relationship of exposure to health effects for workers exposed to organic and inorganic lead compounds (Kehoe, 1963). In 1926, Cook used a differential white cell count to detect excessive benzene exposure in a bitumen-based cable insulating plant (Cook, 1990). The workers who handled rolled cable and had dermal exposure, not those involved with the application of the insulation, where exposure was controlled, were the ones with decreased WBC counts. Yant et al. (1936) monitored benzene exposure using the ratio of inorganic to organic sulfates in urine, as an estimate of conjugated benzene metabolites in urine. Elkins was among the early proponents of the systematic use of biological monitoring in industry; he felt the adoption of biological monitoring into routine hygiene practice was "inevitable" (Elkins, 1967; Waritz, 1979). More recently, Rappaport (1995) stated that health effects for lead exposure were better predicted by biological monitoring, rather than air sampling data.

Information on environmental exposure can be useful to predict certain health effects, particularly with irritants and agents that act on the lung itself. However, for many exposures, this is not sufficient. For example, when exposure is via the dermal, oral, and pulmonary routes, then air sampling measures but a single component of the total exposure. Further, when internal biological processing is necessary in order for a toxic effect to occur, as when the material is metabolized into a more toxic component, then air sampling is only a first approximation of the potential for an effect. N-hexane and methyl-n-butyl ketone must be metabolized into 2,5-hexanedione by the cytochromes P450 and alcohol dehydrogenase before they cause the spinal and peripheral neuropathy with which they are associated (Herskowitz et al., 1971; Yamada et al., 1967). Despite the fact that, in some cases, the internal dose of a compound can be directly predicted by the airborne concentration to which the worker is exposed, it may be simpler and more cost effective to use biological monitoring. This may be the case when breath or urine samples can be collected from a large group of workers at one time and obviate the need of the hygienist to hang and monitor a large number of pumps. Biomarkers also persist for various times after an exposure and can be used to reconstruct exposures when no air sampling was done at the time.

The classic picture of an exposure-disease relationship is given in Figure 8-3. In this model, the group of people who are exposed are a black box that absorbs the exposure; only a fraction of those exposed may get the disease. For example, although lung and urinary bladder

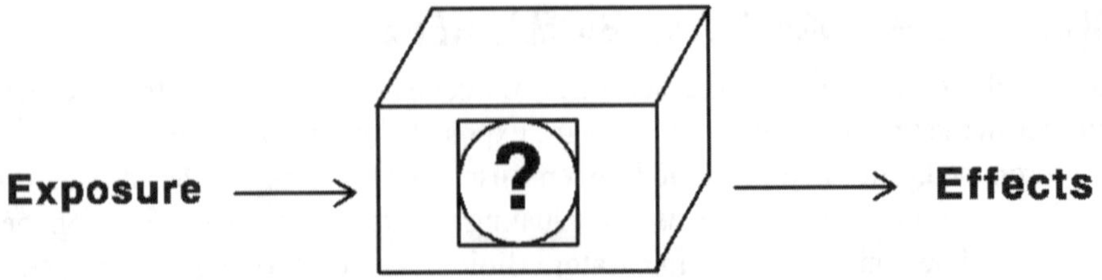

Figure 8-3. The Black Box Model of exposure-effects, i.e., without information of biological dosimetry.

cancers are seven and five times more common in smokers than in nonsmokers, only about 10% of the smokers will get either type of cancer. Why is this? Exposure? Genetics? A combination? The black box model has been useful to identify toxic or carcinogenic agents in cases where the exposure can be accurately assessed, and where the relationship between the exposure and the outcome is strong. This approach has been useful in certain occupational studies when the exposed group can be easily assessed, and the impact of the exposure on the disease is high. Examples would include asbestos, hard wood dust, and vinyl chloride (McDonald and McDonald, 1977, Acheson et al., 1968, Creech and Johnson, 1974).

The black box model fails when exposure cannot be accurately assessed or when internal processing of the exposure is necessary for the toxic effect. These are the conditions where biological monitoring can be most helpful. The advent of biological monitoring has allowed investigations into the black box. This is exciting from a research perspective, because mechanisms can be proposed and investigated. It is also exciting from an exposure assessment perspective because there may be multiple ways exposure and early effects can be estimated. A model of an exposure-disease relationship proceeding through a continuum of biological changes and integrating the range of exposure assessment techniques is given in Figure 8-4. This model suggests that there may be several levels where dose and its effects can be measured within the individual. The separate components between exposure and response are different levels of monitoring that can be done. While these are listed as discrete entities, there is often a considerable amount of overlap between these markers.

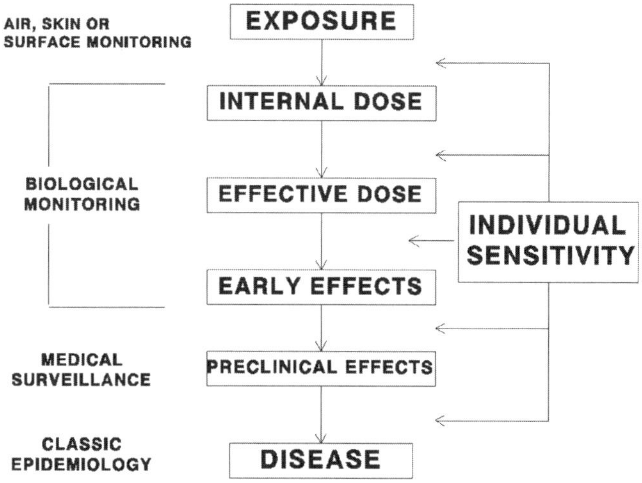

Figure 8-4. Classes of environmental and biological exposure measurements that could be made along the Exposure-Disease Continuum.

Clearly, the closer a measurement is made relative to the disease, the greater the probability that the measurement will predict the disease, i.e., the greater the positive predictive value of the marker. On the other side of the coin, some measurements may be so close to the overt disease themselves that the process may not be reversible, and the marker is one essentially of early disease detection. For example, with a positive sputum cytology, there is a high probability that the person will develop a lung neoplasm. However, the chance of preventing this irreversible disease is correspondingly low at this point, even if exposure was terminated immediately. From the view of the hygiene program, these latter markers are not acceptable because they are not preventive and do not protect the health of the worker. Markers appropriate for industrial hygiene are those associated with exposure and early, reversible effects, or those effects that have a low probability of being associated with disease.

The left column in Figure 8-4 indicates the sampling or surveillance realm of the particular marker. External exposure evaluation includes measurement of skin wipes and related techniques (Van Hemmen and Brouwer, 1995), and surface wipes (McArthur, 1992, Leung and Paustenbach, 1994), as well as air sampling. Biological monitoring includes measurements of internal dose, effective dose, and early effects. Internal dose is a measurement of the material that has actually entered the body. Effective dose is the fraction of the internal dose, which was delivered to, or metabolically converted to the form capable of interacting with the target organ or molecule. Early effect markers are those that measure reversible changes in the body that indicate exposure, but not disease. Distinctions between these biomarkers are sometimes blurred. Internal dose markers can be effective dose markers in some instances. Pre-clinical effects are those with a high positive predictive value, and are diagnostic in nature; these fall under the realm of medical surveillance. Pre-clinical markers would, therefore, include traditional medical markers such as positive sputum and urine cytology, positive spirometry, or chest x-rays for asbestosis or silicosis. In addition, new markers such as detection of activated oncogenes and their proteins may also be included in this category because they are more associated with the disease than exposure and prevention of disease (Brandt-Rauf, 1990). Medical surveillance and classical epidemiology are also concerned with frank disease.

BIOMARKERS OF INTERNAL DOSE

Markers of internal dose estimate the amount of material that the person has absorbed from all routes. So, if a person or persons must work in such a way so as to allow contact of the compound with the skin, or without the benefit of engineering controls or PPE, elevated levels of internal dose markers would indicate that they are exposed. Alternatively, if the physical workload, work temperature or other factors of one or several people were greater than for others, markers of internal dose would be useful to estimate how much more material is absorbed by these persons than by other workers. Fiserova-Bergerova (1995) reported the findings of a study by Volf and Volf of operating room personnel. The highest anesthetic gas air levels were found in the samples taken in the breathing zone of the more sedentary anesthesiologists, while the highest internal dose levels were in the more mobile and active operating room nurses.

Another use of internal dose biological markers in hygiene practice would be to exclude differences in exposure and absorption in the workforce. This approach is used in the OSHA biological monitoring requirements for lead and cadmium. Workers are monitored at infrequent intervals if their results are below certain "trigger" values. To rule out differences, sampling is done at the appropriate time following an exposure and the values of individuals compared to identify outliers. These persons could then be interviewed and/or observed to determine if increased exposure caused the elevated markers. This is the approach Cook (1990) used to determine that the rollers in the insulating plant had an unanticipated exposure while rolling finished product, due to the nature of the task.

Internal dose biomarkers are the most commonly used form of biological monitoring; they include measurement of urinary metabolites and metals, volatile, parent compounds in exhaled air, or parent compounds, metals, or metabolites in blood. The majority of the current Biological Exposure Indices (BEIs®) established by the American Conference of Governmental Industrial Hygienists (ACGIH®) can be considered internal dose markers. Internal dose is linked with exposure and not effect, consequently, there is less ambiguity in their application to industrial hygiene, relative to occupational medicine. There are many examples of the use of internal dose markers in the literature. Lauwerys' and co-workers' (1980) classic use of internal dose markers to determine route of entry and effective interventions for N,N-dimethylformamide will be discussed below. The levels of cadmium in blood are considered to be a reflection of current exposure to the material, while the urinary cadmium levels are thought to reflect the chronic ingestion and total body burden (ACGIH®, 1996).

A good example of an internal dose marker is urinary 1-hydroxypyrene (1HP), which is the major metabolite of the non-carcinogenic polycyclic aromatic hydrocarbon (PAH), pyrene. PAH exposure in industry always involves exposures to complex mixtures including coal tar pitch volatiles, asphalt, coke oven emissions, aluminum electrode emissions. These mixtures contain pyrene, usually as a major component. Pyrene appears to be non-carcinogenic because virtually all of it is metabolized to 1HP and almost none through epoxide formation, which often results in carcinogenicity for other PAH (Jongeneelen et al., 1988a; Strickland, 1994). Pyrene is absorbed like carcinogenic PAH, being similar in lipophilicity and vapor pressure. Thus, detection of 1HP in the urine indicates that there has been absorption of carcinogenic PAH; however, it does not give information regarding individual variability in metabolism of the carcinogenic components of the mixture. Increases in 1HP have been found in such occupations as aluminum plant workers (Petry et al. 1996; Overboe et al., 1994), iron foundry workers (Hansen et al., 1994) road pavers (Zhou et al., 1996; Jongeneelen et al., 1989a), coke oven workers (Van Rooij et al., 1994) and creosote workers (Van Rooij, 1993c). The data obtained in these studies indicated that the dermal route accounted for the majority of the total internal dose for pavers, creosote and coke oven workers, while no comparison was made in the others. Significantly, implementation of interventions including laundering gloves and clothes daily and hand and face washes prior to each break measurably reduced internal dose in coke oven and creosote workers (Van Rooij et al., 1994). 1HP is eliminated from the body with a half-life of 5 to 10 hours following an ingested dose (Buckley and Lioy, 1992).

While it may be difficult to assign an index value for 1HP applicable to all industrial exposures, the determinant could be very useful to track individual exposures during the work-week. Environmental sampling to determine the relative amount of pyrene in the mixture, in the air, and on surfaces, would help relate 1HP levels to carcinogen absorption. Studies should also be done to determine the relationship between 1HP levels and other biomarkers, particularly effective dose and early effect markers.

Hippuric acid, the major metabolite of toluene, is a more commonly used internal dose marker. Hippuric acid measurement has been well-validated in human exposure situations. It has been used not only to estimate the exposure to toluene, but as we have seen, to evaluate the effect of exercise on toluene absorption, and to test exposure interactions between toluene and xylene (Baelum et al., 1987; Tardif et al., 1991, 1995).

Effective Dose Biomarkers

Effective dose biomarkers are those that estimate the level of the compound or metabolite in the form that causes the biological effect. Because the active form of the compound is measured, these markers integrate for activating metabolism and should be more closely related to the effect in question than measurement of internal dose. In the case where the parent compound produces the effect, as with the acute narcosis induced by solvents, the internal dose <u>IS</u> the effective dose. This may also be true for some metals such as cadmium. However, in many cases the metabolite of an absorbed compound is toxic and affects a specific target system. For example, n-hexane is metabolized to 2,5-hexanedione, which induces a chronic neuropathy; the measurement of this metabolite is an estimate of the effective dose that takes into consideration individual differences in metabolism. Similarly, 2-ethoxyacetic acid and 2-methoxyacetic acid are the active metabolites of 2-ethoxyethanol and 2-ethoxymethanol responsible for the reproductive and teratogenic effects of these agents (Foster et al., 1987).

Most chemical carcinogens must be metabolized to exert their effect. Therefore, an exciting area in biological monitoring is the development of effective dose markers for these compounds. Measurement of carcinogen-DNA or protein adducts indicates that the active metabolites of carcinogens have been formed, and, in the case of DNA adducts, have bound to critical target macromolecules (Talaska et al., 1992). Many occupational studies have been performed indicating that groups with carcinogen exposure will have significantly higher levels of adducts than those non-exposed. For example, Schoket et al. (1991) reported that aluminum plant workers had significantly increased levels of carcinogen-DNA adducts in blood leukocytes and Rothman et al. (1996) reported significantly higher levels of benzidine-DNA adducts in exfoliated urothelial cells of benzidine exposed workers relative to workers exposed only to benzidine based dyes, or those not exposed. The adduct levels were, on average, 24 times higher than the controls, and there was a dose response related to recent absorption of benzidine. There were three persons (of 15 in the exposed group) whose adduct levels were 90 or more times higher than the controls. As with 1HP measurements, DNA adduct analysis has not yet been validated sufficiently to have recommended limits. The particular problems of these measurements include difficulty in obtaining chemical standards, the expense and skill required to perform the analysis, and a large study-to-study variability. As a result, adduct analyses still appear to be experimental and not ready for routine use (even after 15 years of development) (Aitio, 1994). The relationship between adduct levels and the disease is not known with surety, yet a prospective study has been done that showed that adducts are associated with liver cancer in a population exposed to aflatoxin B_1 (Ross et al., 1992). The promise of these effective dose markers is that they will identify populations at potential risk due to exposure, absorption, metabolism, and repair differences. Even if absolute quantitation of adduct levels is not available, adduct analysis could certainly have a place in hygiene practice to indicate where significant problems exist and whether remediation is effective (Rothman et al., 1996).

It would be preferable to measure the effective dose at the site of the effect; the level of cadmium in the kidneys, lead in the brain or bone marrow, or DNA adducts in the lung or urinary bladder, for example (Dorian et al., 1992; Talaska et al., 1992). Obtaining these types of tissue samples is usually too invasive for routine monitoring. Instead, surrogate matrices and tissues are monitored. Blood leukocytes have been used as a surrogate for measurements in the target organ (Talaska et al., 1992). These cells are relatively easy to obtain and will be exposed to any active carcinogen circulating in the blood. On the other hand, it is not known whether the leukocytes metabolize carcinogens in a fashion similar to the target organ. Few studies have examined the relationship between adduct levels in the surrogate and in the target organs. Animal and human studies with tobacco smoke suggest that there might not be such a relationship (Reddy et al., 1990). Recently, DNA adducts levels in both blood lymphocytes and in exfoliated urothelial cells of benzidine exposed workers were studied and a significant correlation between the two measurements was reported (Zhou et al., 1997). Whether this relationship will also be seen with other compounds and exposures should be the focus of a significant research effort.

Often, it is not possible to collect air samples when exposures are accidental or unforeseen. Since biomarkers can persist following an exposure, they can be used to reconstruct the exposures that occurred during these events (Que Hee, 1997). A worker was accidentally sprayed with liquid methylene bis-2-chloroaniline (MOCA), a compound that is readily absorbed through the skin. This was an inadvertent event and air samples were not collected. However, urine samples were collected at intervals following the exposure and levels of urinary MOCA and MOCA-DNA adducts in urothelial cells were measured to estimate and reconstruct exposure and recovery (Osario et al., 1990, Kaderlik et al., 1993). Documentation of these levels may be useful to determine the relationship between exposure and health effects.

Early Effect Biomarkers

Early effect biomarkers estimate the level of reversible or repairable effects that occur relative to exposure. There are some well-known examples of this class of biomarker in common use in hygiene. Carbon monoxide has a 200- to 250-fold higher affinity for hemoglobin than does oxygen and decreases the efficacy of oxygen transfer in the tissues. The levels of carboxyhemoglobin in the blood can be measured by analysis using either visible spectrophotometry or gas chromatography (Dubowski and Luke, 1973; Baretta et al., 1978). The effects of CO exposure are fully reversible provided significant tissue hypoxia has not occurred. The levels of carboxyhemoglobin in the blood can be predicted by the following equation:

$$\%HbCO = \%CO \text{ (in air)} \times \text{exposure time (min)} \times K \qquad \text{[Equation 8-1]}$$

K = 3 @ rest; 5 @ light work; 8 @ moderate work;
11 @ heavy work

The exposure-related effects are as follows:

(1) 8% COHb@ TLV (25 ppm) (8% is BEI)

(2) @ 30% COHb fainting occurs

(3) 50-80% COHb has caused death

There are two other effect markers that have found use in hygiene practice. Methemoglobinemia (MetHb) involves the oxidation of heme iron from ferrous to ferric (+2 to +3) that eliminates the ability of hemoglobin to convey oxygen to the tissues. Methemoglobinemia occurs following exposure to specific aromatic amines, nitrates and nitro aromatic compounds, with toxicity by the latter being considered the most hazardous because of the delay in the onset of symptoms following exposure (ACGIH®, 1996). The normal range of methemoglobin levels in unexposed populations is up to about 1.5%. Cyanosis becomes readily apparent at 10-15% and symptoms of weakness, shortness of breath and increased heart rate occur at levels from 30-50%. Levels higher than 50% are often fatal without treatment. Methemoglobin measurement is nonspecific in the sense that it occurs following exposure to a variety of agents, yet provides a good marker of the effects of these agents. Many of the compounds that produce methemoglobin also cause other toxic effects, including anemia, urinary bladder cancer, and neurological damage, all of which overshadow the toxicity of Methemoglobinemia (ACGIH®, 1996). For these compounds (anisidine, perchlorylfluoride, cyclohexylamine, tetranitromethane, [beta] naphthylamine, 2,4,6 trinitrotoluene [TNT] and 2 nitropropane) MetHb should be considered an exposure, rather than an effect, marker.

Measurement of acetylcholinesterase (AchE) inhibition in the red blood cell (RBC) is an effect marker for exposure to any of the organophosphate or carbamate compounds that inhibit brain acetylcholinesterase. It is interesting that the inhibition of an isozyme in the red blood cell that obviously cannot be involved in nerve conduction is correlated to the acute effects in the CNS. However, the same is not true for the chronic effects of certain organophosphate compounds, such as triorthocresol phosphate, which produce a delayed neurotoxicity. Therefore, acetylcholinesterase inhibition is not an effect marker for any of the so-called "delayed" effects of these exposures. There is great inter-individual variability in the unexposed levels of RBC AchE. For this reason, baseline measurements of activity are required prior to exposure. The inhibition of AchE following exposure is protracted, with a half life of about 50 days. While substantial acute OP exposures will rapidly inhibit AchE and bring about symptoms, smaller chronic doses can cause the same degree of inhibition. However, with chronic exposure, some acclimatization or adaptation occurs; workers with a history of recent exposure may have 70-80% diminished activity without overt symptoms, while a person without prior exposure may report symptoms with 20% inhibition. A 30% inhibition is considered an action level and is the recommended BEI®.

SAMPLING MATRICES AND SAMPLE COLLECTION TIMING

The selection of an appropriate sampling matrix and sample collection timing are the most complex issues that must be addressed before a biological monitoring program can begin. The key to understanding biological monitoring is an understanding of biology of an exposure, i.e., toxicokinetics and toxicodynamics: the movement and transformation of absorbed compounds within the body. A simple theoretical model can be constructed containing the conceptual information needed to understand — to the first approximation — what is going on within the body. These models are hardly more complex than diagrams that indicate how responsibilities flow within a corporation. More complex models are required to more accurately estimate the quantitative movement of specific compounds within the body (Fiserova-Bergerova, 1990). These models are beyond the needs of this chapter, and are not really necessary to begin a biological monitoring program. A simple model suffices to illustrate principles. The movement of an organic solvent from the ambient environment to inside the body and its elimination will be followed. After this example, the reader will appreciate readily how similar concepts can be applied to materials as solid particles and liquids.

Systems tend to come to steady state. A common example, familiar to all industrial hygienists, is the vaporization of a liquid solvent into the air of a closed room. The concentration of the solvent vapor will increase until it is at equilibrium with the liquid phase. The equilibrium concentration in the air is governed by the ambient temperature and the vapor pressure of the solvent (see Volume 1, page 172). For example, our model system involves a closed system without ventilation and a material with a vapor pressure of 0.076 mm Hg. At equilibrium, the concentration of the compound would be $0.076 \text{ mm} \times 10^6/760 = 100$ ppm. Now consider what happens when a human is added to this system. The material will tend to move from an area of high concentration (100 ppm) to an area of low concentration, i.e., into the person. The first factor in absorption of a compound from the environment and into various tissues is governed by the relative concentration of the material outside the body, in the blood and the various tissues. Compounds differ in their affinity for the vapor phase (vapor pressure), and they also differ in their affinity for blood and other tissues. Since the human body is largely water, and because a vaporized material will come in intimate, if not physical, contact with the blood water at the alveoli, the water solubility of the compound is an excellent predictor of the affinity of the material for blood water (Fiserova-Bergerova, 1995). If there is a steep concentration gradient and the water solubility of the material is high, then a large amount of material will be absorbed. Given enough time, an equilibrium will be established between the concentrations in the air and in the body. Compounds that have high water solubility are readily absorbed by the blood at the gas exchange region. The limiting factor for these compounds is the amount of contaminated air that contacts the blood. Water soluble compounds are said to be "ventilation-limited" for this reason. The compounds that are included as water soluble include those that have a water: gas (or blood: air) partition coefficient of greater than 5. The majority of organic solvents including benzene, toluene, acetone and the volatile alcohols are included in this group. The absorption of compounds, which are not very water soluble into the body, depends on how much uncontaminated blood comes to the lung. These compounds are said to be "perfusion-limited." At a blood: air coefficient of 1, ventilation and perfusion have about an equal effect, as the coefficient increases, the relative importance of ventilation increases, and vice versa. When exposure from air or skin is occurring, the absorption of the compounds will continue to increase until an equilibrium is reached. Not surprisingly, water soluble compounds will reach this equilibrium point slowly because a relatively large amount of material can be dissolved into the blood. Water insoluble compounds reach equilibrium rapidly since the blood can only "hold" a relatively small amount of the material before it becomes saturated. Rate constants can be determined for the movement of material into the body based upon the water solubility. These are used to develop the mathematical models needed to predict the quantitative movement of material.

Thus far, only the absorption of compounds into the body has been considered. There are also two metabolic activities within the body that remove the material from the blood and tend to disrupt the equilibrium. Distribution of the material into body tissues for storage, metabolism, and elimination, reduces the blood level, and reestablishes a concentration gradient at the site of absorption, allowing more material to be absorbed. Figure 8-5 is a flow chart for the absorption of a compound into the body and the possible fates following absorption. In one process, pulmonary exposures enter the arterial blood and are rapidly distributed throughout the body. Dermal exposures enter the local venous blood, which mixes in the heart and lungs before general distribution. Oral exposures are still different in that compounds are delivered to the liver via the portal vein before reaching the general circulation; the liver is given a "first pass" at metabolism before the rest of the body is exposed. Ultimately, all exposures enter the blood, the central compartment (1. in Figure 8-5.). Not surprisingly, distribution from arterial blood to any particular tissue depends upon how well the tissue is perfused. The lungs receive 100% of the cardiac output, for example, while the kidneys receive 25%. Body fat receives a small amount of the cardiac output. While compounds are distributed to all organs (numbers 2,

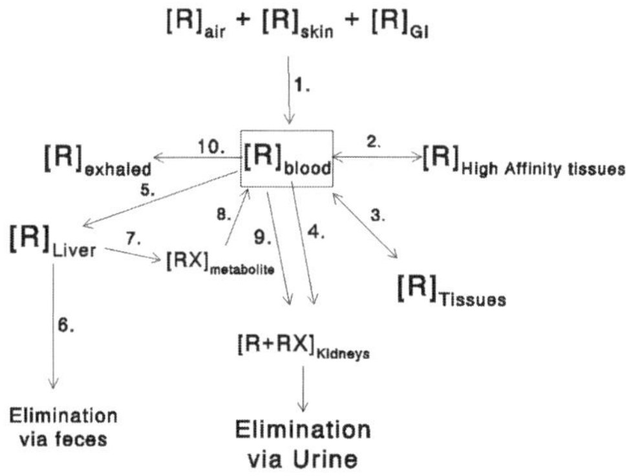

Figure 8-5. Schematic of the biological fate of toxicants. ([R] is the tissue concentration of the compound) entering the body by any route enter the blood for distribution to the tissues (arrow 1). Distribution tends to reduce the blood levels, reestablishing concentration gradients and permitting more absorption during the active exposure phase. High affinity tissues avidly take up compound from the blood (see text). However, when exposure ceases and the concentration in the blood decreases, compounds reenter the blood from these tissues; reentry rate is inversely related to the affinity of the compound for the tissue. The differential return of compounds from these tissues is responsible for the multiple slopes seen following exposure to lipophilic compounds like benzene. For this reason, arrows 2 and 3 are double headed. Compounds inhaled or absorbed through the skin enter the liver through general circulation. However, those ingested enter the liver directly via the hepatic portal system. The liver is said to be given a metabolic first pass at the material in this case. Hepatic metabolites (RX) (arrow 7) reenter the blood (arrow 8) for distribution to other tissues and for elimination (arrow 9). Most metabolites, because they are less volatile, and more water soluble, are eliminated via the kidneys. Volatile compounds may also be exhaled (arrow 10) as described in the text. It should be noted that this is a simplification. For example, the lung and skin also have an opportunity for "first pass" metabolism for materials absorbed via these routes, and metabolism also likely occurs to some extent in each tissue.

3, 4, and 7), retention in particular organs depends upon the relative concentrations of the compound in the blood and the tissue, the affinity of the compound for that organ and the metabolism that occurs within that organ. Tissue affinities are very chemical-specific. Lead has a high affinity for bone because it is chemically similar to calcium. Compounds with a high oil- to-water partition coefficient prefer to be in body fat than in blood, and have a high affinity for this tissue and other tissues with a high fat content (brain, adipose tissue, and peripheral nerves). Compounds like hexane, toluene, PAH, and benzene may reach concentrations in the fat several hundred or thousand times that of blood. Still, because fat is poorly perfused, it may take several hours for lipophilic substances to reach maximal levels in the fat. When a lipophilic compound reaches the fat it may remain until the blood level drops to the point where redistribution occurs, or until the fat is utilized for energy purposes.

Compounds are sometimes initially distributed into one set of organs based on perfusion and then redistributed into other organs based on affinity. After allowing time for distribution and redistribution to occur, the highest concentration of lead will be in the bone, and lipophilic solvents are concentrated in the adipose tissues. This is in spite of the fact that each of these tissues receives only a small fraction of cardiac output. Overall distribution rate constants can be assigned to each tissue and used to predict the amount of material in the tissue; these are the bases of many available toxicokinetic models.

The second process of compound removal from blood is through elimination. It should be remembered that elimination occurs simultaneously with absorption and distribution. There are three distinct elimination processes, all of which are occurring simultaneously (Monster et al., 1979, Imbriani et al., 1986). Compounds can be eliminated unchanged through either the kidneys or the lungs following distribution to these organs. Pulmonary elimination is based upon the relative concentration in blood and air and the blood: air partition coefficient. In other words, the same mechanisms that work to move compounds into the body will also allow compounds to leave when the concentration gradients are reversed. To be eliminated by exhalation, compounds must be somewhat volatile at body temperature. Since metabolites of organic chemicals are almost always less volatile, only the parent compound is eliminated through the lungs. Poorly metabolized, poorly water soluble compounds like perchloroethylene are eliminated predominately via the lungs, while rapidly metabolized and/or water soluble compounds like xylene, methanol, and furfural are eliminated via the kidneys. Compounds like trichloroethylene, hexane, benzene, and others are eliminated by both mechanisms (Opdam and Smolders, 1989; ACGIH®, 1996). Some fraction of the parent compound is eliminated directly by the kidneys following filtration.

The third mechanism of elimination involves metabolism of the compound by the liver and the metabolite's redistribution and elimination via the kidney. Metabolites of molecular weight of more than 300 are eliminated by the liver through the bile and feces. If the half-life of the biomarker in the body is long enough, absorption and elimination reach a steady state and the level in the body plateaus.

When exposure stops, elimination predominates. Arrows 2 and 3 are double headed because compounds distributed to tissues when the concentration in the blood is relatively high (during exposure) will partition back into the blood when the blood concentration falls. As can be expected, the amount that leaves any tissue depends upon the relative concentrations in tissue and blood, the perfusion to and metabolism in the tissue, and the affinity of the material for the tissue. Because of the greatly different perfusion rates and affinities of the different tissues, there can be a range of residence times in a given tissue after an exposure. Compounds with high affinity for poorly perfused tissues will take a very long time to completely leave the tissue.

When compounds have similar residence times in groups of tissues, these tissues are said to act as "compartments" for that compound. Grouping tissues with similar behavior into compartments makes it easier to model the movement of the material in the body. The term half-life is used for the period of time it takes for the concentration in a compartment (or the body as a whole) to decrease by half following exposure. Some compounds seem to disappear into and appear from several different compartments. For example, benzene has a half-life in blood of about 25 minutes; in lean tissues, about 2.5 hours; and in adipose tissue about 30 hours (Berlin et al., 1980; Sherwood, 1972a, 1972b). This can been seen graphically from a single elimination curve because there will be three distinct slopes in the elimination (see Figure 8-6). The curve represents the distribution and redistribution of compounds that have different affinities for different compartments.

The models that have been developed appear to work to the first or second approximation when applied under controlled experimental conditions. However, the behavior of an agent in a particular individual on a given day will likely be much different than predicted by a model, or even an average value for a group. It is at this point that the differences in the individual, the agent, and the environment come into play. A person who is gaining weight will be altering their compartments, as will be a person losing weight. So will a person who has just eaten a large or fatty meal. In addition, we are in our infancy in estimating how exposure to multiple chemicals interacts to produce effects. How can these data be interpreted for occupational health? It is likely that human experience, *viz.* the careful collection of biological monitoring data and health outcome, will be useful to answer these questions.

BIOLOGICAL MONITORING MATRICES

Exhaled breath, blood, and urine are the most common biological monitoring matrices.

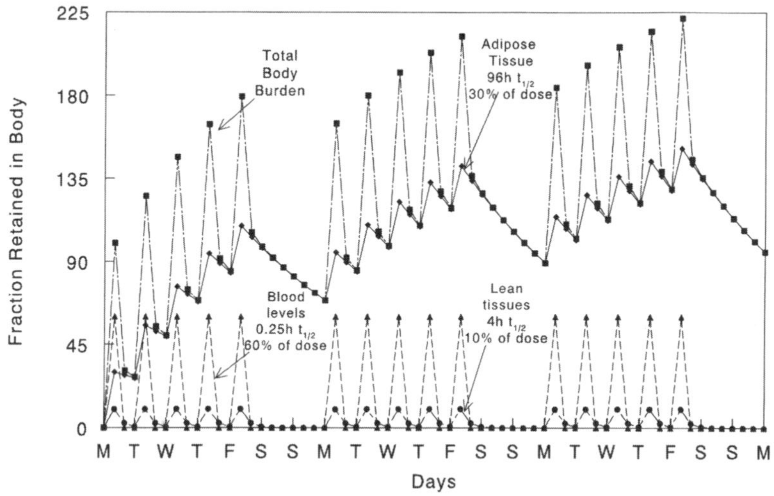

Figure 8-6. Elimination Curve for a compound distributed into three compartments having half-lives of 0.25, 3.0 and 30.0 hours, respectively. The influence of the first compartment is strongest between 0 and 1 hour after cessation of exposure. The slope at this time is about -30 units/hour. Between 1 and 6 hours, the rate of loss decreases to about -5 units per hour. After 6 hours, when material from the first two compartments is largely gone, the rate decreases further to about -1.5 units per hour. This figure gives the overall picture of what one would see by sampling every hour. Also see Figure 8-10 where the kinetics of each compartment are modeled individually.

Breath Analysis

What do industrial hygiene students returning from a field trip to a chemical plant, a patient waking from a surgical procedure, and workers exposed to many solvents have in common? Among other things, they report the experience of tasting or smelling the material to which they were exposed. Often the realization that the reason one can "taste" the material that was once in their bodies is because it is being off-gassed from their lungs, is a new and rudely awakening experience. Many lipophilic compounds that are readily absorbed through the skin have a significant vapor pressure, are only sparingly water soluble, and are poorly metabolized. These compounds can be measured in exhaled breath using the appropriate collection system. The advent of sensitive analysis of volatile compounds, specifically, gas chromatography, has made breath sampling relatively straightforward. The primary requirement of compound candidates for breath analysis is volatility at body temperature. This excludes almost all metals and metabolites of any kind. Compounds that have high water solubility will not likely be found significantly in exhaled air (alcohols, many ketones), neither will those that are metabolized extensively be found significantly.

When the concentration of a volatile material in the ambient air decreases, blood and exhaled breath rapidly equilibrate because of lung perfusion. Breath samples are a reflection of the instantaneous venous blood levels. The fraction that is exhaled depends upon volatility, water solubility (water: air partitioning), the rate of breathing, and lung perfusion. This can be estimated as:

$$\%\,\text{retention} = \frac{\lambda \times 100}{\lambda + \dfrac{V}{Q}} \qquad \text{[Equation 8-2]}$$

Where λ is the water: air partition coefficient, V is the ventilation rate in LPM, and Q is the lung perfusion rate in LPM. As ventilation increases relative to blood perfusion, as during exercise, the denominator will become larger and the fraction retained in the body smaller (Fiserova-Bergerova, 1995).

From a functional viewpoint, there are two regions of the human lung, the conducting and respiratory regions. The conducting region consists of the nasal passages, the oral pharynx, and the bronchi (see Chapter 1). The volume of this region is about 150 ml and stays constant with changes in breathing rate. No gas exchange takes place within the conducting region. The respiratory, or gas-exchange region consists of the respiratory bronchioles and alveoli — the sites of gas exchange. Under normal, quiet breathing the volume of the respiratory region is about 300-350 ml, about 2/3 of the total. As demand increases, this volume can rise at least 10-fold. Since breath samples are obtained with persons at low activity levels, the tidal volume is of concern here. As air is inhaled and the lungs are filled, the last 150 ml fills the conducting regions and does not enter the alveoli. Then with expiration, air from the conducting passages leaves the respiratory system first, followed by the air that was in the respiratory region. Since no gas exchange takes place in the conducting regions, it is not surprising that there is very little exhaled compound in this anatomical "dead space." Figure 8-7 shows an idealized curve for concentration of a volatile gas in the exhaled breath as a function of time from the beginning of exhalation. The figure suggests that there would be very little mixing between the two regions, and that appearance of the analyte in the air would be very discrete (Wilson, 1986).

Generally, air from the conducting passages should not be used to determine the internal dose. More useful collection methods exclude the initial 150-200 ml of a breath sample. This can be done manually, by asking the person to exhale to the atmosphere and then, on a signal, to begin to collect the sample. Alternatively, a valve can be used to shunt the breath at the appropriate time. It can be appreciated from Figure 8-7 that once air from the conducting region has passed, the concentration in the air from the respiratory region will be relatively constant, and a sample can be taken from any portion. Sampling techniques have been developed to sample all, or some fraction of the air from the respiratory region (Wilson, 1986). The collection of samples remains somewhat of an art form, as there are no techniques that have been adopted by ACGIH®, AIHA, or OSHA. Unfortunately, there has not been a recent review of breath collection techniques; the last in industrial hygiene journals appeared in 1986 (Wilson). Although several new approaches have been described in the more recent literature, they are based upon techniques described in the Wilson review.

The procedure for collection of mixed exhaled air is somewhat simpler. Techniques have been described and widely used to collect multiple samples (Echeverria et al., 1995). The BEIs® are given in terms of end-exhaled air, so there is no direct comparison with mixed air

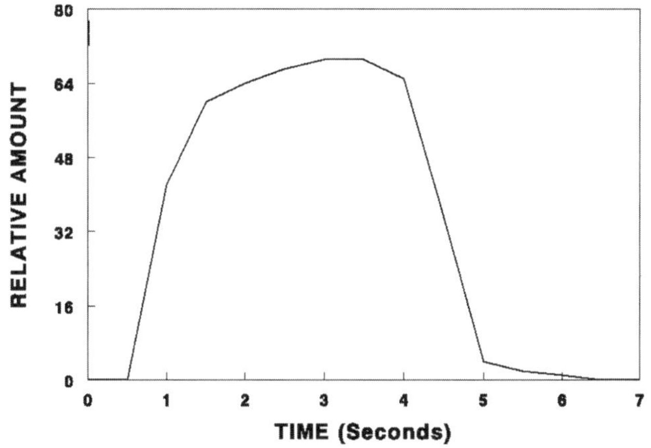

Figure 8-7. Instanteneous levels of a hypoethetical compound in expired air during respiration. Only the expiration phase is shown. In the first phase, the air in the conducting passages is being cleared; there is no compound in this air as it never reached the alveoli. The concentration rapidly increases and plateaus within 3 seconds, which is the best time for sampling. Figure adapted from Wilson (1986).

samples. However, as long as the data are collected and interpreted appropriately, mixed air samples can be very useful for exposure assessment. If all the data are collected in the same way, the results will be internally consistent and will indicate relative exposure of different individuals. Imbriani et al. (1988), Petreas et al. (1992), and Solet et al. (1990) found that airborne perchloroethylene levels were significantly correlated with the levels in mixed exhaled air. Petreas et al. showed that breath samples could predict the summation of the exposure during the workweek. They recommended biological monitoring in this case because fewer samples would be needed to assess exposure.

Several general considerations and recommendations for the collection of breath samples can be made:

1. The collection device should be designed to minimize breathing resistance; persons should not be asked to breathe into solvent collection tubes that would present a severe constriction.
2. Collection of multiple breaths will tend to increase the precision of the exposure estimate; collection of multiple breaths will also increase method sensitivity as more of the analyte is available.
3. Exhaled air is saturated with water; if the analyte is water soluble, care must be taken to prevent or minimize condensation in the collection apparatus, which would tend to decrease the concentration in the vapor phase. This can be done by including an impinger, a hygroscopic agent such as Drier-Rite® in the sampling train, or by warming the collection device to over 40°C.
4. The samples must be collected in clean air, particularly where a preshift sample is obtained to sample the most stable compartment. The concentration in these samples can be a small fraction of the workplace concentration. Allowing several breaths of pure, humidified air from a cylinder may increase method sensitivity (Wallace and Pellizzari, 1995) because the background level of any interfering substance is presumably lower in this than in room air.

The recent literature suggests that the use of Tedlar bags for sample collection is increasing. This method permits the use of a concentration phase to increase sensitivity. Obviously, a sample containing 10 end-exhaled air samples will contain as much as 10 times the material available as from any single breath sample. Air from a collection bag can be metered into a concentrating device such as a charcoal tube, or other specific sorbent, then the concentrated analyte can be desorbed and analyzed. Such an approach has been taken by the U.S. EPA (Wallace et al., 1986; Wallace and Pellizzari, 1995) and others (Newman, 1997) to monitor very low levels of exposure.

Exhalation can be a very effective means of removing substances from the body if the substance is volatile and only slightly water soluble. The initial half-lives of volatile compounds can be very short; the half-life of n-hexane in blood is 15 minutes. A very short half-life exacerbates sample collection problems. With a 15-minute half-life, a significant amount of compound can be exhaled and lost if there is an interval of as little as 5 to 10 minutes between the end of exposure and sampling. This could easily occur if the workers had to walk to a medical facility or if there were a number of workers waiting to have samples collected at the end of a shift. If the half-life of the compound and the time of last exposure are known, a correction could be applied to the results. However, because of the variability of half-lives in different individuals, it is often better to employ other sampling strategies. Fortunately, many volatile compounds are also lipophilic and partition into several compartments. Their loss from these other compartments can be monitored after exposure ceases, as will be discussed below.

The recommended breath sampling time for some other lipophilic solvents is much longer than their initial blood half-life. Benzene should be sampled 16 h post-exposure, and perchloroethylene prior to the last shift of the workweek, due to long half-lives in the adipose compartment. Figure 8-8 provides the rationale in graphical form. A hypothetical compound is determined to be distributed differentially to three compartments that have greatly different half-lives. The blood half-life is 0.25 h and 60% of the total dose is distributed to this compartment. The lean tissue half-life is 4 hours, but only 10% of the total dose is distributed there. The adipose tissue half-life is 96 hours since the compound is lipophilic and 30% of the total dose is distributed to the fat because of the relatively poor perfusion of this compartment. The total body burden is also shown. Although this model is very simple, it gives a great deal of informa-

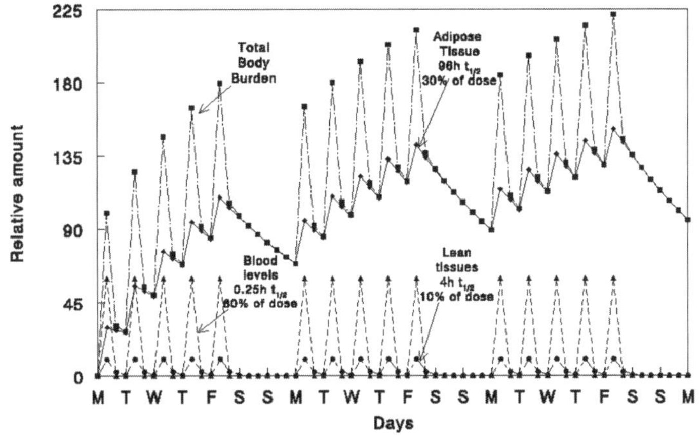

Figure 8-8. Model of levels in each of three tissue compartments with differing rates of perfusion and affinities for the compound. The total body burden is also shown. The levels in compartments with half-lives of 0.25 and 4 hours rapidly fluctuate during exposure. The total body burden also reflects these changes. The levels in the 96 hour compartment increase during the workweek. Within four hours after exposure ceases, these levels and the total body burden essentially coincide. After steady state has been reached (by the end of the third 40-hour week), a sample collected before the shift Friday morning would reflect increases during the week, but avoid the fluctuations that would result if the samples were collected at the end of the exposure. The exact time of sampling is obviously less critical with the pre-shift sample (half-life = 96 hours) than the post shift sample (half-life @ 0.25 hours).

tion about the approximate kinetics of the compound. The level in the blood rises rapidly and falls off quickly after each exposure. The lean tissue level is much lower because of the low affinity of the compound for it even though the half-life in this compartment is eight times that of the blood. Because of high affinity, the levels in the adipose tissue rise steadily with each exposure. By the end of the first week, the adipose tissue levels are twice as high as the peak blood levels. There is also a week-to-week accumulation in the adipose tissue. Steady state levels are reached after three weeks of 8 hour per day, 5 day per week exposure. The amount of compound in the adipose tissue rapidly constitutes the largest, and, due to the long half-life, the most stable body fraction. The total body levels rapidly fluctuate due to the influence of the short-lived compartments. A preshift Friday measurement would be the best choice to capture the accumulation that occurs during the workweek. A sample collected then would be the least influenced by fluctuations in short-lived compartments, and thus be relatively insensitive to minor differences in sample collection time. This is the course that the BEI® Committee selected in their recommendations for benzene, perchloroethylene, methyl chloroform, and trichloroethylene. Petreas et al. (1992) reported that mixed exhaled perchloroethylene samples from both the end of last shift and morning following the last shift of the workweek predicted the average air level of exposure about equally well. There are important shortcomings of a sampling regime that focuses on measuring the adipose compartment. While sampling the adipose tissue compartment minimizes the influences of events within a day and day-to-day variability, the ability to determine the maximal exposure caused by these events is lost. These events can be of great interest, for example, a significant fraction of the total exposure of a dry cleaner may occur during the addition and removal of clothes. A special protocol may have to be designed to capture these events, with breath samples obtained at specified times after each event and preshift samples taken at the prescribed time. Alternatively, air sampling of average and peak exposures could also be performed.

A drawback to most hygiene measurements, including biological monitoring, is the interval between sampling and when the results return from the laboratory. Recall of events involving exposures becomes less clear with time, even if good notes are taken, making it difficult to reconstruct events. If you cannot recall what you did, it's very difficult to determine what you could do better, limiting the ability to intervene. Until recently, there were severe limitations on collecting and analyzing samples in "real-time." Real-time analyses are available with dedicated meters and area samples. While there are many types of meters (see Volume 1), area samples are not generally predictive of individual exposures. Ideally, both short term exposures as well as integrated total dose and steady state, should be captured in real time. The possibilities of real time analysis are very exciting for hygiene practice. The ability to give quick feedback following an exposure can be extremely effective as a training tool. Many minor, unnecessary exposures are tolerated by workers because they do not feel a significant effect with their senses. However, repeated small exposures may produce chronic effects. With real time biological monitoring, a worker exposed to a solvent on bare skin could immediately see that the "accident" had an impact on exposure, and take steps to prevent recurrence. If analysis were rapid and inexpensive, a larger number of samples could be collected to better estimate exposure variability. A single hygienist can monitor only about 5 full shift air samplers. But, if samples could be rapidly collected and analyzed in real time, then the number of persons sampled could include each person on a line or process. The hygienist might observe the monitored workers so that exposure issues could be discussed as they arose. The data could be prepared in a report within a few days and discussed with the worker relative to specific, noted work practices while recall of these practices is fresh.

There is a possibility that real time breath analysis may be used for this purpose in the near future. An individual would breathe into a device connected directly to an analytical instrument. For example, techniques are being developed using portable gas chromatography - mass spectrometry (GC-MS). Wilson and co-workers were among the first to use MS to determine the breath levels of volatile solvents (Wilson, 1981). Campbell et al. (1985) measured pre- and post-shift carbon disulfide breath levels in viscose rayon workers. Not surprisingly, they reported that workers with jobs with higher average airborne exposures had higher levels of CS_2 in post shift samples. However, they also reported that there was some accumulation of CS_2 in end of week preshift samples, suggesting that there are multiple compartments for the compound in the body. A wide variation in values was noted, and was attributed to the short initial half-life of the compound and to the poorly characterized exposure of workers near the end of the shift. These findings illustrate the importance of hygiene observations in a biological monitoring program.

The major problems with using mass spectroscopy, the analytical techniques employed in these studies, included the size of the vacuum pumps and their energy consumption. Advances have been made in both of these fronts, and portable equipment with real-time capabilities is available. However, with any portable GC-MS there will be a very large initial dollar outlay and the necessity for maintenance by a trained person. Somewhat simpler instrumentation can now provide near-real time analysis. Breath samples can be collected in Tedlar® or other bags, then analyzed directly by injection of an aliquot into a regular GC or GC-MS. Alternatively, an increase in sensitivity can be gained by concentrating multiple breath samples from the same person on a sorbent tube or other device. The material can be desorbed in a small volume and then analyzed. Using the former method, Sweet and co-workers (1997) saw that certain events, like loading and unloading of clothes, were associated with peak levels of perchloroethylene exposure in dry cleaners. Rappaport and co-workers (Rappaport, 1991; Petreas, 1992) collected total mixed exhaled breath samples on charcoal tubes, desorbed perchloroethylene with toluene, and analyzed the samples using GC with an electron capture detector. Wallace et al. (1986) collected breath samples on Tenax tubes prior to analysis with GC-MS.

Once cost and analytical problems are solved, breath sampling has the potential to make biological monitoring commonplace in industry.

Blood Analysis

Blood is the central compartment (see Figure 8-5) and some fraction of all absorbed compounds will enter the blood. Theoretically then, blood can be used as a matrix for any determinant. Practically speaking, however, there are limitations to which determinants can be measured in the blood because the half-life of many determinants in the blood is very short due to exhalation or metabolism. Grossly, blood is more than 50% water as plasma, the remainder being red and various white cells, which are also water-containing. It is not surprising, therefore, that the water solubility of compounds predicts their absorption into the blood to the first approximation. Blood also contains considerable lipid; cell membranes are largely fatty acids, and the lipid content of the blood in a fasting individual ranges from 450-850 mg/dl (Dorland, 1988). As noted above, this can rise dramatically following meals containing fats and oils. The blood also contains soluble and structural proteins. Blood can be collected as serum, packed cells or whole blood; thus, advantage can be taken of the fact that determinants distribute differently between the different blood components. Care must be taken to collect blood in the appropriate fashion for the component and determinant of interest.

There are differences between venous, arterial, and capillary blood aside from the content of oxygen and carbon dioxide. Arterial blood returning from the lungs during an ambient air exposure will have a higher concentration of an air contaminant. When the exposure ceases, the same arterial blood will have a lower concentration proportionate to the amount of material eliminated through exhalation. All biological monitoring determinations are made on venous blood, usually obtained from an arm. Venous blood will have a somewhat lower concentration of the determinant than arterial blood, proportionate to the amount absorbed and/or metabolized by the tissue. Capillary blood, obtained from a finger stick, consists of a mixture of venous and arterial blood and will reflect the concentration in the two sources. It is very difficult to determine the relative proportion of venous and arterial blood in the mixture and, therefore, finger sticks are not recommended for biological monitoring. There is also the potential problem of recent dermal exposure on the hands or arms influencing the level in the blood taken from that arm. This will be discussed under biological monitoring weaknesses.

As is obvious from earlier discussion, the concentration of any determinant in the blood may change rapidly due to absorption, distribution, metabolism, and elimination. Therefore, blood samples are "snapshots" of the instant they were taken. It is important to be able to anticipate the behavior of a determinant in the blood based on its biological half-life and collect the sample at the proper time and interpret the data appropriately.

Blood components and cells are also the source of many early effect biomarkers. Measurements of carboxyhemoglobin and methemoglobin are determined relative to blood hemoglobin levels; the inhibition of red blood cell acetylcholinesterase is made in whole blood (ACGIH®, 1996). Red blood cell hemoglobin is also used to monitor the level of carcinogen-protein interactions as will be discussed below (Skipper and Tannenbaum, 1994). In addition, the RBC are used as a sentinel tissue to monitor mutation frequencies of glycophorin A, an RBC cell membrane surface marker (Langlois et al., 1994). Since circulating RBC do not possess a nucleus, or DNA, the white blood cells have been used as a surrogate to monitor the degree to which carcinogens cause changes in the chromosomal material of target tissues (Hemminki et al. 1988, 1990; Herbert et al., 1990; Rothman et al., 1995; Zhou et al., 1997).

These same cells have also been used to monitor for chromosomal damage and mutations caused by industrial agents (Compton et al., 1991; Perera et al., 1993).

Blood lead levels were used by Kehoe (1963) to identify workers having significant exposures in various industries. OSHA now requires that blood lead levels be determined in all workers who have exposure to air lead levels of greater than 30 $\mu g/m^3$ for more than 30 days in the year (Dept. of Labor, 1978, 1982). Other more recent investigators have used blood lead levels to identify individuals with high exposure, to implement and monitor the effectiveness of intervention programs. For example, Kononen et al. (1989) studied the changes in blood lead levels in several types of automotive plants where monitoring programs were in place for at least five years. They reported that there was a general downward trend in blood lead levels in all facilities over the time of the program. Ulenbelt (1991) reported that the air lead levels in a secondary smelter were inversely correlated with the blood lead levels of the workers, but the principal reason is that the workers with highest exposure were most likely to use airstream helmets, gloves and other hygienic interventions, which likely decreased absorption. Measurements of blood lead, chromium, and cadmium were performed on ceramic glazers, and it was determined that only the blood lead levels were elevated (Arai et al., 1994). Blood sampling is also required by OSHA for cadmium exposure (Fed. Reg., 1992).

Blood sampling may also be used for solvent exposure. Ghittori et al. (1987) saw that there was a significant correlation between carbon tetrachloride air and blood levels ($r=0.87$, $p<0.001$) in workers exposed to the compound as a vapor only. A head space technique was developed for this study where blood was warmed to 37°C for 2 h and then 2 ml of the head space was removed for GC-MS analysis.

Blood sampling is physically invasive and requires that trained personnel be used to collect samples. Workers sometimes object to blood sampling because of the relatively invasive nature. Certainly this procedure is too invasive for daily or weekly monitoring. The potential for sample degradation is high, particularly for labile determinants. Hemolysis can also occur, which can invalidate the results for determinants based on hemoglobin, or if the lysed red blood cells release an interfering substance. Sample work up for blood samples tends to be complex because of the potential interferences from proteins and other blood constituents. Blood sampling has the virtue of being very selective of the particular compound, and to reflect relatively recent exposures.

Urine

Water soluble compounds, metals, and metabolites of low molecular weight are filtered by the kidneys and excreted in the urine. The urine also contains epithelial cells from the kidneys, ureters, urinary bladder, and urethra. The concentration of a determinant in the urine is a function of the blood concentration during the interval since the last urination and the volume of the urine. Both urinary volume and the interval between urination vary widely due to how individuals maintain their hydration status. The "normal" range of 24-hour urine volume is from 600 to 2500 ml (Dorland, 1988). More extreme values are common, particularly when exertion or heat causes water loss through sweating, or when copious amounts of water are ingested. An obvious method of controlling these variables and estimating daily exposure is collection of a 24-hour urine sample. The volume of the total sample is measured and an aliquot is analyzed to determine the concentration. The data will tell you how much material a person absorbed in the interval since the last urination before collection began. A variant of this approach would be to collect all urine during a work shift. Only the material that was excreted after the exposure stopped would be missed. However, neither of these approaches is practical for biological monitoring purposes; workers cannot be expected to take sampling containers home with them or cart them around on the job. Consequently, spot samples must be collected at the appropriate time following an exposure and adjustments must be made to control for differences in urinary volume. While there is a greater than four-fold difference in the normal volume of urine excreted, the mass of solid materials dissolved in urine has only a two-fold variability (range 3-7 g). Therefore, adjustments that are based on the amount of either total or specific dissolved materials provide a more reliable means of concentration adjustment. Acceptable methods of adjustments are based upon expected, normal observed values for two urinary parameters, specific gravity, and creatinine concentration. Each method of urinary adjustment has pros and cons, but theoretically, all the adjusted values are equivalent. In practice, it is important to realize that often one type of conversion may be more appropriate because a medication or physiological condition may rule out the use of a certain method of concentration adjustment. In many cases, it is important to include the unadjusted values in statistical analysis for correlation and regression. The current BEIs® are expressed as creatinine adjusted values.

Adjustment using specific gravity has the longest history. Because the total amount of dissolved solids is excreted at a relatively constant rate, urinary specific gravity increases as urine becomes more concentrated (less water volume), and decreases as urine becomes more diluted with water. A person who drinks relatively little water, or who loses a significant fraction of body water through sweating, may excrete all of their metabolites in a very small volume of urine, while a person who drinks ample water and does not sweat for evaporative cooling may excrete the same amount of metabolite in a large volume. The correction equation below simply normalizes the measured amount of the determinant to an average urinary specific gravity of 1.020. Although the use of 1.020 as the normalizing value is arbitrary, it is a value not uncommon in workplace samples.

$$C\,(\text{mg}/\text{liter}) = \frac{(\text{measured amount})\,(0.020)}{(\text{sp. grav.} - 1)} \qquad \text{[Equation 8-3]}$$

If the measured concentration of a determinant was 50 mg/l of urine, and the measured specific gravity was 1.010, the adjusted concentration would be (50*0.020)/(0.010) or 100 mg/l.

This denotes that a relatively dilute urine sample with specific gravity of 1.010 would have twice the level of determinant per unit volume if it were concentrated to the "normal" specific gravity of 1.020. The BEI® Committee recommends that samples in the range of 1.003 to 1.030 are valid for this adjustment. Obviously, the adjustment factor can be very large for very dilute samples. Specific gravity measurement is easy, fast, and inexpensive. Almost all medical units possess a refractometer for this purpose. This makes it easy to conduct measurements of specific gravity in the field.

Specific gravity measurements are problematic at the extremes of very dilute (large multiplication factor) and very concentrated samples. There are several potential reasons for samples with high specific gravity, only one being low water intake. A diabetic might have a very high specific gravity if the disease is not controlled. Elevated urinary protein levels, for whatever reason, would also cause high specific gravity. These reasons can be confirmed or ruled out by further analysis of glucose and protein. It is also recommended that measurement of creatinine be made on these samples.

Creatinine is an excretory product of normal muscle metabolism and is a reflection of muscle mass. Under usual conditions, it is excreted at a relatively stable rate, 1-1.6 g per day. Adjustment of determinant values using creatinine is based on:

$$C\ (\mathrm{mg}/\mathrm{g}_{\mathrm{creatinine}}) = \frac{C\ (\mathrm{mg}/\mathrm{l})}{(\mathrm{grams\ creatinine}/\mathrm{l})} \qquad \text{[Equation 8-4]}$$

A separate analysis of urinary creatinine concentration must be made, obviously, but this is relatively straightforward and can be performed very inexpensively by clinical laboratories. Creatinine, most xenobiotics and their metabolites are eliminated from the blood by glomerular filtration into the kidneys and urine. Therefore, the elimination of these compounds can be normalized relative to creatinine. Drexler et al. (1994) reported that the levels of the carbon disulfide glutathione-conjugated metabolite, 2-thiothiazolidine-4-carboxylic acid, adjusted by creatinine excretion were better correlated with exposure in 362 workers than was the excretion of the unbound parent compound. There are data suggesting that the same is true for many of the compounds for which there are BEIs®.

Creatinine excretion is increased, unfortunately, by factors such as physical exertion and extreme weight loss, which alter muscle mass. In addition, it has been shown that creatinine may not be excreted uniformly at low urine volumes (Araki, 1986). In this case, several alternative adjustments have been proposed (Araki, 1986; Greenberg, 1989). When these conditions are present, it would be advisable to utilize another method of normalization or to compare the results with specific gravity adjustment by performing both.

There are specific cases where creatinine adjustment is not appropriate because the material is eliminated from the body by a mechanism different than that of creatinine. However, certain compounds like nitrous oxide, acetone, and methanol are eliminated primarily by diffusion from the capillary bed of the distal convoluted tubules into the urine. This process may be due to association of the material to proteins and/or lipids that limit their filtration. Therefore, urinary acetone levels are not creatinine adjusted (ACGIH®, 1996).

The BEI® Committee recently recommended that measurements of creatinine and/or specific gravity be made only to identify samples outside of the normal range that should not be analyzed; another sample should be collected when samples are outside the range of 1.010 and 1.030 for specific gravity and 0.3 g/l and 3.0 g/l for creatinine concentration. Correction of sample concentration inside of these ranges is not necessarily recommended and the BEI® should be consulted in each case. This decision was based on an analysis of the precision of values with and without the correction.

Urine samples are relatively easy to collect and aside from the privacy issues discussed below, noninvasive. Analyses of compounds found in urine follow techniques used from the corresponding air samples after cleanup. Sample cleanup is generally required since urine may contain several hundred compounds normally; thus, there is a potential for interference. The potential for a problem is greater for metabolite analysis than for metals or parent solvents. Metals are analyzed by atomic absorption or inductively coupled plasma; interfering substances are relatively rare with these compounds, but background levels may still pose a problem. Similarly, parent solvents can be measured in the head space above the urine sample by GC or GC-MS, which often eliminates interferences. Metabolites or metals cannot be analyzed this way because of their low volatility. Most metabolites are analyzed using HPLC following a clean-up procedure that is often specific for a single compound, or by chemically forming a volatile derivative that can be analyzed by gas chromatography.

Urinary measurements in occupational settings have been made for many compounds, from acetone to zinc. It is not possible to catalog here all the studies that have been conducted, but several will be mentioned in the course of following discussions.

MULTIPLE EXPOSURES

Multiple exposures to various compounds are more often the rule than the exception in occupations. Exposure can be simultaneous or sequential and there may be an impact. Interpretation is difficult because interactions can occur to either increase of decrease the effects relative to each compound alone. There are four potential interactions with exposure to mixtures. We will discuss each of these interactions in the context of effects and how they are dealt with using air sampling before switching into how biological monitoring can be used for exposure assessment.

Independent effects indicate no interaction. The effects of carbon monoxide and xylene, for example, appear to be completely independent, although it might be possible that an oxygen-starved brain may be more sensitive to the narcotic effects of a solvent. The majority of occupational health guidelines assume independent effects and no interaction.

Additive interactions occur when the effect of an exposure to two (or n, where n is the number of compounds in the mixture) compounds results in effects that are twice (or n) as great as when exposure is to a single compound. Because compounds have different toxicity, the air levels must be normalized to take this into account. The TLV® levels are usually used for this normalization. The ACGIH® TLV® Committee guidelines recommend that an effective TLV® for each component in a mixture be reduced by 1/n (cf. Volume 1, pp. 217-219); the summation of all fractional exposures should not exceed one. The guidelines recommend that effects are to be assumed to be additive only when each of the agents has similar toxicological properties. Mechanisms of additive effects are straightforward; for example, if two materials compete for the same receptor or enzyme active site, and these sites have similar affinity and turnover numbers for each substrate, then the net effect will be equivalent to adding twice as much of either compound. Seiji et al. (1989) reported that co-exposure to trichloroethylene and tetrachloroethylene had an additively increased narcotic effect because they competed with each other for detoxication sites on enzymes and slowed down each others metabolism. For biological monitoring purposes, it should be borne in mind that the measured effect will likely be decreased formation of a metabolite of one compound when there are multiple exposures.

Antagonism is a reduction in effect due to the interaction. There are four mechanisms of antagonism. *Functional antagonism* occurs when the antagonist has the reverse effect on the same system as the toxicant, as with the anti-convulsant, diazepam. *Chemical antagonism* occurs when another agent chemically reacts with the toxicant and reduces its effect. Chelating agents, such as dimercaptopropanol (British anti-lewisite, or BAL), and EDTA, which bind metals and facilitate their excretion, antivenoms, and even the metabolic phase II molecules, glutathione and glucuronic acid are examples of chemical antagonists. *Receptor antagonism* occurs when an agent competes for the same target molecule as the toxicant. Pure oxygen is used to counteract the effects of carbon monoxide as both compete for binding sites on hemoglobin and other heme proteins. *Dispositional antagonism* is a broad category that refers to the ability to alter the absorption, distribution metabolism or excretion of a toxicant and thus reduce its toxicity. Dispositional antagonists include agents like charcoal or ipecac given orally to retard absorption of ingested materials. Other agents reduce or increase metabolism of materials making them less toxic. Ethanol is a dispositional antagonist for methanol and ethylene glycol poisoning because it reduces their metabolism to the toxins, formaldehyde and oxalic acid, respectively. Ethanol has also been shown to decrease the toxicity of xylene by increasing its metabolism to methyl hippuric acid (Sato et al. 1990). BEIs® are not adjusted for antagonisms, and extreme care must be taken in interpreting data when the possibility of antagonism exists.

Potentiation is an interaction where a relatively nontoxic component of a mixture enhances the toxicity of another component. Isopropanol significantly increases the hepatotoxicity of carbon tetrachloride, although it is not hepatotoxic itself (Folland et al., 1976).

Synergism involves increasing the toxicity of all components of a mixture; often the effect of the interaction is multiplicative. The carcinogenic risk of persons exposed to both asbestos and cigarette smoke is increased to 50 times that of persons not exposed to either (Selikoff, 1977).

Used appropriately, biological monitoring is a powerful tool to estimate exposures from all routes, mixture interactions and the impact of exposures outside of the workplace. One of the possible interactions between solvents is that one component of the mixture will inhibit the metabolism of the other. If the toxic effects of the material are due to the parent compound, then toxicity will be increased. Tarkowski (1982) experimentally exposed humans for four hours to 25% of the TLV® for xylene and a mixture of alcohols and ketones (butanol, MEK, MIBK) at about 25% of their combined TLV®. The major effect of xylene is general solvent narcosis caused by the parent compound. Tarkowski reported that the metabolism of xylene was inhibited by 60%, which would cause more intensified and prolonged solvent effects. These data suggest that the interaction between these solvents would be more than additive.

In some cases, a threshold for the interaction has been reported or hypothesized. Tardif and co-workers (1991, 1993) reported that the interaction between toluene and xylene was dose-dependent. They saw that at 50 and 40 ppm for toluene and xylene, respectively, there was no interaction. However, when exposure was raised to 95 and 80 ppm, respectively for four hours, metabolism of xylene was diminished and blood levels of both compounds were elevated. On the other hand, Engstrom et al. (1984) reported that the metabolism of both xylene and ethyl benzene were inhibited when volunteers were exposed to levels of each at 50% above their respective TLVs® for four hours. At much lower levels, Wallen et al. (1985) exposed volunteers to 25 ppm xylene and 55 ppm toluene and found that the metabolism of both compounds was diminished. These data suggest that a dose threshold for metabolic interaction does not exist for these mixtures. The discordance between the two sets of data may be due to individual variability.

As the Tardif group studied a relatively small number of volunteers, their sample may not be representative of the population response. In addition, the studies were conducted in countries with different ethnic backgrounds; the metabolism of the different ethnic groups can vary significantly. It is known, for example, that about 15% of Japanese persons are of the slow acetylator phenotype, a trait possessed by 50% of the persons in North America.

Inhibition of metabolism can decrease toxicity if the metabolites are more toxic than the parent compound. Shibata and coworkers (1990) reported that co-exposing rats to hexane and various concentrations of methyl ethyl ketone led to a decrease in the formation of 2,5-hexanedione, the active neurotoxic metabolite (see Figure 8-9). This suggests that this particular co-exposure would antagonize the chronic neuropathic effects of n-hexane, although chronic studies were not conducted to determine if this occurred. Whether or not this co-exposure would lessen the toxicity in humans has not been determined. Ethanol has long been known to be an antidote for methanol poisoning, as it successfully competes with alcohol and aldehyde dehydrogenases and inhibits the formation of formaldehyde and formic acid, the toxic metabolites of methanol (Schecter, 1978). The toxicity of the methanol is decreased because more of it is excreted un-metabolized. Carcinogens in complex mixtures have been both potentiated

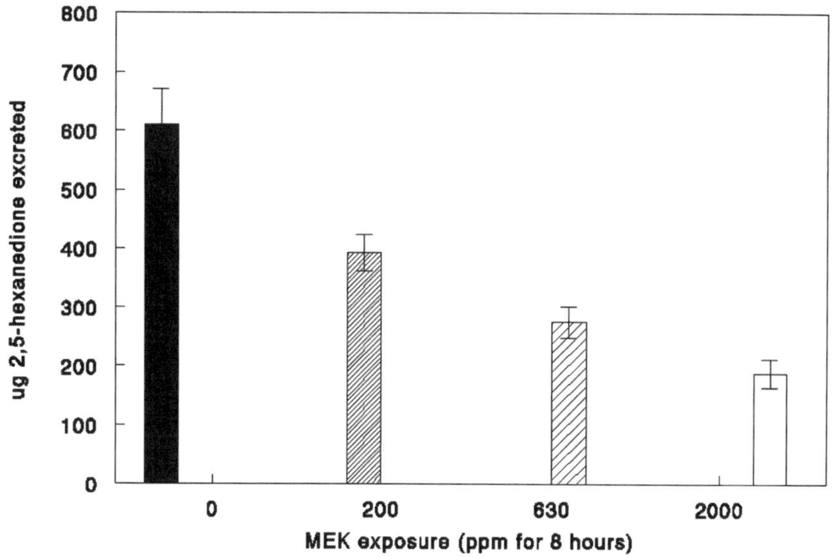

Figure 8-9. Effect of increasing 8-hour exposure to methyl ethyl ketone (MEK) on the excretion of 2,5-hexanedione in hexane exposed rats. Animals were exposed to a binary mixture of hexane at 2000 ppm and MEK at the indicated level. Levels of total hexane metabolites are shown. Adapted from Shibata et al. (1990).

and antagonized by other components of the mixture (DiGiovanni et al., 1982). If the parent compound is more toxic than metabolites, then increasing the metabolism can decrease toxicity. Sato et al. (1990) reported that persons with prior exposure to ethanol excreted methyl hippuric acid (xylene exposure) at a faster rate than did persons without ethanol exposure.

EFFECTS OF UNUSUAL WORK SCHEDULES AND "MOONLIGHTING" (OUTSIDE EMPLOYMENT)

Exposures that occur outside the workplace may also have an impact on an individual's health; these include hobbies, habits, and outside employment. When the outside exposure is to the same compounds as in the workplaces, continuation of exposure into the day or weekend confounds the normal, expected elimination of the material from the body and may allow toxic levels to be reached. It is not possible for measurements of workplace air to take these exposures into account. However, biological monitoring can be used to determine whether such exposures are occurring and if there is a potential problem. Figures 8-10 a and b are very simple models: they assume equal exposure each workday and single phase (one compartment) elimination. If the exposure is the same during the week and on weekends, the effect of weekend "moonlighting" is dependent upon the half-life of the marker. With a 24 hour half-life (see Figure 8-10a), steady state exposure is reached within 2 weeks with or without weekend work.

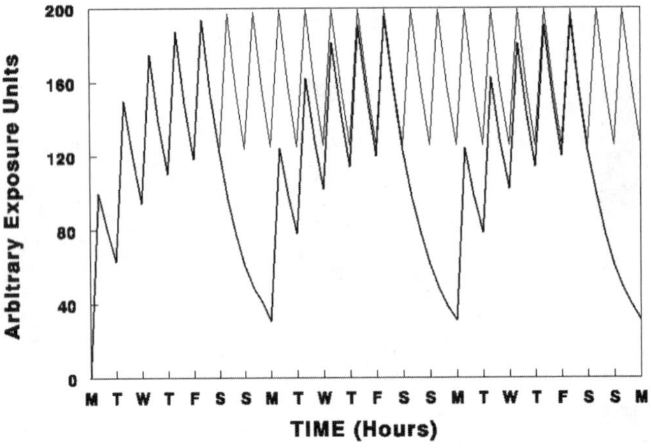

Figure 8-10a. Model of the body burden of a compound with a 24-hour half-life during an idealized exposure 8 hours per day and either 5 days per week, or 7 days per week to demonstrate the effect of working such a schedule. The heavy line shows the 5 day per week curve and the light line shows the 7 day per week curve. (See text for discussion.)

The preshift levels on Monday morning would be substantially different (126 versus 30) and would alert the hygienist to the fact that an additional exposure is occurring. However, note that the differences would tend to disappear during the usual workweek and by Friday afternoon there would only be a 2.5% difference (200 vs. 195). If health effects are related to the cumulative area under the exposure curve, rather than the steady state levels, then the "moonlighter's" total exposure would be increased 35% relative to working five days per week. If the biomarker half-life was less than 24 hours, then the difference between the steady state levels would be smaller and coincidence of the lines would occur earlier in the week. With a half-life of 48 hours (see Figure 8-10b), steady state is reached within three weeks, and the differences between the Monday morning preshift levels would again be substantial. However, the Friday post shift level would be only 10% higher (341 vs. 308). The total area under the exposure curve would be proportional to the extra days worked per week (33% higher after 2 weeks if both weekend days were worked). These models would suggest that the effect of working weekends would not have a dramatic effect on steady state levels, but would proportionally increase the total exposure of the person, assuming that all exposures are the same. If the exposure was higher on weekends, they would be indicated by the preshift Monday sample. If the intent of the program is to protect the health of the worker, then biological monitoring can be used to advise workers of the potential for overexposure during the weekends and the necessity for the proper controls to limit exposure as much as possible. The potential for weekend exposures underscores the importance of obtaining pre- and post-exposure samples; no difference would be seen with a Friday post-shift sample.

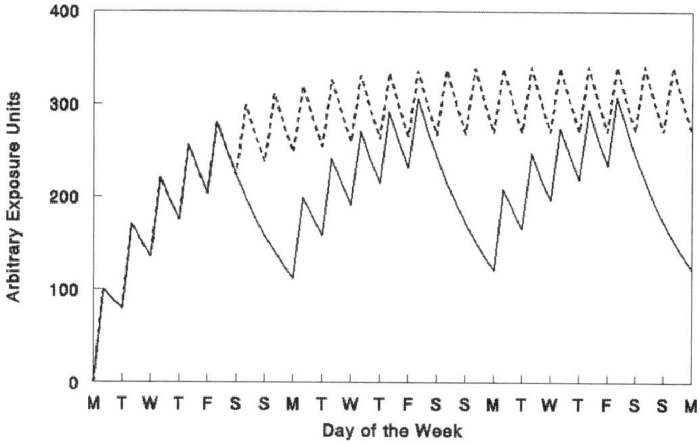

Figure 8-10b. Model similar to Figure 8-10a, but the compound has a 48-hour half-life. The dotted line is the 7 day per week schedule.

WHAT SHOULD BE CONSIDERED BEFORE ADDING BIOLOGICAL MONITORING TO AN OCCUPATIONAL HEALTH PROGRAM?

It is simply not wise to begin a program of biological monitoring without thorough preparation and attention to detail. The pressure to "do something" when a potential exposure occurs occasionally results in damaging incidents due to incomplete preparation. Attention to detail is important so that the proper marker and the sampling time are selected to increase the sensitivity and specificity of the program, i.e., the ability to detect exposures when they exist and to rule them out when they do not. One should be prepared for the eventuality that some individuals — and maybe not those with the highest ostensible exposure — will have a response that indicates exposure or damage. One should also be aware of potential confounders and alert the monitored population to the possibility beforehand that certain foods or drugs may cause false positive results. Consider the following example.

Urinary arsenic levels were determined in a production facility. Production employees, as well as office workers, who were to act as negative controls, were sampled. When the results became available, the levels in the office employees were a significant fraction of those in the production workers. This finding led to a high level of anxiety in the offices as it appeared that arsenic contamination was widespread. What had occurred was that many in the office had gone to a local seafood restaurant for a banquet just prior to the sampling. Seafood, and shellfish in particular, contain high levels of organic arsenic, which the employees were excreting. Unfortunately, the analysis requested did not include speciation of the arsenic as organic or inorganic. When speciation was included, the levels of inorganic arsenic in the office workers were as low as might be anticipated.

Why Is This Sampling Being Done?

The first question to ask is, "why is this sampling being done?" This is a doubled-edged question. The first response should deal with the ethical reasons for the sampling, e.g., biological monitoring is to be used to increase understanding of employee exposure in the workplace and to protect worker health. At a symposium some years ago, a speaker described a drug testing program that he had brought to a particular company. The speaker suggested that the plant had some trouble with a safety record, which was ascribed to a widespread use of drugs in the workplace. The program was described in great detail including the assays used, the number of people who failed the initial screening, and then the secondary screening, the number of employees who were counseled, and the number discharged. However, when the question was asked, "What was the impact of your program upon the company's safety record?," the speaker replied that the data were not available. Was this program put in place to contribute to health and safety, or simply to gain control over the employees? Programs with the latter goal that use biological monitoring are doomed to failure even if they gain initial compliance. Workers will eventually rebel over this level of control when there is no direct benefit to health and safety.

In the utilitarian sense of the question, one might want to embark on biological monitoring to rule out alternate exposure routes, or to determine if PPE is working as expected. Response to this question will help guide the selection of determinant, matrix, sample timing, and participants.

Costs Versus Benefits?

The next question is whether the costs of doing biological monitoring outweigh the benefits of the sampling. Included in the costs of biological monitoring are the increased training of staff and the time it will take to provide education regarding the results and their interpretation. The benefits of having a more well educated workforce should also be included in the equation, however. Other benefits include a better understanding of workplace exposures, and greater employee participation in the health and safety program.

What Markers Should Be Sampled, and in What Matrix?

The next important questions are, "for what marker should I sample, and, in what matrix?" Assay specificity relates to the ability of the test to respond only to a certain specific stimulus. From the example above, a specific marker would be inorganic arsenic in urine, while a less specific marker would measure both inorganic and organic arsenic (total arsenic). As was seen from the example, selection of the nonspecific marker caused a considerable ambiguity in the interpretation of the results. However, there are cases where a less specific marker is perfectly acceptable; if one is sure that there is no potential for confounding exposures. Without any exposure to ethyl benzene, or several related chemicals, levels of mandelic acid in the urine can be assumed to result from exposure to styrene. Selection of the proper marker requires an understanding of the biology of the marker, its half life, and possible interferences. If it is anticipated that there will be individual differences in the response, then effective dose or early effect markers are the best selection. If the effects of concern are simply due to the amount of material inside the body, then internal dose markers are sufficient. Because markers are available for many exposures, factors such as ease of sample collection, cost, and the availability of analytical procedures also become important considerations. Generally, the best marker to choose will be the one that integrates exposure for the longest period. A marker with a long biological half-life will be less affected by day-to-day variation in exposure and require a smaller number of samples to estimate the mean exposure (Droz et al., 1991). However, when peak exposure predicts effects, then exposure estimates that can capture these events, air sampling or biological monitoring markers with a short half-life or integration period, would be more appropriate.

For certain markers, it is necessary to obtain baseline samples prior to exposure because there may be considerable variability in the normal values of the population, due, for example, to common sources outside the workplace, or endogenous production in the body. This is essential for determination of markers like acetylcholinesterase activity, where there is significant person-to-person variability in activity before exposure, and urinary zinc, where dietary intake can have a major effect on individual levels. The BEI® *Documentation* for acetylcholinesterase activity recommends that two pre-exposure samples, at least three days apart, be taken from each person (ACGIH®, 1996).

There are *Documentation* available from ACGIH® (1996) for biological monitoring of over 35 industrial agents. Baselt (1997) has published methods used in the literature for 101 agents, including several carcinogens and other compounds for which the available methods and procedures are less well validated than those for which there are BEIs®. These sources can aid in the selection of determinants, the proper collection procedures, and analytical methods (see Appendix 8-1 for a list of agents covered). The same sources will provide information regarding when and where the samples are to be collected. These precautions will minimize the effects of collection and storage artifacts and decomposition. Arrangements must be made in advance to have samples collected by medical personnel (for blood samples particularly) and for coordination of efforts between the medical and industrial hygiene personnel.

Prior consideration must also be given to selection of participants. The purposes of the program will often dictate participant selection. If the question is to determine whether there are individuals with high internal doses for whatever reason, the primary consideration should be to select those persons with the potential for the highest exposure by all routes. This requires the *recognition* skills of the hygienist. The medical staff may also be aware of individuals who have medical conditions that may cause them to absorb more of a compound, or predispose them to the effects of exposure, and who should, therefore, also be included in the program. Basic epidemiological principles should also be followed in the selection of individuals for a program. This would include selection of individuals with low exposure as a comparison group, particularly in the early phases of a program. If resources exist, sampling of persons with presumed intermediate exposure is useful, at least initially. For one, a dose response can be observed only if there are differences in dose. In addition, the hygienist assigns persons into groups based on previous data and experience and the data from biological monitoring would reinforce those selections and judgments. If biological monitoring is to be included in a routine exposure monitoring program, then efforts should be made to select participants from all possible exposure categories as would be done in any program.

In-house analysis is less common today and generally contract laboratories conduct analyses. The key requirement for the analysis of biological monitoring samples is that the laboratory should participate in proficiency testing procedures. Sources of approved laboratories for performing biological monitoring tests are given in Appendix 8-2.

The issue of timing of sample collection has been discussed. Decisions about when to sample must be made prior to the beginning of a program. If no protocol is available, then studies of the kinetics of the marker should be made in order to interpret results reliably. While this is not as difficult as it may appear, it would be wise to contact persons with prior experience for assistance in the design (The *Documentation* of the BEIs® contains the references upon which the guideline was based). Essentially, the focus of the study would be to determine the half-life of the marker's loss from the body. Models can be found in the literature (Monster et al., 1979; Mraz and Nohova, 1992) and a reasonable protocol can be adapted. Clues as to the potential behavior of the compound/metabolite in question could be garnered from analogous chemicals studied previously. This is a research study, and exposed workers would be recruited as volunteers. They may be asked to allow that multiple blood and/or breath samples be collected, to collect and store urine samples over the weekend, and to provide documentation of non-exposure. Interpretation of kinetic data is made more difficult by interindividual variability. Instead of a single curve, it is more likely that a range of kinetic patterns will be found with different people. There may also be day-to-day variations within the same person due to the impact of such factors as diet, general health, and body weight. The findings of Drexler et al. (1995) are especially germane in this regard. They found that the elimination of a CS_2 metabolite was delayed until after the end of the work-shift in overweight workers. Because CS_2 is rapidly metabolized and excreted, the level in these workers in the next morning pre-shift sample was no higher than those of other workers. Drexler et al. (1995) estimated that each 10% increase in a specific body mass index would reduce the level in a sample taken immediately at the end-of-shift by 0.35 mg/g creatinine, about 7% of the BEI®. Therefore, kinetic studies should include samples taken after workers have been off work several hours.

When groups are sampled, a pattern of elimination should be expected instead of a single curve; individuals *will* differ from the overall pattern. See Skarping et al. (1991) and Sakai et al., (1995) for examples. The last real exposure of the persons (as opposed to the scheduled end-of-shift) should also be carefully documented during kinetic studies.

Who Gets the Results?

Once the samples are analyzed, several issues arise regarding their interpretation and dissemination. The results of any workplace sampling are required by law (Code of Federal Regulations, 1989) to be given to the worker. The information should be sufficient so the worker can make an informed judgment as to its meaning. This would include specific results in light of standards and guidelines and the results of other workers (but without identification) in the workplace. The hygienist does not have the same access to individual medical records as does the occupational physician, therefore, medical and health information should be provided to the worker by the physician or nurse. The hygienist has access to biological monitoring data as exposure data. However, hygienists may not be privy to the individual identity of the persons sampled; rather, they may obtain information relative to the classes of workers exposed. In order to provide accurate information to the worker, medical persons must also be aware of the biology of the markers and the relevance of the marker to exposure. This is information that the hygienist can provide.

Not being medical personnel *per se*, the hygienist is not privy to individual medical records, including biological monitoring results, without specific release by the individual. The same is true for other employer representatives. On the other hand, the hygienist must be able to utilize abstracts of the results, i.e., the results without personal identifiers, to determine whether the work environment is safe. Transfer of abstracted biological monitoring data from occupational medicine to hygiene must be included in the biological monitoring program from its inception. Abstracts of the same data may also be made available to union representatives and others in the company.

BIOLOGICAL MONITORING AND THE INTERACTION BETWEEN OCCUPATIONAL HEALTH PROFESSIONALS

The advent of biological monitoring programs will require closer interaction between the occupational nurse, the physician, and the industrial hygienist. A clear distinction must be made between diagnostic tests where the health status of the worker is ascertained and biological monitoring that examines exposure and reversible effects associated with exposure. The hygienist is charged with protecting the health of the workers by ensuring that they are employed in a healthy work environment. As such, the hygienist understands the plant operations and work schedules and is in the position to determine how, when, and where exposure monitoring samples should be collected, both environmental and biological. To apply this knowledge to a biological monitoring program, the hygienist must understand how work practices and schedules may impact on biological monitoring data, i.e., the toxicokinetics of the biological marker. The hygienist is uniquely qualified for this role. On the other hand, the hygienist is generally not a health practitioner and, therefore, the actual collection of the biological samples will be most often left to medical personnel.

Breath, blood, and urine samples can be obtained by any trained person, but it is often best to do this within the medical unit of the facility. (In some states venipuncture requires certification.) The role of Occupational Medicine and Nursing personnel is medical surveillance and health intervention, which includes obtaining and making health performance recommendations based upon biological data. The role of the hygienist in interpretation of the data has less to do with individuals and more to do with the environment in which the individuals are employed. There is considerable overlap between these activities with the *species differentia* being that the physician works with individuals, symptoms and diagnoses, while the hygienist works with environments and controls exposure and absorption within those environments. To complete the distinction, the hygienist may not be privy to biological data that contains individual identifiers. An example may illustrate the role of the hygienist in such a program. During a discussion with the medical staff, an increase in blood lead levels is noted in a subgroup of exposed workers. Before the individuals reach blood lead levels consistent with medical removal, the hygienist should investigate the exposure of the subgroup. A work process could be identified and corrective procedures implemented so that levels do not increase to the point of medical removal. In this way, medical removal or diagnosis of occupational disease can be viewed as the failure of the hygiene program to control the environment (King, 1990), or failure of standards and guidelines to protect health; very important information in either case.

BIOLOGICAL MONITORING DATA AND THE HEALTH OUTCOME

The ultimate test of whether an exposure estimation, be it air sampling or biological monitoring, protects health is to determine the association with disease at various marker levels. While BEI® levels are thought to be protective for most workers, in most cases, this has not been proven; BEIs® are often based on TLVs® and the TLVs® are revised periodically. Therefore, it is important that data collection programs involving sampling and health outcome are designed to facilitate evaluation. The expertise of epidemiologists, particularly those with experience using biomarkers will be essential to the design of data collection summaries and archives.

Biological monitoring is important in epidemiology to reduce exposure misclassification, i.e., to help classify exposure status accurately. In order to contribute to this process the biological monitoring program must be prepared to answer two important questions: "What exposure period is critical for the effects?" and "Can the selected biomarker estimate the exposure during that period?" Very few biomarkers can estimate exposures that occurred 20 or more years ago. Alternatively, historical exposures may have to be reconstructed (with a great loss in precision) using current marker and exposure levels and interview data or other records that document changes in processes and techniques upon which exposure reconstructions can be based (Rice, 1991). Regardless, it becomes very important that the biological monitoring program be designed to be interactive with exposure and health data.

Compounds like dioxin and polychlorinated biphenyls are metabolized very slowly, accumulate in adipose tissue, and have a very long biological half-life. These are exceptions. A problem with markers with a long biological half-life is that they do not respond quickly to changes in exposure, but measure cumulative exposure. Once steady state is reached (see Figure 8-11), decreases in exposure will be difficult to detect until considerable time has passed, making it more difficult to evaluate interventions. As a result, a 50% reduction in daily exposure would result in a slow reduction in the marker levels that might not be readily apparent for several weeks. Recognition of this potential problem with cadmium led to use of urinary cadmium levels as indicators of chronic exposure (half-life 10 to 30 years, Lauwerys, 1983) and

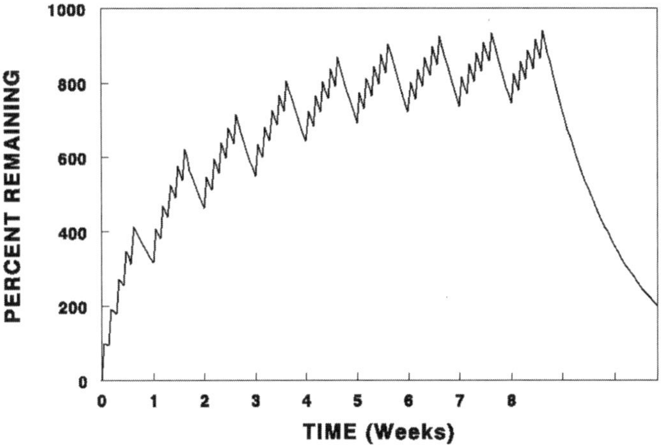

Figure 8-11. Accumulation of a compound with a 168-hour half-life in the body with a 5 day per week, 8 hour per day idealized exposure. A compound with this biological behavior would have both day to day and week to week accumulation until the steady state is established after 6 weeks of employment. The end of the curve declines sharply as there is no exposure. This might be expected during a vacation or other prolonged absence from exposure.

blood cadmium levels, because they change with a shorter biological half-life (initial phase 75-130 days Jarup et al., 1983), as markers of more recent exposure. Even with these, the effects of interventions will not be readily apparent for several weeks or months.

STANDARDS, ADVISORY AND LEGAL

Biological monitoring is being accepted worldwide. As indicated above, biological monitoring has been relatively slow to catch on in the U.S. In Europe, Japan, and other countries where litigation is less common and the standard setting process more dynamic, acceptance of biological monitoring has been very widespread.

BEIs® As Standards

While two OSHA standards, those for lead and cadmium, currently require biological monitoring, there are currently 38 "biological exposure indices" or "BEIs®," set by ACGIH® to assist the hygienist in development of a biological monitoring program. ACGIH®'s BEI® Committee was formed in 1982. The charge and approach of the BEI® Committee is very similar to that of the TLV® Committee. BEIs® are reference levels of biological determinants; the levels most likely observed when the person is exposed via all routes at the equivalent of the TLV®-TWA. To obtain these values, the Committee must estimate the amount of material that would be absorbed from the lungs at TLV®-TWA exposures and convert this to the expected level of the determinant based on either experimental data or what is known about the rate of metabolic conversion. Alternatively, the levels observed in studies in which there is reasonable certainty that exposure is limited to only the inhalation route are considered directly. The intended uses of these values are to evaluate exposure from all sources, to account for individual differences in response and to evaluate the effectiveness of protective measures.

There are several potential problems that may occur if there is a requirement that BEIs® be based on the TLV®-TWAs. Obviously, there can then be no BEI® for compounds for which there is no TLV®. Several compounds, benzo(a)pyrene, benzo(a)fluoranthrene, β-naphthylamine, and 4-aminobiphenyl for example, have no TLVs® because they are recognized human carcinogens, yet they are important components of mixtures of coal tar and other combustion products to which workers are commonly exposed. There is precedent for development of a BEI® not based on the TLV®, but on the health effects, namely, lead. The development of more BEIs® based on observed health effects obviously require that data on biological monitoring and health effects be collected. There is currently a paupacy of this type of data, and its collection is strongly encouraged. Another related approach to BEI® development may be to provide guidance as to the background variability of a particular index and suggest levels above the background that can be reasonably caused by occupational exposure. For example, if the mean level of a urinary metabolite in a series of studies in an unexposed population is 25, with a standard error of 12, then there is a 95% certainty that levels above 50 are not due to background exposures, or endogenous sources, but to some other factor, namely, occupational exposure. If health effects data are collected at the same time, then the protection afforded by the guidance level can be evaluated. This approach was recently presented to the industrial hygiene community by Roy Albert, recipient of the 1999 Stokinger Award (Albert, 1999).

The TLVs® are based upon a qualitative risk assessment that protects "most" of the exposed workers. In practical terms, does "most" mean that one or two persons per plant so exposed would be affected, or about one out of 500 or 1000? These levels of assessed risk are relatively high when compared to those developed by other agencies such as the U.S. EPA. Some difference between workplace and environmental exposure levels is probably warranted because there is the possibility for recovery during periods when workers are not working and being exposed. Should standards protect against any risk above that seen in the unexposed population? This seems reasonable and points to the application of the approach proposed by Albert for standard setting in occupations.

Basing the BEIs® on the TLVs® also means that they are not necessarily associated with health outcome. Often data relating a given BEI® level to outcome are completely absent, so it is not really known whether the BEI® levels are protective. In several cases, it would seem that they are protective because even dietary or endogenous sources of agents like phenol or formaldehyde interfere with the levels seen in occupations. Exposure to toluene at the current TLV® induces a level of hippuric acid in exposed persons that cannot be readily discerned from background levels in the population. Thus, persons excreting materials at or below these BEIs®

should have health effects from these exposures not different than those seen in the unexposed population.

As discussed above, while the TLVs® and BEIs® are not intended to be fine lines between harmful and safe levels, they are often used in that light. The evolution of the TLVs® is clearly an iterative process based on open reporting of exposures and effects. Earlier versions of the TLV® *Documentation* show that the process by which TLVs® were established depended upon the observation of hygienists that no, or slight, effects were noted at exposure level "A," but clear effects were seen at level "B." As the ability to resolve effects improved, the acceptable levels tended to decrease. The BEI® Committee may be moving toward this approach as evidenced by their recommendation of a "no value" or non-quantitative BEI® for methylene-bis-2-chloroaniline (MOCA). By suggesting that biological monitoring be done for this compound, but not providing an "acceptable" exposure value, the BEI® Committee hoped to foster the collection of exposure and effect information so that the current industrial experience may lead to an understanding of what levels of the metabolite are acceptable. Industrial hygiene professionals should expect to be able to exercise professional judgment and not rely completely upon government, physicians, or professional agencies to determine what is a "safe" level of exposure. By the same token, is it reasonable for hygienists to make decisions based upon individual experiences without the input of professional peers? Better decisions will be made as more information is available from the collective experience. The adoption of "no-value" BEIs® for certain compounds should encourage critical evaluation of data and experience for other compounds, markers and exposures heretofore "untouchable," and is a positive step for the field.

German Biological Tolerance Values for Working Materials, BATs

"Biological tolerance values for working materials" are set by a scientific committee appointed by the German government (Deutsche Forschungsgemeinschaft, 2000). The philosophy behind the BATs is similar to that offered by the BEI® Committee, except that the BAT values are to be observed as *ceiling* levels, never to be exceeded. One might anticipate, therefore, that the BEIs® would generally be lower than the BATs, but this is not uniformly true. BATs are also available for several occupational carcinogens currently without BEIs®. These include ethylene oxide, hydrazine, and vinyl chloride.

BIOLOGICAL MONITORING: APPLICATIONS AND EXAMPLES

Metals

Metalworking was among the first concerted activities of human society. The hazards of metalworking were well known in antiquity (see Volume 1, Chapter 2). It is not surprising that metals analyses were among the first biological monitoring tools developed and used. Since different metals affect different targets, there is no single pathway for the biological activity of metals. Some, like lead, are sequestered in bone and are remobilized when the bone is restructured, or as the blood level falls. Cadmium and mercury reach highest concentrations in their target organs, kidney and brain. Chromium affects tissues at the site of exposure/absorption. Many metals are essential to life and a daily intake is required, others mimic essential metals and have toxic effects at very low doses. Organometallic compounds can be particularly dangerous because they are often well absorbed through the skin. The recent case of a cancer researcher who was exposed to dimethylmercury (almost certainly less than 0.5 ml!) through her gloved hands, which resulted in her death 80 days later, illustrates how toxic these materials can be (ACGIH®, 1997). Metals often bind to proteins within cells, and hence may have longer biological half-lives than other classes of compounds. They present the problem of accumulation from day to day, week to week, month to month, or even year to year given occupational exposure. Because certain metals are commonly found in food, a pre-exposure sample is often necessary for correct interpretation of the data.

The first biological monitoring of metals was blood lead analysis. As indicated above, Kehoe was among the pioneers in the development of the field when he began the study of the effects of tetraethyl lead in the 1920s for the Ethyl Corporation (Harris, 1997). OSHA requires that blood lead measurements be made if the exposure to lead in the workplace exceeds a threshold of 30 $\mu g/m^3$ for 30 days in a year. (The PEL for lead is 50 $\mu g/m^3$.) Biological monitoring for lead includes blood lead as a marker of internal dose, and d-aminolevulinic acid as an effective dose (marker of inhibition of heme synthesis). Analysis of zinc protoporphyrin (ZPP) has also been used to evaluate the effect of lead on heme synthesis. OSHA currently mandates medical removal when blood lead levels exceed 50 µg/dl. The affected worker may return to work when their levels are reduced to 40 µg/dl.

OSHA also requires biological monitoring when exposure to cadmium is greater than 2.5 $\mu g/m^3$ for more than 30 days in any year. The standard is complex, requiring that blood and urine be analyzed for cadmium concentration and effects. Blood cadmium is a measure of the current exposure to the metal; urinary cadmium reflects chronic exposure and β_2-micro globulin is a measure of the cumulative damage to the kidneys. β_2-micro globulin is a low molecular weight blood protein that is usually filtered into the glomerulus, but reabsorbed in the proximal renal tubules. Its presence in the urine indicates that there is damage to the normal function of the tubules. OSHA requires a staged response based upon the individual biological monitoring results. Medical removal is indicated if blood levels exceed 15 µg/l, or urine levels are greater than 15 µg/g creatinine, or if urinary β_2-microglobulin levels exceed 1500 µg/g creatinine. OSHA has developed a computer-based program called "GoCad" to assist in interpretation of results and the necessary follow-up. This program can be obtained by contacting the OSHA website (www.osha.gov/oshasoft/gocad.html). The BEI® Committee considers β_2-micro globulin an indicator of early renal pathology and a diagnostic marker of pre-clinical effects and not strictly in the realm of biological monitoring.

BEIs® currently exist for several other important metals including arsenic, chromium, cobalt, mercury, and vanadium. While the BEIs® for lead and cadmium suggest that timing is not critical in the analysis, it is important to remember that metals with long biological half-lives may accumulate for many weeks or months before maximal levels are reached given a regular exposure (see Figure 8-12). Detection methods for metals in body fluids are very sensitive. For example, Gao et al. (1994) reported that while a significant difference could be detected in urinary chromium levels between exposed workers and non-exposed controls, no difference could be detected in lymphocyte DNA strand breaks, or 8-hydroxyguanine adducts in the same workers.

Solvents

Many organic compounds are lipophilic and able to readily pass through the skin when contact is allowed. Many are also often volatile. These qualities of solvents make biological monitoring an important consideration for exposure assessment. The toxic effect common to organic solvents is their narcotic effect on the CNS. Different solvents differ in the dose required to produce narcotic effects; halogenated compounds (the current anesthetics) being the most potent. The mechanism of this effect remains to be positively established, but it may have to do with the solvent changing the fluidity of the neuronal cell membranes, altering their function. Most persons are familiar with the sequence of increasing solvent-induced narcosis through ingestion of alcohol, or during solvent exposure. These effects can be detected at very low exposure levels by an early effect marker that detects changes in postural sway, i.e., quantitative

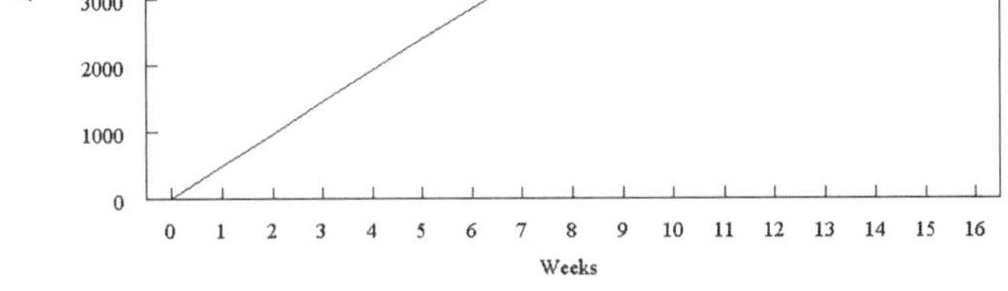

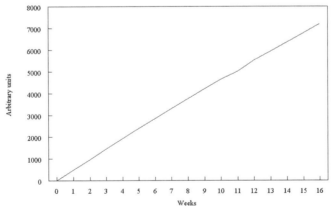

Figure 8-12. Accumulation of a compound with a biological half-life of 120 days under same idealized workweek as in Figure 8-11. Steady state would not be attained until after almost two years of employment. The level at 40 weeks would be over 250 times the level on the first day of exposure. While samples collected before the steady state would underestimate the ultimate levels significantly, they can be very useful to evaluate whether or not the person is accumulating the material at a rate that would ultimately exceed some index value. Then, the exposure could be controlled and the shape of the curve at the peak level could be altered.

posturography (Bhattacharya et al., 1987). The general toxicity of a solvent is acute, due to the activity of the solvent itself, and is reversible, provided paralysis of respiratory centers has not occurred. Monitoring for the parent compound in the exhaled breath, blood, or urine provides an estimate of the internal dose and the general narcotic effect.

In addition to narcosis, a significant fraction of solvents also possess toxicity not directly related to narcosis effect, but to effects caused by their metabolites. The target organ affected by this toxicity varies. For example, ethanol causes cirrhosis; methanol, ocular toxicity; benzene, leukemia; n-hexane and methyl butyl ketone, peripheral neuropathy; ethylene glycol, renal damage. There is evidence that the damage caused by the toxic metabolites of these compounds is cumulative and irreversible. Because these effects are caused by specific metabolites, measurements of effective dose (the internal dose of the metabolite) will likely have a greater predictive value than simple measurements of dose. Since these metabolites are not as volatile as the parent compound, they are not detectable in the breath, so they must be monitored in the blood or, more commonly, the urine. Metabolite formation generally occurs some time after exposure, so care must be taken to understand the kinetics and sample at the appropriate time.

GENETIC SUSCEPTIBILITY AND BIOLOGICAL MONITORING

Introduction

In discussing Figure 8-4, factors of individual susceptibility were noted to play significant roles in the transitions between exposure and effect biomarkers and disease. These factors can include differences in work, work habits, or physiology, which increase exposure, absorption, or retention of the compound. There may also be differences in cellular uptake and processing of the toxicant, its metabolism, and/or the repair of damage, all related to the expression of genetic traits relative to the exposure.

To understand how genetics and exposure can interact, a quick refresher of genetics and protein expression is helpful. DNA contains the genetic code. The DNA bases are G (guanine), A (adenine), C (cytidine), and T (thymidine). To make proteins, a messenger RNA molecule is synthesized or "transcribed" from the DNA template. The RNA molecule is then "translated" into a chain of amino acids, a polypeptide, or protein. Some of these proteins make up the structure of cells and tissues. Others, enzymes, have catalytic activity, facilitating the thousands of reactions that occur to sustain life. After an initiation point, each consecutive series of three DNA bases determines which amino acid will be incorporated into the new polypeptide. For example, the DNA series ACC codes for the amino acid glycine. If the DNA code is changed, or "mutated," a different amino acid could be inserted into the protein, which could change its function.

The effect of mutations is measured by how much they change the activity of the enzyme; different mutations can have greatly different effects on enzyme activity. Some mutations, frame shift mutations that involve the loss or gain of bases in the DNA sequence, change the genetic code from the point where they occur to the end of the gene. These usually turn the protein into gobbilldy-gook, with complete loss of enzyme function. At the other extreme are those mutations that do not change the amino acid sequence at all. This can occur because the DNA code is "degenerate," i.e., in some cases, more than one set of three nucleotides will code for the same amino acid. ACC and GCC both code for glycine, for example. So, a mutation that would cause the replacement of the initial A to a G would have no effect on which amino acid was inserted into the protein. This type of mutation is termed a silent mutation. Mutations that cause amino acids with only slightly different structure or chemical function might also have little impact on the activity of the protein. For example, if glycine was replaced by alanine (ACC to AGC) in a non-critical part of the enzyme, then there might likely be little effect on the protein activity. However, a mutation in the catalytic or recognition sites of the protein may cause a significant change in the enzyme's activity.

Mutations that occur in germ cells can be inherited. There are two copies, or "alleles," of most genes (see Figure 8-13). One copy is inherited from each parent. Each allele may be translated into protein, so that a person with no mutations in either copy is said to be the "wildtype." A heterozygote is a person who has received a normal allele from one parent and an allele from the other parent that contains a mutation (indicated by a crossbar in Figure 8-13). The homozygous recessive has at least one mutation on each copy. If the enzyme is critical to the survival of the organism and the mutation removes most or all of the activity, the organism will not survive. If, however, the gene is not critical to survival under usual conditions, or the mutations do not eliminate activity, then the mutation can be inherited and passed along through

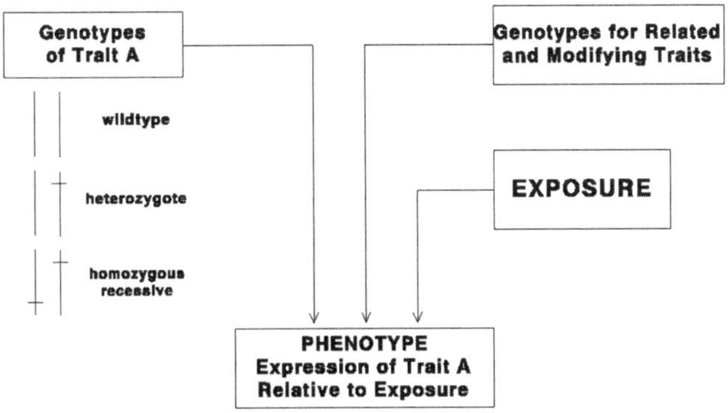

Figure 8-13. The relationship between genotype and phenotype for a person. The final expression of a trait (phenotype) is a function of the DNA coding for the specific gene (genotype) and is also influenced by other genes that compete for the same substrate, influence the expression of the gene, or otherwise affect is ultimate disposition of the material. In addition, the exposure to the compound and other compounds also impact the expression of phenotype.

the generations; this is what occurs in almost all cases of susceptibility markers involved with environmental exposures.

All this said, mutations are a positive force in evolution and population variability. A mutation that slightly alters an enzyme function might increase the chance for survival under some other stress. Sickle cell anemia is a case in point. A person who has sickle cell anemia (not the trait) is a homozygous recessive with mutations on each copy of the gene. Their red blood cells readily sickle under mild anoxia. However, carriers of the sickle cell trait (one mutated gene), do not seem to have a noticeable reaction to most stressors. The evolutionary advantage to keep the gene is that the red blood cells of the heterozygotes are resistant to attack by malaria parasites, *Plasmodium spp*, a very common disease in equatorial Africa. One of 10 African-Americans are heterozygotes and carry the trait, but fewer than 1 in 3000 are homozygote recessives and have sickle cell disease (Lehninger, 1975).

There are other cases where the homozygous recessives are rare in the population because having two mutations limits the ability of the person to survive and reproduce. The frequency of persons without methemoglobin reductase (an enzyme that repairs the damage caused by nitrocompounds and aromatic amines) in the population is about one in 40,000 (Elder, 1949; Omenn, 1982). The heterozygotes for this trait might be at increased risk following occupational exposure to nitrocompounds and aromatic amines because they have one mutation and some fraction of activity compared to the wildtype. However, since heterozygotes for this trait make up less than 1% of the population, no studies demonstrating an effect have been conducted.

Certain mutations in other genes are well tolerated in the population and persons with at least one mutation can be a large part of the population. For example, persons with at least one mutation in the N-acetyltransferase-2 gene (NAT2) appear to make up about one-half of the U.S. population. The important consideration in these cases is that the persons are normal in many, most, or all ways with the exception of their response to exposure or overexposure to certain chemical agents. It is possible, but not known, whether there is any selective advantage to having a mutation in this gene.

Figure 8-13 illustrates some of the important issues of gene mutations and their expression relative to exposure to compounds. Genotype is currently determined by obtaining DNA from a person and hybridizing it with known sequences of complementary DNA. Phenotype is the manifestation of the genotype relative to an exposure. As in Figure 8-13, other genes may also act on the same compound and alter the effect of the exposure. For example, aromatic amines are metabolized by the CYP450 enzymes of the liver and other tissues. One specific CYP450, 1A2 was thought to be responsible for the majority of the activation of these compounds to carcinogenic metabolites. However, Butler et al. (1989a, 1989b) have shown that many CYP450 enzymes can perform this activation, and which one is the most active depends on the particular aromatic amine. The formation of 4-aminobiphenyl carcinogen-DNA adducts were studied in mice genetically engineered to be without CYP450 1A2 activity. The DNA adduct levels in these mice were from 0 to 66% lower than in the wildtype, depending on the tissue (Underwood et al., 1996). However, adducts were formed in all tissues tested, indicating that activating metabolism is taking place despite the missing gene. Apparently some other CYP450 was able to perform the metabolism. This example illustrates the difference between genotype and phenotype. The genotype of the mice was that they were completely lacking CYP 4501A2. However, phenotypically they were able to carry out the same reactions to almost the same degree as the wildtype. Genotype is determined by rote genetics, the linear sequence of DNA bases, while phenotype is more flexible and can be modified by differences in affinity for substrate and the interaction of the substrate with other enzymes that can "compensate" for the loss of gene function.

Perhaps the major controversial areas of biological monitoring are genetic screening and the biological markers that may be useful to predict whether an individual may be sensitive. The controversies center on the use of genetic screening in the pre-employment phase to determine how a potential employee might respond if exposed and the dissemination of this information to others, such as insurance companies (Murray, 1983). There have been several monographs on this topic and the serious reader should refer to these for more extensive discussions of the issues (Vineis and Schulte, 1995; Ashford et al., 1990; Ashford, 1986). The discussion below is based upon issues raised in these papers.

It has been long appreciated that there might be a genetic component to occupational disease. J.B.S. Haldane had this to say in 1938 regarding potters and bronchitis . . .

> "While I am sure that our standards of industrial hygiene are shamefully low, it is important to realize that there is a side of the question which so far has been completely ignored. The majority of potters do not die of bronchitis. It is quite possible that if we really understood the causation of this disease we should find out that only a fraction of potters are of a constitution that renders them liable to it. If so, we could eliminate potters' bronchitis by regulating entrants into the potters industry who are congenitally exposed to it" (Haldane, 1938 as cited in Omenn, 1982).

Have there been any advances on this front in the intervening 60 years? Are we any closer to finding the elements of constitution that cause some people to be more susceptible? Can (and should) we prevent those persons from entering the workplace? What would happen to disease rates if we were able to limit entry based on putative genetic predisposition?

In the case of the British potters, the level of silica exposure was the main culprit, and those with the highest exposure, generally involved with the polishing of the china and earthenware, developed the disease. Kehoe (1963) was of the opinion that the same was true for lead exposure and that biological monitoring could be used to determine those highly exposed . . .

> "It is important to realize and have the means of demonstrating that, in a generally safe series of industrial operations, there may be specific opportunities for excessive exposure to lead. The failure to recognize this simple fact in many a lead-using industry is the primary reason

for the frequent belief and statement that an "unusually susceptible" workman contracted lead poisoning under conditions that were "well-controlled" and safe for other workmen."

Screening to Identify Fitness for Duty

Screening for job fitness has long been a part of pre-employment procedures in many industries. The need to screen to visual acuity for professions like drivers and pilots can be readily appreciated and is non-controversial. Such tests are quantitative and are unambiguously related to fitness for duty. Similarly, screening for the rare homozygous mutations associated with severe response to chemical exposure might be indicated. Hereditary methemoglobinemia involves mutations in both alleles of the methemoglobin reductase gene. High exposures to compounds like nitro and aromatic amines cause an increase in methemoglobin in all persons. In persons without mutations, methemoglobin levels increase, but then rapidly decline as methemoglobin reductase reverses the damage. However, since the homozygous recessive person completely lacks the ability to repair the damage, their methemoglobin levels increase almost irreversibly. However, screening for this deficiency has never been done, because it is so rare in the population (1 in 40,000). In addition there are no data indicating that heterozygotes are at any increased risk.

Screening to Identify Fitness for Exposure

Inherited mutations more common in the population necessarily have less dramatic effects on individuals. Genetic screening for these traits is more controversial because the question becomes not fitness for duty, but fitness for exposure. The criterion to determine whether a susceptibility marker is important include association with disease, i.e., whether the biomarker really is a risk factor, unambiguous determination of the trait, test accuracy, and the rate of false positive and false negative findings.

SCREENING FOR GENOTYPE OR PHENOTYPE

Not surprisingly, screening tests can be *genotypic* or *phenotypic*. *Genotype* tests look for differences at the DNA level, i.e., for specific mutations in the genes of interest. An accurate test of this type would have the advantage of being very specific for the trait in question and would not vary from day to day. Since samples are usually screened against a known DNA sequence, if the individual possesses a new or novel mutation not in the mutation library of the lab, then a false negative may result. Expression of enzymes in response to an exposure is a complex process associated with previous exposure history, dose, and the structure of the specific compound (Nebert, 1989). As noted above, loss of the CYP450 1A2 enzyme is compensated by other genes. Tests of susceptibility *phenotype* investigate the expression of a genetic trait relative to an exposure or challenge. Because they measure the result of metabolism, phenotype measures can often ameliorate some of the problems associated with the genotypic tests measured above.

GENETIC SCREENING IN OCCUPATIONAL AND ENVIRONMENTAL STUDIES

Four examples of genetic traits that may have an impact upon occupational exposures will be discussed. Mutations in glucose-6-phosphate dehydrogenase (G6PD), a gene on the X chromosome have been associated with hemolytic anemia in persons with exposure to dapsone, primaquine, and other therapeutic oxidizing drugs (Omenn, 1982). The possibility exists that persons with these mutations may also be more sensitive to the hemolytic effects of certain compounds including arsine, phenylhydrazine, and various nitroaromatic and aromatic amines. There has been one report of a small group of G6PD-deficient workers in a trinitrotoluene plant developing severe hemolytic anemia. However, this is the only report to date of the effect of G6PD mutation within occupations (Omenn, 1982).

α-Antitrypsin deficiency is associated with greatly increased rates of emphysema in those with two mutant alleles (genes). This enzyme is apparently responsible for the degradation of proteases that are able to destroy the structure of the alveoli and pulmonary bronchioles. Heterozygotes (those with one mutant and one normal copy of the gene) represent about 3-5% of the total population, and have enzyme activities of about one-half that of those without any mutant genes. Smokers who are heterozygotes may be at greater risk of emphysema. Heterozygotes are also thought to be over represented in workers with impaired lung functions (Lappe, 1988). However, it does not seem that the heterozygotes are at an extreme risk of effects, and programs that focus on exposure reduction and early detection of symptoms may be more useful than general screening programs.

N-ACETYLTRANSFERASE

A good example of a susceptibility marker and the issues that surround its use in the workplace is the N-acetyltransferase phenotype and aromatic amine induced urinary bladder cancer. Many epidemiological studies indicate that slow N-acetyltransferase status may be a risk factor for urinary bladder cancer in persons with exposure to aromatic amines. The relative risk, i.e., the excess risk in the slow acetylators relative to their fast counterparts has been estimated from one (no difference) to about 16 in various studies. While the effect appears reproducible (12 of 14 studies found that slow acetylators were at higher risk for urinary bladder cancer), the magnitude of the effect is not dramatic with the average relative risk for bladder cancer in all studies being about two-fold greater in the slow acetylators relative to their fast counterparts.

The test for acetylation phenotype is fairly simple, involving the measured administration of caffeine and collection of timed urine samples to determine the relative amounts of two caffeine metabolites, 5-acetylamino-6-formylamino-3-methyluracil and 1-methylxanthine. A reasonable correspondence has been shown between genotype and phenotype by this assay (Vineis et al., 1994). However, the results of the phenotypic assay are expressed as a continuous distribution of values, so that while fast and slow acetylators may be very distinct at the extremes of the distribution, there is overlap between the two groups in the intermediate values. This will result in some misclassification of persons as slow or fast acetylators depending on thc variability in the assay and where the cutoff line between the "fast" and "slow" phenotypes is drawn, as will be seen below.

Two studies involving biomarkers and acetylation phenotype have been performed and have provided some very interesting data regarding this susceptibility marker and the probability and extent of DNA and protein damage. Figure 8-14 shows the influence of acetylation status on the level of 4-aminobiphenyl-hemoglobin adducts in smokers and non-smokers (Bartsch et al., 1991). Exposure is clearly the main effect; adduct levels were significantly higher in smokers than non-smokers. However, in both smokers and non-smokers, the adduct levels were higher in the slow acetylators. The magnitude of the effect is consistent with the increased risk of urinary bladder cancer reported for slow acetylators. Thus, the effect of acetylation on the rate of urinary bladder cancer and carcinogen-hemoglobin and DNA adduct levels does not seem to be more than about two-fold. However, acetylation phenotype is clearly an effect modifier of 4-ABP exposure; adduct levels are much higher in smokers than in non-smokers and only moderated by acetylation status.

Benzidine (BZ) is a potent human urinary bladder carcinogen that has not been widely used in the U.S. since the 1970s. However, exposure to this compound did have dramatic and devastating effects in certain American workplaces. Bingham and co-workers (1973) studied the effects of exposure in one plant and reported a prevalence of urinary bladder cancer of 50% in the workers (Dr. Zavon has since told the author that the prevalence has risen to 70%, the expected prevalence would be less than 2%). However, BZ exposure remains a problem in the world as the technology to produce, but not control, BZ has been exported. A recent epidemiological study of Chinese workers was conducted by the National Cancer Institute (Bi et al.,

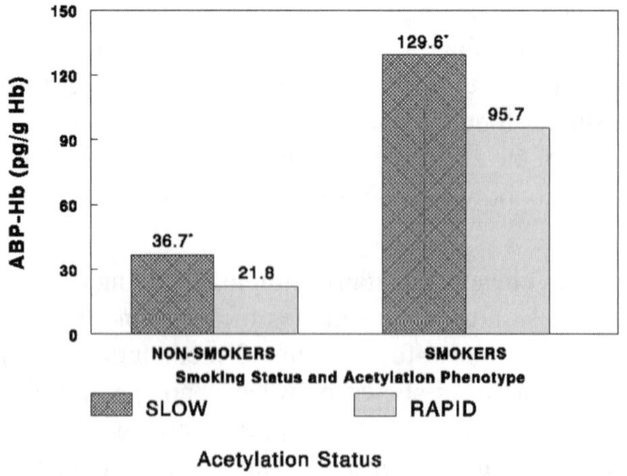

Figure 8-14. The levels of 4-aminobiphenyl-hemoglobin adducts by smoking status and acetylator phenotype. The data are from Bartsch et al. (1991).

1992). The authors reported that the risk of urinary bladder cancer was 25 times higher overall in the group with reported BZ exposure. When reported exposure was categorized as either low, medium, or high, the increased risk of urinary bladder cancer was 5, 36 or 160 times, respectively. However, there was no effect of acetylation phenotype or genotype. In fact, fast acetylators appeared to be at a slightly higher risk. Recent studies of benzidine-exposed workers in India led to remarkably similar conclusions (Rothman et al., 1996). Urine and blood samples were obtained from volunteers without exposure and from groups exposed to benzidine and benzidine-based dyes. Levels of excreted urinary metabolites, mutagens, and DNA adducts in exfoliated urothelial cells were obtained as well as DNA adducts in blood lymphocytes. Figure 8-15 shows that the benzidine DNA adduct levels in the exfoliated urothelial cells in the combined exposed groups were 12 times higher on average than in the controls. However, when the exposed workers were split out into benzidine workers and dyeworkers, the average adduct levels were 24 times higher in the BZ group than in the controls, while the levels in workers exposed to only BZ-dyes were about two times higher (see Figure 8-16). In addition, when the individual values were plotted (see Figure 8-17), a wide range of adduct values was found even in the benzidine group. Some workers appeared to have very low exposure while others had levels more than 100 times that in the controls. This range might be unexpected if all workers are considered equally "exposed," however, there was strong evidence that the biomarker data controlled for exposure misclassification. The DNA adduct data was highly correlated with the excretion of urinary metabolites and mutagens, suggesting that the adduct levels reflected real differences in exposure within the exposed group (DeMarini et al., 1997).

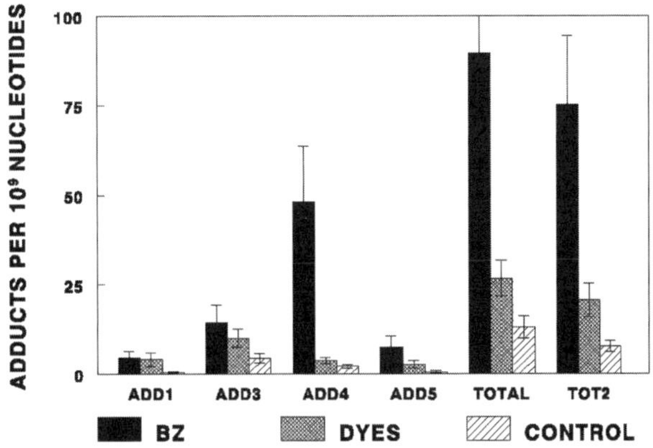

Figure 8-15. The level of DNA adducts in exfoliated urothelial cells of workers and non-exposed controls. In this figure, the exposed group includes both benzidine and benzidine-based dye workers. Adduct 4 (Add4) was tentatively identified as N-(deoxyguanosin-8-yl)-N'-acetylbenzidine. Other adducts could not be identified. Total is the sum of all adducts for each person and TOT2 is the sum of those that were higher in the exposed persons relative to the controls.

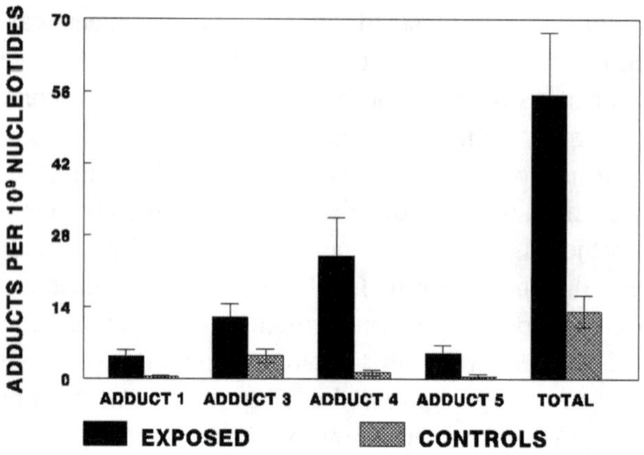

Figure 8-16. The levels of benzidine-DNA adducts in exfoliated urothelial cells of workers and non-exposed controls. The levels for benzidine production and benzidine-based dye workers are shown separately. The adduct identifiers are as in Figure 8-15.

In addition, they were very similar to the report of Bi et al. However, there was no correlation with acetylator status. It appears that the effect of acetylator phenotype is very compound-specific, significantly modifying the effect of 4-aminobiphenyl, but not of benzidine.

The potential for susceptibility misclassification was shown in the same study. Persons with two mutations were 57 times more likely to be grouped with the slow acetylators, but those with one mutation were only about five times as likely to be classified as slows (Rothman et al., 1996). In epidemiological terms, the assay had 91% sensitivity (the ability to predict slow acetylator status when a mutation was present) but only 83% specificity (ability to predict fast acetylator status in the absence of mutations). These data would suggest that although there is a small probability that persons would be classified as slow acetylators in the absence of mutations, there is a 50% chance that persons with mutations would be classified as fast acetylators and thus be permitted to work, defeating the premise of the testing for medical exclusion.

Accepting that the relative risk associated with the slow acetylator phenotype is two-fold, and that the specificity of the assay is 100%, the impact that perfect genetic screening would have on the number of urinary bladder cancer cases can be calculated:

no screening; number of cases = (case rate slows × proportion slows) + (case rate fasts × proportion fasts)

$$= (4 \times 0.5) + (2 \times 0.5)$$
$$= (2 + 1)$$
$$= 3$$

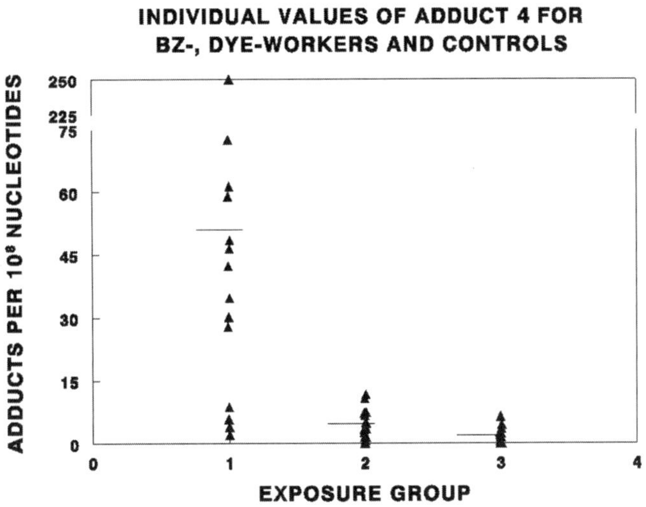

Figure 8-17. Individual values for benzidine-DNA adducts (adduct 4) in benzidine and benzidine-dye exposed workers and controls.

$$\begin{aligned}\text{screening; number of cases} &= (\text{case rate fasts} \times \text{proportion fasts})\\ &= (2 \times 1)\\ &= 2\end{aligned}$$

Operating under the best case of zero misclassification, there would be a maximum reduction in disease of 33% if all slow acetylators were removed from the workforce. Were this the case, the industrial hygienists would have to be certain that the employers or the employees themselves did not come to view themselves as resistant to urinary bladder cancer and treat exposure to aromatic amines in a cavalier fashion. Fast acetylators are not resistant to the disease; with an equal exposure, they merely have a lower propensity to form adducts in their urinary bladders and a corresponding lower rate of urinary bladder cancer. However, increasing exposure may negate any protective effect of the rapid phenotype and must be guarded against. In addition, the potential for misclassification of slow acetylators as rapids would also tend to diminish the difference in cancer rate from 33% to 16.5%.

In addition, the effect of the acetylator phenotype is also compound-specific, as was shown above. There seems to be an increased risk for slow acetylators regarding 4-aminobiphenyl, but slow acetylators do not seem to be at increased risk from benzidine. Skarping et al. (1991) reported that acetylator phenotype had no impact on the elimination of amine metabolites of toluene diisocyanates, although these authors did not make genotoxicity measurements.

Because the effects of susceptibility factors are chemical-specific, screening programs would have to be re-evaluated as the chemicals used in them are changed. Against certain agents, like BZ, screening for acetylator phenotype cannot be protective, because no group is at higher risk. Screening would then be a waste of resources as well as a source of divisiveness.

HLA-DPb1 GLU 69

HLA-DPb1 is a genetic sequence within the major histocompatability region associated with the activation of T-cells. A mutation in this genetic region, the HLA-DPb1 Glu 69 marker has been strongly associated with an increased rate of berylliosis in exposed workers (Kreiss et al., 1996). This marker is positive in about 1/3 of the total workforce. Kreiss and coworkers found that a positive marker and exposure acted independently and at least additively to produce the disease; there was a significant 11.8-fold increased risk in berylliosis in workers positive for the marker while exposure alone was associated with a 10-fold increased risk. In machinists with a median exposure of 0.9 μg/m^3 overall, 10.6% of the workforce developed berylliosis. However, there was a strong case for genetic predisposition: 3.2% of the workers without the marker developed berylliosis, while 25% of those positive for the marker developed the disease (Richeldi, 1997). In this case, screening to remove those persons positive for the marker would reduce the rate of disease from 10.6% to 2.8%. However, to bring about this reduction, 36 workers would have been denied employment. The authors noted another approach. Reduction of exposure from 0.9 μg/m^3 to 0.3 μg/m^3 (the median exposure level for workers other than machinists) reduced the disease rate to 1.3% in all workers (4% of those with exposure and positive for HA-DPb1 Glu 69 presented symptoms). Are genetic screening and exposure control necessarily mutually exclusive or, can they be used in hygiene practice to "protect the health of the worker"? Perhaps the model to be used is that suggested by Richeldi et al., "However, the berylliosis model indicates that genetic studies are useful in determining exposure-related risk levels for the large genetically susceptible segment of the population. By means of genetic epidemiology research studies, exposure control may be designed to protect all (sic) workers, thereby preventing the large majority of disease cases."

Although it may be useful in rare cases to avoid potentially dangerous situations, and in other cases to alert the employer to the degree of control needed, genetic screening for exclusion has a basic conflict with the focus of occupational safety and health programs: to provide a safe workplace.

FRONTIERS OF BIOLOGICAL MONITORING

Carcinogens

Another controversial area is that of biological monitoring methods for chemical carcinogens; their biological significance and the resulting legal-medical implications. As mentioned above, carcinogens pose a particular monitoring problem for industrial hygienists; many have low vapor pressures, but many are readily absorbed through the skin. Dermal contact with carcinogens is relatively common, particularly as components of mixtures. It appears that for mixtures such as asphalt, coke ovens, roofing tars, and used gasoline engine oils, the skin is the major exposure route (Talaska et al., 1996; Van Rooij et al. 1993a; Zhou et al., 1997).

MONITORING CARCINOGEN METABOLITES

Techniques are available to monitor carcinogens at each level of the biological monitoring continuum (see Table 8-1). Analysis of 1-hydroxypyrene (1-HP), the predominate metabolite of the non-carcinogenic PAH, pyrene, has been discussed as an internal dose marker. Pyrene is a common component of many PAH-containing mixtures. With a corroborating exposure history, the presence of 1HP in the urine indicates that the person absorbed some of the mixture to which they were exposed. Pyrene is a four ring PAH. Other metabolites of carcinogens can be readily monitored to determine the internal dose. Zenser and co-workers have developed and used a sensitive GC-based method for analysis of benzidine, N-acetylbenzidine and N,N'-diacetylbenzidine in human urine. They reported a significant increase in the excretion of these compounds in workers exposed to benzidine and benzidine-based dyes (Rothman et al., 1996). In addition, the levels were significantly correlated with the levels of carcinogen-DNA adducts in exfoliated urothelial cells and blood lymphocytes (Zhou et al., 1997).

NO VALUE BEIs®

The proposed BEI® for urinary levels of 4,4'-methylene-bis-2-chloroaniline (MOCA) does not list an index value. MOCA is a confirmed animal and suspected human urinary bladder carcinogen (ACGIH®, 1997). MOCA appears to bind to DNA in animals with about the same potency as benzidine (Segerback et al., 1993). Dermal exposure appears to be particularly important for this compound, as it is for many aromatic amines. Because biological monitoring can provide a much better estimate of internal dose than can air sampling, and because the method of monitoring urine for MOCA and metabolites is well-validated, the BEI® Committee proposed a BEI® even in the absence of reliable data demonstrating the internal dose-health relationship (Cocker et al., 1990). Employers will have to take greater responsibility for interpreting the data and using it to control exposures. This is a positive step in the use of biomarkers. Data may be generated that will document effective control strategies. On the other hand, the results will need to be interpreted for the employee in terms of exposure, rather than health effects, until a sufficient data base is developed.

URINARY MUTAGENICITY

Often the identity of the compounds is not known with complex mixtures, yet exposure may produce significant effects. A non-specific measure of carcinogen exposure is the urinary *Salmonella typhymurium* assay that measures the bacterial mutagenicity of a given sample. This assay is relatively simple and has been used to detect both occupational and environmental exposures to mutagens (Yamasaki and Ames, 1977; Legator et al., 1975). Unlike 1HP, which measures the excretion of a non-mutagen, an estimate of active mutagens is obtained. On the other hand, the routine method of performing this assay includes mechanisms to activate mutagens that may be excreted in a non-mutagenic form. This strategy is useful to increase the sensitivity of the assay. Thus, without other information, this assay is best viewed as a measure of internal dose of potential mutagens.

Table 8-1
Spectrum of sampling tehcniques for carcinogens

Factor	Techniques	Examples
External Dose	Skin Wipes Air Sampling Surface Wipes	Please see earlier chapters
Inernal Dose	Urinary metabolites and excretory products	1-hydroxypyrene (Jongeneelen, 1992); acetylated aromatic amines (Rothman et al., 1996); methylene bis-2-chloroaniline; urinary mutagenicity (Demarini et al., 1997)
Effective Dose	Hemoglobin Adducts DNA Adducts	Vineis et al. (1994) Skipper and Tannenbaum (1994) Rothman et al. (1996)
Early Effects	DNA Adducts Cytogenetics Mutation Assays	Artuso et al. (1995) Gonzalez et al. (1990) Albertini et al. (1990) Sarto et al. (1990)
Preclinical Effects	Oncogene Activation Cytology	Brant-Rauf (1990)

MUTATION ASSAYS

Basepair mutations occur when DNA strands containing DNA adducts are replicated. Currently, there are two mutations assays used to study human populations. Both assays involve "reporter" genes, i.e., genes that are not involved with carcinogenesis, but are such that mutations in the genes can be readily measured, not a general property of all genes. Mutations in the hypoxanthine guanine phosphoribosoyl transferase (HGPRT) gene decrease gene activity and are detected by the ability of the cell clones to grow in media that would be lethal if the gene were active. Mutations can then be confirmed by sequencing the gene in the cell clones (Albertini et al., 1990). The glycophorin A assay (GPA) detects mutations in the gene coding for the M,N blood group, a red blood cell surface marker. Only cells from individuals who are heterozygous for the gene (i.e., those who co-dominantly express both M and N) can be assayed as the system detects loss of expression of the M allele (Langlois et al., 1994). Since only 50% of any population can be expected to be MN heterozygotes, the sample size needed to detect changes in this marker is of some concern, as only half of the workers can be monitored, on average.

Occupational exposure to ethylene oxide, butadiene and cyclophosphamide and several others have been associated with increased frequency of mutations in the locus (Ward et al., 1994). Perera and coworkers (1993) published an informative study of foundry workers as they measured and investigated the correlation between air levels and several biomarkers including DNA adducts and mutations in HGPRT and GPA. They reported that leukocyte benzo(a)pyrene-DNA adduct levels better predict mutations in HGPRT than did the air levels of benzo(a)pyrene. This would certainly be expected if adducts were causal for most of the mutagenic effects, and/or the skin was the major route of exposure.

CARCINOGEN-PROTEIN AND DNA ADDUCTS

Most chemical carcinogens are electrophiles either as they are absorbed, like ethylene oxide, or after metabolism, as with benzo(a)pyrene diol epoxide. Electrophiles react with lone pair electrons on such elements as sulphur, oxygen, and nitrogen; sulphur being the most reactive (most nucleophilic). Protein contains all three reactive elements, while only nitrogen and oxygen are found in DNA and RNA. Covalent bonds are formed between the electrophile and the macromolecule, and there is a permanent addition of the carcinogen to the molecule. Whether the electrophile binds to protein, DNA or RNA is a matter of relative concentration, the reactivity of the reactants (the more selective electrophiles tend to bind to stronger nucleophiles, all other things being equal) and the size of the molecules. Reaction of an electrophile with a protein or RNA can lead to a toxic response if it occurs to a sufficient degree to halt metabolism. However, from the point of view of carcinogenicity, these are detoxication reactions because neither protein nor RNA is the genetic material. A notable example of this is the detoxication reaction of glutathione with electrophiles; a reaction catalyzed by the enzyme, glutathione-S-transferase. It is important to realize that the same activated form of a carcinogen binds to either protein, RNA or DNA. In fact, most carcinogens probably bind to all three in varying proportions. The measurement of effective dose can be made by determining the covalent binding of a carcinogen to any of these. RNA binding is rarely used because RNA is more difficult to prepare and handle than the others. RNA is also less relevant to carcinogenesis than DNA. On the other hand, proteins such as hemoglobin or albumin, while not relevant to carcinogenesis, are very simple to isolate and handle. These molecules have shown some utility in the measurement of carcinogen effective dose. In fact, once the proportionality between the relative protein and DNA binding is determined, the effect on DNA can be estimated from the protein binding. Techniques which use radiolabelled compounds to determine DNA and protein binding are not useful in human studies, but can be used to determine the ratio between DNA and protein binding (Meier and Warshawsky, 1994). Measurement of carcinogen-hemoglobin adduction has been the most successful. Adducts to specific carcinogens in tobacco smoke and such occupational exposures as ethylene oxide, ethylene, butadiene, and styrene have been studied (Skipper and Tannenbaum, 1994; Osterman-Golkar and Bergmark, 1988; Osterman-Golkar et al., 1996; Sorsa et al., 1996; Christakopoulos et al., 1993; Tornqvist et al., 1989).

CARCINOGEN-DNA ADDUCTS

The measurement of DNA adducts is probably the most direct estimate of effective dose. Adducts are often measured in surrogate tissues like blood leukocytes because these are abundant and relatively easy to obtain. The targets of many occupational carcinogens (lung, urinary bladder, liver) are internal organs and are more difficult to sample non-invasively, although some strides have been made (see below). Increased DNA adduct levels were reported in the leukocytes of workers exposed to PAH including foundry and coke oven workers, roofers, aluminum reduction workers, and to styrene (Herbert et al., 1990; Hemminki et al., 1988, 1990; Overbo et al., 1994; Horvath et al., 1994). Hemminki and co-workers have been particularly active in this area. In many cases, there was a clear dose-response with increasing exposure and the effects of occupation could be separated statistically from those of exposure.

The controversy surrounding the use of carcinogen-protein and -DNA adducts comes from the fact that they are neither necessary nor sufficient for carcinogenesis, although it is abundantly clear that carcinogen-DNA adducts are the ultimate mutagenic form of most chemical carcinogens (Broyde et al., 1985, Reardon et al., 1990). Put in the perspective of disease prevention, these markers can be extremely useful. The probability that any given DNA adduct will cause a tumor is extremely small owing to the enormous number of nucleotide bases in DNA and the fact that only a very few base pairs are targets that might result in a tumor (Talaska et al., 1992). The presence of elevated levels of adducts suggests that exposure and metabolic factors have allowed the activated form of the carcinogen to interact with the critical molecule. However, selecting and identifying reproducibly one altered DNA base (adduct) within a sea of 10^6 to 10^{10} normal unadducted bases in a cell is not a simple task. In addition, DNA adduct analysis is very expensive and requires highly trained personnel to perform reproducibly (Is this a scientific word?).

DNA adduct analysis is still not a routine laboratory procedure 15 years after it was first proposed. Inter-laboratory differences in measurements, the lack of positive controls and specificity remain problems so that levels involving any two methods, or even different labs are not comparable. Nonetheless, by controlling day-to-day variability and repeat sampling, meaningful data can be obtained within a single study. In the laboratory, the effectiveness of exposure interventions such as washing and barrier creams with small groups of animals exposed to as little as 25-50 μl of used gasoline engine oil per day for five days has been demonstrated (Talaska et al., 1996).

Carcinogenic PAH occur in many complex mixtures, including tobacco smoke and other home products such as driveway sealers and tars. Other carcinogens like aromatic amines, aflatoxin, etc. are found in food sources and other mixtures. In the absence of industrial hygiene studies documenting exposure, it is not possible to assign responsibility for the effect to occupational exposure, smoking or diet for any individual. Controlling for confounding exposures can, of course, still be done on a group or individual basis. As with all other markers, values above a certain point must be followed up by the IH to determine the extent the work environment can contribute to values. The presence of confounding factors should not rule out the use of a potentially valuable tool of exposure assessment, nor of sensible controls.

CHROMOSOMAL DAMAGE ASSAYS

Oncogenes are activated by at least two mechanisms: basepair mutations, where the DNA sequence is altered by changes of one or a relatively few bases; and chromosome mutations where the positioning of DNA sequences is altered by their movement to an inappropriate place on the same or a different chromosome (Battey et al., 1983; Tabin et al., 1982). These mutations result from breaks in both strands in the DNA molecule causing loss of material. Occasionally, when damage is extensive, the broken off piece of DNA re-anneals at the site of another break and is thereby translocated. Obviously, the probability of this occurring is normally very small, but increases with increasing damage. Since the translocations that activate oncogenes are rare events, the assays that have been developed to detect this type of genetic damage measure chromosome breakage either by examining each chromosome in a number of cells (chromosome aberration assay), or by more grossly looking for loose pieces of DNA (micronucleus assay). These damages have been shown to be related to the levels of carcinogen-DNA adducts in animal models (Van De Poll et al., 1990; Talaska et al., 1987). As with DNA adducts and mutations in the reporter genes discussed above, detection of chromosome breaks in a person's DNA can be used as a tool in disease prevention if exposure is substantially reduced, because there is such a low probability that any particular break will initiate the cancer process. Chromosome damage assays are nonspecific in that many agents can produce indistinguishable damage. However, they are very sensitive. Chromosomal assays have proven useful to detect exposure to many occupational genotoxins including ionizing radiation, benzene, styrene, chromium, mercury, leather tanning, formaldehyde, uranium, ethylene oxide, among many others (Yager et al., 1990; Artuso et al., 1995; Sarto et al., 1990; Suruda et al., 1993; Gonzalez et al., 1990; Loomis et al., 1990; Anwar and Gagbal, 1991).

BIOLOGICAL MONITORING IN HYGIENE PRACTICE

Vincent et al. (1993) investigated the exposure routes and internal dose of 2-butoxyethanol, a widely used solvent for lacquers, varnishes, and latex paint. This material is also a major component of many cleaning products and has been associated with bone marrow, kidney, liver, and blood cell damage. Workers in full-service carwashes, and in new and used auto preparation and detail areas may be exposed to the material as a mist, vapor, and as liquid on the skin. Vincent et al. found that personal 8-hour air levels only poorly predicted internal dose of auto detailers (r^2=0.36); however, an index including the duration of use and the concentration of 2-butoxyethanol in the cleaner predicted 93% of the variability in the samples. This, the authors concluded, indicated the degree to which the dermal route contributes to this exposure. Thus, the standard epidemiological approach of classifying exposure status based on years worked and concentration of the material in the solvent was predictive, while air sampling was not.

One of the major advantages of biological monitoring is that feedback that can be given to workers relative to interventions and decreased or increased exposures. Lauwerys and co-workers (1980) examined the urinary levels of N-methylformamide following exposure of workers to N,N-dimethylformamide. Measurements of the metabolites were made daily pre- and post-shift. Workers were asked to vary the personal protective equipment each week for three weeks. During the first week, they wore the usual PPE, gloves. During the second week, they were asked to apply barrier creams in place of the gloves. During the third week, they were asked not to wear gloves or barrier creams, but to wear the recommended air-purifying respirator. The results are shown in Figure 8-18. During the first week, baseline levels were established and although there is variability, it can be seen that the levels in the post shift samples are higher than the preshift, indicating that exposure is occurring and that some DMF must be entering the workers. The data from the second week indicate that the barrier creams are less effective than gloves in reducing exposure. The third week experiment lasted only one day. The levels of NMF increased dramatically, suggesting that the skin was the major route of exposure. Barrier creams, while less effective, were better than the recommended respirator. During the remainder of the third week, the workers used gloves again and the NMF levels were markedly reduced the second day, in agreement with the relatively short half-life of the metabolites in the body. In fact, the levels for the rest of the week were lower than during the first week, indicating that the workers were now looking for ways to reduce their exposure. The data from studies of this type allow workers to understand how they are exposed, and what steps are needed to reduce exposure. This is the real promise of biological monitoring to reduce exposure and effect.

Weaknesses of Biological Monitoring

GENERAL WEAKNESSES AND EMPLOYEE CONCERNS

Biological monitoring also has a unique set of weaknesses associated with its use. As mentioned above, not all exposures are better monitored using biological monitoring. These include exposure to irritants and compounds that produce chronic effects in the lungs without altering other physiological parameters. The other weaknesses of biological monitoring stem from the fact that there has been only a limited tradition for its use in U.S. industry. There are only two standards that require biological monitoring, lead and cadmium. The lack of government requirements has several major impacts on the use of biological monitoring in industry. Edelman (1990) noted that there are psychological and social factors involved in doing biological monitoring. Workers and employers often operate under the illusion that when contact with industrial materials is allowed, there may be no absorption. Biological monitoring proves that exposure and absorption did occur. If not explained appropriately, this knowledge may cause the worker psychological harm, particularly with agents like carcinogens because of the belief that any exposure to these agents may produce a latent and irreversible disease. Workers also express a feeling of being treated as "guinea pigs" because they are being exposed, and the

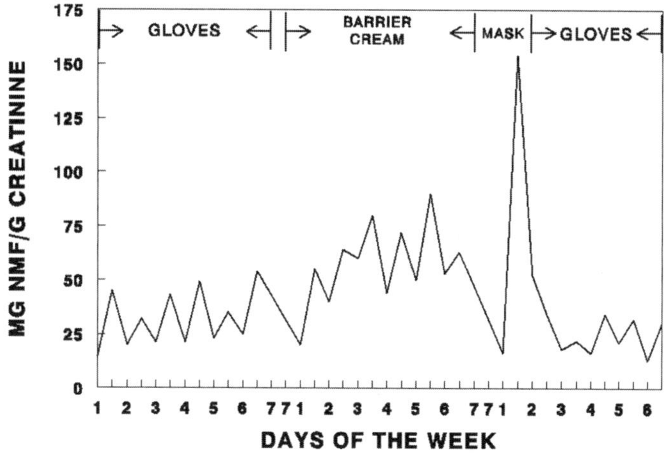

Figure 8-18. Average levels of N-methylformamide in workers exposed to dimethylformamide while using different forms of protective equipment. Adapted from Lauwerys et al., Int. Arch. Occup. Environ. Health, 45:189-203 (1980).

effects of the exposures and the marker levels are not known with certainty. There is no quick cure for these potential problems. If workers have the experience that the sampling is done with the interest of understanding and controlling exposure, the program will ultimately gain acceptance. The hygienist and other health professionals must cooperate to present the program to the workers, what is known about effects and risks to the workers, as well as provide them with information as to what the results of testing mean. Interpretation of the data relative to their co-workers and others in similar industries is often valuable. In many cases, this will require the hygienist to understand the biology of the marker and its place in the exposure-disease pathway. The situation regarding DNA adducts is useful for illustration.

EMPLOYER/EMPLOYEE CONCERNS

Employers fear that the same information may be used against them in court, particularly in the absence of standards of acceptable exposure. This problem is compounded because advisory guidelines are available for only a few carcinogens. Biological monitoring is also more personally invasive than air sampling. Workers may feel threatened by their being identified as having poor work or hygienic practices and may feel at jeopardy of being investigated. Workers have often expressed concerns as being used as guinea pigs for research or exposure and effects assessment with biological monitoring. In addition, workers are also concerned that biological monitoring data may be used against them for job or insurance exclusions. This is particularly true of the susceptibility markers discussed above. Privacy issues are also expressed relative to the collection of urine samples because urine has become the matrix of choice for drug testing. The major concern of most industries is that the results of biological monitoring may be used in court to prove that a worker's condition was related to an exposure. As Edelman (1990) noted, "no expert can testify with certainty that the measured biological level is 'safe.'" Finally, since the results of biological, or any other monitoring may be used in court, standardization, laboratory proficiency, quality assurance, and control are issues for both workers and employers.

SAMPLE CONTAMINATION, DEGRADATION AND EXPENSE

There is an increased probability relative to air sampling that biological samples can be contaminated, or will degrade with improper handling before results can be reported. Guarding against these possibilities often makes the collection and analysis of biological samples more expensive. Contamination is of particular importance when the analysis will be done on the parent compound, especially urine samples and metals. Consider the BEI® for urinary cadmium, which is 5 μg/g creatinine. If the average worker excretes 200 mg creatinine in a three hour urine sample, then 1 μg cadmiun in the sample will be equivalent to the BEI® level. Since cadmium has a density of 0.75 that of lead, only a very small amount of contamination from clothes or hands during the collection would be necessary to give a sample in excess of the recommendation. Therefore, extreme care must be used to avoid contamination from workplace sources. On the other hand, most intentionally, or accidentally spiked samples will be readily evident. The problem of external contamination is less important when metabolites will be analyzed because these are unlikely to be formed outside of the body. Blood sample collection devices can also contain certain metals or compounds like styrene that can absorb into rubber stoppers and give artificially low values. Another potential problem for blood samples is the "exposed arm effect." If an individual is exposed to a lipophilic solvent on one arm or hand prior to the sampling of blood from that hand, then because the venous blood in that hand will be draining the exposed area, the level will be higher than the whole body level. This can be dealt with simply by asking when the last exposure was, and if any exposure was on the sampled arm.

LACK OF PROFESSIONAL TRAINING

Few professionals have training or experience with biological monitoring. Currently, formal graduate level training in biological monitoring is available in a small number of universities. Several of the NIOSH Educational Resource Centers offer short courses on biological monitoring and the American Industrial Hygiene Association Biological Monitoring Committee offers a Professional Development Course, "Biological Monitoring," at the American Industrial Hygiene Conference and Exposition (AIHce). The Biological Monitoring Committee of the AIHA recently polled university-based NIOSH Educational Resource Centers and determined that although few programs had biological monitoring courses in place, there was considerable interest in the subject, and most were interested in adding material regarding biological monitoring to existing courses.

NONSPECIFIC BIOMARKER RESPONSE: INTERFERENCES

In some cases, there are weaknesses in the biomarkers themselves. Biomarkers (or any other measurements) are specific if one, and only one, exposure type leads to a positive finding. This is akin to the well-known problem with interfering substances on colorimetric indicator tubes (Vol. 1, p. 694). Trichloroacetic acid is a metabolite common to exposures to trichloroethylene, methyl chloroform, and perchloroethylene. Similarly, exposure to either ethyl benzene or styrene can lead to the production of mandelic acid. Fortunately, other markers are available for each of these substances (the parent compounds in end-exhaled air), so that strategies can be developed to determine the relative contribution of each component of a mixture. Other nonspecific biomarkers include carboxyhemoglobin (carbon monoxide and methylene chloride), inhibition of acetylcholinesterase (carbamate and organophosphate pesticides), methemoglobin (aromatic amines, nitroaromatics, nitrites), urinary phenol (benzene and phenol), and chromosome damage tests (any clastogenic agent). If the source of exposure is well characterized and processes and chemicals are well-known, then the effect of these confounders is limited. If not, potential co-exposure must be investigated with more traditional hygiene tools, walkthroughs and air sampling.

The body produces other biological determinants during the normal course of metabolism. Formic acid, a metabolite of methanol, formaldehyde, certain halomethanes, and acetone, is formed endogenously, and is also available in certain foods. Yasugi et al. (1992) reported that the background levels of formic acid in urine were 13 times higher than the background levels of methanol itself. These background levels reduce the sensitivity to detect exposure-related changes. Consequently, the BEI® Committee of ACGIH® has removed the BEI® for urinary formic acid and replaced it with a measurement of urinary methanol (ACGIH®, 1996). The best way to deal with background level is to obtain pre-exposure samples. The basal, or pre-exposure level of acetylcholinesterase activity has considerable individual variability and requires that a pre-exposure sample be collected from each person in order to determine effect.

Another source of interference with biological measurements of occupational exposure include those materials which are found in foods and patent medicines (e.g., phenol and aniline in phenacetin), the ambient environment (carbon monoxide), tobacco and other combustion products (carbon monoxide, pyrene and other PAHs) (Fishbeck et al., 1975; Rosenberg, 1990). The exposure levels from these sources may greatly exceed the total occupational dose. For example, Rosenberg indicated that the urinary phenol levels from ingestion of a phenol containing throat lozenge would exceed the occupational internal dose levels permitted for benzene (phenol is a metabolite of benzene) or phenol. Baseline measurements are recommended to detect non-workplace exposures.

AREAS OF FURTHER RESEARCH

Biological monitoring is an emerging field and the validation of determinants for compounds is not complete for the vast majority of important industrial chemicals. The conduct of research into development and validation of new markers will not be possible without the support, direct or indirect, of those who produce and/or use the compounds. Direct support by companies, or industry groups can come in the way of in-house data collection programs. Support of university or institute-based research and development is the indirect option.

Along with extending biomarkers, much work needs to be done on understanding the sources of variability in markers. Even with uniform exposure, markers can be expected to vary between persons, and within the same person when sampled on several different days. Very little research has been done to investigate these sources of variability.

Reviewers approach biological monitoring from different perspectives. Several other reviews are cited for the interested reader (Bernard and Lauwerys, 1987; Decaprio, 1997; Droz et al., 1991; Elkins, 1967; Fiserova-Bergerova, 1990, 1995; Kennedy, 1990; Lauwerys, 1983; Linch, 1974; Que Hee, 1993, 1997).

CONCLUSIONS

Biological monitoring for exposure is of little value for primary irritants and materials that act at the site of injury, like silicosis.

Biological monitoring provides an important tool for exposure assessment. It is most valuable as part of an occupational health program that includes air sampling, medical surveillance, and epidemiology. Biological monitoring should be considered very strongly when it appears that the dermal and/or the oral routes contribute to worker exposure. Biological monitoring can be used to rule out or discover the relative contribution of these routes. It can be a cost-effective alternative to collecting multiple air samples from the same individual. Biological monitoring is not a trivial pursuit, and to get the most from a biological monitoring program, the biology of the exposure must be understood as this influences every subsequent aspect of the program from marker selection to interpretation of results. Biological monitoring methods are available in the literature for over 100 agents; there are ACGIH® Biological Exposure Indices (BEIs®) for more than 30 compounds. Most BEIs® are given as values that are related to the amount that would be absorbed if the air levels were at the TLV® or OEL. Indices without a reference value may still be used to identify excessive exposures and to reduce exposures by evaluation of interventions. In this case, industrial hygiene observations play an important role in the evaluation of the data and explanation of unusual values.

Technical expertise and advice regarding legal and advisory guidelines is now available to assist with the implementation of these programs in any industry. Both ACGIH® and AIHA have volunteer committees interested in biological monitoring who are experts in relevant areas.

ACKNOWLEDGMENT

This Chapter is dedicated to the memory of a quite remarkable man, Jerry Sherwood, CIH. Mr. Sherwood was also a member of the Biological Exposure Indices (BEI®) Committee of ACGIH®.

QUESTIONS

8.1. What would be the effect of exercise at 50 watts on the retention in the body of two compounds, one with a water air partition coefficient of 15 and the other with one of 5?

8.2. Under what conditions can biological monitoring be expected to give a more accurate estimate of exposure than air sampling?

8.3. Biological monitoring is less useful for which class of chemical agents?

8.4. A spill occurred yesterday involving 3 workers and no air sample was taken until the end of the 2 hour cleanup (Levels at that time were 1.4 times the TLV®). Respirators were being worn, yet you suspect that there may have been some dermal exposure or respirator leakage because the workers complained of being dizzy and weak at the end of the cleanup. Today you collect a urine sample from one of the involved workers and measure the level of a metabolite of the compound, finding it to be 1.1 g/ g creatinine. You know that the BEI® is 1.5 g/ g creatinine and, researching the *Documentation* of the BEIs® for the compound, you find the half life is 12 hours. Estimate the peak level of the material and determine whether a significant overexposure occurred during the cleanup of the spill.

8.5. A Monday morning pre-shift urine sample for a determinant suggests that a worker was being exposed during the weekend. Aside from documenting the fact, what action can be taken?

Appendix 1: Part 1: Compounds with BEIs® Adopted, Proposed, or Under Study

OSHA Biological Indices Associated with PELs

Lead	Cadmium		

Compounds with ACGIH® BEIs®

acetone[2]	aniline[4,6]	arsenic compounds[4]	benzene[3,4]
cadmium[1,2]	carbon disulfide[4]	carbon monoxide[3,6]	chlorobenzene[4]
chromium IV compounds[2]	cobalt[1,2]	N,N-dimethylacetamide[2]	
N,N-dimethyl-formamide[4]	2-ethoxyethanol[4]	ethyl benzene[3,4]	
fluorides[2]	furfural[4]	n-hexane[3,4]	lead[1,2,6]
mercury[1,2]	methanol[2]	methemoglobin inducers[6]	
methy ethyl ketone[2]	4,4'-methylene bis-2 -chloroanailine (MOCA)	methyl isobutyl ketone[2]	methyl chloroform[3,4,5]
nitrobenzene[4,6]	pesticide acetylcholinesterase inhibitors[6]	parathion[4,6]	
pentachlorophenol[1,2]	perchloroethylene[1,3,4]	phenol[2]	styrene[1,4]
toluene[7]	trichloroethylene[1,3,4,5]	vanadium pentoxide[2]	xylenes[4]

Other Compounds Under Study to Establish BEIs®

acrylonitrile	aluminum	antimony	beryllium
1,3-butadiene	2-butoxyethanol	carbon disulfide	cyclohexane
2-ethylhexanoic acid	methyl t-butyl ether	Methyl formate	nickel

pentachlorophenol	polynuclear aromatic compounds	2-propanol	pyrethroids
uranium	vinyl chloride		

Issues Under Study

Genetic markers of Exposure	Quality control Issues
Biological monitoring strategies	BEI® notations

1. parent compound in blood; 2. parent compound in urine; 3. exhaled breath; 4. metabolite in urine; 5. metabolite in blood; 6. Effect marker; 7. BEIs® for this compound withdrawn by the committee.

Part 2: Compounds for which biological monitoring methods are published

acetaldehyde	dichlóromethane	methylenebis(2-chloroaniline)
acetone	dichlorophenoxyacetic acid (2,4-D)	methyl ethyl ketone
acetonitrile	dieldrin	methyl isobutyl ketone
acrylonitrile	diethylstilbestrol	nickel
aldrin	dimethylformamide	nickel carbonyl
aluminum	dinitro-o-cresol	nicotine
aniline	dioxane	nitrobenzene
antimony	diquat	nitroglycerin
arsenic	endrin	oxalate
benzene	ethanol	paraquat
benzidine	ether	parathion
beryllium	ethylbenzene	pentachlorophenol

borate
n-butanol
cadmium
camphor
carbaryl
carbon disulfide
carbon monoxide formaldehyde
carbon tetrachloride
chlordane
chlordecone
chlorobenzene
chloroform
chromium
cobalt
copper

cresol
cumene
cyanide
cyclohexane
DDT
diazinon
p-dichlorobenzene

ethylene glycol
ethylene glycol dinitrate
ethylene glycol monoethyl ether
ethylene glycol monomethyl ether
ethylene oxide
fluoride
tetrachloroethylene
hexachlorobenzene
hexachlorophene
hexane
hydrogen sulfide
isopropanol
lead
lindane
lithium

malathion
manganese
mercury
methanol
methyl bromide
methyl n-butyl ketone

phenol
polybrominated biphenyls
polychlorinated biphenyls
selenium
silver
styrene

tetraethyl lead
thallium
tin
toluene
toluene diisocyanate
trichloroethane
trichloroethylene
trichlorophenoxyacetic acid (2,4,5-T)
uranium
vanadium
vinyl chloride
warfarin
xylene
zinc

Appendix 2: Sources for Approved Laboratories

OSHA has approved laboratories for the analysis of blood lead, and for cadmium. These lists are available from the Salt Lake Technical Center, QC division, 1781 South 300 West, Salt Lake City, UT 84115-1802. The Program Director, William E. Babcock, can be reached at (801) 487-0073 ext. 271, or at www.osha-slc.gov.

AIHA also accredits laboratories based on their ability to meet or exceed proficiency tests. The lists of approved laboratories are published in April and September in the *AIHA Journal*. The list may also be accessed on the Internet, http://www.aiha.org.

Appendix 3: Interferences for BIOMONITORING Determinants*

Markers	Potential Interferences
Acetone	Product of fat metabolism, felt pens, nail polish remover, 2-propanol exposure
aniline	Some household dyes; a metabolite of phenylhydroxylamine, nitrobenzene, acetanilide, phenacetin, phenazopyridine
arsenic	Seafood (total arsenic), smoking
benzene	gasoline (up to 1.5%)
cadmium	smoking, seafood, some contaminated foods
carbon disulfide	captan, disulfiram, dithiocarbamates (may be metabolized to CS2)
carbon monoxide	space heaters, auto exhaust, smoking, methylene chloride
chromium	smoking
cobalt	metallic surgical implants
ethyl benzene	styrene, styrene glycol, styrene oxide, phenylglyoxylic acid, and phenylaminoacetic acid are likewise metabolized to mandelic acid
fluorides	rust removers and many foods
hexane	gasoline, certain glues
lead	some paints, leaded gasoline, some pottery
mercury	dental amalgams, fly ash
methanol	gasoline, certain acetates
methyl chloroform	spot removers, some adhesives
methyl ethyl ketone	certain glues and coatings
perchloroethylene	wearing fresh dry cleaned fabrics
phenol	antiseptics, throat lozenges
styrene	auto body repairs
toluene	pain, paint strippers, glues, gasoline
xylenes	paints, varnishes, certain glues

REFERENCES

Acheson, E.D.; Cowdell, R.H.; Hadfield, E.; et al.: Nasal cancer in woodworkers in the furniture industry, Br. Med. J., 2, 587 596 (1968).

Aitio, A.: Biological monitoring today and tomorrow, Scan. J. Work Environ. Health 20, 46-58 (1994).

Albert, R.A.: Unifying the standard setting process for carcinogens and non-carcinogens, Appl. Occup. Environ. Hyg. 14, 742-747 (1999).

Albertini, R.J.; Nicklas, A.J.; O'Neill, J.P.; et al.: In vivo somatic mutations in humans: measurements and analysis, Annu. Rev. Genet. 24, 305-326 (1990).

American Conference of Governmental Industrial Hygienists: Documentation of the Biological Exposure Indices, 6th ed., CD ROM version, ACGIH®, Cincinnati, OH (1996).

American Conference of Governmental Industrial Hygienists: Death from mercury poisoning results in inquiry, American Conference of Governmental Industrial Hygienists: ACGIH® Today! Newsletter, Issue 5, (1997).

Anwar, W.A.; Gabal, M.S.: Cytogenetic study in workers occupationally exposed to mercury, Mutagenesis, 6, 189-192 (date?).

Arai, F.; Yamamura, Y.; Yoshida, M.; et al.: Blood and urinary levels of metals (Pb, Cr, Cd, Mn, Sb, Co and Cu) in cloisonne workers, Ind. Health 32, 67 78 (1994).

Araki, S.; Aono, H.; Murata, K.: Adjustment of urinary concentration to urinary volume in relation to erythrocyte and plasma concentrations: an evaluation of urinary heavy metals and organic substances, Arch. Environ. Health, 41, 171-77 (1986).

Artuso, M.; Angotzi, G.; Bonassi, S.; et al.: Cytogenetic biomonitoring of styrene-exposed plastic boat builders, Arch. Environ. Contam. Toxicol., 29, 270-274 (1995).

Ashford, N.A.: Policy considerations for human monitoring in the workplace, J. Occup. Med., 28, 563-568 (1986).

Ashford, N.A.; Hattis, D.B.; Spadafor, C.J.; et al.: Monitoring the Worker for Exposure and Disease: Scientific, Legal and Ethical Considerations in the Use of Biomarkers, Baltimore, MD The Johns Hopkins Press (1990).

Autrup, H.: Carcinogen metabolism in cultured human tissues and cells, Carcinogenesis, 11, 707-712(1990).

Baelum, J.; Dossing, M.; Hansen, S.H.; et al.: Toluene metabolism during exposure to varying concentrations combined with exercise, Int. Arch. Occup. Environ. Health, 59, 281 294 (1987).

Baretta, E.D.; Stewart, R.D.; Graff, S.A.: Methods developed for the mass sampling analysis of CO and carboxyhemoglobin in man, Am. Ind. Hyg. Assoc. J. 39, 202 209 (1978).

Bartsch, H.; Caporaso, N.; Coda, M.; et al.: Carcinogen-hemoglobin adducts, urine mutagenicity and metabolic phenotype in active and passive cigarette smokers, J. Natl. Cancer Inst., 82, 1826-1831 (1991).

Baselt, R.C.: Biological Monitoring Methods for Industrial Chemicals, 3rd Ed. PSG Publishing Co., Littleton, MA (1997).

Battey, J.; Moulding, C.; Taub, R.; et al.: The human myc oncogene: consequence of tranlocation to the IgH locus in Burkitt lymphoma, Cell, 34, 779-787 (1983).

Berlin, L.; Wass, U.; Audunsson, G.; et al.: Amines: Possible Causative Agents in the Development of Bronchial Hyperactivity in Workers Manufacturing Polyurethane From Isocyanates. Br. J. Ind. Med., 40, 251 257 (1983).

Bernard, A.; Lauwerys, R.: General principles for biological monitoring of exposures to chemicals, In: Ho, M.H. and Dillon, H.K. (eds.) Biological Monitoring of Exposure to Chemicals, Wiley and Sons, New York, pp. 1-16 (1987).

Bhattacharya, A.; Morgan, R.; Shukla, R.; et al.: Noninvasive estimation of afferent inputs for postural stability under low levels of alcohol, Ann. Biomed. Engin., 15, 533-550 (1987).

Bi, W.; Hayes, R.B.; Feng, P.; et al.: Mortality and cancer incidence of bladder cancer of benzidine-exposed workers in China, Am. J, Ind. Med., 21, 481-489 (1992).

Brandt-Rauf, P.: Oncogene proteins as molecular epidemiologic markers of cancer risk in hazardous waste workers, Occup. Med., 5, 59-65 (1990).

metabolites and urothelial DNA adducts, Carcinogenesis, 18, 981-988 (1997).

Broyde, S.; Hingerty, B.E.; Srinavasan, A.R.: Influence of the carcinogen 4 aminobiphenyl on DNA conformation, Carcinogenesis, 6, 719 725 (1985).

Buckley, T.J.; Lioy, P.J.: An examination of the time course from human dietary exposure to polycyclic aromatic hydrocarbons to urinary elimination of 1-hydroxypyrene, Br. J. Ind. Med., 49, 113-124 (1992).

Butler, M.A.; Ivasaki, M.; Guengerich, F.P.; et al.: Human cytochrome P-450paP-4501A2, the phenacetin-O-deethylase, is primarily responsible for the hepatic 3-demethylation of caffeine and N-oxidation of carcinogenic arylamines, Proc. Natl. Acad. Sci. USA, 86,7696-7700 (1989a).

Butler, M.A.; Guengerich, F.P.; Kadlubar, F.F.: Metabolic oxidation of the carcinogens 4 aminobiphenyl and 4,4' methylenebis-2 chloroaniline by human hepatic microsomes and by purified rat hepatic cytochrome P 450 monooxygenases, Cancer Res., 49, 25 31 (1989b).

Campbell, L.; Jones, A.H.; Wilson, H.K.: Evaluation of occupational exposure to carbon disulfide by blood, exhaled air and urine analysis, Am. J. Ind. Med., 8, 143-153 (1985).

Caporaso, N.E.; Tucker, M.A.; Hoover, R.N.; et al.: Lung cancer and the debrisoquine metabolic phenotype, J. Natl. Cancer Inst., 82, 1264-1272 (1990).

Cardona, A.; Marhuenda, D.; Marti, J.; et al.: Biological monitoring of occupational exposure to n-hexane by measurement of 2,5-hexanedione, Int. Arch. Occup. Environ. Health, 65, 71-73 (1993).

Christakipoulos, A.; Bergmark, E.; Zorcec, V.; et al.: Monitoring occupational exposure to styrene through hemoglobin adducts and metabolites in blood, Scan. J. Work Environ. Health, 19, 255-263 (1993).

Clayson, D.B.; Lawson, T.A.: In: Carcinoma of the Bladder, Connolly, J.G., ed.. Raven Press, New York, 91-100, (1987).

Cocker, J.; Boobis, A.R.; Wilson, H.K.; et al.: Evidence that a â-N-glucuronide of 4,4'-methylenebis(2-chloroaniline)(MbOCA) is a major urinary metabolite in man: implications for biological monitoring, Br. J. Ind. Med., 47, 154-161 (1990).

Commission for the Investigation of Health Hazards of Chemical Compounds in the Work Area: List of MAK and BAT values, 1993, Deutsche Forschungsgemeinschaft, VHC Verlagsgesellschaft, Weinheim, Germany (1993).

Compton, J.E.; Hooper, K.; Smith, M.T.: Human somatic mutation assays as biomarkers of carcinogenesis, Environ. Health Perspect., 94, 135-141 (1991).

Cook, W.: Introduction: Supplemental values of atmospheric and biologic monitoring, in, Biological Monitoring of Exposure to Industrial Chemicals, Fiserova-Bergerova, V. and Ogata, M. (eds) American Conference of Governmental Industrial Hygienists, Cincinnati, OH, pp.xi-xii (1990).

Creech, J.L.; Johnson, M.N.: Angiosarcoma of Liver in the Manufacture of Polyvinyl Chloride, J. Occup. Med. 16, 150 151 (1974).

Decaprio, A.P.: Biomarkers: coming of age for environmental health and risk assessment, Environ. Sci. Technol., 31, 1837-49 (1997).

DeMarini, D.M.; Brooks, L.; Bhatnagar, V.K.; et al.: Urinary mutagenicity as a biomarker in benzidine-exposed workers: Correlation with urinary metabolites and urothelial DNA adducts, Carcinogenesis, 18, 981-988 (1997).

Deutsche Forschungsgemeinschaft: List of MAK and BAT values, 2000, Commission for the Investigation of Health Hazards of Chemical Compounds in the Work Area, Report No. 36, Wiley-VCH, Germany (2000).

DiGiovanni, J.; Rymer, J.; Slaga, T.J.; et al.: Anticarcinogenic and co-carcinogenic effect of benzo[e]pyrene and dibenz[a,c]anthracene on skin tumor initiation by polycyclic hydrocarbons, Carcinogenesis, 3, 371-375 (1982).

Dorian, C.; Gattone, V.H.; Klaassen, C.D.: Renal cadmium deposition and injury as a result of accumulation of cadmium-metallothionein (CdMT) by proximal convoluted tubules- a light microscopic, autoradiographic study with ^{109}CdMT, Toxicol. Appl. Pharmacol., 114, 173-181 (1992).

Dorland, W.A.N.: Dorland's Illustrated Medical Dictionary, W.B. Saunders and Co., New York, p. 1879, (1988).

Drexler, H.; Goen, T.; Angerer, J.; et al.: Carbon disulphide: external and internal exposure to carbon disulphide of workers in the viscose industry, Int. Arch. Occup. Environ. Health, 65, 359-365 (1994).

Drexler, H.; Goen, T.; Angerer, J.: Carbon disulphide: investigations on the uptake of CS_2 and the excretion of its metabolite 2-thiothiazolidine-4-carboxylic acid after occupational exposure, Int. Arch. Occup. Environ. Health, 67, 5-10 (1995).

Droz, P.O.; Berode, M.; Wu, M.M.: Evaluation of concomitant biological and air monitoring results, Appl. Occup. Environ. Hyg, 6, 465-474 (1991).

Drummond, L.; Luck, R.; Safacan, A.; et al.: Biological monitoring of workers exposed to benzene in the coke oven industry, Br. J. Ind. Med., 45, 256-61 (1988).

Dubowski, K.M.; Lukc, J.L.: Measurement of carboxyhemoglobin and carbon monoxide in blood, Ann. Clin. Lab. Sci. 3, 53 65 (1973).

Echeverria, D.; White, R.F.; Sampaio, C.: A Behavioral Evaluation of PCE Exposure in Patients and Dry Cleaners, J. Occup. Environ. Med., 37:667-680 (1995).

Edelman, P.A.: Social factors hindering implementation of biological monitoring in the USA, in, Biological Monitoring of Exposure to Industrial Chemicals, Fiserova-Bergerova, V. and Ogata, M. (eds) American Conference of Governmental Industrial Hygienists, Cincinnati, OH, pp.193-94 (1990).

Elder, H.A.; Finch, C.; McKee, R.W.: Congenital Methemoglobinemia: A clinical and biochemical study of a case, J. Clin. Invest., 39, 265-272 (1949).

Elkins, H.B.: Excretory and biological threshold limits, Am. Ind. Hyg. Assoc. J., 28, 305-317 (1967).

Elovaara, E.; Heikkilae, P.; Pyy, L.; et al.: Significance of dermal and respiratory uptake in creosote workers: Exposure to polycyclic aromatic hydrocarbons and urinary excretion of 1 hydroxypyrene, Occup. Environ. Med., 52, 196 203 (1995).

Engstrom K.; Riihimaki, V.; Laine, A.: Urinary disposition of ethylbenzene and m-xylene in man following separate and combined exposure, Int. Arch. Occup. Environ. Health, 54, 355-63 (1984).

Fenske, R.A.: Dermal exposure assessment techniques, Ann. Occup. Hyg., 37, 687-706 (1993).

Fiserova-Bergerova, V.: Application of toxicokinetic models to establish biological exposure indicators, Ann. Occup. Hyg., 34, 639-651 (1990).

Fiserova-Bergerova, V.: Introduction: Development of Biological Exposure Indices and their Implementation, IN: Fiserova-Bergerova, V., Ogata, M. (eds.) Topics in Biological Monitoring, American Conference of Governmental Industrial Hygienists, Cincinnati, OH, XII-XXIV (1995).

Fishbeck, W.A.; Langner, R.R.; Kociba, R.J.: Elevated urinary phenol levels not related to benzene exposure, Am. Ind. Hyg. Assoc. J., 36, 820-824 (1975).

Folland, D.S.; Schaffner, W.; Grinn, H.E.; et al.: Carbon tetrachloride toxicity potentiated by isopropanol, J. Am. Med. Assoc., 236, 1853-56 (1976).

Foster, P.M.D.; Lloyd, S.C.; Blackburn, D.M.: Comparison of the in vivo and in vitro Testicular Effects Produced by Methoxy , Ethoxy , and n Butoxy Acetic Acids in the Rat, Toxicology 43, 17 30 (1987).

Gao, M.; Levy, L.S.; Faux, S.P.; et al.: Use of molecular epidemiology techniques in a pilot study on workers exposed to chromium, Occup. Environ. Med., 51, 663-668 (1994).

Ghittori, S.; Imbriani, M.; Pezzagno, G.; et al.: The urinary concentration of solvents as a biological indicator of exposure: Proposal for the biological equivalent exposure limit for nine solvents, Am. Ind. Hyg. Assoc. J. 48, 786 790 (1987).

Gonzalez, C.M.; Lora, D.; Vilensky, M.; et al.: Leather tanning workers: chromosomal aberrations in peripheral lymphocytes and micronuclei in exfoliated cells in urine, Mutat. Res., 259, 197-201 (1990).

Greenberg, G.N.; Levine, R.J.: Urinary creatinine excretion is not stable: a new method for assessing urinary toxic substance concentration, J. Occup. Med. 31, 832-838 (1989).

Hansen, A.M.; Omland, O.; Poulsen, O.M.; et al.: Correlation between work process related exposure to polycyclic aromatic hydrocarbons and urinary levels of alpha naphthol, beta naphthylamine and 1 hydroxypyrene in iron foundry workers, Int. Arch. Occup. Environ. Health, 65, 385 94 (1994).

Harris, R.L.: The Public Health Roots of Industrial Hygiene, Am. Ind. Hyg. Assoc. J., 58, 176-180 (1997).

Hemminki, K.; Perera, F.P.; Phillips, D.H.; et al.: Aromatic deoxyribonucleic acid adducts in white blood cells of foundry and coke oven workers, Scand. J. Work Environ. Health, 14, 55-56(1988).

Hemminki, K.; Randerath, K.; Reddy, M.V.; et al.: Postlabeling and immunoassay analysis of polycyclic aromatic hydrocarbons-adducts of deoxyribonucleic acid in white blood cells of foundry workers, Scand. J. Work Environ. Health, 16, 158-162 (1990).

Herbert, R.; Marcus, M.; Wolff, M.S.; et al.: Detection of adducts of deoxyribonucleic acid in white blood cells of roofers by ^{32}P-postlabelling, Scan. J. Work Environ. Health, 16, 135-143 (1990).

Herskowitz, A.; Ishii, N.; Schaumburg, H.: n Hexane Neuropathy A Syndrome Occurring as a Result of Industrial Exposure. N. Engl. J. Med. 285, 82 85 (1971).

Horvath, E.; Pongracz, K.; Rappaport, S.; et al.: ^{32}P-Post-labeling detection of DNA adducts in mononuclear cells of workers occupationally exposed to styrene, Carcinogenesis, 15, 1309-1315, (1994).

Imbriani, M.; Ghittori, S.; Pezzagno, G.; et al.: Toluene and styrene in urine as biological exposure indices, Appl. Ind. Hyg., 1, 172-178 (1986).

Imbriani, M.; Ghittori, S.; Pezzagno, G.; et al.: Urinary excretion of tetrachloroethylene (perchloroethylene) in experimental and occupational exposure, Arch. Environ. Health, 43, 292-298 (1988).

Jarup, L.; Rogenfelt, A.; Elinder, C. G.: Biological Half Time of Cadmium in the Blood of Workers after Cessation of exposure, Scand. J. Work Environ. Health, 9, 327 331 (1983).

Jawetz, E.; Melnick, J.L.; Adelberg, E.A.: Review of Medical Microbiology, 13th edition, Lange Medical Publications, Los Altos, CA, pp. 134-141 (1978).

Jongeneelen, F.J.; Anzion, R.B.M.; Henderson, P.T.: Determination of hydroxylated metabolites of polynuclear hydrocarbons in urine, J. Chromatog., 413, 227-232 (1988).

Jongeneelen, F.J.; Scheepers, P.J.T.; Groenedijk, A.; et al.: Airborne concentrations, skin contamination and urinary metabolite excretion of polycyclic aromatic hydrocarbons among paving workers exposed to coal tar derived road tars, Am. Ind. Hyg. Assoc. J., 49, 600-607 (1988).

Jongeneelen, F.J.: Biological exposure limit for occupational exposure to coal tar pitch volatiles at coke ovens, Int. Arch. Occup. Environ. Health, 63, 511 6 (1992).

Kaderlik, K.R.; Talaska, G.; DeBord, D.G.; et al.: 4,4'-methylene-bis(2-chloraniline)-DNA adduct analysis in human exfoliated urothelial cells by ^{32}P-postlabelling, Cancer Epidemiol. Biomarkers, Prevention, 2, 63-69 (1993).

Kehoe, R.A.: Industrial lead poisoning IN: F.A. Patty (ed.) Industrial Hygiene and Toxicology, 2nd Edition, Volume II, Intersceince Publishers, New York, pp. 941-986 (1963).

Kennedy, G.L. Jr.: Biological monitoring in the American Chemical Industry, in Biological Monitoring of Exposure to Industrial Chemicals, Fiserova-Bergerova, V. and Ogata, M. (eds) American Conference of Governmental Industrial Hygienists, Cincinnati, OH, pp. 63-67 (1990).

King, E.: Occupational hygiene aspects of biological monitoring, Ann. Occup. Hyg., 34, 315-322 (1990).

Kononen, D.W.; Kintner, H.J.; Bivol, K.R.: Air and Blood lead exposures within a large automobile manufacturing workforce, Arch. Environ. Health, 44, 244-251 (1989).

Kotsonis, F.N.; Burdock, G.A.; Flamm, W.G.: Food Toxicology, IN: Klaassen, C.D., ed. Toxicology: The Basic Science of Poisons, McGraw-Hill, New York, pp.909-949 (1996).

Kreiss, K.; Mroz, M.M.; Newman, L.S.; et al.: Matching risk fo beryllium disease and sensitization with median exposures below 2 ìg/m^3, Am. J. Ind. Med., 30, 16-25 (1996).

Langlois, R.G.; Bigbee, W.L.; Kyoizumi, S.; et al.: Evidence for increased somatic cell mutations at the glycophorin A locus in atomic bomb survivors, Science, 236, 445-448 (1994).

Lapare, S.; Tardif, R.; Broduer, J.: Effect of various exposure scenarios on the biological monitoring of organic solvents in alveolar air, Int. Arch. Occup. Environ. Health, 64, 569-580 (1993).

Lappe, M.: Ethical issues in genetic screening for susceptibility to chronic lung disease, J. Occup. Med., 30, 493-501 (1988).

Lauwerys, R.R.: Industrial Chemical Exposure Guidelines for Biological Monitoring, pp. 17 21. Biological Publications, Davis, CA, (1983).

Lauwerys, R.R.; Kivits, A.; Lhoir, M.; et al.: Biological surveillance of workers exposed to dimethylformamide and the influence of skin protection on its percutaneous absorption, Int. Arch. Occup. Environ. Health, 45, 189-203 (1980).

Legator, M.S., Connor, T.H. and Stoeckel, M.: Detection of mutagenic activity of metronidazole and niridazole in body fluids of humans and mice, Science, 188, 1118-1119 (1975).

Lehninger, A.L.: Biochemistry 2nd Edition, Worth publishers, Inc., New York, pp115-116, (1975).

Leung, H-W.; Paustenbach, D.J.: Techniques for estimating the percutaneous absorption of chemicals due to occupational and environmental exposure, Appl. Occup. Environ. Hyg., 9, 187-197, (1994).

Linch, A.L.: Biological Monitoring for Industrial Chemical Exposure, CRC Press, Cleveland, OH (1974).

Loomis, D.P.; Shy, C.M.; Allen, J.W.; et al.: Micronuclei in epithelial cells from sputum of uranium workers, Scan. J, Work Environ. Health, 16, 355-362 (1990).

McArthur, W.R.: Dermal measurements and wipe sampling measurements: a review, Appl. Occup. Environ. Hyg., 7, 599-606 (1992).
McDonald, J.C.; McDonald, A.D.: Epidemiology of mesothelioma from estimated incidence, Prev. Med., 6, 426 446 (1977).
Monster, A.C.; Boersma, G.; Steenweg, H.: Kinetics of 1,1,1-trichloroethane in volunteers: Influence of exposure concentration and work load, Int. Arch. Occup. Environ. Health, 42, 293-301 (1979).
Mosbech, J.; Acheson, E..D.: Nasal Cancer in Furniture Makers in Denmark, Danish Med. Bull., 18, 34 35 (1971).
Mraz, J.; Nohova, H.: Absorption, metabolism and elimination of N,N-dimethylformamide in humans, Int. Arch. Occup. Environ. Health, 64, 84-92 (1992).
Murray, T.H.: Warning: Screening workers for genetic risk, scientific uncertainty and moral questions, Hastings Center Report, Institute of Society, Ethics and the Life Sciences, 360 Broadway, Hastings-on-Hudson, N.Y. USA, 10706, (1983).
Nebert, D.W.: The Ah locus: Genetic differences in toxicity, cancer, mutation and birth defects, Cr. Revs. Toxicol., 20, 153-174 (1989).
Newman, C.: Development of a highly-sensitive transportable method for on-site analysis of benzene in exhaled air, Masters Thesis, Department of Environmental Health, the University of Cincinnati, Cincinnati, OH, 45267-0056, (1997).
"Occupational Health Standards," Code of Federal Regulations, Title 1910, part 20 (1989).
Occupational Safety and Health Administration: Original Lead Standard. Fed. Reg. 43:52952 53014 (1978); Revised Title 29, Code of Federal Regulations, Part 1910.1025. Fed. Reg. 47:51117 51119, Department of Labor, OSHA, Washington, D.C. (1982).
OSHA: The Cadmium Standard, Fed. Reg. 57, 178: 42102-42463, U.S. Government Printing Office, Washington, D.C., Sept. 14, (1992).
Omenn, G.S.: Predictive identification of hypersusceptible individuals, J. Occup. Med., 24,369-374 (1982).
Opdam, J.J.G.; Smolders, J.F.J.: Alveolar sampling and fast kinetics of tetrachloroethene in man, Br. J. Ind. Med., 43, 814-824 (1986).
Opdam, J.J.G.; Smolders, J.F.J.: A method for the retrospective estimation of the individual respiratory intake of a highly and a poorly metabolising (sic) solvent during rest and exercise, Br. J. Ind. Med., 46, 250-260 (1989).
Osario, A.M.; Clapp, D.; Ward, E.; et al.: Biological monitoring on a worker acutely exposed to MBOCA, Am. J. Ind. Med., 18, 577-589 (1990).
Osterman-Golkar, S.; Bergmark, E.: Occupational exposure to ethylene oxide, Scan. J. Work Environ. Health, 14, 372-377 (1988).
Osterman-Golkar, S.; Peltonen, K.; Anttinen-Klemetti, T.; et al.: Haemoglobin adducts as a biomarker of occupational exposure to 1,3-butadiene, Mutagenesis, 11, 145-149 (1996).
Overboe, S.; Haugen, A.; Fjeldstad, P.E.; et al.: Biological monitoring of exposure to polycyclic aromatic hydrocarbon in an electrode paste plant, J. Occup. Med., 36, 303 310 (1994).
Overboe, S.; Haugen, A.; Phillips, D.H.; et al.: Detection of polycyclic aromatic hydrocarbons-DNA adducts in white blood cells from coke oven workers: correlation with job categories, Cancer Res., 52, 1510-1514 (1992).
Pang, Y.C.; Reid, P.E.; Brooks, D.E.: Solubility and distribution of halothane in human blood, Br. J. Anaesth., 52, 851-62 (1980).
Perera, F.P.; Dickey, C.; Santella, R.; et al.: Carcinogen-DNA adducts and gene mutation in foundry workers with low level exposure to polycyclic aromatic hydrocarbons, Carcinogenesis, 15, 2905-2910 (1993).
Petry, T.; Schmid, P.; Schlatter, C.: Airborne exposure to polycyclic aromatic hydrocarbons (PAHs) and urinary excretion of 1 hydroxypyrene of carbon anode plant workers, Ann. Occup. Hyg., 40, 345 57 (1996).
Pezzagno, G.; Imbriani, M.; Ghittori, S.: Urinary elimination of acetone in experimental and occupational exposure, Scand. J. Work Environ. Health, 12, 603 608 (1986).
Petreas, M.; Rappaport, S.M.; Materna, B.L.; et al.: Mixed-exhaled air measurements to assess exposure to tetrachloroethylene in dry cleaners, J. Exp. Anal. Environ. Epidemiol., Suppl. 1, 25-39, (1992).
Piotrowski, J.: Quantitative estimation of aniline absorption through the skin in man, J. Hyg. Epidemiol. Microbiol. Immunol., 1:23 33, (1957).

Que Hee, S.: Biological Monitoring, IN: DiNardi, S.R., ed., The Occupational Environment: Its Evaluation and Control, American Industrial Hygiene Association, Fairfax, VA, USA,262-283 (1997).

Que Hee, S.: Teaching biological monitoring to Physicians and Industrial hygienists in the United States, in Biological Monitoring of Exposure to Industrial Chemicals, Fiserova-Bergerova, V. and Ogata, M. (eds) American Conference of Governmental Industrial Hygienists, Cincinnati, OH, pp. 79-81 (1990).

Que Hee, S.S.: Biological monitoring: An Introduction, Van Nordstrand Rheinhold (1993).

Rappaport, S.M.; Kure, E.; Petreas, M.; et al.: A field method for measuring solvent vapors in exhaled air: application to styrene exposure, Scan. J. Work Environ. Health, 17, 195-204 (1991).

Rappaport, S.M.; Symanski, E.; Yager, J.W.; et al.: The relationship between environmental monitoring and biological markers in exposure assessment, Environ. Health Perspect., 103, 49-54 (1995).

Rappaport, S.M.: Biological monitoring and standard setting in the USA: a critical appraisal, IN: Mutti, A., Chambers, P.L. and Chambers, C.M. (eds.) Mechanisms of Toxicity and Biomarkers to Assess Adverse Effects of Chemicals, 77, 171-182 (1995).

Reardon, D.B.; Bigger, C.A.H.; Dipple, A.: DNA polymerase action on bulky deoxyguanosine and deoxyadenosine adducts, Carcinogenesis, 11, 165-168 (1990).

Reddy, M.V.; Randerath, K.: A comparison of DNA adduct formation in white blood cells and internal organs of mice exposed to benzo[a]pyrene, dibenzo[c,g]carbazole, safrole and cigarette smoke condensate, Mutat. Res. 241, 37-48, (1990).

Rice, C.H.: Retrospective exposure assessment: a review of approaches and directions for the future, IN: Exposure assessment for Epidemiology and Hazard Control, S.M. Rappaport and T.J. Smith (eds.) Lewis Publishers, Chelsea, MI, pp. 185-197 (1991).

Richeldi, L.; Kreiss, K.; Mroz, M.M.; et al.: Interaction of genetic and exposure factors in the prevalence of berylliosis, Am. J. Ind. Med., 32, 337-340 (1997).

Riihimäki, V.; Pfaffli, P.; Savolainen, K.; et al.: Kinetics of m-xylene, Scan. J. Work Environ. Health, 5, 217-231 (1979).

Rosenberg, J.: Concomitant exposures to medications and industrial chemicals, IN: Fiserova-Bergerova, V., Ogata, M. (eds.) Topics in Biological Monitoring, American Conference of Governmental Industrial Hygienists, Cincinnati, OH, 91-99 (1990).

Ross, R.K.; Yuan, J-M.; Yu, M.C.; et al.: Urinary aflatoxin biomarkers and risk of hepatocellular carcinoma, Lancet, 339, 943-946 (1992).

Rothman, N.; Bhatnagar, V.K.; Hayes, R.B.; et al.: The impact of interindividual variability in N-acetyltransferase activity on benzidine urinary metabolites and urothelial DNA adducts in exposed workers, Proc. Natl. Acad. Sci. (USA), 93, 5084-89 (1996).

Rothman, N.; Shields, P.G.; Poirier, M.C.; et al.: The impact of glutathione S transferase M1 and cytochrome P450 1A1 genotypes on white blood cell polycyclic aromatic hydrocarbon DNA adduct levels in humans, Mol. Carcinog., 14, 63 68 (1995).

Sack, D.; Linz, D.; Shukla, R.; et al.: Health status of pesticide applicators: postural stability assessments, IN: Biological Monitoring, S.Que Hee, Ed, Van Nostrand Reinhold, 1196-1202 (1993).

Sakai, T.; Kageyama, H.; Araki, T.; et al.: Biological monitoring of workers exposed to N,N-dimethylformamide by determination of the urinary metabolites N-methylformamide and N-acetyl-S-(N-methylcarbamoyl)cysteine, , Int. Arch. Occu. Environ. Health, 67, 125-129 (1995).

Saltzman, B.E.: Assessment of health effects of fluctuating concentrations using simplified pharmacokinetic algorithms, J. Air Waste Manage. Assoc., 46, 1022-1034 (1996).

Sarto, F.; Tomanin, R.; Giacomelli, L.; et al.: The micronucleus assay in human exfoliated cells of the nose and mouth: application to occupational exposures to chromic acid and ethylene oxide, Mutat. Res., 233, 345-351(1990).

Sato, A.; Endoh, K.; Kaneko, T.: Effects of ethanol on uptake, distribution and elimination of inhaled vapors of organic solvents, IN: Fiserova-Bergerova, V., Ogata, M. (eds.) Topics in Biological Monitoring, American Conference of Governmental Industrial Hygienists, Cincinnati, OH, 155-158 (1990).

Schecter, P.G. (Ed.): The Merck Index, Merck and Company, Rahway, NJ (1978).
Schoket, B.; Phillips, D.H.; Hewer, A.; et al.: ^{32}P Postlabelling detection of aromatic DNA adducts in peripheral blood lymphocytes from aluminum production plant workers, Mutat. Res., 260, 89 98 (1991).
Schulte, P.A.; Sweening, M.H.: Ethical considerations, confidentiality issues, rights of human subjects, and uses of monitoring data in research and regulation, Environ. Health Perspect., 103, 69-74 (1995).
Segerbäck, D.; Kaderlik, K.R.; Talaska, G.; et al.: ^{32}P-postlabelling analysis of DNA adducts of 4,4'-methylenebis(2-chloraniline) in target and nontarget tissues in the dog and their implications for human risk assessment, Carcinogenesis, 14, 2143-2147 (1993).
Seiji, K.; Inoue, O.; Jin, C.; et al.: Dose-excretion relationship in tetrachloroethylene exposed workers and the effect of tetrachloroethylene co-exposure on trichloroethylene metabolism, Am. J. Ind. Med., 16, 675-84 (1989).
Selikoff, I.J.: Perspectives in Preclinical Management of Cancer: Initiation of a Mesothelioma Therapy Research Program. Mount Sinai J. Med., 44:645-647 (1977).
Sherwood, R.J.: Benzene: The Interpretation of Monitoring Results, Ann. Occup. Hyg.,15, 409 421 (1972a).
Sherwood, R.J.: Comparative Methods of Biologic Monitoring of Benzene Exposures (One Man's Elimination of Benzene C[sub 6]H[sub 6]). Proceedings of the 3rd Annual Conference of Environmental Toxicology, pp. 29 52. AMRL TR 72 130 Wright Patterson Air Force Base, Dayton, OH (1972b).
Shibata, E.; Huang, J.; Ono, Y.; et al.: Changes in urinary n-hexane metabolites by co-exposure to various levels of methyl ethyl ketone, and fixed n-hexane levels, Arch. Toxicol., 64, 165-168 (1990).
Skarping, G.; Brorson, T.; Sango, C.: Biological monitoring of isopcyanates and related amines, test chamber exposure of humans to toluene diisocyanates, Int. Arch. Occup. Environ. Health, 63, 83-88 (1991).
Skipper, P.L.; Tannenbaum, S.R.: Molecular dosimetry of aromatic amines in human populations, Environ. Health Perspect., 102, 17 21 (1994).
Snyder, R.; Andrews, L.S.: Toxic effects of solvents and vapors, IN: Klaassen, C.D., ed. Toxicology: The Basic Science of Poisons, McGraw-Hill, New York, pp.737-772 (1996).
Solet, D.; Robins, T.G; Sampaio, C.: Perchloroethylene exposure assessment among dry cleaning workers, A.m Ind. Hyg. Assoc. J., 51, 566-574 (1990).
Sorsa, M.; Peltonen, K.; Anderson, D.; et al.: Assessment of environmental and occupational exposures to 1,3-butadiene and a model for risk estimation of petrochemical emissions, Mutagenesis, 11, 9-17 (1996).
Strickland, P.T.; Kang, D.H.; Bowman, E.D.; et al.: Identification of 1-hydroxypyrene and the major pyrene metabolite in human urine by synchronous fluorescence spectroscopy and gas chromatography-mass spectrometry, Carcinogenesis, 15, 483-87 (1994).
Suruda, A.; Schulte, P.A.; Boeniger, M.; et al.: Cytogenetic effects of formaldehyde exposure in students of mortuary science, Cancer Epidemiol. Biomarkers Preven., 2, 453-460 (1993).
Suskind, R.: Industrial Dematoses, Lecture in Environmental Hygiene and Safety Technology at the University of Cincinnati, Department of Environmental Health, Cincinnati, OH, USA (1997).
Sweet, N.D.; Burroughs, G.E.; Ewers, L.; et al.: A field method for near-real time analysis of perchloroethylene in end-exhaled breath, Appld. Occup. Environ. Hyg., (submitted).
Tabin, C.J.; Bradley, S.M.; Bargmann, C.I.; et al.: Mechanisms of activation of a human oncogene, Nature, 300, 143-149 (1982).
Talaska, G.; Au, W.W.; Ward, J.B. Jr.; et al.: The Correlation between DNA Adducts and Chromosome Aberrations in the Target Organ of benzidine exposed, partially hepatectomized mice. Carcinogenesis, 8, 1895 1903 (1987).
Talaska, G.; Roh, J-H.; Getek, T.: ^{32}P-Postlabelling and mass spectrometric methods for analysis of bulky, polyaromatic carcinogen-DNA adducts in humans, J. Chromatog., 580, 293-323, (1992).
Talaska, G.; Jaeger, M.; Reilman, R.; et al.: Chronic, topical exposure to benzo[a]pyrene induces relatively high steady state levels of DNA adducts in target tissues and alters kinetics of adduct loss, Proc. Natl. Acad. Sci. (USA), 93, 7789-7793, (1996).
Talaska, G.; Cudnick, J.; Jaeger, M.; et al.: Development of non-invasive biomarkers for carcinogen-DNA adduct analysis in occupationally exposed humans: Exposure monitoring of chemical dye workers and monitoring the effectiveness of interventions to dermal exposure of used gasoline engine oils, Toxicol., 110, 1-6. Invited Manuscript (1996).

Talaska, G.; Al-Juburi, A.Z.S.S.; Kadlubar, F.F.: Smoking-related carcinogen-DNA adducts in biopsy samples of human urinary bladder: Identification of N-(deoxyguanosin-8-yl)-4-aminobiphenyl as a major adduct, Proc. Natl. Acad. Sci. (USA), 88, 5350-5354 (1991).

Tardif, R.; Lapare, S.; Plaa, G.I.; et al.: Effect of simultaneous exposure to toluene and xylene on their respective biological exposure indices in humans, Int. Arch. Occup. Environ. Health, 63, 279-84 (1991).

Tardif, R.; Lapare, S.; Charest-Tardif, G.; et al.: Physiologically-based pharmacokinetic modeling of a mixture of toluene and xylene in humans, Risk Anal., 15, 335-342 (1995).

Tardif, R.; Goyal, R.; Brodeur, J.: Assessment of occupational health risk from multiple exposure: review of industrial solvent interactions and implication for biological monitoring of exposure, Toxicol. Ind. Health, 8, 37-52 (1992).

Tarkowski, S.: Excretion of methyl hippuric acid in the urine of persons exposed to xylene and a mixture of solvents, IN: Combined exposure to chemicals (interim document 11) pp. 269-82, Organization mondale de la sante, Copenhagen (1982).

Tornqvist, M.; Almberg, J.; Bergmark, E.; et al.: Ethylene oxide exposures in ethylene exposed fruit store workers, Scan. J. Work Environ. Health, 15, 436-438 (1989).

Tsuruta, H.: Dermal Absorption, IN: Biological Monitoring of Exposure to Industrial Chemicals, Fiserova-Bergerova, V. and Ogata, M. (eds) American Conference of Governmental Industrial Hygienists, Cincinnati, OH, pp.131-36 (1990).

Ulenbelt, P.; Lumens, M.E.G.L.; Geron, H.M.A.; et al.: An inverse lead air to lead blood relation: the impact of air stream helmets, Int. Arch. Occup. Environ. Health, 63, 89-95 (1991).

Underwood, P.M.; Zhou, Q.; Jaeger, M.; et al.: Chronic, topical administration of 4-aminobiphenyl induces tissue-specific DNA adducts in mice, Toxicol. Appl. Pharmacol., 144, 125-131 (1997).

Underwood, P.M.; Zhou, Q.; Jaeger, M.; et al.: Tissue-specific 4-aminobiphenyl DNA adduct formation in the cyp1a2(-\-) knockout mouse, Fund. Appld. Toxicol., 30, s1, 281 (1996).

Van Hemmen, J.J.; Brouwer, D.H.:Assessment of dermal exposure to chemicals, Sci. Total Environ., 168, 131-141, (1995).

Van Rooij, J.G.M.; Bodelier-Bade, M.M.; Jongeneelen, F.J.: Estimation of individual dermal and respiratory uptake of polycyclic aromatic hydrocarbons in 12 coke oven workers, Br. J. Ind. Med., 50, 623-632 (1993a).

VanRooij, J.G.M.; De Roos, J.H.C.; Bodelier-Bade, M.M.; et al.: Absorption of polycyclic aromatic hydrocarbons through the human skin: differences between anatomical sites and individuals, J. Toxicol. Environ. Health, 38, 355-368 (1993b).

Van Rooij, J.G.M.; Van Lieshout, E.M.A.; Bodelier-Bade, M.M.; et al.: Effect of the reduction of skin contamination on the internal dose of creosote workers exposed to polycyclic aromatic hydrocarbons, Scan. J. Work Environ. Health, 19, 200-207 (1993c).

Van Rooij, J.G.M.; Bodelier-Bade, M.M.; Hopmans, P.M.J.; et al.: Reduction of urinary 1-hydroxypyrene excretion in coke oven workers exposed to polycyclic aromatic hydrocarbons due to improved hygienic skin protective measures, Ann. Occup. Hyg., 38, 247-256 (1994).

Vineis, P.; Schulte, P.A.: Scientific and ethical aspects of genetic screening of workers for cancer risk: The case of N-acetyltransferase phenotype, J. Clin. Epidemiol., 48, 189-197 (1995).

Vineis, P.; Bartsch, H.; Caporaso, N.; et al.: Genetically-based N-acetyltransferase metabolic polymorphism and low level environmental exposure to carcinogens, Nature (London), 369, 154-156 (1994).

van de Poll, M.L.M.; van der Hulst, D.A.M.; Tates, A.D.; et al.: Correlation between clastogenicity and promotion activity in liver carcinogenesis by N-hydroxy-2-acetylaminofluorene, N-hydroxy-4'-fluoro-4-acetylaminobiphenyl and N-hydroxy-4-acetylaminobiphenyl, Carcinogenesis, 11, 333-339 (1990).

Volf, J. Sr.; Volf, J. Jr.: Determination of halothane exposure of operating room personnel. Presented at 1st Teisinger's Day on Industrial Toxicology, Prague (as cited in Fiserova-Bergerova (1995).

Wall, K.L.; Gao, W.; Koppele, J.M.; et al.: The liver plays a central role in the mechanism of chemical carcinogenesis due to polycyclic aromatic hydrocarbons, Carcinogenesis, 12, 783-786 (1991).

Wallace, L.A.; Pellizzari, E.; Hartwell, T.; et al.: Concentrations of 20 volatile organic compounds int the air and drinking water of 350 residents of New Jersey compared with concentrations in their exhaled breath, J. Occup. Med., 28, 603-608 (1986).

Wallace, L.A.; Pellizzari, E.D.: Recent advances in measuring exhaled breath ad estimating exposure and body burden for volatile organic compounds(VOCs), Envirn. Health Perspect., 103, 95-97 (1995).
Wallen, M.; Holm, S.; Nordgvist, M.B.: Co-exposure to toluene and p-xylene in man: uptake and elimination, Br. J. Ind. Med., 42, 111-116 (1985).
Ward, J.B. Jr.; Ammenheuser, M.M.; Bechtold, W.E.; et al.: Mutant lymphocyte frequencies in workers at a 1,3-butadiene production plant, Environ Health Perspect., 102, 79-85 (1994).
Waritz, R.S.: Biological indicators of chemical dosage and burden, IN, Cralley, L.J. and Cralley, L.V. Industrial Processes, Wiley and Sons, New York, 257-318 (1979).
Weinberg, R.A.: Oncogenes, antioncogenes and the molecular bases of multistep carcinogenesis, Cancer Res., 49, 3713 3721 (1989).
Wilson, H.K.; Ottley, T.W.: The use of transportable mass spectrometer for the direct measurement of industrial solvents in breath, Biomed. Mass Spectrom., 8, 606-610 (1981).
Wilson, H.K.: Breath Analysis: Physiological basis and sampling techniques, Scan. J. Work Environ. Health, 12, 174-192 (1986).
Yager, J.W.; Eastman, D.A.; Robertson, M.L.; et al.: Characterization of micronuclei induced in human lymphocytes by benzene metabolites, Cancer Res., 50, 393-399 (1990).
Yamada, S.: Polyneuritis in Workers Exposed to n Hexane, Its Cause and Symptoms. Jpn. J. Ind. Health 9, 651 659 (1967).
Yamasaki, E.; Ames, B.N.: Concentration of mutagens from urine by adsorption on the non-polar resin XAD-2: cigarette smokers have mutagenic urine, Proc. Natl. Acad. Sci., 74, 3555-3559 (1977).
Yant, W.P.; Schrenk, H.H.; Sayers, R.R.; et al.: Urine sulfate determinations as a measure of benzene exposure. J. Ind. Hyg. Toxicol., 18, 69 90 (1936).
Yasugi, T.; Kawai, T.; Mizunuma, K.; et al.: Formic acid excretion in comparison with methanol excretion in urine of workers, occupationally exposed to methanol, Int. Arch. Occup. Environ. Health, 64, 329 337 (1992).
Zavon, M.; Hoegg, U.; Bingham, E.: Benzidine exposure as a cause of bladder tumors, Arch. Environ. Health, 27, 1 7 (1973).
Zhou, Q.; Schulte, P.A.; Connally, B.; et al.: Polycyclic aromatic hydrocarbon-DNA adducts in exfoliated urothelial cells and urinary 1-hydroxypyrene in road paving workers, Fund. Appld. Toxicol., 30, s1, 235 (1996).
Zhou, Q.; Talaska, G.; Jaeger, M.; et al.: Benzidine-DNA adducts levels in human peripheral white blood cells correlate significantly with levels in exfoliated urothelial cells, Mutat. Res., 393, 199-205 (1997).

9

DERMAL EXPOSURE ASSESSMENT

Richard A. Fenske, PhD

The purpose of this chapter is to provide occupational hygienists and other public health professionals with information about the recognition, evaluation, and control of dermal exposures. Key elements include the identification of important sources and exposure pathways, and quantitative evaluation of dermal exposures. The field of dermal exposure assessment remains relatively undeveloped when compared with other aspects of occupational and environmental hygiene. While research efforts have increased in this area recently, there remains a paucity of practical guidance and validated measurement techniques. This chapter reviews available assessment methods, noting their strengths and limitations, and discusses current strategies for reducing dermal exposures in the workplace. Boeniger reviews skin anatomy, physiology, and permeation theory, as well as the management of occupational dermatoses, (see Chapter 3).

THE DERMAL ROUTE OF EXPOSURE

The skin is considered to be an organ system, providing the body primary protection from environmental insults. The epidermis has been described as "the cellular investment of the whole organism . . . a continuous stratified epithelial sheet with mucocutaneous junctions at all major body orifices." (Odland and Short, 1968) The ability of chemical substances to permeate intact skin was generally accepted by 1900, but a clear understanding of the barrier properties of the skin did not emerge until the 1950s (Scheuplein and Blank, 1971).

In industrial settings, percutaneous absorption was recognized as a contributor to phenol poisonings as early as 1880 (Deichmann, 1949). In the early 20th century, compounds such as tetra-ethyl lead and aromatic amines were discovered to enter the body through the dermal route (Hamilton, 1920, 1925). A 1951 report of benzidine exposure during chemical manufacturing, in which the importance of dermal uptake was clearly documented and controls were introduced to reduce skin contamination, remains a classic case study for occupational hygiene investigations of multi-route exposures (Meigs et al., 1951).

It has been well documented since the 1950s that workers exposed to pesticides receive most of their dose via the skin (Durham and Wolfe, 1962). Poisonings and deaths from dermal exposure to such chemicals as organophosphorus insecticides are well documented. Regulatory agencies such as the U.S. Environmental Protection Agency (EPA) have focused on reducing skin exposure in their efforts to prevent pesticide-related illness.

Despite these many recognized instances of chemical uptake via the skin, there remains no systematic approach to the evaluation and control of dermal exposures within occupational hygiene science and practice. In the 1960s the relevance of the skin route received formal recognition by professional societies such as the American Conference of Governmental Industrial Hygienists (ACGIH®) through establishment of a "skin notation." The list of chemicals with a skin notation continues to grow, and now constitutes about one quarter of all chemicals with an assigned ACGIH® Threshold Limit Value (TLV®) (ACGIH®, 2000). Yet most current discussions of exposure assessment and sampling strategies in occupational hygiene focus exclusively on measurements of air concentrations. It is encouraging to note that dermal exposure is now accorded recognition at such national meetings as the American Industrial Hygiene Conference and Exposition (AIHce), and that the development of dermal occupational exposure limits is debated in the scientific literature.

RATIONALE FOR DERMAL EXPOSURE ASSESSMENT

Since exposure of the skin to chemical substances can contribute significantly to total dose in a variety of workplace situations, knowledge of dermal exposure is fundamental to hazard evaluation and control (Table 9-1). Measurements of skin exposure can help to define the importance of sources and exposure pathways, evaluate the effectiveness of controls designed to reduce skin exposure, and determine the contribution of the dermal route to total dose. Dermal exposure measurements and biological monitoring can often play complementary roles in defining occupational exposures.

Several rationales can be offered for more careful evaluation of the skin exposure route. First and foremost, current airborne standards are not protective for workers who receive a substantial proportion of their internal dose through dermal absorption. This has been shown for pesticides, polycyclic aromatic hydrocarbons, and many other chemicals. Second, new toxicological findings, coupled with refinements in risk assessment methods, have led to proposed decreases in many airborne exposure limits (NIOSH, 2001). If such reductions take place without parallel efforts to reduce skin exposure, then the relative importance of the dermal route is magnified. In the case of benzene, for instance, the contribution of skin uptake to daily dose among tire workers in the United States was estimated by OSHA to increase from 4% to 30% with a reduction in air concentration from 10 ppm to 1 ppm (Susten et al., 1985). If the benzene standard were further reduced to 0.1 ppm, and skin exposure remained constant, then dermal absorption could be responsible for more than 60% of the remaining internal dose.

Table 9-1
Rationale for dermal exposure assessment

Dermal Exposure Assessment Objectives	Goals	Measurement Requirements	Alternative Methods
Hazard evaluation	Define sources and pathways responsible for dermal exposure	Qualitative or semi-quantitative	Visual observation
Hazard control	Test effectiveness of controls designed to reduce dermal exposure	Quantitative or semi-quantitative	Biological monitoring
Dose estimation	Define contribution of dermal exposure to dose	Quantitative	Biological monitoring

Source: Adapted from Fenske, 1993

Furthermore, the substitution of less volatile compounds in the workplace may result in a greater fraction of dose attributable to dermal absorption (see Chapter 3). Finally, primary prevention of skin disorders and dermal exposures in the workplace has a positive feedback loop effect, in that intact skin provides a defense against many potential assaults, whereas damaged or chemically exposed skin can lead to increased hazards from subsequent exposures.

DETERMINANTS OF DERMAL EXPOSURE

Theoretical Considerations

An effective exposure assessment requires an understanding of source-receptor interactions. In traditional occupational hygiene assessments, source strength is characterized by measuring workplace air concentrations. A typical assessment of air contaminant exposure involves sampling in a worker's breathing zone. This measurement is assumed to capture variability in both the source and receptor, since it determines concentration at the point of entry to the lungs. Thus, the most common mathematical definition of exposure is the product of air contaminant concentration and exposure time. Threshold limit values, permissible exposure limits, and other occupational exposure limits are based on this definition. Dermal exposure assessment requires characterization of concentrations and exposure time, but must also evaluate exposure distribution over the entire skin surface. Worker behavior can thus have a significant impact on the magnitude of exposures, and the assessment requires a careful analysis of job task and worker activities. Figure 9-1 illustrates this difference in air and skin exposure assessment.

In the upper portion of this figure, air concentrations vary over time due to changes in source strength or work task. In the lower portion, dermal exposures are shown to vary over time, but exposures to particular body regions vary as well. In this hypothetical example, hand exposure increases over the workday due to exposure through work gloves, while exposure is reduced periodically by hand washing; arm contamination reflects this trend to a lesser degree (less contamination, but also less thorough washing). If workpants are not changed and penetration occurs, leg exposure gradually increases over the workday; exposure is negligible for

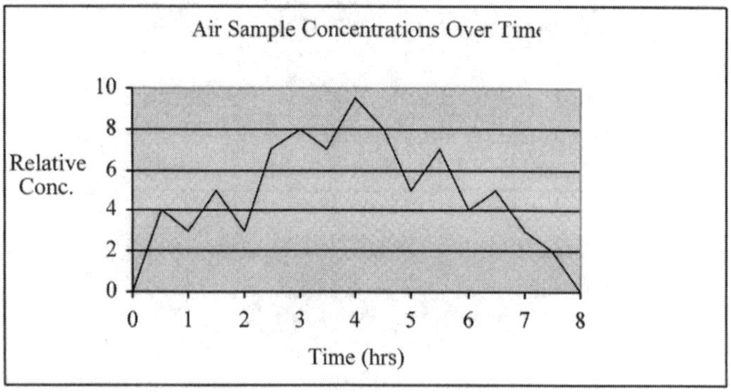

Figure 9-1a. Plot of air concentration over time as a means of monitoring inhalation exposure.

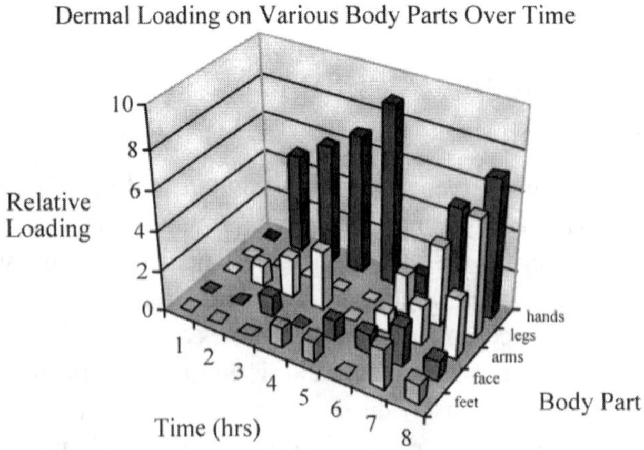

Figure 9-1b. Plot of dermal loading over time by body part as a means of monitoring dermal exposure. Concentration and loading scales are arbitrary.

the feet and face. This three-dimensional nature of dermal exposure makes it impossible to use a single measurement to characterize exposure in most cases. In fact, even the presentation of concentration distributions over specific body regions is a simplification of the dermal exposure process, since exposure can vary within such regions.

A recently proposed dermal exposure model (Zartarian et al., 1997) views exposure as contact between an agent (e.g., a chemical contaminant) and a target (i.e., the skin), as illustrated in Figure 9-2. When a portion of the skin, referred to here as the "contact boundary," intersects with a "contact zone" containing a contaminant, the interaction results in an instantaneous point concentration that can be considered dermal exposure. This definition allows a more rigorous mathematical treatment of dermal exposure, and is consistent with mathematical expressions of exposure for the inhalation route. A conceptual model of occupational dermal exposure has been proposed recently by European researchers (Schneider et al., 1999).

Physical States and Exposure Pathways

The physical state of the material of concern, whether liquid, vapor, or solid, will influence the opportunity for worker-chemical contact. Dermal exposure pathways include (1) direct contact with bulk (solid) contaminants, (2) splashes and immersion in liquids, (3) skin deposition of liquid or solid aerosols and contact with vapors, and (4) contact with contaminated items and surfaces. Multiple physical states and exposure pathways may exist for a given set of workplace conditions for a particular chemical compound. The conceptual model pro-

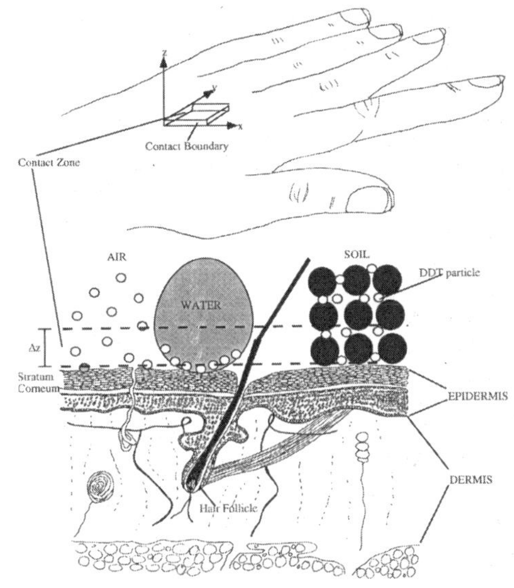

Figure 9-2. Contact boundary. (from Zartarian, 1997).

posed by Schneider and colleagues (1999) delineates these various pathways. A recent study of the rubber manufacturing industry provides a particularly useful analysis of dermal exposure pathways (Vermeulen et al., 2000).

A simplified schematic of the skin is presented in Figure 9-3 to indicate the various processes that can occur at the skin surface. Chemical substances that contact the skin may permeate the stratum corneum and enter the epidermis, and ultimately the dermis. The skin may serve as a reservoir for chemicals, or as a site of metabolism; chemicals that move into the capillary-rich dermis can enter systemic circulation, and be transported throughout the body. Alternatively, chemicals may be removed from the surface of the skin by perspiration, volatilization, mechanical action, or washing. The dynamic nature of skin-contaminant contact makes the assessment of dermal exposure particularly challenging.

DIRECT CONTACT WITH BULK CONTAMINANTS

Many workplace activities involve handling materials in bulk. Materials that need to be mixed or transferred, for example, may reach the workplace in bags or barrels. During handling such materials can vaporize or aerosolize, but they may also be contacted directly. For dry materials, the exposure pathway is not unlike contact with contaminated soils. Research in this area has been aimed at deriving soil adherence factors and soil contact rates (see Chapter 11) to estimate skin exposure (U.S. EPA, 1992a, 1997; Kissel et al., 1996a, 1996b, 1998). Such values are not normally available for workplace materials.

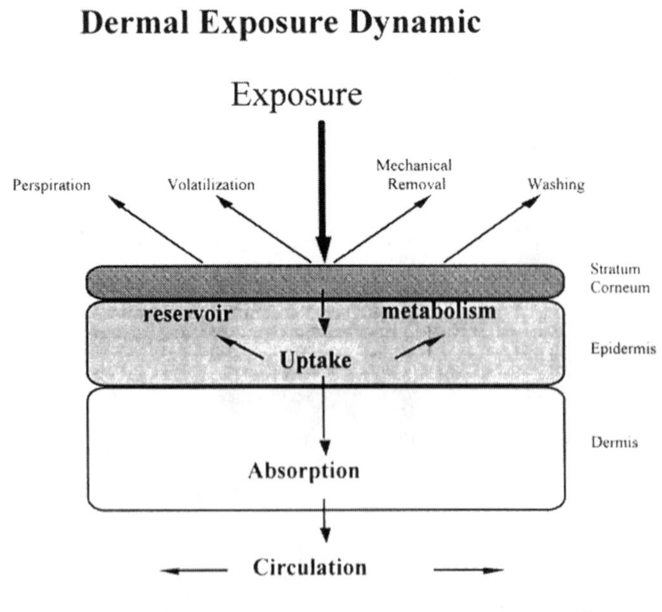

Figure 9-3. Schematic of the dermal exposure dynamic.

IMMERSION AND SPLASHES

Immersion of the skin in liquids is perhaps the most hazardous exposure pathway. Immersion enhances exposure and absorption because there is normally complete surface contact of the liquid with the exposed skin, and an essentially infinite source of contaminant available for uptake. In some cases, solvent carriers can increase absorption. Splashes, spills to skin, and contact with wet surfaces can all be considered variations of the immersion pathway. In the evaluation of liquid contact hazards, chemical protective clothing cannot be considered *a priori* to be an impermeable barrier, as discussed in sections that follow.

DEPOSITION AND VAPOR CONTACT

Airborne material, including sprays, other aerosols, and vapor, may come into contact with skin. Potentially important occupational exposures by this route include solvents in paint spray operations, metalworking fluids, spray of isocyanate-containing urethanes and or other allergens. Spray contact is a major pathway for pesticide exposure in outdoor agricultural and greenhouse environments. As with other exposures to liquids, the carrier solvent may enhance exposures to substances in it (e.g., allergenic metals in metalworking fluids).

Vapor uptake by the skin has often been considered to be a minor contributor to exposure compared to respiratory uptake, but the validity of this assumption remains in question. The logic that has been applied is that in the absence of effective respiratory protection, the inhalation route of vapor uptake will generally predominate because the lungs present far more surface area and a much more permeable barrier to most substances than does the skin. However, transdermal absorption of vapor has been demonstrated to be competitive with respiratory uptake for some chemicals, but for most chemicals, no investigations have been conducted on this question. One compound that has received considerable study in this regard is 2-butoxyethanol (Johanson and Boman, 1991; Corley et al., 1994; 1997). (For further discussion, see Chapter 11).

CONTACT WITH CONTAMINATED SURFACES AND ITEMS

Films, powders, particles, and other solid, viscous, or liquid materials may be present on the surfaces contacted in the workplace. The wide variety of contaminated surfaces that may be contacted include work materials, instrument handles, tools, shop rags, work shoes and clothing, and chemically treated foliage. Repeated contact with a contaminated surface results in a complex pattern of dermal loading and unloading. This process is discussed in more detail in the next section. The long-recognized dermal exposure to pesticides experienced in agricultural hand harvesting is an important example of surface-to-skin contact with contaminants (Popendorf and Leffingwell, 1982). Workers' dermal contact with just-sprayed, wet foliage can pose an acute systemic hazard, whereas contact with nearly the same quantity of active ingredient after sprays have dried is far less hazardous.

Chlorophenols have also been monitored on workplace surfaces, and provide a good example for skin exposure. Pentachlorophenol and tetrachlorophenol are high molecular weight, low volatility compounds. Chlorophenols in timber mill treatments have also been shown to pose a hazard that is primarily dermal, as air concentrations are low, but urinary concentrations are high (Fenske et al., 1987; Bentley et al., 1989).

Behavioral Aspects of Dermal Exposure

Workplace behaviors can have a significant impact on skin exposure. Some of these behaviors may be associated with requirements of the job or task, while others involve personal hygiene and the extent to which workers choose to engage in risk-taking activities. Five topics will be addressed briefly here: skin contact and transfer, duration of contact, use of personal protective equipment, washing, and the skin-to-ingestion pathway.

SKIN CONTACT AND TRANSFER

A central determinant of dermal exposure is the extent to which work tasks require contact between skin and workplace contaminants. A number of variables influence the importance of work activity for skin exposure, including the proportion of the body's skin surface area in direct contact with contaminants, and the duration of contact. The transfer of material from a contaminated surface to the skin is particularly challenging to quantify. Contaminant transfer is dependent on factors including contact motion and force at the point of interaction, surface concentration of contaminant, efficiency of contaminant transfer to the skin, and the presence of sweat, oil, and other substances on the skin surface (Brouwer et al., 1999). Repeated surface contact can result in contaminant loading and unloading from skin in ways that have not been characterized empirically (Zartarian et al., 1997).

These concepts have been developed most fully in the study of occupational pesticide exposure. Researchers characterizing dermal contact with pesticide-treated plants in agricultural settings have observed that faster workers are more exposed (Fenske et al., 1989). In addition, the total exposure varies by the amount of skin surface that is in contact with treated foliage, which may be only the hands, or may include the arms, torso, and legs.

Dermal transfer coefficients for agricultural pesticides in harvesting operations have been calculated from empirical data for a variety of work tasks (Nigg et al., 1984; Zweig et al., 1985). The coefficient is constructed from a measurement of surface concentration, exposure, and exposure duration. In this case, surface concentration is referred to as the "dislodgeable foliar residue." This value is defined as the amount of pesticides that can theoretically be transferred to the worker during foliar contact. Dermal exposure is measured by one of several available methods (see the following section on personal monitoring of dermal exposure). Exposure duration is considered to be the length of the work shift. This relationship is contained in the following equation:

$$TC = DE/DFR \times T$$

where:

TC	=	transfer coefficient, in foliar area per unit time (e.g., cm^2/hr)
DE	=	dermal exposure, in mass units (e.g., μg)
DFR	=	dislodgeable foliar residue, in mass per unit of foliar area (e.g., $\mu g/cm^2$)
T	=	time, or length of the workshift (e.g., hr)

This empirically calculated transfer coefficient does not specifically characterize surface-to-skin transfer as separate from the activity component of contaminant transfer, since the so-called transfer coefficient incorporates both transfer and activity.

DURATION OF CONTACT

The duration of contact with contaminated surfaces can be an important predictor of total dermal exposure. In most exposure scenarios, it is assumed that dermal exposure accrues linearly, although, in many cases, such exposures are intermittent over the course of the work shift. Other hazards, such as splashes, may be conceptualized in terms of probability over time, and thus, the likelihood of their occurrence increases with time. It should also be considered that, unlike inhalation exposure, the absorption of contaminant from the skin following workplace dermal exposure may continue after the work shift if material is retained on the skin or clothing (Kissel and Fenske, 2000).

PERSONAL PROTECTIVE EQUIPMENT

The selection, use, and maintenance of gloves, coveralls, aprons, face shields, head and foot coverings, and other chemical protective clothing (CPC) and equipment influence the amount of contaminant that is available for dermal absorption, although not always in the way that might be expected. Most combinations of chemical protective clothing and a particular workplace chemical have a permeation versus time curve such that the amount of chemical passing through the clothing material is relatively negligible for some period after exposure begins, and then increases rapidly. The extent of "low" permeation and its time period may vary over several orders of magnitude depending on the materials. Garment openings can have a major influence on both the total quantity and the distribution of dermal exposure (Fenske, 1988a). Large neck and sleeve openings, for example, can compromise the real effectiveness of chemical protective coveralls during agricultural spraying operations. Ordinary clothing and items such as cotton gloves can, in some cases, become both a source and a means of occlusion for a contaminant on skin, thus actually increasing total dermal absorption.

WASHING

Washing can remove various contaminants from the skin and reduce the total time-integrated contaminant loading on skin. Nonetheless, the effectiveness of washing should be carefully considered within the context of the exposure, for example, compounds that tend to bind to or to rapidly penetrate the skin. Laboratory evaluations of handwash removal efficiency of pesticides have shown that over one-half of on-skin chlorpyrifos and one-third of on-skin captan remained sorbed to skin after vigorous hand washing (Fenske and Lu, 1994; Fenske et al., 1998). In addition, skin cleansers that are abrasive, drying, or chapping to the skin should be avoided.

SKIN-TO-INGESTION PATHWAY

Materials deposited on skin may present an occupational hazard due to incidental ingestion of skin contaminants through eating, drinking, smoking, nail-biting, or any other hand-to-mouth contact, even when the skin contact and transdermal hazards are themselves minimal (Ulenbelt et al., 1990). This pathway is recognized in occupational regulations such as the U.S. lead and cadmium standards (OSHA, 2001a; 2001b), which require that employers provide washing facilities and lunch areas that are separate from work areas. These administrative controls are designed to prevent incidental ingestion of these highly toxic metals while eating. Any film, dust, aerosol, or bulk material that is hazardous via ingestion is potentially of concern for this pathway, including heavy metals, radionuclides, antineoplastic pharmaceuticals, pesticides, PCBs, and contaminated soil. Recognition of hazards from this pathway must simultaneously involve identification of those contaminants that are toxic by absorption through the gastrointestinal tract, and that also have the potential for skin contact and hand-to-mouth transfer.

PERSONAL MONITORING OF DERMAL EXPOSURE

Techniques for measurement of chemical exposure to the skin have been available since the 1950s, and method development has increased in the past 20 years as the importance of dermal exposure has been recognized. Several articles have reviewed these methods (McArthur, 1992; Fenske, 1993). One important aspect of skin exposure measurements is area of the skin surface exposed. Skin surface area is difficult to measure directly, and so is usually estimated from models using height and weight parameters. The U.S. Environmental Protection Agency conducted an extensive review of the literature on this subject for their Exposure Factors Handbook (U.S. EPA, 1997), and have published these data recently. Table 9-2 is drawn from the EPA analysis, and provides surface area estimates for various body regions. Table 9-2a provides values for men, and Table 9-2b provides values for women.

Surrogate Skin Techniques

These methods involve placing a chemical collection medium against the skin and subsequently analyzing it for chemical content. Two general approaches have been used: patch samplers covering small skin surface areas, and garment samplers covering entire anatomical regions. These methods have been reviewed recently (Soutar et al., 2000).

Table 9-2a
Surface area and percentage of total body surface area by body part for adult men

Body Part	N[a]	Surface Area (cm^2) Mean ± sd[b]	5th percentile	95th percentile	Percentage of Total Body Surface Area Mean ± sd[b]	5th percentile	95th percentile
Head	32	1180 ± 160	900	1610	7.8 ± 1.0	6.1	10.6
Trunk	32	5690 ± 1040	3060	8930	35.9 ± 2.1	30.5	41.4
Upper extremities	48	3190 ± 461	1690	4290	18.8 ± 1.1	16.4	21.0
Arms	32	2280 ± 374	1090	2920	14.1 ± 0.9	12.5	15.5
Upper arms	6	1430 ± 143	1220	1560	7.4 ± 0.5	6.7	8.1
Forearms	6	1140 ± 127	945	1360	5.9 ± 0.3	5.4	6.3
Hands	32	840 ± 127	596	1130	5.2 ± 0.5	4.6	7.0
Lower extremities	48	6360 ± 994	2830	8680	37.5 ± 1.9	33.3	41.2
Legs	32	5050 ± 885	2210	6560	31.2 ± 1.6	26.1	33.4
Thighs	32	1980 ± 1470	1280	4030	18.4 ± 1.2	15.2	20.2
Lower legs	32	2070 ± 379	930	2960	12.8 ± 1.0	11.0	15.8
Feet	32	1120 ± 177	611	1560	7.0 ± 0.5	6.0	7.9
TOTAL		19400[c] ± 37.4[d]	16600	22800			

[a]number of observations
[b]standard deviation
[c]median
[d]standard error
Source: Adapted from U.S. EPA, 1997

Table 9-2b
Surface area and percentage of total body surface area by body part for adult women

		Surface Area (cm^2)			Percentage of Total Body Surface Area		
Body Part	**N[a]**	**Mean ± sd[b]**	**5th percentile**	**95th percentile**	**Mean ± sd[b]**	**5th percentile**	**95th percentile**
Head	57	1100 ± 62.5	953	1270	71.1± 0.6	5.6	8.1
Trunk	57	5420 ± 712	4370	8670	34.8 ± 1.9	32.8	41.7
Upper extremities	57	2760 ± 241	2150	3330	17.9 ± 0.9	15.6	19.9
Arms	13	2101 ± 129	1930	2350	14.0 ± 0.6	12.4	14.8
Upper arms	--	----	----	----	----	----	----
Forearms	--	----	----	----	----	----	----
Hands	12	746 ± 51.0	639	824	5.1 ± 0.3	4.4	5.4
Lower extremities	57	6260 ± 675	4920	8090	40.3 ± 1.6	36.0	43.2
Legs	13	4880 ± 515	4230	5850	32.4 ± 1.6	29.8	35.3
Thighs	13	2580 ± 333	2580	3600	19.5 ± 1.1	18.0	21.7
Lower legs	13	1940 ± 24.0	1650	2290	12.8 ± 1.0	11.4	14.9
Feet	13	975 ± 90.3	834	1150	6.5 ± 0.3	6.0	7.0
TOTAL		16900[c] ± 37.4[d]	14500	20900			

[a]number of observations
[b]standard deviation
[c]median
[d]standard error
Source: Adapted from U.S. EPA, 1997

Patch samplers arose initially in the context of hazard evaluation for acute intoxications among users of organophosphorus insecticides (Durham and Wolfe, 1962). Patches provided a first approximation of risk to health, as well as demonstrating the predominant role of dermal exposure under these conditions. The technique has since become recognized as a standard quantitative exposure method for pesticides (WHO, 1986; U.S. EPA, 1987), and its use has been extended to assessment of such occupational hazards as polycyclic aromatic hydrocarbons and dichlorobenzidene (Jongeneelen et al., 1988; London et al., 1989).

Its validity as a means of assessing exposure rests on one of two critical assumptions. Typically, patch sampling assumes uniform exposure; i.e., deposition on the patch is representative of deposition over that part of the body. Alternatively, this method assumes worst-case exposure; i.e., the patch has been located at the point of highest exposure potential for that part of the body. These assumptions have not been investigated systematically, so that the accuracy of exposure estimates derived from this technique is open to question (Franklin et al., 1981; Fenske, 1990).

Patches are essentially spot or grab samples of the skin, whereby dermal exposure is calculated by extrapolating the patch loading level to the surface area of an entire anatomical region. Table 9-3 indicates the percent of the skin covered by typical 25 cm^2 samplers for three anatomical regions.

In the case of the chest and stomach, for example, two patch samplers represent less than 1% of the skin surface, a ratio of sampled surface area to total surface area analogous to air sampling for four minutes to characterize an eight hour exposure. The likelihood that such a limited sample will be representative of true exposure would appear low. Furthermore, expo-

Table 9-3
Percent of body surface areas represented by dermal patch samplers for specific regions — Values are compared to the length of sampling time represented by the corresponding percentage of an eight hour air sample*

Body Region	Patch Surface Area	Percentage as Portion of an 8 hour air sample
Chest & Stomach	0.73 %	4 minutes
Thighs	1.4 %	7 minutes
Forearms	4.3 %	21 minutes

* *Source*: Fenske, 1993

sure patterns for different work activities can differ systematically, and patch samplers placed at one site can produce misleading exposure estimates. A study of pesticide handlers illustrates differences in exposure estimates derived from wearing a head patch during pesticide mixing and application (Fenske, 1990). The normal patch location (front) overestimated the average exposure during application by 35%, but underestimated average exposure during mixing by 75%. These problems suggest that patch sampling cannot inherently provide accurate estimates of dermal exposures.

If these limitations are borne in mind, patch sampling can serve as a simple and cost-effective method for hazard evaluation and control through comparative studies, and within a given work task can provide a useful index of exposure. Machado Neto et al. (1992), for example, demonstrated a significant reduction in exposure by increasing the distance between workers and spray nozzles during pesticide applications in staked tomato crops. Similarly, Methner and Fenske (1994a) reported that unidirectional ventilation in greenhouses reduced dermal exposure substantially, and confirmed patch sampling data with fluorescent tracer/video imaging analysis. Numerous investigators have quantified protective clothing penetration by placing patch samplers inside and outside of the fabric barrier (Gold et al., 1982; Nigg et al., 1986; 1992; Keeble et al., 1988; Perkins and Knight, 1989).

Limitations inherent in patch sampling can be overcome by sampling entire anatomical regions with garments. Absorbent gloves (e.g., cotton) have been used frequently to estimate hand exposure during contact with equipment or materials, and in harvesting of crops treated with pesticides (Davis et al., 1983; Fenske et al., 1989; Brouwer et al., 1992). Whole-body garment samplers have been used most frequently to assess pesticide exposure (Bonsall, 1985; Abbott et al., 1987), and have been proposed recently as a standard method for measuring the exposure of users (Chester, 1993), and worker exposure to antifungal products during commercial wood treatment (Teschke et al., 1992; 1994). In a study aimed at estimating children's exposure to pesticides indoors, subjects were asked to wear whole body garments while doing an exercise routine on a treated carpet (Ross et al., 1990).

The major assumption underlying all surrogate skin techniques is that the collection medium captures and retains chemicals in a manner similar to that of skin, but none of the garment samplers in common use has been systematically tested for retention efficiency. Concerns have been raised regarding potential overestimation of exposure, since sampling media may be selected primarily for their absorbent properties. Several studies found that cotton gloves collected more of azinphosmethyl residues than did handwashing, but the removal efficiency of the handwash technique was not known (Davis et al., 1983; Fenske et al., 1999). Fenske et al. (1989) found that differences between glove and handwash measurements of captan residues decreased with increasing sample time, as indicated in Table 9-4. Thus, the representativeness of glove and other garment samplers remains an open question.

Garment sampling has several advantages: distributional assumptions are not required, a standard sampling approach can be applied to virtually all body regions (e.g., hands, feet, arms, legs), and the sampling of work activities with different skin exposure patterns is comparable.

Table 9-4
Glove and hand wash derived exposure rate estimates by sampling time interval during harvesting of captan-treated strawberries[1]

Time (hr)	Mean Glove Exposure Rate (mg/hr)	CV (%)	Mean Handwash Exposure Rate (mg/hr)	CV (%)	Glove to Handwash Ratio
0.5	43.6[2]	64	18.0	50	2.4[4]
1.0	32.5	55	15.5	52	2.4[4]
1.5	23.2[3]	34	16.6	48	1.4
3.0	21.0[3]	43	14.7	61	1.4

[1] 8 samples for each time period and sampling method
[2,3] glove rate at 0.5 hours significantly different than rate at 1.5 and 3.0 hours (ANOVA, $p<.05$)
[4] glove rates significantly different than handwash rates (ANOVA, $p<.05$)
Source: Fenske et al., 1989

Several disadvantages can also be noted: putting garments on and taking them off can be cumbersome, extraction requires large volumes of solvents, and garments are susceptible to permeation and penetration, and may require changing during the workshift. A major obstacle to widespread use is the lack of standard garments for sampling. In studies to date, garments have been selected for convenience rather than according to scientific criteria, and garment characteristics have not been well described. Fiber content, construction, finish, weight and thickness should be specified. Ideally, garments employed as dermal samplers would be pre-tested for their ability to absorb and retain the particular chemical under study. Volatile chemical sampling presents special challenges, requiring charcoal gloves in some cases (Perkins and Rainey, 1997).

Chemical Removal Techniques

Chemicals deposited on skin can be removed by washing or wiping. A review of current methods has been published recently (Brouwer et al., 2000). Water/surfactant mixes or water/alcohol wash solutions are generally used only to assess hand exposure, while wiping techniques can, in theory, be applied to larger and more heterogeneous skin surfaces. Handwash sampling procedures can normally be standardized to ensure that they are operator-independent, so that studies can be compared (Davis, 1980), whereas skin-wiping relies on procedures that are inherently operator-dependent, and thus include an unknown component of variability.

Keenan and Cole (1982) demonstrated the feasibility of washing skin contaminated with coal liquids with a spray delivery system; however, the technique was designed to sample only 5 cm^2 of skin, and thus does not provide an integrated measure of dermal exposure over large skin surfaces.

Measurements of chemical removal represent only what can be removed from the skin at the time of sampling rather than the actual skin loading. Yet investigations employing these techniques have often reported measured values as "exposure," on an implicit assumption that removal was 100%. Fenske and Lu (1994) proposed a standard laboratory procedure for assessing removal efficiency of handwash techniques. Evaluation of several handwash techniques indicated a removal efficiency for the pesticide chlorpyrifos of less than 50% when skin was washed immediately after exposure, and less than 25% one hour post-exposure. Furthermore, removal efficiency decreased with decreased skin loading. These findings suggest that exposure data estimated by chemical removal techniques are likely to be difficult to interpret, and will require appropriate laboratory removal efficiency studies as a part of quality assurance.

A recent occupational exposure study explored dermal exposure sampling systematically (Brouwer et al., 1998). The study examined the correspondence of glove and handwash measurements of 4,4'-methylene dianiline (MDA) exposure during plastic pipe manufacturing. Measurements were conducted over an entire week for four winders and two liners. Figure 9-4 presents a scatter plot of these data.

Each worker had from 18 to 23 discrete measurements of each type (glove and handwash); plotted values are geometric means. Five of the six worker measurements fell close to the unity line, but for one worker the handwash exposure estimate was three times greater than the glove estimate.

A comparison of dermal exposure and biological monitoring values was also made for two workers exposed to 4,4'-methylene dianiline (MDA) while conducting the same task during plastic pipe manufacturing (Figure 9-5). Glove and handwash values are geometric means based on 18-20 measurements for each worker and each sampling method. MDA urine values are arithmetic means of seven complete 24-hour urine samples. Note that glove and handwash measurements differ by a factor of two for worker 2, and dermal exposure estimates for the two workers are dissimilar, regardless of sampling method, yet MDA excretion is virtually identical. This study illustrates the complexity of dermal exposure measurement, and the value of biological monitoring as a complementary assessment method.

Fluorescent Tracer Techniques

Dermal exposure can be quantified directly and non-invasively by measuring fluorescent materials. The natural fluorescence of polycyclic aromatic hydrocarbons has been evaluated with a small scanning device (luminoscope), allowing identificaton of skin contamination patterns (Schuresko, 1980; Vo-Dinh and Gammage, 1981; Vo-Dinh, 1987). The luminoscope can provide spot measurements of contamination and allows rapid identification of skin surfaces

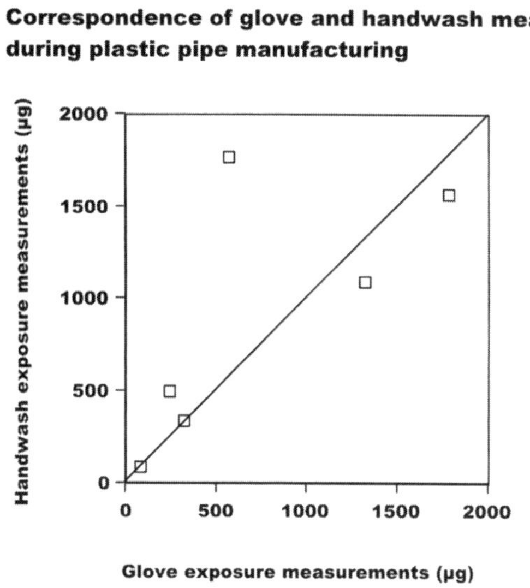

Figure 9-4. Correspondence of glove and handwash measurements of 4,4'-methylene dianiline (MDA) exposure during plastic pipe manufacturing. Measurements were conducted over an entire work week for four winders and two liners. Each worker had from 18 to 23 discrete measurements of each type; plotted values are geometric means. Five of the six worker measurements fell close to the unity line, but for one worker, the handwash exposure estimate was three times greater than the glove estimate (from Brouwer et al., 1998, Table 1).

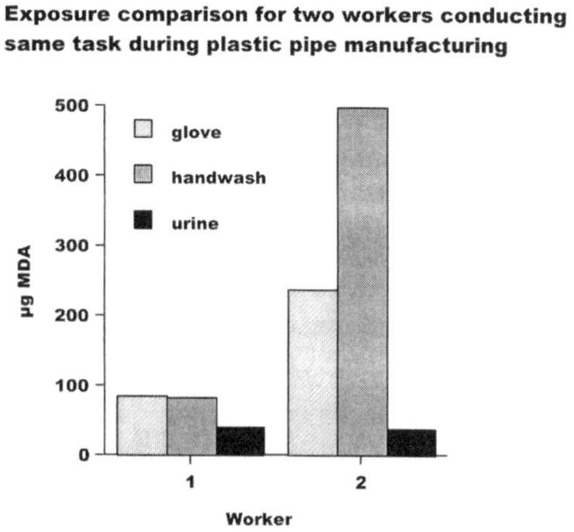

Figure 9-5. Correspondence of glove and handwash measurements of 4,4'-methylene dianiline (MDA) exposure during plastic pipe manufacturing. Measurements were conducted over an entire work week for four winders and two liners. Each worker had from 18 to 23 discrete measurements of each type; plotted values are geometric means. Five of the six worker measurements fell close to the unity line, but for one worker, the handwash exposure estimate was three times greater than the glove estimate (from Brouwer et al., 1998, Table 1).

requiring decontamination. This instrument was developed to monitor worker exposure to synthetic fuels in the United States, but has not found widespread usage since the decline of this industry.

FLUORESCENT WHITENING AGENTS AS TRACERS OF EXPOSURE

Visualization of skin exposure patterns with fluorescent tracers is a relatively new assessment method. Compounds known as fluorescent whitening agents (FWAs) were first reported to be useful tools for characterizing skin deposition of pesticide sprays in the early 1980s (Franklin et al., 1981; Fenske et al., 1985). FWAs, also known as optical brighteners, were developed in the 1950s as additives to various commercial processes (Fenske et al., 1986b). They absorb longwave ultraviolet light energy (UV-A, 340-400 nm) and emit visible light (400-440 nm).

IMAGING ANALYSIS

Quantification of fluorescent whitening agents on human skin was first achieved with a prototype video imaging technique for assessing exposure, or VITAE system (Fenske et al., 1985; 1986a). A second generation VITAE system was assembled in the late 1980s to take advantage of advances in computer-based imaging analysis technology (Fenske and Birnbaum, 1997). Figure 9-6 illustrates the design and layout of this system.

This system has been deployed in a number of field investigations, including evaluation of chemical protective clothing during airblast applications of ethion in citrus orchards (U.S. EPA, 1993; Fenske et al., 2002), analysis of factors affecting pesticide exposure during greenhouse applications (Methner and Fenske, 1994a, 1994b), and measurement of skin exposure among children contacting chemically treated lawns (Black, 1993).

The technique has been adopted by several other laboratories, and has been employed to estimate dermal exposures among golf course workers (Kross et al., 1996), and greenhouse applicators (Archibald et al., 1995; Bierman et al., 1998). The Health and Safety Executive in the United Kingdom has produced a novel lighting system to improve quantitative accuracy (Roff, 1994; 1997). The dodecahedral lighting system, illustrated in Figure 9-7, illuminates all skin surfaces evenly, reducing the number of correction factors that must be employed in the quantification procedures. A review of fluorescence techniques for dermal exposure assessment was published recently (Cherrie et al., 2000).

FLUORESCENT TRACER APPLICATIONS

Video imaging techniques can produce exposure estimates over virtually the entire body (Fenske and Birnbaum, 1997). This requires pre- and post-exposure images of skin surfaces under longwave ultraviolet illumination, development of a standard curve relating dermal fluorescence to skin-deposited tracer, and chemical residue sampling to quantify the relationship between the tracer and the chemical substance of interest as they are deposited on skin.

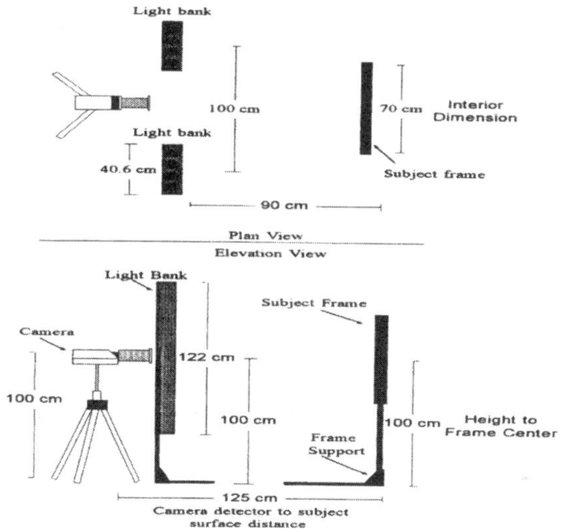

Figure 9-6. Comparison of dermal exposure and biological monitoring values for two workers exposed to 4,4'-methylene dianiline (MDA) while conducting the same task during plastic pipe manufacturing. Glove and handwash values are geometric means based on 18-20 measurements for each worker and each sampling method. MDA Urine values are arithmetic means of 7 complete 24 hour urine samples. Note that glove and handwash measurements differ by a factor of two for worker 2, and dermal exposure estimates for the two workers are dissimilar, regardless of sampling method, yet MDA excretion is virtually identical (from Brouwer et al., 1998, Tables 1 and 2).

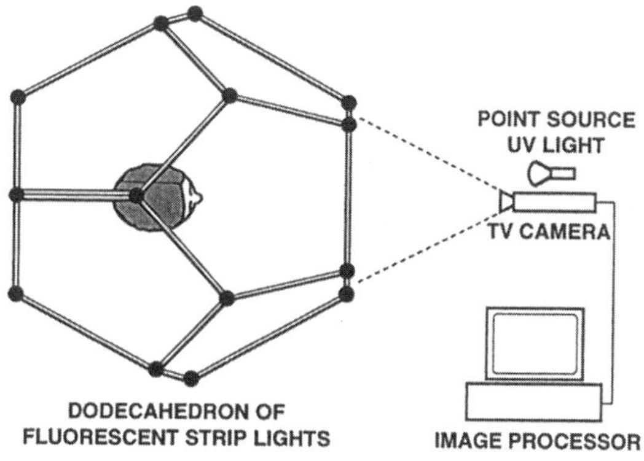

Figure 9-7. FIVES.

Ideally, this method can provide improved accuracy over other methods (e.g., the patch technique) in dermal exposure assessment, since it measures actual skin loading levels, requires no assumptions regarding exposure distribution over skin surfaces, and can identify hitherto unrecognized exposure pathways. In practice, however, it has several important limitations: First, use of a tracer requires the introduction of a foreign substance into the production system. In most agricultural settings this has not posed a problem, since the tracer compounds used to date do not appear to be phytotoxic or otherwise incompatible with pesticide use. However, for many industrial processes, addition of a foreign chemical substance could not be tolerated without extensive pre-testing. Second, the relative transfer of the tracer and chemical substance of interest must be demonstrated during field investigations. The need for ancillary studies involving chemical extraction and analysis to some extent undermines the primary advantages of video imaging; i.e., samples are collected and analyzed rapidly, and stored in digital form. Third, additional quality assurance steps may be required during field studies, including determination of the instrument's quantitative range, and evaluation of potential tracer degradation due to sunlight. Fourth, when protective clothing is worn, separate studies may be required to determine the relative fabric penetration of the tracer and the chemical substance of interest.

Qualitative use of fluorescent tracers can indicate dermal exposure patterns and identify differential exposure potential among worker groups, and is also valuable for worker education and training (Fenske, 1988c; Bentley et al., 1989; Allen et al., 2000). Quantitative studies are likely to be most successful when the amount of tracer deposited on or transferred to the skin is predictable as with controlled spraying or routine surface contact activities rather than spills and splashes.

Workplace Area Monitoring

As an alternative to personal monitoring, exposure may be assessed through sampling of surfaces that workers are likely to contact. This approach is analogous to area sampling for air contaminants in that it requires assumptions concerning worker location and behavior. Contaminated surfaces have long been recognized as an occupational health hazard. Surface sampling has been addressed most systematically in the field of radioactive contaminants (Dunster, 1962; IAEA, 1970). The U.S. Occupational Health and Safety Administration has published a standard wipe sampling procedure for workplace surfaces (OSHA, 2001c), and several investigators have tested or modified such procedures (Chavalitnitkul and Levin, 1984; Lichtenwalner, 1992; Fenske et al., 1990; McArthur, 1992). These studies indicate that wiping a discrete area is critical to reducing sampling variability, and that physical surface characteristics, contaminant surface loading, sampling material, and wipe sampling procedures all influence the accuracy and precision of measurements.

There is a clear need to develop a surface sampling technique which employs standard materials and procedures, samples a defined surface area, and is operator-independent; i.e., pressure is not produced by hand motion. The goal of such a technique is to provide a reproducible measure of what Royster and Fish (1967) called "removeable residues," which will be referred to here as "transferable" residues. Thus, it is not necessary or even desirable to remove 100% of surface residues, but to measure those residues which are likely to be transferred to human skin. This sampling objective is exemplified in the development of the "dislodgeable residue" concept as applied to fieldworkers exposed to pesticides. This approach assumes that the hazardous component of residues found on foliar surfaces can be dislodged from the leaf surface due to worker contact. This is in contrast to the total extractable foliar residue (Iwata et al., 1977).

Several sampling instruments incorporating standard features have been developed recently for estimating surface pesticide residues in homes. Hsu et al. (1990) have used a polyurethane foam roll attached to a roller that delivers a known pressure to a measured surface independent of the operator. Ross et al. (1991) placed sampling material of known dimensions on a surface, covered it with a plastic sheet, and then rolled a weighted roller across the surface according to a standard protocol. Recent studies indicate that these techniques remove surface contaminants with comparable reproducibility (U.S. EPA, 1998), but that such measurements may not be representative of actual surface-to-skin transfer (Lu and Fenske, 1999). Housedust has also been sampled in residential environments to estimate children's exposures to pesticides (Lewis et al., 1994; Simcox et al., 1995). Recent work has focused on the development of novel methods for assessing contamination of workplace surfaces (Dost, 1996; Byrne, 2000).

Recent Advances in Dermal Exposure Assessment Methods

Recent advances in occupational exposure assessment with direct reading instruments such as attenuated total reflectance fourier transform infrared spectroscopy (ATR-FTIR) have led to new methods of skin exposure assessment (Doran et al., 2000). ATR-FTIR spectra of non-volatile compounds on the skin surface can be obtained rapidly (within a few seconds) with minimal sampling preparation. ATR-FTIR spectra provide the same characteristic infrared absorption features as other types of infrared analysis. The shape and location of absorption features indicate the structure and can be used to qualitatively identify an unknown compound. A quantitative measure of the chemical on skin can be obtained by relating the magnitude of the absorption features to spectral standards with known surface loading for the target chemicals. ATR-FTIR can also simultaneously identify and quantify multiple compounds (mixtures or formulations) on the skin surface *in vivo* using multivariate analysis methods such as partial least squares or classical least squares. Such approaches overcome the inherent limitations of tracer techniques by the direct measurement of workplace contaminants on skin.

OTHER INDUSTRIAL HYGIENE ASPECTS OF DERMAL EXPOSURE

Sampling Strategies

Sampling strategy design should be based on a recognition that exposures are likely to vary widely over time and between workers (see Volume 1, Chapters 14-16). A preliminary sampling strategy for dermal exposure has been proposed by van Hemmen and Brouwer (1995), and is outlined in Table 9-5.

Several issues can be highlighted. First, sampling must be *representative*; that is, all skin surfaces with potential for exposure need to be identified and monitored. In practice, the exposed skin surfaces may be obvious and quite limited (e.g., hands). Under many circumstances, however, dermal exposure can extend over substantial portions of the body, and be unpredictable, for example, as the result of accidents or of unusual conditions. Second, *sample duration* should be tailored to the sampling technique selected. Removal techniques require sampling immediately after exposure events, but frequent sampling over a workshift may alter the barrier properties of the skin (e.g., use of alcohols for washing or wiping may drive compounds into the skin). When garment samplers are employed, care must be taken to avoid saturation and penetration. Exposure intensity can vary substantially over regions of the body, and double-layered samplers may be needed for regions with high penetration potential. Losses may also occur from garment samplers, so that samplers may have to be changed periodically. Fluorescent tracer evaluation requires range-finding studies to determine the appropriate tracer concentration for specific exposure scenarios.

Assessment of occupational exposure to antifungal agents during commercial wood production illustrates how a dermal exposure sampling strategy might be implemented across an industry. Exposure assessment guidelines to support registration of antisapstain agents were prepared in British Columbia, and included a substantial discussion of the statistical basis for the proposed sampling strategy (Teschke et al., 1992; 1994). Based on previous work in this industry (Kauppinen and Lindroos, 1985; Fenske, et al., 1987) and an occupational hygiene survey conducted at the outset of the project, it was concluded that dermal exposure would occur primarily by one of two pathways: skin deposition for users of formulated products and for workers in the vicinity of dipping and spraying operations; surface contact for many of these workers and for all other workers at the facility. Qualitative use of fluorescent tracers was recommended to characterize exposed populations and exposure pathways at particular worksites, and surrogate skin techniques (garment samplers) were selected to measure dermal exposure. The proposed study design is outlined in Table 9-6.

The sampling guidelines proposed included four points. First, standardized sampling garments should be used; i.e., uniform fabric, construction, weight, thickness. Second, sampling garments should include gloves, shirt, hood, pants, and socks to measure exposure to all regions of the body. Third, sampling garments should be worn beneath protective clothing or normal work clothing; clothing worn normally by workers should not otherwise be altered. Fourth, samples should be collected over the entire work shift, including clean-up of equipment and materials at the end of the shift. Fifth, sampling garments should be changed at regular intervals (e.g., gloves four times per shift), and more frequently if there is evidence of saturation or if the compound under study is volatile. Repeat sampling of individual workers was also included in the strategy. This study design could be expected to produce exposure data sufficient to compare work activities and work sites, and suitable for use in risk assessment models.

Table 9-5
Dermal exposure sampling strategy[a]

Exposure Pathway	Sampling Method	Body Area Sampled	Time of Sample[b] Collection	Quality Assurance Requirements	Additional Requirements[c]
Immersion	Chemical removal	Contact area	End of immersion	Removal efficiency studies	Concentration and duration
Deposition	Chemical removal	Contact area	End of deposition	Removal efficiency studies	Identify exposed areas
	Surrogate skin	Whole body	End of shift	Avoid garment saturation & losses	
	Tracer	Whole body	End of shift	Range-finding studies	Relative deposition studies
Surface Contact	Chemical removal	Contact area	End of contact	Removal efficiency studies	Identify exposed areas Surface sampling
	Surrogate skin	Contact area	End of shift	Avoid garment saturation & losses	Identify exposed areas Surface sampling
	Tracer	Contact area	End of shift	Range-finding studies	Relative transfer studies

[a] adapted from van Hemmen and Brouwer (1995)
[b] may require consecutive samples within shift
[c] in all cases estimate effect of personal hygiene on skin loading; e.g., handwashing

Table 9-6
Dermal exposure sampling strategy for commercial wood treatment[A]

Sampling Garment	Body Region	Number of Changes (times per day)	Sampling Time[B] (hr)
Gloves	Hands	4 (at breaks/lunch)	2
Shirt	Upper Torso	2 (at lunch)	4
Hood	Head	2 (at lunch)	4
Socks	Feet	1	8
Pants	Lower Torso/Legs	1	8

[A] from Teschke et al. (1992)
[B] assumes 8 hour workshift

Biological Monitoring and Dermal Exposure Assessment

Biological monitoring, which typically includes the sampling of blood or urine, can provide additional information on exposure to some chemicals. Biological monitoring does not specifically address the dermal route of exposure. However, when used to supplement air monitoring, it can help determine the relative importance of the inhalation and dermal exposure routes. It is important that the chemicals of concern have been characterized in terms of metabolism and excretion over time. (see Volume 2, Chapter 8).

Dermal Exposure Control Strategies

As with any occupational exposures, permanent controls that do not place the burden on the worker are favored in the hierarchy of controls. Where process modification and engineering controls can be applied, these are preferred over types of controls that require administrative vigilance or the use of protective equipment, which can be burdensome, inconvenient, and less reliable than other means, and can create heat stress issues in some situations. Industrial hygiene practice favors changes in processes or products to reduce dermal exposure risks; for example, substitution of solvents that penetrate the skin easily. Administrative controls may also be used to reduce dermal exposures. The restricted entry period in agricultural workplaces is the primary control for pesticide health risk among fieldworkers (U.S. EPA, 1992b).

In many circumstances, however, protective clothing is often the only solution considered when dermal exposure risks are present. Clothing may not always provide complete or adequate protection from dermal exposure to chemicals. Indeed, under some circumstances, ordinary or protective clothing can facilitate exposure and absorption by holding liquid materials near the skin, or trapping and occluding contaminant on the skin, often in a warmer, moister micro-environment (such as inside gloves) that is conducive to absorption (see Chapter 3 for further details). Substantial research has been conducted to understand glove permeability.

Regulatory Aspects of Dermal Exposure Assessment

In the United States, the Occupational Safety and Health Administration (OSHA), National Institute for Occupational Safety and Health (NIOSH), and the U.S. Environmental Protection Agency (EPA) have all promulgated regulations that address dermal exposures. These regulations are largely limited to addressing pesticides in agricultural occupations, and transdermal exposures in a few other industries. Risks of dermatoses are not specifically addressed. The best known guidance related to dermal exposure risks — the skin notation — was developed by ACGIH®, and has been adopted by OSHA. However, this innovation has not led to quantitative guidelines.

Recent analyses of chemical hazards have recognized the importance of the dermal route of exposure in occupational settings, but have failed to mandate the evaluation of skin exposures. NIOSH, for example, has found that skin exposure to glycol ethers is an important exposure route, but concluded that direct measurement of such exposures was not practical (NIOSH, 1991). Similarly, the United States (OSHA) standard on methylene dianiline (MDA) recognized that skin exposure is the overwhelming contributor to dose for most workers, but promulgated a permissible air exposure limit (Chalk, 1989). These cases illustrate how skin exposure is viewed as a confounder in the context of current regulatory requirements. The most often cited explanation is a lack of methods for measuring skin exposure and the complications associated with such measurements. While this argument accurately reflects the status of current methods, there is no inherent barrier to measuring skin exposures with acceptable accuracy.

The establishment of dermal occupational exposure limits has been proposed recently (Bos et al., 1998; Brouwer et al., 1998), and presents several challenges. First, there is a need for an operational definition of "potential dermal exposure." Those workers whose activities bring them into routine contact with skin-permeable compounds need to be identified systematically, as do the compounds themselves. Second, improved methods of measuring chemical deposition on the skin are essential. Substantial resources will be needed for development, testing and validation of such methods under laboratory and field conditions. Third, more study is required of hygienic behavior. There is a need to characterize how personal behavior affects exposure, and how it may be constrained by the demands of particular work tasks. Fourth, the quantitative relationship between skin exposure and uptake must be determined for individual compounds, involving validation of percutaneous absorption models, and a greater undersatnding of how physiological paramaters relevant to work activity (e.g., skin temperature, perspiration) impact chemical flux through the skin. Finally, the setting of dermal occupational exposure limits must be actively pursued. Such limits could be based on workplace contamination levels in ambient air or on contacted surfaces, if sufficient information were available to generate exposure models for relevant scenarios. Alternatively, biological measures of exposure, such as the BEI® approach taken by ACGIH®, might be most effective.

Whatever method is employed, however, it must be based on an estimate of dose attributable to the dermal route. Thus, in addition to the toxicological data and risk assessment procedures that serve as the basis for airborne exposure limits, extensive knowledge of dermal absorption processes and/or detailed human pharmacokinetic information will be essential. Ultimately, a combination of airborne and dermal occupational exposure limits can lead to more complete appraisal of workplace hazards, and thus greater protection of exposed worker populations.

QUESTIONS

9.1. Name three ways that dermal exposure can occur in the workplace, and give an example of an exposure event for each.

9.2. How does personal behavior affect skin exposure? Provide two examples of how workers can reduce skin contact through hygienic behavior.

9.3. What are the strengths and weaknesses of the three major methods of dermal exposure monitoring?

9.4. When workplace area monitoring, rather than personal monitoring, is used to assess dermal exposure, what information is needed besides environmental concentrations?

9.5. Is the creation of dermal occupational exposure limits a desirable and/or practical goal?

REFERENCES

Abbott, I.M.; Bonsall, J.L.; Chester, G.; et al.: Worker Exposure to a Herbicide Applied with Ground Sprayers in the United Kingdom. Am Ind Hyg Assoc J 48:167-175 (1987).

ACGIH®: Threshold Limit Values. American Conference of Governmental Industrial Hygienists, Cincinnati, OH (2000).

Allen E.; Fenske R.; Foss C.; et al.: Evaluating a fluorescent tracer demonstration as a dermal pesticide exposure educational technique for non-agricultural pesticide applicators in western Washington State. Abstract. International Society for Exposure Analysis Annual Meeting, Monterey, CA, October 24-27 (2000).

Archibald, B.A.; Solomon, K.R.; Stephenson, G.R.: Estimation of Pesticide Exposure to Greenhouse Applicators Using Video Imaging and Other Assessment Techniques. Am Ind Hyg Assoc J 56:226-235 (1995).

Bentley, R.K.; Horstman, S.W.; Morgan, M.S.: Reduction of Sawmill Worker Exposure to Chlorophenols. Appl Ind Hyg 469-74 (1989).

Bierman, E.P.B.; Brouwer, D.H.; VanHemmen, J.J.: Implementation and evaluation of the fluorescent tracer technique in greenhouse exposure studies. Ann Occup Hyg 42:467-475 (1998).

Black, K.G.: An Assessment of Children's Exposure to Chlorpyrifos from Contact with a Treated Lawn. Ph.D. Dissertation, New Jersey Experiment Station, Rutgers University, New Brunswick, NJ (1993).

Bonsall, J.: Measurement of Occupational Exposure to Pesticides. In: Occupational Hazards of Pesticide Use, G.J. Turnbull, Ed. Taylor & Francis, London (1985).

Bos, P.M.; Brouwer, D.H.; Stevenson, H.; et al.: Proposal for the assessment of quantitative dermal exposure limits in occupational environments: part 1. Development of a concept to derive a quantitative dermal occupational exposure limit. Occup Environ Med 55:795-804 (1998).

Brouwer, D.H.; Brouwer R.; DeMik, G.; et al.: Pesticides in the cultivation of carnations in greenhouses: Part I. Exposure and concomitant health risk. Am Ind Hyg Assoc J 53:575-581 (1992).

Brouwer, D.H.; Hoogendoorn, L.; Bos, P.M.J.; et al.: Proposal for the assessment of quantitative dermal exposure limits in occupational environments: part 2. Feasibility study for application in an exposure scenario for MDA by two different dermal exposure sampling methods. Occup Environ Med 55:805-811 (1998).

Brouwer, D.H.; Kroese, R.; van Hemmen, J.J.: Transfer of contaminants from surface to hands: experimental assessment of linearity of fixed pressure and repeated contact with surfaces contaminated with a powder. Appl Occup Environ Hyg 14:231-239 (1999).

Brouwer, D.H.; Boeniger, M.F.; van Hemmen, J.: Hand wash and manual skin wipes. Ann Occup Hyg 44(7):501-510 (2000).

Byrne, M.A.: Suction methods for assessing contamination on surfaces. Ann Occup Hyg 44(7):523-528 (2000).

Chalk: 4,4'-Methylene Dianiline: Industrial Hygiene Experiences. In: Proceedings of the Conference on Occupational Health Aspects of Advanced Composite Technology in the Aerospace Industry, pp. 60-63. App Ind Hyg [Special Issue] (1989).

Chavalitnitikul, C.; Levin, L.A.: A Laboratory Evaluation of Wipe Testing Based on Lead Oxide Surface Contamination. Am Ind Hyg Assoc J 45:311-317 (1984).

Cherrie, J.W.; Brouwer, D.H.; Roff, M.; et al.: Use of qualitative and quantitative fluorescence techniques to assess dermal exposure. Ann Occup Hyg 44(7):519-522 (2000).

Chester, G.: Evaluation of worker exposure to, and absorption of, pesticides during occupational use and reentry. Ann Occup Hyg (1993).

Corley, R.A.; Markham, D.A.; Banks, C.; et al.: Physiologically based pharmacokinetics and the dermal absorption of 2-butoxyethanol vapor by humans. Fundam Appl Toxicol 39(2):120-30 (1997).

Corley, R.A.; Bormett, G.A.; Ghanayem, B.I.: Physiologically based pharmacokinetics of 2-butoxyethanol and its major metabolite, 2-butoxyacetic acid, in rats and humans. Toxicol Appl Pharmacol 129(1):61-79 (1994).

Davis, J.E.: Minimizing Occupational Exposure to Pesticides: Personnel Monitoring. Residue Rev 75:33-50 (1980).

Davis, J.E.; Stevens, E.R.; Staff, D.C.: Potential Exposure of Apple Thinners to Azinphosmethyl and Comparison of Two Methods for Assessment of Hand Exposure. Bull Environ Contam Toxicol 31:631-638 (1983).

Deichmann, W.B.: Local and systemic effects following skin contact with phenol: a review of literature. J Indust Hyg Toxicol 31(3):146-154 (1949).

Doran, E.; Yost, M.; Fenske, R.A.: Measuring dermal exposure to pesticide residues with attenuated total reflectance fourier transform infrared (ATR-FTIR) spectroscopy. Bull Environ Contam Toxicol 64 666-672 (2000).

Dost, A.: Monitoring surface and airborne inorganic contamination in the workplace by field portable x-ray fluorescence spectrometer. Ann Occup Hyg 40(5):589-610 (1996).

Dunster, H.J.: Surface Contamination Measurements as an Index of Control of Radioactive Materials. Health Phys 8:353-356 (1962).

Durham, W.F.; Wolfe, H.R.: Measurement of the Exposure of Workers to Pesticides. Bull World Health Organ 26:79-91 (1962).

Fenske, R.A.: Comparative Assessment of Protective Clothing Performance by Measurement of Dermal Exposure During Pesticide Applications. Appl Ind Hyg 3:207-213 (1988a).

Fenske, R.A.: Visual Scoring System for Fluorescent Tracer Evaluation of Dermal Exposure to Pesticides. Bull Environ Contam Toxicol 41:727-736 (1988c).

Fenske, R.A.: Nonuniform Dermal Deposition Patterns during Occupational Exposure to Pesticides. Arch Environ Contam Toxicol 19:332-337 (1990).

Fenske, R.A.; Black, K.G.; Elkner, K.P.; et al.: Potential Exposure and Health Risks of Infants Following Indoor Residential Pesticide Applications. Am J Public Health 80:689-693 (1990).

Fenske, R.A.: Dermal Exposure Assessment Techniques. Ann Occup Hyg 37:687-706 (1993).

Fenske, R.A.; Leffingwell, J.T.; Spear, R.C.: Evaluation of Fluorescent Tracer Methodology for Dermal Exposure Assessment. In: Dermal Exposure Related to Pesticide Use – ACS Symposium Series 273, pp. 377-393, R.C. Honeycutt; G. Zweig; N.N. Ragsdale, Eds. American Chemical Society, Washington, D.C. (1985).

Fenske, R.A.; Leffingwell, J.T.; Spear, R.C.: A Video Imaging Technique for Assessing Dermal Exposure – I. Instrument Design and Testing. Am Ind Hyg Assoc J 47:764-770 (1986a).

Fenske, R.A.; Wong, S.M.; Leffingwell, J.T.; et al.: A Video Imaging Technique for Assessing Dermal Exposure – II. Fluorescent Tracer Testing. Am Ind Hyg Assoc J 47:761-775 (1986b).

Fenske, R.A.; Horstman, S.W.; Bentley, R.K.: Assessment of Dermal Exposure to Chlorophenols in Timber Mills. Appl Ind Hyg 2:143-147 (1987).

Fenske, R.A.; Birnbaum, S.G.; Methner, M.M.; et al.: Methods for Assessing Fieldworker Hand Exposure to Pesticides During Peach Harvesting. Bull Environ Contam Toxicol 43:805-813 (1989).

Fenske, R.A.; Lu, C.: Determination of handwash removal efficiency: incomplete removal of the pesticide, chlorpyrifos, from skin by standard handwash techniques. Am Ind Hyg Assoc J 55:425-432 (1994).

Fenske, R.A.; Birnbaum, S.G.: Second generation video imaging technique for assessing dermal exposure (VITAE System). Am Ind Hyg Assoc J 58(9):636-645 (1997).

Fenske, R.A.; Schulter, C.; Lu, C.; et al.: Incomplete removal of the pesticide captan from skin by standard handwash exposure assessment procedures. Bull Environ Contam Toxicol 61:194-201 (1998).

Fenske, R.A.; Simcox, N.J.; Camp, J.E.; et al.: Comparison of three methods for assessment of hand exposure to azinphos-methyl (guthion) during apple thinning. Appl Occup Environ Hyg 14:618-623 (1999).

Fenske, R.A.; Birnbaum, S.G.; Methner, M.M.; et al.: Fluorescent tracer evaluation of chemical protective clothing during pesticide applications in central Florida citrus gorves. J Agric Sfty Hlth 8(3):319-331 (2002).

Franklin, C.A.; Fenske, R.A.; Greenhalgh, R.; et al.: Correlation of Urinary Pesticides Metabolite Excretion with Estimated Dermal Contact in the Course of Occupational Exposure to Guthion. J Toxicol Environ Health 7:715-731 (1981).

Gold, R.E.; Leavitt, J.R.C.; Holsclaw, T.; et al.: Exposure of Urban Applicators to Carbaryl. Arch Environ Contam Toxicol 11:63-67 (1982).

Hamilton, A.: Industrial poisoning by compounds of the aromatic series. J Ind Hyg 1:200-212 (1920).

Hamilton, A.; Reznikoff, P.; Burnham, G.M.: Tetra-ethyl lead. J Am Med Assoc 84(20):1481-1486 (1925).

Hsu, J.P.; Camann, D.; Schattenberg, H.; et al.: New Dermal Exposure Sampling Technique. In: Measurement of Toxic and Related Air Pollutants, pp. 489-498. Air and Waste Management Association, Publication VIP-17, Pittsburgh, PA (1990).

IAEA: Monitoring of Radioactive Contamination of Surfaces. Technical Report Series No 120, International Atomic Energy Agency, Vienna (1970).

Iwata, Y.; Spear, R.C.; Knaak, J.B.; et al.: Worker Reentry into Pesticide-Treated Crops – I. Procedure for the Determination of Dislodgeable Pesticide Residues on Foliage. Bull Environ Contam Toxicol 18:649-655 (1977).

Johanson, G.; Boman, A.: Percutaneous Absorptions of 2-Butoxyehtanol Vapour in Human Subjects. Br J Ind Med 48(11):788-792 (1991).

Jongeneelen, F.J.; Scheepers, P.T.J.; GroeneNdijk, A.; et al.: Airborne Concentrations, Skin Contamination, and Urinary Metabolite Excretion of Polycyclic Aromatic Hydrocarbons among Paving Workers Exposed to Coal Tar Derived Road Tars. Am Ind Hyg Assoc J 49:600-607 (1988).

Kauppinen, T.; Lindroos, L.: Chlorophenol Exposure in Sawmills. Am Ind Hyg Assoc J 46:34-38 (1985).

Keeble, V.B.; Dupont, R.R.; Doucette, W.J.; et al.: Guthion Penetration of Clothing Materials During Mixing and Spraying in Orchards. In: Performance of Protective Clothing: Second Symposium, STP 989, pp. 573-583, S.Z. Mansdorf; R. Sagar; A.P. Nielson, Eds. American Society for the Testing of Materials, Philadelphia, PA (1988).

Keenan, R.R.; Cole, S.B.: A Sampling and Analytical Procedure for Skin Contamination Evaluation. Am Ind Hyg Assoc J 43:473-476 (1982).

Kissel, J.C.; Richter, K.Y., Fenske, R.A.: Field measurement of dermal soil loading attributable to various activities: implications for exposure assessment. Risk Analysis 16:115-125 (1996a).

Kissel, J.C.; Richter, K.Y.; Fenske, R.A.: Investigation of factors affecting soil adherence to skin using a hand-press technique. Bull Environ Contam Toxicol 56:722-728 (1996b).

Kissel, J.C.; Shirai, J.H.; Richter, K.Y.; et al.: Investigation of dermal contact with soil in controlled trials. J Soil Contamination 7:737-752 (1998).
Kissel, J.C.; Fenske, R.A.: Improved estimation of dermal pesticide dose to agricultural workers upon reentry. Appl Occup Environ Hyg 15:1-7 (2000).
Kross, B.C.; Nicholson, H.F.; Ogilvie, L.K.: Methods Development Study for Measuring Pesticide Exposure to Golf Course Workers using Video Imaging Techniques. Appl Occup Environ Hyg 11(11):1346-1350 (1996).
Lewis, R.G.; Fortmann, R.C.; Camaan, D.E.: Evaluation of Methods for Monitoring the Potential Exposure of Small Children to Pesticides in the Residential Environment. Arch Environ Contam Toxicol 26:37-46 (1994).
Lichtenwalner, C.P.: Evaluation of Wipe Sampling Procedures and Elemental Surface Contamination. Am Ind Hyg Assoc J 53:657-659 (1992).
London, M.A.; Boiano, J.M; Lee, S.A.: Exposure assessment of 3,3'-dichlorobenzidine (dcb) at two chemical plants. Appl Ind Hyg 4:101-104 (1989).
Lu, C.; Fenske, R.A.: Dermal transfer of pesticide residues from treated surfaces: comparison of hand press, hand drag, wipe and PUF roller measurements following residential flea control applications. Environ Health Perspect 107:463-467 (1999).
Machado Neto, J.G.; Matuo, T.; Matuo, Y.K.: Dermal Exposure of Pesticide Applicators in Staked Tomato (Lycopersicon Esculentum Mill) Crops: Efficiency of a Safety Measure in the Application Equipment. Bull Environ Contam Toxicol 48:529-534 (1992).
McArthur, B.: Dermal Measurement and Wipe Sampling Methods: A Review. Appl Occup Environ Hyg 7:599-606 (1992).
Meigs, J.W.; Brown, R.M.; Sciarini, L.J.: A study of exposure to benzidine and substituted benzidines in a chemical plant. Am Med Ass Archs Ind Hyg Occup Med 4:533-540 (1951).
Methner, M.M.; Fenske, R.A.: Pesticide Exposure during Greenhouse Applications, Part I. Dermal Exposure Reduction due to Directional Ventilation and Worker Training. Appl Occup Environ Hyg 9:560-566 (1994a).
Methner, M.M.; Fenske, R.A.: Pesticide Exposure during Greenhouse Applications, Part II. Chemical Permeation through Protective Clothing in Contact with Treated Foliage. Appl Occup Environ Hyg 9:567-574 (1994b).
Nigg, H.N.; Stamper, H.H.; Queen, R.M.: Dicofol Exposure to Florida Citrus Applicators: Effects of Protective Clothing. Arch Environ Contam Toxicol 15:121-134 (1986).
Nigg, H.N.; Stamper, H.H.; Queen, R.M.: The Development and Use of a Universal Model to Predict Tree Crop Harvest Pesticide Exposure Am Ind Hyg Assoc J 45:182-186 (1984).
Nigg, H.N.; Stamper, J.H.; Easter, E.; et al.: Field Evaluation of Coverall Fabrics: Heat Stress and Pesticide Penetration. Arch Environ Contam Toxicol 23:281-288 (1992).
NIOSH: Criteria for a Recommended Standard: Occupational Exposure to Ethylene Glycol Monomethyl Ether, Ethylene Glycol Monoethyl Ether, and Their Acetates. National Institute for Occupational Safety and Health, U.S. Department of Health and Human Services, Cincinnati, OH (1991).
NIOSH: Pocket Guide to Chemical Hazards. U.S. Department of Health and Human Services, Public Health Services, Centers for Disease Control and Prevention, National Institute for Occupational Safety and Health, Washington, DC (2001).
Odland, G.F.; Short, J.M.: Structure and development of the skin. Chapter 3 in Biology and Pathphysiology of Skin, pp. 39-48 (1968).
OSHA: Cadmium Standard. U.S. Occupational Safety and Health Administration website, accessed April 6, 2001. http://www.osha-slc.gov/OshStd_data/1910_1027.html (2001a).
OSHA: Lead Standard. U.S. Occupational Safety and Health Administration website, accessed April 6, 2001. http://www.osha-slc.gov/OshStd_data/1910_1025.html (2001b).
OSHA: Technical Manual, Section II, Chapter 2. U.S. Occupational Safety and Health Administration website, accessed April 6, 2001. http://www.osha-slc.gov/dts/osta/otm/otm_ii/otm_ii_2.html (2001c).
Perkins, J.L.; Knight, V.B.: Risk Assessment of Dermal Exposure to PCBs Permeating a PVC Glove. Am Ind Hyg J 50:A-171 (1989).
Perkins, J.L.; Rainey, K.: The Effect of Glove Flexure on Permeation Parameters. Appl Occ and Env Hygiene 12:206-210 (1997).

Popendorf, W.J.; Leffingwell, J.T.: Regulating OP Pesticide Residues for Farmworker Protection. Residue Rev 82:125-201 (1982).

Roff, M.W.: A Novel Lighting System for the Measurement of Dermal Exposure Using a Fluorescent Dye and Image Processor. Ann Occup Hyg 38:903-919 (1994).

Roff, M.W.: Accuracy and reproducibility of calibrations on the skin using the FIVES fluorescence monitor. Ann Occup Hyg 41:313-324 (1997).

Ross, J.; Thongsinthusak, T.; Fong, H.R.; et al.: Measuring Potential Dermal Transfer of Surface Pesticide Residue Generated from Indoor Fogger Use: An Interim Report. Chemosphere 20:349-360 (1990).

Ross, J.; Fong, H.R.; Thongsinthusak, T.; et al.: Measuring Potential Dermal Transfer of Surface Pesticide Residue Generated from Indoor Fogger Use: Using the CDFA Roller Method, Interim Report II. Chemosphere 22:975-984 (1991).

Royster, G.W.; Fish, B.R.: Techniques for Assessing "Removable" Surface Contamination. In: Surface Contamination, pp. 201-208, B.R. Fish, Ed. Pergamon Press, New York, NY (1967).

Scheuplein, R.J.; Blank, I.H.: Permeability of the skin. J Invest Dermatol 67:31-38 (1971).

Schneider, T.; Vermeulen, R.; Brouwer, D.H.; et al.: Conceptual model for assessment of dermal exposure. Occup Environ Med 56:765-773 (1999).

Schuresko, D.D.: Portable Fluorometric Monitor for Detection of Surface Contamination by Polynuclear Aromatic Compounds. Anal Chem 52:371-373 (1980).

Simcox, N.J.; Fenske, R.A.; Wolz, S.; et al.: Pesticides in housedust and soil: exposure pathways for children of agricultural families. Environmental Health Perspectives 103:1126-1134 (1995).

Soutar, A.; Semple, S.; Aitken, R.J.; et al.: Use of patches and whole body sampling for the assessment of dermal exposure. Ann Occup Hyg 44(7):511-518 (2000).

Susten, A.S.; Dames, B.L.; Burg, J.R.; et al.: Percutaneous penetration of benzene in hairless mice: an estimate of dermal absorption during tire-building operations. Am J Ind Med 7(4):323-35 (1985).

Teschke, K.; Fenske, R.A.; van Netten, C.; et al.: Generic Guidelines for Assessing Worker Exposure to Antisapstain Chemicals in the Lumber Industry. Report to the Health Monitoring Sub-Committee of the B.C. Stakeholder Forum on Sapstain Control, Vancouver, British Columbia, Canada (1992).0

Teschke, K.; Marion, S.A.; Jin, A.; et al.: Strategies for determining occupational exposures in risk assessments: a review and a proposal for assessing fungicide exposures in the lumber industry. Am Ind Hyg Assoc J 55:443-449 (1994).

Ulenbelt, P.; Lumens, M.E.G.L.; Geron, H.M.A.; et al.: Work Hygienic Behavior as Modifier of the Lead Air-Lead Blood Relation. Int Arch Occup Environ Health 62:203-207 (1990).

USEPA: Pesticide Assessment Guidelines, Subdivision U, Applicator Exposure Monitoring (PB87-13328). U.S. Department of Commerce, National Technical Information Service, Springfield, VA (1987).

USEPA: Dermal Exposure Assessment: Principles and Applications. EPA/600/891/011B. Office of Research and Development, Washington, D.C. (1992a).

USEPA: Worker Protection Standards for Agricultural Pesticides. Federal Register 57(163):38102-38176. 21 August (1992b).

USEPA: Fluorescent Tracer Evaluation of Protective Clothing Performance, by R.A. Fenske (EPA /ORD Pub No EPA/600/R-93/143). Risk Reduction Engineering Laboratory, Cincinnati, OH (1993).

USEPA: Exposure Factors Handbook, Volume I: General Principles. EPA/600 /P-95/002Fa. Office of Research and Development, Washington DC (1997).

USEPA: Round Robin Testing of Three Methods for Measurement of Pesticides on Surfaces. Final Report, Office of Research and Development, Washington DC (1998).

Van Hemmen, J.J.; Brouwer, D.H.: Assessment of dermal exposure to chemicals. Sci Total Environ 168:131-141 (1995).

Vermeulen, R.; Heideman, J.; Bos, R.P.; et al.: Identification of dermal exposure pathways in the rubber manufacturing industry. Ann Occup Hyg 44(7):533-541 (2000).

Vo-Dinh, T.: Evaluation of an Improved Fiberoptics Luminescence Skin Monitor with Background Correction. Am Ind Hyg Assoc J 48:59-598 (1987).

Vo-Dinh, T.; Gammage, R.B.: The Lightpipe Luminoscope for Monitoring Occupational Skin Contamination. Am Ind Hyg Assoc J 42-112-120 (1981).

WHO: World Health Organization Fields Surveys of Exposure to Pesticides Standard Protocol. Toxicol Lett 33:223-235 (1986).

Zartarian, V.G.; Ott, W.R.; Duan, N.: A quantitative definition of exposure and related concepts. J Exposure Anal Environ Epidemiol 7(4):411-437 (1997).

Zweig, G.; Leffingwell, J.T.; Popendorf, W.J.: The Relationship between Dermal Pesticide Exposure by Fruit Harvesters and Dislodgeable Foliar Residues. J Environ Sci Health B 20:27-59 (1985).

10

EPIDEMIOLOGY

Elizabeth Delzell, DSc

INTRODUCTION

This chapter provides information on epidemiology that should be useful for industrial hygienists. It discusses the definition, scope, and uses of epidemiology; defines basic measures of disease frequency and of association that are estimated in epidemiologic research; describes the main study designs used to estimate these measures; compares the main study designs in terms of their advantages and disadvantages; briefly describes several other study designs used in epidemiologic research; and summarizes issues of causal inference and interpretation in epidemiologic research. The emphasis throughout is on occupational epidemiology and on the epidemiology of chronic diseases, rather than acute disorders. Acute disorders due to an occupational factor typically develop soon after an exposure has occurred (e.g., a musculoskeletal injury sustained as a result of workplace activity or accident). The epidemiologic investigation of the effect of occupational factors on acute disorders is relatively simple and usually is confined to actively working persons. Chronic diseases, in contrast, develop years after exposure has occurred. Thus, the epidemiologic investigation of the etiology of chronic diseases is complex, requiring observation of exposed and unexposed groups over long periods of time, including follow-up of people after they leave active employment. The term "disease" is used in this chapter as a synonym for any medical outcome evaluated in epidemiologic research, including fatal and nonfatal conditions, physiologic dysfunction, and accidental injuries.

DEFINITION, SCOPE, AND USES

Epidemiology is the study of determinants of disease in human populations. The ultimate purpose of epidemiology is disease prevention. Epidemiologic research pursues this goal by describing the occurrence of disease in groups of people and by identifying the factors that explain disease patterns and cause or prevent disease in human populations. Causal factors typically are thought of as physical, chemical, and biological agents that produce or protect against an adverse physiological response or that increase or decrease the risk of disease (see Volume 1, Chapter 4).

Epidemiology has a number of special subject areas. These are specified on the basis of methodology (descriptive or analytical) or on the basis of orientation towards particular diseases or exposures.

Descriptive epidemiology develops and assesses hypotheses about possible causes by evaluating trends in disease rates by personal characteristics such as age, gender, race or ethnicity, and occupation; by place, typically, geographic regions such as countries, states or counties within countries; and by calendar time period (i.e., assessing increases or decreases over time). Such evaluation may identify disease patterns that are correlated with the distribution of causal agents in the populations of interest. For example, descriptive epidemiology was used to document a pattern of relatively high lung cancer rates in coastal areas of the United States in the 1950s-1970s (Blot et al., 1979). This pattern led to the development of the hypothesis that employment in shipyards during World War II might have increased the risk of lung cancer because of concomitant exposure to asbestos (Blot et al., 1980). Typically, descriptive epidemiologic studies provide estimates of measures of occurrence, such as incidence or mortality rates.

Analytic epidemiology evaluates hypotheses about causal factors by quantifying the relationship between exposure to a specific agent (e.g., asbestos) and disease or by studying the relationship between factors correlated with a possible causal agent (e.g., shipyard worker occupation) and disease. Analytic epidemiologic studies provide estimates of measures of association, including rate ratios, risk ratios, and rate or risk differences. The emphasis of etiologic research in epidemiology tends to be on the identification of modifiable causes of disease, namely factors that are encountered as external exposures (occupational exposures, diet, smoking, pharmaceutical agents), rather than on intrinsic personal characteristics. However, the latter factors must always be considered as potentially explaining differences in disease patterns between exposed and unexposed persons.

Disease- and exposure-specific subject areas in epidemiology include reproductive, cancer, cardiovascular disease, injury, occupational, infectious disease, genetic, molecular, and nutritional epidemiology. Occupational epidemiology, one of the most relevant subject areas for the industrial hygienist, deals with the identification of work-related etiologic exposures or correlated factors (Checkoway et al., 1989; Monson, 1980). Work-related factors include agents directly used or produced in industries, contaminants of materials being used as part of a production process, and physical or mental stress due to the ergonomic or psychological requirements of work. The etiologic significance of such factors may be studied either directly, or indirectly, through the evaluation of correlates such as job title or industry.

Occupational epidemiology interacts primarily with the fields of industrial hygiene and toxicology in providing the scientific basis for primary disease prevention in the occupational setting. In its focus on quantifying etiologic relationships between workplace exposures and disease in populations, occupational epidemiology is distinct from occupational medicine, which concentrates on diagnosing, treating, and to some extent, preventing disease in individual workers.

The aim of most epidemiologic research is to identify causes of disease, with the broad goal being disease prevention. Occupational epidemiologic research focuses on detecting causal relationships between work-related factors and disease (Wegman, 1992), although estimation of the impact of lifestyle factors on the health of workers in various occupations and industries also may be an important objective. The measurements obtained from epidemiologic research may be used for many other purposes, including:

- to set work place and general environmental regulatory standards.
- to carry out disease surveillance in occupational and other groups.
- to develop risk assessments for occupational and general community groups.
- to develop interventions (reduction in exposure; screening).
- to evaluate interventions.
- to determine the focus of future research.

MEASURES OF DISEASE FREQUENCY

Most epidemiologic studies use one of two basic methodologic designs: 1) the follow-up study design, or 2) the case-control study design. In a follow-up study, a group of individuals who have a characteristic in common (the "subjects" or the "study group") and who are initially free of the disease of interest is identified. Each person in the study group is monitored ("followed up") over time to determine changes in exposure status and to determine if he or she develops the disease of interest. The subsequent analysis determines if the disease experience of exposed persons is higher, lower, or the same as that of unexposed persons. Case-control studies evaluate the relationship between an agent and a disease by identifying a group of people who have developed the disease of interest (the "cases") over a particular time period and a group of people who do not have the disease (the "controls"), and comparing these two groups with regard to their history of exposure to one or more agents of interest. Thus, one distinction between the follow-up and the case-control study designs lies in the initial approach used to select subjects: the follow-up study identifies subject groups on the basis of their exposure status, whereas the case-control study identifies and selects subjects on the basis of their disease status. The "cross-sectional" study design, used extensively in the past but now recognized as having severe limitations, uses yet another approach to identify and select subjects. This design identifies a group of people who have had a particular exposure history at some point in time and a group of people who are unexposed; measures the disease status of people in the two groups at a single point in time; and compares a measure of disease occurrence in the exposed and unexposed groups.

The three epidemiologic study designs mentioned above are discussed in more detail in later sections. For the present, it is useful to understand that all epidemiologic studies are measurement instruments having as their central aims the quantification of disease frequency and the estimation of measures of association. All three designs estimate one of three basic measures of disease frequency in the population(s) under study during a particular time period (Rothman and Greenland, 1998, pp. 29-46). The three measures are the incidence or mortality rate; incidence proportion, cumulative incidence, or risk; and the prevalence. Terms essential for understanding these measures are defined below:

Study base: the exposure and disease experience of a group of people over a specified period of time. This experience is the basis for selecting study subjects and for measuring their disease rates (Miettinen, 1985, pp. 46-68).

Study period: the total calendar time period covered by an epidemiologic investigation.

Observation period: the calendar time period during which an individual subject is observed in a study; the observation period may the all or part of the study period.

Person-time: this refers to the amount of time each person is observed in a population under study (i.e., the length of the observation period of each subject, measured in terms of person-days, -weeks, -months, or -years).

Population at risk: persons who are candidates for developing a disease or other outcome condition at the beginning of their observation period (i.e., they do not already have the disease); individuals contributing person-time to a study base and, hence, to the denominator of an incidence or mortality rate.

Open population: a study group which may accrue new subjects during the study period and which subjects may leave if they die, emigrate, or are lost. In studies of open populations the amount of observation time may vary from subject to subject.

Closed population: a study group to which no new subject is added during the study period, and which subjects leave only if they die. In studies of closed populations the amount of observation time should be approximately the same for each subject.

Incidence and Mortality Rates

An incidence or mortality rate (IR or MR) is the number of cases of (or deaths from) a disease, divided by the total person-time experience giving rise to the cases. IRs and MRs are estimated in follow-up studies and in some case-control studies, but not in cross-sectional studies.

IRs and MRs have a numerator (the number of cases or deaths) and a denominator (person-time). The numerator may contain only those events that occur among persons contributing to the denominator at the time of the events. IRs and MRs are measured in units of reciprocal time (years^{-1}, months^{-1} or days^{-1}) and range from 0, with no upper bound. A typical formula for computing an IR is:

$$IR = c/PY$$

where c is the total number of cases and PY is the total number of person-years of observation, summed over all subjects and all observation periods. The formula for computing an MR is:

$$MR = d/PY$$

where d is the total number of deaths and PY is the total number of person-years, again summed over all subjects and all observation periods.

Figure 10-1 displays the person-years and mortality experience of a hypothetical workforce of 10 persons in a study of an open population. The study period is 40 years, 1955 through 1994. The length of the observation period (number of person-years of observation) varies from five years (subjects 1 and 10) to 35 years (subject 7), and the total amount of person-years is 200 (PY=200). During the period of 1955-1995, two deaths occur among the 10 subjects (d=2), and the MR is two deaths/200 person-years, or 10/1,000 years.

Incidence Proportion

Incidence proportion (IP) is defined as the proportion of a closed population at risk that develops disease during a specified study period. "At risk" implies that at the beginning of the study period, no member of the study group has the disease of interest. The term "cumulative incidence" may be used as a synonym for IP. Also, IP often is used interchangeably with the term "risk." However, risk refers to the probability that an individual will develop disease in a specified time period, whereas IP is a measure of average risk, computed for a group of individuals. It is dimensionless and ranges in value from 0 to 1. IP may be used as a measure of occurrence in epidemiologic investigations of spontaneous abortions and other medical conditions when the observation period for each subject is about the same and is short in duration.

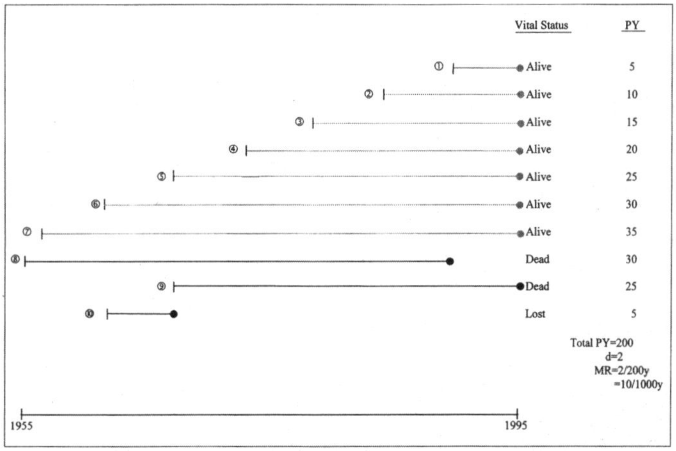

Figure 10-1. Observation period, person-years (PY), number of deaths (d), and mortality rate (MR) among subjects during a 40-year study period of 1955 through the end of 1994.

Also, IP is often used as a measure of occurrence in studies of acute medical conditions. IP is used as a disease occurrence measure in some follow-up and case-control studies but not in cross-sectional studies.

IP may be computed as follows:

$$IP = c/N,$$

where *c* is the total number of cases of illness occurring among all subjects during the study period and *N* is the total number of subjects at risk of developing the illness at the beginning of the study period.

To illustrate the computation of IP, consider a hypothetical study that includes 100 pregnancies conceived among women in a workforce during a five-year interval. The observation period for each pregnancy is about nine months. Among the 100 pregnancies (N=100), 20 spontaneous abortions (SABs) occur (c=20). The IP of SAB is c/N or 20/100 (20%).

Prevalence Proportion

Prevalence proportion (or prevalence) refers to the proportion of a population or study group that has a disease or other medical outcome at a particular point in time and is used in cross-sectional studies. Prevalence proportion at a particular point in time is influenced both by the incidence rate of a disease and by the probability of surviving with disease. Most nonexperimental epidemiologic research focuses on identifying factors that cause disease, rather than on factors that influence the progression of the disease or survival once the disease has developed. Because prevalence reflects both determinants of incidence (development of disease) and determinants of progression/survival, this occurrence measure is of less interest in epidemiologic research than are measures of incidence (Eisen, 1995). However, prevalence proportions (and the cross-sectional study design) have been used for convenience as measures of occurrence under certain conditions when:

- the disease or medical condition of interest is nonfatal, so that mortality rates cannot be measured
- the disease or medical condition must be detected by clinical examination or a laboratory test or symptom report
- the point of onset of the disease or medical condition is difficult to establish retrospectively or prospectively.

Prevalence proportion is dimensionless and ranges in value from zero to one. This measure will not be discussed further.

ESTIMATION OF THE EFFECT OF A FACTOR ON A DISEASE: MEASURES OF ASSOCIATION

The ultimate goal of epidemiologic research, disease prevention, requires the identification of factors that cause disease and the quantification of the effect of a causal factor. The term "effect" refers to a causal relationship and is defined as the amount of change in the frequency of a disease brought about by a causal exposure. In concept, an effect measure contrasts the frequency of the outcome of interest in a single population under two different conditions (e.g., exposed and unexposed) (Rothman and Greenland, 1998, p. 64). In practice, epidemiology evaluates effects by measuring and interpreting associations. Any association, unlike an effect, may be causal or noncausal. Quantifying and interpreting an association involves:

- computing a point estimate of a measure of association; a point estimate is the single best value of the measure, computed from study data
- evaluating the validity of the estimate
- evaluating the precision of the estimate.

Point Estimates of Measures of Association

Measures of association may be absolute or relative. All measures of association compare disease frequencies (IR, MR, IP, prevalence proportion) or exposure frequencies in two or more different study groups or populations.

Absolute Measures of Association

An absolute measure of association is the difference in IR, MR, IP or prevalence proportion between an exposed and an unexposed group of individuals. An IR difference ranges in value from minus infinity to plus infinity and has the dimension time.$^{-1}$ An IP difference or risk difference and a prevalence difference range in value from minus one to plus one and are dimensionless. If a factor is accepted as causal, IR and IP differences are referred to as attributable rates or risks. These are useful measures of the public health impact of a causal exposure, in that they indicate the number of cases (or deaths) that would be prevented if exposure were removed.

Relative Measures of Association

A relative measure of association refers to the ratio of two IRs, MRs, IPs or prevalence proportions, specifically the ratio of the occurrence measure among the exposed to the occurrence measure among the unexposed. A relative measure of association also may refer to the ratio of two exposure odds, specifically the ratio of the exposure odds in a group of cases (persons with a particular disease or medical condition) to the exposure odds in a group of controls (persons without the disease). In this chapter, the emphasis is on relative measures of association. This is largely because relative measures may provide a clearer indicator of the strength of an association or causal role than do absolute measures (Miettinen, 1985, p. 18). Thus, relative measures are used more often in etiologic research in epidemiology. Commonly used relative measures of association include:

- the incidence rate ratio (IRR) or the IR among exposed subjects, divided by the IR among unexposed subjects
- the mortality rate ratio (MRR) or the MR among exposed subjects, divided by the MR among unexposed subjects
- the incidence proportion ratio (IPR) or the IP among exposed subjects, divided by the IP among unexposed subjects
- the odds ratio (OR) or the odds of exposure in diseased subjects, divided by the odds of exposure in nondiseased subjects; the OR is used to estimate the IRR or the IPR in case-control studies, discussed in a later section, and may also be used when prevalence is measured (Zocchetti et al., 1997).

The "null" value (corresponding to the situation in which there is no association) of an IRR, MRR, IPR, or OR is 1.0, and each of the ratio measures has a possible range of 0 to infinity. Formulas for computing each measure are:

$$IRR = [c_1/(PY)_1] \div [c_0/(PY)_0]$$

where c_1 and c_0 are the numbers of cases among exposed and unexposed subjects, respectively, and $(PY)_1$ and $(PY)_0$ are the numbers of person-years among the exposed and the unexposed, respectively.

$$MRR = [d_1/(PY)_1] \div [d_0/(PY)_0]$$

where d_1 and d_0 are the numbers of deaths among exposed and unexposed subjects, respectively, and $(PY)_1$ and $(PY)_0$ are the numbers of person-years among the exposed and the unexposed, respectively.

$$IPR = [c_1/N_1] \div [c_0/N_0]$$

where c_1 and c_0 are the numbers of cases among exposed and unexposed subjects, respectively, and N_1 and N_0 are the numbers of exposed and the unexposed subjects, respectively.

$$OR = (a/b) \div (c/d) = (ad)/(bc)$$

where *a* is the number of people with both disease and exposure (exposed cases), *b* is the number of unexposed people with disease (unexposed cases), *c* is the number of exposed people without disease (exposed controls), *b* is the number of unexposed people without disease (unexposed controls).

Figure 10-2 and Table 10-1 illustrate the computation of an MRR. The hypothetical data in Figure 10-2 depict the person-years and disease experience of a workforce of 10 persons included in a study. The study period is 40 years, 1955 through 1994. Subjects 4 (20 person-

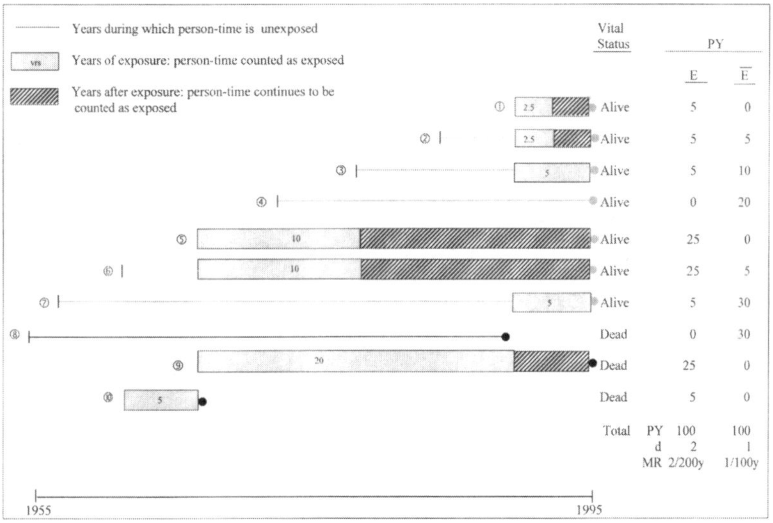

Figure 10-2. Observation period, person-years (PY), number of deaths (d), and mortality rate (MR) among exposed (E) and unexposed (U) subjects during a 40-year study period of 1955 through the end of 1994.

Table 10-1
Computation of a mortality rate ratio (MRR)

	Exposed	Unexposed
Number of deaths	$d_1 = 2$	$d_0 = 1$
Number of person-years	$(PY)_1 = 100$	$(PY)_0 = 100$
Mortality rate, MR	$(MR)_1 = d_1/(PY)_1$ $(MR)_1 = 2/100$ years	$(MR)_0 = d_0/(PY)_0$ $(MR)_0 = 1/100$ years
Mortality rate ratio, MRR	$MRR = (IR)_1 \div (IR)_0$ $MRR = [d_1/(PY)_1] \div [d_0/(PY)_0]$ $MRR = (2/100\ y)/(1/100\ y) = 2.0$	

years) and 8 (30 person-years) are unexposed for all of their observation period. All person-years of subjects 1, 5, 9, and 10 are classified as exposed. (Once a subject is exposed, all subsequent person-years are considered "ever exposed."). Subject 2 has 5 unexposed and 5 exposed person-years, subject 3 has 10 unexposed and 5 exposed person-years, subject 6 has 5 unexposed and 25 exposed person-years and subject 7 has 30 unexposed and 5 exposed person-years. Among the exposed, there are two deaths (subjects 9 and 10) ($d_1 = 2$), a total of 100 person-years [$(PY)_1 = 100$] and a MR_1 of 2/100 years. Among the unexposed, there is one death (subject 8) ($d_0 = 1$), 100 person years [$(PY)_0 = 100$], and an MR_0 of 1/100 years. The data are summarized in Table 10-1. The MRR is [(2/100 years) ÷ (1/100 years)], or 2.0, indicating that the exposed have a MR two times higher than the MR of the unexposed.

Table 10-2 illustrates the computation of an IPR. The data are from a study of pregnancies among women who worked as telephone operators during the period 1983-1986 (Schnorr et al., 1991). Of the 882 pregnancies among these women during the study period, 366 were among women who used visual display units as directory assistance operators (exposed pregnancies, N_1), and 516 were among women who did not use visual display units and were general operators (unexposed pregnancies, N_0). Spontaneous abortion was the outcome of 54 (c_1) of the 366

Table 10-2
Computation of an incidence proportion ratio (IPR)

	Exposed	**Unexposed**
Number of spontaneous abortion cases	$c_1 = 54$	$c_0 = 82$
Number of pregnancies	$N_1 = 366$	$N_0 = 516$
Incidence proportion, IP	$(IP)_1 = c_1/N_1$	$(IP)_0 = c_0/N_0$ $(IP)_1 = 54/366$ $(IP)_0 = 82/516$ $(IP)_1 = 0.148$ (14.8%) $(IP)_0 = 0.159$ (15.9%)
Incidence proportion ratio, IPR	$IPR = (IP)_1 \div (IP)_0$ $IPR = (c_1/N_1) \div (c_0/N_0)$ $IPR = 0.148 \div 0.159 = 0.93$	

Source: Schnorr et al. (1991)

exposed pregnancies and 82 (c_0) of the 516 unexposed pregnancies, yielding IPs of 54/366 (c_1/N_1) or 14.8% among the exposed and 82/516 (c_0/N_0) or 15.9% among the unexposed and a IPR of 0.93 (0.148/0.159) for exposed compared to unexposed pregnancies. The IPR of 0.93 indicates that the risk of spontaneous abortion was 7% lower among exposed than among unexposed pregnancies.

Table 10-3 illustrates the computation of an OR. The data come from a study of fatal injuries in motor vehicle accidents among workers at a large Brazilian steel plant (Barreto et al., 1997). The investigation included 145 cases who died in motor vehicle accidents while actively employed and 551 controls who also were working at the plant when the cases died. Having a hearing defect was one of the "exposures" evaluated as possibly contributing to the risk of fatal motor vehicle-related injuries. Plant records containing data from employee medical examinations conducted every two years indicated that 13 of the 145 cases, and 22 of the 551 controls, had a hearing defect. The exposure odds were 13/132 (a/b) among the cases and 22/529 (c/d) among the controls, and the OR, the ratio of these exposure odds, was [(13/132) ÷ (22/529)], or 2.37. This OR is interpreted as meaning that workers dying of a motor vehicle injury were about 2.4 times more likely than other workers to have had a hearing defect.

VALIDITY

Validity in an epidemiologic study refers to lack of systematic error due to selection and/or information bias and lack of confounding in the measurement of an association. Systematic error and confounding can lead to the misleading appearance of an effect when in fact the

Table 10-3
Computation of an odds ratio (OR)

	Hearing defect, YES	Hearing defect, NO
Cases	a = 13	b = 132
Controls	c = 22	d = 529

Odds ratio, OR = (a/b) ÷ (c/d) = (ad) ÷ (bc)
OR = (13/132) ÷ (22/529) = (13 × 529) ÷ (22 × 132)
OR = 2.37

Source: Barreto et al. (1998)

exposure under investigation does not cause the disease of interest. Considerations of validity are, therefore, essential in interpreting the results of an epidemiologic study.

Selection Bias

Selection bias is a systematic error that arises because of the manner in which study subjects and their person-time are selected for inclusion in a study. Specifically, selection bias occurs when procedures used to recruit or choose subjects and their person-time for a study result in an actual study group with disease rates or exposure frequencies that differ from the disease rates or exposure frequencies in the larger population consisting of all persons who in theory are eligible to be in the study. Selection bias arises when participation is less than 100% among all persons or person-time theoretically eligible for a study. The proportion of eligible person-time included, or of persons who actually participate, may be determined by the investigator, by the potential subjects themselves, or by both. Typically, concern about possible selection bias is greatest when potential subjects select themselves into a study; that is, when participation is voluntary, as is often the case in epidemiologic research.

Table 10-4 illustrates selection bias. In part one of this example, the OR of 2.0 is computed using data selected in an unbiased manner from the total study base. This version of the study includes all 100 cases that occurred in the study base and a sample of 200 controls, chosen without bias from the total study base. Because all cases are included, and because the controls have the same exposure frequency (22/200 or 11%) as in the total study base, the OR is bias-free. In part two of the example, participation by some of the subjects included in part one is less than 100%. Specifically, 100% of exposed cases participate, but only 80% of other subjects participate (i.e., 80% of unexposed cases, 80% of exposed controls and 80% of unexposed controls). The resulting OR is 2.5, indicating selection bias away from the null; that is, the OR is farther from the OR's null value of 1.0 than the true OR of 2.0.

Selection bias is avoided by enumerating all subjects or person-time eligible for a study by including all eligible subjects or person-time, and by achieving 100% participation. In practice, these selection goals cannot be achieved. Feasibility considerations may require sampling from among all eligible subjects to obtain a manageable study size. Sampling may be random, from a roster of all eligible subjects, or may consist of obtaining a convenience sample. If all eligible subjects are enumerated and random selection is carried out, selection bias may still occur by chance. Participation often is voluntary and will not be 100% of those asked to be in the study. Participants may differ from nonparticipants in terms of exposure and disease relationships, and such differences may distort the exposure-disease association. Evaluation of potential selection bias is difficult, and it usually is not possible to determine whether selection bias has occurred in a particular study.

Information Bias

Information bias can happen only after subjects are selected for a study, during the course of measuring study participants' exposure and disease status. This type of bias occurs because

Table 10-4
Selection bias in an epidemiologic study

Part 1. Unbiased Data

	Exposed	Unexposed	
Cases	20	80	Odds Ratio = (20/80) ÷ (22/178) **OR = 2.0**
Controls	22	178	

Part 2. Selection Bias: Participation is 100% for exposed cases, 80% for unexposed cases, and 80% for both exposed and unexposed controls

	Exposed	Unexposed	
Cases	20	64	Odds Ratio = (20/64) ÷ (18/142) **OR = 2.5**
Controls	18	142	

of errors in classifying subjects (or person-time) with regard to exposure, or with regard to disease. The effect of misclassification depends on whether errors in classification are differential or nondifferential. Differential errors occur if the extent of misclassification of subjects according to disease status varies by exposure category, or if the extent of misclassification of subjects by exposure status differs by disease category (diseased, not diseased). Error in disease classification is referred to as nondifferential when the extent of error is equal for exposed and unexposed subjects, and error in exposure classification is nondifferential when the extent of error in exposure classification is equal for diseased and nondiseased subject.

It is virtually certain that misclassification will occur in epidemiologic studies. This is particularly true of classification of subjects with regard to exposures that occurred in the past (i.e., before the investigators begin the study) and that depend on subjects' memories (e.g., self-reported history of smoking, of occupational exposures, of nonprescription drug use, of diet, etc.) or that must be inferred from indirect information (e.g., occupational exposure to butadiene inferred from information on particular jobs held by a subject in the synthetic rubber industry). Misclassification with respect to disease status also occurs in epidemiologic research.

Disease information may come from death certificates, which do not always accurately record decedents' medical conditions, or from clinical examinations or subjects' self reports, both of which may be subject to error.

Nondifferential misclassification errors generally are considered to have less grave consequences than differential errors (Rothman and Greenland, 1998, pp. 126-132). This is because nondifferential misclassification of a binary exposure or disease variable either produces no bias, or produces bias in a predictable direction; that is, towards the null (Birkett, 1992). For example, an IRR for a causal association, when measured in a study with nondifferential misclassification of subjects relative to a binary exposure variable (exposed v not exposed), will appear to be closer to the null value of 1.0 than it truly is. Differential misclassification, on the other hand, can either exaggerate or reduce the magnitude of an association. To minimize the potential for differential misclassification, investigators typically use objective procedures to determine subjects' exposure and disease status. That is, instead of obtaining information from subjects' unsubstantiated reports of exposure or disease, investigators preferably would use industrial hygiene measurements, laboratory tests, historical exposure or disease records, and would review and process all such data in a "blind" fashion (see later discussion).

An objective approach to exposure and disease ascertainment is, of course, highly desirable. However, objectivity merely ensures that classification errors will **tend** to be nondifferential. In small studies, it is reasonable to assume that misclassification of just a few subjects could introduce either differential or nondifferential error.

Table 10-5 illustrates nondifferential misclassification in a study examining the relationship between a binary exposure variable and mortality from a particular cause of death. Part one of the example shows the correctly classified exposure distribution of decedents and person-years. The unbiased MRR is 2.0. In part two of the example, 5% of exposed decedents and person-years and 10% of unexposed decedents and person-years are incorrectly classified with respect to exposure. The misclassification is dependent on exposure status (i.e., it is greater for the unexposed than for the exposed), but not on disease status (i.e., within each of the two exposure categories, the per cent misclassified is the same for decedents and person-years). The MRR of 1.8 is biased towards the null; that is, the MRR is closer to the null value of 1.0 than the true MRR of 2.0. A similar result would be obtained if 10% rather than 5% of exposed decedents and person-years, and 5% rather than 10% of unexposed decedents and person-years, were incorrectly classified with respect to exposure.

Table 10-6 illustrates differential misclassification in a study comparing the odds of exposure to a particular agent in a group of cases of a given disease with the odds of exposure in a group of controls without the disease of interest. Part one of the example shows the correctly classified data with an unbiased OR of 2.0. In part two, it is assumed that subjects were asked about their exposure history, and that recall of exposure was perfectly correct for 100% of both exposed and unexposed cases and for 100% of unexposed controls. However, only 82% of truly exposed controls remembered their exposure. Thus, subjects' exposure misclassification errors were differential; that is, dependent both on exposure status (more accurate for unex-

Table 10-5
Nondifferential exposure misclassification in an epidemiologic study

Part 1. Correctly classified exposure data for decedents and person-years

	Exposed	Unexposed	
Deaths	200	100	Mortality rate ratio = 200/100 **MRR = 2.0**
Person-years	100,000	100,000	
Mortality rate (MR) (x 10^5 years)	200	100	

Part 2. Biased data due to nondifferential misclassification of exposure: 5% of exposed decedents and person-years are misclassified (counted as unexposed); 10% of unexposed decedents and person-years misclassified (counted as exposed)

	Exposed	Unexposed	
Deaths	200*	100‡	Mortality rate ratio = 190/105 **MRR =1.8**
Person-years	105,000†	95,000§	
Mortality rate (MR) (x 10^5 years)	190	105	

* 200 – (0.05x200) + (0.10x100).
† 100,000 – (0.05x100,000) + (0.10x100,000).
‡ 100 - (0.10x100) + (0.05x200).
§ 100,000 – (0.10x100,000) + (0.05x100,000).

Table 10-6
Differential exposure misclassification in an epidemiologic study

Part 1. Correctly classified exposure data for cases and controls

	Exposed	**Unexposed**	
Cases	20	80	Odds ratio = (20/80) ÷ (22/178) **OR = 2.0**
Controls	22	178	

Part 2. Biased data due to differential misclassification of exposure: 100% of both exposed and unexposed cases and 100% of unexposed controls are correctly classified; 18% (4) of exposed controls are misclassified as unexposed

	Exposed	**Unexposed**	
Cases	20	80	Odds ratio = (20/80) ÷ (18/182) **OR = 2.5**
Controls	18	182	

posed) and on disease status (more accurate for cases than for controls). The biased OR in part two of the example is 2.5, and the bias is away from the null.

Many strategies of study design and data collection and development are used for the purpose of ensuring that classification errors are nondifferential. These include using comparable data collection and development techniques for the two or more groups to be compared in a study; using "blind" procedures for determining exposure and diseases status (i.e., persons responsible for classifying subjects according to exposure are unaware of subjects' disease status, and persons responsible for classifying subjects according to disease status are unaware of subjects' exposure status); and using objective records, rather than self-reports, to determine subjects' exposure and disease status.

Another approach that is controversial but continues to be used involves selecting study groups that are likely to be comparable in terms of the extent of misclassification in their data (Greenland and Robins, 1985). For example, if all cases in a case-control study are deceased, an investigator may choose to require that all controls also be deceased. The rationale is that the two decedent groups will have a similar degree of misclassification in data, such as information on smoking and lifetime job history, that must be furnished by surrogates. If some cases are alive at the time of data collection and some are deceased, an investigator may match cases and controls on vital status, choosing living controls for cases who are alive and deceased controls for cases who are dead.

Confounding

Confounding refers to an error that occurs when the groups being compared in a study differ with respect to disease determinants other than the factor whose possible effect is being assessed. When an association is confounded, the estimated effect of the exposure is distorted because it is mixed with the effect of another factor. In order to be a confounder a factor must be predictive of the occurrence of the disease under study and must be associated with the exposure under study, as explained further in the next example. The impact of confounding can be large or small and can produce an under- or over-estimate of the effect of the exposure of interest.

Table 10-7 illustrates confounding by age in a hypothetical study of the relationship between occupational exposure to bis-chloromethyl ether and the occurrence of lung cancer. In this example, the rate of lung cancer is three times higher among the exposed than among the unexposed **within each age group** (i.e., each of the three age-specific IRRs is 3.0). However, if one ignores age and computes the rate for the entire exposed group (i.e., for the "total") as done in the first part of this example, this rate is 51 cases/64,000 person-years, for an IR of 79.7/100,000 years. Again, if one ignores age and computes the rate for the entire unexposed group, the IR is 13 cases/35,000 person-years, or 37.1/100,000 years. The IRR is 79.7/37.1 or 2.1, indicating erroneously that the rate of lung cancer is only two times higher among the exposed than among the unexposed. In this example, age is a confounder:

- age is "predictive of" the lung cancer IR – the older the age, the higher the IR; and
- age is associated with exposure – the exposed group is younger than the unexposed group, since 70% of the person-years of the exposed are in the youngest age group, whereas only 57% of the person-years of the unexposed are in the youngest age group.

Epidemiologists use several strategies to control for potential confounding. These consist of restricting a study to subjects in one category of a potential confounder (e.g., if a study is restricted to men; gender cannot be a confounder) and adjusting measures of association for potential confounding by using certain analytic techniques, such as standardization, other stratified analysis procedures, or multivariable modeling. The main purpose of these analytic procedures is to remove distortion due to the groups' being compared having different distributions of causal factors other than the particular factor under investigation.

A measure of occurrence or of association that does not take other factors (potential confounders) into consideration is referred to as "crude." For example, the data in Table 10-7 indicate that the crude IRR (i.e., the IRR computed ignoring age) is 2.1. A measure of occurrence or of association that uses stratified or multivariable analysis to consider a factor or fac-

Table 10-7
Confounding by age in a study of bis-chloromethyl ether (BCME) exposure and lung cancer

Age group	Exposed to asbestos: Obs.* cases	Exposed to asbestos: Person-years	Exposed to asbestos: IR (incidence rate)	Unexposed to asbestos: Obs. cases	Unexposed to asbestos: Person-years	Unexposed to asbestos: IR (incidence rate)	Age-specific IRR (rate ratio)
20-39 yrs	24	40,000	$60/10^5$ y	400	2,000,000	$20/10^5$ y	3.0
40-59 yrs	18	15,000	$120/10^5$ y	400	1,000,000	$40/10^5$ y	3.0
60-89 yrs	9	3,000	$300/10^5$ y	500	500,000	$100/10^5$	3.0
Total	51	64,000	79.7	1,300	3,500,000	37.1	**2.1 (crude IRR)**

Standardization of the IRR to obtain a standardized incidence ratio (SIR)

Age group	Obs.* cases	Expected number (in each age group, the person-years of the exposed, multiplied by the rate of the unexposed†)	IRR (Obs/Exp no.)
20-39 yrs	24	40,000 PY x $20/10^5$ y = 8	24/8 = 3.0
40-59 yrs	18	15,000 PY x $40/10^5$ y = 6	18/6 = 3.0
60-89 yrs	9	3,000 PY x $100/10^5$ y = 3	9/3 = 3.0
Total	51	8 + 6 + 3 = 17	**51/17 = 3.0 (age-adjusted IRR= SIR)**

* Obs, observed number of cases of lung cancer.
† Expected, expected number of cases of lung cancer in the exposed, if the exposed had the same lung cancer rate and same age distribution as the unexposed.

tors other than the particular one under investigation is referred to as "adjusted" for the other factor(s).

As shown in part two of the example in Table 10-7, an adjusted IRR would be 3.0, obtained by dividing the observed number of lung cancer cases among the exposed by the expected number, 51/17 = 3.0. Table 10-7 illustrates the computation of an age-adjusted IRR for the exposed compared to the unexposed, obtained using a technique called standardization. The latter technique is a form of stratified analysis. In standardization, one computes an "expected" number of cases of lung cancer for each age group of the exposed. The expected number is the number of cases that would occur in the exposed if they retained their own age distribution but had the same age-specific lung cancer IRs as the unexposed group. Such an IRR is referred to as a standardized IR (SIR) or standardized MR (SMR). In this example, the SIR is 3.0, exactly as one would expect, given that each of the three age-specific IRRs is 3.0. This SIR also can be viewed as a weighted average of the age-specific IRRs, with the weight for each age-specific IRR equal to the person-years of the exposed in the corresponding age group, multiplied by the age-specific rate in the unexposed (see Table 10-7, part 2).

Confounding can be either negative or positive (Rosner, 1995, p. 401). Negative confounding, when the crude is lower than the adjusted measure of association, occurs if the confounder is associated positively with disease and negatively with exposure, or if the confounder is associated negatively with disease and positively with exposure. The above example illustrates negative confounding. The crude IRR of 2.1 is lower than the adjusted IRR of 3.0, and the confounder, age, is associated positively with lung cancer and negatively with exposure. Positive confounding, when the crude is higher than the adjusted measure of association, occurs either if the confounder is positively associated both with disease and with exposure, or if the confounder is negatively associated both with disease and with exposure.

PRECISION

Precision refers to lack of random error in the measurement of an association, and random error refers to that part of an experience that is due to chance (see Volume 1, Chapter 7). Random error can lead to the appearance of an effect, even when the exposure under investigation actually does not cause the disease of interest. In interpreting any association observed in a particular epidemiologic study, it is important to consider, among other things, how likely it is that the association could have occurred by chance, if no causal effect exists. Similarly, when no association is seen, it is important to consider whether a true effect could have been missed simply by chance; that is, simply because the study was too small. Measurement of the precision of IRRs, MRRs, IPRs or rate or risk differences allows one to evaluate the role of chance in producing an apparent relationship or in failing to find a true effect.

Precision depends on the size of the study, on the relative size of the groups being compared, and on the frequency of exposure and of disease in the populations studied. Precision can be improved by increasing the number of subjects or other units of observation to be included in an investigation (Rothman and Greenland, 1998, pp. 135-137). When cost and effort constraints make it infeasible to include larger numbers of subjects, precision can be enhanced by focusing on a study base in which the disease and/or the exposure of interest are relatively common. For example, one might choose to restrict a study according to age, including only older persons in whom disease rates tend to be relatively high; or one might choose to restrict a study according to geographic region, including only those persons living in a region where the exposure of interest is relatively common. Matching (defined briefly in a later section) groups to be compared in a study on the basis of potential confounders may also be used to increase precision.

The precision of estimates of IRRs, MRRs, or IPRs, of rate or risk differences, of ORs, and of other measures of association of interest in epidemiology typically is expressed as a confidence interval or as a standard error. P-values also may be used. Confidence intervals and p-values are discussed briefly below. Computation of these measures is beyond the scope of this chapter, but details are provided in epidemiology text books (Checkoway et al., 1989; Rothman, 1986; Rothman and Greenland, 1998).

Confidence Interval

A confidence interval (CI) specifies the range of values of a true effect with which the data from a particular study are compatible. The width of the CI of a particular measure of association is determined by the extent of random variability in the study and by an arbitrary "confidence level," typically 95% in observational epidemiologic research. For example, in a study of synthetic rubber industry workers, the MRR for leukemia among hourly workers compared to the general population was 1.4, with a 95% CI of 1.0-1.9. Assuming the point estimate of 1.4 is valid, this CI indicates that it is 95% likely that the true value of the MRR lies in the range of 1.0 to 1.9. If a series of similar investigations of synthetic rubber industry workers were conducted, the MRR would be in the 1.0-1.9 range in 95% of the studies.

P-Value

Another measure of random error is the P-value, computed from a statistical hypothesis test. P often is construed as the probability that a test statistic as large or larger than that found in a study would arise by chance alone if the exposure and disease were not, in fact, associated. P-values range from 0 to 1, with small P-values often taken as indicating that the observed results have little compatibility with the null hypothesis. Usually, a P-value of 0.05 or lower is considered statistically significant, implying that the result probably did not occur by chance. These definitions and interpretations of P-values are conventional, but inaccurate, and can be misleading (Rothman and Greenland, 1998, pp. 185-186). CIs and standard errors are considered to be preferable to P-values in expressing the amount of precision in epidemiologic studies (Lang et al., 1998).

Power

Power indicates the adequacy of a given study size (number of subjects or amount of person-time) for detecting an exposure-disease association of a certain minimum magnitude, if in fact a true effect exists. Consideration of power can be useful for research planning purposes. This is because a power calculation indicates whether a study of a particular size will yield statistically significant results, given an outcome of interest of a particular frequency, and given an assumption about the magnitude of the association to be investigated and about the statistical tests to be performed. If, for example, exposure to ethylene glycol ether is assumed to produce a fivefold increase in the IP for congenital malformation, a smaller number of pregnancies can be studied than if ethylene glycol ether is assumed to produce a twofold increase in IP. Similarly, a statistically informative study of a medical condition occurring in occupational groups will require fewer subjects if the condition is common than if the condition is rare. After a study has been conducted, the issue of power has little relevance; instead, the best way to assess precision is to examine CIs (Checkoway et al., 1989, pp. 76-77).

MAJOR EPIDEMIOLOGIC STUDY DESIGNS

Most research in occupational epidemiology is nonexperimental. Because investigators do not manipulate the exposure conditions experienced by study subjects, nonexperimental research is more prone to systematic error and to confounding than are experiments. Each of the two main study designs used in epidemiology (the follow-up and the case-control study designs) has specific uses and areas of application, specific measures of association, and particular advantages and disadvantages. Also, both follow-up and case-control studies have several important variants.

Follow-up Studies

In a follow-up study, a group of individuals who have a characteristic in common (the "study group" or "subjects", sometimes referred to as a "cohort") and who are initially free of the disease of interest is identified. Each subject is monitored ("followed up") over time to determine changes in exposure status. This information may be used to form exposed and unexposed subgroups of subjects or subgroups classified according to level of exposure; alternatively, all members may be considered "exposed." Each person in the study group is followed up over time to determine whether he or she develops the disease of interest, information that is used to measure the groups' morbidity and/or mortality rates. The subsequent analysis determines if the IR, MR, or IP of exposed subjects is higher, lower or the same as that of unexposed subjects.

A follow-up study may involve open or closed populations (subjects). In studies of open groups, the unit of observation usually is a unit of person-time, such as a person-year, and the measure of association is an IRR or an MRR. In follow-up studies of closed groups, the unit of observation may be either person-time or the individual subject, and correspondingly, the measure of association is an IRR or MRR or an IPR. Most follow-up studies in the occupational setting are of open study groups, and most estimate IRRs or MRRs as measures of association.

A follow-up study may be retrospective (historical) or prospective. In retrospective follow-up studies, subjects are identified as of some time in the past and are followed into the present, often over several decades. In prospective follow-up studies, the study group is enumerated currently and followed into the future. Prospective follow-up studies investigating the relationship between exposures and chronic diseases may continue for many decades. Some studies have both retrospective and prospective components.

Retrospective follow-up studies of completely closed study groups are rare in occupational epidemiology, whereas prospective follow-up studies of closed study groups are more common. In occupational epidemiology, follow-up studies are used to investigate the causes of cancer, chronic respiratory disease, and many other chronic diseases. Retrospective follow-up studies have been used particularly often to investigate fatal diseases, because mortality registries, which facilitate such studies, have existed in many countries for many decades.

Retrospective follow-up studies have not been as useful for determining the etiology of nonfatal diseases or physiologic dysfunction (e.g., pulmonary, renal, neurologic dysfunction). These are best suited to prospective follow-up studies. Chronic diseases with long induction times (amount of time required from first exposure to a causal agent to the development or detection of overt, irreversible disease) have most often been studied with retrospective follow-up studies. This is because prospective follow-up study design would require enormous resources to observe large numbers of subjects into the future to determine their exposures and disease experience, with evaluation of hypotheses possible only after much elapsed time.

The key steps required to conduct follow-up studies are:

- Specification of objectives
- Identification of study group members and specification of exposed groups
- Identification of comparison groups
- Follow-up or tracing of subjects to determine their vital status, their disease status and their individual person-years of observation
- Analysis, or comparison of disease rates between exposed and unexposed groups or across groups specified on the basis of level of exposure; control of confounding; examination of the cause-to-effect time sequence (induction time effects)
- Interpretation of results.

Follow-up studies may have a rather broad, descriptive purpose, such as determining if workers in a particular plant, group of plants, or occupation have any unusual disease occurrences. Alternatively, follow-up studies may be done to evaluate a specific hypothesis, such as the hypothesis that exposure to butadiene causes leukemia in humans (Delzell et al., 1996; Macaluso et al., 1996). A clear statement of objectives and rationale helps to guide design, data collection, analysis and interpretative decisions (Epidemiology Task Group, 1991).

In occupational epidemiology subjects may be defined as all or some employees of a particular plant or of a group of facilities in a given industry; subgroups of employees in a plant or set of plants, specified on the basis of exposure to a putative etiologic agent of interest; members of a union or professional organization; persons applying for a license to carry out certain types of work (e.g., pesticide applicators). Eligibility for inclusion may also be determined on the basis of consideration of duration of employment (short-term workers often are excluded), time period of employment, payroll status (wage v. salaried), and other factors. The two main steps in specifying a study group are the identification of individuals employed at the facilities or in the job group(s) under investigation, and the identification of subjects with and without exposure.

In occupational follow-up studies of open study groups, subjects are identified on the basis of membership in an employee or job group at any time during the study period. This typically is done by reviewing plant, union or licensing records. The types of record of interest are those that identify all employees regardless of employment status, vital status or job type, and that have information on name, Social Security number, birth date, race, gender, and employment dates. Eligibility for inclusion as a subject typically requires, at minimum, information on name, Social Security number, birth date, race, gender, and employment dates.

Exposure is defined in a variety of ways in occupational epidemiology. In some studies, exposure may simply be defined as employment in a particular industry or job, with no agent specified. The null hypothesis under investigation will be that employment in that particular industry or job has no effect on a disease or set of diseases of interest. Typically, in such studies all subjects are classified as exposed at the beginning of their observation period, which often coincides with the date of first employment. Amount of exposure is defined as duration of employment in the industry or job group.

In other studies of particular agents, the specification of exposure groups and the measurement of exposure may be considerably more complex. Entry into the study may require employment at a particular facility or in an industry or job that used the agent. The subjects are then examined over time to determine which individuals experience exposure to the agent and to determine among those ever exposed, the timing of their exposures. The study group's person-time experience is then divided into exposed and unexposed subgroups. Alternatively, eligibility for inclusion in a study may be determined after a group of workers is identified and subjects' records are examined to determine if they were ever exposed. Subsequently the study may be restricted to those employees with at least some exposure to the agent of interest. In either case, retrospective exposure estimation must be carried out to determine which subjects were exposed and to determine the timing of exposure and the amount of exposure. Values not only of exposure, but also of other "time-dependent" variables must be determined for each person-year of follow-up of a subject. Examples of time dependent variables that for a given subject may vary from person-year to person-year include the calendar year of observation, age, duration and amount of exposure, duration of employment, and years since first exposure.

Retrospective exposure assessment almost always is problematic in epidemiology and is particularly so when exposure is involuntary (subjects do not know whether or not they were exposed). Exposure is routinely considered to be involuntary in retrospective follow-up studies in occupational settings. Historical industrial hygiene measurements of the workplace are rarely adequate to characterize subjects' exposure. Instead, complex historical assessment procedures, often combining measurements with extrapolation, modeling and expert judgement, are used to construct one or more "job-exposure matrices" (see, for example, Hallock et al., 1994; Hammond et al., 1995; Stewart et al., 1992, 1995, 1996).

Development of a job-exposure matrix begins with an understanding of plant operations; process or work areas; job titles; potential for exposure to one or more agents within a particular work area, job, or combination; and changes in processes, exposure potential, and job characteristics over time. This information may be gained by conducting plant walk-throughs; by examining operations at other, similar plants; by examining industrial hygiene and plant engineering records; by conducting industrial hygiene exposure measurement surveys of current workplace conditions; and by obtaining opinions about past exposure conditions from experts. Next, a matrix specified on the basis of work area, job, or combinations of work area and job on one axis, and time period on the other axis, is constructed. The work area/job axis covers all work areas and/or jobs appearing in subjects' work histories. Each category on this axis consists of one or more work areas, jobs, or combinations assumed to have relatively homogeneous exposure potential within a particular time interval. The time axis covers the entire temporal experience of subjects included in the study, from the earliest to the most recent date of potential exposure at work. Each category on this axis consists of an interval of time during which exposure in a particular work area/job is considered constant. Next, exposure estimates are developed (using expert opinion, historical industrial hygiene measurements, extrapolation or modeling) for each work area/job/time period combination in the matrix. These estimates are linked to each subject's work history to develop cumulative exposure indices for each person-year of observation.

Table 10-8 illustrates a job-exposure matrix and the linkage of a job-exposure matrix to a subject's work history in order to estimate the subject's cumulative exposure to 1,3-butadiene (BD). The data come from a study of the synthetic rubber industry that evaluated the occurrence of leukemia and other diseases among workers exposed to BD and styrene (Macaluso et al., 1996). Part one of Table 10-8 shows a partial job-exposure matrix for BD. The "job" axis specifies three (out of many) work areas at one of the eight plants included in the study. The "time" axis specifies seven time intervals during the overall study period of 1943 through 1991. Each value within the matrix is an estimate of the eight-hour time-weighted average intensity of BD exposure (in parts per million, ppm) for a particular work area/time period combination. For example, in 1943-1947 the BD exposure estimate was 7.8 ppm for a reactor operator.

Part two of Table 10-8 displays the job history of a particular study subject who was employed at the study plant from August 1943 to April 1971. The data include for each of the subject's jobs, the dates of employment during the time periods indicated in the job-exposure

Table 10-8
Example of a partial job-exposure matrix for butadiene (BD) exposure intensity (ppm) and of computation of cumulative exposure (ppm-years) for one subject by linking a job-exposure matrix to a work history

Part 1. Job-exposure matrix

	Time Axis						
Job axis (work area)	**1943-47**	**1948-59**	**1960**	**1961**	**1962-64**	**1965-67**	**1968-72**
Tank farm operator	9.6	9.3	5.6	5.2	5.3	4.2	3.8
Reactor operator	7.8	7.2	4.0	4.0	4.0	2.9	2.4
Recovery operator	7.9	7.9	7.8	7.8	7.8	5.0	3.8

Part 2. Linkage with one subject's work history and estimation of cumulative exposure

	Dates				Cumulative ppm-years	
Job	**Start**	**End**	**Years worked**	**BD intensity (ppm)**	**This job**	**This plus previous jobs**
Tank farm operator	08/43	09/44	1.1	9.6	10.6	10.6
Recovery operator	10/44	12/44	0.25	7.9	2.0	12.6
Reactor operator	01/45	12/47	3.0	7.8	23.4	36.0
Reactor operator	01/48	10/48	0.84	7.2	6.0	42.0
Recovery operator	11/48	12/59	11.2	7.9	88.3	130.3
Recovery operator	01/60	12/64	5.0	7.8	39.0	169.3
Recovery operator	01/65	12/67	3.0	5.0	15.0	184.3
Recovery operator	01/68	04/71	3.3	3.8	12.6	196.9

matrix, the estimated BD intensity from the job-exposure matrix, the years worked in the job from the start to the end date, the cumulative exposure (ppm × years) while working in the job from the start to the end date, and the cumulative exposure (ppm × years) from the beginning of employment through the end date of the job. At the end of employment, the subject's overall cumulative BD exposure is obtained by summing ppm-years over all job/time period combinations. In this example, the cumulative exposure and the end of employment in 1971 is 196.9 ppm-years.

The next step in carrying out a follow-up study is specification of a comparison group. Conceptually, the comparison group provides the background disease rate, the rate that would be expected in the exposed group if exposure had not occurred. In retrospective follow-up studies, comparison groups may be "external" (consisting of subjects from outside the work setting of the exposed subjects; typically, a national, state, or regional general population) or "internal" (consisting of subjects who come from the same occupational setting as the exposed, but who are unexposed, or who have low exposure to the agent of interest). Each of the two types of comparison group, in theory, has advantages and disadvantages over the other type.

External comparison groups are large, provide statistically precise data, and cost little in terms of time, money and effort. However, general population groups differ from occupational groups in several ways. They may have different risk factor distributions, and procedures used to compile information on disease occurrence cannot be assumed to be identical for the general population and a study group. These differences may lead to uncontrolled confounding and to information bias when comparisons are made. The "healthy worker effect" (defined and described later in more detail) is a manifestation of these multiple problems. Internal comparison groups tend to be small and to provide information subject to considerable random statistical variability. However, because they derive from the same occupational setting as the exposed groups, they may provide comparisons less susceptible to confounding and information bias.

Once all subjects in the study group(s) have been identified, they are traced to determine their vital status. Plant records may be used to identify subjects who died while actively working or while retired and receiving pension benefits. Other vital status tracing is done by linking information on the subjects, compiled from their employment records, with data from various federal and state government agencies, including records of:

- the National Death Index (contains data on decedents in the United States who died in 1979 or later)
- the Social Security Administration Death Master File (contains data on decedents in the United States who died in about 1940 or later)
- the Internal Revenue Service
- the Health Care Financing Administration (HCFA) and
- state drivers license offices.

The first four of the above information sources are helpful both with identifying deceased subjects and with confirming that other subjects were alive at the close of the study. Drivers license data usually are useful only for confirming as alive those subjects not identified as deceased. Credit bureau and post office records also may be useful for tracing purposes, if it is necessary to locate and contact subjects to determine if they are alive.

Usually, a complex combination of record linkages and individual tracing is used to identify decedents in the study group and to determine if other persons who have left employment are still alive. Typically it is possible to document the vital status of at least 95% of an occupational study group. This percentage may be lower for women than for men, because surname changes among women make tracing and record linkage more difficult.

The next step is the determination of disease status. In retrospective follow-up studies, disease status most often is determined from death certificates as the underlying cause of death. Medical outcomes other than death from particular diseases are rarely determined because old records do not exist for doing this. For example, it is unlikely that medical records, covering both active employment and post-employment time periods, would be available for identifying cases of nonmalignant respiratory disease, neurologic dysfunction, reproductive dysfunction, musculoskeletal disorders, and the like.

Death certificates may be obtained from states where deaths occurred and evaluated by one or more nosologists. Nosologists use death certificate information on medical conditions leading to death to identify the "underlying" cause of death, as well as contributory causes, and to assign numerical disease codes using a standard classification system, called the International Classification of Diseases (ICD). ICD-coded underlying causes of death are used to compile death rates at the local, state and national levels. The National Death Index includes cause of death ICD codes for persons dying in or after 1979.

Other, rarely used options for determining disease status in retrospective follow-up studies are linkage with disease morbidity registries (e.g., state cancer registries; United States Renal Data System, which contains data on Medicare recipients receiving renal dialysis for end-stage renal disease), and the use of questionnaires sent to living subjects or the next of kin of deceased subjects asking about disease experience. The latter approach requires concomitant retrieval of historical medical records to confirm subject-reported illnesses. Questionnaires are the least used approach, because they require ascertainment of subjects' current addresses and/or telephone numbers, as well as the subjects' written consent to participate, and because medical record retrieval can be extremely difficult for illnesses diagnosed years before data collection is done for a particular epidemiologic study. Regardless of the procedures used to obtain disease information, it is important that determination of disease status be conducted by persons who are unaware of subjects' exposure status in order to avoid information bias.

Once disease status has been ascertained, it is possible to compute the person-years of observation of each subject. Usually the person-years of an individual subject's observation period extend from the subject's date of first employment, date of achieving some minimum duration of employment (e.g., one year) or date of first exposure to a particular agent, until the subject is no longer a candidate for disease. Most often, the end of the observation period is determined by death (in mortality studies), by diagnosis of the disease of interest (in morbidity studies), by loss to follow-up (i.e., the point in time after which the subject cannot be traced to determine vital status), or by the end date of the study chosen by the investigator (as close in time to the present as possible).

Analysis involves the computation of IRs or MRs (these always should be used in follow-up studies of open groups) or IPs (these, or IRs or MRs, may be used in follow-up studies of closed groups) and the estimation of measures of association, usually IRRs, MRRs or IPRs. Control of confounding in the epidemiologic analysis of follow-up studies is accomplished through stratified or multivariable analysis techniques.

The basic approach to stratified analysis (Table 10-7) involves dividing the exposed and referent groups into categories ("strata") of a potential confounder, such as age (e.g., each stratum may consist of a five-year age group). Within each stratum of age, the disease rate of the exposed group is compared with the disease rate of the unexposed group, and a measure of association (the "stratum-specific" measure, or in this example, the "age-specific" measure) is computed. A "summary" measure of association is then obtained by computing a weighted average of the stratum-specific measures. This summary measure is said to be "adjusted" for confounding by age. Commonly used adjusted measures of association obtained by stratified analysis procedures include the standardized mortality ratio (SMR) and the Mantel-Haenszel rate ratio (MHRR). The SMR, illustrated in table 10-7, is used most often for comparing a study group to an external referent group, typically a national or regional general population. The MHRR typically is used for comparing a study group to an internal referent group. In addition to differing in terms of the typical referent group, the SMR and the MHRR differ with respect to their computational formulas (Checkoway et al., 1989, pp. 122-125).

Multivariable analysis procedures for obtaining adjusted summary measures of association include Poisson regression and proportional hazards modeling. These procedures are used primarily in internal analyses. In both a stratified and a multivariable analysis, dose-response is evaluated by examining patterns in measures of association across categories or levels of exposure. Induction time also should be examined in the analysis. This may be done by examining confounder-adjusted risk ratios within discrete categories of time since first exposure or by restricting an analysis to the subject experience 15-20 or more years since first exposure.

Potential confounders usually considered in the analysis of an occupational follow-up study include gender, race, age, and calendar time. Geographic location also may be controlled. Other potential confounders, such as smoking and dietary factors, often are not considered because collection of the pertinent data is difficult, especially in retrospective follow-up studies. Excellent computer software is available for performing stratified and multivariable analysis (Monson, 1974; Marsh et al., 1998; Preston et al., 1993).

Interpretation of a follow-up study begins with an assessment of the quality of the investigation. Quality is judged by efforts made to avoid selection and information biases and to control for confounding. A high-quality follow-up study will have the following attributes relating to the potential for selection bias, information bias, and confounding:

Selection bias: the study should include all eligible subjects and should have a low proportion (<5%) of subjects lost to follow-up; if these conditions are not met, there should be no evidence that inclusion in the study or in subject exposure subgroups was related to disease (e.g., study procedures should provide assurance that the records of deceased employees were not missed in assembling the study group) or that the probability of being lost depended both on exposure and disease (e.g., there should be no evidence that subjects lost to follow-up are more likely to be alive than subjects who were traced).

Information bias: the study should have used comparable procedures to ascertain disease in exposed and unexposed or general population referents and should have used objective data sources and procedures to determine level of exposure(i.e., this should have been done without knowledge of disease status).

Confounding: the study should have controlled for possible confounding by personal characteristics such as age, gender, and race; it should have considered whether extrinsic risk factors for the disease of interest (smoking, other occupational exposures) may be related to the exposure of interest and, if so, should have controlled for potential confounding by these factors; it should have used both internal and external referent groups when possible to estimate measures of association.

EXAMPLE OF A RETROSPECTIVE FOLLOW-UP STUDY

A recent investigation of mortality patterns among industrial workers exposed to acrylonitrile (AN) provides a useful example of a retrospective follow-up study (Blair et al, 1998). AN is an important industrial chemical used to make synthetic fibers and resins. It causes cancer of the brain, mammary gland, Zymbal gland, and forestomach in animals, but it is unknown whether it causes cancer in humans who are occupationally or environmentally exposed. To investigate the potential human carcinogenicity of AN in humans, Blair et al. conducted a retrospective follow-up study of persons employed at any of four plants that produced AN monomer, three plants that produced acrylic fibers, and one plant that made AN resins. All plants were in the United States. Eligibility was restricted to persons who began working at the plants before 1984, when study plans were initially developed, and who were employed at any time after the beginning of AN operations at each plant, ranging from 1952 to1965. There were 25,460 men and women who met these criteria. Plant records were used to obtain data on each subject's name, race, gender, birth date, Social Security number, and work history. The closing date of the study period was January 1, 1990.

Exposure estimates were developed for about 18,000 combinations of plant, department, job title, and time period (Stewart et al., 1992, 1995, 1996). Historical monitoring data, most of which dated from the late 1970s, and extrapolation backwards in time based on engineering and other workplace changes were used to develop estimates of time-weighted average exposure intensities (ppm of AN) for an eight hour period, as well as estimates of cumulative exposure, expressed in ppm-years. Cumulative exposure was estimated for a particular subject as the ppm intensity for each job/time period category in the subject's work history, multiplied by the amount of time spent in that category, and summed over all job/time period combinations in the subject's work history.

Of the total 545,368 person-years included in the study, 348,642 were exposed, and 196,727 were unexposed to AN. The exposed person-years were further divided into AN dose categories of >0-0.13, >0.13-0.57, >0.57-1.5, >1.5-8.0 and >8.0 ppm-years. The unexposed subject experience comprised an internal referent group. In addition, the United States general population served as an external referent group.

Determination of vital status as of January 1, 1990, involved linking the subjects' records with records of the plants, the Social Security Administration, the National Death Index, the Health Care Financing Administration, and credit bureaus. These linkages identified 2,038 subjects as deceased (8% of the study group) and 22,438 (88%) as alive. Vital status could not be determined for 984 (4%).

The unit of analysis was the person-year of observation. In analyses using an external referent group, stratified analysis procedures were used to compute SMRs, adjusted for age, gender, race, and calendar period of observation for the exposed subgroup compared to the United States general population, and for the unexposed subgroup compared to the United States general population. In analyses using an internal referent group, Poisson regression modeling was used to compute MRRs, adjusted for year of birth, plant, and calendar period for the total exposed group and for subgroups of the exposed, specified on the basis of ppm-years of exposure compared to the unexposed.

The overall crude mortality rate was 702 deaths/196,727 person-years (or 3.6/1,000 years) in the unexposed and 1,217 deaths/348,642 person-years (or 3.5/1,000 years) in the exposed (Table 10-9). Both exposed and unexposed subject groups had a favorable overall mortality experience when compared to the United States general population: the SMR of 0.7 for each subgroup indicated that their all-causes mortality rate is 30% lower than the mortality rates of the general population, after adjusting for differences between each subgroup and the general population. Results for all cancers combined, cerebrovascular disease, ischemic heart disease, nonmalignant respiratory disease, and accidents also indicated a favorable mortality experience. Both for the exposed and for the unexposed, the 95% CI of the all-causes SMR was quite narrow, indicating that for all causes of death combined the results were precise, as would be expected in a study this large. Also, the 95% CI of each all-causes SMR had an upper limit that was below the null value of 1.0. This indicated that the overall MRs of the exposed and unexposed subjects were statistically significantly below general population rates at the 0.05 probability level.

These results are illustrative of the healthy worker effect, mentioned earlier. The healthy worker effect refers to the observation that mortality rates among occupational study groups are consistently lower than rates in the general population (Checkoway et al., 1989). The effect

Table 10-9
Study of workers exposed to acrylonitrile (AN): results of analyses using an external referent group*

	Exposed to AN (PY=348642)			Unexposed to AN (PY=196727)		
Cause of death	**OBS**	**SMR**	**95% CI**	**OBS**	**SMR**	**95% CI**
All causes	1217	0.7	0.6-0.7	702	0.7	0.7-0.8
All cancers	326	0.8	0.7-0.9	216	0.9	0.8-1.1
Colon	19	0.6	0.4-0.9	19	1.0	0.6-1.5
Lung	134	0.9	0.8-1.1	59	0.8	0.6-1.1
Central nervous system	12	0.7	0.4-1.3	11	1.3	0.7-2.3
Lymphopoietic system	27	0.6	0.4-0.9	18	0.8	0.5-1.2
Cerebrovascular disease	37	0.5	0.4-0.7	23	0.5	0.4-0.8
Ischemic heart disease	374	0.8	0.7-0.9	186	0.8	0.7-0.9
Respiratory disease	40	0.4	0.3-0.6	36	0.7	0.5-1.0
Cirrhosis of the liver	18	0.3	0.2-0.4	12	0.4	0.2-0.7
Accidents	156	0.8	0.7-1.0	63	0.7	0.5-0.9
Suicide	57	0.8	0.6-1.0	34	1.0	0.7-1.4

* Abbreviations: OBS, observed number of deaths; SMR, standardized mortality ratio, CI, confidence interval.
Source: Blair et al. (1998)

may be due to a combination of selection bias, information bias, and confounding. Because the study by Blair et al. used procedures that probably avoided information bias in the form of underascertainment of deaths (e.g., a high proportion of the study group was successfully traced to establish vital status), the healthy worker effect most likely was due primarily to confounding by unidentified factors associated with health and with fitness for employment in the AN-related industry.

For most specific forms of cancer, SMRs for exposed and for unexposed subjects also were below or close to the null value of 1.0 (Table 10-9). Exposed and unexposed subjects had a lung cancer SMR of 0.9 (95% CI, 0.8-1.1) and 0.8 (95% CI, 0.6-1.1), respectively, and a central nervous system cancer SMR of 0.7 (0.4-1.3) and 1.3 (0.7-2.3), respectively. The internal analysis (Table 10-10) indicated that compared to the unexposed group, the exposed had an MRR of 1.2 (0.9-1.6) for lung cancer overall, an MRR of 1.3 (0.8-2.0) for the subgroup with 20+ years since hire, an MRR of 0.5 (0.2-1.2) for central nervous system cancer overall, and an MRR of 0.6 (0.2-1.8) for the subgroup with 20+ years since hire. The weak positive association (MRR>1.0) between any AN exposure and lung cancer was not statistically significant, nor

Table 10-10
Study of workers exposed to acrylonitrile (AN): results of analyses using an internal referent group (comparing exposed with unexposed)

	All exposed v. unexposed		20+ years since first exposure v. unexposed	
Cause of death	**RR**	**95% CI**	**RR**	**95% CI**
All causes	0.9	0.8-1.0	0.9	0.8-1.0
All cancers	0.8	0.7-1.0	0.8	0.7-1.1
Colon	0.5	0.3-1.0	0.4	0.2-0.9
Lung	1.2	0.9-1.6	1.3	0.8-2.0
Central nervous system	0.5	0.2-1.2	0.6	0.2-1.8
Lymphopoietic system	0.7	0.4-1.4	0.6	0.3-1.5
Cerebrovascular disease	0.9	0.5-1.6	1.0	0.5-1.9
Ischemic heart disease	1.0	0.8-1.2	0.9	0.7-1.2
Cirrhosis of the liver	0.8	0.4-1.7	0.8	0.3-2.2
Accidents	1.1	0.8-1.4	1.3	0.6-2.7
Suicide	0.7	0.4-1.1	0.5	0.2-1.2

Source: Blair et al. (1998)

was the stronger, but inverse, association (MRR<1.0) seen for central nervous system cancer. Table 10-11 shows MRRs for lung cancer and for central nervous system cancer by AN ppm-years. For each MRR, the referent group consisted of the unexposed subject experience. No consistent pattern of increasing MRR with increasing amount of AN exposure was evident for either form of cancer. The highest AN exposure category (>8.00 ppm-years) had a lung cancer MRR of 1.5 (0.9-2.4) and a central nervous system RMR of 0.5 (0.1-2.5).

This study did not find a relationship between AN and any disease that can be interpreted as causal. The 95% CI of the RR for AN-exposed compared to unexposed subjects was 0.9-1.6. This CI indicates that if the study is valid, the data are compatible with, on the one hand a 10% lower rate among the exposed compared to the unexposed, and on the other hand, at most a 60% higher rate among the exposed compared to the unexposed. Thus, the study found an association between AN and lung cancer that was not statistically significant and was weak in magnitude, even at the upper bound of the CI.

Table 10-11
Number of deaths, rate ratios (RRs) and 95% confidence intervals (CIs) for cancer of the lung and of the central nervous system (CNS) by cumulative acrylonitrile (AN) exposure

Cumulative AN exposure (ppm-years)	Lung Cancer			CNS Cancer		
	No. of deaths	RR	95% CI	No. of deaths	RR	95% CI
0	59	1.0	Referent	11	1.0	Referent
0.01-0.13	27	1.1	0.7-1.7	3	0.5	0.1-1.8
0.14-0.57	26	1.3	0.8-2.1	1	0.2	0.1-1.7
0.58-1.50	28	1.2	0.7-1.9	2	0.5	0.1-2.3
1.51-8.00	27	1.0	0.6-1.6	4	0.8	0.1-2.4
>8.00	26	1.5	0.9-2.4	2	0.5	0.1-2.5

Source: Blair et al., 1998

Selection bias is an unlikely explanation for the weak positive result because of an extensive review of plant personnel records to identify eligible subjects. On the other hand, information bias, confounding and/or chance, easily may have produced the weak association or could have obscured a stronger association. Information bias undoubtedly occurred because of difficulties in developing historical exposure estimates. Objective procedures were used for exposure estimation, so any error probably would have been nondifferential. The impact of nondifferential misclassification errors on the dose-response analysis is uncertain. Confounding by age, gender, race, and calendar time was controlled for in most analyses. However, potential confounding by other nonoccupational factors was not addressed. Chance, information bias, and confounding cannot be ruled out as possible explanations for the positive association with lung cancer (RR=1.5) or for the inverse association with central nervous system cancer (RR=0.5) seen in the highest AN exposure category.

To summarize the evidence on the human carcinogenicity of AN, associations reported by Blair et al. were for the most part weak, were not statistically significant, and may have been due to chance, information bias, or confounding. Dose-response patterns were irregular. Other studies in humans do not consistently indicate a relationship between AN and any form of cancer. Toxicologic studies do not indicate that the lung is a target for AN-induced cancer in laboratory animals. On balance, AN does not appear to be a carcinogen in humans exposed to levels encountered in occupational settings in the United States in the 1950s, 1960s and 1970s.

EXAMPLE OF A PROSPECTIVE FOLLOW-UP STUDY

A component of a large investigation of adverse reproductive outcomes among semiconductor industry workers illustrates the prospective follow-up study design. In the late 1980s, Pastides et al. (1988) conducted a retrospective study of spontaneous abortion (SAB) among the pregnancies of women working at a semiconductor manufacturing facility. The study reported an increased risk of SAB among the pregnancies of women working in silicon wafer fabrication. High risk work areas included diffusion, which involved potential exposure to acids and metals, and photolithography, which involved potential exposure to glycol ethers and other solvents. Glycol ethers were widely used in industry and in the general population, and certain classes of these chemicals are known reproductive toxicants in animals. Schenker and coworkers undertook a series of additional studies to evaluate further the relationship between semiconductor industry work and SAB, and other adverse reproductive outcomes (Schenker et al., 1995; Eskenazi et al., 1995; Hammond et al., 1995). One of their investigations was a prospective follow-up study of SAB among pregnancies occurring in women working at any of seven semiconductor industry facilities in the United States (Eskenazi et al., 1995). A total of 3,915 women were initially considered for inclusion in the study. Of these, 2,639 completed questionnaires designed to determine eligibility for the study, with eligibility defined as being 18-44 years of age, as being able to become pregnant, as having a working freezer (to store urine specimens required to identify pregnancy), as having no plans to leave the company for three months, and as speaking English, Vietnamese, Spanish or Tagalog. Of those completing the screening questionnaire, 739 were eligible for the study, and 403 participated (152 fabrication and 251 nonfabrication workers).

Pregnancies occurring over a six-month period (pregnancy was the unit of observation in this study) were identified by analyzing daily urine samples. Job activities and nonoccupational exposure factors (potential confounders) were measured using daily diaries kept by the women and monthly telephone interviews. Workplace exposure measurement used job and exposure information from interviews, plant records, and industrial hygiene assessments. Pregnancies were classified according to fabrication work (≥5 hours versus none), work group, and exposure to ethylene glycol ether and fluoride. Pregnancy outcomes among clinically recognized and unrecognized pregnancies were determined from interviews conducted at the time of clinically recognized SAB or around the time a normal pregnancy was expected to conclude. All comparisons were internal. No external comparison group could be used because data on clinical and unrecognized SABs are not available for the general population of the United States. Cumulative incidence was used as the measure of occurrence, and incidence proportion ratios and odds ratios were used as measures of association. Both crude analyses and analyses adjusting for potential confounding by multiple factors were done. Selected results are displayed in Table 10-12.

The adjusted odds ratio for all SAB for fabrication compared to nonfabrication pregnancies was 1.25 (95% CI, 0.63-1.76), similar to the crude IPR. Four pregnancies were identified among women exposed to ethylene glycol ethers, and all four ended in SAB.

This study required intense efforts to identify eligible pregnancies, to follow-up eligible pregnancies to determine the outcomes, and to obtain data on occupational factors and potential confounders during the study period. The small size of the investigation is not surprising, given the intricacies of the data collection procedures. However, small size and the attendant problems of imprecision and possibly poor control for confounding are the main limitations of the study. None of the results was statistically significant, and the study was inconclusive regarding the possible relationship between fabrication work and SAB.

Case Control Studies

Case-control studies evaluate the relationship between an agent and a disease by addressing the question: Are people with the disease equally likely, more likely, or less likely to have been exposed to the agent than people without the disease? In a case-control study, people with

Table 10-12
Results of a prospective follow-up study of spontaneous abortion (SAB) among pregnancies experienced by women working in the semiconductor industry

	Fabrication workers N (% of all pregnancies)		Nonfabrication workers N (% of all pregnancies)	
Clinical SABs	3	(15.8)	3	(9.1)
Early SABs	9	(47.4)	12	(36.4)
All SABs	12	(63.2)	15	(45.5)
Total pregnancies	19		33	

IPR for fabrication compared to nonfabrication workers:

Clinical SAB: 0.158/0.091 = 1.74 (95% CI, 0.39-7.76)
Early SAB: 0.474/0.364 = 1.30 (95% CI, 0.68-2.51)
All SAB: 0.632/0.455 = 1.39 (95% CI, 0.84-2.31)

Source: Eskenazi et al., 1995.

a particular disease (the "cases") are compared to people who were candidates for the disease (they did not have the disease) at the time when the cases occurred (the "controls"). Specifically, the cases and controls are compared with regard to their history of exposure to one or more agents of interest. Exposure may be defined as ever exposed, cumulative amount of exposure up until the time the case occurs, average intensity of exposure, frequency of exposure, etc. In occupational epidemiology, subjects in a case-control study of a chronic disease may be identified retrospectively (all cases occurred in the past relative to the time when the investigation is being done) or prospectively (cases occur after the investigation has begun), but exposure measurement typically involves retrospective assessment of exposure to one or more agents of primary interest in a particular occupational setting, as well as assessment of potential confounders. Case-control studies have been used to study the effects of occupational factors on many different types of fatal and nonfatal medical conditions, including cancer, congenital malformations, spontaneous abortion, injuries, cardiovascular disease, kidney disease, neurologic dysfunction, etc.

There are two main types of case-control study, the **nested case-control study** and the **registry-based case-control study**. The main distinction between the two designs lies in the approach used to specify the study base from which subjects are chosen. Nested case-control studies use a base consisting of the person-time experience of a defined occupational study group, typically identified in a follow-up study of a particular plant or industry, done before the case-control study. Registry-based case-control studies have a "hypothetical" base that often is conceptualized as the general population residing in a geographic area during a time period, but that is not comprehensively identified before or during the study. The key elements of both types of case-control study are:

- Specification of objectives and hypotheses
- Specification of the study base
- Identification of all eligible cases
- Specification of controls
- Recruitment of cases and controls
- Measurement of exposures and of potential confounders
- Analysis
- Interpretation

NESTED CASE-CONTROL STUDIES

The main objective of a nested case-control study is always to evaluate in detail the relationship between a disease and a specific exposure, set of exposures, or other job characteristics **within a specific workplace setting**. A nested case-control study is usually done after the completion of an initial follow-up study that evaluated disease patterns in the overall study group, but not in subgroups specified on the basis of detailed exposure information. A secondary objective may be to obtain information on and to analyze lifestyle and other factors that are potential confounders and that are not recorded in documents (e.g., personnel records) from the specific workplace under study. Often, the initial follow-up study's evaluation of the impact of work-related factors on disease has been limited to an assessment of disease patterns by job or work area categories, without estimating exposure to particular occupational agents and without controlling for possible confounding by nonoccupational factors. A nested case-control study provides an approach that is more efficient than follow-up or registry-based case-control studies for carrying out detailed retrospective exposure measurement and for obtaining data on potential confounders. Thus, a sequential approach — an initial follow-up study, then a nested case-control study — is used.

In a nested case-control study the study base is the person-time experience of a defined group of individuals in a population of interest, e.g., employees of a particular plant or group of plants, workers in a particular occupation, the general population of a particular community. The study base is defined and completely enumerated before or as part of the case-control study, usually as part of conducting the initial, or "parent," follow-up study.

Both cases and controls are selected from among persons included in a study group assembled for the parent follow-up study. Every member of the study group is identified, and the subjects are traced over time to identify cases. Cases in a nested case-control study are all subjects diagnosed with or dying from a particular disease within the study period. As in follow-up studies, cases may be identified from death certificates, cancer registries or other ad hoc procedures used to ascertain disease occurrence in the parent follow-up study. A distinguishing, and problematic, feature of nested case-control studies is that cases typically have occurred over a long period of time, equal to the entire study period of the parent retrospective follow-up study, often decades in length. The long length of the case accrual period makes retrieval of information, that must be obtained by questionnaire, difficult.

Controls for each case are chosen randomly from a roster consisting of all study group members who were candidates for disease at the time of the case's diagnosis or death. If it is desired that ORs estimate IRRs or MRRs, one or more controls for each case will be selected by incidence density matching (Flanders and Louv, 1986). Incidence density matching requires the identification of a "risk set" of subjects for each case and the random selection of one or more subjects from among those in the risk set. A risk set for a particular case consists of all subjects who were at risk of disease occurrence at the time the case was diagnosed, and who "match" the case with regard to year of birth, race, and gender. Matching involves the pairing of one or more controls to each case (individual matching) or a group of controls to a group of cases (frequency matching) on the basis of one or more potential confounders (Wacholder et al., 1992c). The purpose of matching is to achieve efficient statistical control of confounding. When incidence density matching is used, controls for a case may later (i.e., subsequent to the case's occurrence date) become cases themselves. Although incidence density matching is desirable for nested case-control studies of chronic disease, other forms of matching are often used (Flanders and Louv, 1986).

Enrollment of subjects into the study may be automatic, if the only sources of information on work history and potential confounders are records assembled for the parent retrospective follow-up study or other historical personnel and work records that are kept by the facilities where the subjects worked and that are readily available to the investigator. However, if external sources of information are to be used (for example, questionnaire information on lifestyle factors that are to be considered as potential confounders), interviews must be conducted with subjects or with their next-of-kin or fellow workers, and only those eligible, initially selected subjects for whom an interview can be carried out or who complete a mail questionnaire, will be included in the study. Some subjects (or surrogates) may be untraceable; others may refuse to participate in an interview or may refuse to complete a mail questionnaire. In any case, participation may not be 100% for either cases or controls. Because cases and controls may have died or left work many years before the nested case-control study is done, participation is often problematic and may be as low as 70%. In some studies, initially selected controls who do not participate are replaced with additional controls. This practice increases the potential for selection bias.

Exposure to the agent(s) of interest is measured using existing historical work records and job-exposure matrices developed for the study. To avoid observation bias, exposure measurement for individual subjects is carried out "blind," that is, without referring to a subject's status as a case or a control. Exposure to potential confounders, in contrast, usually involves using responses to questionnaires administered in person, by telephone, or by mail. Thus, measurement of potential confounders is less objective than exposure assessment in the nested case-control study, as it involves memory and is often not blind. In developing measures of workplace exposure and of potential confounders, the exposures of controls who continue to live or to work after the case occurs are truncated as of the case's death or diagnosis date.

In the analysis of a nested case-control study, cases and controls are compared with regard to each group's odds of exposure to agents of interest and/or with regard to each group's cumulative exposure. The OR typically is the measure of association. Several approaches are available for estimating ORs in case-control studies, and the procedures are similar for nested and registry-based case-control studies. These include crude estimation, an example of which was provided earlier; matched or unmatched analysis using stratified analysis procedures (e.g., the Mantel-Haenszel procedure) (Checkoway et al., 1989, pp. 184-188) and multivariable analysis using conditional logistic regression (for designs using matching) or unconditional logistic regression (for designs not using matching). The stratified analysis and logistic regression procedures allow for control of confounding by the matching factors and by other factors, as well as for the evaluation of effect modification. Dose-response is also assessed using these procedures. Induction time and latency may be considered by incorporating terms for time since first exposure into the analyses, and by excluding exposures occurring within various time periods preceding death or diagnosis.

In assessing the quality of a nested case-control study, crucial issues of selection bias, information bias, and confounding are as follows:

- Selection bias: Subject participation rates should be high; all, a random sample, or a high proportion of eligible cases should have been included; a high proportion of initially selected eligible controls should be included, and there should be no evidence of major differences between participating and nonparticipating subjects
- Information bias: Sources of work history and occupational exposure data should have been objective (i.e., based on records established before disease occurrence); exposure assessment should have been carried out blindly; if exposure measurement was based on self-reports, results should be reasonably specific (e.g., positive associations for multiple exposure-disease relationships might indicate differential errors in recalling or reporting exposure)
- Confounding: Exposure to potential confounders should have been measured using objective data; adjustment for confounding should have been carried out using categorizations of confounders that were as detailed as possible; analytic procedures should have taken any matching into account.

Example of a Nested Case-Control Study

An investigation of malignant and nonmalignant respiratory disease among employees of a fiberglass manufacturing facility illustrates the nested case-control study design. Chiazze et al. (1992, 1993) conducted a nested case-control study of respiratory cancer and nonmalignant respiratory disease deaths among subjects employed at a large glass fiber manufacturing plant. All subjects had been members of a retrospective follow-up study, which included workers employed for at least one year between 1940 and 1963 and followed up through the end of 1982.

The nested case-control study included two case series, one consisting of 164 persons who died of lung cancer, and the other consisting of 112 persons who died of nonmalignant respiratory diseases other than pneumonia or influenza during the parent follow-up study. Controls were other study group members. Each case was matched with up to four controls whose years of birth were within two years of the case's year of birth and who had lived at least as long as two years before the case's year of death. There were 379 controls for the lung cancer cases and 270 controls for the nonmalignant respiratory disease cases.

The investigators used historical plant engineering records, process descriptions, material ad product specifications, laboratory notebooks, expert opinion, and other information to develop time period-specific estimates of exposure to various substances (eight-hour time-weighted average exposure concentrations for respirable fibers, asbestos, talc, formaldehyde, respirable silica, asphalt fumes and total particulate) for particular departments and jobs (Chiazze et al., 1993). These job-exposure matrices were linked to subjects' work histories to develop cumulative exposure estimates for each person in the study.

The investigators attempted to trace and to interview each subject or a surrogate respondent to obtain information, via in-person and telephone interviews, on potential confounders, including smoking, marital status, and years of education. Logistic regression procedures were used to estimate ORs, the measures of association used in the study.

Interview data were available for 88% of lung cancer cases and 79% of their controls and for 91% of nonmalignant respiratory disease cases and 74% of their controls. Cumulative exposure to respirable fibers was not associated meaningfully with lung cancer or with nonmalignant respiratory disease (Table 10-13). Smoking-adjusted ORs for subjects with high fiber exposure (300+ fibers/ml-days) compared to subjects with low fiber exposure (<100 fibers/ml-days) were 0.6 (95% CI, 0.2-1.7) for lung cancer and 1.5 (95% CI, 0.6-4.1) for nonmalignant respiratory disease. Dose-response patterns for fiber exposure were irregular. No result was statistically significant. For example, the nonmalignant respiratory disease OR of 1.5 for the high compared to low fiber exposure had a wide 95% CI that included the null value of 1.0. On the other hand, cigarette smoking was strongly associated with lung cancer, with a statistically significant OR for smoking 6+ months of 26.2, and was moderately associated with nonmalig-

Table 10-13
Selected results from a nested case-control study of lung cancer and nonmalignant respiratory disease among fiberglass manufacturing workers

Exposure variable	Lung cancer			Nonmalignant respiratory disease		
	Ca/Co*	OR†	95% CI	Ca/Co*	OR†	95% CI
Respirable fibers (fibers/ml × days)						
<100	98/236	1.0		70/159	1.0	
100-<300	37/60	1.7	0.8-3.9	23/62	0.9	0.4-2.1
300+	27/67	0.6	0.2-1.7	18/41	1.5	0.6-4.1
Smoking						
Never	4/47	1.0		10/41	1.0	
Smoked 6+ months	139/209	26.2	3.3-207	88/135	2.6	1.1-6.1
Smoking data not available	20/80	-		10/69	-	

* Number of exposed cases (Ca) and exposed controls (Co).

† Odds ratios for respirable fibers are adjusted for smoking and for several other variables. Odds ratios for smoking are adjusted for respirable fibers and for several other variables.

Source: Chiazze et al., 1992, 1993

nant respiratory disease, with a statistically significant OR for smoking 6+ months of 2.6. The results of this study are consistent with results of other investigations, which indicate that fiberglass exposure is unassociated, or at most weakly associated, with lung cancer and nonmalignant respiratory disease.

REGISTRY-BASED CASE-CONTROL STUDIES

A registry-based (population-based) case-control study is typically done when a large exposed study group is difficult to specify and assemble (e.g., farmers, mechanics), when data on occupation are collected as part of routine registry procedures (death certificates), when data on occupation are collected as part of a study of nonoccupational factors, when an objective is to assess the public health impact (to estimate the population attributable risk for an occupational characteristic or exposure), or when the disease of interest is extremely rare.

In a registry-based case-control study, the study base is defined secondarily and is not enumerated over time (Wacholder et al., 1992a). That is, the study base is hypothetical, not explicit. Most often, the study base consists of the population of a particular geographic region during a particular time period: for example, all persons living in a state or group of states during 1990-1995. Cases and controls are selected from this study base.

Cases may be identified in an existing or ad hoc disease registry, such as a population-based death registry or cancer or other disease registry, or from a registry consisting of hospital records (Wacholder et al., 1992b). Cases are defined as all persons with a particular disease diagnosed or otherwise identified in the registry during a relatively short time period, either retrospectively (that is, over a several-year period in the past) or concurrently with the conduct of the study (that is, some time period starting with the commencement of the study and continuing for several years into the future). Unlike the situation in most nested case-control studies, the study period for identifying cases and controls is relatively short, often two to five years, rather than decades. The short case accrual study period offers some advantages over the long case accrual period typical of nested case-control studies. Subjects are more often still living and are relatively easy to trace. Thus, participation in interviews and the quality of information obtained in interviews or from mail questionnaires, particularly information on potential confounders, tend to be better in registry-based than in nested case-control studies.

Controls for registry-based case-control studies are selected from among persons who were at risk of developing the disease and of being included in the registry at the time of the case's diagnosis. Typically, a roster of all such persons does not exist, and many procedures may be used to address this deficiency. Some examples of commonly used sources of controls in registry-based case-control studies in the United States are (Wacholder et al., 1992b):

- persons recruited from the general population using random digit telephone dialing
- neighbors or friends of cases
- persons selected from lists of registered voters

- persons selected from among Medicare recipients
- persons included in the registry from which the cases are chosen but having diseases other than the case disease (if cases are persons who died of aplastic anemia in the United States during 1990-1994 and who have aplastic anemia recorded as their cause of death on their death certificate, controls may be selected from among other persons dying of other diseases in the same time period).

Controls may be individually or frequency matched to cases or may be selected without matching. The average control to case subject ratio varies considerably in registry-based case-control studies, as in nested case-control studies. If the case series is large, the control to case ratio may be 1:1 or occasionally less. If controls are selected from the same registry as the cases, and if all exposure and confounder data already exist in the registry and do not have to be actively collected, the control to case ratio may be much larger. If controls are to be actively recruited from the study base and interviewed or otherwise contacted for data collection, it is unusual to encounter a control to case ratio above 4:1.

Recruitment of subjects may be passive (for example, if all data for the study, including data on occupational exposures and potential confounders, come from a death registry for both cases and controls) or active (if data come from questionnaires or interviews). If recruitment is passive, then participation may not be an issue, as it may be possible to include all persons initially selected. Selection bias will be a problem only if the registry is incomplete in terms of its identification of cases, if some subjects must be eliminated because of missing registry data on exposures or confounders, or if the registry controls have an exposure frequency different from that in the study base in which the cases arose. The use of controls selected from other conditions included in the disease registry has been particularly controversial, but this procedure continues to be used for convenience. If controls are persons with "other" diseases, a strategy used to reduce the potential for selection bias is the elimination as potential controls persons with other diseases thought or known to be related to the exposure of interest.

If recruitment is active, then all initially selected, eligible subjects must be contacted and asked to participate. Sometimes, before this is done, the attending physicians must be contacted to obtain permission to contact their patients. Thus, subjects and physicians self-select themselves to be included in the study, and the potential for selection bias is enhanced. It is usually impossible to judge whether selection bias has occurred. Thus, studies with participation rates below 90% are suspect.

Work history or exposure data, as well as data on potential confounders, come either from existing registry records (if both cases and controls are chosen from among people included in the registry) or from interviews or questionnaires. In the latter instances, exposure and confounder data are, of course, self-reported by the subjects or if the subject is deceased or otherwise cannot be interviewed, by a surrogate (spouse, child, etc.). Compared to data based on historical work records, self-reported data on work histories and occupational exposures tend to be more error prone. This is because they are based on memory and are reported after disease occurrence and thus potentially influenced by disease status. Standard procedures used by epidemiologists to minimize information bias include blinding interviewers to study hypotheses and subjects' status as cases or controls; providing subjects with information on general, rather than specific, study hypotheses; using deceased controls for deceased cases; and attempting to validate exposure data obtained from subjects or surrogates with objective historical data (rarely feasible).

Typical occupational data obtained in registry-based case-control studies consist of a list of all jobs a subject held during the working lifetime, and for each job, the time period, industry, tasks, and potential agents encountered. These data have a number of severe limitations: recall of jobs (and exposures) held in the distant past may be quite poor; it is usually infeasible to elicit very detailed job/exposure histories; and exposure conditions are heterogeneous in jobs in different industries or at different plants in the same industry. Despite these limitations, much effort has been devoted to developing job-exposure matrices for use in registry-based case-control studies to estimate exposures based on self-reported occupational histories (Goldberg et al., 1986; Gomez et al., 1994).

The analysis of registry-based case-control studies is similar to that of nested case-control studies. Both stratified and multivariable modeling procedures are used to control for confounding, and both may be matched or unmatched, depending on design features of the study. The Mantel-Haenszel procedure and logistic regression are the typical stratified analysis and multivariable approaches used to estimate ORs.

In evaluating the validity of a registry-based case-control study, important considerations are:

- Selection bias: The disease registry used to identify cases should have been complete (evaluation of this can be done by comparing rates reported by the registry with rates reported by other registries); if not, it is important to consider whether the exposure frequency is likely to be higher, lower, or the same in cases included in the registry compared to cases not included; procedures used to choose controls should have yielded an initially selected group having an exposure frequency similar to that in the study base giving rise to the cases; the participation rate of cases and of controls should have been high; if not, there should be no evidence that low participation affected exposure frequencies
- Information bias: objective procedures should have been used to collect data on exposures and potential confounders; self-reported exposures should have been confirmed with objective records, where possible
- Confounding: Analytic procedures to control for confounding should have been appropriate; matched analyses should have been done if individual matching was used in selecting controls; adjustment for confounding should have been carried out using categorizations of confounders that were as detailed as possible.

Example of a Registry-Based Case-Control Study

An investigation of occupational factors and congenital malformations serves as an example of a registry-based case-control study. Cordier et al. (1997) evaluated the relationship between maternal occupational exposure to glycol ethers and other agents during pregnancy and the occurrence of various types of congenital malformation. Cases were 984 children with one or more major congenital malformations detected prenatally, at birth, or within the first week of life among induced abortuses, still births, and live births. They were identified at six European centers during the period 1989 to 1992. The participating centers were registries included in the European Registration of Congenital Anomalies (EUROCAT) surveillance system and were located in Italy, France, Scotland, and the Netherlands. One or two controls for each case were selected from among live births without malformations occurring at the same hospitals or in the same broad geographic region as the case. Controls were matched to cases on approximate birth date. If initially selected controls left the delivery hospital before an interview could be done, could not be traced, or declined to participate in the study, they were replaced. Interviews were carried out to obtain information on maternal occupations, occupational exposures, hobbies, medical and obstetrical history, drug and vitamin use, alcohol and tobacco use, and other factors. Information obtained for each occupation included a description of the industrial activity, tasks, products handled, and frequency of use of various products.

There were four periods of interest for ascertaining exposure: one month before conception and each trimester of pregnancy. Exposure to glycol ethers and to other chemicals encountered in the occupational setting was determined by industrial hygienists at each center. The industrial hygienists were not aware of subjects' status as cases or controls. They reviewed job descriptions, specified likely workplace exposures for each job and rated each exposure according to four attributes: likelihood of actual exposure (possible, probable, certain); level (low, medium, high); frequency (<5%, 5-50%, >50%); and exposure route (inhalation, dermal, both). An "overall exposure" index combined likelihood, level, and frequency of exposure into four categories: none, very low, low, and medium-to-high. Analyses computed ORs for the association of glycol ethers with all congenital malformations combined, with eight major categories of malformations, and with 18 subcategories. The investigators computed ORs adjusted for maternal age, socioeconomic status, area of residence, country of origin, and center.

The study found a positive association between any exposure to glycol ethers and all malformations combined (OR=1.4, 95% CI=1.1-1.9) (Table 10-14). An OR above the null value of 1.0 was also present for all of the eight major malformation categories (the eight ORs ranged from 1.3 to 2.0) and for 17 of the 18 subcategories of malformations (the 17 ORs ranged from 1.2 to 2.6). The OR was statistically significant at the 0.05 probability level or was of borderline statistical significance for all cardiac malformations (OR=1.5, CI=1.0-2.1), all central nervous system malformations (OR=1.9, CI=1.2-3.0), neural tube defects (OR=1.9, CI=1.2-3.2), spina bifida (OR=2.4, CI=1.2-4.6), cleft lip/palate (OR=2.0, CI=1.2-3.25), cleft lip with or without cleft palate (OR=2.0,CI=1.1-3.7), and multiple anomalies (OR=2.0, CI=1.2-3.2).

Further analyses were done for endocardial cushion/septal defects, neural tube effects, oral clefts, and multiple anomalies. There was no evidence of increasing ORs with increasing

Table 10-14
Selected results from a registry-based case-control study of occupational exposure to glycol ethers and congenital malformations

Type of malformation	Number of cases: Total	Number of cases: Exposed	Odds ratio*	95% CI
Cardiac	249	56	1.45	0.99-2.13
Endocardial cushion defects and septal defects	148	32	1.52	0.93-2.49
Defect of cardiac chambers	56	13	1.17	0.59-2.32
Hypoplastic left heart	22	6	1.73	0.62-4.84
Anomalies of valves	41	9	1.53	0.67-3.52
Other	20	5	1.71	0.56-5.21
Musculoskeletal	158	38	1.32	0.85-2.05
Diaphragmatic defects	43	10	1.44	0.65-3.17
Omphalocele	32	9	1.98	0.85-4.60
Gastroschisis	20	4	N/A†	
Limb reductions	50	7	0.75	0.32-1.77
Other	20	8	2.55	0.93-6.97
Central nervous system defects	140	40	1.91	1.23-2.96
Neural tube defects	94	28	1.94	1.16-3.24
Anencephaly	29	7	1.34	0.53-3.37
Spina bifida	51	18	2.37	1.22-4.62
Hydrocephalus	34	9	2.20	0.92-5.31
Other	12	3	N/A	
Cleft lip/palate	100	34	1.97	1.20-3.25
Cleft palate without cleft lip	36	11	1.68	0.75-3.76
Cleft lip with or without cleft palate	64	23	2.03	1.11-3.73
Digestive system	70	16	1.40	0.74-2.62
Esophageal defects	32	6	1.18	0.45-3.13
Atresia & stenosis of intestine	35	10	1.92	0.84-4.36
Other	5	1	N/A	
Genital and urinary	61	14	1.32	0.67-2.59
Genital defects	18	4	1.34	0.39-4.58
Urinary defects	49	11	1.25	0.59-2.63
Other anomalies	28	7	1.64	0.64-4.18
Total number of cases	648	158	1.44	1.10-1.90
Total number of controls	751	137	1.0	Referent

*Odds ratio is adjusted for maternal age, socioeconomic status, area of residence, country and center.
†N/A, data not provided by investigators.
Source: Cordier et al., 1997

exposure likelihood, level, frequency, or overall exposure index for endocardial cushion/septal defects, neural tube defects or multiple anomalies. A positive trend with exposure likelihood and with the overall exposure index was present for cleft lip with or without cleft palate but was based on small numbers of exposed subjects. ORs for cleft lip with or without cleft palate were 1.5 (CI=0.6-3.8, 7 exposed cases, 59 exposed controls), 2.2 (CI=1.0-4.4, 14 exposed cases, 73 exposed controls) and 5.9 (CI=0.9-39, 2 exposed cases, 5 exposed controls) for possible, probable, and "certain" exposure, respectively. There was no consistent trend for the overall exposure index, with ORs of 1.4 (CI=0.5-3.6), 3.1 (CI=1.4-6.9) and 1.8 (CI=0.7-4.75), respectively, for very low, low, and medium-to-high exposure. No cleft lip case was classified as having a high level or a high frequency of exposure. Cordier et al. concluded that their results were consistent with previous epidemiologic research reporting a positive association between maternal exposure to solvents other than glycol ethers and cardiac defects, neural tube defects, and oral clefts.

An unusual feature of the results of this study is the elevated OR for all but one category of congenital malformation. This pattern has two possible interpretations. Such a pattern could indicate that glycol ethers cause most major malformations. Alternatively, the nonspecific positive associations could be due to bias. The latter interpretation is plausible because of several important technical weaknesses that compromise the validity of the study's results. The most important of these are possible bias in selecting cases and controls, possible bias in ascertaining exposure, possible confounding by factors such as alcohol and socioeconomic status, small numbers of subjects in malformation subcategories, misinterpretation of trend data, and poor documentation of exposure.

Selection bias is possible because the physicians, midwives, and nurses who selected the cases may have had access to at least some information on occupational history and may have used that information in case selection. Low participation rates, both among potential cases and among potential controls, may also have contributed to selection bias. Participating cases may have had a higher exposure frequency than nonparticipating cases, and a similar, balancing tendency may not have been present among controls. Controls who declined to participate or could not be traced were replaced with more accessible controls. Replacement controls may have differed from initially selected controls with regard to exposure frequency.

Information bias is also a plausible explanation of the positive results. It appears that, although the industrial hygienists who evaluated job histories were blind, the interviewers who obtained occupational data from the mothers of cases and controls were not blind. It is not known whether subjects were told about the hypothesis under investigation. Case mothers may have been queried in more detail about jobs and exposures or may have been more thorough than control mothers in reporting jobs and in providing job descriptions that led to classification of a pregnancy as exposed. This problem would not have been corrected by the presumably objective approach used by industrial hygienists in reviewing the questionnaire data and assigning exposure level, frequency, and likelihood.

Confounding may not have been adequately controlled. For example, alcohol and vitamin use appear not to have been considered. Cordier et al. state that alcohol has been linked to oral clefts, and folic acid has been shown to protect against neural tube defects. On the other hand, it is unlikely that confounding by these factors could have been responsible for the nonspecific pattern of positive results.

The study was large overall. However, the number of exposed cases in many of the congenital malformation subcategories was small, precluding a meaningful analysis of trends in OR with exposure likelihood, frequency, and level. In particular, the number of cleft lip cases and controls with "certain" glycol ether exposure were inadequate for drawing conclusions.

It is difficult to judge whether or not the classification of subjects as exposed to glycol ethers was credible or to identify the type of glycol ethers involved. Few subjects had "certain" exposure to glycol ethers, and no information was presented on the jobs or job descriptions of, or products handled by, any subject in this exposure category. The investigators indicate that service workers (cleaners, cooks, waiters, hairdressers, beauticians, nursing aides, launderers) and sales workers accounted for a high proportion of exposed subjects among controls. These job titles suggest considerable uncertainty and diversity in exposure potential.

The study found a positive association between glycol ethers and virtually all congenital malformations. Such a relationship has not been reported previously, and for this reason alone, the result does not warrant a causal interpretation. Moreover, several features of the methods and results suggest that the associations reported are more likely to be due to a general bias than to a series of causal relationships. First, the high potential for selection and observation bias suggests that the results could be due to systematic technical errors. Second, the observed statistical associations are, for the most part, weak and, thus, could easily be due to bias. Where the associations are moderate to strong in magnitude, they are also quite imprecise and could be due to chance. Third, human teratogens generally have not been shown to increase the occurrence of such a broad spectrum of congenital malformations. The biologic plausibility of the results is low for this reason and, further, because of the lack of information on the types of glycol ether involved in each of the observed associations. The lack of information on specific types of glycol ethers is also an important concern because teratogenicity, as determined from animal experiments, is not the same for all glycol ethers. Ethylene glycol methyl and ethyl ethers are teratogenic, whereas ethylene glycol butyl, propyl, isopropyl and phenyl ethers, and propylene glycol ethers are not. Ethylene glycol methyl and ethyl ethers have been removed progressively from consumer products since the early 1980s.

ADVANTAGES AND DISADVANTAGES OF FOLLOW-UP AND CASE-CONTROL STUDIES

This section summarizes the main advantages or strengths and the main disadvantages or limitations of follow-up studies compared to case-control studies and of each major type of case-control study compared to other major study designs.

Follow-up Studies

ADVANTAGES

All eligible subjects are included or a sample of eligible subjects, selected before disease occurrence, is included; thus, selection bias should be nonexistent in a follow-up study, whereas it is a major potential concern in case-control studies, particularly in registry-based case-control studies.

Many different diseases may be assessed in relation to various employment factors and exposures.

Study groups may be selected in a way that optimizes the frequency of exposure; this enhances precision and is a major advantage of a follow-up study over a registry-based case-control study.

Exposure assessment is typically based on objective records (retrospective follow-up studies), objective industrial hygiene measurements or biomonitoring data (prospective follow-up studies); this decreases the possibility of observation bias.

Measurement of potential confounders should be relatively good in a prospective follow-up study, as this is done concomitantly with the conduct of the study and is entirely under the control of the investigator; however, in retrospective follow-up studies, lack of data on potential confounders can be a major deficiency.

The prospective follow-up study is the design of choice for the epidemiologic investigation of nonfatal diseases and other medical conditions.

DISADVANTAGES

Many years are required to obtain reasonably precise data on the occurrence of rare diseases; thus, follow-up studies tend to be imprecise and uninformative when study periods are short (under two decades) and/or when very rare diseases (e.g., specific forms of cancer) are evaluated.

Prospective follow-up studies tend to be expensive and effort-intensive. Even though prospective measurement of exposure and of potential confounders is a potential advantage, this requires frequent active monitoring of the study group to collect the information throughout the time period preceding the onset of disease.

Prospective follow-up studies often require the active participation of subjects to measure exposure and to measure the disease outcome; losses to follow-up may be a severe problem.

Nested Case-Control Studies

ADVANTAGES

Compared to retrospective follow-up studies, they are inexpensive, because they require the collection of data on exposure and potential confounders on fewer subjects.

Compared to retrospective follow-up studies, they tend to have better control of confounding by lifestyle factors because it is feasible to obtain data on potential confounders and because analyses use internal referent groups.

Compared to registry-based case-control studies, they have less potential for selection bias, because of random selection of controls from completely enumerated risk sets.

Compared to registry-based case-control studies, they have less potential for information bias affecting the assessment of occupational exposure, since this is typically done using objective records rather than questionnaire responses.

Compared to registry-based case-control studies, they have a higher exposure frequency in the study base; therefore, they tend to be more precise and informative.

DISADVANTAGES

Compared to retrospective-follow-up studies, they have a greater possibility of error due to selection bias, particularly if participation in study components involving active data collection is low.

Compared to retrospective follow-up studies, they are less precise.

Compared to registry-based case-control studies, they have poorer information on potential confounders because data on such factors must be obtained, in many instances, years after disease occurrence.

Compared to registry-based case-control studies, the number of cases may be small.

Registry-Based Case-Control Studies

ADVANTAGES

The design can be specified in a way that enlarges the size of the case series.

They may be used to study occupations and diseases for which historical data are not available for compiling retrospective follow-up studies (e.g., farm work).

Because cases are typically identified close in time to study commencement or prospectively, participation may be higher than in nested case-control studies, and information on confounders may be better.

DISADVANTAGES

The occupational exposure of interest may be quite low in the hypothetical study base; thus, the study may be relatively imprecise.

Information on occupational exposures of interest is usually self-reported; hence, the possibility of information bias is relatively high.

The potential for selection bias is high if the disease registry is incomplete and/or if participation of initially selected eligible cases and controls is low.

OTHER STUDY DESIGNS

Case-cohort studies are similar to nested case-control studies (Wacholder, 1991; Wacholder et al., 1992c). In a case-cohort study, cases that occur in a defined study group during the study period are compared, with respect to exposure, to controls who are selected at random from the total study group, without specifying risk sets, as in the typical nested case-control study. Like the nested case-control study, the case-cohort study provides an efficient and feasible approach to obtaining detailed data on exposure and on potential confounders. The research on AN workers by Blair et al. (1998) included a case-cohort study of lung cancer, in which the investigators obtained data on smoking and performed analyses of AN exposure and lung cancer, adjusting for smoking as a potential confounder.

Cross-sectional studies measure disease prevalence at a single point in time and compare prevalences in exposed and unexposed persons. Such studies are typically done when it is desired to investigate factors related to nonfatal, persistent diseases, symptoms (e.g., symptoms of respiratory disorders), or biomarkers of physiologic dysfunction, and typically require the voluntary participation of subjects in medical examinations or interviews. In occupational epidemiology, cross-sectional studies are nearly always limited to subjects who are actively working, and these investigations have other, serious methodologic weaknesses, for the most part dealing with selection and information biases and unclear time order or cause-to-effect temporal sequence (Eisen, 1995). Specific disadvantages of cross-sectional studies are:

- absence of data on retired and other separated workers who, in many situations, are the most likely to have experienced the disease or medical condition of interest
- inability to establish that the exposure of interest predated the medical condition of interest (the required temporal sequence if an agent is a causal factor)
- measurement of disease prevalence, which is a blend of disease incidence and disease duration, so that observed associations may be due to factors that cause the disease, factors that influence survival after the disease has occurred or both
- inability to determine if selective migration of diseased workers out of exposed jobs has occurred in studies where only current exposure is considered
- likelihood of information bias in studies using self-reports of medical conditions or symptoms
- likelihood of selection bias in studies with poor participation.

Prospective follow-up studies, although more difficult to carry out than cross-sectional studies, are the preferable design.

Proportional mortality ratio (PMR) studies compare exposed and unexposed groups of decedents with respect to their relative frequency of one or more specific causes of death among all deaths (Kupper et al., 1978). This design was used in the past when it was infeasible to conduct retrospective follow-up studies, and has sometimes been used to supplement conventional follow-up study analyses (Park et al., 1991). However, the PMR study design is recognized as methodologically inferior because the measure of association may reflect an increase in one cause of death or a deficit in other causes of death (Wong and Decoufle, 1982; Wong et al. 1985). The PMR study is now considered a variant of the case-control study (Miettinen and Wang, 1981; Robins and Blevins, 1987).

Ecologic or correlational studies are investigations in which the unit of observation is a group of people, rather than an individual subject or person-year of experience (Morgenstern, 1995). Typically, groups are defined by residence in a particular geographic unit, such as a county, and the analysis examines the correlation between exposure prevalences in county populations (e.g., proportion of the county employed in transportation equipment manufacturing) and county population disease rates (e.g., bladder cancer incidence rates). Data sources often consist of readily available census information on exposure prevalence and equally easily obtained mortality rates by county of residence at the time of death. Ecologic studies may be difficult to interpret because they do not directly examine disease rates among exposed and unexposed persons. In a given unit of observation, such as a county, some individuals will be exposed and others will be unexposed, yet all are classified according to the exposure category assigned to the county. An interpretative error that can occur when one makes an inference on the basis of such data — that is, when one makes an inference about individuals on the basis of data pertaining to groups — is called an "ecologic fallacy." Ecologic studies are also difficult to interpret because they do not adequately control for possible confounding factors.

Meta-analysis refers to "methods used for contrasting and combining results of different studies" (Greenland, 1987; Dickersin and Berlin, 1992). Meta-analysis is used in observational epidemiology to summarize the results of a group of studies dealing with the same exposure-disease relationship. Summarization involves obtaining the single, "best" quantitative estimate of the magnitude of an association between a particular exposure and disease. The unit of analysis is the individual study. The best quantitative estimate derived from a meta-analysis and its CI are typically computed as a weighted average of estimates extracted from individual studies. The weight assigned to a study usually depends on its precision and may also depend on a quantitative estimate of the quality of the study.

Bhatia et al. (1998) carried out a meta-analysis of studies on occupational diesel exhaust exposure and lung cancer. They identified 23 follow-up or case-control studies with useful information on diesel exhaust and lung cancer. Of these, 21 had RRs greater than 1.0, and the summary RR, weighted for precision, was 1.3 (95% CI, 1.2-1.4). The summary RR was similar for follow-up and case-control studies and for studies that controlled for possible confounding by smoking and studies that did not. Meta-analysis, like that of Bhatia et al. (1998), has become a commonly used tool for examining a body of published studies. Much has also been written about the pitfalls of the approach. These include publication bias (studies with null results may tend not to be published, leading to summary results for an exposure-disease relationship that are unduly positive) and a tendency to emphasize the quantitative aspects of meta-analysis, which enhance precision, rather than an appropriate and thorough evaluation of the validity of the individual studies that contribute to the meta-analysis (Shapiro, 1994).

INTERPRETATION OF EPIDEMIOLOGIC RESEARCH

Several statisticians and epidemiologists have proposed and refined guidelines for interpreting the results of epidemiologic studies (Hill, 1965; Rothman and Greenland, 1998, pp. 24-28). Cole (1997) has suggested three contexts for examining interpretive issues: context 1 – evaluation of the validity of the findings of a single study; context 2 – assessment of a causal hypothesis using evidence from all available epidemiologic studies, as well as pertinent information from other fields (general context); and context 3 – determination of the causal role of a particular agent in the illness of an individual person. Contexts 1 and 2 are discussed below.

Assessing the Validity of a Single Study (Context 1)

In developing an interpretation of an individual epidemiologic study, it is important to first recognize that any association, whether positive or negative, observed in such a study has several alternative explanations: it may be due to chance, to bias or to confounding, or it may reflect a valid relationship between the exposure and the disease under investigation. In discriminating among these interpretations, Cole (1997) suggests considering the following criteria:

- Minimal random error
- Minimal systematic error
- Minimal confounding
- Strength
- Internal consistency
- Biologic plausibility

It is never possible to completely rule out chance, selection bias, information bias or confounding as explanations of an observed association. Rather, it is essential to evaluate the credibility of each as part of the interpretation of any study; guidelines for doing this were provided earlier for each of the major epidemiologic study designs.

Strength is an important criterion for judging whether an observed association is causal or noncausal. Typically, epidemiologic studies are subject to many sources of selection, information bias, and confounding. Thus, it is difficult to rule out systematic error as an explanation of apparent associations that are weak (RR<2.0) or even moderate (RR 2.0-5.0) in strength. Strong associations (RR>=5.0), in contrast, are less likely to be due to systematic error unless clearly inappropriate subject selection and information gathering procedures were used, or unless a potent confounding risk factor for disease is highly correlated with the exposure of interest. It is also important to consider the precision of the estimated measure of association in individual studies. Even a strong association may be uninterpretable if it has a wide 95% confidence interval. For example, an RR of 5.0 with a 95% confidence interval of 0.6-8.0 in a single study is uninformative in the context of the single study in which it is measured. This is because the data from the study are compatible with a range of possible effects of exposure on disease that is too broad: the data in this example are consistent with protective effect of exposure (RR<1.0), with no effect of exposure (RR=1.0) and with a very strong effect (RR=8.0). Even a weak association can be causal. Strength is not an invariable biologic property of a causal relationship between a particular exposure and a particular disease. Rather, strength may vary according to the prevalence of other causes in the population under study. Even an association that is not statistically precise can be causal.

Internal consistency is an extremely important criterion for evaluating whether an observed association is valid. Associations that are found both in female and male subjects or in all race groups are more likely to be causal than associations that are limited to one gender or one race, unless there is a credible biologic basis for differences observed among the gender or race groups. If the disease rate tends to increase as exposure level increases (positive dose-response), this is another internally consistent feature of the results that supports a causal interpretation. A time sequence in which exposure clearly precedes disease onset and in which there is an appropriate time lag between first exposure and observation of an increased rate of disease is also an important aspect of internal consistency. Relationships that are not internally consistent may, nonetheless, be valid. Dose-response may be absent despite a causal connection between exposure and disease because of exposure or disease measurement errors, uncontrolled confounding, statistical imprecision, or because exposure does not vary in a biologically meaningful way among the subjects studied.

Biologic plausibility involves the description of a potential mechanism by which the exposure could have caused the disease of interest. Results of experimental research in toxicology or other fields may support biologic plausibility. When pertinent experimental mechanistic data are lacking, an assessment of biologic plausibility may simply involve a subjective determination of whether the exposure of interest could have reached the target organ for disease occurrence.

Evaluating Causality (Context 2)

Perhaps most important in establishing causality in epidemiologic research is the criterion that a relationship must be observed in more than one high-quality study. Because of the many potential sources of error in epidemiologic studies, findings from a single study are not sufficient to generate a consensus among scientists that a causal relationship exists. Instead, an association must be seen in multiple studies before a causal effect of exposure is presumed to exist. This is true even when supportive toxicologic and other data exist.

In assessing the consistency of epidemiologic information pertaining to a hypothesized causal effect, it is important to consider studies with null results, as well as studies showing an effect of exposure on disease. Null results from a study or set of studies cannot be used to argue strongly **against** the existence of an effect if the null results come from imprecise investigations. Rather, such investigations are simply uninformative: they will contribute little to the overall evaluation of causality.

Strong associations observed in multiple studies are more readily accepted as causal than are weak associations. Nonetheless, weak associations may become established as causal if found consistently in a large number of well-conducted studies and if they have a high degree of biologic plausibility.

Finally, a causal interpretation of an association seen in multiple epidemiologic studies must be consistent with all available epidemiologic, toxicologic, and other biologic information. There should not be explicit evidence that argues plausibly against causality. Under these circumstances, the body of evidence is likely to be judged as being "coherent," and the association accepted as causal.

Health and safety officials and regulators may often be confronted with data that suggest an association between workplace exposure and disease but that are not sufficient for establishing causality. In such situations, it may be appropriate to base decisions about control of possible hazards on prudent policy, rather than on science.

CONCLUDING REMARKS

Epidemiologic research plays a crucial role in identifying the causes of disease and in providing a basis for disease prevention. Its particular strength rests in its focus on evaluating relationships between exposure and illness in humans, rather than in experimental animals. However, this attribute also leads to numerous potential imperfections in the epidemiologic study as a measurement instrument. An understanding of sources of error, of the strengths and limitations of various study designs, and of the process of evaluating validity and causality is essential to the proper interpretation of epidemiologic research.

QUESTIONS

10.1. Define epidemiology and list its uses in the occupational setting.

10.2. List and define four measures of disease frequency used in epidemiology.

10.3. List four relative measures of association used in epidemiology, and indicate the epidemiologic study designs in which each measure is used.

10.4. Identify and define three problems that threaten the validity of an epidemiologic study. List standard procedures used in epidemiologic research to minimize each of these problems.

10.5. A study of the relationship between occupation and stomach cancer found an OR of 1.5, with a 95% confidence interval of 0.9-2.4 for mechanics. What is the meaning of the confidence interval?

10.6. Define the terms "open study group" and "closed study group," and indicate the study design in which each type of study group is usually found.

10.7. Describe the main steps required to conduct a follow-up study.

10.8. List the main advantage and two potential disadvantages of using an external comparision group in a follow-up study.

10.9. List at least three data sources commonly used to determine subjects' vital status in occupational follow-up studies.

10.10. A follow-up study includes 10,000 subjects, a total of 200,000 person-years of observation, and a total of 100 incident cases of lung cancer. a) What is the lung cancer incidence rate in this study group? The exposed subgroup has 2,000 subjects, 50,000 person-years and 50 cases. b) What is the crude IRR for exposed compared to unexposed subjects?

10.11. List two types of procedures used to obtain confounder-adjusted summary rate ratios in follow-up studies of open study groups.

10.12. Explain why prospective follow-up studies are not often done in evaluating relationships between occupational factors and cancer.

10.13. What are the two main types of case-control study, and what features distinguish the two types?

10.14. Explain how controls are usually selected in a nested case-control study.

10.15. A case-control study of children born with cleft lip/cleft palate includes a total of 100 cases and 400 controls. Thirty of the cases and 60 of the controls have mothers who were occupationally exposed to ethylene glycol ethers. What is the odds ratio for the association between ethylene glycol ethers and cleft lip/cleft palate?

10.16. List the main advantages and disadvantages of the three main study designs used in occupational epidemiology.

10.17. Explain how validity is assessed in evaluating the quality of an epidemiologic study.

10.18. Explain how causality is evaluated in epidemiologic research.

REFERENCES

Barreto, S. M.; Swerdlow, A.J.; Smith,P.G.; et al.: Risk of death from motor-vehicle injury in Brazilian steelworkers: A nested case-control study. Int. J. Epidemiol. 26:814-821 (1997).

Bhatia, R.;Lopipero P.; Smith, A.H.: Diesel exhaust exposure and lung cancer. Epidemiology 9:84-91 (1998).

Birkett, N. J.: Effect of nondifferential misclassification on estimates of odds ratios with multiple levels of exposure. Am. J. Epidemiol. 136:356-362 (1992).

Blair, A.; Stuart, P.A.; Zaebst, D.D.; et al.: Mortality study of industrial workers exposed to acrylonitrile. Scand. J. Work Environ. Health 24:25-41 (1998).

Blot, W.J.; Stone, B.J.; Fraumeni, Jr., J.F.; et al.: Cancer mortality in U.S. Counties with shipyard industries during World War II. Environ. Res. 18:281-290 (1979).

Blot, W.J.; Morris, L.E.; Stroube,R.; et al.: Lung and laryngeal cancers in relation to shipyard employment in coastal Virginia. J. Natl. Cancer Inst. 65:571-575 (1980).

Checkoway, H.; Pearce, N.E.; Crawford-Brown, D.J.: Research Methods in Occupational Epidemiology. Oxford University Press, New York (1989).

Chiazzc, Jr., L; Watkins, D.K.; Fryar, C.: A case control study of malignant and non-malignant respiratory disease among employees of a fibreglass manufacturing facility. Br. J. Ind. Med. 49:326-331 (1992).

Chiazze, L. Jr.; Watkins, D.K.; Fryar, C.; et al.: A case-control study of malignant and non-malignant respiratory disease among employees of a fibreglass manufacturing facility. II. Exposure assessment. Am. J. Ind. Med. *50:*717-725 (1993).

Cole, P.: Causality in epidemiology, health policy, and law. Environ. Law Reporter 27:10279-10285 (1997).

Cordier, S.; Bergeret, A.; Goujard, J.; et al.: Congenital malformations and maternal occupational exposure to glycol ethers. Epidemiology. 8:355-362 (1997).

Delzell, E.; Sathiakumar, N.; Hovinga, M.; et al.: A follow-up study of synthetic rubber workers. Toxicology. 113:182-189 (1996).

Dickersin, K.; Berlin, J.A.: Meta-analysis: state-of-the-science. Epidemiol. Rev. 14:154-176 (1992).

Eisen, E.A.: Healthy worker effect in morbidity studies. Med. Lav. 86:125-138(1995).

Epidemiology Task Group: Guidelines for good epidemiology practices for occupational and environmental epidemiologic research. Epidemiology Resource and Information Center (ERIC) Pilot Project. Chemical Manufacturers Association (1991).

Eskenazi, B.; Gold, E.B.; Lasley, B.L.; et al.: Prospective monitoring of early fetal loss and clinical spontaneous abortion among female semiconductor workers. Am. J. Ind. Med. 28:833-846 (1995).

Flanders, W.D.; Louv, W.C.: The exposure odds ratio in nested case-control studies with competing risks. Am. J. Epidemiol. 124:684-692 (1986).

Goldberg, M.S.; Siemiatycki, J.; Gerin, M.: Inter-rater agreement in assessing occupational exposure in a case-control study. Br. J. Ind. Med. 43:667-676 (1986).

Gomez, M.R.; Dosemeci, M.; Stewart, P.A.: Occupational exposure to chlorinated aliphatic hydrocarbons: Job exposure matrix. Am. J. Ind. Med. 26: 171-183 (1994).

Greenland, S.: Quantitative methods in the review of epidemiologic literature. Epidemiol. Rev. 9:1-30 (1987).

Greenland, S.; Robins, J.M.: Confounding and misclassification. Am J Epidemiol 122:495-506 (1985).

Hallock, M.F.; Smith, T.J.; Woskie, S.R.; et al.: Estimation of historical exposures to machining fluids in the automotive industry. Am. J. Ind. Med. 26:621-634 (1994).

Hammond, S.K.; Hines, C.J.; Hallock, M.F.; et al.: Tiered exposure-assessment strategy in the semiconductor health study. Am. J. Ind. Med. 28:661-680 (1995).

Hill, A.B.: The environment and disease: association or causation? *Proc.* R. Soc. Med. 58:295-300 (1965).

Kupper, L.L.; McMichael, A.J.; Simmons, M.J.; et al.: On the utility of proportional mortality analysis. J. Chronic Dis. 31:15-22 (1978).

Lang, J. M.; Rothman, K.J.; Cann, C.I.: That confounded *p*-value. Epidemiology 9:7-8 (1998).

Macaluso, M.; Larson, R.; Delzell, E.; et al.: Leukemia and cumulative exposure to butadiene, styrene and benzene among workers in the synthetic rubber industry. Toxicology 113:190-202 (1996).

Marsh, G.; Youk, A.O.; Stone, R.A.; et al.: OCMAP-Plus: A program for the comprehensive analysis of occupational cohort data. J. Occup. Environ. Med. 40:351-362 (1998).

Miettinen, O.S.; Wang, J.D.: An alternative to the proportionate mortality ratio. Am. J. Epidemiol. 114:144-148 (1981).

Miettinen, O.S.: Theoretical Epidemiology. Principles of Occurrence Research in Medicine John Wiley and Sons, New York (1985).

Monson, R.R.: Analysis of relative survival and proportional mortality. Comput. Biomed. Res. 7:325-332 (1974).

Monson, R.R.: Occupational Epidemiology. CRC Press, Boca Raton, Florida (1980).

Morgenstern, H.: Ecologic studies in epidemiology. Annu. Rev. Public Health 16:61-81 (1995).

Parent, M.E.; Siemiatycki, J.; Fritschi, L.: Occupational exposures and gastric cancer. Epidemiology 9:48-55 (1998).

Park, R.M.; Maizlish, N.A.; Punnet, L.; et al.: A comparison of PMRs and SMRs as estimators of occupational mortality. Epidemiology 2:49-59 (1991).

Pastides, H.; Calabrese, E.J.; Hosmer, D.W., Jr.; et al.: Spontaneous abortion and general illness symptoms among semiconductor manufacturers. J. Occup. Med. 30: 543-551 (1988).

Pastides, H.; Calabrese, E.J.; Hosmer, D.W., Jr.; et al.: Spontaneous abortion and general illness symptoms among semiconductor manufacturers. J. Occup. Med. 30:543-551 (1988).

Preston, D.L.; Lubin, J.H.; Pierce, D.A.; et al.: EPICURE User's Guide. HiroSoft International Corporation, Seattle, WA (1993).

Robins, J.M.; Blevins, D.: Analysis of proportionate mortality data using logistic regression models. Am. J. Epidemiol. 125:524-535 (1987).

Rosner, B.: Fundamentals of Biostatistics. Wadsworth Publishing Co., Belmont, CA (1995).

Rothman, K.J.; Greenland, S.: Modern Epidemiology. Lippincott-Raven Publishers, Philadelphia, PA (1998).

Rothman, K.J.: Lessons from John Graunt. Lancet 347:37-39 (1996).

Rothman, K.J.: Modern Epidemiology. Little, Brown and Company, Boston (1986).

Schenker, M.B.; Gold, E.B.; Beaumont, J.J.; et al.: Association of spontaneous abortion and other reproductive effects with work in the semiconductor industry. Am. J. Ind. Med. 28:639-659 (1995).

Schnorr, T.M.; Grajewski, B.A.; Hornung, R.W.; et al.: Video display terminals and the risk of spontaneous abortion. N. Engl. J. Med. 324:727-733 (1991).

Shapiro, S.: Meta-analysis/shmeta-analysis. Am. J. Epidemiol. 140:771-778 (1994).

Stewart, P.A.; Lemaski, D.; White, D.; et al.: Exposure assessment for a study of workers exposed to acrylonitrile. l. Job exposure profiles: A computerized data management system. Appl. Occup. Environ. Hyg. *7:*820-825 (1992).

Stewart, P.A.; Triolo, H.; Zey, J.: et al.: Exposure assessment for a study of workers exposed to acrylonitrile. ll. A computerized exposure assessment program.. Appl. Occup. Environ. Hyg. 10:698-706 (1995).

Stewart, P.A.; Zey, J.N.; Hornung, R.; et al.: Exposure assessment for a study of workers exposed to acrylonitrile. lll. Evaluation of exposure assessment methods. Appl. Occup. Environ. Hyg. *11:*1312-1321 (1996).

Stewart, W.; Hunting, K.: Mortality odds ratio, proportionate mortality ratio, and healthy worker effect. Am. J. Ind. Med. 14:345-353 (1988).

Wacholder S.; McLaughlin, J.K.; Silverman, D.T.; et al.: Selection of controls in case-control studies, l: principles. Am. J. Epidemiol. 135:1019-1028 (1992a)

Wacholder, S.: Practical considerations in choosing between the case-cohort and nested control design. Epidemiology 2:155-158 (1991).

Wacholder, S.; Silverman, D.T.; McLaughlin, J.K.; et al.: Selection of case-control studies, ll: types of controls. Am J. Epidemiol. 135:1029-1041 (1992b).

Wacholder, S.; Silverman, D.T.; McLaughlin, J.K.; et al.: Selection of controls in case-control studies. lll: design options. Am. J. Epidemiol. 135:1042-1050 (1992c).

Wegman, D.H: The potential impact of epidemiology on the prevention of occupational disease. Am. J. Public Health 82:944-954 (1992).

Wong, O.; Decoufle, P.: Methodological issues involving the standardized mortality ratio and proportionate mortality ratio in occupational studies. J. Occup. Med. 24:299-304 (1982).

Wong, O.; Morgan, R.W.; Kheifets, L.; et al.: Comparison of SMR, PMR, and PCMR in a cohort of union members potentially exposed to diesel exhaust emissions. Br. J. Ind. Med. 42:449-460 (1985).

Zocchetti, C.; Consonni, D.; Bertazzi, P.A.: Relationship between prevalence rate ratios and odds ratios in cross-sectional studies. Int. J. Epidemiol. *26:*220-223 (1997).

11

HEALTH RISK ASSESSMENT AND THE PRACTICE OF INDUSTRIAL HYGIENE

Dennis J. Paustenbach, PhD

INTRODUCTION

Risk assessment has been broadly defined as the methodology that attempts to predict the likelihood of numerous unwanted events, including industrial explosions, workplace injuries, failures of machine parts, natural catastrophes (e.g., earthquake, tidal waves, hurricanes, volcanic eruptions, tornadoes, blizzards), injury or death due to an array of voluntary activities (e.g., skiing, football, sky diving, flying, hunting), diseases (e.g., cancer, developmental toxicity caused by chemical exposures), death due to natural causes (e.g., heart attack, cancer, diabetes), death due to lifestyle (e.g., smoking, alcoholism, diet), and others (Paustenbach, 1995, 2000).

Health risk assessment is the process wherein toxicology data from animal studies and human epidemiology are evaluated, a mathematical formula is applied to predict the response at low doses, and then information about the degree of exposure to the disease agent is quantitatively used to predict the likelihood that a particular adverse response will be seen in a specific human population (Paustenbach, 1995, 2000). The risk assessment process has been used by regulatory agencies for almost 40 years, most notably within the United States Food and Drug Administration (FDA) (Lehman and Fitzhugh, 1954). However, the difference between assessments performed in the 1950s and 1960s versus those performed in the 1980s and 1990s is that dose-extrapolation models, quantitative exposure assessments, and quantitative descriptions of uncertainty have been added to the process. Due to an increase in understanding, the availability of desk top computers and better quantitative methods for predicting the low dose response, such as physiologically-based pharmacokinetic (PB-PK) models, risk assessments conducted in the future should provide more accurate risk estimates than in the past.

Since 1980, most environmental regulations and some occupational health standards have, at least in part, been based on health risk assessments (Paustenbach, 1995, 2000). These include standards for pesticide residues in crops, drinking water, ambient air, and food additives, as well as exposure limits for contaminants found in indoor air, consumer products, and other media. Risk managers increasingly rely on risk assessment to decide whether a broad array of risks are significant or trivial; an important task since, for example, more than 400 of the approximately 2,000 chemicals routinely used in industry have been labeled carcinogens in various animal studies. In theory, in the United States, the results of risk assessments should influence virtually all regulatory decisions involving so called "toxic agents."

The emergence of "modern-day" health risk assessment can be traced to the mid-1970s with the publication by Crump et al. (1976), which described an approach for predicting cancer rates at low doses using a mathematical model. Since then, at least six dozen guidance documents (about 10,000 pages) on how risk assessments should be conducted have been written by the U.S. Environmental Protection Agency (EPA) and other agencies (Paustenbach, 1995, 2002). During much of the past 20 years, when the United States chose to standardize the process through regulation, the guidance for conducting risk assessment was relatively inflexible and unable to keep pace with our ability to characterize the true or most likely risks. Due to this inflexibility, most attempts by regulatory agencies to standardize health risk assessment methods introduced several levels of conservatism. The end result was that most risk assessments did, for the sake of public safety, significantly overestimate rather than underestimate the true risks (Nichols and Zeckhauser, 1978; Maxim and Harrington, 1989). Fortunately, this changed in the late 1990s as agencies were more comfortable in dealing with complex biological issues (EPA 1999a, b).

In the United States, the popularity of human and environmental risk assessment as a policy instrument has been due, in large part, to public concern regarding the relationship between cancer and chemical exposure, as well as other adverse health effects, and frustration with the perceived inability of the government to adequately regulate these chemicals (Paustenbach, 1995, 2000). At the same time, industry has insisted, through the court system, that an adequate scientific rationale be provided to support potentially restrictive regulatory decisions. Fortunately, risk assessment has generally been successful in satisfying these often-conflicting mandates. As a result, over the past 20 years in the United States, quantitative risk assessment has evolved into an accepted methodology for identifying a range of options for tackling environmental and occupational health issues (Hartwell and Graham, 1997; EPA, 2000a). Ideally, when developing each of these options, specific cost and benefit information was provided to risk managers, policy makers, and the public so that they could make reasoned and measured decisions. Although risk assessment has played a prominent role in the environmental arena since about 1980, scientists have only recently begun to apply risk assessment methods to the practice of industrial hygiene or the setting of occupational exposure limits (Paustenbach, 1990a, 1997a; Jayjock et al., 2000; Rozman and Doull, 2002).

FUNDAMENTALS OF HEALTH RISK ASSESSMENT

The risk assessment process can be divided into four parts: hazard identification, dose-response assessment, exposure assessment, and risk characterization (see Figure 11-1). Hazard identification is the first and most easily recognized step in a risk assessment. It is the process

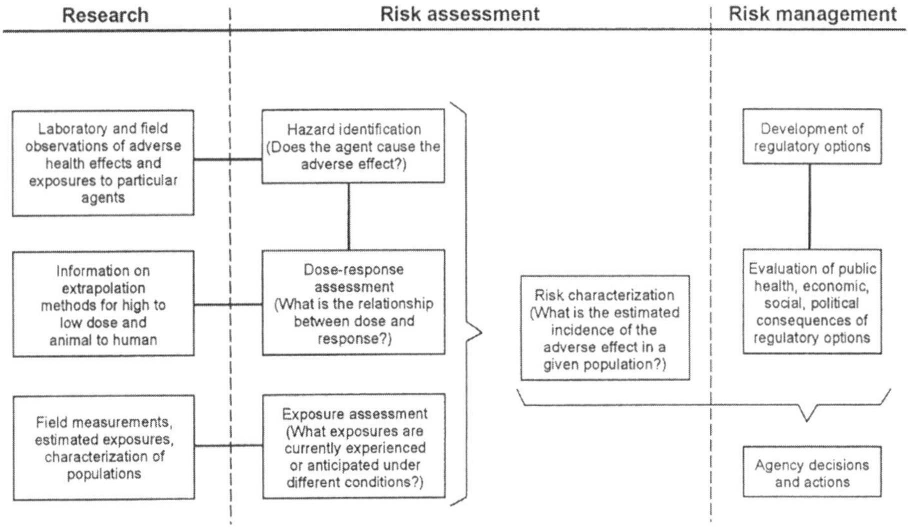

Figure 11-1. A schematic illustration of the four components of a risk assessment (NAS, 1983).

of determining whether exposure to an agent could (at any dose) cause an increase in the incidence of adverse health effects (cancer, birth defect, etc.) in humans or wildlife. Dose-response evaluations define the relationship between the dose of an agent and the probability of a specific adverse effect in laboratory animals. Exposure assessment addresses the probability of uptake (systemic absorption) of a chemical in the environment by any combination of oral, inhalation, and dermal routes of exposure. Perhaps the most important part of an assessment, risk characterization, summarizes and interprets the information collected during the previous activities and identifies the limitations and the uncertainties in the risk estimates (National Academy of Sciences [NAS], 1983; 1994).

Risk assessments conducted to evaluate an occupational hazard are much different than those used to characterize an environmental hazard. For example, as many as 20 or more factors need to be considered in an exposure assessment of most environmental situations while most occupational assessments focus only on dermal and inhalation exposure (Paustenbach, 1995; Paustenbach et al., 1992).

Hazard Identification

Hazard identification involves gathering and evaluating data on the types of health injury or disease that may be produced by a chemical and on the conditions of exposure under which injury or disease may occur. It may also involve characterization of the behavior of a chemical within the human body and its interactions with organs, cells, or macromolecules.

Fundamentally, hazard identification is a scientific, qualitative evaluation of all available toxicological, epidemiological, biological, and structural analogy data. Its principal function is to identify the predominate type(s) of adverse health effect that a particular toxicant might produce as a result of human exposure. It is inclusive of all health endpoints from irritation to cancer, and uses all toxicological information including mechanistic work like PB-PK study. This biological information can be useful in determining if the kinds of adverse effects known to be produced by a chemical agent in one population group or in a particular experimental setting are likely to be produced in the human population group of interest (NAS, 1983).

Hazard identification is an expert scientific judgment process that typically involves groups or panels of multi-disciplinary experts from the fields of toxicology, epidemiology, medicine, biology, and other disciplines. Governmental hazard identifications often result in complex documents generated by contractors, further refined by agency scientists, and ultimately reviewed by external scientific review committees, such as the EPA Science Advisory Board (SAB). The SAB is a group of about 200 professionals who are considered among the very best in the field, who conduct outside peer review on the most significant scientific decisions of EPA (about 60% are professors).

To a large degree, the hazard identification process is dominated by data provided by toxicologists, epidemiologists, and occupational physicians on whether or not a certain dose of a chemical can cause adverse effects within a given population. However, risk (i.e., the probability of occurrence) is not assessed at this stage of the assessment. Instead, the hazard identification is conducted to determine whether and to what degree it is scientifically correct to infer that toxic effects observed in one setting will occur in another (e.g., are chemicals found to be carcinogenic, teratogenic, or neurotoxic in experimental animals likely to produce similar effects at the same dose in humans?) (NAS, 1983; 1994).

Data on the toxicity of a chemical may come from several sources. The most relevant data come from studies in humans. During the early years of toxicology, from 1920 to 1970, a large body of data was derived from the observation of adverse effects in workers (Hamilton, 1928; Irish, 1959). More recently, however, most toxicological information comes from experiments with laboratory animals. Although it is preferable to learn from animal studies, one must be cautious in extrapolating animal toxicology data to humans because many assumptions about the biological similarity of mammalian species to humans will often not apply to all chemicals. The assumptions concerning the applicability of toxicity test conditions (e.g., exposure route, doses) to human exposure conditions, as well as some quantitative understanding about the appropriateness of the scale-up procedure, need to be understood and accounted for (Andersen et al., 1987a, 1995; NAS, 1994; Krishnan and Andersen, 2001) during the risk assessment process.

Although there is a substantial body of evidence showing that the results of animal studies are, with the appropriate qualifications, generally applicable to humans, there are important exceptions to this broad assertion (Ottoboni, 1991; Krishnan and Andersen, 2001). Nonetheless, in an effort to err on the side of safety, unless there are data from human studies that refute a specific finding in animals, or unless there are other biological reasons to consider certain types of animal data irrelevant to humans (e.g., $\propto_{2u}$-globulin mediated nephropathy in male rats), it is generally assumed that the human response can be inferred from observations in experimental animals (Albert, 1994). The assumption that all forms of adverse health effects observed in experimental animals could also be observed in humans is a conservative one which is accepted, in part, for reasons of prudent public health policy or regulatory mandate (Paustenbach, 1995, 2000). However, the bulk of the evidence indicates, and most experts believe, that animal data alone are not sufficient to conclude that an agent will cause specific types of effects in humans exposed to a particular dose, although there is generally good concordance that an effect in an animal study can be seen in humans at some dose.

Having recently celebrated the 25th year in the current-day era of the practice of health risk assessment (post-1976), we now have enough experience to understand the strengths and weaknesses of the methodology, and, in particular, the hazard identification portion. We have also learned a lot about the likely human health risks of exposure to low doses of various classes of chemicals (e.g., neurotoxicants, carcinogens, developmental toxicants). For example, it has become clear that most animal carcinogens (at some dose) will probably pose a human cancer hazard, however, the risk at very low doses may be very small and, often, may be zero (Abelson, 1993). We know that criteria by which a risk assessor should determine that a chemical could pose a significant carcinogenic or developmental threat to humans based on animal studies requires that at least six factors be considered. For carcinogens, these factors include the number of animal species affected, the number and types of tumors occurring in the animals, the dose (relative to the acute toxic dose) at which the animals are affected, the dose/response relationship, and the genotoxicity of the chemical (Squires, 1981; EPA, 1999a). For developmental toxicants, the key issues in the hazard identification step are similar to those for carcinogens and include the number of species affected, the severity of the effect, and the relationship between the dose which affects the mother compared to that which affects the offspring (Wang and Schwartz, 1987).

Over the past few years, the scientific community has learned that, in the hazard identification process, it should not place an equal weight on all data (EPA, 1996, 1999a). Unlike the early years, regulatory agencies now attempt to resist the tendency to place an emphasis on any piece of data that suggests that a chemical might pose a particular hazard, and little weight on data that suggest that the chemical failed to cause these effects. This approach was once considered prudent and health protective from a public policy standpoint (Tengs et al., 1995). Today, most scientists have come to accept that only data of similar quality should be judged equally. In the United States, two basic approaches have emerged on the subjective weighting of data. One is called the "strength of the evidence" and the second is called the "weight of the evidence." Upon initial review, these appear similar, but in fact, they are very different. One's conclusion about the hazards posed by a chemical can differ significantly depending on which approach is being used. "Strength of the evidence" approaches tend to focus on evaluating how strongly positive the study or studies are (how many tumors, across what sites, how many positive studies, etc.). It tends to ignore or downplay the importance or meaning of negative studies. "Weight of evidence" evaluations factor in both positive and negative studies and seek to integrate positives and negatives into a scientific judgment about the inherent hazard of the substance (Jayjock et al., 2000).

Different organizations use different approaches to evaluate data. For example, the International Agency for Research on Cancer (IARC) and the National Toxicology Program (NTP) systems tend to be "strength of the evidence" approaches. If there are several positive studies, these approaches drive one to assign a cancer classification whether there are negative studies or not. These systems are meant to be conservative or err on the side of safety. Obviously, it is important that the *user* of the system's output understand the basis of the classification.

In contrast, ACGIH® tends to have a "weight of the evidence" approach in its classification system, scientifically integrating all adequate quality studies. Similarly, the EPA alleges that they use a "weight" approach, but in practice, to date, it tends to operate most like a "strength" approach although this is under revision (EPA, 1999a). This document will most likely undergo additional changes in 2003 and 2004.

In any of these systems, the experts usually score the animal evidence and rank the human epidemiological data on a continuum of categories: "sufficient or clear" evidence, "limited or mixed" evidence, "inadequate" evidence, "no data available," or "no evidence" of the effect in studies. This scoring is performed based on comprehensive review of all qualified studies, and it is done separately according to animal evidence and the human evidence. They are usually combined in a scheme to yield a carcinogen classification. Mechanistic information such as genetic toxicity studies is typically integrated as modifying information or to support the initial classification (Jayjock et al., 2000).

The 1999 *EPA Draft Guidelines for Carcinogenic Risk Assessment* represent the first time a United States regulatory agency has been required to use this approach. It is an important change in policy from that used from 1960 to 1990 when the study, which produced the lowest NOEL (highest risk), was given the greatest weight. The use of a weight-of-evidence approach is applicable not only to the hazard identification segment of risk assessment, but also the exposure and dose-response evaluations (EPA, 1999a, 1999b). One benefit of using this approach is that it minimizes the possibility that huge sums of money will be spent to conduct several high quality toxicity studies simply to refute the results of one or two poorly conducted ones. The test case regarding the amount of information needed by an agency to call a chemical a "threshold carcinogen" involves chloroform and it is still being debated. Specifically, chloroform is present in drinking water in some towns at concentrations that would pose a theoretical cancer risk as high as 1 in 1,000, however, few toxicologists believe these predictions to be accurate. The reason is that the chemical is a weak mutagen, only produces tumors in animals at doses that produce toxicity, and because the epidemiologic data doesn't suggest a risk due to chlorinated water.

Dose-Response Assessment

A dose-response assessment describes the quantitative relationship between the amount of chemical that is administered (applied dose) and the incidence of toxic injury or disease (NAS, 1983; 1994). Data are usually collected in animal studies or, less frequently, from studies of exposed human populations. There may be many different dose-response relations for a chemical agent depending on the conditions of exposure (e.g., single versus repeated exposures), the route of administration (oral, dermal, or inhalation), and the response (e.g., cancer, systemic toxicity) being considered.

When applying data derived from experiments with animals to estimate the risks to humans, at least two major extrapolations are required: 1) interspecies adjustments for differences in size, life span and basal metabolic rate, and 2) extrapolation of the dose-response relation observed at doses used in animal experiments to the lower doses to which humans are likely to be exposed. Models have been developed that attempt to estimate the results because, in general, the responses at these low doses (e.g., those which produce less than a 10% response) cannot be observed or measured (Food Safety Council, 1980; Krewski et al., 1989; International Life Sciences Institute [ILSI], 1995) (see Figure 11-2).

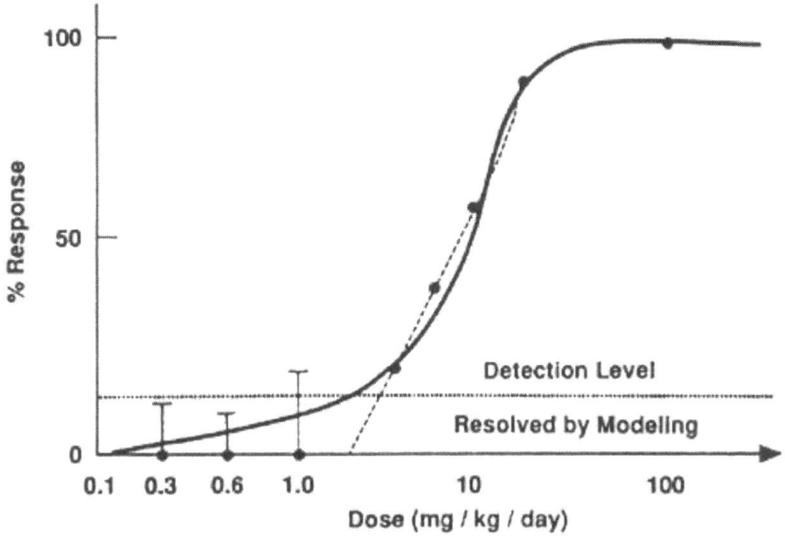

Figure 11-2. A dose response curve from an exceptionally thorough (8 dose groups) carcinogenicity study. The solid line is a best fit of the eight data points identified in the test. The three lowest data points indicate that at these doses, no increased incidence in tumors was observed in the test animals. The error bars on the three lowest doses indicate the statistical uncertainty in the test results since a limited number of animals were tested (n = 100). In an effort to derive risk estimates that are unlikely to underestimate risk, the models usually derive risk estimates based on the estimated upper bound of the plausible response, rather than the best estimate (from Paustenbach, 1989).

Interspecies adjustments of dose, often called interspecies scaling factors, are necessary to account for differences between humans and laboratory animals in size, life span, and basal metabolic rate (Andersen et al., 1987a, 1995; Allen et al., 1988; Freedman and Sisal, 1988; Travis and White, 1988; Travis et al., 1990). Because most studies involve oral dosing, the most commonly used measures of dose are milligrams of chemical per kilogram of body weight of the animal per day (mg/kg-day) and milligrams of chemical per square meter of body surface area per day (mg/m^2-day). There are numerous other measures of dose (i.e., dose metrics) that can be used, such as area-under-the-blood/time curve (AUC), time above a particular tissue concentration of a toxicant, peak circulating blood concentration, etc. In recent years, it has become clear that one dose metric alone cannot be used as the basis for interspecies extrapolation for all adverse effects; e.g., liver toxicity, neurotoxicity, developmental effects, irritation, or cancer. Rather, the proper dose metric will depend on the mechanism of action, the specific biological effect, the pharmacokinetics of the chemical and other factors (Andersen et al., 1987a; Aylward et al., 1996).

For carcinogenic compounds, several different scaling factors have been used by different risk assessors and by different federal agencies to adjust the dose administered to animals to predict the equivalent human dose. For example, the EPA originally used mg/m^2-day, which is the more conservative scaling factor because it tends to give higher risk estimates per unit of dose. In contrast, the FDA has historically used mg/kg-day. Later, EPA suggested that mg/kg-day to the 0.66 or 0.75 power might be a better predictor of risk (EPA, 1996). For some carcinogens, like the persistent organohalogens, which often produce tumors through non-genotoxic mechanisms, peak concentrations in a tissue times the amount of time above that concentration will probably be the most relevant dose metric (Aylward et al., 1996). Comparison of risk estimates generated from human data and animal data on a number of chemical carcinogens indicates that the use of body weight provides good agreement for some of them, while surface area is better for others. A significant amount of research effort is needed to identify the appropriate scaling factor approach and dose metric for specific chemicals and toxic effects.

With rare exception, for any given chemical and route of exposure, the severity and frequency of the effect (i.e., the response) decreases with decreasing dose. In addition, the type of adverse effect may change with dose. For example, anesthetic gases cause death at high doses, induce sleep at moderate doses, produce headaches or lethargy at low doses, and produce no detectable effect at even lower doses (Ottoboni, 1991). Clearly, an understanding of any dose-response relationship and its effect is critical to describing the human health hazard. To the extent that this relation is not well defined, uncertainty about the estimated risk at various doses will exist and, accordingly, the uncertainty will be greater as one attempts to predict effects far below the doses tested.

NON-CARCINOGENS

For purposes of risk assessment, chemicals are usually divided into two categories — carcinogens and non-carcinogens — because different methods are used for estimating the risks at various levels of exposure to each of these classes of chemicals. All carcinogens, by definition, can produce non-carcinogenic effects, but for virtually all chemicals the doses needed to minimize the cancer hazard are so low that the non-carcinogenic effects are not a concern. Therefore, the term "non-carcinogen" is strictly operational, i.e., a term that is useful for describing a situation rather than an inherent biological property (Environ, 1994). For example, it is used to describe either chemicals that have not been shown to be carcinogenic by epidemiological or experimental studies, or chemicals that have not been evaluated for carcinogenic potency. The term is used only for convenience and it is recognized that future studies may suggest carcinogenic activity. Non-carcinogenic toxic effects include such diverse responses as sensory irritation, damage to specific organs (e.g., kidney, heart, liver, or nervous system), and birth defects.

To identify so-called "safe" levels of exposure for the non-carcinogens, one of four methods are typically used. They are the no observed effect level (NOEL) safety factor approach, mathematical models with thresholds, the benchmark dose method, and the PB-PK model. Each has its own strengths and weaknesses and, at this time, one is not considered superior to the other in all circumstances.

The NOEL-safety factor approach is the original and still most frequently used method for identifying acceptable doses. Basically, the NOEL from the best animal study is selected and an uncertainty factor is divided into it to identify a "safe" level of exposure. This has been the most frequently used approach in safety assessment since the 1940s (Lehmann and Fitzhugh, 1954). As shown in Table 11-1, the uncertainty factors are used to account for our lack of understanding about such issues as the difference in susceptibility between rodents and humans, differences in inter-individual susceptibility among persons, and other factors (Dourson and Stara, 1983; Calabrese, 1978; Renwick, 1995). This approach is also used to identify reference doses (RfD) for exposure to chemicals in our diet so it is sometimes called the RfD method.

There are several limitations in the use of the RfD method. This approach focuses only on the dose that is the no observed adverse effect level (NOAEL) and does not incorporate information on the slope of the dose-response curve at higher doses where effects are observed, or the variability in the data (Allen et al., 1994a). Since data variability is not taken into account, the NOAEL from a study with few animals will likely be higher than the NOAEL from a similar study but with more animals of the same species. Additionally, the NOAEL is usually one of the experimental doses, and the number and spacing of doses in a study can influence the dose that is chosen for the NOAEL. The NOAEL is defined as a dose that does not produce an observable change in adverse responses from control levels. The NOAEL is dependent on the power (and thus sample size) of the study (Allen et al., 1994b). Theoretically, the risk associated with it may fall anywhere between zero and an incidence just above that detectable from control levels. This detectable level is usually in the range of 7 to 10% for quantal data (Ottoboni, 1991; Rodricks, 1994; Jayjock et al., 2000).

Table 11-1
Criteria and guidelines for the application of uncertainty factors

Type	Magnitude	Comments
Interindividual variation	3-5	• Intended to account for variation in susceptibility among the human population (i.e., high risk groups).
Interspecies variation	3-5	• Intended to account for uncertainty in extrapolating results obtained in animals to the general human population. • Applied in all cases where the NOAEL[1] and LOAEL[2] were derived from an animal study.
Subchronic to chronic of less-than-lifetime to lifetime	5 or 10	• Intended to account for the uncertainty in extrapolating from less-than-lifetime to lifetime exposure or subchronic to chronic exposure. • UF[3] of 10 applied whenever study duration is 90 days or less in rodent species or approximately 1/10 to 1/5 life span in other species.
LOAEL to NOAEL	5 or 10	• Intended to account for uncertainty inherent in extrapolating downward from a LOAEL to a NOAEL.

Source: Adapted from Calabrese and Kenyon (1991)

[1] NOAEL – No-Observed-Adverse-Effect-Level. That exposure level at which there are no statistically and biologically significant increases in frequency or severity of adverse effects (adverse effects are defined as any effect resulting in functional impairment and/or pathological lesions that may affect the performance of the whole organism, or that reduce an organism's ability to cope with an additional challenge) between the exposed populations and its appropriate control. Effects are produced at this level, but they are not considered to be adverse (EPA,1994).

[2] LOAEL – Lowest-Adverse-Effect-Level. The lowest exposure level in a study or group of studies that produces statistically and biologically significant increases in frequency or severity of adverse effects between the exposed population and its appropriate control (EPA, 1994).

[3] UF – uncertainty factor

The second approach for identifying the "likely" safe level of exposure to a non-carcinogen involves use of a simple curve-fitting model. The dose-response data from a toxicity study is input into a "best fit" computer program known as a probit. The probit method relies upon a statistical assumption about variability of data. It has been used in toxicology to identify the LD_{50} (lethal dose for 50% of the exposed population) for many years. The probit approach has been discussed elsewhere in detail (Ottoboni, 1991; Klaassen, 2000, Hayes, 2001).

The third approach, the benchmark dose methodology, relies upon a combination of both low dose extrapolation and the application of uncertainty factors. The benchmark dose is a model-derived estimate of the upper confidence limit on the effective dose that produces a certain increase in incidence above control levels, such as 1, 5, or 10% (Barnes et al., 1995). As mentioned, most toxicity studies are designed to detect changes from control on the order of 10% to 20%. Because the application of the model is to derive an estimate of dose for a given incidence that is likely to fall within or very close to the experimental dose range, it does not require extrapolation to estimates far below the experimental dose range and the uncertainty in the estimates is lessened.

As shown in Figure 11-3, the benchmark dose is derived by modeling the data in the observed range, calculating the upper confidence limit on the dose-response curve, and select-

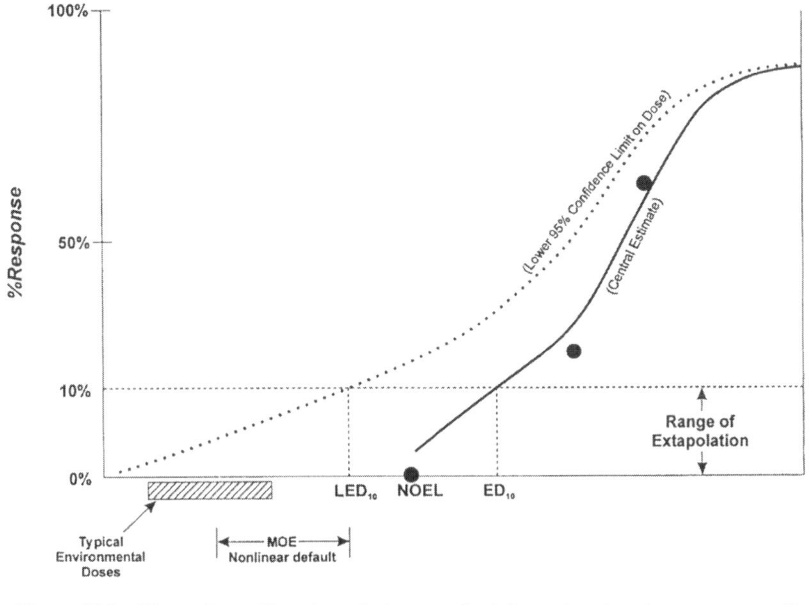

Figure 11-3. Illustration of benchmark dose method for estimating dose corresponding to specified level of increased response; ED_{10} and LED_{10} (the benchmark dose) are central estimate and lower confidence limit of dose corresponding to 10% increase in response, respectively. These are obtained from statistical fitting of dose-response model to dose-response data, and MOE is margin of exposure (safety factor) that can be applied to the benchmark dose (from Paustenbach, 2002).

ing the point on the upper confidence curve corresponding to, e.g., a 10% increase in incidence (ED_{10}) of an effect. Using this approach, a benchmark dose can be calculated for each toxicological response (adverse effect) that has adequate data. The adverse effect with the lowest value is considered the critical or driving effect and it becomes the one that is considered "safe." In some cases, the data may be adequate to also estimate benchmark doses at lower incidence levels, e.g., the 1% level, which may be closer to a true no-effect dose.

Several models may be used to calculate the benchmark dose, but choice of the model may not be critical since estimation is often within the observed dose range. Thus, any model that fits the empirical data well is likely to provide a reasonable estimate of the benchmark dose. For example, similar results are obtained from the Weibul, the multi-stage, or the probit models. If, however, there is some biological reason to incorporate particular factors in the model (e.g., intralitter correlation for developmental toxicity data), these should be included to account as much as possible for variability in the data. The exposure limit is then derived using an uncertainty factor as described above applied to the benchmark dose (e.g., 10% response dose) as a starting point (Jayjock et al., 2000).

Another possible use of the 10% benchmark dose is to assume a linear/no threshold model to estimate the proportion of the population protected at lower doses. Applying a mild extrapolation (a factor of 10 to 100) to this 10% benchmark yields exposure limits that should protect about 99.0 to 99.9% of the exposed population. It is noteworthy that the resulting exposure limit is identical to applying a "safety factor" of between 10 and 100 to the "benchmark" of the NOAEL. In this way, the "nearly all" level of worker protection is quantified under the assumptions of this model.

Physiologically-Based Pharmacokinetic Models

Many published papers have described the use of PB-PK modeling in human health risk assessment (Ramsey and Andersen, 1987; Clewell, 1992; Paustenbach et al., 1988; Andersen, 1995; Leung, 2000). PB-PK modeling tries to simulate the biological processes in an attempt to convert the administered dose to internal dose in enough detail and with enough accuracy to yield more realistic estimates of the internal doses. Early generations of PB-PK models subdivide the body into distinct physiological compartments such as alveolar air, arterial blood, fatty tissue, poorly perfused tissues (muscles and skin), richly perfused tissues (brain, heart, kidney, endocrine glands, liver and other metabolizing tissue groups), and mixed venous blood (Andersen et al., 1987a, 1987b) (see Figure 11-4). In more recent times, even portions of specific target tissues like the liver have been mathematically described, and then this information was used to predict target doses to critical cells (Andersen et al., 1995).

The most promising method for identifying the most biologically reasonable human response from rodent data is the PB-PK model. These models quantitatively account for the various differences between the test species and humans by considering body weight, metabolic capacity and products, respiration rate, blood flow, fat content, and a number of other parameters. Although confidence in the results of PB-PK models often relies on some frequently untestable assumptions such as the delivered dose of an unstable metabolite to a target organ, it represents one of the most important advances in toxicology and health risk assessment. To date, PB-PK models have been developed and validated for carbon tetrachloride, styrene, methylene chloride, chloroform, trichloroethane, dioxane, tetrachlorodibenzo-p-furan,

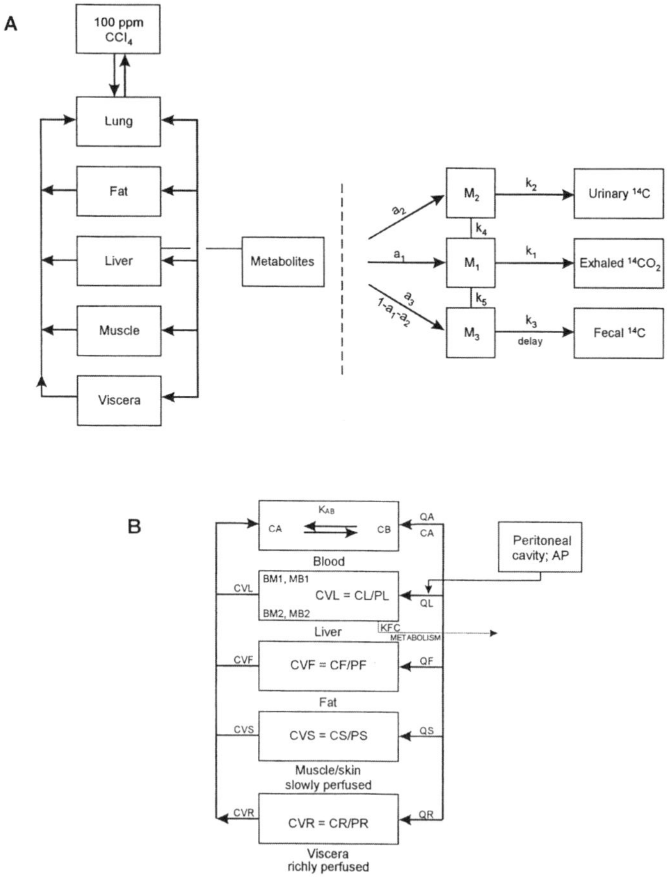

Figure 11-4. Illustration depicting a PB-PK model for inhalation exposure to carbon tetracholoride (CCl_4) (Panel A) (Paustenbach et al., 1988) and a PB-PK model for uptake following intraperitoneal dosing (Panel B) of a non-volatile chemical. In the figure, a_1, a_2, a_3 = proportion of total metabolites excreted in urine, feces, and CO_2; M_1, M_2, M_3 = corresponding metabolic compartment for urine, feces, and CO_2; k_1-k_5 = first-order rate constants; ^{14}C and $^{14}CO_2$ = radiolabeled carbon and carbon dioxide, respectively; C_A, C_B = concentration in arterial blood that is free and bound, respectively; K_{AB} = dissociation constant; CV = compartment volume for the liver (L), fat (F), slowly perfused (S), and richly perfused (R) compartments; C = concentration in L, F, S, and R compartments; P = partition coefficient in L, F, S, and R compartments; Q = flow rate in L, F, S, R compartments; AP = amount in peritoneal cavity; BM_1, BM_2 = binding capacities; and MB_1, MB_2 = binding constants.

tetrachlorodibenzo-p-dioxin (TCDD), benzene, trichloroethylene, vinyl-chloride, and about 50 other agents (Leung and Paustenbach, 1995). Perhaps the best example of the predictive power of the PB-PK model involved vinyl chloride where the number of human cancers estimated from the modeled animal data was virtually identical to that observed in a major epidemiology study (Reitz et al., 1996).

PB-PK models are used as "front ends" to dose-response models to solve some of the problems that arise when parametric statistical risk models are fit to raw administered dose response data (Jayjock et al., 2000). Specifically, they represent a quantitative approach that attempts to account for the following issues:

(1) Inter-individual variability

PB-PK models allow different individuals to have different administered dose-response functions based on inter-individual variability in their pharmacokinetic parameters. Thus, inter-individual variability in uptake metabolism and elimination (although not in pharmacodynamics) can be accounted for by using PB-PK models in which parameter values for individuals are drawn from a population joint distribution.

(2) Low-dose extrapolation

PB-PK models provide a basis for adjusting administered dose response data at high doses to correct for the effects of metabolic saturation. This yields a set of adjusted dose-response pairs to which the multistage model (or other biologically motivated statistical risk model) might reasonably be expected to apply better than they do to the raw administered dose-response data. It also provides a basis for extrapolating from high-concentration dose-response observations to predict the low-concentration dose response relation (Andersen et al., 1995).

(3) Extrapolation across routes of administration

One of the most common uses of PB-PK models is to calculate the amount of chemical administered by one route (e.g., inhalation concentration and duration) that is "equivalent" to a given amount administered by another route (e.g., oral gavage). Equivalence is defined by equality of internal doses produced at the target organs. If a measure of internal dose has been defined that truly explains or predicts response probability, the route of administration is no longer relevant once the internal dose has been calculated (McDougal et al., 1990).

(4) Interspecies extrapolation

Administered dose-response functions can be extrapolated across species using PB-PK models for the different species (Paustenbach et al., 1988; Clewell, 1995). Once a measure of internal dose has been defined and the internal dose-response relation has been quantified in one species, it can be used to estimate the relationship between administered and internal dose in other species. Combining the administered-to-internal dose function for a species with the internal dose response function (assuming that the internal dose-response function is the same across species) then gives an estimate of the administered dose-response function for that species.

These clear advantages over classic extrapolation approaches have given most agencies the confidence to adopt, or at least consider, the results of the PB-PK models in quantitative risk assessment.

CARCINOGENS

Non-carcinogenic effects are generally thought not to occur until some minimum (threshold) level of exposure (absorbed dose) is reached (Klaassen, 2000). In contrast, genotoxic carcinogens are thought, by many scientists, to pose a finite risk at nearly any absorbed dose. This "no-threshold" assumption has been adopted by federal agencies for most carcinogens as a prudent practice for protecting public health. However, it stems from our experience with radiation, which is a known genotoxic carcinogen. Further, the administered dose of ionizing radiation can be directly converted to a delivered dose to the target organ; a characteristic unlike that of a chemical carcinogen. In some instances, scientists believe there may well be a threshold for some classes of carcinogens, because they appear to act through mechanisms that require a threshold dose to be exceeded prior to initiation of the carcinogenic process (Butterworth and Slaga, 1987; Butterworth, 1990, 1993; Williams and Weisburger 1991). For example, chemicals like chloroform (as discussed previously) are not likely to pose a cancer hazard at very low doses for a number of biological reasons (Butterworth et al., 1998).

Functionally, most toxicologists believe that, in light of the fact that our diets are abundant with carcinogens, there must be a practical threshold below which we would call the doses "safe" (Ames et al., 1987; Williams and Weisburger, 1994; Gold et al., 2002). Further, there seems to have been growing agreement in the scientific community that not all carcinogens pose a hazard at all doses. That is, some chemicals act through genotoxic mechanisms while others act through non-genotoxic mechanisms. This is reflected in the recent EPA draft carcinogen assessment guidelines (EPA, 1999a), which acknowledge that many chemicals that are carcinogens at high doses in animals may not pose a risk to humans at much lower doses (e.g., they have a practical threshold).

As noted previously, scientists have developed several mathematical models to estimate low-dose carcinogenic risks from observed high-dose responses measured in animal studies (Crump et al., 1976; Krewski et al., 1984; Holland and Sielken, 1993). Such models describe the expected quantitative relation between risk and dose at doses too small to resolve experimentally. These models have historically been applied by regulatory agencies to both genotoxic and non-genotoxic chemicals. They are used to estimate a value for the risk at the doses that are relevant to exposures in our diet or the environment, which is often 1,000-fold lower than the lowest animal dose tested. For example, if it is known that there was a 20% increased incidence of tumors in rodents exposed to 100 μg/kg-day, and a 50% increased incidence at 1,300 μg/kg-day, they attempt to describe the risk to humans who ingest only 1.0 mg/kg-day.

The accuracy of the risk estimates at the dose of interest is a function of how accurately thc mathcmatical model describes the true, but unmeasurable, relation between dose and risk at the low-dose levels. While most of the available models fit the experimental, high-dose data, the predicted risks at low doses may vary significantly (see Figure 11-5). In general, because of the "spread" in the various risk estimates from the various models at low doses, and because science cannot be sure which model is likely to be most accurate, most U.S. regulatory agencies rely upon the most conservative method (the one suggesting the lowest dose for a given risk) — the linearized multi-stage model, developed by Crump et al. (1976). However, it is often useful to apply other models and to compare the results. If they vary significantly, it can be worthwhile to understand why the various models are predicting risks significantly different from one another.

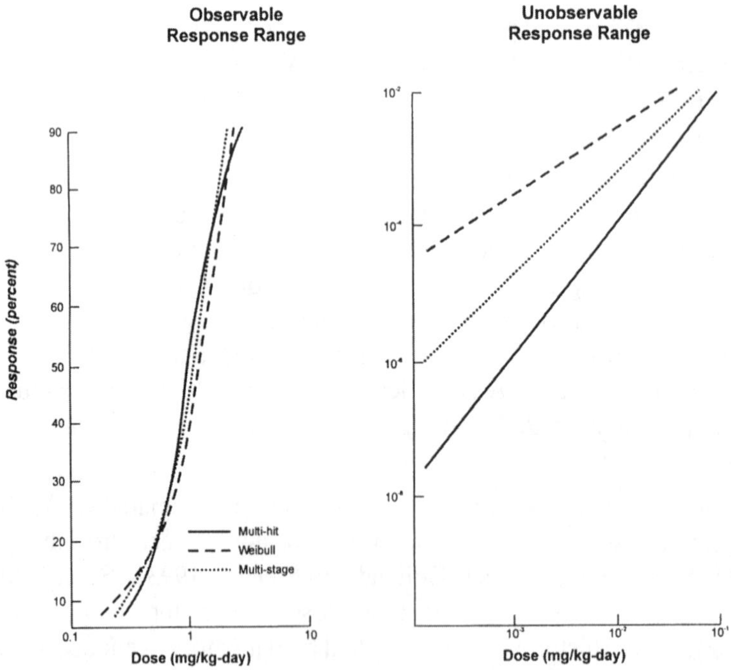

Figure 11-5. The fit of most dose-response models to data in the observable range is generally similar (left plot), as shown in this example for DDT. However, due to the differences in the assumptions upon which the equations are based, the risk estimates at low doses can vary dramatically between the different models (right plot) (from Paustenbach, 2002).

SOME SUBTLETIES ABOUT DOSE-RESPONSE EXTRAPOLATIONS

It is reasonably well accepted that the most uncertain portion of the risk assessment of chemicals, especially carcinogens, is the low-dose extrapolation. Most toxicologists would agree that science has a limited ability to estimate the risks associated with typical levels of environmental exposure based on the results of standard rodent tests, including lifetime tests or bioassays. There are many reasons why this is so. First, animals are usually dosed at levels at least 1,000-fold above those to which humans will be exposed. Second, we do not fully understand all of the various possible mechanisms of action for carcinogens (Williams and Weisburger, 1991; Butterworth, 1993). So it is difficult to estimate how humans will respond at such low doses. For example, if it is known that ingesting four beers a day increases the rate of liver cancer, how could we estimate the cancer risk for four beers every year? Third, the doses under which we conduct the animal tests are so high that they often produce effects that would not occur at the doses to which people are exposed (Ames, 1998). Fourth, there are usually significant differences between animals and humans with respect to the rate at which chemicals are metabolized, distributed, and excreted (Paustenbach, 2000). Fifth, the delivered dose to specific target tissues in animals for a given oral dose will frequently produce a different delivered organ dose in humans even at the same oral dose (Andersen et al., 1995). Sixth, for a given target tissue dose, the toxicity may also vary among species. Accordingly, scientists must rely on models or theory to estimate the human response at doses often 1,000-fold below the lowest animal dose tested. The most scientifically rigorous (and likely most valid) models for estimating safe levels of human exposure based on animal data are the PB-PK models (Leung and Paustenbach, 1995; Reitz et al., 1997).

Low-dose extrapolation models (often erroneously called a quantitative risk assessment [QRA]) are the backbone of dose-response assessments for carcinogens. Because these models have played such a dominant role in the regulatory process, it is important to understand some of their characteristics. First, the most routinely used models will usually fit the rodent data in the dose region used in the animal tests. Second, the different models usually yield very different estimates of risk at the doses to which humans are typically exposed, as exemplified by the analysis of DDT (see Figure 11-5). Third, the results of these six models usually vary in a predictable manner because they use different mathematical equations for predicting the chemical's carcinogenic potency (see Table 11-2). Although not in all cases, the one-hit and linearized multi-stage models will usually predict the highest risk and the probit model will predict the lowest. To date, most regulatory decisions in the United States used to identify safe or acceptable levels of exposure to air, water, and soil contaminants have been based on these statistical models, as opposed to biologically based models like the PB-PK analyses or the Moolgavkar-Venson-Knudson (MVK) model (Moolgavkar et al., 1988). Based on EPA's *1999 Draft Guidelines for Carcinogenic Risk Assessment*, it can be expected that, in the coming years, much greater weight will be placed on those assessments that attempt to quantitatively account for biological phenomenon.

Table 11-2

The results of low-dose extrapolation models usually vary in the following predictable manner:

MODEL	PREDICTED RISK
Linear	highest ↑
One-Hit	
Multistage	
Weibull	
Multi-Hit	
Logit	
Probit	lowest

It is noteworthy that the lifetime rodent studies (bioassays) now used to quantitatively predict the magnitude of the human cancer risk were never intended for that purpose. Instead, these studies were designed to qualitatively identify potential human hazards, not to quantitatively estimate the human risk at low levels of exposure. For this reason, and others, there has been a movement away from strict adherence to regulatory guidance for conducting dose-response assessments that have traditionally required that only a single mathematical model be used. Such requirements have constrained the analysis and limited the valuable biological information (that could dramatically alter the results) that can be obtained from these studies.

We have learned a good deal about the biology of cancer during risk assessment's formative years (1975-2000) and, where possible, this information should be incorporated into the dose-response assessment. For example, we now know that there are at least three broad classes of mechanisms by which chemicals may produce a carcinogenic response in rodents: repeated cytotoxicity, promotion, and initiation. Some have suggested that there may be at least eight different classes of carcinogens (Butterworth, 1993). These distinctions are important since the appropriate approach to estimating the cancer risk for humans exposed to low doses of a cytotoxicant or promotor should be different than that for an initiator. Because at relatively high doses nongenotoxicants may produce repeated cytotoxicity, the primary reason for excessive cell turnover, many scientists expect them to possess a threshold dose below which no cancer risk would be present. This is in contrast with genotoxicants that may pose some risk, albeit small, even at very low doses (Andersen, 1991).

In general, the scientific underpinnings of the dose-response models used for assessing carcinogens are based on our understanding of ionizing radiation and genotoxic chemicals (Hoel,

1987). Both types of agents may well have a linear, or a nearly linear, response in the low dose region. In contrast to radiation and chemical initiators, promoters and cytotoxicants need not have a linear dose-response curve. Scientific data increasingly suggest that the non-genotoxic chemicals would be expected to have a very non-linear response at low doses and, as importantly, probably have a genuine or practical threshold (Sielken, 1985). The increased acceptance of this postulate is evidenced by EPA's position that the linearized multi-stage model is inappropriate for chloroform, thyroid type carcinogens, nitrilotriacetic acid, d-Limolene, and presumably, similar non-genotoxic chemicals.

Historically, when presenting the results of the low-dose models, most scientists and regulatory agencies have generally only presented the upper bound risk from these models. However, in recent years, it has become customary to present the best estimates (maximum likelihood estimates), as well as the upper and lower bounds on the risk estimates. The objective of the bounding techniques is to attempt to account for the statistical uncertainty in the results of the animal tests, however, the degree of potential conservatism within the bounding procedure can be significant (Sielken, 1985). Historically, most risk managers have not been fully aware of the breadth of equally plausible risk estimates. For example, the cancer risk associated with exposure to dioxin in our food has been reported to be as high as one in 1,000 using the upper bound risk estimate of the multi-stage model. However, using the same model, the best or maximum likelihood estimate of the risk is about one in 100,000 and the lower bound estimate is virtually zero. Therefore, the plausible range of increased lifetime cancer risk is as high as one in 1,000 and as low as zero. When biological factors are considered, such as its weak genotoxicity and the human experience, the carcinogenic risk associated with low concentrations of dioxin (below 2 pg/kg-day TEQ) is most likely to be quite small or negligible (Greene et al., in press). The message here is simply that one should not focus exclusively on the upper bound, or q*, risk estimate when attempting to predict the actual cancer risk.

ROLE OF EPIDEMIOLOGY

Although rarely given proper consideration, epidemiology studies should be an important component of a risk assessment, especially when the art is applied to the practice of industrial hygiene. Critics often claim that these studies are almost never as statistically robust as the animal studies and, therefore, are not very useful in the dose-response assessment (Silbergeld, 1988). Others contend that total acceptance of this assertion is inappropriate because epidemiological studies can, at the least, establish the degree of confidence that should be placed in the results of low-dose extrapolation models (Layard and Silvers, 1989; Federal Focus, 1995). However, in this author's opinion, even less-than-perfect epidemiology studies coupled with retrospective exposure assessments should be able to yield much more defensible estimates of the likely human health hazard than statistical models based only on animal studies. Although in the United States there continues to be a struggle between those who want to quantitatively adjust risk estimates from animal studies by using epidemiology data, there appears to be more receptivity for this practice in the European community (Federal Focus, 1995).

Exposure Assessment

An exposure assessment describes the nature and size of the various populations exposed to a chemical agent and the magnitude and duration of their exposures (EPA, 1988, 1996b). Exposure assessments can address past, current, or future anticipated exposures (NAS, 1983, 1994; Ripple, 1992; Paustenbach, 2000).

Exposure assessments determine the degree of contact a person has with a chemical and estimate the magnitude of the absorbed dose. Several factors influence how to estimate the absorbed dose, including the duration of the exposure, route of exposure, bioavailability of the chemical from the contaminated media (e.g., soil) and, sometimes, the unique characteristics of the exposed population (e.g., hairless mice absorb a greater percent of chemical than haired mice). By definition, "duration" is the period of time over which the person is exposed. An "acute" exposure generally involves one contact with the chemical, usually for less than a day. An exposure is considered "chronic" when it covers a substantial portion of the person's lifetime. Exposures of intermediate duration are called "subchronic."

Knowledge of the concentration of a chemical in an environmental medium is essential to determine the magnitude of the absorbed dose. This information is usually obtained by analytical measurements of samples of the medium. Estimates can also be made using mathematical models, such as models relating air concentrations at various distances from a point of release (e.g., a smoke stack), to factors including release rate, weather conditions, distance, and stability of the agent (Scott et al., 1997).

In general, exposure assessments will contain less uncertainty than other steps in the risk assessment (Paustenbach, 1995, 2000). Admittedly, there are a large number of factors to consider when estimating exposure and it is a complicated procedure to understand the transport and distribution of a chemical which has been released into the environment. Nonetheless, the available data indicate that scientists can do an adequate job of quantifying the concentration of the chemicals in the various media and the resulting uptake by exposed persons if they account for all the factors that should be considered.

The primary routes of human exposure to chemicals in the ambient environment are inhalation of dusts and vapors, dermal contact with contaminated soils or dusts, and ingestion of contaminated foods, water, or soil. In the workplace, the predominant route of exposure is usually inhalation, followed by dermal uptake and, to a lesser extent, ingestion of dust due to hand-to-mouth contact (Paustenbach and Rothrock, 1998). The uncertainty in an occupational exposure assessment is often less than in an assessment of environmental exposure. Initial efforts to quantitatively estimate the uptake of environmental contaminants by humans were first conducted by scientists in the field of radiological health (Romney et al., 1963; Martin, 1964; Russell, 1966; Baes et al., 1985), and their work can be a source of valuable information when conducting assessments of chemical contaminants. Numerous methodologies for estimating the human uptake of environmental contaminants have been described and refined in recent years (McKone and Bogen, 1991; Fries and Paustenbach, 1990; Paustenbach, 2000). Not long ago, the EPA released the second edition of its *Exposure Factors Handbook*, a text containing 300 pages of information on exposure assessment (EPA, 1997).

MONTE CARLO TECHNIQUES

Beginning in the late 1970s, it became regulatory policy to use conservative approaches to conduct exposure assessments. This was considered prudent because it guaranteed that risk would not be underestimated. By about 1985, a concern evolved that the repeated use of conservative exposure assumptions of exposure factors was producing unrealistically high estimates of exposure (Paustenbach et al., 1986; Maxim and Harrington, 1989; Nichols and Zeckhauser, 1986). Although it took until about 1990 to learn how to use Monte Carlo techniques to properly account for this phenomenon, their application to exposure assessment has dramatically improved our level of understanding and their use has altered the field permanently (Thompson and Burmaster, 1991; Paustenbach and Finley, 1994; Burmaster and Harris, 1996; Cullen and Frey, 1999).

The probabilistic or Monte Carlo model accounts for the uncertainty in select parameters used in evaluating the range and probability of plausible exposure levels. Instead of the input parameters being specified as single values, this model allows for consideration of the probability distributions. The Monte Carlo statistical simulation is a statistical model in which the input parameters to an equation are varied simultaneously (Portier and Kaplan, 1989). The values are chosen from the parameter distributions with the frequency of a particular value being equal to the relative frequency of the parameter in the distribution (Figure 11-6). The simulation involves the following three steps:

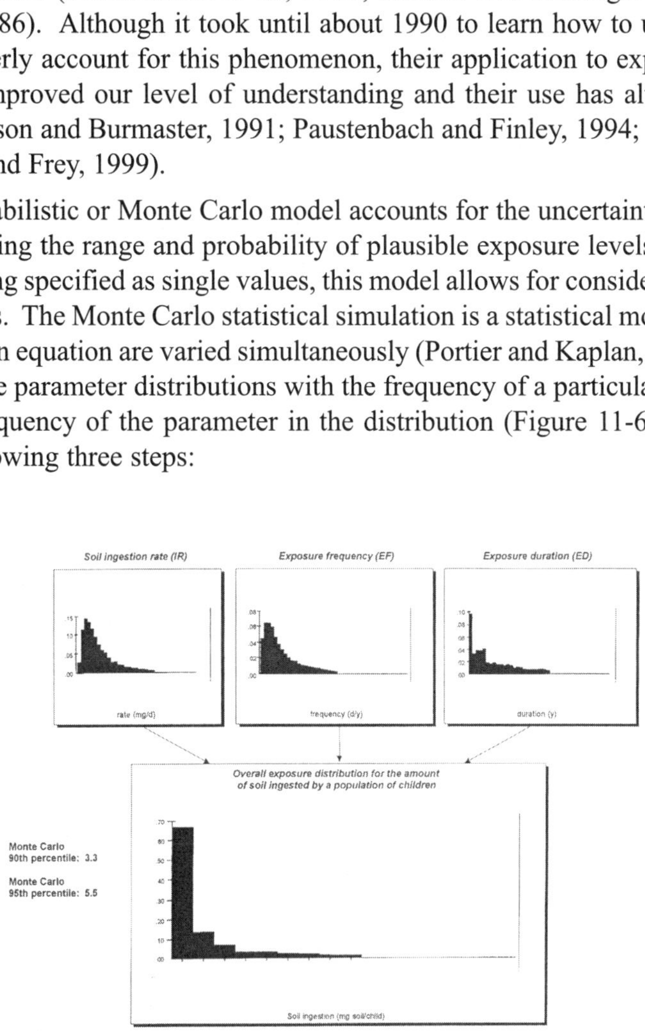

Figure 11-6. An example of how probability density functions (distributions) for three different related exposure factors are combined to form a distribution for the amount of soil ingested by a population of children. The Monte Carlo technique allows the risk assessor to account for the variability in many exposure parameters within a population and then produce a distribution that characterizes the entire population (from Paustenbach, 2002).

1. The probability distribution of each equation parameter (input parameter) is characterized, and the distribution is specified for the Monte Carlo simulation. If the data cannot be fit to a distribution, the data are "boot-strapped" into the simulation, meaning that the input values are randomly selected from the actual data set without a specified distribution.
2. For each iteration of the simulation, one value is randomly selected from each parameter distribution, and the equation is run. Many iterations are performed, such that the random selections for each parameter approximate the distribution of the parameter. Five thousand iterations are typically performed for each dose equation.
3. Each iteration of the equation is evaluated and saved; hence a probability distribution of the equation output (possible doses) is generated (see Figure 11-6).

This technique generates distributions that provide a description of the uncertainty associated with the risk estimate (resultant doses). The predicted dose for every 5th percentile up to the 95th percentile of the exposed population and the true mean dose are estimated. Using these models, the assessor is not forced to rely solely on a single exposure parameter or the repeated use of conservative assumptions to identify the plausible dose and risk estimates. Instead, the full range of possible values and their likelihood of occurrence is incorporated into the analysis to produce the range and probability of expected exposure levels (Finley and Paustenbach, 1994; Jayjock et al., 2000).

The methodology can be illustrated through an example. This Monte Carlo analysis is applied to understand the distribution of time needed to go shopping based on the three activities involved in shopping. Time spent shopping each month (minutes) is estimated by the product of two parameters: the number of trips per month and the total time spent in the store (minutes). Total time spent in the store is the sum of time spent shopping and time spent in line. Using Monte Carlo techniques, a distribution of likely values is associated with each of these parameters. These distributions are dependent upon the detail of information available to characterize each parameter. For example, the distribution uses all of the information such as those days when the line at the check-out counter is short, as well as those when it is long. It is noteworthy that each parameter has a different distribution: log-normal, gaussian, and square. Total time spent shopping is then calculated repeatedly by combining parameter values that are randomly selected from these distributions. The result is a distribution of likely time spent shopping each month.

A significant advantage of Monte Carlo analysis is that it shares a great deal more information with the risk manager. Instead of presenting a single point estimate of risk, probabilistic analyses characterize a range of potential risks and their likelihood of occurrence. In addition, those factors which most affect the results can be easily identified. For example, in a probabilistic analysis, one can present the risk manager with the following type of information: "the plausible increased cancer risks for the 50th, 95th, and 99th percentiles of the exposed population are 1×10^{-8}, 5×10^{-7}, and 1×10^{-6}, respectively."

With the introduction of Monte Carlo approaches, it has become possible to begin to understand and characterize the statistical confidence in our estimates of risk. Risk managers and the public were eager to have access to this information so, not unexpectedly, uncertainty and variability analysis began to be conducted on the exposure assessment on a routine basis around 1994. Many papers in recent years have illustrated how these can be conducted (McKone and Bogen, 1991, Sheehan et al., 1991; Finley and Paustenbach, 1994; Paustenbach et al., 2002). Applying Monte Carlo techniques to evaluations of the occupational setting, industrial hygienists have the tools to characterize risk. Although there are several software applications for doing Monte Carlo analysis, the most popular is Crystal Ball (Decisioneering, Inc., Denver, CO, USA).

BIOLOGICAL MONITORING

Before closing a discussion about exposure assessment in the occupational environment, it is important to reiterate that biological monitoring can often be an exceptional tool for estimating exposure.

Over the past five to 10 years, analytical chemists have increased their ability to detect very small quantities of non-natural chemicals in blood, urine, hair, feces, breath, and fat (Lynch, 1994). Measurement of parts per trillion and parts per quadrillion is now possible. For many chemicals, the results represent a direct indicator of either recent or chronic exposure to a chemical, and these measurements can serve as a validation for the accuracy of the risk assessment calculations. For example, the uptake of dioxin by Vietnam veterans exposed to 2,4,5-trichlorophenoxyacetic acid (2,4,5-T) was recently evaluated by analyzing the amount of dioxin in their blood (Aylward et al., 1996). This and other studies, conducted more than 20 years after the last day of service in Vietnam, allowed epidemiologists to conclude that the vast majority of veterans had only a modest degree of exposure to this chemical that has been alleged to have produced numerous adverse health effects in field soldiers.

Risk Characterization

Risk characterization involves integration of the data and analyses from the other three steps of risk assessment to determine the likelihood that the human population of interest will experience any of the various forms of toxicity associated with a chemical under the known or anticipated conditions of exposure (Figure 11-7) (NAS, 1983, 1994; EPA 2000b).

It has been said that the most carefully conducted hazard identification, exposure assessment and dose-response assessment will not be useful if the resulting information is not fully considered in forming conclusions about risk, i.e., risk characterization (Jayjock et al., 2000). In addition, the risk assessment document must be clear, transparent, and objective. The goal is that the stakeholders (environmental regulators in the state or federal government, people living near an industrial facility or hazardous waste site, the lawyers, your employer, the workers at your facility) who read the document will understand it. These issues are so important to the

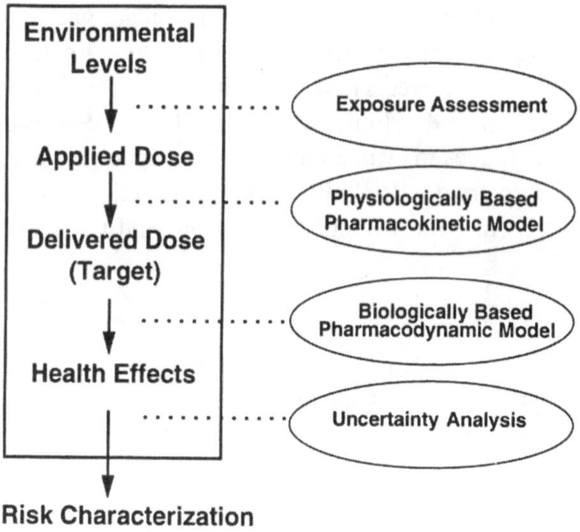

Figure 11-7. A general description of the kinds of issues that need to be addressed in a high quality risk characterization.

EPA that they have published guidance to ensure that risk characterizations utilize critical information from each stage of the process, and communicate this information to risk managers (EPA, 1995).

The EPA primarily intended their guidance to be used to assist human health or ecological risk assessors writing documents for review by the agency. However, they contain useful information that can guide our efforts to write industrial hygiene risk assessments. The pertinent messages are that risk characterization should be based on reliable scientific information from many different sources. The need for the assessment to be transparent is perhaps the most recent challenge presented to assessors. By "transparency" we mean any conclusion drawn from the science should be clearly separated from policy judgments, and that all assumptions should be carefully stated. In their specific guidelines for how to write a risk characterization, EPA (1995) has suggested the following:

> "The risk assessment report must discuss key scientific concepts, data, and methods, from each stage of the process. It also must include acknowledgment and analysis (qualitative or quantitative) of uncertainties, and an estimate of confidence in the risk characterization. Well-balanced risk assessments present risk conclusions along with strengths and limitations of the analysis."

The risk assessment should present several types of risk information. A range of exposure scenarios and multiple risk descriptors (central tendency, high end of individual risk, population risk, important subgroups) should be included. Use of several descriptors, rather than a single descriptor (such as the reasonable maximal exposed [RME] individual's risk), enables presentation of a fuller picture of risk. Situation-specific information adds perspective on possible future events or regulatory options. Presentation of a numerical risk estimate is not complete risk characterization; the numbers must be supported by documentation of data, concepts, assumptions, models, and conclusions.

When writing a risk characterization, it is customary to discuss the non-cancer and cancer hazards separately.

NON-CARCINOGENS

The risk of non-carcinogenic adverse health effects is generally characterized by comparing the predicted dose with established regulatory criteria. It is generally agreed that non-carcinogenic health effects are not expected to occur until after a threshold dose is reached; as opposed to genotoxic carcinogens. Thus, doses which should protect everyone from adverse effects can be identified. As discussed previously, the EPA has established criteria for acceptable exposure known as RfDs for most "threshold chemicals" (Jarabek et al., 1990). This figure is normally set near a value about 1/100th to 1/1000th of the NOAEL reported in animal studies.

To evaluate the acceptability of a particular dose relative to the reference criterion, the data are analyzed using the "hazard quotient" approach (EPA, 1989a). The hazard quotient (HQ) is the ratio of the predicted average daily dose (ADD) to the RfD:

$$\text{Hazard Quotient} = \frac{\text{ADD}}{\text{RfD}} \qquad \text{[Equation 11-1]}$$

If the potential exposure is through ingestion, the value for the "oral RfD" is used to calculate the HQ. If the potential exposure is through inhalation, risk is characterized using the "inhalation reference concentration (RfC)." The current value for the RfD or RfC for a chemical can be found on-line by using the Integrated Risk Information System (IRIS), which is administered by the National Library of Medicine (IRIS, 1996). Although inhalation exposure is usually evaluated using the inhalation RfCs, when these are not available, one can convert the oral RfD to an inhaled dose to yield an approximation of the safe level of exposure (if a route-to-route extrapolation of this type is assumed appropriate). When characterizing workplace hazards, it is often more appropriate to use occupational exposure limits like TLVs® to assess risk. Conceptually, the RfD or the occupational exposure limit (OEL) could be used in HQ calculations and the two results can be compared.

The EPA has not established reference doses for dermal exposure; however, in general, the risks due to dermal uptake can usually be estimated using the oral RfD. Oral toxicity criteria can be adjusted by incorporating appropriate dermal absorption factors (ABF) into the dose equation. Because most substances are typically absorbed less efficiently via the dermal route than the oral route, dermal ABFs are used to account for the likelihood that a smaller fraction of the substance will be absorbed following dermal contact than if the substance were ingested. Guidance (California Department of Toxic Substances Control, 1994) is available for applying dermal ABFs.

An HQ less than or equal to one indicates that adverse health effects are not expected to occur. An HQ greater than one indicates only that further evaluation is required since the dose cannot be readily considered trivial; however, it does not mean that adverse effects will occur.

The likelihood of adverse health effects due to potential exposure to multiple chemicals is addressed by calculating a hazard index (HI). This is a screening approach wherein all substances are conservatively assumed to act in an additive manner. While this approach, by design, overstates the potential for non-carcinogenic health effects, it can be used as a rapid screening tool to eliminate those exposure scenarios that do not pose a signature non-carcinogenic hazard. For the non-carcinogens, only chemicals that act through a similar mechanism of action or act on the same target organ should be assessed together.

The HI is calculated as follows:

$$\text{Hazard Index} = \sum \text{Hazard Quotients} = \sum \frac{\text{ADD}_i}{\text{RfD}_i} \qquad \text{[Equation 11-2]}$$

The following example illustrates how one can determine whether exposure to a group of agents is acceptable.

Example 11-1: ***Non-carcinogenic Hazard from Multiple Chemical Exposure***

A female construction worker is exposed to chlorobenzene and polychlorinated biphenyls (PCBs) due to dermal contact with subsurface soils and it is predicted through exposure calculations that the ADD is 4.2×10^{-7} and 1.5×10^{-4} mg/kg-day, respectively. How might one characterize the risk of exposure? [Chlorobenzene has an RfD of 2×10^{-2} mg/kg-day (IRIS, 1997) and PCBs have an RfD of 5×10^{-5} mg/kg-day (IRIS, 1996; Health Effects Assessment Summary Tables (HEAST), 1997).] Assume the TLV® for chlorobenzene is 10 ppm (50 mg/m^3), and for the PCBs it is 0.5 mg/m^3. Conduct the assessment in both ways; e.g., one that uses the RfD as risk criterion and one that uses the TLV®.

Approach A (RfD Approach)

Using Equation 11-1:

$$\begin{aligned} \text{HQ}_{\text{chlorobenzene}} &= \text{ADD/RfD} \\ &= (4.2 \times 10^{-7} \text{ mg/kg-day})/(2 \times 10^{-2} \text{ mg/kg-day}) \\ &= 2.1 \times 10^{-5} \end{aligned}$$

$$HQ_{PCBs} = ADD/RfD$$
$$= (1.5 \times 10^{-4} \text{ mg/kg-day})/(5 \times 10^{-5} \text{ mg/kg-day})$$
$$= 3$$

Using Equation 11-2:

$$HI = HQ_{chlorobenzene} + HQ_{PCBs}$$
$$= 2.1 \times 10^{-5} + 3$$
$$= 3$$

NOTE: Therefore, based on this calculation, and assuming that one wants to meet EPA criteria, because the HI is greater than one, it would be recommended that exposure to the PCBs be reduced until the total HI is less than one.

Approach B (TLV® Approach)

In the industrial hygiene community, it is more common to attempt to comply with ACGIH® TLVs® or Occupational Safety and Health Administration (OSHA) permissible exposure limits (PELs) than to EPA RfDs. Using the information that has been provided, an airborne concentration of chlorobenzene and PCB equivalent to the dermal exposure can be back calculated from the skin contact data. Assume that persons inhale about 10 m³ of air every 8-hour workday.

$$\frac{(10 \text{ m}^3/\text{day})(\text{x concentration in air})}{70 \text{ kg}} = 4.2\times10^{-7} \text{mg/kg} - \text{day}$$

x = 2.8×10^{-6} mg/m³ [Chlorobenzene airborne equivalent exposure]

$$\frac{(10 \text{ m}^3/\text{day})(\text{x concentration in air})}{70 \text{ kg}} = 4.2\times10^{-7} \text{mg/kg} - \text{day}$$

x = 3.5×10^{-4} mg/m³ [PCB airborne equivalent exposure]

These values are then applied to Equation 11-1.

$$\frac{2.8\times10^{-6} \text{ mg/m}^3}{50 \text{ mg/m}^3} = 5.6\times10^{-8} \qquad \frac{3.5\times10^{-4} \text{ mg/m}^3}{0.5 \text{ mg/m}^3} = 7.0\times10^{-4}$$

Assuming that both chlorobenzene and PCB work through similar mechanisms of action, and that the TLVs® were designed to protect against the same adverse effects (e.g., liver toxicity), then one could add the two ratios (the mixture formula) to determine acceptability and a hazard index (Equation 11-2).

$$(5.6 \times 10^{-8}) + (7.0 \times 10^{-4}) = 7.0 \times 10^{-4}$$

NOTE: It is apparent from these calculations that if one uses the EPA RfD approach, then the exposure is not acceptable. If one uses the alternative method, which relies on TLV® values, the exposure would clearly be of no concern. There are several reasons for this result. First, the margin of safety inherent in regulatory standards meant to protect the general public (that is, EPA criteria) is much larger than that used to protect workers. Second, in its RfD, the EPA based its value for PCB at a level that would not pose a very low cancer risk while the rationale for the TLV® was to prevent liver toxicity. In virtually all cases, EPA criteria will be more strict than either OSHA or ACGIH® criteria used to protect workers.

In those situations where a large group of chemicals is being added together, and the HI is greater than 1.0, then a slightly more complex analysis should be conducted. Specifically, it is suggested that the chemicals be grouped into categories based on the target organ affected or the mechanism of action. Lists of primary target organs and/or mechanism of action have been assembled for most chemicals to which workers are exposed (Federal Register, 1991; ACGIH®, 2001).

CARCINOGENS

For carcinogens, the plausible incremental cancer risk is estimated by multiplying the Lifetime Average Daily Dose (LADD) by the risk per unit of dose that is predicted from one of the dose-response models. A range of risks will generally be produced from different models and assumptions about the dose-response curve. The predicted risks will also wary with information on the relative susceptibilities of humans and animals.

In risk characterizations, the excess cancer risk is based on the total LADD for the chemical of interest (the exposure averaged over a 70-year lifetime). Once the dose has been calculated, the estimated cancer risk can be predicted as follows by:

$$\text{Inhalation Route} = \text{Inhalation LADD}_i \times Sf_{\text{inhalation, i}} \qquad \text{[Equation 11-3]}$$

$$\text{Oral Route} = \text{Oral LADD}_i \times Sf_{\text{oral, i}} \qquad \text{[Equation 11-4]}$$

$$\text{Dermal Route} = \text{Dermal LADD}_i \times Sf_{\text{oral,i}} \qquad \text{[Equation 11-5]}$$

$$\text{TOTAL RISK} = \text{Inhalation} + \text{oral} + \text{dermal} \qquad \text{[Equation 11-6]}$$

Where Sf = slope factor or cancer potency factor. This value is derived from the dose-response assessment and can be found in IRIS.

The total cancer risk estimate is calculated by summing the predicted carcinogenic risk for each chemical (i) across all routes and media (e.g., workplace, ingestion of dust, indoor air). This approach is expected to overestimate the true risk because different substances generally have different mechanisms of action and target organs. In addition, this approach assumes that exposure to carcinogens at any dose will present some degree of risk. Although it is not known to be true for numerous categories of carcinogenic substances, additivity is assumed unless adequate data are available to show otherwise.

Example 11-2: ***Calculating the Cancer Risk from Inhalation Exposure***

A utility worker is exposed to PCB particles in soils and is estimated to absorb via inhalation, on average, over his/her 70 year lifetime, an LADD of 6.4×10^{-7} mg/kg-day. What is his/her cancer risk, assuming that PCBs have an inhalation slope factor of 2 $(\text{mg/kg-day})^{-1}$?

Using Equation 11-3:

$$\begin{aligned}\text{Cancer Risk} &= \text{Inhalation LADD} \times Sf_{\text{inhalation}} \\ &= (6.4 \times 10^{-7}\ \text{mg/kg-day})\,(2\ (\text{mg/kg-day})^{-1}) \\ &= 1.3 \times 10^{-6}\end{aligned}$$

In short, the above calculation indicates that if humans respond similarly to rodents who were exposed in a lifetime bioassay, then the increased lifetime cancer risk for this person would probably be no greater than 1.3 in 1,000,000. Since the background lifetime cancer mortality rate for cancer for humans is about 25%, the exposure to the PCBs in this setting may raise his/her risk from 25.000% to 25.0001%.

Example 11-3: ***Estimating Risk from Exposure via All Three Routes***

Assume that the worker in Example 2 has chosen not to carefully adhere to what he learned during training. Consequently, having forgotten to wash his hands before lunch, he ingests about 10 mg of soil or dust. In addition, he allowed moist soil to be in contact with his hands and forearms for about one hour during the day. The resulting dermal contact has produced an uptake into the blood of about 0.005 mg of PCBs. Assume the concentration of PCB in the soil was 10 ppm (10 mg/kg). What was the total cancer risk due to inhalation, ingestion, and skin contact. Assume, for the sake of this calculation, that this occurs every day for 70 years.

$$\begin{aligned}\text{Ingestion} &= (10\ \text{mg PCB/kg soil})(10\ \text{mg soil})(\text{kg}/1{,}000{,}000\ \text{mg}) \\ &= 0.0001\ \text{mg [PCB]} \\ \text{Dose} &= \text{Inhalation} + \text{Ingestion} + \text{Skin}\end{aligned}$$

$$= \left(\frac{6.4 \times 10^{-7}\, mg}{kg - day}\right) + \left(\frac{0.0001\, mg}{70\, kg - day}\right) + \left(\frac{0.005\, mg}{70\, kg - day}\right)$$

$$= (6.4 \times 10^{-7}) + (14 \times 10^{-7}) + (7 \times 10^{-5})$$

$$= 7 \times 10^{-5} \text{ mg/kg-day}$$

Total Risk $= [7 \times 10^{-5} \text{ mg/kg-day}][2(\text{mg/kg-day})^{-1}]$

$$= 1.4 \times 10^{-4}$$

NOTE: This calculation shows that the predominant exposure pathway is dermal contact, and that the incremental cancer risk, if exposure occurred every day for 70 years, would be about 1.4 in 10,000.

However, if exposure is assumed to occur for only 10 days during the worker's lifetime, then an adjustment in the risk **estimate** is needed.

$$\text{Risk} = (\text{lifetime risk})(\text{time correction})$$

$$\text{Time Correction} = \frac{(10 \text{ days})}{(365 \text{ days})(70 \text{ years})} = 3.9 \times 10^{-4}$$

$$(1.4 \times 10^{-4})(3.9 \times 10^{-4}) = 5.48 \times 10^{-8}$$

Therefore, the increased lifetime cancer risk for having been exposed under these conditions for 10 days is only 5.5 in 10,000,000, an insignificant level.

Over the past 20 years, risk characterizations have consistently been the weakest component of the assessment. The reason is that, when writing the characterization, the assessor must draw on his/her understanding of science, as well as regulatory policy to describe whether a significant human health hazard exists in a specific setting (Paustenbach et al., 2002). For example, a thorough characterization should discuss background concentrations of the chemical in the environment and in human tissue, pharmacokinetic differences between the animal test species and humans, the impact of using a PB-PK or biologically-based model to extrapolate between animals and humans, the nuances of dose-response models, the effect of selecting specific exposure parameters (a sensitivity analysis), uncertainty and variability analyses, as well as other factors that can influence the magnitude or confidence in the estimated risks (EPA, 1995). An extensive and thorough review of the optimal conduct of risk characterization was recently published (Williams and Paustenbach, 2002).

Uncertainty

As discussed in the EPA's 1999 *Exposure Assessment Guidelines*, it is important to identify variability and uncertainty in risk assessment (especially in risk characterizations). Hygienists should be specific whether they are addressing variability, uncertainty, or a combination of the two. There are many sources of variability and uncertainty in exposure factors. Variability is the imprecision that occurs due to actual differences among members of a population. Variability is not reduced, but is only more accurately characterized with additional data. Variability in a factor may occur as the result of inter-individual variation [e.g., one adult is taller than the next], age related variation [e.g., adults eat more fish than children], or spatial variation [e.g., pollutant concentration is greater at location A than at location B] (Finley and Paustenbach, 1994; Finley, 1994; Abdel-Rahman and Kadry, 1995; Dourson et al., 1998).

In contrast, uncertainty is the imprecision in an estimate of an exposure factor that results from the limitations in the thoroughness of our measurement or estimate of the truth. In short, as more information is obtained, uncertainty is reduced. There are two major forms of uncertainty. The first is intrinsic to the available data and arises from limitations of study design and analytical techniques. The second type of uncertainty arises from the application of the data to non-sampled populations. For example, while distributions for human activity patterns developed from national surveys might be shown to adequately represent activity patterns among individuals living in California, these same distributions may not be applicable to native populations in the Yukon or in the Amazon rainforest. This second type of uncertainty will vary from assessment to assessment and will be a function of the differences between the sampled population and the assessment's target population.

Although exposure distributions largely reflect interpersonal variability, all the distributions represent varying amounts of uncertainty and variation. For example, the distributions for body weight and surface area reflect almost entirely inter-individual variation, with only a minor uncertainty component. In contrast, soil adherence estimates are dominated by analytical and measurement uncertainty, with inter-individual variability comprising only a minor component of the distribution (Finley et al., 1994).

A good risk characterization will present the best estimates and upper bounds on risk. All of the key assumptions and uncertainties within the dose extrapolation model, for example, should be discussed. Further, a risk manager should be told about the scientific plausibility or certainty of the estimate of risk. As noted previously, discussion of the foundation for the risk estimated, its merit, and the uncertainties in the risk estimates is called "transparency" (NAS, 1994). This movement toward sharing all of the "hidden" parts of the assessment began in the early 1990s and it is now the hallmark of a high quality risk characterization (NAS, 1996).

One of the biggest problems with most risk characterizations historically performed by regulatory agencies, from a scientific perspective, is that the default assumptions and methods that are often used are scientifically plausible for some chemicals but not for others. That is, the "plausible upper bounds" of carcinogenic potency may provide reasonable estimates for some compounds and wild overestimates for the rest.

For example, the default, conservative methods of risk assessment used by most regulatory agencies assume a dose-response function that is linear in the low-dose region and has no threshold. There is evidence that some agents, like certain types of radiation and directly mutagenic chemicals, may indeed have this type of dose-response relationship. However, many scientists believe the linear, no-threshold approach to risk estimation is inappropriate for most other chemical carcinogens, especially those that are not direct mutagens (Butterworth, et al., 1995).

This means that when one applies standard procedures to all chemicals without giving serious thought to the characteristics of a given substance, the amount of conservatism in a risk estimate can vary greatly. For example, a risk estimate for a powerful direct mutagen may be quite close to the calculated "plausible upper bound," while for a nonmutagenic compound the estimate may be an extreme overestimate of the actual risk. Two "plausible upper bound" risk estimates that are generated through consistent procedures may have very different levels of scientific plausibility. Presentation of multiple estimates of risk and the conditions which would make one more probable than another will help to avoid the trap of implying there is more precision in the risk estimates than can currently be provided.

The significance of model-estimated cancer risks needs to be communicated in a thorough and an understandable fashion. For example, the goal of some environmental standards, such as the maximum contaminant levels (MCLs) for drinking water, is to keep the maximum plausible cancer risk between 1 in 100,000 and 1 in 1,000,000. However, it might be useful to communicate to all involved that about 30% of all Americans will eventually develop cancer and about 25% will eventually die from it. Accordingly, a 1×10^{-6} risk is equivalent to ensuring that the lifetime cancer risk for persons exposed to this level of contamination will be no greater than 250,001 in 1,000,000 (25.0001%), rather than the background rate of 250,000 in 1,000,000 (25%). If society demands this standard of care, that is its choice; however, both society and its risk managers should be aware of the relative magnitude of various risks before deciding to spend money on one hazard versus another. For example, at least one analysis predicts that if the EPA were to regulate all carcinogens in air, soil, and water to levels generally considered insignificant, the decrease in the cancer incidence in the United States would probably be negligible (a reduction of between 0.25% to 1.3% of the annual cancer mortality) (Gough, 1993).

Cost Benefit Analyses

One reason that transparency and uncertainty in risk assessment has become pivotal to the long term viability of the process is that there is a need to compare and prioritize the many public health risks facing our country. Since statistics for many other public health threats, such as motorcycle accidents or AIDS cases, are not deliberately inflated, health risk assessments about chemicals must go beyond single "plausible upper bound" risk characterization to ensure meaningful comparisons (Tengs et al., 1995). If we can put all risk assessment on a level playing field, then it should be possible for risk managers to make holistic decisions about how America should use its limited financial resources.

Risk characterizations are much more informative if they contain cost/benefit analyses (Williams et al., 2002). That is, to the extent possible, the risk manager should understand how much the risks are likely to be reduced for various portions of the population for each dollar spent. Although relatively few have been conducted thus far, most risk scholars and risk managers believe that these risk/benefit versus cost analyses are the primary justification for performing health risk assessments. For these applications the use of Monte Carlo techniques could be invaluable.

Numerous papers in recent years have presented so called "cost-benefit" and "cost-effectiveness" analyses. These analyses have, to a large extent, made regulatory agencies, legislators, and the public reflect on the wisdom of some of our decisions. Table 11-3 presents examples of cost effectiveness relationships.

APPLYING RISK ASSESSMENT TO THE OCCUPATIONAL SETTING

Virtually all of the research and published papers addressing the practice of health risk assessment have been directed at evaluating environmental, rather than workplace, hazards. In fact, perhaps the first paper which presented a description of risk assessment to industrial hygienists was published about 15 years after the methodology had been used to assess hazardous waste sites, contaminated water, and airborne air (Paustenbach, 1990). There are a number of reasons why this occurred. First, historically, it has been widely held that the primary route of

Table 11-3

Examples of cost effectiveness relationships

Benzene emission control at chemical manufacturing process vents	$530,000,000
Benzene emission control at rubber tire manufacturing plants	$20,000,000,000
Chlorination of drinking water	$3,100
Coal-fired power plants emission control through high stacks, etc.	≤$0
Chloroform reduction at 33 worst pulp and paper mills	$57,000,000
Chloroform private well emission standard at 48 pulp mills	$99,000,000,000
Ban urea-formaldehyde foam insulation in homes	$11,000
Formaldehyde exposure standard of 1 (vs. 3) ppm in wood industry	$6,700,000
1,3-butadiene exposure standard of 2 (vs. 1000) ppm emissions from polymer plants	$770,000

*Costs are in units of per life saved.

Source: Tengs et al., 1994

exposure to occupational hazards was via inhalation rather than dermal or oral uptake. Therefore, comparing airborne concentrations to an ACGIH® TLV® seemed to be an adequate characterization of risk. Second, there were techniques available to measure airborne concentrations of chemicals whereas it was much more difficult to assess dermal uptake; the other likely route of workplace exposure to chemicals. Third, it was generally accepted that persons in the workplace could protect themselves from the dermal or ingestion hazard via training and through the use of protective clothing and gloves. So there was no reason for industrial hygienists to go beyond comparing airborne concentrations to OELs.

However, industrial hygienists today are asked to evaluate settings other than those seen in the traditional indoor environment. As such, they should be capable of assessing all three routes of exposure, and it would be useful if they could establish a risk criterion from basic toxicology data. Applying risk assessment methods to the field of industrial hygiene is relatively straightforward (Jayjock, et al., 2000). For example, the hazard identification, dose response assessment, and risk characterization are performed in the same manner in the occupational environment as is performed to assess chemicals in the ambient environment. The exposure assessment of workplace scenarios is generally much less complicated than those associated with assessment of such complex situations as the risk of airborne emissions from hazardous waste combustors. In the workplace, there are only three routes of entry to consider and only three media. That is, one can inhale gases/vapors or particles, ingest particles, or allow solvents to penetrate the skin. As such, calculations for estimating the uptake (dose) via inhalation and uptake due to dermal exposure are usually all that need to be performed.

Occupational Exposure to Airborne Chemicals

Humans can be exposed to chemicals via inhalation of vapors, gases, and particles (dust, fumes, etc). To estimate uptake via inhalation, the calculation is straightforward. Only three pieces of information are needed: the average airborne concentration of the chemical, the duration of exposure, and the average amount of air breathed during the period of exposure. This calculation is illustrated in the following example:

Example 11-4: ***Estimating Uptake Via Inhalation***

Assume that a person works in a paint factory and is exposed to solvent vapors. An industrial hygienist obtains representative measurements of the airborne concentration of methylene chloride and concludes that the 8-hour, time-weighted average concentration is 25 ppm (40 mg/m^3).

What is the ADD via inhalation for a typical worker who is exposed for 8 hours? Assume that a typical worker breathes 10 m^3 of air in 8 hours and body weight is 70 kg.

$$\text{ADD} = \frac{(40\ \text{mg/m}^3)(10\ \text{m}^3)}{70\ \text{kg}} = 5.7\ \text{mg/kg-day}$$

Slope factors are generally derived based on 365 days of exposure for 70 years. Thus, when exposure is less than this, the LADD is calculated with an adjusted slope factor. What is the LADD if the person works 200 days/year and has a 25 year employment (assume airborne concentration is stable)?

$$\text{LADD} = \frac{(200\ \text{days/yr})(25\ \text{yr})(5.7\ \text{mg/kg} - \text{day})}{(365\ \text{days/yr})(70\ \text{yr})}$$

$$\text{LADD} = 1.1\ \text{mg / kg-day}$$

The methodology for calculating the uptake of inhaled particles is the same as for gases and vapors.

Uptake Via the Skin

Although the uptake of chemicals via the skin has been generally overlooked in most assessments of workplace exposure, it probably represents a substantial portion of the exposure for many occupations. Although the use of gloves is much greater now than in years past, and training regarding the possible hazards of dermal exposure has increased, there is still ample evidence to indicate that, to conduct a complete exposure assessment, this route of exposure deserves attention (Fenske, 1993; Paustenbach et al., 1992; Leung and Paustenbach, 1994; EPA, 1999c).

In addition to the risks associated with systemic toxicity due to uptake via the skin, there is often a need to evaluate the allergic contact dermatitis hazard (ACD). In recent years, techniques have been developed to quantitatively predict the likelihood of illicitation and induction of ACD (Nethercott et al., 1994).

In the workplace, a worker's skin frequently comes into contact with solvents or chemicals mixed in water (aqueous materials). Fortunately, a good deal of research has been conducted to understand the rate at which chemicals pass through the skin. Percutaneous absorption of neat chemicals (e.g., the pure liquid) had historically been studied in humans until the late 1970s (Stewart and Dodd, 1964; Dutkiewicz and Tyras, 1967; Feldmann and Maibach, 1969, 1974; Piotrowski, 1977). Due to the potential toxicity of many chemicals and improved laboratory techniques, *in vivo* human studies have been largely supplanted by experiments with laboratory animals or athymic rodents grafted with human skin (Klain and Black, 1990). Historical research has shown that, in general, the penetration of chemicals through the human skin is similar to that of the pig or monkey, and much slower than that of the rat and rabbit (Bartek et al., 1972).

Starting in the 1980s, *in vitro* studies using human skin began to be conducted on a more routine basis. In these studies, a piece of excised skin is attached to a diffusion apparatus with a top chamber to hold the applied chemical and a temperature-controlled bottom chamber containing saline or other fluids (plus a sampling port to withdraw fractions for analysis) (Frantz, 1990). Although human forearm skin is optimal, it is difficult to obtain, so abdominal or breast skin is commonly used. Generally, a properly conducted *in vitro* test can be a reasonably good predictor of the absorption rate *in vivo* (Bronaugh et al., 1982; Scott et al., 1992). However, due to the fragile nature of the technique, these studies must be carefully interpreted (Barber et al., 1992). Often, depending on the conditions of the test, the results are not applicable to humans.

Aside from neat liquids, dermal exposures in the workplace can also occur through contact with dust or dirt on surfaces and by way of contact with soil or dust-bound contaminants (Paustenbach et al., 1997b). Few studies (Lepow et al., 1975; Que Hee et al., 1985; Driver et al., 1989; Sheppard and Evenden, 1994; Finley et al., 1994a; Kissel et al., 1996a) have directly estimated soil loading on human skin and only one of them attempted to measure occupational contact with equipment (Marlow et al., 1992). Although the available studies probably contain sufficient data to generate point estimates of soil adherence and perhaps can provide a reasonable probability density function (PDF) for most persons exposed to contaminated soils, the degree of representatives of the data to the general population is unknown (Finley et al., 1994a). A recent study measured the adherence of soil to multiple skin surfaces under ambient and recreational conditions (Kissel at al., 1996b). The hands, forearms, lower legs, faces, and feet were measured. Dermal loading on the hands was found to vary over five orders of magnitude and to be dependent on the type of activity. Differences between pre- and post-activity adherence demonstrated the episodic nature of dermal contact with soil. However, due to the activity-dependent nature of soil exposure, data from these studies must be interpreted for their relevance to the type of activity, frequency, duration, and otherwise site-specific nature of exposure. These various studies involving contaminated soil are informative for giving an estimate of exposure which is a couple orders of magnitude greater than what might be expected in a chemical plant but it is nonetheless a "starting point" for bracketing exposure.

Recently, there has been a reasonable level of research investigating exposure to house dust. The basis for the concern has been increasing evidence that perhaps controlling exposure to house dust, especially in homes that have been located in proximity to sites with considerable surface soil contamination, is more important for reducing the health hazard than remediating the soil (Paustenbach et al., 1997b, 1997c, 1997d). Of particular interest is the recent work to develop standardized approaches for wipe sample collection and for estimating the amount of dust loading on the palm of the hand (Lioy et al., 1993, 1998). Although dermal absorption of toxicants due to house dust will almost always pose an insignificant dermal uptake hazard, the research is of interest to occupational dermatologists who may want to draw on these procedures to assess dermal contact with a number of media.

QUANTITATIVE DESCRIPTION OF DERMAL ABSORPTION

For the purposes of risk assessment, percutaneous absorption is defined as the transport of an externally applied chemical through the cutaneous structures and the extra-cellular medium to the bloodstream. The simplest way to describe the rate of skin absorption is to apply Fick's First Law of Diffusion at steady-state (Wepierre and Marty, 1979):

$$J = dQ / dt = Dk\Delta C / e \approx K_p C \qquad \text{[Equation 11-7]}$$

Where:

J	=	chemical flux or rate of chemical absorbed (mg/cm²/h),
D	=	diffusivity in the stratum corneum (cm²/h),
K	=	stratum corneum/vehicle partition coefficient of the chemical (unitless),
ΔC	=	concentration gradient (mg/cm³),
e	=	thickness of the stratum corneum (cm),
K_p	=	permeability coefficient (cm/h), and
C	=	applied chemical concentration (mg/cm³).

The concentration gradient is equal to the difference between the concentration above and below the stratum corneum. Because the concentration below is small compared to the concentration above, C can be approximated as equal to the applied chemical concentration. From the above equation, it can be seen that the rate of absorption is directly proportional to the applied concentration. The diffusivity represents the rate of migration of the chemical through the stratum corneum. Since the stratum corneum has a nonnegligible thickness, there is a period of transient diffusion (lag time) during which the rate of transfer rises to reach a steady state. In these studies, the steady state is thereafter maintained indefinitely, provided the system remains constant. Depending on the type of chemical, the lag time can range from minutes to days (Leung and Paustenbach, 1994). From an exposure assessment standpoint, if the exposure duration is shorter than the lag time, it is unlikely that there will be any significant systemic absorption (Gargas et al., 1989; Flynn, 1990).

The partition coefficient in Equation 11-7 illustrates the importance of solubility characteristics for a chemical to penetrate the skin (Anderson et al., 1988; Surber et al., 1990). Fatty chemicals tend to accumulate in the stratum corneum. Conversely, the stratum corneum is an effective barrier for hydrophilic substances that tend to have low skin absorption rates. Because stratum corneum/vehicle partition coefficients are difficult to measure, the three parameters, D, k, and e, are combined to give an overall permeability coefficient, K_p. It must be emphasized that Equation 11-7 only approximates most *in vivo* exposure situations where true steady-state conditions are rarely attained. In spite of its limitations, this equation has yielded satisfactory estimations of the actual absorption rates of chemicals for many situations.

RECOGNIZED SYSTEMIC DERMAL HAZARDS

Almost all OELs, such as the ACGIH® TLVs® and the OSHA PELs, are acceptable exposure concentrations to chemicals through inhalation. However, it is well known that skin contact is also an important portal of entry for many industrial chemicals used in the workplace (Scansetti et al., 1988; Grandjean et al., 1988; Fiserova-Bergerova, 1993; Leung and Paustenbach, 1994; Hakkert et al., 1996; McDougal et al., 1997). For certain chemicals that are judged to have a significant potential to be absorbed following skin contact, a skin notation is appended to the OEL. For example, the ACGIH® *TLVs® and BEIs® Book* (2001) states:

"Listed substances followed by the designation 'skin' refer to the potential significant contribution to the overall exposure by the cutaneous route, including mucous membranes and the eyes, either by contact with vapors or, of probable greater significance, by direct contact with the substance. Vehicles present in solutions or mixtures can also significantly enhance potential skin absorption. It should be noted that while some materials are capable of causing irritation, dermatitis, and sensitization in workers, these properties are not considered relevant when assessing a skin notation."

Clearly, this definition is rather loose and lacks a precise criterion for deciding if a chemical warrants a skin notation. The inclusion or exclusion of the skin notation is largely a process of subjective judgment by the deliberating standard-setting committees, resulting in many inconsistencies (Grandjean et al., 1988; Scansetti et al., 1988; Fiserova-Bergerova et al., 1990; Wilschut et al., 1995). To bring some consistency to the skin notation process, a quantitative criterion for the biological significance of dermal absorption needs to be established, and several of these approaches are described below.

Estimating the Potential for Skin Uptake of Chemicals in Aqueous Solution

The first such criterion to identify chemicals that deserve a skin notation was proposed by Hansen (1982), who based his proposal on the swelling of psoriasis scales. More recently, Fiserova-Bergerova et al. (1990) proposed another scheme that is toxicologically based and is more consistent with ACGIH®'s definition. In this approach, the dermal absorption potentials of the industrial chemicals, as defined by fluxes predicted from physicochemical properties, are compared with certain defined reference values, referred to as critical fluxes. Berner and Cooper (1987) proposed the following model for calculating the fluxes:

$$J = (C/15)\,(0.038 + 0.153\,K_{ow})\,e^{-0.016\,MW} \qquad \text{[Equation 11-8]}$$

Where:

J = flux

C = saturated aqueous solution concentration of the chemical

K_{ow} = octanol-water partition coefficient

MW = molecular weight of the chemical

Critical fluxes are determined by comparing the skin uptake rate under specified exposure conditions with the inhalation uptake rate during exposure to the time-weighted average TLV® at steady-state conditions. The specified exposure condition for the critical flux can be calculated assuming exposure of 2% of the body surface area (equivalent to the stretched palms and fingers) to a saturated aqueous solution of the chemical. Two reference values have been recommended as criteria for using the skin notation (Fiserova-Bergerova, 1993): dermal absorption potential and dermal toxicity potential. If the dermal absorption (flux) of a chemical for the specific exposure exceeds the critical flux by 30%, the chemical should be classified as possessing "dermal absorption potential." If the flux of a chemical exceeds three times the critical flux, the chemical should be classified as possessing "dermal toxicity potential," and should carry a skin notation. Based on this scheme, a dermal absorption potential was predicted for 122 chemicals, of which only 43 currently (in 1991) carry a skin notation with their TLV® (Fiserova-Bergerova, 1993). A significant dermal toxicity potential was indicated for 77 chemicals, of which only 40 currently have skin notations (see Table 11-4).

In general, whenever dermal uptake is potentially significant, biological monitoring of the chemicals involved is recommended. For chemicals classified as possessing dermal toxicity potential, biological monitoring should be instituted because measurement of airborne concentrations alone may not yield enough information to protect potentially exposed workers. Physicians and others should be aware that, generally, a skin notation is not used for chemicals whose TLVs® are based on skin irritation. This is because the threshold for systemic toxicity is usually higher than that for irritation, and the critical fluxes for chemical irritants are likely underestimated.

The percutaneous permeability coefficients for some common industrial chemicals in aqueous media have been determined experimentally in humans. The corresponding K_p values for the chemicals were calculated using Equation 11-7 above (see Table 11-5). Comparison of the observed and calculated permeability coefficients suggests that there is a general concordance, i.e., the lower the observed value, the lower the predicted value and vice versa. However, the accuracy of the calculated values was rather poor. Specifically, the model equation grossly over-predicted the dermal penetration rates. For example, of the 43 chemicals, the permeability coefficients of only 12 were predicted within a factor of 5. Thus, while this model is useful as a qualitative screening tool to classify organic chemicals in aqueous media according to their skin absorption potential, it should not be used for precise determination of dermal absorption rates.

Table 11- 4
Chemicals with dermal absorption potential

Acetone	Isopropyl alcohol
n-Amyl acetate	Methoxychlor
Atrazine	Methyl acetate
1,3-Butadiene	Methyl n-amyl ketone
n-Butyl acetate	Methyl chloroform
p-ter-Butyltoluene	Methyl ethyl ketone
Cumene	Methyl isoamyl ketone
Cyclohexanone	Methyl isobutyl ketone
o-Dichlorobenzene	Methyl propyl ketone
1,1-Dichloroethane	Metribuzin
1,2-Dichloroethylene	Naphthalene
Diethyl ketone	Nitromethane
Diuron	Pentaerythritol
Enflurane	n-Propyl acetate
Ethanol	Propylene dichloride
Ethyl acetate	Strychnine
Ethyl benzene	Styrene
Ethyl ether	Toluene
Ethyl formate	1,2,4-Trichlorobenzene
Halothane	Trichloroethylene
n-Heptane	Trimethyl benzene
Hexachloroethane	o-Xylene
n-Hexane	m-Xylene
Isoamyl alcohol	p-Xylene

Source: Leung and Paustenbach, 1994

Table 11-5
Human cutaneous permeability coefficient values for some common organic industrial chemicals in aqueous medium

	MW[A]	K_{ow}[B]	Observed[C]	Calculated[D]	Reference
Organic Chemicals					
2-Amino-4-nitrophenol	154.13	21.38	0.00066	0.019	Bronaugh and Congdon, 1984
4-amino-2-nitrophenol	154.13	9.12	0.0028	0.0081	Bronaugh and Congdon, 1984
Aniline	93.12	7.94	0.041[B]	0.019	Baranowska-Sutkiewicz, 1982
Benzene	78.11	134.9	0.11	0.39	Blank and McAuliffe, 1985
p-Bromophenol	173.02	389.05	0.036	0.25	Roberts et al., 1977
Butane-2,3-diol	90.12	0.12	<0.00005	0.0009	Blank et al., 1967
n-Butunol	74.12	7.59	0.0025	0.024	Scheuplein and Blank, 1973
2-Butanone	72.10	1.94	0.0045	0.007	Blank et al., 1967
Carbon disulfide	76.14	100	0.54[B]	0.3	Baranowska-Sutkiewicz, 1982
Chlorocresol	142.58	1258.93	0.055	1.31	Roberts et al., 1977
S-Chlorophenal	128.56	147.91	0.033	0.19	Roberts et al., 1977
p-Chlorophenal	128.56	257.04	0.036	0.34	Roberts et al., 1977
Choroxylenal	156.61	1621.81	0.059	1.35	Roberts et al., 1977
m-Cresol	108.13	100	0.015	0.18	Roberts et al., 1977
o-Cresol	108.13	100	0.016	0.18	Roberts et al., 1977
p-Cresol	108.13	85.11	0.018	0.15	Roberts et al., 1977
Decanol	158.28	37153.52	0.08	30.11	Scheuplein and Blank, 1973
2,4-Dichlorophenol	163.01	1995.26	0.06	1.5	Roberts et al., 1977
1,4-Dioxane	88.10	0.38	0.00043	0.0016	Bronaugh, 1982
Ethanol	46.07	0.49	0.0008	0.0036	Scheuplein and Blank, 1973
2-Ethoxyethanol	90.12	0.29	0.0003	0.0013	Blank et al., 1967
Ethylbenzene	106.16	1412.54	1.215[B]	2.65	Sutkiewicz and Tyras, 1967
Ethylether	74.12	6.76	0.016	0.022	Blank et al., 1967
p-Ethylphenol	122.17	549.54	0.035	0.79	Roberts et al., 1977
Heptanol	116.20	257.04	0.038	0.41	Blank et al., 1967
Hexanol	102.17	107.15	0.028	0.21	Bond and Barry, 1988

Table 11-5 (Cont'd)
Human cutaneous permeability coefficient values for some common organic industrial chemicals in aqueous medium

	MW[A]	K_{ow} [B]	Observed[C]	Calculated [D]	Reference
Methanol	32.04	0.17	0.0016	0.0026	Southwell et al., 1988
Methyl hydroxybenzoate	152.15	91.2	0.0091	0.082	Roberts et al., 1977
b-naphthol	144.16	691.83	0.028	0.7	Roberts et al., 1977
3-Nitrophenol	139.11	100.00	0.0056	0.11	Roberts et al., 1977
4-Nitrophenol	139.11	81.28	0.0056	0.09	Roberts et al., 1977
Nitrosodiethanolamine	134.13	0.13	0.0000055	0.0005	Bronaugh et al., 1981
Nonanol	144.26	2951.21	0.06	2.99	Scheuplein and Blank, 1973
Octanol	130.22	933.25	0.061	1.19	Southwell et al., 1988
Pentanol	88.15	36.31	0.006	0.091	Scheuplein and Blank, 1973
Phenol	94.11	32.36	0.0082	0.074	Roberts et al., 1977
Propanol	60.09	2.00	0.0017	0.0088	Blank et al., 1967
Resorcinol	110.11	6.03	0.00024	0.011	Roberts et al., 1977
Styrene	104.14	891.25	0.635[D]	1.72	Dutkiewicz and Tyras, 1968a
Thymol	150.21	1995.26	0.053	1.84	Roberts et al., 1977
Toluene	92.13	489.78	1.01	1.15	Dutkiewicz and Tyras, 1968a
2,4,6-Trichlorophenol	197.46	2344.23	0.059	1.02	Roberts et al.,1977
3,4-Xylenol	122.16	169.82	0.036	0.25	Roberts et al., 1977

[A] MW = molecular weight
[B] K_{ow} = octanol-water partition coefficient
[C] All the observed permeability coefficients were obtained by using *in vitro* techniques, except those denoted with superscript D, which were determined *in vivo*.
[D] Permeability coefficients calculated with Equation 11-7.

Source: Leung and Paustenbach, 1994

Pharmacokinetic Models for Estimating the Potential Uptake of Chemicals

Aqueous Solution

A four-compartment pharmacokinetic model to describe the percutaneous absorption of chemicals has been proposed (Guy et al., 1982). This model describes, using first-order rate constants, a chemical moving through the compartments representing the various skin structures. The model has been used successfully to predict the disposition of chemicals in the skin and plasma as a function of their physicochemical properties, and when an input rate constant to the skin surface is added to the model, it can be used to assess vehicle effects. A similar model that treats the barrier membrane as a series of spaces filled with immiscible liquids has also been developed (Zatz, 1985); its advantage is that it allows the examination of nonsteady-state conditions where Fick's Law does not apply.

Under an infinite dose situation where the amount of a chemical lost by penetration is too small to alter the applied concentration (e.g., where one is swimming), the rate of absorption is essentially linear once steady-state has been reached. In the finite dose system, however, the chemical solution is applied as a thin film and the concentration decreases as penetration proceeds (e.g., a splash). All other model parameters being the same, penetration is reduced under finite dose conditions. This is because the chemical concentration is continuously reduced over time, resulting in a decrease in the gradient across the stratum corneum. These modeling results indicate that the mechanism by which fluxes are affected must be considered when extrapolating to nonsteady-state conditions.

Although classic pharmacokinetic modeling like that described by Guy et al. (1982) can provide a good mathematical description of the disposition of chemicals, it does not depict exactly the biological processes in the intact animal. Fortunately, due to recent increases in the availability of computer hardware and software, pharmacokinetic methods based on physiological principles are now feasible alternatives for analysis of *in vivo* skin penetration studies. These PB-PK approaches realistically describe the disposition of the chemical in the intact animal in terms of rates of blood flow, permeability of membranes, and partitioning of chemicals into tissues (Andersen et al., 1987a; McDougal, 1996). Characterizing dermal absorption in terms of actual anatomical, physiological, and biochemical parameters facilitates extrapolations to the real species of interest, humans.

In 1991, a PB-PK model was developed to describe the percutaneous absorption of volatile organic contaminants in dilute aqueous solutions (Shatkin and Brown, 1991). The exposure scenario modeled was either hand or full-body immersion into a vessel of solute-contaminated water. Modeling results suggest that the uptake of chemicals in aqueous solutions is most markedly influenced by epidermal blood flow rates followed by epidermal thickness and lipid content of the stratum corneum. In general, thicker and fattier skin provides a better barrier to dermal penetration of chemicals. This model also predicts that the dose of some volatile organic chemicals in water absorbed through the skin during a 20-minute bath may be equivalent to the amount inhaled (Shatkin and Brown, 1991).

The most complex and perhaps the most well validated of the various models for dermal uptake of liquids is that developed by McDougal et al. (1996). This group has successfully predicted dermal uptake rates for humans for nearly a dozen chemicals based on animal and physical chemistry data. One advantage of dermal PB-PK models over traditional *in vivo* methods is their ability to accurately describe nonlinear biochemical and physical processes. For example, describing skin penetration based on blood concentrations or excretion rates as "percent absorbed" assumes that all processes have a simple linear relationship with the exposure concentration. This is often not the case. The kinetics become non-linear when the absorption, distribution, metabolism, or elimination of a chemical is saturated at high exposure concentrations. Their model and some of those that have been developed since then address this phenomenon in a reasonable manner.

Soil

Fortunately, one of the most frequently occurring exposure situations facing industrial hygienists who evaluate environmental exposures involves contaminated soil (Paustenbach et al., 1986). Unfortunately, the dermal uptake of chemicals found on soil has been infrequently evaluated experimentally (Leung and Paustenbach, 1994). A model to estimate the amount of a chemical in soil that crosses the stratum corneum into the underlying tissue layer has been developed (McKone, 1990). To differentiate this absorptive process from bioavailability, which also includes transport into blood, McKone refers to the percentage of the available chemical as an uptake fraction. The approach is based on the concept of fugacity which measures the tendency of a chemical to move from one phase to another. Because the skin has a fat content of about 10 percent and soil has an organic carbon content of one to four percent, a chemical in soil placed on the skin will move from the soil to the underlying adipose layers of the skin. However, this transfer depends on the period of time between deposition on the skin and removal by evaporative processes. It is the mass-transfer coefficients of the soil-to-skin layer and the soil-to-air layer that define the rate at which these competing processes occur.

Results of this model suggest that the uptake fraction of a chemical in soil varies with the duration of exposure, soil deposition rate, and physical properties of the chemical, and is particularly sensitive to the values of the K_{ow}, and the mass or depth of soil deposited on the skin. When the amount of soil on the skin is low (<1 mg/cm^2), a high uptake fraction, approaching unity in some cases, is predicted. With higher soil loading (20 mg/cm^2), an uptake of only 0.5 percent is predicted. Because of the diverse variations of the uptake fraction with soil loading, results obtained from experiments with a single soil loading should be applied with caution to human soil-exposure scenarios.

The dermal uptake of chemicals in soil is a complex process, but its behavior is predictable if the controlling factors are accounted for and quantified (McKone, 1990; Leung and Paustenbach, 1994). In situations involving a relatively thin layer of a chemical on the skin, a few generalizations can be made. First, for chemicals with a high K_{ow} and a low air:water partition coefficient for risk assessment purposes, it is reasonable to assume 100 percent uptake in 12 hours. Second, for chemicals, with an air:water partition coefficient greater than 0.01, the uptake fraction is unlikely to exceed 40 percent in 12 hours. Third, for chemicals with an air:water partition coefficient greater than 0.1, one can expect less than 3 percent uptake in 12 hours. In most occupational settings, contaminated soil will rarely be in contact with the skin for greater than 4 hours before it is washed off. Consequently, this should be accounted for when attempting to predict systemic uptake.

FACTORS USED TO ESTIMATE DERMAL UPTAKE

At least 10 different factors need to be quantitatively accounted for in order to estimate the likely systemic uptake of a chemical that comes into contact with the skin, either as a liquid or when present in soil or dust.

Bioavailability

The typical media of concern for assessing cutaneous contact to environmental chemicals, in contrast with occupational exposure, are house dust, soil, fly ash, and sediment. In the workplace, dermal uptake is due to direct contact with liquids and contact with surfaces contaminated with dirt, condensed solids, or liquids. A number of parameters can influence the degree of cutaneous bioavailability of chemicals in complex matrices. These may include aging (time following contamination), soil type (e.g., silt, clay, and sand), type and concentration of co-contaminants (e.g., oil and other organics), and the concentration of the chemical contaminant in the media (Shu et al., 1988). The bioavailability of a chemical in soil will usually be affected by its physicochemical properties. Large molecular weight chemicals tend to bind to soil/dust and be less water soluble, while smaller molecules will frequently be water soluble, less tightly bound, and relatively bioavailable (McKone, 1990). The cutaneous bioavailability of various chemicals in soils has been determined in animals (Lucier et al., 1986; Umbreit et al., 1986a, 1986b; Shu et al., 1988; Skrowronski et al., 1988; Wester et al., 1993; Hrudey et al., 1996; Ruby et al., 1999). These studies show that different media and different chemicals can yield dramatically different cutaneous bioavailabilities. The results of these studies, for example, range from 0.01% to 3% uptake for chemicals from soil.

Skin Surface Area

The "rule of nines" may be used for estimating the surface area of the human body (Snyder, 1975): the head and neck are 9 percent; upper limbs are each 9 percent; lower limbs are each 18 percent; and the front or back of the trunk is 18 percent. The EPA (1996b) has estimated an exposed surface area (arms, hands, legs, and feet) of 2,900 cm^2 for children 0 to 2 years old; 3,400 cm^2 for children 2 to 6 years old; and 2,940 cm^2 for adults (an adult is assumed to wear pants, an open-neck short-sleeve shirt, shoes, and no hat or gloves). When assessing exposure to chemicals in the ambient environment, most of the necessary surface area information can be found in the EPA *Exposure Factors Handbook* (1996). When estimating exposure in the occupational environment, one should rely on professional judgment regarding how to extrapolate the data from studies of the ambient environment and site specific exposure information. Table 11-6 presents the skin surface areas commonly used when conducting exposure assessments (Snyder, 1975).

Soil Loading on the Skin

Although not frequently needed to assess occupational exposure, soil-to-skin adherence rates of 0.5 to 0.6 mg/cm^2 and 0.2 to 2.8 mg/cm^2 have been reported for adults and children, respectively (Que Hee et al., 1985; Paustenbach et al., 1986; Colorado Department of Human Services [CDHS], 1986; Driver et al., 1989). Recent work by Finley et al. (1994a) and Kissel et al. (1996a) has built on those studies and has shown that dermal loading can vary significantly among different activities and different people. Based on data collected prior to recent studies, the EPA (1996) has suggested a default soil-to-skin adherence rate of 0.2 mg/cm^2 (median) and 1.0 mg/cm^2 (95th percentile) for an adult.

DERMAL UPTAKE OF CONTAMINANTS IN SOLUTION

To estimate the uptake of a chemical, one needs to know the percutaneous absorption rate, the area of exposed skin, the concentration of the chemical, and the duration of exposure (Paustenbach, 1988). One scenario is that of a thin film of chemical on the skin. For this finite dose scenario:

$$\text{Uptake (mg)} = (C)(A)(x)(f)(t) \qquad \text{[Equation 11-9]}$$

Where:

C = concentration of the chemical (mg/cm^3),

A = skin surface area (cm^2),

x = thickness of the film layer (cm)

f = absorption rate (percent per hour)

t = duration of exposure (hour).

Another scenario is where there is an excess amount of a chemical on the skin (i.e., infinite dose). In this case, the thickness of the chemical layer is not calculated and steady-state kinetics are assumed. For a chemical in an aqueous or gaseous media:

$$\text{Uptake (mg)} = (C)(A)(K_p)(t) \qquad \text{[Equation 11-10]}$$

Table 11-6
Representative surface areas of the human body (adult male)*

Body Portion	Area (cm^2)
Whole body	18,000
Head and neck	1620
Head	1260
Back of head	320
Neck	360
Back of neck	90
Torso	6480
Back	2520
Chest	2520
Sides	1440
Upper limbs	3240
Upper arms (elbow-shoulder)	1440
Lower arms (elbow-wrist)	1080
Hands	720
Palms	360
Upper arms (back of)	360
Lower arms (back of)	270
Lower limbs	6480
Thighs	3240
Lower legs (knee-ankle)	2160
Feet 1080	
Soles of feet	540
Thighs (back of)	810
Lower legs (back of)	540
Perineum	180

* Data adapted from Snyder (1975)

Where:

- C = concentration of chemical (mg/cm^3)
- A = area (cm^2)
- K_p = permeability coefficient (cm/h)
- t = time (h)

For a neat liquid chemical:

$$\text{Uptake (mg)} = (A)(J)(t)(C) \qquad \text{[Equation 11-11]}$$

Where:

- A = area (cm^2)
- J = flux of chemical (mg/cm^2/h).
- t = time (h)
- C = concentration of chemical (mg/cm^3)

$$\text{Uptake} = \left(\frac{250 \text{ ng dioxin}}{1 \text{ g soil}}\right)\left(\frac{0.2 \text{ mg soil}}{\text{cm}^2 \text{ skin}}\right)\left(\frac{1 \text{ g}}{10^3 \text{ mg}}\right)(1{,}800 \text{ cm}^2 \text{ skin})(0.01)$$

The EPA has suggested using the following equation for estimating the percutaneous absorption of chemicals in soil (EPA, 1996; EPA, 1989a):

$$\text{Uptake (mg)} = (C)(A)(r)(B) \qquad \text{[Equation 11-12]}$$

Where:

- C = concentration of the chemical in soil (mg/g),
- A = skin surface area (cm^2),
- r = soil-to-skin adherence rate (g/cm^2),
- B = cutaneous bioavailability or absolute transfer (unitless).

Example 11-5: *Skin Uptake of a Chemical in Soil*

A man gardens with soil contaminated on average with 250 ng dioxin per gram of soil (250 ppb). Assuming that both hands and lower arms are in contact with the soil and that the cutaneous bioavailability of dioxin in soil is one percent (Shu et al., 1988), what is the plausible uptake of dioxin by this person (using Equation 11-12). From the EPA Exposure Factors Handbook we estimate that 0.2 mg soil/cm^2 skin adhere per 4 hours and 1800 cm^2 are exposed.

Assume that he washes his hands every four hours. From Equation 11-12, where C = 250 ng/g, A = 1,800 cm² (from EPA, 1996), and B = 0.01 (Shu et al., 1988):

$$\text{Uptake} = \left(\frac{250 \text{ ng dioxin}}{1 \text{ g soil}}\right)\left(\frac{0.2 \text{ mg soil}}{\text{cm}^2 \text{ skin}}\right)\left(\frac{1 \text{ g}}{10^3 \text{ mg}}\right)(1{,}800 \text{ cm}^2 \text{ skin})(0.01)$$

$$= 0.9 \text{ ng dioxin}$$

UPTAKE OF CHEMICALS IN AN AQUEOUS MATRIX

Although most published estimates of dermal uptake of chemicals in water were not directed at understanding workplace exposure, the methodologies used provide some insight to industrial hygienists. For example, if interested in the possible uptake of a chemical present in water, the amount of chlordane absorbed through the skin by a man swimming for four hours in water containing 1 ppb chlordane has been estimated (Scow et al., 1979). Along similar lines, the amount of chloroform absorbed by a boy swimming for three hours in water containing 0.5 ppm chloroform has been calculated to be about 1.65 mg (Beech, 1980). Approximately 0.5 μg and 3.3 μg, respectively, can be absorbed through the skin during a 10-minute shower and a 20-minute bath with water containing 1 ppb 1,1,1-trichloroethane (Byard, 1989).

About 10 years ago, it was recognized that in the indoor environment, sometimes an obvious dermal exposure did not necessarily represent the vast majority of the risk. Specifically, it has been found that the inhalation exposure due to the release of vapors from liquids to which people were in close contact could be relatively high. For example, comparisons have been made of the chloroform concentration in exhaled breath after a shower to that after an inhalation-only exposure (Jo et al., 1990). The concentration after showering was about twice that after inhalation-only exposure, indicating that the absorbed dose from skin absorption was approximately equivalent to that from inhalation absorption.

Example 11-6: *Skin Uptake of a Chemical from Water*

A person has filled his swimming pool with shallow well water contaminated with 0.002 mg/ml (2 ppb) toluene. What is the plausible dermal uptake of toluene while swimming in the contaminated water for half an hour? K_p for toluene in water is 1.01 cm/h. From the EPA Exposure Factors Handbook, totaol body surface is 18,000 cm². From Equation 11-10 and using information from Tables 11-3 and 11-6:

$$\text{uptake} = (0.002 \text{ mg/ml})\ (18{,}000 \text{ cm}^2)\ (1.01 \text{ cm/h})\ (0.5 \text{ h})\ (1 \text{ ml water}/1 \text{ cm}^3)$$

$$\text{uptake} = 18 \text{ mg}$$

PERCUTANEOUS ABSORPTION OF LIQUID SOLVENTS

While the percutaneous absorption of chemical solutes generally proceeds by simple diffusion, the skin uptake of neat chemical liquids is not necessarily exclusively governed by Fick's Law. Consequently, the uptake of neat liquid through the skin needs to be estimated using direct *in vivo* skin contact techniques. Table 11-7 presents the percutaneous absorption rates of some neat industrial liquid solvents that have been determined for humans.

Example 11-7: *Skin Uptake of a Neat Liquid Chemical*

Due to carelessness or a leak, the inside of a glove gets contaminated with 2-methoxyethanol. How much can be absorbed if a worker wears the contaminated glove on one hand for half an hour? J = 2.82 mg/cm^2-h (see Table 11-10). From Equation 11-11:

$$\text{uptake} = (360\ \text{cm}^2)\ (2.82\ \text{mg/cm}^2\text{-h})\ (0.5\ \text{h})$$

$$\text{uptake} = 508\ \text{mg}$$

To understand the relative hazard from skin exposure versus inhalation exposure, the dose of 2-methoxyethanol absorbed by the same worker via inhalation for eight hours (10 m^3 of air inhaled) at the TLV® of 16 mg/cm^3 can be estimated. Assume an 80 percent inhalation uptake efficiency.

$$\text{Inhalation uptake} = (16\ \text{mg/m}^3)(10\ \text{m}^3)(0.8)$$

$$= 128\ \text{mg}$$

Thus, the uptake of 2-methoxyethanol following 30 minutes of skin exposure of a single hand can be as much as four times that from inhalation for eight hours at the TLV® concentration; a presumably safe level of exposure. From this example, it is clear that the cutaneous route of entry can, in some situations, significantly contribute to the total absorbed dose, especially in the occupational setting.

PERCUTANEOUS ABSORPTION OF CHEMICALS IN THE VAPOR PHASE

Until the 1990s, it was generally assumed that the plausible dose resulting from vapors absorbed through the skin was too low to pose a hazard. Only a few studies have examined this issue (Krivanek et al., 1978; Mraz and Nohova, 1992; Kezic et al., 1997). A few clinical reports have encouraged some limited *in vitro* research to evaluate the absorption of several chemicals in the gaseous phase through the human skin (Table 11-8). A chamber system to measure the whole-body percutaneous absorption of chemical vapors in rats has been described by McDougal et al. (1990), and this work has produced some interesting results (Mattie et al., 1994). In this system, the flux of a chemical across the skin is determined from the chemical concentration in

Table 11-7
Absorption rates of some neat industrial liquid chemicals in human skin *in vivo*

Chemical	Absorption Rate (mg/cm²-h)	Reference
Aniline	0.2 - 0.7	Piotrowski, 1957
Benzene	0.24 - 0.4	Hanke et al., 1961; Maibach and Anjo, 1981
2-Butoxyethanol	0.05 - 0.68	Johanson et al., 1988
2-(2-Butoxyethoxy) ethanol	0.035	Dugard et al., 1984
Carbon disulfide	9.7	Baranowska-Dutkiewicz, 1968
Dimethylformamide	9.4	Mraz and Nohova, 1992
Ethylbenzene	22 - 23	Dutkiewicz and Tyras, 1967
2-Ethoxyethanol	0.796	Dugard et al., 1984
2-(2-Ethoxyethoxy) ethanol	0.125	Dugard et al., 1984
Methanol	11.5	Dutkiewicz et al., 1980
2-Methoxyethanol	2.82	Dugard et al., 1984
2-(2-Methoxyethoxy) ethanol	0.206	Dugard et al., 1984
Methyl butyl ketone	0.25 - 0.48	Divincenzo et al., 1987
Nitrobenzene	2 1960	Salmowa and Piotrowski,
Styrene	9 - 15	Dutkiewicz and Tyras, 1968a
Toluene	14 - 23	Dutkiewicz and Tyras, 1968b
Xylene (mixed)	4.5 - 9.6	Dutkiewicz and Tyras, 1968a
m-Xylene	0.12 - 0.15	Engstrom et al., 1977; Riihimaki, 1979

Source: Leung and Paustenbach, 1994

Table 11-8
Percutaneous absorption rates for chemical vapors *In vivo**

Chemical	Skin Uptake in a Combined Exposure** (%)	Permeability Coefficient K_p (cm/h)	
		Rat	Human
Styrene	9.4	1.75	0.35 - 1.42
m-Xylene	3.9	0.72	0.24 - 0.26
Toluene	3.7	0.72	0.18
Perchloroethylene	3.5	0.67	0.17
Benzene	0.8	0.15	0.08
Halothane	0.2	0.05	
Hexane	0.1	0.03	
Isoflurane	0.1	0.03	
Methylene chloride		0.28	
Dibromomethane		1.32	
Bromochloromethane		0.79	
Phenol			15.74 -17.59
Nitrobenzene			11.1
1,1,1-Trichloroethane			0.01

* Rat data adapted from McDougal et al. (1990) and human data from Rioihimaki and Pfaffli (1978)

** Combined exposure is one in which rats are simultaneously absorbing chemical vapors byinhalation exposure as well as through whole-body absorption through the skin.

blood during exposure by using a PB-PK model. In most cases, the absorption of vapors through the skin amounts to less than 10% of the total dose received from a combined skin and inhalation exposure. While there is good agreement between the rat and human in the relative ranking of the permeability coefficients among the chemicals studied, for an individual chemical, the rat skin appears to be two to four times more permeable than the human skin. This observation is consistent with previously reported data (Piotrowski, 1977; Bronaugh et al., 1982).

It is generally not necessary to account for the contribution due to percutaneous uptake of vapors when the OEL is used as a guideline for acceptable exposure, because uptake via this route is usually inherent in the data used to derive the OEL; that is, in the studies of animals or humans from which the data were collected exposure via inhalation (whole body) and dermal uptake occurred. However, although good work practices and the law require that these situations not occur in the workplace, sometimes in emergency situations, airline (supplied air) respirators or self-contained breathing apparatus are worn in environments containing chemical concentrations 10-fold to 1,000-fold greater than the TLV®. In these cases, it can be important to account for the uptake of vapor through either exposed or covered skin.

Although nearly all data on the absorption of vapors involve bare skin, the role of clothing in preventing skin uptake has occasionally been evaluated. For example, a study of workers wearing denim clothing indicated no decreased uptake of phenol vapors (Piotrowski, 1971), but found a 20% and a 40% reduction in the uptake of nitrobenzene (Piotrowski, 1967) and aniline vapor (Dutkiewicz and Piotrowski, 1961), respectively. Although standard clothing may slightly decrease the amount of a chemical that is transferred from air through the skin, it can be a significant source of continuous exposure if the clothing has been contaminated.

Example 11-8: ***Skin Uptake of a Chemical Vapor***

Assuming that a person wears an airline respirator for a half hour in a room containing 500 mg/m^3 nitrobenzene (100 times the current TLV®), how much nitrobenzene might be absorbed through the skin?

The head, neck, and upper limbs are assumed to be exposed (surface area = 4,860 cm^2), and the rest of the body (surface area = 13,140 cm^2) is covered with clothing. Assume the percutaneous K_p of nitrobenzene in air is 11.1 cm/h, and clothing reduced the skin uptake rate of vapors by about 20 percent (Piotrowski, 1967). Equation 11-10 is applied twice.

Uptake = $(C)(A)(K_p)(t)$

Uptake through exposed skin = $(500\ \text{mg/m}^3)\ (4{,}860\ \text{cm}^2)\ (11.1\ \text{cm/h})\ (0.5\ \text{h})\ (1\ \text{m}^3/10^6\ \text{cm}^3)$ = 13.5 mg

Uptake through clothing = $(500\ \text{mg/m}^3)\ (13{,}140\ \text{cm}^2)\ (11.1\ \text{cm/h})\ (1-0.2)\ (0.5\ \text{h})\ (1\ \text{m}^3/10^6\ \text{cm}^3)$ = 29 mg

Total uptake = 13.5 + 29

= 42.5 mg

From this example, it is clear that if one enters an environment that contains an exceedingly high concentration of airborne contaminant, even if a supplied-air respirator is worn, the degree of skin uptake of the vapor may be worthy of evaluation to insure that the worker is protected. Uptake following one day of inhalation exposure at the TLV® (5 mg/m^3) results in 50 mg uptake [(10 m^3)(5 mg/m^3)].

Interpreting Wipe Samples

In some workplaces, wipe sampling has historically been conducted to assess the degree of surface contamination. Hospitals were among the first occupational settings, as long ago as 1940, to rely on this method to determine microbial levels in operating rooms. In pharmaceutical manufacturing, wipe sampling has been used as an indicator of hygienic conditions since the 1960s. The health physics profession has utilized wipe samples extensively as an indicator of the need for better housekeeping and decontamination, and has performed most of the early work in quantifying the relationship of wipe samples to inhalation, dermal, and oral uptake.

Over the years, only a few papers have discussed how to collect and interpret wipe samples (European Centre for Ecotoxicology and Toxicology of Chemicals, 1993a, 1993b; Caplan, 1993; McArthur, 1992; Brouwer et al., 1992; Fenske, 1993). When the primary effect of a chemical is skin discoloration, ACD, or chloracne, wipe sampling has nearly always been the preferred approach for assessing the acceptability of the workplace (rather than rely on air samples). Beginning in the 1980s, a substantial number of wipe samples were collected within office buildings contaminated with dioxins and furans following electrical transformer fires to estimate the potential human exposure (Michaud et al., 1994). The interpretation of these data was often mishandled and a number of decisions of risk managers would have been improved had better exposure assessment been conducted. In more recent times, as in the evaluation of the health hazards posed by dusts in office buildings surrounding the World Trade Center disaster, the importance of wipe sample data and how to interpret it has been elevated.

Although wipe sampling data have been used generally as an indicator of cleanliness (Caplan, 1993), these data can also be used to estimate systemic uptake of a contaminant if the degree of skin contact with the contaminated surfaces is known. While this exposure assessment methodology is rather imprecise, it is useful in obtaining a rough estimate of the possible exposure that can be refined later with other means such as biological monitoring. Again, such approaches have been found useful by those who work in the nuclear industry.

If one knows that wipe sampling results are representative of what comes into contact with the hands, i.e., actually able to be absorbed, then the procedures for converting wipe sample data to estimates of systemic uptake are straightforward. For example, if one knows the number of times a surface (e.g., valve handle, instrument controller, or drum) is touched, the surface area of the hand touching these items (usually the palm), and the percutaneous absorption rate of the chemical, then the uptake can be estimated using wipe sample information. One of the most thorough evaluations that relies on wipe sample data and time-motion studies was conducted by the National Institute for Occupational Safety and Health (NIOSH) to examine the amount of dioxins absorbed by chemical operators in a 2,4,5-T manufacturing plant. Regrettably, the study was never published (Marlow et al., 1990). In general, most of the historical wipe sample data within the chemical industry over the past 40 years has not been shared in the literature nor has it been used in retrospective risk assessments. Perhaps the most robust data sets within the chemical industry have been those collected by Dow Chemical and Monsanto during the 1950s when they were trying to understand the cause of chloracne, which had been observed in chemical workers.

Until recently, no standardized approaches existed for conducting wipe sampling. Differences in the use of wetting agents (acetone, methylene chloride, water, saline, isopropanol, and ethanol) and sampling media (paper, cotton, and synthetic fibers) produced drastically different results. In some procedures, especially those that used methylene chloride (where the paint was concurrently stripped by the solvent), the chemical in the paint matrix was assumed to be bioavailable (a completely unreasonable assumption). Clearly, much of the previous work that measured the amount of chemical released following the aggressive scrubbing of the contaminated surfaces with detergent or solvent did not reflect a realistic exposure scenario. Thus, there has been a need for standard techniques that attempt to mimic the conditions in which a hand comes into contact with a contaminated surface (McArthur, 1992). Virtually all of the best work to-date has been conducted by hygienists involved in agricultural assessment (Knaak et al., 1989, 2002).

In an attempt to fill this need, fairly sophisticated work to standardize these procedures has been conducted by researchers at Rutgers University, and their wipe sampling procedure and device have been patented (Lioy et al., 1993). They have also developed a dry contact sampling device (Lioy et al., 1998) that offers promise for understanding the hazard due to surface dusts. The implications of the recent wipe sampling research are that: (1) a minimum number of samples is needed to have statistical confidence; (2) the pressure applied to the cloth during sample collection should be standardized; (3) neat solvent should not be used as a collection media; (4) the size of the sample area needs to be sufficient to collect enough contaminant for quantification; and (5) the technique should be validated by using glove analyses.

Example 11-9: ***Using Wipe Sample Data to Estimate Skin Uptake***

Wipe samples have been collected in office buildings where electrical transformers containing PCBs have caught fire and the smoke has been distributed throughout the building in the ventilation system. What dose of dioxin due to skin contact with contaminated building surfaces might be possible if the dioxin concentration in the wipe samples is 10 pg/cm^2 (use Equation 11-12).

In applying Equation 11-12 to wipe sampling, C = concentration of chemical in contaminated surface (mg/cm^2), A = surface area of one palm (cm^2), r = removal efficiency of chemical from contaminated surface by skin (unitless), and B = cutaneous bioavailability (unitless). The chemical concentration in the wipe samples (C_{wipe}) is a measure of the surface contamination as it is affected by the removal efficiency (R_{wipe}) of the particular wiping procedure, i.e., $C = C_{wipe}/R_{wipe}$. Thus, based on values for removal efficiencies below,

$$\text{Uptake} = (C_{wipe})(A)(r)(B)/R_{wipe}$$

where: C_{wipe} = 10 pg/cm^2

A = 180 cm^2 (Table 11-6)

r = 10 percent

B = 1 percent (Shu et al., 1988)

R_{wipe} = 50 percent

Using substitution,

$$\text{uptake} = (10\ \text{pg/cm}^2)\ (180\ \text{cm}^2)\ (0.10)\ (0.01)/(0.50)$$

$$= 3.6\ \text{pg}$$

RETROSPECTIVE EXPOSURE ASSESSMENT

Estimating exposures is an essential component of the risk assessment process and is necessary in epidemiological studies to evaluate the dose-response relationship of an agent (Paustenbach, 2000; Williams and Paustenbach, 2002). Although modern era exposure assessments rely primarily on available industrial hygiene measurements, evaluations performed using historical datasets (i.e., pre-1970) often require the reconstruction of exposures based on incomplete or inadequate sampling data. In such instances, retrospective exposure analyses can be used, in conjunction with chemical or biological data, to estimate applied or internal doses under a variety of scenarios.

In a retrospective exposure assessment (more recently referred to as "dose-reconstruction"), available records, monitoring data, interview notes, expert judgment, exposure factors, and new information are used to generate a best estimate of probable workplace or community exposures that occurred in the past. More recent monitoring data are also typically used to extrapolate backward in time (i.e., to estimate historical conditions), taking into account potential changes in plant operations, engineering controls, job functions, and other factors that may affect likely exposure levels.

A popular method for conducting occupational retrospective exposure assessments is to use a "job-exposure matrix" approach, in which exposures are estimated for specific job categories (rather than individuals) based on specific tasks or job functions. A number of retrospective exposure assessments have been conducted over the last 20 years for a variety of chemicals, and these have involved the use of historic industrial hygiene monitoring data, mortality and morbidity data, and state-of-the-art knowledge about exposure scenarios in similar settings (see Table 11-9).

It is important to recognize, however, that most retrospective exposure analyses lack high quality industrial hygiene monitoring data (e.g., full-shift, time-weighted average, personal exposure measurements from typical operations). This is because these types of data were not often available until the early 1970s. One method for recreating exposure conditions for specific tasks or scenarios when few or no data are available is to conduct an exposure simulation. For example, Madl and Paustenbach (2002) recently described an occupational exposure simulation study in which airborne concentrations of benzene and other chemicals were measured in an enclosed "roundhouse" where diesel train engines were serviced. Specifically, workplace exposures were recreated inside a turn-of-the-century roundhouse. This study found that simulation methods provided a reasonable estimate of historic benzene air concentrations that workers could have been exposed to under various conditions.

Each retrospective exposure assessment is a unique investigation that takes advantage of the available information and relies on creative and innovative approaches to exploit that information. Clearly the lack of historical data, however, presents a challenge to quantitatively characterize the degree of chemical exposure that may have occurred over time. In addition, the evaluation of pertinent data such as company archives, plant processes, and current and retired employee recollections, can be an expensive and time-consuming process. Despite these difficulties, major improvements in the quality of retrospective exposure assessments have been made in the past decade. The use of retrospective exposure assessment will undoubtedly continue to grow in occupational and other settings. Further, ongoing research will continue to improve current methods for conducting these analyses.

CLOSING THOUGHTS

As described in this chapter, the field of health risk assessment has evolved significantly over the past 20 years. Nearly all of the classic risk assessments that have been conducted over those years have addressed exposure to chemical contaminants in the ambient environment. Specifically, many people studied exposure to contaminated soils (Kimbrough et al., 1984; Paustenbach et al., 1986; Sheehan et al., 1988), contaminated water (McKone and Bogan, 1992), contaminated foods, and ambient air. Perhaps as many as 50-100 risk assessments have been published in peer reviewed journals over the past 15 years and certainly several thousand environmental risk assessments have been conducted by regulatory agencies, consulting groups, universities, and others. They have become the basis for most decisions about what to do about contaminated media.

However, the conduct of comprehensive risk assessments has just begun to become a part of the practice of industrial hygiene. There are good reasons why this has occurred, and one could argue that, in most cases, it is unnecessary to invest the time and money to perform

Table 11-9
Selected retrospective exposure assessments

Site or Population Studied	Chemical	Citation
Petroleum marketing and distribution workers	Benzene, total hydrocarbons	Armstrong et al., 1996
Workers	Formaldehyde	Blair and Stewart, 1992
Paving workers	Bitumen and PAHs	Burstyn et al., 2000
Workers in the North Carolina Dusty Trades (1935- 1980)	Silica	Checkoway and Rice, 1992
Selected plant locations in Tyler, Texas	Asbestos	Corn, 1992
Painesville, Ohio chromate production workers	Hexavalent chromium	Crump et al., 2002 Luippold et al., 2002 Proctor et al., 2002a, b
Asbestos workers	Asbestos	Esmen and Corn, 1999
Automative parts industry	Machining fluids and aerosols	Hallock et al., 1994
Embalmers from funeral homes	Formaldehyde	Hornung et al., 1996
Indoor wood use	Isothiazolone-treated wood	Jayjock et al., 1995
Manufacturersand sprayers	Phenoxy herbicides, chlorophenols, dioxin	Kaupinnen et al., 1994
Synthetic rubber industry	Butadiene, styrene, benzene	Macaluso et al., 1996
Rocky Flats nuclear weapons plant	Plutonium	Mongan et al., 1996 a, b Ripple, 1992

Table 11-9 (Cont'd)
Selected retrospective exposure assessments

Site or Population Studied	Chemical	Citation
Rubber hydrochloride (Pliofilm) workers	Benzene	Paustenbach et al., 1992
Fiberglass and rock or wool slag plant workers	Man-made vitreous fiber (MMVF)	Quinn et al., 2001 Smith et al., 2001
Rocky Flats nuclear weapons plant	Plutonium and uranium	Ripple et al., 1996
Coal miners	Respirable dust	Seixas et al., 1993
Workers loading biocide into brominators	Halogen gas from bromine-based biocide	Shade and Jayjock, 1997
Silicon carbide production workers in Canada	Silicon carbide	Smith et al., 1984
	Respirable dust, hexane	Smith, 1987
Gasoline distribution workers, truck and marine vessels	Gasoline	Smith et al., 1993
Workers in acrylonitrile producing and using factories	Acrylonitrile	Stewart et al., 1995 Stewart et al., 1998
Workers from 300 industries	Formaldehyde	Stewart et al., 2000
Men in fertility clinics or seeking medical advice regarding infertility	Organic solvents, chromium	Tielemans et al., 1999
	Gasoline	Thomas et al., 1993
European carbon black manufacturing industry	Inhalable and respirable dust	Van Tongeren et al., 1997

Table 11-9 (Cont'd)
Selected retrospective exposure assessments

Site or Population Studied	Chemical	Citation
Workers in formaldehyde producing and using factories	Formaldehyde	Vetter et al., 1993
Oak Ridge Reservation	Radionuclides, radioiodine, mercury, PCBs	Widner et al., 1996
Rubber hydrochloride (Pliofilm) workers (1936-1976)	Benzene	Williams and Paustenbach, 2003
Hospital sterilizer workers	Ethylene oxide	Yager, 1991

holistic assessments like those conducted by environmental professionals. The reason industrial hygienists have not had to rely on risk assessment methods is due primarily to the existence of the TLVs® and other occupational exposure limits. While the environmental community tried to understand the health hazard posed by one of 2,000 different chemicals in one or more different environmental media (soil, water, dust, air, food), the industrial hygiene community had already conducted their own quantitative risk assessment when the TLVs® were developed. By assuming that dermal contact was negligible, hygienists have inferred that if the airborne concentrations were lower than the TLV®, then the risks to the worker were negligible.

To further understand why classic risk assessment was not a pressing issue to most hygienists, one needs to reflect on the four parts of a classic analysis. The first step, hazard identification, is simply identifying the chemical to which persons are exposed and then deciding if any adverse effects could occur under foreseeable conditions of exposure. This step is obvious in the practice of industrial hygiene and the *Documentation of the TLVs®* tells the professional which adverse effect has been observed in experimental animals and in workers. Thus, the first step was completed unless the chemicals did not have a TLV® or other criteria. The second step, dose-response assessment, also did not need to be conducted by most practitioners. Again, during its deliberations, the TLV® Committee assembled all of the relevant data, identified the no-observed-effect levels for animals and humans, then identified a dose which they considered acceptable. Thus, the dose-response assessment was completed. The third step, exposure assessment, was also functionally conducted because when the TLV® was established, it was assumed that exposure would occur only through inhalation and that exposure would occur about 8 to 10 hours/day for five days per week for about 40 years. No exposure assessment of dermal uptake was needed because contact was assumed to be negligible. The fourth step, risk characterization, was also completed for the hygienist as this information was conveyed in the *Documentation of the Threshold Limit Values and Biological Exposure Indices*.

In short, it is not surprising that most old and young industrial hygienists may ask themselves "Why do I care about a methodology that was developed to fill a major need in the environmental sciences?" Although there is some merit to embracing this position, there appears to be ample evidence why risk assessment methods should be well understood by occupational health professionals.

First, only about 800 chemicals currently have TLVs® and another 300 or so have had some type of OEL set for them. Thus, many hygienists, occupational physicians, and/or toxicologists have some responsibility to identify safe levels of exposure for the 1,000 other agents that can be found commonly in the workplace and perhaps for some of the other 80,000 chemicals that are less common. The risk assessment process is ideally suited to serve this purpose. Second, over the past 15 years industrial hygienists have become more and more involved in environmental, rather than classic occupational issues. If they are to function effectively in this arena, it will be very helpful for them to understand and be proficient at risk assessment. Third, retrospective exposure assessments will continue to evolve into an important adjunct to epidemiology studies (Ripple, 1992). Thus far, about 10 chemicals have been evaluated using dose-reconstruction or retrospective risk assessment (see Table 11-9) but medical and exposure information for as many as 300 chemicals now resides in computer files in corporations across the United States, which may be unloaded and analyzed over the next 10 years. As evidenced in the assessments conducted thus far of the pliofilm workers (benzene) (Paustenbach et al., 1992, 2002), those who worked at The Rocky Flats Nuclear Arsenal (Ripple, 1992), those exposed to butadiene (Cole et al., 1997), the stone workers (Smith et al., 1994), and by others, it is clear that risk assessment methods will be the basis for interpreting the vast amounts of historical exposure and medical data that have been collected and accumulated by the private sector over the past 30 to 45 years.

In short, health risk assessment is quite likely to continue to evolve into an integral part of the practice of health risk assessment. Undoubtedly, the profession will be better off as a result (Jayjock et al., 1999, 2000).

QUESTIONS

11.1. What are the four components of a health risk assessment?

11.2. Please describe in a sentence or two the kind of information that one should expect to find in each of the above mentioned components.

11.3. In the practice of industrial hygiene, what routine activity could be classified as being part of the "hazard identification process"?

11.4. In your daily activities as an industrial hygienist, how do you see the dose-response assessment being conducted when you assess the health risks to workers exposed to a particular scenario?

11.5. It has been said that exposure assessment is the specialty of industrial hygienists. Why do you think this was said and explain how the hygienist might improve the quality of his exposure assessment of workers?

11.6. It has been said that, "Dose-response is the most uncertain step in health risk assessment while exposure assessment has evolved into a rather complex art form." Please explain why our confidence in many dose-response assessments is so limited.

11.7. Risk characterization is generally considered to be the most important part of the assessment, yet it is the most difficult to perform. Why?

11.8. Physiologically-based pharmacokinetic (PB-PK) models are considered one of the most important improvements in the practice of toxicology that has occurred over the past 40 years. What are they and why are they thought to pose so many advantages over traditional approaches?

REFERENCES

Abdel-Rahman, M.S.; Kadry, A.M.: The use of uncertainty factors in deriving RfDs: Overview. Hum. Ecol. Risk Assess. 1(5): 614-625 (1995).

Abelson, P.H.: Exaggerated Carcinogenicity of Chemicals. Science 256:160 (1992).

Abelson, P.H.: Health Risk Assessment. Regul Toxicol Pharm 17(2 Pt. 1):219-223 (1993).

Ackerman, A.B.: Histologic Diagnosis of Inflammatory Skin Disease, p. 233-236. Lea and Febiger, Philadelphia, PA (1978).

Adams, R.M.: Allergic Contact Dermatitis. In: Occupational Skin Disease, 2nd ed., p. 26-31, R. Adams, Ed. WB Saunders Company, Philadelphia, PA (1990).

Agency for Toxic Substances and Disease Registry (ATSDR): Public Health Assessment Guidance Manual. Lewis Publishers, Atlanta, GA (1992).

Albert, R.E.: Carcinogen Risk Assessment in the United States. Crit Rev Toxicol 24:75-85 (1994).

Allen, B.C.; Crump, K.S.; Shipp, A.M.: Correlation between Carcinogenic Potency of Chemicals in Animals and Humans. Risk Anal 8:531-544 (1988).

Allen, B.C.; Kavlock, R.J.; Kimmel, C.A.; et al.: Dose response assessments for developmental toxicity: II. Comparison of genetic benchmark dose estimates with NOAELs. Fundam. Appl. Toxicol. 23: 487-495 (1994a)

Allen, B.C.; Kavlock, R.J.; Kimmel, C.A.; et al.: Dose response assessments for developmental toxicity: III. Statistical models. Fundam. Appl. Toxicol. 23: 496-509 (1994b).

Allen, B.; Kimmel, C.; Kavlock, R.; et al.: Dose-response Assessment for Development Toxicity II: Comparison of Generic Benchmark Dose Estimates with NOAELs. Fund Appl Toxicol 23:487-495 (1994).

American Conference of Governmental Industrial Hygienists: Threshold Limit Values for Chemical Substances and Physical Agents. ACGIH®, Cincinnati, OH (2001).

Ames, B.N.; Magaw, R.; Gold, L.S.: Ranking Possible Carcinogenic Hazards. Science 236:271-273 (1987).

Ames, B.N. (1987). Six Common Errors Relating to Environmental Pollution. Regul. Toxicol. Pharm. 7:379-383 (1987).

Ames, B.N.; Gold, L.S.; Willett, W.C.: The Causes and Prevention of Cancer. Proc Natl Acad Sci USA 92: 5258-5265 (1995).

Andersen, M.E.; MacNaughton, M.G.; Clewell, H.J.; et al.: Adjusting Exposure Limits for Long and Short Exposure Periods Using a Physiological Pharmacokinetic Model. Am Ind Hyg Assoc J 48(4):335-343 (1987a).

Andersen, M.E.; Clewell, H.J.; Gargas, M.L.; et al.: Physiologically-based Pharmacokinetics and the Risk Assessment for Methylene Chloride. Toxicol Appl Pharmacol 185-205 (1987b).

Andersen, M.E.: Quantitative Risk Assessment and Chemical Carcinogens in Occupational Environments. Appl Ind Hyg 3:267-273 (1991).

Andersen, M.E.; Clewell, H.J.; Gargas, M.L.; et al.: Physiologically-based Pharmacokinetic Modeling with Dichloromethane, Its Metabolite, Carbon Monoxide, and Blood Carboxyhemoglobin in Rats and Humans. Toxicol Appl Pharmacol 108:14-27(1991).

Andersen, M.E.; Clewell, H.J.; Krishnan, K.: Tissue Dosimetry, Pharmacokinetics Modeling, and Interspecies Scaling Factors. Risk Anal 15:533-537 (1995).

Anderson, B.D.; Higuchiand, W.I.; Raykar, P.V.: Heterogeneity Effects on Permeability: Partition Coefficient Relationships in Human Stratum Corneum. Pharm Res 5:566-573 (1988).

Anderson, E.: Quantitative Approaches in Use to Assess Cancer Risk. Risk Anal 3:277-295 (1983).

Anderson, F.E.: Cement and Oil Dermatitis: The Part Played by Chrome Sensitivity. Br J Dermatol 72:108-117 (1960).

Armstrong, T.W.; Pearlman, E.D.; Schnatter, A.R.; et al.: Retrospective Benzene and Total Hydrocarbon Exposure Assessment for a Petroleum Marketing and Distribution Worker Epidemiology Study. Am Ind Hyg Assoc J 57: 333-343 (1996).

Aylward, L.L.; Hays, S.M.; Karch, N.J.; et al.: Relative Susceptibility of Animals and Humans to the Cancer Hazard Posed by 2,3,7,8-Tetrachlorodibenzo-p-dioxin Using Internal Measures of Dose. Environ Sci Technol 30(12):3534-3543 (1996).

Bagdon, R.E.; Hazen, R.E.: Skin Permeation and Cutaneous Hypersensitivity as a Basis for Making Risk Assessments of Chromium as a Soil Contaminant. Environ Health Perspect 92:111-119 (1991).

Barber E.D.; Teetsel, N.M.; Kolberg, K.F.; et al.: A Comparative Study of the Rates of in Vitro Percutaneous Absorption of Eight Chemicals Using Rat and Human Skin. Fund Appl Toxicol 19:493-497 (1992).

Barnes, D.G.; Dourson, M.: Reference Dose (RfD): Description and Use in Health Risk Assessments. Regul Toxicol Pharm 8:471-486 (1988).

Barnes, D.G.; Daston, G.P.; Evans, J.S.; et al.: Benchmark Dose Workshop: Criteria for Use of a Benchmark Dose to Estimate A Reference Dose. Regul Toxicol Pharmacol 21:296-306 (1995).

Bartek, M.J.; LaBudde, J.A.; Maibach, H.I.: Skin Permeability in Vivo: Comparison in Rat, Rabbit, Pig, and Man. J Invest Dermatol 58:114-123 (1972).

Beech, J.A.: Estimated Worst-Case Trihalomethane Body Burden of a Child Using a Swimming Pool. Med Hypoth 6:303-307 (1983).

Begley, R.: Risk-based Remediation Guidelines: Take Hold. Environ Sci Technol 30(10):438A-441A (1996).

Berner, B.; Cooper, E.R.: Models of Skin Permeability. In: Transdermal Delivery of Drugs, Chapter 2, A.F. Kydonieu and B. Berner, Eds. CRC Press, Boca Raton, FL (1987).

Black, K.G.; Fenske, R.A.: Dislodgeability of Chlorpyrifos and Fluorescent Tracer Residues on Turf: Comparison of Wipe and Foliar Wash Sampling Techniques. Arch Environ Contam Toxicol 31:563-570 (1996).

Blair, A.; Stewart, P.A.: Do Quantitative Exposure Assessments Improve Risk Estimates in Occupational Studies of Cancer? Am. J. Ind. Med. 21: 53-63 (1992).

Bronaugh, R.L.; Stewart, R.F.; Congdon, E.R.; et al.: Methods for In Vitro Percutaneous Absorption Studies I.: Comparison with In Vitro Results. Toxicol Appl Pharmacol 62:474-480 (1982).

Brouwer, D.H.; van Hemmen, J.J.: Elements of a Sampling Strategy for Dermal Exposure Assessment (Abstract). First International Scientific Conference. International Occupational Hygiene Association, Brussels, Belgium (December 7-10, 1992).

Burmaster, D.E.; Maxwell, N.I.: Time and Loading—Dependence in the McKone Model for Dermal Uptake of Organic Chemicals from a Soil Matrix. Risk Anal 11:491-497 (1991).

Burmaster, D.E.; von Stackelberg, K.: Using Monte Carlo Simulations in Public Health Risk Assessments: Estimating and Presenting Full Distributions of Risk. J Expos Anal Environ Epid 1:491-521 (1991).

Burmaster, D.E.; Anderson, P.D.: Principles of Good Practice for the Use of Monte Carlo Techniques in Human Health and Ecological Risk Assessments. Risk Anal 14(4):477-481 (1994).

Burmaster, D.E.; Harris, R.: The magnitude of compounding conservatisms in Superfund risk assessments. Risk Anal. 13: 131-134 (1993).

Burrows, D.; Adams, R.M.: Metals. In: Occupational Skin Disease, 2nd ed., p. 349-386, R.M. Adams, Ed. W.B. Saunders Company, Philadelphia, PA (1990).

Burrows, D.; Calnan, C.D.: Cement Dermatitis II: Clinical Aspects. Trans St John Hosp Dermatol Soc 51:27-39 (1965).

Burrows, D.: Adverse Chromate Reactions on the Skin. In: Chromium: Metabolism and Toxicity, D. Burrows, Ed. CRC Press, Boca Raton, FL (1983).

Burstyn, I.; Kromhout, H.; Kauppinen, T.; et al.: Statistical Modelling of the Determinants of Historical Exposure to Bitumen and Polycyclic Aromatic Hydrocarbons Among Paving Workers. Ann. Occup. Hyg. 44: 43-56 (2000).

Butterworth, B.E.: Nongenotoxic Mechanisms in Carcinogenesis. Banbury Report 25. Cold Spring Harbor Press, New York, NY (1987).

Butterworth, B.E.: Consideration of Both Genotoxic and Nongenotoxic Mechanisms in Predicting Carcinogenic Potential. Mutat Res 239:117-132 (1990).

Butterworth, B.; Conolly, R.B.; Morgan, I.J.: A Strategy for Establishing Mode of Action of Chemical Carcenogens as a Guide for Approaches to Risk Assessments. Cancer Letters 93:129-146 (1995)

Butterworth, B.; Kideris, G.; Conolly, R.: The Chloroform Cancer Risk Assessment: A Mirror of Scientific Understanding. CIIT Activities 18(4), Research Triangle Park, NC (1998).

Byard, J.: Hazard Assessment of 1,1,1-Trichloroethane in Groundwater. In: The Risk Assessment of Environmental and Human Health Hazards: A Textbook of Case Studies, p. 331-334, D.J. Paustenbach, Ed. John Wiley and Sons, New York, NY (1989).

Calabrese, E.J.: Methodologic Approaches to Deriving Environmental and Occupational Health Standards. John Wiley and Sons, New York, NY (1978).

Calabrese, E.J.: Pollutants and High Risk Groups: The Biological Basis of Increased Human Susceptibility to Environmental and Occupational Pollutants. John Wiley and Sons, New York, NY (1978).

Calabrese, E.J.: Principles of Animal Extrapolation. John Wiley and Sons, New York, NY (1983).

Calabrese, E.J.; Baldwin, L.A.: Hormesis as a default parameter in RfD derivation. Hum. Exp. Toxicol. 17: 444-447 (1998).

Calabrese, E.; Kenyan, E.M.: Air Toxics and Risk Assessment. Lewis Publishers, Ann Arbor, MI (1991).

Calabrese, E.J.; Gilbert, C.E.: Lack of Total Independence of Uncertainty Factors (UF): Implications for Size of Total Uncertainty Factor. Regul Toxicol Pharm 17:44-51 (1993).

Calabrese, E.J.; Stanek, E.J.: A guide to interpreting soil ingestion studies II: Qualitative and quantitative evidence of soil ingestion. Regul. Toxicol. Pharmacol. 13: 278-292 (1991).

Calabrese, E.J.; Stanek, E.J.; Barnes, R.: Methodology to estimate the amount and particle size of soil ingested by children: Implications for exposure assessment at waste sites. Regul. Toxicol. Pharmacol. 24: 264-268 (1996).

Calabrese, E.J.; Stanek, E.J.; Gilbert, C.E.; et al.: Preliminary adult soil ingestion estimates: Results of a pilot study. Regul. Toxicol. Pharmacol. 12: 88-95 (1990).

Calabrese, E.J.: The frequency of U-shaped dose responses in the toxicological literature. Toxicol. Sciences 62: 330-338 (2001).

Calabrese, E.J.; Baldwin, L.A.: Scientific foundations of hormesis. Crit. Rev. Toxicol. 31(4,5) (2001).

California Department of Health Services (CDHS): Development of Applied Action Levels for Soil Contact: A Scenario for the Exposure of Humans to Soil in a Residential Setting. CDHS, Sacramento, CA (1986).

California Department of Toxic Substance Control: Guidance on Dermal Absorption Factors. Environmental Protection Agency, Sacramento, CA (1994).

Caplan, K.: The significance of wipe samples. Am Ind Hyg Assoc J 53(2): 70-75 (1993).

Caplan, K.; Lynch, J.: A Need and an Opportunity: AIHA Should Assume a Leadership Role in Reforming Risk Assessment. Amer Ind Hyg Assoc J 57:231-237 (1996).

Carnegie Commission on Science, Technology, and Government: Risk and the Environment, Improving Regulatory Decision Making. The Carnegie Corporation, New York, NY (1993).

Carson, R.: Silent Spring. Houghton Mifflin, Boston, MA (1962).

Center for Risk Analysis: Historical Roots of Health Risk Assessment. Harvard University School of Public Health, Cambridge, MA (1994).

Checkoway, H.; Rice, C.H.: Time-Weighted Averages, Peaks, and Indices of Exposure in Occupational Epidemiology. Am. J. Ind. Med. 21: 25-33 (1992).

Chemical Manufacturers Association (CMA): Chemicals in the Community: Methods to Evaluate Airborne Chemical Levels. CMA, Washington, DC (1988).

Cherrie, J.W.; Robertson, A.: Biologically Relevant Assessment of Dermal Exposure. Ann Occup Hyg 39:387-92 (1995).

Chester, G.: Revised Guidance Document for the Conduct of Field Studies of Exposure to Pesticides in Use. In: Methods of Pesticide Exposure Assessment, P.B. Curry, S. Iyengar, P. Maloney, and M. Maroni, Eds. Plenum Press, New York and London (1995).

Clewell, H.J.: The Application of Physiologically-based Pharmacokinetics Modeling in Human Health Risk Assessment of Hazardous Substances. Toxicol Letters 79:207-217 (1995).

Clewell, H.J.; Gentry, P.R.; Gearhart, J.M.; et al.: Considering Pharmacokinetics and Mechanistic Information in Cancer Risk Assessments for Environmental Contaminants: Examples with Vinyl Chloride and Trichlorethylene. Chemosphere 31:2561-2578 (1995).

Clewell, H.J.; Lee, T.S.; Carpenter, R.L.: Sensitivity of Physiologically-based Pharmacokinetics Models to Variation in Model Parameters: Methylene Chloride. Risk Anal 14:533-554 (1994).

Cole, P.; et al.: The Epidemology of 1,3 Butadiene. Conference on Butadiene and Isoprene, Blaine, WA (1997).

Commission on Risk Assessment and Risk Management (CRAM): Framework for Environmental Risk Management. Presidential/Congressional Commission on Risk Assessment and Risk Management. Vol. 1. Washington, DC (1997a).

Commission on Risk Assessment and Risk Management (CRAM): Risk Assessment and Risk Management in Regulatory Decisionmaking. Vol. 2. Washington, DC (1997b).

Copeland, T.L.; Holbrow, A.H.; Otani, J.M.; et al.: Use of Probabilistic Methods to Understand the Conservatism in California's Approach to Assessing Health Risks Posed by Air Contaminants. JAWMA 44:1399-1413 (1994).

Copeland, T.L.; Paustenbach, D.J.; Harris, M.A.; et al.: Comparing the Results of a Monte Carlo Analysis with EPA's Reasonable Maximum Exposed Individual (RMEI): A Case Study of a Former Wood Treatment Site. Regul Toxicol Pharm 18:275-312 (1993).

Corn, J.K.: Response to Occupational Health Hazards: A Historical Perspective. Van Nostrand Reinhold, New York, NY (1992).

Corn, M.: Historical Perspective on Approaches to Estimation of Inhalation Risk by Air Sampling. Am. J. Indust. Med. 21: 113-123 (1992).

Crump, C.; Crump, K.S.; Hack, E.; et al.: Dose-response analyssi of hexavalent chromium and lung cancer mortality. Risk Anal (2002).

Crump, K.S.; Hoel, D.G.; Langler, C.H.; et al.: Fundamental Carcinogenic Processes and Their Implications for Low Dose Risk Assessment. Cancer Res 36:2973-2979 (1976).

Crump, K.S.: Risk of Benzene-Induced Leukemia: A Sensitivity Analysis of the Pliofilm Cohort with Additional Follow-up and New Exposure Estimates. J Toxicol Environ Hlth 42:219- 2421 (1994).

Crump, K.: Calculation of Benchmark Doses from Continous Data. Risk Anal 15:79-85 (1995).

Crump, K.S.: The Linearized Multistage Model and the Future of Quantitative Risk Assessment. Human and Experimental Toxicol 15:787-798 (1996).

Crump, K.S.; Hoel, D.G.; Langley, C.H.; et al.: Fundamental Carcinogenic Processes and Their Implications for Low Dose Risk Assessment. Cancer Res 36:2973-2979 (1976).

Crump, K.S.; Clewell, H.J.; Andersen, M.E.: Cancer and non-cancer risk assessment should be harmonized. Hum. Ecol. Risk Assess. 3(4): 495-499 (1997).

Cullen, A.C.: Measures of Compounding Conservatism in Probabilistic Risk Assessment. Risk Anal 14(4):389-393 (1994).

Cullen, A.C.; Frey, H.C.: Probabilistic Techniques in Exposure Assessment. Plenum, NY (1999).

Doll, R.; Peto, R.: The Causes of Cancer: Quantitative Estimates of Avoidable Risks of Cancer in the United States Today. J Natl Cancer Inst 66:1191-1308 (1981).
Dorland, I.; Newman, W.A.: Dorland's Illustrated Medical Dictionary. W.B. Saunders Company, Philadelphia, PA (1994).
Dourson, M.L.; Felter, S.P.; Robinson, D.: Evolution of science-based uncertainty factors for noncancer risk assessment. Regul. Toxicol. Pharmacol. 24: 108-120 (1998).
Dourson, M.L.; Stara, J.F.: Regulatory History and Experimental Support of Uncertainty (Safety) Factors. Regul Toxicol Pharm 3:224-238 (1983).
Dourson, M.L.; Herizberg, R.C.; Hartung, R.; et al.: Novel Methods for the Estimation of Acceptable Daily Intake. Toxicol Ind Hlth 1:23-42 (1985).
Driver, J.H.; Konz, J.J.; Whitmyre, G.K.: Soil Adherence to Human Skin. Bull Environ Contam Toxicol 17(9):1831-1850 (1989).
Dupuis, G.; Benezra, C.: Allergic Contact Dermatitis to Simple Chemicals: A Molecular Approach. Marcel Dekker, New York, NY (1982).
Dutkiewicz, T.; Piotrowski, J.: Experimental Investigations on the Quantitative Estimation of Aniline Absorption in Man. Pure Appl Chem 3:319-323 (1961).
Dutkiewicz, T.; Tyras, H.: A Study of the Skin Absorption of Ethylbenzene in Man. Br J Ind Med 24:330-332 (1967).
European Center for Ecotoxicology and Toxicology of Chemicals (ECETOC): Strategy for Assigning a "Skin Notation." Revised Document No. 31. ECETOC, Brussels, Belgium (1993a).
European Center for Ecotoxicology and Toxicology of Chemicals (ECETOC): Percutaneous Absorption. Monograph No. 20. ECETOC, Brussels, Belgium (1993b).
Elias, P.M.; Feinghold, K.R.; Menon, G.K.; et al.: The Stratum Corneum Two-Compartment Model and Its Functional Implications. In: Pharmacology and the Skin: Skin Pharmacokinetics, vol. 1, B. Shroot and F. Schaefer, Eds. Karger Publishing, Basel, Switzerland (1987).
Environmental Protection Agency (EPA): Unfinished Business. EPA Science Advisory Board, Washington, DC (1988).
Environmental Protection Agency (EPA): Rationale for Air Toxics Control in Seven State and Local Agencies. EPA-450/5-86-005, NATICH. EPA, Washington, DC (1988).
Environmental Protection Agency (EPA): Risk Assessment Guidance for Superfund, vol. 1, part A: Human Health Evaluation Manual. Interim Final. Publication 540/1-89/002. EPA Office of Emergency and Remedial Response, Washington, DC (1989).
Environmental Protection Agency (EPA): NATICH Database Report on State, Local, and EPA Air Toxics Activities. EPA-450/3-89-29. EPA, Washington, DC (1989b).
Environmental Protection Agency (EPA): Reducing Risk: Setting Priorities and Strategies for Environmental Protection. EPA Science Advisory Board, Washington, DC (1990).
Environmental Protection Agency (EPA): Dermal Exposure Assessments: Principles and Applications. Publication EPA/600/8-91/011B. Exposure Assessment Group, EPA Office of Health and Environmental Assessment, Washington, DC (1992).
Environmental Protection Agency (EPA): Dermal Exposure Assessment: Principles and Applications. EPA 600/8-91/011B. EPA, Washington, DC (1992).
Environmental Protection Agency (EPA): Guidance for Risk Characterization. EPA Science Policy Council, Washington, DC (Feb 1995).
Environmental Protection Agency (EPA): The Use of the Benchmark Dose Approach in Health Risk Assessment. EPA/630/R-94//007. EPA Office of Research and Development, Washington, DC (Feb 1995).
Environmental Protection Agency (EPA): Proposed Guidance for Risk Assessment of Carcinogens. EPA, U.S. Government Printing Office, Washington, DC (1996).
Environmental Protection Agency (EPA): Draft Guidelines for Carcinogen Risk Assessment. Federal Register 61(79):17960-18011 (1996a).
Environmental Protection Agency (EPA): Exposure Factors Handbook, vol. 1 of 3: General Factors. Review Draft. EPA/600/P-95/002B. National Center for Environmental Assessment, EPA Office of Research and Development, Washington, DC (Aug1996b).

Environmental Protection Agency (EPA): IRIS: Integrated Risk Information System. DataBase, Washington, DC (1998).

Environmental Protection Agency (EPA): Guidelines for Carcinogen Risk Assessment (SAB review copy, July 1999). EPA, Washington, DC. http://www.epa.gov/ncea/raf/crasab.htm (1999a).

Environmental Protection Agency (EPA): Risk Assessment Guidance (RAGS3A) for Conducting Probabilistic Risk Assessment. EPA, Washington, DC (1999b).

Environmental Protection Agency (EPA): Risk Assessment Guidelines for Dermal Assessment. EPA, Washinogton, DC (1999c).

Environmental Protection Agency (EPA): Toward Integrated Environmental Decision-Making. EPA-SAB-EC-95-007, EPA, Washington, DC (2000a).

Environmental Protection Agency (EPA): Risk Characterization Guidance, http:/www.epa.gov/ord/spc/rchandbk.pdf., EPA, Washington, DC (2000b).

Esmen, N.A.: Analysis of strategies for reconstructing exposures. Appl Occup Environ Hyg 6(6): 488-494 (1991).

Esmen, N.A.; Corn, M.: Airborne Fiber Concentrations During Splitting Open Boxes and Boxing Bags of Asbestos. Toxicol. Indust. Health 14(6): 843-56 (1999).

Evans, J.S.; Graham, J.D.; Gray, G.M.; et al.: Distributional Approach to Characterizing Low-Dose Cancer Risk. Risk Anal 14(2):25-34 (1994a).

Evans, J.S.; Gray, G.M.; Sielken, R.L., Jr.; et al.: Use of Probabilistic Expert Judgment in Uncertainty Analysis of Carcinogenic Potency. Regul Tox Pharmacol 20:15-35 (1994b).

Environ: Fundamentals of Risk Assessment. Environ Corporation, Washington, DC (1994).

Federal Focus: Integrating Epidemiology with Risk Assessment. Washington, DC (1995).

Federal Register: Updating Permissible Exposure Limits (PELs) for Air Contaminants. 61(16):1948-1950. U.S. Government Printing Office, Washington, DC (January 24, 1996).

Feldmann, R.J.; Maibach, H.I.: Percutaneous Absorption of Steroids in Man. J Invest Dermatol 52:89-94 (1969).

Feldmann, R.J.; Maibach, H.I.: Percutaneous Penetration of Some Pesticides and Herbicides in Man. Toxicol Appl Pharmacol 28:126-132 (1974).

Fenske, R.A.: Dermal exposure assessment techniques. Ann. Occup. Hyg. 37:687-706 (1993).

Finkel, A.M.: Is Risk Assessment Really Too Conservative? Revising the Revisionists. Columbia J Env Law 14:427-467 (1989).

Finkel, A.M.: Edifying Presentation of Risk Estimates: Not As Easy As It Seems. J Policy Anal Manage 10:296-303 (1991).

Finkel, A.M.: A Second Opinion on an Economical Misdiagnosis: The Risky Prescriptions of Breaking the Vicious Circle. NYU Law J 3:295-381 (1995).

Finley, B.L.; Scott, P.; Paustenbach, D.J. Using an Uncertainty Analysis of Exposure to Contaminated Tapwater to Evaluate Health-Protective Cleanup Goals. Regul Toxicol Pharm 18:438-455 (1993).

Finley, B.L.; Scott, P.K.; Mayhall, D.A.: Development of A Standard Soil-to-Skin Adherence Probability Density Function for Use in Monte Carlo Analyses of Dermal Exposure. Risk Anal 14:555-569 (1994).

Finley, B.L.; Mayhall, D.A.: Airborne Concentrations of Chromium Due to Contaminated Interior Building Surfaces. Appl Occup Environ Hyg 9:433-441 (1994).

Finley, B.L.; Paustenbach, D.J.: The Benefits of Probabilistic Exposure Assessment: Three Case Studies Involving Contaminated Air, Water, and Soil. Risk Anal 14(1):53-73 (1994).

Finley, B.L.; Proctor, D.; Scott, P.; et al.: Recommended Distributions for Exposure Factors Frequently Used in Health Risk Assessment. Risk Anal 14(4):533-553 (1994).

Finley, B.L.; Paustenbach, D.J.: Using Applied Research to Reduce Uncertainty in Health Risk Assessment: Five Case Studies Involving Human Exposure to Chromium in Soil and Groundwater. J Soil Contamination 6:649-705 (1997).

Finley, B.L.; Kerger, B.D.; Katona, M.W.; et al.: Human Ingestion of Chromium (VI) in Drinking Water: Pharmacokinetics Following Repeated Exposure. Toxicol Appl Pharmacol 142:151-159 (1997).

Finley, B.L.; Scott, P.: Response to Letter by Kissel. Risk Anal 18G:9-12 (1998).

Fiora, G.; Specht, P.G.: Cost-Benefit Analysis and Risk: In the Hands of the Supreme Court. Professional Safety 37(2):24-28 (April 1, 1992).

Fischer, T.I.; Maibach, H.I.: Easier Patch Testing with TRUE-Testä. J Am Acad Dermatol 20:447-453 (1989).
Fischer, T.I.; Maibach, H.I.: The Thin Layer Rapid Use Epicutaneous Test (TRUE-Testä), a New Patch Test Method with High Accuracy. Br J Dermatol 112:63-68 (1985).
Fischer, T.I.; Maibach, H.I.: Amount of Nickel Applied with A Standard Patch Test. Contact Derm 11:285-287 (1984).
Fiserova-Bergerova V.; Pierce, J.T.; Droz, P.O.: Dermal Absorption Potential of Industrial Chemicals: Criteria for Skin Notation. Am J Ind Med 17:617-635 (1990).
Fiserova-Bergerova, V.: Relevance of Occupational Skin Exposure. Ann Occup Hyg 37:673-85 (1993).
Flynn, G.L.: Physicochemical Determinants of Skin Absorption. In: Principles of Route-to-Route Extrapolation for Risk Assessment, p. 93-127, T.R. Gerrity and C.J. Henry, Eds. Elsevier Science Publishing, New York, NY (1990).
Food Safety Council: Quantitative Risk Assessment. In: Food Safety Assessment, Chapter 11. Food Safety Council, Washington, DC (1980).
Fosbroke, D.E.; Kisner, S.M.; Myers, J.R.: Working Lifetime Risk of Occupational Fatal Injury. Am J Ind Med 31:459-467 (1997).
Frantz, S.W.: Instrumentation and Methodology for in Vitro Skin Diffusion Cells. In: Methods for Skin Absorption, p. 35-59, B.W. Kemppainen and W.G. Reifenrath, Eds. CRC Press, Boca Raton, FL (1990).
Freedman, D.A.; Zeisel, H.: From Mouse to Man: The Quantitative Assessment of Cancer Risk. Stat Sci 3:3-56 (1988).
Fries, G.F.; Paustenbach, D.J.: Evaluation of Potential Transmission of 2,3,7,8-tetrachlorodibenzo-p-dioxin-contaminated Incinerator Emissions to Humans via Foods. J Toxicol Environ Hlth 29:1-43 (1990).
Frosch, P.; Wissing, C.: Cutaneous Sensitivity to Ultraviolet Light and Chemical Irritants. Arch Dermatol Res 272:263-278 (1992).
Galer, D.M.; Leung, H.W.; Sussman, R.G.; et al.: Scientific and Practical Considerations for the Development of Occupational Exposure Limits (OELs) for Chemical Substances. Regul Toxicol Pharm (15):291-306 (1992).
Gargas, M.; Burgess, R.J.; Voisaro, G.E.; et al.: Partition Coefficients of Low Molecular Weight Volatile Chemicals in Various Liquids and Tissues. Toxicol Appl Pharm 98:87-99 (1989).
Gargas, M.L.; Norton, R.L.; Paustenbach, D.J.; et al.: Urinary excretion of chromium by humans following ingestion of chromium picolinate: Implications for biomonitoring. Drug Metab. Dispos. 22: 522-528 (1994a).
Gargas, M.L.; Norton, R.B.; Harris, M.A.; et al.: Urinary excretion of chromium following ingestion of chromite-ore processing residues in humans: Implications for biomonitoring. Risk Anal. 14: 1019-1024 (1994b).
Geiser, J.D.; Jeanneret, J.P.; Delacretaz, T.: Eczema Au Ciment et Sensibilization au Cobalt. Dermatologica 121:1-7 (1960).
Gold, L.S.; Ames, B.N.; Sloan, T.H.: Misconceptions About the Causes of Cancer. In D.J. Paustenbach, Ed., Human and Ecological Risk Assessment. Theory and Practice. John Wiley: New York, NY 1415-1460 (2002).
Gold, L.W.; Backman, G.M.; Hooper, N.K.; et al.: Ranking the Potential Carcinogenic Hazards to Workers from Exposures to Chemicals that are Tumorigenic in Rodents. Environ Hlth Perspect 76:211-219 (1987).
Goldstein, B.D.: The Maximally Exposed Individual: An Inappropriate Basis for Public Health Decision Making. Environ Forum Nov/Dec (13):13-16 (1989).
Gough, M.: How Much Cancer Can EPA Regulate Anyway? Risk Anal 10:1-6 (1990).
Graham, J.D.; Green, L.; Roberts, M.J.: In Search of Safety: Chemicals and Cancer Risks, p. 80-114. Harvard University Press, Cambridge, MA (1988).
Grandjean, P.A.; Berlin, M.G.; Penning, W.: Preventing Percutaneous Absorption of Industrial Chemicals: The "Skin" Notation. Am J Ind Med 14:97-107 (1988).
Grandjean, P.A.: Skin Penetration. In: Hazardous Chemicals at Work. Taylor and Francis Ltd., London (1990).
Greene, J.; Hays, S.; Paustenbach, D.: Basis for a Proposed Reference Dose (RfD) for dioxin of 1-5 pg/kg-day: A Weight of Evidence Evaluation of the Human and Animal Studies. J Toxicol Environ Health. In press.
Guy, R.H.; Hadgraft, J.; Maibach, H.I.: A Pharmacokinetic Model for Percutaneous Absorption. Int J Pharm 11:119-129 (1982).
Habicht, F. H.: Guidance on Risk Characterization for Risk Managers and Risk Assessors. Memorandum on Behalf of the Environmental Protection Agency's Risk Assessment Council. EPA Deputy Administrator to Assistant and Regional Administrators. EPA, Washington, DC (February 26, 1992).

Haber, L.; Dollarhide, J.; Maier, A.; et al.: Noncancer risk assessment: Principles and practice in environmental and occupational settings. Patty's Toxicology, Vol. 9, Ch. 5. E. Bigham, B. Cohrssen, and C. Powell, Eds., John Wiley and Sons, New York, NY (2001).

Hakkert, B.C.; Stevenson, H.; Bos, P.J.M.; et al.: Methods for the Establishment of Health-Based Recommended Occupational Exposure Limits for Existing Substances. Zeist: TNO-Report 96,463;1-16 (1996).

Hall, K.J.; Helman, G.: In Vitro Percutaneous Absorption in Mouse Skin: Influence of Skin Appendages. Toxicol Appl Pharm 94:93-103 (1988).

Hallock, M.F.; Smith, T.J.; Woskie, S.R.; et al.: Estimation of Historical Exposure to Machining Fluids in the Automobile Industry. Am. J. Ind. Med. 26: 621-634 (1994).

Hamilton, A.: Industrial Toxology. Little, Brown, and Company, Boston, MA (1928).

Hansen, C.M.: The Absorption of Liquids into the Skin. Report T3-82. Scandinavian Painting Ink Research Institute, Horsholm, Denmark (1982)

Hart, W.L.; Reynolds, R.C.; Krasavage, W.J.; et al.: Evaluation of Developmental Toxicity Data: A Discussion of Some Pertinent Factors and A Proposal. Risk Anal 8:59-70 (1988).

Hartwell, J.; Graham, J.: The Greening of Industry. Harvard University Press, Cambridge, MA (1997).

Hattis, D.; Burmaster, D.: Assessment of variability and uncertainty distributions for practical risk analyses. Risk Anal. 14: 713-729 (1994).

Hayes, A.W.: Principles of Toxicology, 4th Ed. Taylor and Francis, Philadelphia, PA (2001).

Health Effects Assessment Summary Tables (HEAST): F.Y. 1997 Update. EPA-540-R-97-036. Office of Solid Waste and Emergency Response, U.S. Environmental Protection Agency, Washington, DC (1997).

Henschler, D.: The Concept of Occupational Exposure Limits. Sci Tot Environ 101:0-16 (1991).

Hewett, P.; Ganser, G.H.: Simple procedures for calculating confidence intervals around the sample mean and exceedance fraction derived from lognormally distributed data. Appl. Occup. Environ. Hyg. 12(2): 132-142 (1997).

Hoel, D.: Cancer Risk Models for Ionizing Radiation. Environ Hlth Perspect 76:121-124 (1987).

Holland, C.D.; Sieklen, R.L., Jr.: Quantitative Cancer Modeling and Risk Assessment. Prentice Hall, Englewood Cliffs, NJ (1993).

Holmes, K.K.; Kissel, J.C.; Richter, K.Y.: Investigation of the Influence of Oil on Soil Adherence to Skin. J Soil Contam 5:301-308 (1996).

Hornung, R.W.; Herrick, R.F.; Stewart, P.A.; et al.: An Experimental Design Approach to Retrospective Exposure Assessment. Am. Ind. Hyg. Assoc. J. 57(3): 251-256 (1996).

Horowitz, S.B.; Finley, B.L.: Using Human Sweat to Extract Chromium from Chromite Ore Processing Residue: Applications to Setting Health-Based Cleanup Levels. J Toxicol Environ Hlth 40:585-599 (1994a)

Horowitz S.B.; Finley, B.L.: Setting Health Protective Soil Concentrations for Dermal Contact Allergens: A Proposed Methodology. Regul Toxicol Pharm 19:31-47 (1994b).

Hrudey, S.E.; Chen, W.; Rousseaux, C.: Bioavailability. CRC-Lewis Publishers, New York, NY (1996).

Hueber, F.; Wepierre, J.; Schaefer, H.: Role of Transepidermal and Transfollicular Routes in Percutaneous Absorption of Hydrocortisone and Testosterone: In Vivo Study in the Hairless Rat. Skin Pharm 5:99-107 (1992).

Huff, J.; Haseman, J.; Rall, D.: Scientific Concepts, Value, and Significance of Chemical Carcinogenesis Studies. Ann Rev Pharm Toxicol 31:621-652 (1991).

Integrated Risk Information System (IRIS): National Library of Medicine. Bethesda, MD (1996).

Irish, D.: Patty's Industrial Hygiene and Toxicology, vol. 2. John Wiley and Sons, New York, NY (1959).

Jarabek, A.M.; Menache, M.G.; Overton, J.H., Jr.; et al.: The U.S. Environmental Protection Agency's Inhalation RFD Methodology: Risk Assessment for Air Toxics. Toxicol Ind Hlth 6:279-301 (1990).

Jayjock, M.A.: Back pressure modeling of indoor air concentrations from volatilizing sources. Am Ind Hyg Assoc J 55(3):230–-235 (1994).

Jayjock, M.A.; Doshi, D.R.; Nungesser, E.H.; et al.: Development and evaluation of source/sink model of indoor air concentrations from isothioazolone-treated wood used indoors. Am Ind Hyg Assoc J 56(6):546–557 (1995).

Jayjock, M.A.; Hawkins, N.C.: A proposal for improving the role of exposure modeling in risk assessment. Am Ind Hyg Assoc J 54(12):733–741 (1993).

Jayjock, M.A.; Lynch, J.R.; Nelson, D.I.: Risk Assessment: Principles for the Industrial Hygienist. ACGIH®, Cincinnati, OH (2000).

Jo, W.K.; Weisel, C.P.; Lioy, P.J.: Routes of Chloroform Exposure and Body Burden from Showering with Chlorinated Tap Water. Risk Anal 10:575-580 (1988).

Kauppinen, T.P.; Pannett, B.; Marlow, D.A.; et al.: Retrospective Assessment of Exposure Through Modeling in a Study on Cancer Risks Among Workers Exposed to Phenoxy Herbicides, Chlorophenols, and Dioxins. Scand J Work Environ Hlth 20:262–271 (1994).

Kerger, B.D.; Schmidt, C.E.; Paustenbach, D.J.: Assessment of airborne exposure to trihalomethanes from tap water in residential showers and baths. Risk Anal. 20: 637-651 (2000).

Kezic, S.; Mahieu, K.; Monster, A.C. et al.: Dermal Absorption of Vapors and Liquid 2-Methoxyethanol and 2-Ethoxyethanol in Volunteers. Occup Environ Med 54:38-43 (1997).

Kimbrough, R.D.; Falk, H.; Stehr, P.; et al.: Health Implications of 2,3,7,8-Tetrachlorodibenzo-P-Dioxin (TCDD) Contamination of Residential Soil. J Toxicol Environ Hlth 14:47-93 (1984).

Kimmel, C.A.; Kavlock, R.J.; Allen, B.C.; et al.: Benchmark Dose Concept Applied to Data from Conventional Developmental Toxicity Studies. International Congress of Toxicology - VII. Seattle, WA (July 2-6, 1995).

Kissel, J.C.; Richter, K.Y.; Fenske, R.A.: Field Measurement of Dermal Soil Loading Attributable to Various Activities: Implications for Exposure Assessment. Risk Anal 16(1):115-125 (1996a).

Kissel, J.C.; Richter, K.Y.; Fenske, R.A.: Factors Affecting Soil Adherence to Skin in Hand-Press Trails. Bull Environ Contam Toxicol 56:722-728 (1996b).

Klain, G.J.; Black, K.E.: Specialized Techniques: Congenitally Athymic (nude) Animal Models. In: Methods for Skin Absorption, p. 165-174, B.W. Kemppainen and W.G. Reifenrath, Eds. CRC Press, Boca Raton, FL (1990).

Klaassen, C.D.: Casarett and Doull's Toxicology: The Basic Science of Poisons, 6th ed. McGraw-Hill, New York, NY (2000).

Knaak, J.B.; Iwata, Y.; Maddy, K.T.: The Worker Hazard Posed by Reentry Into Pesticide-Treated Foliage: Development of Safe Reentry Times, with Emphasis on Chlorhiophos and Carbosulfan. In: The Risk Assessment of Environmental Hazards: A Textbook of Case Studies, p. 797-842, D.J. Paustenbach, Ed. John Wiley and Sons, New York, NY (1989).

Knaak, J.B.; Dary, C.C.; Patterson, G.; et al.: The worker hazard posed by reentry into pesticide-treated foliage: Reassessment of reentry intervals using foliar residue transfer-percutaneous absorption PB-PK/PD models, with emphasis on isofenphos and parathion. In: Human and ecological risk assessment: Theory and practice, D.J. Paustenbach, Ed. John Wiley and Sons, New York, NY (2002).

Kodell, R.L.; Chen, J.J.: Reducing Conservatism in Risk Estimation for Mixtures of Carcinogens. Risk Anal 14(3):327-332 (1994).

Kodell, R.L.; Howe, R.B.; Chen, J.J.; et al.: Mathematical Modeling of Reproductive and Developmental Toxic Effects for Quantitative Risk Assessment. Risk Anal 11:583-590 (1991).

Krewski, D.; Brown, C.; Murdoch, D.: Determining Safe Levels of Exposure: Safety Factors or Mathematical Models. Fund Appl Toxicol 4:383-394 (1984).

Krewski, D.; Zhu, Y.: A Simple Data Transformation for Estimation Benchmark Doses in Development Toxicity Experiment. Risk Anal 15:29-39 (1995).

Krewski, D.; Murdoch, D.; Withey, J.R.: Recent Developments in Carcinogenic Risk Assessment. Health Physics 57(1):313-325 (1989).

Krewski, D.; Thorslund, T.; Withey, J.: Carcinogenic Risk Assessment of Complex Mixtures. Toxicol Ind Hlth 5:851-867 (1989).

Krishnan, K.; Anderson, M.: Physiologically-based pharmacokinetic modeling in toxicology. Principles and Methods in Toxicology, Ch. 5, A.W. Hayes, Ed. Taylor and Francis, Philadelphia, PA (2001).

Krivanek, N.; et al.: Monomethylformide Levels in Human Urine After Repetitive Exposure to Dimethylformamide. J Occup Med 20:179-187 (1978).

Lammintausta, K.H.; Maibach, H.I.: Contact Dermatitis Due to Irritation. In: Occupational Skin Disease, 2nd ed., p. 1-3, R.M. Adams, Ed. W.B. Saunders Company, Philadelphia, PA (1990).

Lavy, T.; Shepard, J.; Bouchard, D.: Field Worker Exposure and Helicopter Spray Pattern of 2,4,5-T. Bull Environ Contam Toxicol 24(1):90-96 (1980).

Lavy, T.; Walstad, J.; Flynn, R.; et al.: (2,4-Dichlorophenoxy) Acetic Acid Exposure Received by Aerial Application Crews During Forest Spray Operations. J Agric Food Chem 30:375-361 (1982).

Lehmann, A.J.; Fitzhugh, O.G.: 100-Fold Margin of Safety. Q Bull Assoc. Food Drug Office U.S. 18:33-35 (1954).

Lepow, M.L.; Bruckman, L.; Gillette, M.; et al.: Investigations Into Sources of Lead in the Environment of Urban Children. Environ Res 10:415-426 (1975).

Leung, H.W.; Paustenbach, D.J.: Setting Occupational Exposure Limits for Irritant Organic Acids and Bases Based on Their Equilibrium Dissociation Constants. Appl Ind Hyg 3:115-118 (1988).

Leung, H.W.: Development and Utilization of Physiologically-based Pharmacokinetic Models for Toxicological Applications. J Toxicol Environ Hlth 32:247-267 (1991).

Leung, H.W.; Paustenbach, D.J.: Techniques for Estimating the Percutaneous Absorption of Chemicals due to Environmental and Occupational Exposure. Appl Occup Environ Hyg 9(3):187-197 (1994).

Leung, H.W.; Paustenbach, D.J.: Physiologically-based Pharmacokinetic and Pharmacodynamic Modeling in Health Risk Assessment and Characterization of Hazardous Substances. Toxicol Letters 79:55-65 (1995).

Leung, H.W.; Auletta, C.S.: Evaluation of Skin Sensitization and Cross-Reaction of Nine Alkyene Amines in the Guinea Pig Maximization Test. Cut Ocular Toxicol 16:189-195 (1997).

Leung, H.W.: Physiologically-based pharmacokinetic models. In: General and Applied Toxicology, 2nd ed., pp. 141-154. Ballantyne, Marrs, and Syverson, Eds. MacMillan, London (2000).

Levin, H.M.; Brunner, N.J.; Ratner, L.T.: Lithographer's Dermatitis. JAMA 169:566-569 (1959).

Lewis, S.C.; Lynch, J.R.; Nikiforov, A.I.: A New Approach to Deriving Community Exposure Guidelines from No Observed Effect levels. Regul Toxicol Pharm 11:314-330 (1990).

Lioy, P.J.; Wainman, T.; Weisel, C.: A Wipe Sampler for the Quantitative Measurement of Dust on Smooth Surfaces: Laboratory Performance Studies. J Exp Anal Environ Epidemiol 3:315-320 (1993).

Lioy, P.J.; Yiin, L.M.; Adgate, J.; et al.: The Effectiveness of a Home Cleaning Intervention Strategy in Reducing Potential Dust and Lead Exposures. J Exp Anal Environ Epidem 8(1): 17-35 (1998).

Lioy, P.; Lippmann, M.: Toxic Air Pollutant Guidelines. ACGIH®, Cincinnati, OH (1988).

Luippold, R.S.; Mundt, K.A.; Panko, J.P.; et al.: Lung cancer mortality among chromate workers for risk assessment. Occ and Environ Med (submitted) (2002).

Lynch, J.: Biological Monitoring. Wiley Publishing, New York, NY (1994).

Macaluso, M.; Larson, R.; Delzell, E.; et al.: Leukemia and cumulative exposure to butadiene, styrene and benzene among workers in the synthetic rubber industry. Toxicol 113:190–202 (1996).

Maddaloni, M.; Lolacono, N.; Manton, W.; et al.: Bioavailability of soilborne lead in adults, by stable isotope dilution. Environ Hlth Persp 106 Suppl 6:1589–1594 (1998).

Madl, A.; Paustenbach, D.J.: Airborne concentration of benzene due to diesel locomotive exhaust in a roundhouse. J Toxicol Environ Hlth (in press) (2002).

Maibach, H.I.; Feldmann, R.J.; Milby, T.H.; et al.: Regional Variation in Percutaneous Penetration in Man. Arch Environ Hlth 23:208-211 (1971).

Mantel, N.; Bryan, W.R.: "Safety" Testing of Carcinogenic Agents. J Natl Cancer Inst (U.S.) 27:455-460 (1961).

Mantel, N.; Schneiderman, M.A.: Estimating "Safe" Levels of Hazardous Undertaking. Cancer Res 35:1379-1386 (1975).

Marks, J.G.; DeLeo, V.A.; et al.: North American Contact Dermatitis Group Standard Tray Patch Test Results (1992-1994). Am J Contact Derm 6(3):160-165 (1995).

Marks, J.G.; DeLeo, V.A.: Contact and Occupational Dermatology. Mosby Year Book, St. Louis, MO (1992).

Marks, R.M.; Barton, S.P.; Edwards, C., Eds.: The Physical Nature of the Skin. MTP Press, Lancaster, PA (1988).

Marlow, D.; Sweeney, M.H.; Fingerhut, M.: Estimating the Amount of TCDD Absorbed by Workers Who Manufactured 2,4,5-T. Presented at the 10th Annual International Dioxin Meeting, Bayreuth, Germany (1990).

Marzulli, F.N.; Tregear, R.T.: Identification of a Barrier Layer in the Skin. J Physiol 157:52-53 (1961).

Mason, J.W.; Dershin, H.: Limits of Occupational Exposure in Chemical Environments Under Novel Work Schedules. J Occup Med 18:603-607 (1976).

Mattie, D.R.; Bates, G.D.; Jepson, G.W.; et al.: Determination of Skin: Air Partition Coefficients for Volatile Chemicals: Experimental Method and Applications. Fund Appl Toxicol 22:51 (1994).

Maxim, D.: Problems Associated with the Use of Conservative Assumptions in Exposure and Risk Analysis. In: The Risk Assessment of Environmental and Human Health Hazards: A Textbook of Case Studies, pp. 526-560. D.J. Paustenbach, Ed. John Wiley and Sons, New York, NY (1989).

McArthur, B.: Dermal Measurement and Wipe Sampling Methods: A Review. Appl Occup Environ Hyg 7:599-606 (1992).

McDougal, J.N.; Grabau , J.H.; Dong, L.; et al.: Inflammatory Damage to Skin by Prolonged Contact with 1,2-Dichlorobenzene and Chloropentafluorobenzene. Microsc Res Tech 37(3): 214-220 (1997)

McDougal, J.N.; Jepson, G.W.; Clewell, H.J.; et al.: Dermal Absorption of Organic Chemical Vapors in Rats and Humans. Fund Appl Toxicol 14:299-308 (1990).

McKone, T.E.: Dermal Uptake of Organic Chemicals from a Soil Matrix. Risk Anal 10:407-419 (1990).

McKone, T.E.; Bogen, K.T.: Predicting the Uncertainties in Risk Assessment. Environ Sci Technol 25:16-74 (1991).

McKone, T.E.; Bogen, K.T.: Uncertainties in Health-Risk Assessment: An Integrated Case Study Based on Tetrachloroethylene in California Groundwater. Regul Toxicol Pharm 15:86-103 (1992).

Methner, M.M.; Fenske, R.A.: Pesticide Exposure during Greenhouse Applications, Part 1: Dermal Exposure Reduction due to Directional Ventilation and Worker Training. Appl Occup Environ Hyg 9:560-566 (1994a).

Methner, M.M.; Fenske, R.A.: Pesticide Exposure during Greenhouse Applications, Part II: Chemical Permeation Through Protective Clothing in Contact with Treated Foliage. Appl Occup Environ Hyg 9:567-574 (1994b).

Michaels, A.S.; Chandrasekaran, S.K.; Shaw, J.E.: Drug Permeation Through Human Skin: Theory and In Vitro Experimental Measurement. AIHce 19:985-996 (1975).

Michaud, J.M.; Huntley, R.A.; Paustenbach, D.J.: PCB and Dioxin Re-entry Criteria for Building Surfaces and Air. J Exposure Anal Environ Epidemiol 4:197-227 (1994).

Mongan, T.R.; Ripple, S.R.; Brorby, G.P.; et al.: Plutonium releases from the 1957 fire at Rocky Flats. Hlth Phys Soc 71(4):510–521 (1996a).

Mongan, T.R.; Ripple, S.R.; Winges, K.D.: Plutonium releases from the 903 Pad at Rocky Flats. Hlth Phys Soc 71(4):522–531 (1996b).

Moolgavkar, S.H.; Dewangi, A.; Venson, D.J.: A Stochastic Two-Stage Model for Cancer Risk Assessment: The Hazard Function and the Probability of Tumor. Risk Anal 8:383-392 (1988).

Morgan, D.L.; Cooper, S.W.; Carlock, D.L.; et al.: Dermal Absorption of Neat and Aqueous Volatile Organic Chemicals in the Fischer 344 Rat. Environ Res 55:51-63 (1991).

Morgan, M.D.; Henrion, M.: Uncertainty: A Guide to Dealing with Uncertainty in Quantitative Risk and Policy Analysis. Cambridge Univ Press, Cambridge, MA (1990).

Mraz, J.; Nohova, M.: Percutaneous Absorption of N,N-Dimethylformamide in Humans. Int Arch Occup Environ Hlth 64:79-83 (1992).

National Academy of Sciences (NAS): Risk Assessment in the Federal Government: Managing the Process. National Academy Press, Washington, DC (1983).

National Research Coucil (NRC): Science and Judgement in Risk Assessment. National Academy Press, Washington, DC (1994).

National Research Council (NRC): Understanding Risk: Informing Decisions in a Democratic Society. National Academy Press, Washington, DC (1996).

Nethercott, J.R.: Practical Problems in the Use of Patch Testing in Evaluation of Patients with Contact Dermatitis. In: Current Problems in Dermatology, p. 101-123, W. Westin, Ed. Mosby Year Book, St. Louis, MO (1990).

Nethercott, J.; Finley, B.; Horowitz, S.; et al.: Safe Concentrations of Dermal Allergens in the Environment. New Jersey Med 91(10):694-697 (1994).

Nethercott, J.; Paustenbach, D.J.; Adams, R.; et al.: A Study of Chromium-Induced Allergic Contact Dermatitis with 54 Volunteers: Implications for Environmental Risk Assessment. Occup Environ Med 51(6):371-380 (1994).

New Jersey Department of Environmental Protection (NJDEP): Derivation of a Risk-based Chromium Level in Soil Contaminated with Chromite-Ore Processing Residue in Hudson County. NJDEP, Trenton, NJ (1990).

New Jersey Department of Environmental Protection (NJDEP): Risk Assessment of the ACD Potential of Hexavalent Chromium in Contaminated Soil Derivation of an Acceptable Soil Concentration. NJDEP, Trenton, NJ (1992).

North American Contact Dermatitis Group (NACDG): Preliminary Studies of the TRUE-Testä Patch Test System in the United States. J Am Acad Dermatol 21:841-843 (1984).

Nichols, A.L.; Zeckhauser, R.J.: The Perils of Prudence: How Conventional Risk Assessments Distort Regulations. Regul Toxicol Pharm 8:61-75 (1998).

Occupational Safety and Health Administration (OSHA): U.S. Congress (91st) S.2193. Public Law 91-596. Washington, DC (1970).

Occupational Safety and Health Administration (OSHA): OSHA Compliance Officers Field Manual. Department of Labor, Washington, DC (1979).

Occupational Safety and Health Administration (OSHA): Air Contaminants: Final Rule. Federal Register 54:2332-2983, Washington, DC (1989).

Occupational Safety and Health Administration (OSHA): Updating Permissible Exposure Limits (PELs) for Air Contaminants. Federal Register 61(16):1948-1950, Washington, DC (January 24, 1996a).

Office of Technology Assessment (OTA): Researching Health Risks. OTA-BBS-570. U.S. Congress, Office of Technology Assessment, Washington, DC (1993).

Ottoboni, M.A.: The Dose Makes the Poison: A Plain Language Guide to Toxicology, 2nd ed. Van Nostrand Reinhold, New York, NY (1991).

Park, C.N.; Snee, R.D.: Quantitative Risk Assessment: State-of-the-Art for Carcinogensis. Fund Appl Toxicol 3:320-333 (1983).

Paustenbach, D.J.: The history and biological basis of occupational exposure limits for chemical agents. In: Patty's Industrial Hygiene, 5th Ed., Vol. 3, R.L. Harris, Ed., pp. 1903-2000. John Wiley and Sons, New York, NY (2000).

Paustenbach, D.J.: Human and ecological risk assessment: Theory and practice. John Wiley and Sons, New York, NY (2002).

Paustenbach, D.J.: Assessment of the Developmental Risks Resulting from Occupational Exposure to Select Glycol Ethers Within the Semiconductor Industry. J Toxicol Environ Hlth 23:29-75 (1988).

Paustenbach, D.J.: Shortcomings in the Traditional Practice of Health Risk Assessments. Columbia J Environ Law 14:411-427 (1989a).

Paustenbach, D.J.: A Survey of Health Risk Assessment. In: The Risk Assessment of Environmental and Human Health Hazards: A Textbook of Case Studies, Chapter 1, p. 27-124, D.J. Paustenbach, Ed. John Wiley and Sons, New York, NY (1989b).

Paustenbach, D.J.: The Risk Assessment of Environmental Hazards: A Textbook of Case Studies. John Wiley and Sons, New York, NY (1989c).

Paustenbach, D.J.: Health Risk Assessment and the Practice of Industrial Hygiene. Am Ind Hyg Assoc J 51(7)339-351 (1990a).

Paustenbach, D.J.: Occupational Exposure Limits: Their Critical Role in Preventive Medicine and Risk Management-A Guest Editorial. Am Ind Hyg Assoc J 51:332-336 (1990b).

Paustenbach, D.J.: Jousting with Environmental Windmills. Letter to the Editor. Risk Analysis 13(1):13-15 (1993).

Paustenbach, D.J.: The Practice of Health Risk Assessment in the United States (1975-1995): How the U.S. and Other Countries Can Benefit from that Experience. Human Ecolog Risk Assess 1(1):29-79 (1995).

Paustenbach, D.J.: OSHA's Program for Updating the Permissible Exposure Limits (PELs): Can Risk Assessment Help "Move The Ball Forward?" Risk in Perspective 5(1):1-6 Harvard's Center for Risk Analysis, Boston, MA (1997).

Paustenbach, D.J.: Methods for Setting Limits for Acute and Chronic Toxic Ambient Air Contaminants. Appl Occup Environ Hyg 12(6):418-428 (1997).

Paustenbach, D.J.; Shu, H.P.; Murray, F.J.: A Critical Examination of Assumptions Used in Risk Assessment of Dioxin Contaminated Soil. Regul Toxicol Pharm 6:284-307 (1986).

Paustenbach, D.J.; Langner, R.R.: Corporate Occupational Exposure Limits: The Current State of Affairs. Am Ind Hyg Assoc J 47:809-818 (1986).

Paustenbach, D.J.; Clewell, H.J.; Gargas, M.L.: et al.: A Physiologically-based Pharmacokinetic Model for Carbon Tetrachloride. Toxicol Appl Pharm 96:191-211 (1988).

Paustenbach, D.J.; Meyer, D.M.; Sheehan, P.J.; et al.: An Assessment and Quantitative Uncertainty Analysis of the Health Risks to Workers Exposed to Chromium-Contaminated Soils. Toxicol Ind Hlth 7:159-196 (1991a).

Paustenbach, D.J.; Rinehart, W.E.; Sheehan, P.J.: The Health Hazards Posed by Chromium-Contaminated Soils in Residential and Industrial Areas: Conclusions of an Expert Panel. Regul Toxicol Pharm 13:195-222 (1991b).

Paustenbach, D.J.; Sheehan, P.J.; Lau, V.; et al.: An Assessment and Quantitative Uncertainty Analysis of the Health Risks to Workers Exposed to Chromium-Contaminated Soils. Toxicol Ind Hlth 7:159-196 (1991c).

Paustenbach, D.J.; Sheehan, P.J.; Paul, J.M.; et al.: Review of the ACD Hazard Posed by Chromium-Contaminated Soil: Identifying a Safe Concentration. J Toxicol Environ Hlth 37:177-207 (1992a).

Paustenbach, D.J.: Methods for Setting Limits for Acute and Chronic Toxic Ambient Air Contaminants. Appl Occup Environ Hyg 12(6): 418-428 (1997).

Paustenbach, D.J.: The U.S. EPA Science Advisory Board Evaluation (2001) of the EPA Dioxin Reassessment. Regul Toxicol Pharmacol 36:211-219 (2002).

Paustenbach, D.J.; Wenning, R.J.; Lau, V.; et al.: Recent Developments on the Hazards Posed by 2,3,7,8-Tetrachlorodibenzo-P-Dioxin in Soil: Implications for Setting Risk-Based Cleanup Levels at Residual and Industrial Sites. J Toxicol Environ Hlth 36:103-148 (1992b).

Paustenbach, D.J.; Price, P.E.; Bradshaw, R.D.; et al.: Re-evaluation of Benzene Exposure for the Pliofilm Workers (1939-1976). J Toxicol Environ Hlth 36:177-232 (1992).

Paustenbach, D.J.; Finley, B.L.; Long, T.F.: The Critical Role of House Dust in Understanding the Hazards Posed by Contaminated Soils. Internatl J Toxicol 16:339-362 (1997b).

Paustenbach, D.J.; Bruce, G.; Chrostowski, P.: Current Views on the Oral Bioavailability of Inorganic Mercury in Soil: Implications for Health Risk Assesments. Risk Anal 27(5):533-544 (1997d).

Paustenbach, D.J.; Jernigan, J.; Bass, R.; et al.: A Proposed Approach to Regulating Contaminated Soil: Identify Safe Concentrations for Seven of the Most Frequently Encountered Exposure Scenarios. Regul Toxicol Pharm 16:21-56 (1992).

Paustenbach, D.J.; Hays, S.; El-Sururi, S.; et al.: Comparing the Estimated Uptake of TCDD Using Exposure Calculations with the Actual Uptake: A Case Study of Residents of Times Beach, Missouri. In: Organohal. Comp. 34: 25-31. Proceedings of 17th International Symposium on Chlorinated Dioxins and Related Compounds. Indianapolis, IN (1997c).

Paustenbach, D.J.; Williams, P.R.D.; Warmerdam, J.: Reconstruction of Benzene Exposure for the Pliofilm Cohort (1936-1976) Using Monte Carlo Techniques. J. Tox. Environ. Health (2003).

Paustenbach, D.J.; Leung, H.W.; Rothrock, J.: Risk Assessment and the Practice of Occupational Medicine. In: Occupational Skin Disease, 2nd ed., Chapter 18, pp. 291-323. R. Adams, Ed. Taylor and Francis, New York, NY (1998).

Piotrowski, J.K.: Further Investigations on the Evaluation of Exposure to Nitrobenzene. Br J Ind Med 24:60-65 (1967).

Piotrowski, J.: Evaluation of Exposure to Phenol: Absorption of Phenol Vapor in the Lungs and Through the Skin and Excretion of Phenol in Urine. Br J Ind Med 28:172-178 (1971).

Piotrowski, J.: Exposure Tests for Organic Compounds in Industrial Toxicology. National Institute for Occupational Safety and Health (NIOSH), Cincinnati, OH (1977).

Pirila, V.: On the Role of Chromium and Other Trace Elements in Cement Eczema. Acta Derm Venereol 34:137-143 (1954).

Proctor, D.M.; Panko, J.P.; Liebig, E.W.; et al.: Exposure reconstruction for hexavalent chromium in a chromate production plant: 1940–1973. Appl Occup Environ Hyg (submitted) (2002a).

Que Hee, S.S.; Peace, B.; Scott, C.S.; et al.: Evolution of Efficient Methods to Sample Lead Sources, Such as House Dust and Hand Dust, in the Homes of Children. Environ Res 38:77-95 (1985).

Quinn, M.M.; Smith, A.O.; Youk, G.M.; et al.: Historical Cohort Study of US Man-Made Vitreous Fiber Production Workers: VIII. Exposure-Specific Job Analysis. J Occup Environ Med 43:824–834 (2001).

Raykar, P.V.; Fung, M.; Anderson, B.D.: The Role of Protein and Lipid Domains in the Uptake of Solutes by Human Stratum Corneum. Pharm Res 5:140-150 (1988).

Rayner, S.; Cantor, R.: How Fair is Safe Enough? The Cultural Approach to Sociental Technology Choice. Risk Anal 7(1):3-9 (1987).

Reitz, R.H.; Gargas, M.L.; Andersen, M.E.; et al.: Predicting Cancer Risk from Vinyl Chloride Exposure with a Physiologically-based Pharmacokinetic Model. Toxicol Appl Pharm 137:253-267 (1996).

Renwick, A.G.: The use of an additional safety or uncertainty factor for nature of toxicity in the estimation of acceptable daily intake and tolerable daily intake values. Regul. Toxicol. Pharmacol. 22: 250-261 (1995).

Rice, C.: Retrospective Exposure Assessment: A Review of Approaches and Directions for the Future. In: Industrial Hygiene Science Series: Exposure Assessment for Epidemiology and Hazard Control, pp. 185–197. S.M. Rappaport, T.J. Smith, Eds. Lewis Publishers, Chelsea, MI (1991).

Rietschel, R.L.; Marks, J.G.; Adams, R.M.: Preliminary Studies of the TRUE-Test Patch Test System in the United States. J Am Acad Dermatol 21:841-843 (1989).

Rietschel, R.L.; Fowler, J.F.: Fisher's Contact Dermatitis, 4th ed. Williams and Wilkins, Baltimore, MD (1995).

Ripple, S.R.: Looking Back at Nuclear Weapons Facilities: The Use of Retrospective Health Risk Assessments. Environ Sci Technol 26(7):1270–1277 (1992).

Ripple, S.R.; Widner, T.W.; Morgan, T.R.: Past Radionuclide Releases from Routine Operations at Rocky Flats. Hlth Phys 71(4):502–509 (1996).

Roach, S.A.; Rappaport, S.M.: But They Are Not Thresholds: A Critical Analysis of the Documentation of the Threshold Limit Values. Am J Ind Med 17:727-753 (1990).

Robinson, M.K.; Stotts, J.; Danneman, P.J.: A Risk Assessment Process for Allergic Contact Sensitization. Food Chem Toxicol 27:479-489 (1989).

Robinson, J.C.; Paxman, D.G.; Rappaport, S.M.: Implications of OSHA's Reliance on TLVs in Developing the Air Contaminants Standard. Am J Ind Med 19:3-13 (1991).

Rodricks, J.V.; Brett, S.M.; Wrenn, G.C.: Significant Risk Decisions in Federal Regulatory Agencies. Regul Toxicol Pharm 7:307-320 (1987).

Rook, A.; Wilkinson, D.S.; Ebling, F.J.; et al.: Textbook of Dermatology, p. 350-450. Blackwell Scientific, Oxford, England (1986).

Rosenthal, A.G.; Gray, M.; Graham, J.D.: Legislating Acceptable Cancer Risk from Exposure to Toxic Chemicals. Ecology Law Quarterly 190:269-362 (1992).

Rozman, K.K.; Doull, J.: Derivation of an Occupational Exposure Limit (OEL) for n-Propyl Bromide Using an Improved Methodology. App. Occup. Environ. Hyg. 17(10): 711-716 (2002).

Ruby, M.V.; Schoof, R.; Brattin, W.; et al.: Advances in evaluating the oral bioavailability of inorganics in soil for use in human health risk assessment. Environ. Sci. Technol. 33:21): 3697-3705 (1999).

Ruby, M.V.; Fehling, K.A.; Paustenbach, D.J.: Estimation of the oral bioaccessibility of dioxins/furans in weathered soil. US-Vietnam Scientific Conference on Human Health and Environmental Effects of Agent Orange/Dioxin, Hanoi, Vietnam, March 3–6, 2002 (2002).

Schwartz, L.; Tulipan, L.; Peck, S.M.: Occupational Diseases of the Skin. Lea and Febinger, Philadelphia, PA (1947).

Scott, P.K.; Harris, M.A.; Rabbe, D.E.; et al.: Background Air Concentrations of Cr(VI) in Hudson County, New Jersey: Implications for Setting Health-based Standards for Cr(VI) in Soil. J Air Waste Mgmt 47:592-600 (1997).

Scott, P.K.; Sung, H.; Finley, B.L.; et al.: Identification of an Accurate Soil Suspension/Dispersion Modeling Method for Use in Estimating Health-based Soil Cleanup Levels of Hexavalent Chromium in Chromite-Ore Processing Residues. J Air Waste Mgmt 47(7): 753-765 (1997).

Scott, R.C.; Batten, P.L.; Clowes, H.M.; et al.: Further Validation of an In Vitro Method to Reduce the Need for In Vivo Studies for Measuring the Absorption of Chemicals Through Rat Skin. Fund Appl Toxicol 19:484-492 (1992).

Scow, K.; Wechsler, A.E.; Stevens, J.; et al.: Identification and Evaluation of Waterborne Routes of Exposure from Other than Food and Drinking Water. EPA-440/4-79-016. Environmental Protection Agency (EPA), Washington, DC (1979).

Seidenari, S.: Skin Sensitivity, Interindividual Factors: Atopy. In: The Irritant Contact Dermatitis Syndrome, P.G.M. van der Valk, H.I. Maibach, P. van der Valk, Eds. CRC Press, New York, NY (1996).

Seixas, N.S.; Robbins, T.G.; Becker, M.: A Novel Approach to the Characterization of Cumulative Exposure for the Study of Chronic Occupational Disease. Am J Epi 137(4):463–471 (1993).

Shade, W.D.; Jayjock, M.A.: Monte Carlo uncertainty analysis of a diffusion model for the assessment of halogen gas exposure during dosing of brominators. Am Ind Hyg Assoc J 58(6):418–424 (1997).

Shatkin, J.A.; Brown, H.S.: Pharmacokinetics of the Dermal Route of Exposure to Volatile Organic Chemicals in Water: A Computer Simulation Model. Environ Res 56:90-108 (1991).

Sheehan, P.; Meyer, D.M.; Sauer, M.M.; et al.: Assessment of the Human Health Risks Posed by Exposure to Chromium-Contaminated Soils at Residential Sites. J Toxicol Environ Hlth 32:161-201 (1991).

Sheppard, S.C.; Evenden, W.G.: Contaminant Enrichment and Properties of Soil Adhering to Skin. J Environ Qual 23:604-613 (1994).

Shoaf, C.R.: Current Assessment Practices for Noncancer End Points. Environ Hlth Perspect 95:111-119 (1991).

Shu, H.P.; Teitelbaum, P.; Webb, A.S.; et al.: Bioavailability of Soil-Bound TCDD: Dermal Bioavailability in the Rat. Fund Appl Toxicol 10:648-654 (1988).

Sielken, R.L.: Some Issues in the Quantitative Modeling Portion of Cancer Risk Assessment. Regul Toxicol Pharm 5:175-181 (1985).

Silbergeld, E.K.: Risk Assessment: The Perspective and Experience of the U.S. Environmentalists. Environ Hlth Perspect 101:100-104 (1993).

Skog, E.; Wahlberg, J.W.: Patch Testing with Potassium Dichromate in Different Vehicles. Arch Dermatol 99:697-700 (1969).

Skrowronski, G.A.; Turkall, R.M.; Abdel-Rahman, M.S.: Soil Absorption Alters Bioavailability of Benzene in Dermally Exposed Male Rats. Am Ind Hyg Assoc J 49:506-511 (1988).

Smith, C.M.; Christian, D.C.; Kelsey, K.T.: Chemical Risk Assessment and Occupational Health. Auburn House,Westport, CT (1994).

Smith, R.L.: Use of Monte Carlo Simulation for Human Exposure Assessment at a Superfund Site. Risk Anal 14(4):433-439 (1994).

Smith, R.W.; Sahl, J.D.; Kelsh, M.A.; et al.: Task-based exposure assessment: analytical strategies for summarizing data by occupational groups. Am Ind Hyg Assoc J 58(6):402–412 (1997).

Smith, T.J.: Exposure assessment for occupational epidemiology. Am J Ind Med 12:249–268 (1987).

Smith, T.J.; Hammond, S.K.; Laidlaw, F.; et al.: Respiratory Exposures Associated With Silicon Carbide Production: Estimation of Cumulative Exposures for an Epidemiology Study. Br J Ind Med 41:100–108 (1984).

Smith, T.J.; Hammond, S.K.; Hallock, M.; et al.: Exposure Assessment for Epidemiology: Characteristics of Exposure. Appl Occup Environ Hyg 6(6):441–447 (1991).

Smith, T.J.; Hammond, S.K.; Wong, O.: Health effects of gasoline exposure. In: Exposure assessment for U.S. distribution workers. Environ Hlth Persp Suppl 101(Suppl 6):13–21 (1993).

Smith, T.J.; Quinn, M.M.; Marsh, G.M.; et al.: Historical Cohort Study of US Man-Made Vitreous Fiber Production Workers: VII. Overview of the Exposure Assessment. J Occup Environ Med 43(9):809–823 (2001).

Snyder, W.S.: Report of the Task Group on Reference Man. International Commission on Radiological Protection, Pub No 23. Pergamon Press, New York, NY (1975).

Squires, R.: Ranking Animal Carcinogens. Science 214:877-880 (1981).

Stern, A.H.; Freeman, N.C.G.; Plesan, P.; et al.: Residential Exposure to Chromium Waste—Urine Biological Monitoring in Conjunction with Environmental Exposure Monitoring. Environ Res 58:147-162 (1992).

Stewart, R.D.; Dodd, H.C.: Absorption of Carbon Tetrachloride, Trichloroethylene, Tetrachloroethylene, Methylene Chloride, and 1,1,1-Trichloroethane Through the Human Skin. Am Ind Hyg Assoc J 25:439-446 (1964).

Stewart, P.A.; Herrick, R.F.: Issues in Performing Retrospective Exposure Assessment. Appl Occup Environ Hyg 6: 421-427 (1991).

Stewart, P.A.; Triolo, H.; Zey, J.; et al.: Exposure Assessment for a Study of Workers Exposed to Acrylonitrile. II. A Computerized Exposure Assessment Program. Appl Occup Environ Hyg 10(8):698–706 (1995).

Stewart, P.A.; Lees, P.S.J.; Francis, M.: Quantification of historical exposures in occupational cohort studies. Scand J Work Environ Hlth 22:405–414 (1996).

Stewart, P.A.; Zaebst, D.; Zey, J.N.; et al.: Exposure Assessment for a Study of Workers Exposed to Acrylonitrile. Scand J Work Environ Hlth 24(Suppl 2):42–53 (1998).

Stewart, P.: Challenges to retrospective exposure assessment. Scand J Work Environ Hlth 25(6):505–510 (1999).

Stewart, P.A.; Carel, R.; Shairer, C.; et al.: Comparison of industrial hygienists' exposure evaluations for an epidemiologic study. Scan J Work Environ Hlth 26(1):44–51 (2000).

Stokinger, H.E.: The Case for Carcinogen TLVs Continues Strong. Occup Hlth Sfty 46:54-58 (1977).

Storrs, F.J.; Rosenthal, L.E.; Adams, R.M.; et al.: Prevalence and Relevance of Allergic Reactions in Patients Patch Tested in North America, 1984 to 1985. J Am Acad Dermatol 20:1038-1044 (1989).

Surber, C.; Wilhelm, K.P.; Maibach, H.I.; et al.: Partitioning of Chemicals into Human Stratum Corneum: Implications for Risk Assessment Following Dermal Exposure. Fund Appl Toxicol 15:99-107 (1990).

Tengs, T.O.; Adams, M.E.; Pliskin, J.S.; et al.: Five Hundred Life-Saving Interventions and Their Cost-Effectiveness. Risk Anal 15(3):369-390 (1995).

Thomas, J.S.; Hammond, S.K.; Wong, O.: Health Effects of Gasoline Exposure. 1.Exposure Assessment for U.S. Distribution Workers. Environ Hlth Perspect 101(Suppl 6):13–21 (1993).

Thompson, K.M.; Burmaster, D.E.; Crouch, E.A.C.: Monte Carlo Techniques for Quantitative Uncertainty Analysis in Public Health Risk Assessments. Risk Anal 12(1):53-63 (1992).

Tielemans, E.; Heederik, D.; Burdorj, A.; et al.: Assessment of Occupational Exposures in a General Population: Comparison of Different Methods. Occup Environ Med 56:145–151 (1999).

Travis, C.C.; Richter, S.A.; Crouch, E.A.C.; et al.: Cancer Risk Management: A Review of 132 Federal Regulatory Decisions. Environ Sci Technol 21:415-420 (1987).

Travis, C.C.; White, R.K.: Interspecies Scaling of Toxicity Data. Risk Anal 8:119-125 (1988).

Travis, C.C.; White, R.K.; Ward, R.C.: Interspecies Extrapolation of Pharmacokinetics. J Theoret Biol 142:285-304 (1990).

Treffel, P.; Muret, P.; Muret-Daniello, P.; et al.: Effect of Occlusion on In Vitro Percutaneous Absorption of Two Compounds with Different Physicochemical Properties. Skin Pharm 5:108-113 (1992).

U.S. Office of Science and Technology Policy (DDHS): Researching Health Risks. Office of Technology Asessment, Washington, DC (1993).

U.S. Superintendent of Documents: Home Page on the U.S. Government Printing Office Website: http://www.access.gpo.gov./su_doc/. (August 1996).

Umbreit, T.H.; Hesse, E.J.; Gallo, M.A.: Acute Toxicity of TCDD-Contaminated Soil from an Industrial Site. Science 232:497-499 (1986a).

Umbreit, T.H.; Hesse, E.J.; Gallo, M.A.: Comparative Toxicity of TCDD-Contaminated Soil from Times Beach, Missouri and Newark, New Jersey. Chemosphere 15:121-2124 (1986b).

Upadhye, M.; Maibach, H.I.: Influence of Area of Application of Allergen on Sensitization in Contact Dermatitis. Contact Derm 27:281-286 (1992).

van Hemmen, J.J.; Brouwer, D.H.: Assessment of Dermal Exposure to Chemicals. Sci Total Environ 168:131-41 (1995).

Van Tongeren, M.; Gardiner, K.; Calvert, I.; et al.: Efficiency of Different Grouping Schemes for Dust Exposure in the European Carbon Black Respiratory Morbidity Study. Occup Environ Med 54:714–719 (1997).

Vetter, R.; Stewart, P.A.; Dosemeci, M.; et al.: Validity of Exposure in One Job as a Surrogate for Exposure in a Cohort Study. Am J Ind Med 23(4):641–651 (1993).

Vickers, C.F.H.: Existence of Reservoir in the Stratum Corneum. Arch Dermatol 88:20-23 (1963).

Wahlberg, J.E.: Clinical Overview of Irritant Dermatitis. In: The Irritant Contact Dermatitis Syndrome, P.G.M. van der Valk, H.I. Maibach, P. van der Valk, Eds. CRC Press, New York, NY (1996).

Wang, R.; Schwetz, B.: An Evaluation System for Ranking Chemicals with Teratogenic Potential. Teratogen Carcinogen Mutagen 7:133-190 (1987).

Watanabe, K.; Bois, F.Y.; Zeise, L.: Interspecies Extrapolation: A Reexamination of Acute Toxicity Data. Risk Anal 12:301-310 (1992).

Weil, C.S.: Statistics versus Safety Factors and Scientific Judgment in the Evaluation of Safety for Man. Toxicol Appl Pharm 21:454-463 (1972).

Wepierre, J.; Marty, J.P.: Percutaneous Absorption of Drugs. Trends Pharm Sci 1:23-26 (1979).

Wester, R.C.; Maibach, H.I.; Sedik, L.; et al.: Percutaneous Absorption of Pentachlorphenol from Soil. Fund Appl Toxicol 20:68-71 (1993).

Wester, R.C.; Noonan, P.K.; Maibach, H.I.: Percutaneous Absorption of Hydrocortisone Increases with Long-Term Administration. Arch Dermatol 116:186-188 (1980).

Wester, R.C.; Noonan, P.K.: Relevance of Animal Models for Percutaneous Absorption. Internatl J Pharm 7:99-110 (1980).

Widner, T.E.; Ripple, S.R.; Buddenbaum, J.E.: Identification and Screening Evaluation of Key Historical Materials and Emission Sources at the Oak Ridge Reservation. Hlth Phys 71(4):457–469 (1996).

Williams, G.M.; Weisburger, J.H.: Chemical Carcinogenesis. In: Casarett and Doull's Toxicology: The Basic Science of Poisons, 4th ed., Chapter 5, p. 127-236, M.D. Amdur, J.D. Doull, and C.D. Klaassen, Eds. Pergamon Press, New York, NY (1991).

Williams, P.; Paustenbach, D.J.: Risk Characterization. In: Human and Ecological Risk Assessment. Theory and Practice. D.J. Paustenbach, Ed., pp. 293-366. John Wiley and Sons, New York, NY (2002).

Wilschut, A.; TenBerge, W.F.; Robinson, P.J.; et al.: Estimating Skin Permeation: The Validation of Five Mathematical Skin Permeation Models. Chemosphere 30:1275-1296 (1995).

Wilson, R.; Crouch, E.A.C.: Risk-Benefit Analysis, 2nd ed. Harvard University Press (2001).

Wilson, J.D.: Assessment of Low Exposure Risk From Carcinogens: Implications of The Knudson-Moolgavkar Two-Critical Mutation Theory in Biologically-based Models for Cancer Risk Assessment, p.275-288, C. Travis, Ed. Pergamon Press, New York, NY (1989).

Wright, R.W.: Evaluation of Contact Dermatitis using the TRUE-Testä Patch Test. J Ark Med Soc 88:271-272 (1991).

Wyzga, R.E.: The Role of Epidemiology in Risk Assessments of Carcinogens. Adv Mod Environ Toxicol 15:189-208 (1988).

Yager, J.: Development of Biological Monitoring as a Tool for Exposure Assessment. In: Industrial Hygiene Science Series: Exposure Assessment for Epidemiology and Hazard Control, pp. 115–130. S.M. Rappaport, T.J. Smith, Eds. Lewis Publishers, Chelsea, MI (1991).

Young, F.A.: Risk Assessment: The Convergence of Science and the Law. Regul Toxicol Pharm 7:179-184 (1987).

Zatz, J.L.: Computer Simulation Using Multicompartmented Membrane Models. In: Percutaneous Absorption: Mechanisms, Methodology, and Drug-Delivery, vol. 6, p. 165-181, R.L. Bronaugh and H.I. Maibach, Eds. Marcel Dekker, New York, NY (1985).

Zelger, J.: Zur klinik und pathogenses dis chromate ekzems. Arch Clin Exper Dermatol 18:499-542 (1964).

12

THE HISTORY AND BIOLOGICAL BASIS OF OCCUPATIONAL EXPOSURE LIMITS*

Dennis J. Paustenbach, PhD

INTRODUCTION

Over the past 50 years, many organizations in numerous countries have proposed occupational exposure limits (OEL) for airborne contaminants. The limits or guidelines that have gradually become the most widely accepted both in the United States and in most other countries are those issued annually by the American Conference of Governmental Industrial Hygienists (ACGIH®) known as Threshold Limit Values (TLVs®) (LaNier, 1984; Cook, 1986; ACGIH®, 2002).

The usefulness of establishing OELs for potentially harmful agents in the working environment has been repeatedly demonstrated since their inception (Stokinger, 1970; Cook, 1986; Doull, 1994). It has been claimed that whenever these limits have been implemented in a particular industry, no worker has been shown to have sustained serious adverse effects on his health as a result of exposure to these concentrations of an industrial chemical (Stokinger, 1981). Although this statement is arguable with respect to the acceptability of OELs for those chemicals established before 1980 and later found to be carcinogenic, there is little doubt that many individuals have avoided serious effects of workplace exposure due to their existence.

The contribution of OELs to the prevention or minimization of disease is now widely accepted, but for many years such limits did not exist, and even when they did, they were often not observed (Cook, 1945, 1986; Smyth, 1956; Stokinger, 1981; LaNier, 1984). It was, of course, well understood as long ago as the 15th century that airborne dusts and chemicals could bring about illness and injury, but the concentrations and lengths of exposure at which this might be expected to occur were unclear (Ramazinni, 1700).

* Portions of this chapter were published as Chapter 41 of *Patty's Industrial Hygiene*, 5th Edition, Volume 3. Edited by Robert Harris. John Wiley and Sons, New York, NY (2000).

As reported by Baetjer (1980), "early in this century when Dr. Alice Hamilton began her distinguished career in occupational medicine, no air samples and no standards were available to her, nor were they necessary. Simple observation of the working conditions and the illnesses and deaths of the workers readily proved that harmful exposures existed. Soon however, the need for determining standards for safe exposure became obvious."

Cook has reported that the earliest efforts to set an OEL were directed towards carbon monoxide, the toxic gas to which more persons are occupationally exposed than to any other. The work of Max Gruber at the Hygienic Institute at Munich was published in 1883. The paper described exposing two hens and 12 rabbits to known concentrations of carbon monoxide up to 47 hours over three days. Gruber stated that "the boundary of injurious action of carbon monoxide lies at a concentration on all probability of 500 parts per million, but certainly (not less than) 200 parts per million." In arriving at this conclusion, Gruber had also exposed himself. He reported no symptoms or uncomfortable sensations after three hours on each of two consecutive days at concentrations of 210 parts per million (ppm) and 240 ppm (Cook, 1986).

According to Cook (1986), the earliest and most extensive series of animal experiments on exposure limits were those conducted by K.B. Lehmann and others under his direction at the same Hygienic Institute where Gruber had done his work with carbon monoxide. The first publication in the series entitled *Experimental Studies on the Effect of Technically and Hygienically Important Gases and Vapors on the Organism* was a report on ammonia and hydrogen chloride gas. The report was 126 pages in length within Volume 5 of Archiv für Hygiene (Lehmann, 1886). This series of reports on animal experimentation with a large number of chemical substances by Lehmann and associates continued through Part 35 in Volume 83 (1914) followed by a final comprehensive paper of 137 pages on chlorinated hydrocarbons (Volume 116 (1936) of the German Archive). These reports became the standard against which others would be compared for nearly 30 years.

Kobert (1912) published one of the earlier tables of acute exposure limits. Concentrations for 20 substances were listed under the headings: (1) rapidly fatal to man and animals, (2) dangerous in 0.5 to one hour, (3) 0.5 to one hour without serious disturbances, and (4) only minimal symptoms observed (Cook, 1986). In his paper on *Interpretations of Permissible Limits*, Schrenk (1947) noted that the "values for hydrochloric acid, hydrogen cyanide, ammonia, chlorine, and bromine as given under the heading 'only minimal symptoms after several hours' in the foregoing Kobert paper agree with values as usually accepted in present-day tables of Maximum Allowable Concentrations [MACs] for reported exposures." However, values for some of the more toxic organic solvents, such as benzene, carbon tetrachloride and carbon disulfide, far exceeded those currently in use (Cook, 1986).

One of the first tables of exposure limits to originate in the United States was that published by the U.S. Bureau of Mines (Fieldner et al., 1921). Although its title does not reflect the content, the 33 substances listed are those encountered in workplaces. Cook (1986) also noted that most of the exposure limits through the 1930s, except dusts, were based on rather short animal experimentation. A notable exception was the study of chronic benzene exposure by

Greenburg of the U.S. Public Health Service conducted under the direction of a committee of the National Safety Council (NSC, 1926). From this, an acceptable OEL was derived based on long-term animal experimentation.

According to Cook (1986), for dust exposures, permissible limits established before 1920 were based on exposures of workers in the South African gold mines where the dust from drilling operations was high in crystalline free silica. The effects of the dust exposure were followed by periodic chest x-ray examination and the dust concentrations were monitored by an instrument known as a konimeter that collected a nearly instantaneous sample. In 1916, based on a correlation of these two sets of findings, an exposure limit of 8.5 million particles per cubic foot of air (mppcf) with an 80 to 90% quartz content was set for the dust (Report of Miners, 1916). Later, the level was lowered to 5 mppcf. Cook (1986) also reported that, in the United States, standards for dust (also based on exposure of workers) were recommended by Higgins et al. following a study at the southwestern Missouri zinc and lead mines in 1917. The initial level established for high quartz dusts was 10 mppcf, appreciably higher than was established by later dust studies conducted by the U.S. Public Health Service.

The most comprehensive list of OELs up to 1926 was that for 27 substances which was published in Volume 2 of International Critical Tables (Sayers, 1927). Sayers and Dalle Valle (1935) published a table giving physiological response to five levels of concentrations of 37 substances. The first four refer to acute effects, but the fifth is for the maximum allowable concentration for prolonged exposure. In 1930, the USSR Ministry of Labor issued a decree that included the first actual approval of workplace maximum allowable concentrations for USSR with a list of 12 industrial toxic substances. About this time, Lehmann and Flury (1938) and Bowditch et al. (1940) published papers that presented tables with a single value for repeated exposures to each substance.

As noted by Cook (1986), many of the exposure limits developed by Lehmann were included in the Henderson and Haggard monograph (1943) initially published in 1927 and a little later in Flury and Zernik's Schadilche Gase (1931). According to Cook (1986), this book acted as the bible on effects of injurious gases, vapors, and dusts in industrial exposures until Volume II of Patty's Industrial Hygiene and Toxicology (1949) was published.

Baetjer (1980) has reported that the first list of standards for chemical exposures in industry in the United States was called MAC, which was prepared in 1939 and 1940. They represented a consensus of the American Standard Association (ASA) and a number of industrial hygienists who had formed ACGIH® in 1938. These "suggested standards" were published in 1943 by James Sterner.

A committee of ACGIH® met in early 1940 to begin the task of identifying safe levels of exposure to workplace chemicals. They began by assembling all the data they could locate related to the degree of exposure to a toxicant and the likelihood of the exposure producing an adverse effect (Stokinger, 1981; LaNier, 1984). This task, as might be expected, was a formidable one. After much painstaking research and labor intensive documentation, the first set of values issued by ACGIH® was released in 1941 by this committee which was composed of Warren Cook, Manfred Boditch (reportedly America's first hygienist employed by industry), William Fredrick, Philip Drinker, Lawrence Fairhall, and Alan Dooley (Stokinger, 1981).

In 1941, a committee, designated as Z-37, of the American National Standards Institute (ANSI) (then known as the American Standards Association), developed its first Standard - carbon monoxide, with an acceptable value of 100 ppm. The committee issued separate bulletins through 1974 including exposure standards for 33 toxic dusts and gases.

At the Fifth Annual Meeting of ACGIH® in 1942, the newly appointed Subcommittee on Threshold Limit Values presented in its report a table of 63 toxic substances with the "maximum allowable concentrations of atmospheric contaminants" from lists furnished by the various state industrial hygiene units. The report contained the statement, "The table is not to be construed as recommended safe concentrations. The material is presented without comment" (Cook, 1986).

In 1945, a list of 132 industrial atmospheric contaminants with MACs was published by Cook. This is considered a landmark publication since it was thorough, included references on the original investigations, and provided the rationale leading to the values. The table included the then current values for six states, California, Connecticut, Massachusetts, New York, Oregon, and Utah, values presented as a guide for occupational disease control by the U.S. Public Health Service, and 11 other standards. In addition, Cook included a list of MACs that appeared best supported by the references to original investigations (Cook, 1986).

At the 1946 Eighth Annual Meeting of ACGIH®, the Subcommittee on Threshold Limit Values presented their second report with the values for 131 gases, vapors, dusts, fumes, mists, and 13 mineral dusts. As stated in the report, the values were "compiled from the list reported by the subcommittee in 1942, from the list published by Warren Cook in *Industrial Medicine* (1945), and from published values of the Z-37 Committee of ANSI. The committee emphasized that the "list of MAC values is presented ... with the definite understanding that it be subject to annual revision."

The overall impact of these efforts to develop quantitative limits to protect humans from the adverse effects of workplace air contaminants and physical agents could not have been anticipated by the early TLV® Committees. To their credit, even though toxicology was then only a fledgling science, their approach to setting limits has been shown to be generally correct, even by today's standards. For this reason, most of the techniques for setting limits established by this committee are still in use today (Cook, 1945; Smyth, 1956; LaNier, 1984; Stokinger, 1970; World Health Organization (WHO), 1977; Dourson and Stara, 1983; Doull, 1994; Leung, 2002). The principles they used to set OELs were similar to those later used to identify safe doses of food additives and pharmaceuticals (Lehmann and Fitzhugh, 1954).

From the perspective of the hygienist, engineer, and businessperson, there have been many benefits of setting OELs. The establishment of limits, by their very nature, implies that at some concentration or dose, exposure to a toxicant can be expected to be safe and pose no significant risk of harm to exposed persons. The key to the success of limits has not only been that they were established on solid scientific principles; rather, the setting of any goal gives a sense of purpose and direction to industrial, occupational, or medical programs, which prior to the TLVs®, had been difficult to evaluate. The setting of goals, such as controlling workplace concentrations below an occupational exposure limit, establishes an objective that can then be mutually pursued by the occupational health team, engineers, and management. History has shown that by introducing the concept of "safe level of exposure" and by establishing a type of "management by objectives," occupational health programs will establish and pursue a systematic program for reducing exposure (Paustenbach, 1990c).

Intended Use of OELs

The ACGIH® TLVs® and most other OELs used in the United States and some other countries are limits that refer to airborne concentrations of substances and represent conditions under which "it is believed that nearly all workers may be repeatedly exposed day-after-day without adverse health effects" (ACGIH®, 1997). In some countries, which will be discussed later, the OEL is set at a concentration that will protect virtually everyone. It is important to recognize that unlike some exposure limits for ambient air pollutants, contaminated water, or food additives set by other professional groups or regulatory agencies, exposure to the TLV® will not necessarily prevent discomfort or injury for everyone who is exposed (Adkins et al., 1990). ACGIH® recognized long ago that because of the wide range in individual susceptibility, a small percentage of workers may experience discomfort from some substances at concentrations at or below the threshold limit value and that a smaller percentage may be affected more seriously by aggravation of a preexisting condition or by development of an occupational illness (Cooper, 1973; ACGIH®, 1994). This is clearly stated in the introduction to ACGIH®'s annual booklet *Threshold Limit Values for Chemical Substances and Physical Agents and Biological Exposure Indices* (ACGIH®, 2002).

This limitation, although perhaps less than ideal, has been considered a practical one since airborne concentrations so low as to protect hypersusceptibles have traditionally been judged infeasible due to either engineering or economic limitations. This shortcoming in the TLVs® has, until about 1990, not been considered a serious one. In light of the dramatic improvements of the past 10 years in our analytical capabilities, personal monitoring/sampling devices, biological monitoring techniques, and the use of robots as a plausible engineering control, we are now technologically able to consider more stringent OELs.

The background information and rationale for each TLV® are published periodically in the *Documentation of the TLVs® and BEIs®* (ACGIH®, 2001). Equally high quality documentation is available for the limits set by the German Maximum Arbeitsplatz Konzentration (MAK) Commission. Some type of documentation is occasionally available for OELs set in other countries. The rationale or documentation for a particular OEL, as well as the specific data which were considered in establishing it, should always be consulted before interpreting or adjusting an exposure limit (ACGIH®, 2002).

Threshold Limit Values, like OELs used in most other countries, are based on the best available information from industrial experience, experimental human and animal studies and, when possible, from a combination of the three (Smith and Olishifski, 1988; ACGIH®, 2002). The rationale for each of the values differs from substance to substance. For example, protection against impairment of health may be a guiding factor for some, whereas reasonable freedom from irritation, narcosis, nuisance, or other forms of stress may form the basis for others. The age and completeness of the information available for establishing most OELs also varies from substance to substance; consequently, the precision of each particular TLV® is not constant. The most recent TLV® and its *Documentation* should always be consulted in order to evaluate the quality of the data upon which that value was set.

Even though all of the publications that contain OELs emphasize that they were intended for use only in establishing safe levels of exposure for persons in the workplace, they have been used at times in other situations. It is for this reason that all exposure limits should be interpreted and applied only by someone knowledgeable of industrial hygiene and toxicology. The TLV® Committee did not intend that they be used, or modified for:

1) Evaluation or control of community air pollution nuisances;
2) Estimating the toxic potential of continuous, uninterrupted exposures or other extended work periods;
3) Proving an existing disease or physical condition;
4) Adoption or use by countries whose working conditions or cultures differ from those of the United States and where substances and processes differ (ACGIH®, 2002).

It is noteworthy to remember that the ACGIH® TLV® Committee has repeatedly stated that "these limits are not fine lines between safe and dangerous contaminations." In short, without knowing the toxic endpoint to be avoided, and the physical properties of the agent, they can't even be used as relative indices of toxicity.

The TLV® Committee and other groups that set OELs warn that these values should not be "directly used" or extrapolated to predict safe levels of exposure for other exposure settings. However, if one understands the scientific rationale for the guideline and the appropriate approaches for extrapolating data, considering the pharmacokinetics and mechanism of action of the chemical, they can be used to predict acceptable levels of exposure for many different kinds of exposure scenarios and work schedules (Hickey and Reist, 1979; Paustenbach, 2000b).

The reason that ACGIH® has stated that the TLVs® should not be used for other purposes, and that they should be used only by professionals trained in the field is because of the history of misuse. For example, there are dozens of examples where lawyers, physicians, and others have erroneously concluded that if a worker was exposed above a TLV® concentration, then the person was at "real" danger, or worse, that they may have been harmed. Often, such interpretations were self-serving and persons were adversely effected by such unfounded opinions. In addition, the TLVs® have often been inappropriately used by regulatory agencies as a basis for establishing "temporary standards" for everything from ambient air guidelines to emergency evacuation criteria. Often, the group issuing the draft criteria knew that this was not a proper use of the TLVs®, but such action was considered justifiable because it was "science forcing." That is, it made professionals in the regulated community go about the task of doing the more detailed work necessary to develop proper standards since the regulatory agency was simply too understaffed to handle such a large task. Because no one has been able to anticipate all of the various ways these values could be misused, ACGIH® decided to issue a "blanket disclaimer" many years ago. That is "the TLVs should not be directly used to predict safe levels of exposure for non-occupational settings." Obviously, the *Documentation of the TLVs® and BEIs®* (ACGIH®, 2001) provides a good deal of important information to the health scientist. From this, if all of the appropriate factors are considered, a professional should be able to derive other criteria.

Philosophical Underpinnings of TLVs® and Other OELs

TLVs® were originally prepared to serve only for the use of industrial hygienists who could exercise their own judgment in their application. They were not to be used for legal purposes (Baetjer, 1980). However, in 1968 the Walsh-Healey Public Contract Act incorporated the 1968 TLV® list, which covered about 400 chemicals. Then, in the United States, when the Occupational Safety and Health Act (OSHA) was passed (1970), it allowed OSHA, for a period of two years, to adopt national consensus standards or established federal standards, such as Walsh-Healey.

Exposure limits for workplace air contaminants are based on the premise that, although all chemical substances are toxic at some concentration when experienced for a period of time, a concentration (e.g., dose) does exist for all substances at which no injurious effect should result no matter how often the exposure is repeated. This premise applies to substances whose effects are limited to irritation, narcosis, nuisance or other forms of stress (Stokinger, 1981).

This philosophy thus differs from that applied to physical agents such as ionizing radiation, and for some chemical carcinogens, since it is possible that there may be no threshold or dose at which zero risk would be expected (Stokinger, 1981). Even though many would say that this position is too conservative in light of our current understanding of the mechanisms by which cancer occurs, there are some data on some genotoxic chemicals that support this theory (Bailer et al., 1988; Finkel, 1994, 1998). On the other hand, a large number of toxicologists believe that a threshold dose exists for those chemicals that are carcinogenic in animals, but that act through a non-genotoxic (sometimes called epigenetic) mechanism (Watanabe et al., 1980, Stott et al., 1981; Butterworth and Slaga, 1987; Butterworth et al., 1995; Park and Hawkins, 1995). Still others maintain that a practical threshold exists for even genotoxic chemicals, although they agree that the threshold may be at extremely low doses (Seiler, 1977; Stott et al., 1981; Wilkinson, 1988; Bus and Gibson, 1994; Gold et al., 2002).

With this in mind, some OELs for carcinogens proposed by regulatory agencies in the early 1980s were set at levels that, although not completely without risk, posed risks which were no greater than classic occupational hazards such as electrocution, falls, etc. (about one-in-1,000) (Graham and Hartwell, 1997). Although a clear description of this risk level or the rationale for the criteria has rarely been presented or discussed, it is now apparent that it was used to justify these limits (Rodricks et al., 1987; Travis et al., 1987, Hewett, 1996).

Occupational Exposure Limits in the United States

A comprehensive listing of the various OELs used throughout the world can be found in one of two references. One is the *Occupational Exposure Limits For Airborne Toxic Substances*, 4th Edition, published by International Labour Office of the World Health Organization (WHO, 1995), and the other is *Occupational Exposure Limits-Worldwide* published by the American Industrial Hygiene Association (AIHA) (Cook, 1986).

The philosophical underpinnings for the various OELs vary between the organizations and countries that develop them. For example, in the United States, at least six groups recommend exposure limits for the workplace. These include the TLVs® of ACGIH®, the Recommended Exposure Limits (RELs) suggested by the National Institute for Occupational Safety and Health (NIOSH) of the U.S. Department of Health and Human Services, the Workplace Environment Exposure Limits (WEEL) developed by AIHA (AIHA, 2000), standards for workplace air contaminants suggested by the Z-37 Committee of ANSI, the proposed workplace guides of the American Public Health Association (APHA, 1991), and finally, recommendations that have been made by local, state, or regional government. In addition to these recommendations or guidelines, Permissible Exposure Limits (PELs), which are regulations that must be met in the workplace in the United States because they are law, have been promulgated by the Department of Labor and are enforced by OSHA.

In addition to the OELs established by professional societies and regulatory bodies, guidance has also been provided by many corporations who handle or manufacture specific chemicals (Paustenbach and Langner, 1986; Naumann et al., 1996). For example, beginning in the 1960s, Dow Chemical, DuPont, Kodak, Union Carbide, and some other large firms began to establish internal OELs, which were intended to protect their workers, as well as their customers who purchased the chemicals. Later, due to the fact that only their workers would be exposed (without a prescription) to the chemicals that they manufactured, the pharmaceutical industry began to set limits for some of their intermediate and final products. Workers in the drug industry needed these guidelines because the doses to which they could be exposed each day had the potential to be several fold greater than the therapeutic dose. Today, perhaps more than 40 firms in the United States set internal OELs.

Outside the United States, as many as 50 other countries or groups of countries have established OELs (Stokinger, 1970; Cook, 1986; WHO, 1995). Many, if not most, of these limits are nearly or exactly the same as the ACGIH® TLVs® developed in the United States (Magnusson, 1964; Cook, 1986; WHO, 1995). In some cases, such as in the East European countries, including former Soviet bloc countries, and in Japan, the limits can be dramatically different than those used in the United States. Differences among various limits recommended by other countries can be due to a number of factors:

1) Difference in the philosophical objective of the limit and the untoward effects they are meant to minimize or eliminate,
2) Difference in the predominant age and sex of the workers,
3) The duration of the average workweek,
4) The economic state of affairs in that country, or
5) A lack of enforcement (therefore the OEL serves only as a guide).

For example, limits established in what is now the former Soviet Union are often based on a premise that they will protect "everyone, rather than nearly everyone, from any (rather than most) toxic or undesirable effects of exposure" (Elkins, 1961; Lyublina, 1962; Stokinger, 1963, 1970; Magnusson, 1964; LaNier, 1984, WHO, 1995).

In recent years, ACGIH® has worked closely with the European community (in particular, the German MAK Committee) in an attempt to harmonize the various approaches used to set OELs. For example, in 1997 a joint meeting between the TLV® and the German MAK Committees was held in Germany to move this initiative forward. The meeting went well and differences of opinion were shared freely. These two groups have met every other year since then to exchange information and reach agreement on recommendations for certain chemicals. However, due to a number of differences in their views of various scientific approaches and social considerations, it is unlikely that a single method for dealing with each category of adverse effects will be adopted in the foreseeable future by these two organizations.

APPROACHES USED TO SET OELs

OELs established both in the United States and elsewhere are based on data from a wide number of studies and sources. As shown in Table 12-1, the 1968 TLVs® (those adopted by OSHA in 1972 as federal regulations) were based largely on human experience. This may come as a surprise to many hygienists who have recently entered the profession since it indicates that, in most cases, the setting of an exposure limit was often after it had been found to have toxic, irritational, or otherwise undesirable effects on humans. As might be anticipated, many of the more recent exposure limits for systemic toxins, especially those internal limits set by manufacturers, have been based primarily on toxicology tests conducted on animals, which is in contrast with waiting for observations of adverse effects in exposed workers (Paustenbach and Langner, 1986; Naumann et al., 1996).

Several approaches for deriving OELs from animal data have been proposed and put into use over the past 40 years. The approach used by the TLV® Committee and others is not markedly different from that used by the U.S. Food and Drug Administration (FDA) in establishing Acceptable Daily Intakes (ADI) for food additives as far back as the early 1950s (Lehman and Fitzhugh, 1954; WHO, 1977). An understanding of the FDA approach to setting exposure limits for food additives and contaminants can provide good insight to hygienists who are involved in interpreting OELs (Rodricks and Taylor, 1983; Dourson and Stara, 1983).

Discussions of methodological approaches that can be used to establish workplace exposure limits based exclusively on animal data have also been presented (Weil, 1972; WHO, 1977; Calabrese, 1978, 1983; Zielhuis and van der Kreek, 1979a, 1979b; Dourson and Stara, 1983; Andersen, 1991; Leung et al., 1988; Finley et al., 1992). A review of the general procedures that OEL setting groups use has been published and it warrants evaluation (AIHA, 1996). The OELs derived from the various approaches contain some degree of scientific uncertainty associated with them; that is, a particular OEL may also prevent illness or irritation at 2-3 times the OEL that is selected and conversely, in some cases, symptoms of overexposure may be seen in a few persons exposed at the recommended OEL. Many scientists believe that much of the uncertainty can be reduced by the availability of better epidemiological information or through the use of physiologically-based pharmacokinetic (PB-PK) models, a much better approach than the traditional qualitative extrapolation of animal test results to humans (Reitz et al., 1995; Federal Focus, 1996; Krishnan and Andersen, 2001).

In 1968, approximately 50% of the TLVs® were derived from human data, and approximately 30% were based on animal data (see Table 12-1). By 1998, about 50% of the 700 or so TLVs® were derived primarily from animal data. The criteria used to develop the TLVs® have been classified into four groups: morphologic, functional, biochemical, and miscellaneous (nuisance, cosmetic) (Stokinger, 1970).

Table 12-1
The sources of information upon which TLVs® were derived in 1968

Procedure	Number	Percent Total
Industrial (human) experience	157	38
Human volunteer experiments	45	11
Animal, inhalation—chronic[a]	83	20
Animal, inhalation—acute[a]	8	2
Animal, oral—chronic	18	4.5
Animal, oral—acute	2	0.5
Analogy	101	24

[a]Exclusive of inert particulates and vapors.
Source: Stokinger (1970)

Of those TLVs® based on human data, most are based on the effects observed in workers who were exposed to the substance for many years. In short, about half of the existing TLVs® have been based on the results of workplace monitoring, coupled with qualitative and quantitative observations of the human response (Stokinger, 1970; Paull, 1984). In recent times, TLVs® for new chemicals have been based primarily on the results of animal studies rather than on human experience.

The endpoints upon which most of the historical TLVs® have been established is presented in Table 12-2. It is noteworthy that in 1968 only about 50% of the TLVs® were intended primarily to prevent systemic toxic effects. Roughly 40% were based on irritation and about 2% were intended to prevent cancer. By 1998, about 50% of the TLVs® were meant to prevent systemic effects, 30% to prevent irritation, 5-10 % to prevent cancer, and the remainder to prevent other adverse effects (ACGIH®, 1997).

In the early years of the modern era of industrial hygiene (e.g., post-OSHA and 1970), very little information was available to the public regarding the precise methodology by which TLVs®, MAKs, and other OELs were derived. To Warren Cook's credit, he wisely shared his thoughts about the rationale for selecting specific values in the documentation for the various limits (Cook, 1945), but it is clear from his writings that no set of uncertainty factors or criteria were universally adopted in setting these limits of exposure. Furthermore, over the ensuing 50 years, the TLV® Committee never adopted a standard approach and this has been troublesome to critics of the Committee. For example, many professionals believed that a fairly generic approach to setting OELs could be applied to the various classes of chemicals. On only a few occasions has a prescribed approach to setting OELs been suggested in a published paper (Zielhuis and Van Der Kreek, 1979a, 1979b).

Table 12-2
Various toxicologic endpoints upon which OELs may be based

Systemic effects	Neurologic effects
Irritant effects (eye, nose, throat)	Esthetics (blue skin, yellow eyes)
Odors	Local effects (perforated septum)
Cancer	Reproductive/developmental effects
Nuisance	

One reason that neither occupational or environmental limits cannot be derived in a "cookbook" manner is because of the relatively "soft" nature of the data upon which these limits are based. For example, no animal bioassay can ever be large enough to conclusively assure us that we can identify a precise dose that will not pose some theoretical risk to some individual, and no human epidemiology study can be so strong as to show that a chemical may not produce some type of adverse effect in some person. Nonetheless, in spite of the differences in proving the absence of risk, there is clear evidence that *virtually* safe doses for humans can be identified for all substances. The primary question, which is difficult to answer, is "What is the margin of safety between what we label as a 'safe' dose and the dose that might have some probability of producing some type of adverse effect in some individual?"

The following sections describe, in a general way, some of the various approaches that could be or have been used to establish OELs. In the main, the uncertainty factor approach can be used to establish OELs for nearly all adverse effects (except perhaps genotoxic carcinogens). The primary variable in setting these limits is the size of the uncertainty factor, which will vary with the adverse effect to be avoided, as well as the amount of available data. It must be noted here that, in most cases, the use of uncertainty or safety factors that have been used to establish the TLVs®, for example, has rarely been explicitly described in the *Documentation of the TLVs® and BEIs®*. This is a shortcoming in the process that deserves to be addressed.

Uncertainty Factors

Uncertainty factors (UF) or safety factors (as they were called from 1950-1980) are used in health risk assessment to account for a lack of complete knowledge or uncertainties, about the dose delivered, human variations in sensitivity, and other factors associated with extrapolating animal data to estimate human health effects (Dourson and Stara, 1983). They are applied to animal or human data in an attempt to identify safe levels of exposure for most persons. The uncertainty factor approach has been and continues to be used by FDA, U.S. Environmental Protection Agency (EPA), OSHA, and virtually all agencies and scientific bodies who set acceptable levels of exposure to toxic substances. The strengths and weaknesses of the UF approach and refinements to the process still enjoy considerable debate within the scientific community (Adbul-Rahman, 1999, 2001; Beck and Clewell, 2001).

Within the EPA, for example, UFs are factors used to operationally derive the Reference Concentrations (RfC) or Reference Doses (RfD) from experimental or epidemiological data (Jarabek et al., 1990). RfCs and RfDs are guidelines which attempt to protect nearly all citizens from adverse effects of chemicals in the environment, rather than the workplace. The critical dose is usually defined as the No Observed Effect Level (NOEL) for the most sensitive adverse effect from the best animal study.

The RfC or RfD is the result of dividing the critical dose by one or more uncertainty factors and sometimes a modifying factor (MF).

Uncertainty factors are intended to account for:

1) The variation in sensitivity among the members of the human population;
2) The uncertainty in extrapolating animal data to estimate human health effects;
3) The uncertainty in extrapolating from data obtained in a study that is of less than lifetime exposure;
4) The uncertainty in using Lowest Observable Adverse Effect Level (LOAEL) data rather than the No Observed Adverse Effect Level (NOAEL) data; and
5) An incomplete data base, generally with regards to reproductive or developmental toxicity.

Usually, each of these factors is set at 3 or 10, but a value as low as one may sometimes be appropriate (Calabrese, 1978). MFs are used the same way as UFs are used. However, modifying factors are applied when additional uncertainty exists after accounting for the uncertainty within the five listed categories. Modifying factors would not generally be applied when setting an OEL because some human data is almost always available, thus avoiding the need for additional conservatism.

Additional information on the EPA's justification and rationale for the use of UFs and MFs is presented in several documents, including *Interim Methods for Development of Inhalation Reference Doses* (EPA, 1990), *Methods for Derivation of Inhalation Reference Concentrations and Application of Inhalation Dosimetry* (EPA, 1994), *IRIS Supportive Documentation Volume 1* (EPA, 2000), Dourson and Stara (1983), and Barnes and Dourson (1988). In general, when one sets occupational exposure limits, the same factors are considered. The primary differences in setting an RfC vs. an OEL are the number and size of the uncertainty factors, the difference in the length of exposure (continuous vs. 40-hour workweek), the lack of a recovery period, and the difference in the type of exposed population (old, young, and sickly, versus healthy workers). In short, OELs do not attempt to protect everyone in the general population who could be exposed continually for a lifetime; however, this is the intent of an RfD.

SPECIFIC UNCERTAINTY FACTORS USED TO SET AN RfC

As mentioned previously, the EPA issues guidance regarding safe airborne concentrations of various chemicals. This guidance is known as a reference concentration, or RfC. The formula for deriving the RfC is generally as follows:

$$\text{RfC} = \frac{\text{Critical dose}}{\text{UF} \times \text{MF}} \qquad \text{[Equation 12-1]}$$

Where: Critical Dose = Best estimate of the NOEL from the studies that have evaluated the most sensitive releative toxic effect

MF = Modifying factor

UF = Uncertainty factor

When the EPA sets an RfC, a documentation is presented in their data base for each chemical that specifies the uncertainty factors applied and their rationale. This documentation is available to anyone via Integrated Risk Information System (IRIS), a data base (IRIS, 2002). The EPA's application of uncertainty factors has evolved over time. Until recently, they relied only on UFs representing an "order of magnitude" (factor of 10). Where appropriate, the EPA now uses an intermediate uncertainty factor of 3, rather than 10. Since 3 is the approximate mid-point between 1 and 10 on a logarithmic scale, when 2 successive applications of UF 3 are performed, the combined UF is rounded to 10. Over the years, when setting OELs (rather than RfDs) the TLV® Committee frequently applied an uncertainty or safety factor of 3 rather than 10 (although this was rarely explicitly stated in the documentation) and only one or two, rather than three to five, uncertainty factors were applied to human or animal data to form the overall or aggregate uncertainty factor.

Uncertainty factors have often been applied by the EPA in the following manner when setting ambient air limits. Be reminded that only the conceptual approach is applicable to setting an OEL because a smaller population of healthy persons is the focus of workplace standards and they are only exposed 40 hours a week.

- When setting limits to protect the general population, a 10-fold uncertainty factor is used to protect the sensitive individuals. However, as when they derived the RfC for carbon disulfide, the EPA used a three-fold uncertainty factor because the critical dose was measured directly as an internal measure of dose.
- Where necessary, either a 10-fold or three-fold uncertainty factor is used to account for uncertainties in the extrapolation of data from animal studies to estimating effects on humans. When dosimetric adjustments are used, the smaller (three-fold) uncertainty factor is used.
- The EPA uses a 10-fold or three-fold uncertainty factor to extrapolate from subchronic studies to predict the hazard due to exposure that may last a lifetime. The size of the uncertainty factor depends on whether progression of the adverse condition is expected.
- In addition, EPA uses a 10-fold or three-fold uncertainty factor to extrapolate from an LOAEL to an NOAEL (in those studies where no clear NOAEL was observed). The smaller uncertainty factor is used when the adverse effect is judged to be sufficiently mild.
- Where indicated, a 10-fold or three-fold uncertainty factor for database deficiencies can be applied. The most common deficiencies are lack of developmental or reproductive toxicity studies. This additional uncertainty factor is most often used where there is reason to suspect the effect, but the necessary studies are lacking.

Finally, the modifying factor is used by EPA only when there may be indications from previous studies or from supporting biological data that other effects may occur. The following example illustrates how this approach was used to derive a chronic RfC for methyl ethyl ketone for the general population.

Example 12-1: ***Setting an RfC for Methyl Ethyl Ketone***

Recently, the EPA recommended a chronic RfC for methyl ethyl ketone (MEK). It was based on decreased fetal birth weight. They reported that the NOAEL adjusted to a Human Equivalent Concentration (NOAEL [$NOAEL_{(HEC)}$]) from the best study was 2,978 mg/m^3. From this beginning, they established an RfC in the following manner:

a. A total uncertainty factor of 1,000 was derived by multiplying the following factors:
   - 10 to account for interspecies extrapolation
   - 10 to account for sensitive individuals
   - 10 to account for an incomplete database including the lack of chronic and reproductive toxicity studies

b. A modifying factor of 3 was used to address the lack of unequivocal data for respiratory tract (portal of entry) effects.

c. Critical dose = 2,978 mg/m^3 [the $NOAEL_{(HEC)}$]

$$RfC = \frac{\text{Critical Dose}}{UF \times MF} = \frac{2{,}978 \text{ mg/m}^3}{1{,}000 \times 3} \approx 1 \text{ mg/m}^3$$

Note: This RfC was intended to protect virtually anyone in the population including the very young, the very old, and those with one of a number of illnesses. The approach is unnecessarily conservative for establishing an occupational exposure limit that is intended to protect a much smaller and less diverse population that would *not* be continuously exposed to this concentration in ambient air. In the author's view, the EPA should not have added a modifying factor of 3 to account for possible respiratory effects since developmental effects were the most sensitive endpoint. Further, an aggregate 3,000-fold safety factor appears excessive even for an RfC for this chemical given the large amount of toxicity data and the available information from worker studies.

UNCERTAINTY FACTORS AND OELs

The method by which one chooses the size of each of the specific UFs has been debated for many years (Weil, 1972; Dourson and Stara, 1983; Krewski et al., 1984; Barnes and Dourson, 1988; Velazquez et al., 1996). Not surprisingly, there is no single method that has been embraced by all scientific bodies. For example, from about 1950-1985, the FDA often applied a UF of 10 to account for the possible increased susceptibility of humans versus the animals tested and another factor of 10 to account for the differences in susceptibility across the human population (e.g., a total UF of 100) when establishing limits of exposure for food additives and pesticide residues. Often, if chronic animal data were available, this factor of 100 was considered adequate. On the other extreme, Weil (1972) once suggested that if only acute toxicity information is available, such as an LD_{50}, then an aggregate uncertainty factor of 5,000 is reasonable to predict a chronic risk based only on acute toxicity data.

When evaluating the various views about uncertainty factors, it is worthwhile to note that the size of the uncertainty factor used to account for various "unknowns" has changed slightly over the years. For example, it would not be fair to compare the views of Warren Cook in 1945 to those of Dr. Lisa Brosseau in 2000 (the current Chair of the Chemical Substances TLV® Committee). The reason that the size of the overall UF has changed over time is because the margin of safety thought to be necessary to protect "most workers" in 1945 is different than in 2000. In part, this is because society has begun to encourage scientists to select toxicological endpoints that are more sensitive than the traditional endpoints. In addition, the margin of safety thought to be necessary to insure protection of different persons has changed. Further, our knowledge of the degree of inter-individual variability in response to xenobiotics within the population has improved.

The same four extrapolation factors used to adjust animal data to set an ambient air limit (RfC) can be used to derive an OEL, although their magnitude will often be much smaller. A number of papers that have been published over the years discuss the rationale for selecting the various values (Calabrese, 1978, Krewski and Brown, 1984, ILSI, 1992). The use of UFs to set OELs has been discussed by Zielhuis and Van Der Kreck (1979a,b) and others. Unlike the approach used to set RfCs, aggregate or overall UFs much less than 100 have often been applied to lifetime animal studies to predict safe exposures for humans exposed in the workplace. This is because workers are only exposed as adults, they are generally more healthy than the general population, and the exposure is only for 40 hours per week and only for 30-40 years. Consequently, it is not unusal to see an overall UF as small as 10-100 being applied to chronic animal NOELs to establish an OEL for workers. This is not to say that larger safety factors are not often warranted. In addition, when the slope of the dose-response curve is known, it should influence the size of the uncertainty factor.

Regrettably, the ACGIH® TLV® Committee has not explicitly presented the quantitative basis for setting the vast majority of the TLVs®. That is, the key studies are cited, but the size of the safety factor applied to a particular study is rarely presented. Thus, one must look at the value suggested for a specific chemical and the information in the original study or studies that they seemed to rely upon in order to back-calculate the size of "safety factor" that was used. Nevertheless, if one analyzes the historic TLVs® for many of the systemic toxicants, a few generalizations can be offered. First, if a solid NOEL from a six month to two year animal study was available, it appears that a UF of 10 or greater was applied to the NOEL to establish an OEL for a chronic toxicant that was not carcinogenic, assuming there was no evidence of mutagenicity or carcinogenic potential. This margin of safety was considered adequate. Second, if there was a reasonable amount of human experience with the chemical and no adverse effects had been observed in carefully monitored workers, then uncertainty factors as small as 10 have often been used to establish TLVs® when nearly chronic animal data were available. Third, in recent years, there has been some interest, by those involved in setting OELs, to increase the margin of safety inherent to accommodating the differences in susceptibility among workers.

An interesting analysis of the various implied UFs in the TLVs® has been published (Roach and Rappaport, 1990).

Setting Limits for Systemic Toxicants

By far, the majority of chemicals for which OELs have been established are systemic toxicants. By definition, this class of chemicals brings about their adverse effects at a site or target organ distant from the site of contact (Ottobani, 1993). An example of a systemic toxicant is ethanol, which is usually ingested. Although this chemical is absorbed in the gastrointestinal tract, the adverse effects are on the central nervous system and the liver; thus, it is a systemic toxicant.

Before proceeding to interpret toxicity studies, it is important to evaluate the chemical structure of the substance in an attempt to see if it has characteristics that suggest that it may act like another chemical for which the toxicology is well understood. For example, toxicity data on the dioxins are useful for understanding something about the hazards posed by the furans and some polychlorinated biphenyls. Likewise, information about trichloroethylene is probably useful for understanding some aspects of perchloroethylene. Conversely, although one might expect 1,1,2 trichloroethane to be similar in toxicity to 1,1,1 trichloroethane, the two chemicals have markedly different potency. Understanding why there are differences or similarities among chemicals can be helpful to setting proper OELs for them.

Beyond evaluating the structure activity relationship (SAR), it is also useful to assess the available mutagenicity data. The TLV® Committee has often examined homologous series of chemicals to assure consistency in setting TLVs®. For those systemic toxicants that are slightly positive in the Ames test or other tests for genotoxicity, it is often useful to provide a slightly greater margin of safety than might otherwise be indicated if a lifetime bioassay has not been conducted.

The process of setting either an OEL for an industrial chemical or an ADI for a food additive based on animal data is a relatively straightforward exercise. After evaluating the SAR, the genotoxicity battery, and the human experience, all of the applicable toxicity studies are assembled. An emphasis is placed on those studies of the longest duration for which an adequate number of animals were used.

Since inhalation is generally the primary route of exposure to industrial chemicals, tests which use this route of exposure are favored over those where the animal was exposed via ingestion or gavage. If test results for both ingestion (dietary) and inhalation studies are available, then the results should be compared on a mg/kg basis (or other relevant dose metric) so that "first pass" effects can be evaluated. "First pass" effects are those produced prior to having the chemical metabolized by the liver.

Conceptually, the approach to setting an OEL for the systemic toxicants is much the same as that used by EPA to set an RfC. First, all relevant animal studies are evaluated and are placed in order of strength or quality. Second, the best study involving the most relevant species is selected. Third, the NOEL from this study is identified. One caution is that an NOEL should have an endpoint that has genuine biological significance. For example, it is widely held that transient changes in liver enzymes due to the challenge of dealing with a xenobiotic are not usually considered a significant adverse effect so, in these cases, it should not be used as the basis for deriving an OEL. Fourth, the results of the study are compared with those from other studies to determine if there is a consistent and clear message from all data sets. Fifth, if little else is known, the most simplistic approach to setting an OEL for a systemic chemical is to simply divide any NOEL from a high quality chronic animal study by a factor of 100.

Example 12-2: ***Basic Approach to Setting a Chronic OEL***

Assume that the NOEL for rats exposed for 2 years to 1,1,1- trichloroethane is 1,000 ppm (vapor) and that for mice the NOEL was 1,500 ppm, what is a reasonable OEL? Assume that the primary adverse effect observed in animals is liver hypertrophy (a systemic effect). Lastly, experience indicates that human exposure in the workplace to 100 ppm for 8-hours/day for a lifetime provides no adverse effects.

$$\text{OEL} = \frac{\text{Appropriate NOEL}}{(\text{UF}_1)\ (\text{UF}_2)} \qquad \text{[Equation 12-2]}$$

Where: Appropriate NOEL = no observed effect level in animals (most sensitive) if adverse effect is biologically significant

UF_1 = 5 (to account for animal to human differences)

UF_2 = 5 (to account for differences in sensitivity among humans)

Therefore, given equation for OEL:

$$\text{OEL} = 40 \text{ ppm}$$

Setting OELs for Sensory Irritants

Sensory irritants are those chemicals that produce temporary and undesirable effects on the eyes, nose, or throat. These include such chemicals as ammonia, hydrogen sulfide, formaldehyde, sulfuric acid mist, and dozens of others. The general TLV® policy on irritants can be found in the Introduction to the Chemical Substances in the *TLVs® and BEIs® Book*, "The basis on which the values are established may differ from substance to substance, protection against impairment of health may be a guiding factor for some, whereas reasonable freedom from irritation, narcosis, nuisance or other forms of stress may form the basis for others (ACGIH®, 2002)." Nearly 50% of the published TLVs® are based on avoidance of objectionable eye and upper respiratory tract irritation.

The approach typically used by the TLV® Committee has been to assign ceiling values (CV) to rapidly-acting irritants and to assign short term exposure limits (STEL) where the weight of evidence from irritation, bioaccumulation and other endpoints (e.g., central nervous system depression, increased respiratory tract illness, decreased pulmonary function, impaired clearance) combine to warrant such a limit. As noted by Willhite (1998), the MAK Commission has utilized a five category system based on intensive odor, local irritation, and elimination half-life (MAK, 1997), but this system is being replaced to be consistent with the European Union Scientific Committee for Occupational Exposure Limit Values (SCOEL). It is now likely that the MAK will assign a 15-minute STEL when necessary, feasible, and the underlying data are sufficiently developed to justify a compound-specific excursion factor to control body burden (e.g., carbon monoxide). These changes lead, among other things, to modification of eight hour time-weighted average (TWA) and STELs to accommodate unusual work schedules, where compliance often rests with availability of Biological Exposure Indices (BEI®). In each instance, simplicity and practical application are considered along with the relationship between absorbed dose (e.g., area under curve) and biological/medical endpoint of concern.

A significant debate among health professionals has taken place in recent years because some believe that transient irritation does not constitute material impairment of health (Willhite, 1998), while others contend that the TLVs® should protect against any irritation. Although their later proposal was overturned by the Supreme Court, it is noteworthy that in 1989, in OSHA's effort to promulgate PELs, 79 materials were proposed for regulation based on avoidance of sensory irritation (*Federal Register*, 1989):

> "The recognition of sensory irritation as potentially being 'material impairment of health' is consistent with the current scientific consensus related to health effects of environmental agents. Mucous membrane irritants can cause increased blink frequency and tearing, nasal discharge, congestion, and sneezing and cough, sputum production, chest discomfort, sneezing, chest tightness and dyspnea. Work environments often require levels of physical and mental performance considerably greater than encountered in daily living. Even in the absence of any permanent impairment, the symptoms listed can interfere with job performance and safety. Mucous membrane irritation is associated with respiratory illness, depending on the composition of specific exposure and on the dose, duration and frequency of exposure. No universally applicable conclusion can be drawn at this time regarding the association between irritative symptoms and permanent injury or dysfunction. Where certain individuals show no measurable impairment after an exposure, even when experiencing irritative symptoms, others may develop identifiable dysfunction."

OSHA concluded that exposure to sensory irritants can cause inflammation and increase susceptibility to other irritants and infectious agents, lead to permanent injury or dysfunction, and permit greater absorption of hazardous substances. OSHA (1989) also concluded that, "Exposing workers repeatedly to irritants at levels that cause subjective irritant effects may cause workers to become inured to the irritant warning properties of these substances and thus increase the risk of overexposure." TLV® treatment and interpretation of dose-response relationships for irritants are consistent with the OSHA description. While the TLV® Committee assumes a sigmoid concentration-response relationship for common irritants and believes a NOEL (below which the risk of experiencing irritation is trivial) can be identified for "nearly all workers," the Committee seldom has available a data set sufficiently robust so as to assign a specific level of risk. Even though correlations have been drawn, there is no widely accepted method for extrapolation of animal irritation data to human beings.

Historically, OELs for irritants were based on observations of the response of workers to various airborne concentrations measured by industrial hygienists. Basically, concentrations that did not produce irritation were recorded and this information was sent to the TLV® Committee so they could recommend an appropriate OEL. Armed with whatever information the Committee was sent over the next two years following the "notice of intent," if it was reported that workers were experiencing some irritation at the proposed TLV®, the Committee could then choose to make adjustments to the value. It is because of the importance of the subjective information gathered in the workplace that the TLV® Committee has placed such reliance on the advice offered by the "consulting members" of the Committee; e.g., those toxicologists and industrial hygienists who work in industry.

More recently, methodologies for setting OELs for sensory irritants have been based on the results of eye and skin irritation tests conducted in rabbit or rodent studies. Since sensory irritation is generally considered an acute response, animal tests involving limited durations of exposure are now generally considered to be adequate to identify "interim" safe levels of exposure. Many acute toxicology tests of irritation, however, often don't provide a dose metric since they involve directly applying a liquid or solid to the eye or the skin. For example, how does one convert 0.1 ml of a chemical placed in the eye to an airborne concentration? Thus, unless the animals are exposed via inhalation, toxicologists and physicians have had to make rough estimates about the likely airborne concentration that might prevent irritation. Generally, animal studies need to be conducted with vapors to see if adverse effects are observed. A recent symposia on this topic addressed these issues and also discussed different approaches used by various nations (Paustenbach, 2001).

Historically, uncertainty factors applied to NOELs observed in animal studies (when exposed to vapors) to set an OEL for humans to prevent sensory irritation have been rather small. Often, the overall UF has been as low as 2 to 5. There are a number of reasons why this has been the case. First, rabbits are considered more susceptible to eye irritants than humans, so if only animal data are available, these tests have been considered "worse case." Second, mild eye irritation due to vapors, if it occurs in humans, is generally transient. Third, irritation is often accommodated and, by definition, the effects are reversible. Another reason the margins of safety have been small is that, until recent years, most TLVs® and other OELs were established after, rather than before, workers were exposed to the chemical. Thus, when these committees met to establish OELs there was generally a significant amount of information about the airborne concentrations known to produce eye, nose, throat, or lung irritation in workers. Thus, having some information about human response and some animal data, the TLV® Committee apparently believed that they could set TLVs® at values that did not incorporate large uncertainty factors.

Today, there is a greater expectation that nearly everyone should be protected against even minor sensory irritation. Therefore, in recent years, committees that establish OELs tend to apply uncertainty factors of 5–10 or more to an animal NOEL to set an OEL if human data are not available. Of course, for the period during which a new OEL (such as a TLV®) is "out for comment," it is expected that those who use the chemical and will collect industrial hygiene data will submit comments regarding the reasonableness of the proposed guideline. In some cases, such as for chlorine, formaldehyde, and other widely used chemicals, data from controlled human studies have been available (Paustenbach et al., 1997). In those situations where human chamber studies have been conducted, it has not been uncommon for an OEL setting group to apply a very low uncertainty factor, as low as 1–2, to a well understood human NOEL. Whenever possible, it is recommended that expert panels identify all the studies that are considered to be of good quality and build a comprehensive dose-response curve (Paustenbach et al., 1997). As shown in Figure 12-1, these can be very informative.

Setting OELs for Irritants Using Models

Unlike the period 1940-1975, when OELs for irritants were based primarily on the human experience or simple tests with rabbits, we now have a reasonably reliable capability for predicting safe levels of exposure using models. Two kinds of models are available. One is based on tests that consider the response of rats and/or mice (Kane and Alarie, 1977; Alarie, 1981; Abraham et al., 1990; Nielsen, 1991) to irritants. The second is based on chemical properties (Leung and Paustenbach, 1988). Going forward, it is recommended that these be used to identify OELs that can be used until experience has been gained about the experience with workers.

The first approach was developed at the Hazelton Labs, and was later refined at the University of Pittsburgh (Alaire and Nielsen, 1982). It involves exposing rodents to various concentrations of contaminants and then measuring respiratory parameters. In this method, animals are placed in a device that measures the number of breaths and the depth of breathing. These two endpoints are measured at various increasing concentrations. It is known that as the degree of irritation increases, both parameters will decrease in rodents. Rodents, unlike other species, will decrease their metabolism to near death in order to avoid lung damage due to serious irritation.

In one of their more comprehensive papers, they described the success of their methodology to accurately predict safe levels of human exposure for the benzenes and alkylbenzenes.

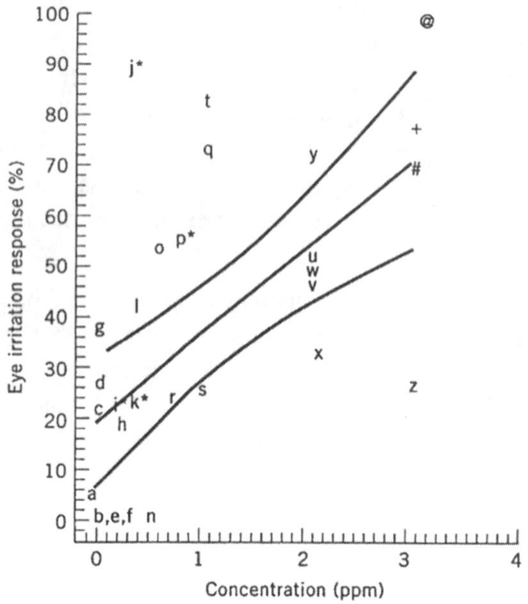

Figure 12-1. Linear concentration-response curve based on the data from various human studies regarding eye irritation due to formaldehyde. Linear least-squares regression analysis of the data presented in Paustenbach et al. (1997), omitting the data for mobile home studies (points i*, j*, k*, and p*). The regression equation is % response = 19.6 + (17.4 x concentrations in ppm): n = 24, $r^2 = 0.45$. The regression, that is positive slope, is significant ($p < 0.001$) and the 95% confidence interval for the regression lines are shown. The data points b, e, and f represent studies with zero response at zero concentration. The fit of the line does not vary appreciably if one fits the line with only the controlled human studies or all of the studies (from Paustenbach et al., 1997).

Their methodology, known as the RD_{50} approach, has been used successfully to predict OELs for many irritants, e.g., as many as 100 different chemicals over the past 15 years. OELs have been predicted with this model for a variety of irritants including ketones, alcohols, alkanes, disulfides, and styrene-like chemicals (Nielsen, 1991; Schaper, 1993).

The second modeling approach to setting OELs has been applied to the organic acids and bases; a class of chemicals that are well known irritants (Leung and Paustenbach, 1988b). A generic method for understanding these chemicals was needed since only a few of the more than 40 of this class of chemicals used in industry had OELs when the approach was developed (Cook, 1986). Although a great structural diversity exists among these chemicals, the primary biological effect produced by exposure to these materials is irritation. These researchers proposed that irritation could be related to their acidity or alkalinity. Since the strength of an organic acid or base is measured by its pK_a, it was shown that this term could be used to identify preliminary acceptable levels of human exposure.

As shown in Figures 12-2 and 12-3, the existing OELs for organic acids and bases correlated well with pK_a. For organic acids and bases for which no OEL has been established, the following equations can be used to set a preliminary limit:

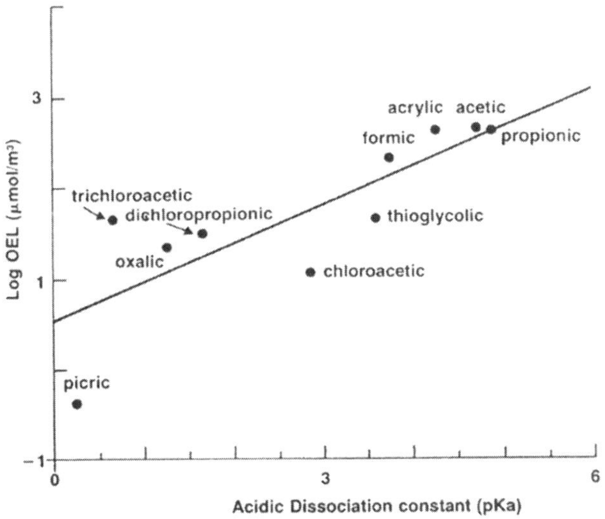

Figure 12-2. Correlation of OELs with equilibrium dissociation constants of organic acids. The regression equation is: log OEL (µmol/m³) = 0.43 pK_a + 0.53 (r = 0.80).

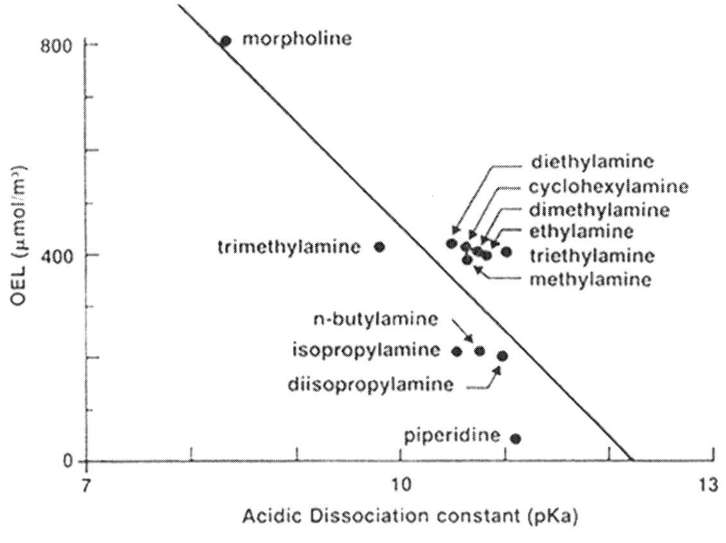

Figure 12-3. Correlation of OELs with equilibrium dissociation constants of organic bases. The regression equation is: OEL (µmol/m³) = - 200 pK_a + 2453 (r = - 0.81).

For organic acids: log OEL (μmol/m^3) = 0.43 pK_a + 0.53

For organic bases: OEL (μmol/m^3) = -200 pK_a + 2453

Table 12-3 presents the OELs calculated with these formulae for a variety of organic acids and bases. A large number of corporate OELs have been adopted based on this approach.

In the coming years, it is quite likely that committees who recommend the TLVs® and other OELs for the sensory irritants will suggest that lower concentrations are needed as society attempts to prevent even transient irritation from occurring in exposed workers.

Setting Limits for Reproductive Toxicants

"Reproductive dysfunction" can be broadly defined to include all effects resulting from paternal or maternal exposure that interferes with the conception, development, birth, and maturation of offspring to healthy adult life (Manson and Kang, 1994). For purposes of this discussion, reproductive effects are those that impair the ability of a male or female to produce offspring (Zenick et al., 1994). The relation between exposure and reproductive dysfunction is highly complex because exposure of the mother, the father, or both may influence reproductive outcome. In addition, exposures may have occurred at some time in the past, immediately prior to conception, or during gestation. For some specific dysfunctions, the relevant period of exposure can be identified, and for others it cannot. For example, chromosome abnormalities detected in the embryo can arise from mutations in the germ cells of either parent prior to conception or at fertilization, or from direct exposure of embryonic tissues during gestation. Major malformations, however, usually occur with exposure during a discrete period of pregnancy, extending from the third to the eighth week of human development (Manson and Kang, 1994).

Although historically the bulk of interest has been on female reproductive function, the precise male contribution to reproductive failure and adverse pregnancy outcomes, although often unknown, is considered to be significant (Zenick et al., 1994). When evaluating males, attention is focused primarily on toxic effects that involve testicular and postspermatogenic processes that are essential for reproductive success. Male reproductive failure resulting from germ cell mutation (i.e., genotoxicity), the role of the endocrine system in the support of reproductive function, and female reproductive toxicity are all important variables. Like the developmental toxicants, there have been few OELs established for chemicals whose primary adverse effect is reproductive toxicity. There are at least two reasons. First, few industrial chemicals have been tested in the classic male or female reproductive toxicity batteries (Zenick et al., 1994; Manson and Kang, 1994). Second, of those chemicals tested, few have been shown to produce adverse effects selectively on reproduction at concentrations or doses lower than those that are known to produce significant adverse effects on other organs, thus, this is rarely the "driving" health effect.

Similar to the historical approaches used to identify OELs that protect against other adverse effects, it appears that uncertainty factors in the range of 10-100 applied to an NOEL from

Table 12-3
Occupational exposure limits for selected high volume organic acids and bases recommended by a mathematical formula based on the disassociation constant[a]

Acid	OEL mg/m^3	OEL ppm	Base	OEL mg/m^3	OEL ppm
Acrylic	16	5	Allylamine	29	12
Butyric	35	10	Dialylamine	58	15
Caproic	49	10	Dibutylamine	43	8
Crotonic	30	8.5	Isobutylamine	21	7
Hepatanoic	55	10	Propylamine	21	7
Isobutyric	35	10	Trialylamines	109	20
Isocaproic	49	10			
Isovaleric	42	10			
Methacrylic	40	8.5			
Pentenoic	32	7.8			
Propiolic	1.5	0.5			
Valeric	42	10			

[a]Exposure limits were calculated by using the equations:
acid: $\log \text{OEL} (\mu\text{mol/m}^3) = 0.43\ pK_a + 0.53$;
base: $\text{OEL} (\mu\text{mol/m}^3) = -200\ pK_a + 2453$.
Source: Leung and Paustenbach, 1988b

well conducted animal studies have been considered adequate to protect humans from reproductive effects. For example, the OSHA PEL for dibromochloropropane is 1 part per billion (ppb), and this appears to be based on applying an uncertainty factor of less than 10 to the NOEL for adverse effects observed in humans. Because there are a limited number of reproductive toxicants for which OELs have been set, and there have been few long term follow-up studies of workers exposed to these agents, it is unclear whether uncertainty factors closer to 10, 30, 100, or slightly higher are the "best ones" for this class of chemicals. As with other adverse effects, the magnitude of the various uncertainty factors used to identify an OEL should be directly related to the strength of the animal and human data, the severity of the adverse effects, the reversibility of the effect, as well as the relationship to doses that cause other toxic effects.

Setting Limits for Developmental Toxicants

Very few of the current OELs that have been set by the ACGIH® TLV® Committee, the MAK, or the AIHA WEEL Committee have the primary objective of preventing developmental effects. One reason is that relatively few of the 1,200 or so chemicals that have OELs have been tested in the standard Segment II test battery for assessing developmental toxicity and, of those chemicals tested, only a fraction have been found to be selectively toxic to the developing fetus (Johnson, 1988; Manson and Kang, 1994). However, in recent years, it has been recognized that OELs that account for the results of developmental toxicity studies need to receive careful consideration. There has been an active dialogue within the toxicology community about the appropriate approach for estimating safe levels of human exposure based on animal data involving agents having developmental and reproductive effects in animals (EPA, 1986; Gaylor, 1989; Johnson, 1989; Kimmel et al., 1995).

For many years, due to the technical difficulties involved in the safety evaluation/extrapolation process used to set OELs for developmental toxicants, toxicologists in industry typically recommended that women of child-bearing age not be placed in locations where exposure to these agents could occur. However, during the 1970s, in an attempt to satisfy federal labor laws in the United States and to give women greater access to higher paying jobs, it was no longer considered acceptable to restrict women from jobs where exposure to the truly significant developmental toxicants was possible. Thus, OELs need to insure that even particularly susceptible women and men need to be considered when the limit is set. The methodologies for setting OELs for these agents continue to evolve as much due to changes in societal values about the required size of the margin of safety, as to the changes in our scientific understanding of the developmental hazard.

By definition, a developmental toxicant is a chemical that can produce an adverse effect on the developing fetus (Hayes, 2000). The range of possible adverse effects on development covers a very broad spectrum spanning small changes in birth weight to gross teratologic effects. The decision to classify a chemical as a developmental toxicant is clouded by the fact that at some dose, virtually all chemicals will produce an adverse effect on developing offspring

(Karnofsky, 1965). To complicate matters further, there are very significant differences in the susceptibility to various developmental effects among the various animal species and humans. There are also differences in the percentage of naturally occurring defects among rodent and non-rodent species, and humans. Lastly, when attempting to identify a reasonable OEL, one must assume that many of these effects are not reversible. For these and other reasons, it is not surprising that a great deal of deliberation needs to occur when attempting to identify safe levels of exposure to developmental toxicants.

In general, the current approach to setting an OEL for these agents is much like that used by FDA to identify acceptable exposure to certain drugs that might pose a developmental hazard. First, the critical NOEL observed in a well conducted Segment II developmental toxicity study is identified. A Segment II test involves exposing a rabbit and a relevant species to three or four doses of a toxicant. Females are first exposed about two weeks prior to pregnancy. Exposures continue throughout the pregnancy. About three days before delivery, the pups are removed via C-section and then examined for detrimental effects (Manson and Kang, 1994). Two species are always evaluated in these tests so two NOELs are produced. If both are similar, the lower of the two is used in the safety evaluation. If they are not similar, then a careful review of the specific adverse effects and the differences in metabolism between the species and human needs to be conducted. If known, the species thought to provide information most relevant to humans should be used. Second, one or more uncertainty factors should be applied to the animal NOEL. Historically, when setting tolerances for food additives and pesticide residues, a UF of 100 has been applied to the NOEL observed in the most sensitive species exposed in a Segment II test to identify the ADI. The size of the uncertainty factor applied to developmental studies (Segment II) used to set OELs appears to have varied over the years from 5 to 50 (Paustenbach, 1988). For some chemicals, the apparent UF incorporated into some TLVs® has been smaller, depending upon the strength of the animal data, the mechanistic data, the warning properties of the chemical, and other factors.

Because developmental effects have a threshold, it is anticipated that the uncertainty factor or benchmark dose approach will continue to be used to identify safe levels of exposure. The benchmark dose approach is a hybrid method (it relies on both low-dose modeling and the safety factor approach), which has certain benefits versus either the modeling or uncertainty factor approaches (Crump, 1995). The methodology is discussed in the chapter on Risk Assessment in this book. Thus far, no OELs have been based on the benchmark dose approach, but it should be one of the methods considered by groups who set limits in the coming years. As different techniques for identifying more sensitive developmental effects evolve, it is likely that the OELs for these chemicals will become smaller (Kimmel et al., 1995).

Setting Limits for Neurotoxic Agents

Chemicals which can produce permanent neurological damage often present significant concern to toxicologists and physicians. This is because many neurotoxins can produce permanent damage at doses that produce no other adverse effects. For this reason, the FDA has traditionally regulated neurotoxic agents that can enter the food chain in a fairly aggressive manner. Many agencies, including the EPA and OSHA, often regulate neurotoxicants nearly as stringently as the carcinogens, except, of course, if a threshold dose is believed to exist. It appears that uncertainty factors in the region of 5 to 100 have been typically applied to animal NOELs to set OELs. As in the setting of other OELs, a large degree of professional judgment is needed to identify the appropriate value. It is noteworthy that, at times, the FDA has applied uncertainty factors as great as 2,000 to identify acceptable levels of exposure to residues of certain pesticides in foods that cause permanent neurotoxicity, like dying-back neuropathies. In contrast, most of the OELs set for the neurotoxins are based on human experience and because of the fairly large database on the human response to these chemicals (like parathion), it appears that this has been the justification for the relatively low uncertainty factors that are inherent in many OELs.

Setting Limits for Aesthetically Displeasing Agents and Odors

The process or approach to setting an OEL for a chemical that tends to have a low odor threshold or for those chemicals that are aesthetically displeasing has been fairly simple. One of two approaches is used. In the first, if the agent has an odor threshold that occurs at an airborne concentration much lower than the concentration that produces even subtle toxic effects, the agent is categorized as one that has "self limiting" exposure characteristics. That is, a worker is not going to allow themselves to be placed at risk of injury due to exposure, unless he/she is exposed in a confined space without easy egress, because the odor is so objectionable. These chemicals generally do not pose much of a concern to toxicologists or hygienists as long as the agent does not cause rapid olfactory fatigue, and the TLV® is set at the concentration where odor causes discomfort.

In the second approach, one usually identifies the airborne concentration at which most persons find the odor of the chemical objectionable, and then divides that concentration by a factor of 2 or 3 to establish a preliminary TLV®. As with other "preliminary TLVs®," during the two year period of comments, the value can be raised or lowered based on the feedback from occupational health professionals.

Historically, human experience in the workplace has been used to identify the concentration at which most persons recognized an odor (generally, one only focuses on the concentration that is found objectionable rather than simply detectable), and then the OEL was established. In recent times, some firms have used odor or irritation panels to identify the concentration at which detection occurs, as well as those concentrations where the odor is considered objectionable. Such panels include men and women of various ages since both sex and age are known to affect the threshold of smell. From the results of the odor panel, a concentration can be identified that is likely to be acceptable to most persons (Dalton, 2001) (See Volume 1, Chapter 13 for more discussion.).

One aspect of setting an OEL for this class of chemicals that requires attention is the phenomenon of "accommodation." Accommodation means that with continuing exposure throughout the day, the objectionable nature of the odor diminishes. For example, many persons who work in factories that use chemicals like amyl alcohol initially find the odor objectionable but within 5 to 10 minutes, and for the remainder of the workday, the workers do not even recognize that it is present. Some have claimed that this fatigue provides an opportunity for chronic cellular irritation, thus, in their view accommodation is not a beneficial response. Although there is no commonly accepted approach for dealing with chemicals with this property, it is important that the concentration that is selected as the OEL be a fraction of that known to produce toxic effects (even if the odor is tolerable).

Setting Limits for Persistent Chemicals

In general, those groups involved in setting OELs have not attempted to quantitatively account for the differences in the pharmacokinetics of chemicals. That is, it has usually been assumed that chemicals with very long biologic half-lives in animals will also have long half-lives in humans, and that this is accounted for in the results of chronic animal studies.

In recent years, more attention has been focused on chemicals with very long half-lives, and toxicologists now know that differences in the elimination between animals and humans can be substantial (even when relative life expectancies are considered). For example, the difference in the biologic half-life for dioxin between rodents and humans is sufficiently great that the steady-state blood concentrations at a given dose are quite different (Leung et al., 1988). For the so-called "long lived or persistent" chemicals, it is prudent to rely upon PB-PK models to account for these differences when deriving OELs (especially those based on animal data). The basis for and the benefits of the PB-PK approach are discussed in Chapter 11.

An approach to setting OELs for persistent chemicals that has been described by Leung et al. (1988) is worthy of evaluation. Basically, the methodology incorporates information on the biologic half-life of the chemical in humans (the pharmacokinetics), as well as the background concentration of the chemical in humans due to contamination of the food chain. The approach assumes that the biological half-life of a chemical in humans can be predicted based on animal data using a PB-PK model when human data are not available. These researchers reasoned that if the amount of chemical absorbed due to workplace exposure is about the amount that every American ingests, or the total uptake (occupational and/or dietary) is much lower than the NOEL, then the occupational contribution almost certainly poses no significant increased health risks.

This approach was applied to setting an OEL for 2,3,7,8-tetrachlorodibenzodioxin (TCDD) because a good deal of toxicology information was available. Since TCDD is highly lipophilic and has a long biologic half-life in humans, it is expected to accumulate in adipose tissue with repeated daily exposure. TCDD levels in the adipose tissue of nonoccupationally exposed persons in the United States was about 7 parts per trillion (ppt) in the mid-1990s and much less today.

Example 12-3: ***Basic Approach to Setting an OEL for a Resistant Chemical***

The steady-state level of TCDD in adipose tissue resulting from occupational exposure to 200 pg/m³ can be estimated by:

$$\text{Steady - state concentration} = \frac{1.44\,(D_t)\,(t_{1/2})}{10.5 + (59.5/10)} \qquad \text{[Equation 12-3]}$$

Where: D_t = daily intake

$t_{1/2}$ = biologic half-life (years)

This calculation assumes that the TCDD concentration in the liver and other tissues is about 1/10 that in adipose tissue, and the average human weighs 70 kg with 10.5 kg (15%) body fat. If the half-life is assumed to be four to eight years, the steady-state TCDD concentration in adipose tissue resulting from occupational exposure at an airborne concentration of 200 pg/m³ will be 89-179 ppt. Thus, exposure to this OEL for 40 years could raise the steady-state body burden well above the 7 ppt background concentration (Leung et al., 1988).

Since the increase in body burden for an OEL of 200 pg/m³ is much greater than due to diet, the dose should not be immediately considered insignificant. The authors then began to evaluate other factors and presented the following rationale for concluding that the OEL was reasonable. First, the concentration of TCDD measured in the adipose tissue of rats exposed for two years to the NOEL of 0.001 µg/kg-day was 540 ppt. Since humans sequester more TCDD in adipose tissue than lower species, which is speculated by some scientists as a protective mechanism, a comparable level in human fat should yield a lesser risk than that suggested in rodent studies. Second, humans exposed to 16 mg TCDD had a theoretical peak adipose tissue level of about 1,300 ppb (16 µg/12.25 kg), yet they only developed chloracne, which resolved within six months. None of those who received 8 µg TCDD developed chloracne, yet their peak adipose tissue levels were about 650 ppt (8 µg/12.25 kg).

The third factor that was considered was the adipose tissue concentration for other persistent chemicals with OELs. Table 12-4 shows that the 26-fold increase over background (179 ppt/7 ppt) for TCDD, when compared with other industrial chemicals following workplace exposure at their corresponding TLVs®, appears to be comparable. Fourth, and most importantly, the risk associated with the proposed OEL was 100-fold below the animal NOEL for carcinogenicity (i.e., 10 pg/kg-day). This was thought to pose no significant human health hazard (Leung et al., 1988). From these data, the risk to humans appeared rather small (especially following a comparison with OELs for other persistent chemicals). The methodology is a useful one for evaluating the hazard posed by persistent chemicals.

Table 12-4
Estimated steady-state adipose tissue level of chemicals following chronic exposure at the OEL

Chemical	OEL		$t_{1/2}$(year)	Adipose Tissue Level				E/B[b]
				Background[a]		Exposed		
DDT[c]	1	mg/m³	1.5	6	ppm	480	ppm	80
Dieldrin	0.25	mg/m³	1.0	0.29	ppm	80	ppm	276
PCB[d]	1	mg/m³	1.5	1	ppm	800	ppm	800
TCDD[e]	200	pg/m³	4-8	7	ppt	89-179	ppt	13-26

[a] Background levels referred to those in non-occupationally exposed general population.
[b] E/B is the ratio of steady-state adipose tissue level in the occupationally exposed to that in the background.
[c] DDT = dichlorodiphenyl trichloroethane
[d] PCB = polychlorinated biphenyl
[e] TCDD = tetrachlorodibenzodioxin

Setting Limits for Respiratory Sensitizers

Respiratory sensitization is an immune status whereas respiratory allergy is a clinical manifestation. Respiratory sensitization results from an immune response to antigen (usually, but not exclusively, exogenous antigen), which may result in clinical hypersensitivity upon subsequent inhalation exposure to the same or similar antigen. An allergic or sensitization response characteristically requires at least two encounters with antigen. Following first exposure, the susceptible individual mounts a primary immune response, which results in sensitization (the induction or sensitization phase). If the sensitized individual subsequently comes into contact with the same antigen, a clinical allergic reaction may be provoked (the elicitation phase). Allergic reactions may be attributable to either antibody or cell-mediated immune responses. Acute allergic reactions in the respiratory tract induced by exposure to exogenous antigens (e.g., some industrial chemicals) are almost invariably associated with specific antibody responses, frequently, but not always, of the IgE class (Briatico-Vangosa, 1994).

Certain chemicals can produce an allergic response as a result of either dermal or inhalation exposure (Sarlo and Karol, 1994). The reason for the interest is that this class of chemicals can, after a sensitizing exposure occurs, produce an adverse effect with subsequent exposures to very small quantities. Respiratory sensitizers can produce asthma in select persons and so-called "attacks or incidents" can be fatal if untreated.

In the late 1990s, respiratory sensitizers received perhaps the most attention of all categories of toxicants with respect to setting OELs. The concern about inhalation sensitizers or allergens came about because researchers reported that the incidence of asthma in children and adults appeared to be increasing in the United States and other Western countries. A few years ago, the German MAK began to identify, with notation, those chemicals that were known or suspect inhalation sensitizers. In 1996, the ACGIH® TLV® Committee chose to pursue the same approach, and has recently made a similar notation to chemicals for which they believe a sensitization hazard exists.

The toxicology community has made significant headway in developing methods for identifying likely dermal and respiratory sensitizers over the past 10 years. Since exposure to most dermal sensitizers is prevented by gloves and other personal protective equipment, the focus of the TLV® and MAK Committees has been on respiratory sensitizers. Fortunately, a model that relies on SAR has been developed for screening groups of chemicals to identify sensitizers (Karol, 1996), thus making the task a manageable one. The SAR model relies upon identifying chemical moiety in a substance, which is known to increase the probability that it will be a respiratory sensitizer, like an aldehyde or cyano group. This model has been applied to nearly 100 different chemicals.

In addition to SAR models, there are *in vitro* and *in vivo* test methods to identify sensitizers. The most common approach includes an *in vitro* assessment of protein binding potential, followed by *in vivo* evaluation using an animal model (Karol, 1996). Diverse species have been used including mice, rats, and guinea pigs, each possessing distinct advantages and disadvantages in representing human disease (Karol, 1994). The guinea pig model (Karol, 1995) assesses several hypersensitivity responses, such as early and late airway reactions, airway hyperreactivity, production of allergen-specific IgE and IgG_1 antibodies, and eosinophilic inflammation. However, the model is costly and requires a high degree of technical skills (Sarlo and Karol, 1994). Mouse models have been described that associate an increase in total IgE with respiratory sensitizers (Kimber and Dearman, 1992) or characterize the cytokines produced following exposure to chemical allergens (Kimber, 1996). Each of the animal models has been tested with only a limited number of chemicals and requires further validation. An excellent paper which reviews current views on toxicology testing of respiratory sensitizers has been published (Briatico-Vangosa et al., 1994).

Of all the tests, the guinea pig sensitization test has been used most frequently and refinements in the procedure have made it a much more powerful and reliable tool for identifying likely human sensitizers (Briatico-Vangosa et al., 1994). Although a related test relies upon administration of an agent via dermal contact (Kimber, 1996), the most reliable way to identify a likely occupational allergen is through inhalation testing. Using the results of the testing, it has been shown that "safe" levels of exposure can be identified by comparing the test results on a new chemical versus the results obtained with a known occupational sensitizer that has a TLV®. Using a simple ratio approach, an OEL for the "new" chemical can be calculated directly from the animal data.

Setting Limits for Chemical Carcinogens

"Carcinogen" is the term applied to a chemical which has been shown to produce a significant increase in the occurrence of tumors (above background) in an appropriately designed and executed animal study or has been shown to produce an increase in the incidence of cancer in a human population. Chemical carcinogens have been the focus of most environmental and occupational regulations for the past 25 years.

In the United States, the impetus to have the TLV® Committee develop a classification scheme for occupational carcinogens began in 1970. At that time, lists were routinely published by numerous agencies and different groups who claimed that a large number of substances were likely to be occupational carcinogens. As noted by Stokinger (1977), substances of purely laboratory curiosity, such as acetylaminofluorene and dimethylaminobenzene, which were found to be tumorigenic in animals, were classified along with known human carcinogens of high potency and individual significance, such as bis-chloromethyl ether. In short, no distinction was made between an animal tumorigen and a likely human carcinogen. Union leaders, workers, and the public would often become equally worried about the positive results of animal bioassays of different chemicals even though for a given dose, the carcinogenic or mutagenic potency could vary by 1,000,000 fold (Stokinger, 1977).

During the 1970s and 1980s, the TLV® Committee believed that the finding of a substance to be tumorigenic, often in a half-dead mouse or rat administered intolerable doses, as was the case for chloroform and trichloroethylene, was not suggestive evidence that it was likely to be carcinogenic in man under controlled working conditions. It is for this reason that the ACGIH® Chemical Substances TLV® Committee, as early as 1972, made a clear distinction between animal and human carcinogens. As time has passed, the TLV® Committee has stood firm that not all carcinogens pose an equal hazard; even when the potency of two chemicals may be equally great. One reason, among others, was that some chemicals are mutagenic or genotoxic while others produce tumors through epigenetic mechanisms (Williams and Weisberger, 1994). By setting exposure limits, ACGIH®, as well as its sister organizations (like the MAK and the Health and Safety Executive of the United Kingdom) adopted the view that all chemical carcinogens should at least have a "practical threshold." This term simply means that some dose of these agents would not be expected to pose a significant cancer risk. The dose is very small compared to that shown to produce cancer in animals, and is probably easily handled by biological progressive mechanisms in humans.

In 1977, Herbert Stokinger, then chairman of the ACGIH® TLV® Committee, summarized the historical philosophy of ACGIH® with respect to TLVs® for carcinogens:

> "Experience and research findings still support the contention that TLVs® make sense for carcinogens. First and foremost, the TLV® Committee recognizes practical thresholds for chemical carcinogens in the workplace, and secondly, for those substances with a designated threshold, that the risk of cancer from a worker's

occupation is negligible, provided exposure is below the stipulated limit. There is no evidence to date that cancer will develop from exposure during a working lifetime below the limit for any of those substances.

Where did the TLV® Committee get the idea that thresholds exist for carcinogens? We have been asked 'Where is the evidence?' ... Well, the Committee thinks it has such evidence, and it takes three forms:

1) Evidence from epidemiologic studies of industrial plant experience, and from well-designed carcinogenic studies in animals,
2) Indisputable biochemical, pharmacokinetic, and toxicologic evidence demonstrating inherent, built-in anticarcinogenic processes in our bodies,
3) Accumulated biochemical knowledge makes the threshold concept the only plausible concept" (Stokinger, 1977).

Although these comments were written about 25 years ago, many industrial hygienists, industrial toxicologists, and occupational physicians generally continue to agree with Dr. Stokinger's position. This has been due, in part, to the work of Ames et al. (1993, 1995, 1996) who have shown that man's diet is abundant with chemical carcinogens and that humans have almost certainly developed adequate mechanisms for dealing with very "low dose" exposures to virtually all classes of carcinogens.

In recent times, the TLV® and MAK Committees have attempted to keep pace with the increased understanding of the hazards posed by chemical carcinogens. For example, beginning in about 1985, they began considering not only the results of mathematical models used to estimate response at low doses but also *in vitro* data, information on the mechanisms of action, case-reports, genotoxicity data, and other information on chemical carcinogens before setting a particular TLV®. Evidence that such discussions occurred is presented in the *Documentation* for the TLV® for trichloroethylene, methylene chloride, 1,1,1-trichloroethane, and others which were revised between 1985 and 1990. Due to the variability in risk estimates between the various statistical or low dose models (e.g., Weibull, multi-stage) and their inability to incorporate biological protective mechanisms, the TLV® Committee has been reluctant to place much emphasis on their results, and thus far, has not set a TLV® based on the results of low dose models.

As noted in the current *TLVs® and BEIs® Book* (ACGIH®, 2002), when deciding on values for chemical carcinogens, the Chemical Substances Group within the TLV® Committee gives greatest weight to epidemiological studies that are based on good quantitative exposure data. When the weight of evidence is convincing, certain chemicals will receive an A1 categorization. These are called Confirmed Human Carcinogens. Next in importance, and more typically available, are positive bioassays involving rats or mice (but human data are lacking). Such substances are given an A2 designation and are called "Suspected Human Carcinogens." In reviewing the key experimental toxicology studies, the Committee considers route of entry (greatest weight given to inhalation studies), dose-response gradient, potency, mechanism of action, cancer site, time-to-tumor, length of exposure, and underlying incidence rate for the type of cancer and species under study. Replication of results is important to the Committee, especially if comparable results are obtained in different species. Other types of information, such as batteries of genetic toxicity studies, are useful in confirming that a substance is a carcinogen, but usually are not helpful in setting a TLV®.

Appendix A of the annual *TLVs® and BEIs® Book* contains a description of categories into which chemical carcinogens have been placed (ACGIH®, 2002). The goal of the Chemical Substances TLV® Committee has been to synthesize the available information in a manner that will be useful to practicing industrial hygienists without overburdening them with needless details. The Committee reviewed current methods of classification used by other groups and in 1991 developed a new procedure for classification (Alvanja et al., 1990). This was generally accepted in 1992 and the following categories for occupational carcinogens are currently used by the TLV® Committee (ACGIH®, 2002):

A1 - Confirmed Human Carcinogen: The agent is carcinogenic to humans based on the weight of evidence from epidemiologic studies.

A2 - Suspected Human Carcinogen: Human data are accepted as adequate in quality but are conflicting or insufficient to classify the agent as a confirmed human carcinogen; OR, the agent is carcinogenic in experimental animals at dose(s), by route(s) of exposure, at site(s), of histologic type(s), or by mechanism(s) considered relevant to worker exposure. The A2 is used primarily when there is limited evidence of carcinogenicity in humans and sufficient evidence of carcinogenicity in experimental animals with relevance to humans.

A3 - Confirmed Animal Carcinogen with Unknown Relevance to Humans: The agent is carcinogenic in experimental animals at a relatively high dose, by route(s) of administration, at site(s), of histologic type(s), or by mechanism(s) that may not be relevant to worker exposure. Available epidemiologic studies do not confirm an increased risk of cancer in exposed humans. Available evidence does not suggest that the agent is likely to cause cancer in humans except under uncommon or unlikely routes or levels of exposure.

A4 - Not Classifiable as a Human Carcinogen: Agents which cause concern that they could be carcinogenic for humans but which cannot be assessed conclusively because of a lack of data. *In vitro* or animal studies do not provide indications of carcinogenicity which are sufficient to classify the agent into one of the other categories.

A5 - Not Suspected as a Human Carcinogen: The agent is not suspected to be a human carcinogen on the basis of properly conducted epidemiologic studies in humans. These studies have sufficiently long follow-up, reliable exposure histories, sufficiently high dose, and adequate statistical power to conclude that exposure to the agent does not convey a significant risk of cancer to humans; OR, the evidence suggesting a lack of carcinogenicity in experimental animals is supported by mechanistic data.

Substances for which no human or experimental animal carcinogenic data have been reported are assigned a no carcinogenicity designation.

Exposures to carcinogens must be kept to a minimum. Workers exposed to A1 carcinogens without a TLV® should be properly equipped to eliminate exposure to the carcinogen to the fullest extent possible. For A1 carcinogens with a TLV® and for A2 and A3 carcinogens, worker exposure by all routes should be carefully controlled to levels as low as possible below the TLV®. Refer to the "Guidelines for the Classification of Occupational Carcinogens" in the Introduction to the 7th Edition of the *Documentation of the Threshold Limit Values and Biological Exposure Indices* for a more complete description and derivation of these designations.

The TLV® Committee continues to evaluate the mechanisms through which various chemical carcinogens act, and they are seeking improved methods for identifying more accurate guidelines (Doull, 1994). In the future, for example, it is possible that the TLV® Committee may place more emphasis on model-derived cancer risk estimates for certain genotoxic agents (Spirtas et al., 1986; Alavanja et al., 1990) rather than on the uncertainty factor approach. As evidenced in their deliberations on benzene, formaldehyde, and vinyl chloride, the Committee has considered PB-PK models, controlled human studies, mechanisms of action data, pharmacokinetic data, and other relevant information when attempting to identify the appropriate TLV®.

TWO APPROACHES FOR IDENTIFYING OELs FOR CARCINOGENS

Even though the ACGIH® TLV® Committee, as well as many other groups that recommend OELs, may believe that there is likely to be a threshold for carcinogens at very low doses, another school of thought is that there is little or no evidence for the existence of thresholds for chemicals that are genotoxic (Crump et al., 1976; Bailer et al., 1988; EPA, 1996). In an attempt to take into account the philosophical postulate that chemical carcinogens do not have a threshold, even though an NOEL can be identified in an animal experiment, and because a test involving several hundred animals cannot describe the large differences among humans in the general population or accurately estimate a NOEL, modeling approaches to estimate the possible cancer risk to humans exposed to very low doses have been developed (Crump et al., 1976; Krewski et al., 1989; Paustenbach, 1990b; Holland and Sielken, 1993).

The rationale for a modeling approach to identify safe levels of exposure is that it is impossible to conduct toxicity studies at doses near those measured in the environment because the number of animals necessary to elicit a response at such low doses would be too great (Crump et al., 1976). Consequently, results of animal studies conducted at high doses are extrapolated by these statistical models to those levels (e.g., doses) found in the workplace or the environment. By the early 1980s, mathematical modeling approaches for evaluating the risks of exposure to carcinogens were relied upon by various regulatory agencies who were attempting to protect the public, these models rapidly identified doses that almost certainly posed no health hazard. Interestingly, the limits derived by these models have rarely been the sole factor on which environmental regulatory limits have been established (Rodricks et al., 1987; Travis et al., 1987; Paustenbach, 1990b).

The most popular models for low dose extrapolation are the one-hit, multi-stage, Weibull, multi-hit, logit and probit. The pros and cons of these models have been discussed in many papers and in Chapter 11 of this book. Since it is usually presumed in these models that at any dose, no matter how small, a response could occur in a sufficiently large population, an arbitrary increased lifetime cancer risk level is usually selected (i.e., from 1 in 1,000 to 1 in 1,000,000) as presenting an insignificant or *deminimus* level of risk. By identifying these *deminimus* levels as virtually safe doses, regulatory agencies do not give the impression that there is an absolutely "safe" level of exposure or that there is a threshold below which no response would be expected. This has historically been considered prudent.

Often the use of these statistical models to help assess risks associated with exposure to carcinogens has been erroneously called "risk assessment" (Paustenbach, 2000). In practice, however, modeling is only one part of the risk assessment process (Park and Snee, 1983). A good dose-response assessment whose purpose is to help identify safe levels of occupational exposure requires exhaustive analysis of all of the information obtained from studies of mutagenicity, acute toxicity, subchronic toxicity, chronic studies in animals and metabolism data, human epidemiology data, and an understanding of the role of dermal uptake (Paustenbach, 1990b).

At this time, in the evolution of our understanding of the cancer process, most scientists would support using quantitative risk modeling as providing only an additional piece of information to consider when setting an OEL. Because there are dozens of shortcomings associated with the models, especially their inability to consider complex biological events which undoubtedly occur at low doses, they have not been used as the sole basis for deriving occupational exposure limits.

Several papers have examined the model predicted upper bound cancer risk for workers exposed at the TLV® concentration of several chemicals (Rodricks et al., 1987; Alvanja et al., 1990). The results are presented in Table 12-5. As shown, the theoretical cancer risk for exposure to many, if not most, occupational carcinogens at the current OSHA PELs is about 1 in 1,000 rather than 1 in 1,000,000 (the goal of many environmental regulations). Some TLVs® for carcinogens have model predicted risks as high as 1 in 100. The exposure to carcinogens in the workplace considered "safe or acceptable" is even more interesting when one considers the estimated steady-state tissue concentration in humans following chronic exposure to the TLV® versus that due to background exposure to that chemical in our diet (see Table 12-4).

The wide disparity between the ambient air guidelines recommended by the EPA (which attempt to limit the model-predicted cancer risk to 1 in 10,000 to 1 in 1,000,000) and the workplace values recommended by the TLV® Committee can be explained primarily by the underlying philosophical principles governing the two organizations and the differences in the exposed population, rather than the technical differences between the two methods for identifying "safe

Table 12-5
Model derived estimates of lifetime risks of death from cancer per 1000 exposed persons associated with occupational exposure at pre-1986 and post-1987 OSHA permissible exposure limits (PELs) for selected substances

Substance	Cases/1000 at Previous PEL	Cases/1000 at Revised PEL
Inorganic arsenic	148-767	8
Ethylene oxide	63-109	1-2
Ethylene dibromide (proposal)	70-110	0.2-6
Benzene (proposal	44-152	5-16
Acrylonitrile	390	39
Dibromochloropropane (DBCP)	—	2
Asbestos	64	6.7

Source: Table reprinted from Rodricks et al. (1987); reprinted with permission of *Regulatory Toxicology and Pharmacology*.

doses." The TLV® Committee is governed by the precept that "Threshold Limit Values refer to airborne concentrations of substances and represent conditions under which it is believed that **nearly all** workers may be repeatedly exposed day after day without adverse effect" (ACGIH®, 2000). In contrast, the EPA's Clean Air Act (CAA) and other environmental regulations are intended to insure that all members of the public are exposed to virtually insignificant risks. For example, the CAA states that air standards must protect the public health with an adequate margin of safety. The requirement for an "adequate margin of safety" is intended both to account for inconclusive scientific and technical information and to provide a reasonable degree of protection against hazards that research has not yet identified. The TLVs® define "adequate margin of safety" differently from EPA since healthy workers allegedly make up the bulk of the workforce, e.g., those who report to work each day must be more healthy than the general population. Environmental exposure is assumed to occur continuously for a 168 hr/week for 70 years. Occupational exposure involves 40 hours per week for about 40 years.

With respect to setting environmental standards, the use of low-dose extrapolation models that are conservative, and the adoption of a 1 in 100,000 or 1 in 1,000,000 risk criterion, have been justified because of a strong desire to protect virtually everyone in the public (e.g., the aged, young, and infirmed); and to account for the fact that the public can be continually exposed for 70 years rather than a portion of a 40 year working lifetime. Due to the very different populations at risk and the fact that workers are sometimes compensated for accepting certain risks, it has been considered reasonable that the approaches used to set various limits (as well as the risk criteria) are different.

Setting Limits For Mixtures

The whole topic of setting OELs for mixtures began to be re-evaluated by the toxicology community in the middle 1990s, and it can be expected that it will receive a good deal of discussion over the next few years. Historically, the ACGIH® TLV® Committee approach has been to consider chemicals that act on the same target organ or act through the same mechanism of action as being additive with respect to their hazard. A number of meetings of experts were held in the 1970s through the 1990s to reassess this approach and, in the main, it was concluded that the methodology described in the *TLVs® and BEIs® Book* was adequate (if not amply health protective) (EPA, 1984, 2002).

The approach currently recommended by the TLV® Committee is as follows (taken from a recent *TLVs® and BEIs® Book*):

> "When two or more hazardous substances which act upon the same organ system are present, their combined effect, rather than that of either individually, should be given primary consideration. In the absence of information to the contrary, the effects of the different hazards should be considered as additive. That is, if the sum of exceeds unity, then the threshold limit of the mixture should be considered as being exceeded.

$$\frac{C_1}{T_1} + \frac{C_2}{T_2} + \ldots \frac{C_n}{T_n} < 1.0 \qquad \text{[Equation 12-4]}$$

C_1 is the observed atmospheric concentration and T_1 the corresponding threshold limit (see Examples in Appendix C of the *TLVs® and BEIs® Book*).

Exceptions to the above rule may be made when there is a good reason to believe that the chief effects of the different harmful substances are not in fact additive, but are independent as when purely local effects on different organs of the body are produced by the various components on the mixture. In such cases, the threshold limit is ordinarily only exceeded when at least one member of the series (C_1/T_1 or C_2/T_2, etc.) itself has a value exceeding unity.

Synergistic action or potentiation may occur with some combinations of atmospheric contaminants. Such cases, at present, must be determined individually since no model is available for predicting when this could occur. Potentiating or synergistic agents are not necessarily harmful by themselves, and dose is a critical factor when attempting to prohibit the health hazard. Potentiating effects of exposure to such agents by routes other than that of inhalation are also possible, e.g., imbibed alcohol and inhaled narcotic agents (trichloroethylene). Potentiation is characteristically exhibited at high concentrations, and less probable at low concentrations.

As described in the ACGIH® *TLVs® and BEIs® Book*, Equation 12-4 only applies when the components in a mixture have similar toxicologic effects. It should not be used for mixtures with widely differing reactivities or mechanisms of toxic action, e.g., hydrogen cyanide and sulfur dioxide. It is essential that the atmosphere be analyzed both qualitatively and quantitatively for each component present in order to evaluate compliance or noncompliance with this calculated TLV®. Examples about how to perform the calculation are presented in Chapter 10 of Volume 1.

Recently, this approach to dealing with mixtures was questioned by the German MAK committee. Specifically, in 1997 the MAK reiterated their view that this simple approach may not be appropriate in some situations. They stated that it was advisable to conduct toxicology tests on the mixture of chemicals to which workers would be exposed rather than rely on equations that attempt to consider only the target organ. In particular, they believed it was essential to evaluate common commercial mixtures, like gasoline or certain commercial all-purpose fluids like mineral spirits, in separate toxicology tests and not to rely on Equation 12-4. Although any group responsible for setting OELs would not take issue with such an approach, most would probably agree that until such data are available, the method recommended by ACGIH® is reasonable. As noted in Chapter 10 of Volume 1, there are several possible scientific shortcomings in the approach, but it will take many years before these are resolved.

DO THE TLVs® PROTECT ENOUGH WORKERS?

Beginning in 1988, concerns were raised by numerous persons regarding the adequacy or health protectiveness of the TLVs® (Castleman and Ziem, 1988; Ziem and Castleman, 1989; Roach and Rappaport, 1990). The key question raised in these papers was "what percent of the working population is truly protected from adverse health effects when exposed to the TLV®?"

In the first of their papers, Castleman and Ziem (1988) claimed that the TLVs® were excessively influenced by corporations and, as a result, suggested that they lacked adequate objectivity. In addition, they indicated that the scientific documentation for many, if not most, of the TLVs® was woefully inadequate. They concluded by suggesting that "an ongoing international effort is needed to develop scientifically based guidelines to replace the TLVs® in a climate of openness and without manipulation by vested interests."

In their second paper, Ziem and Castleman (1989) further discussed their views about the inadequacies of the TLVs®. To a large extent, this paper was a modification and expansion of their 1988 paper. They once again concluded that the TLVs® were not derived with sufficient input from physicians, and that many TLVs® were simply not low enough to protect most workers. They believed that there was more than circumstantial evidence to show that there had been an excessive amount of industrial influence on the TLV® Committee, and that this resulted in TLVs® that were not sufficiently low to protect workers.

The response to these two papers by occupational physicians and hygienists was significant (Finklea, 1988; Paustenbach, 1990a,b). Over the 12 months that followed, more than a dozen letters to the editor were published and editorials appeared in Journal of Occupational Medicine, American Journal of Industrial Medicine, and the American Industrial Hygiene Association Journal. One editorial, written by Tarlau (1990) of the New Jersey Department of Environmental Protection, suggested that hygienists would be better off not relying on the TLVs®. This prompted a rather lengthy response that discussed the historical benefits of the TLVs® and suggested that the papers criticizing the TLVs® had some merit, but that the critics, to a large degree, were applying the social expectations and scientific standards of 1990 on risk decisions that were often performed more than 30 to 40 years ago (Paustenbach, 1990a, 1990b).

During 1988-1990, the claims that the TLVs® were not scientifically sound were, to a large extent, subjective or anecdotal. Although Castleman and Ziem (1988) identified inconsistencies in the margin of safety inherent in various TLVs®, alleged that companies had undue influence on the TLV® Committee, and that objective analysis had not been conducted, the significance of these claims with respect to whether employees were sufficiently protected at the TLV® remained unclear. The situation changed when two professors, one from the University of North Carolina and the other from England, published a rather lengthy paper that analyzed the scientific basis for a large portion of the TLVs® (Roach and Rappaport, 1990). In this paper, the authors showed that for many of the irritants and systemic toxicants, the TLVs® were at or near a concentration where 10 to 50% of the population could be expected to experience some adverse effect. Although for many chemicals the adverse effect might be transient and not very significant, e.g., temporary eye, nose or throat irritation, the authors did offer adequate evidence that there was only a small margin of safety between the TLV® concentration for some chemicals and those concentrations that had been shown to cause some adverse effect in exposed persons.

Roach and Rappaport summarized their work in this manner:

> "Threshold Limit Values (TLVs®) represent conditions under which the TLV® Committee of the American Conference of Governmental Industrial Hygienists (ACGIH®) believes that nearly all workers may be repeatedly exposed without adverse effect. A detailed research was made of the references in the 1976 *Documentation* to data on 'industrial experience' and 'experimental human studies.' The references, sorted for those including both the incidence of adverse effects and the corresponding exposure, yielded 158 paired sets of data. Upon analysis it was found that, where the exposure was at or below the TLV®, only a minority of studies showed no adverse effects (11 instances) and the remainder indicated that up to 100% of those exposed had been affected (eight instances of 100%). Although, the

> TLVs® were poorly correlated with the incidence of adverse effects, a surprisingly strong correlation was found between the TLVs® and the exposures reported in the corresponding studies cited in the *Documentation.* Upon repeating the search of references to human experience, at or below the TLVs®, listed in the more recent 1986 edition of the *Documentation*, a very similar picture has emerged from the 72 sets of clear data which were found. Again, only a minority of studies showed no adverse effects and the TLVs® were poorly correlated with the incidence of adverse effect and well correlated with the measured exposure. Finally, a careful analysis revealed that authors conclusions in the references (cited in the 1976 *Documentation)* regarding exposure-response relationships at or below the TLVs® were generally found to be at odds with the conclusions of the TLV® Committee. These findings suggest that those TLVs® which are justified on the basis of "industrial experience" are not based purely upon health considerations. Rather, those TLVs® appear to reflect the levels of exposure which were perceived at the time to be achievable in industry. Thus, ACGIH® TLVs® may represent guides or levels which have been achieved, but they are certainly not thresholds" (Roach and Rappaport, 1990).

The authors reported the following as their key findings:

> "Three striking results emerged from this work, namely, that the TLVs® were poorly correlated with the incidence of adverse effects, that the TLVs® were well correlated with the exposure levels which has been reported at the time limits were adopted and that interpretations of exposure-response relationships were inconsistent between the authors of studies cited in the 1976 *Documentation* and the TLV® Committee. Taken together these observations suggest that the TLVs® could not have been based purely on consideration of health.
>
> While factors other than health appear to have influenced assignments of particular TLVs®, the precise nature of such considerations is a matter of conjecture. However, we note that one interpretation is consistent with the above results, namely, that the TLVs® represent levels of exposure which were perceived by the Committee to be realistic and attainable at the time" (Roach and Rappaport, 1990).

A number of scientists published comments on the Roach and Rappaport (1990) analysis. One of the more thorough discussion papers was written by the past-chairs of ACGIH® (Adkins et al., 1990). In their letter to the editor, they claimed that the Roach and Rappaport paper was flawed, and that it did not assess the validity of the bulk of the TLVs®. The essence of their criticism was that the:

> "...conclusions which they draw concerning the protection afforded by TLVs® are based on incomplete consideration of all of the data relative to a given substance. The authors present information in their tables as though the effects and exposures are valid and generally accepted by the occupational health community. No single epidemiologic study normally stands by itself. Requirements for inferring a causal relationship between disease and exposure in epidemiological studies are well established and include criteria for temporality, biological gradient with exposure, strength of the association, consistency with other studies, and biological plausibility of the observed effect. Roach and Rappaport present an uncritical analysis of various reports which would lead the uninformed reader to conclude that these criteria have been satisfied. In developing exposure recommendations, the TLV® Committee and most other scientific organizations consider all of the relevant data before drawing conclusions. This includes judgments as to the validity and quality of individual studies in addition to the overall weight of the scientific evidence" (Adkins et al., 1990).

Another rather lengthy counter-analysis of Ziem and Castleman that contained a good deal of historical perspective was written by the ACGIH® Board of Directors (1990). In that paper, the Board stated that:

> "While some criticisms may be valid, these articles do not fairly present the facts concerning historical development of TLVs® nor do they accurately portray procedures followed by the TLV® Committee in developing and reviewing TLV® recommendations. Both articles contain a substantial number of errors and omissions and freely exercise selective quotation and quotation out of context in an effort to make their points. The section of Ziem and Castleman's article which discusses "Origins of TLVs®" is a masterpiece of selective quotation and quotation out of context. This begins with their quoting a statement made by L.T. Fairhall concerning the role of industrial hygienists in setting health standards: "He [industrial hygienist] is in contact with the individuals exposed and therefore soon learns whether the concentrations measured are causing any injury or complaint." The authors use this quote to imply that physicians were excluded from the process of developing exposure guidelines. Taken in context, Fairhall's statement is as follows: 'The industrial hygienist is in contact with not one, but a number of plants, using a given toxic substance. He knows, as no one else knows, the actual aerial concentration of contaminant encountered in practice. And he is in contact with the individuals exposed and therefore soon learns whether the concentrations measured are causing any injury or complaint. His judgement and the combined judgement of this entire Conference group is therefore most valuable in helping formulate maximum allowable concentration values.' Contrary to Ziem and Castleman's comments, Fairhall

> advocated a multi disciplinary approach, including physicians, to making exposure recommendations. This has continued to be the operating philosophy of the TLV® Committee. The conference in its first ten meetings was chaired by five physicians in six of the ten years" (ACGIH® Board of Directors, 1990).

In 1993, Rappaport published a follow-up analysis regarding the adequacy of the TLVs®. He noted that, given the continuing importance of the ACGIH® limits, it was useful to compare the basis of the TLVs® with that employed by OSHA *de novo* in its 12 new PELs. Using benzene as an example, he showed that OSHA's new PELs had been established following a rigorous assessment of the inherent risks and the feasibility of instituting the limit. He concluded that the TLVs®, on the other hand, had been developed by ad hoc procedures and appeared to have traditionally reflected levels thought to be achievable at the time. However, Rappaport noted that this might be changing. Specifically, he said that, "Analysis of the historical reductions of TLVs®, for 27 substances on the 1991–1992 list of intended changes, indicates smaller reductions in the past (median reduction of 7.5-fold between 1989 and 1991). Further analysis suggests a more aggressive policy of ACGIH® regarding TLVs® for carcinogens but not for substances that produce effects other than cancer." He also noted that, "Regardless of whether the basis of the TLVs® has changed recently, it would take a relatively long time for the impact of any change to be felt, since the median age of the 1991–1992 TLVs® is 16.5 years, and 75% of these limits are more than 10 years old" (Rappaport, 1993).

One of the more thought provoking proposals offered by Rappaport was whether the TLV® Committee should consider redefining the definition of the protectiveness of these limits (Rappaport, 1993). Specifically, he suggested, among other things, that ACGIH® "define TLVs® officially as levels that represent guides for purposes of control but that do not necessarily protect 'nearly all' workers." Such a move would be in keeping with what appears to be the traditional basis of TLVs®. This direction could, in time, lead to explicit rules for establishing 'feasibility' and could allow for the direct participation of industry through the submission of data related to levels of exposure in facilities of various types and ages.

It seems clear to most hygienists and toxicologists who have reflected on this issue that, given our increased awareness of the differences in susceptibility of various persons in the workplace, there is a growing lack of confidence that "nearly all workers" are protected against some of the adverse effects at the current TLVs® (such as irritation) unless "nearly all workers" is defined as 80–95% (Newmann and Kimmel, 1998). Whether it is necessary for ACGIH® to ask the TLV® Committee to reduce these values in an attempt to protect an even greater percentage of workers will continue to be a topic of heated discussion.

A related opinion paper/commentary that criticized Rappaport's 1993 analysis was published in 1994 by Castleman and Ziem (Castleman and Ziem, 1994). The authors claimed that the TLV® setting process continued to lack objectivity, and that there was too much opportunity for conflicts of interest to occur. In support of their claims, they presented two tables listing chemicals whose primary TLV® authors were from various chemical companies. Since this paper, little additional debate has occurred and ACGIH® continues to work to improve the process to satisfy its critics.

WHERE THE TLV® PROCESS IS HEADED

Although the merits of the Roach and Rappaport analysis, or for that matter, the opinions of Ziem and Castleman, have been debated for over 10 years, it is clear that the process by which TLVs® and other OELs will be set will never again be as it was between 1945 and 1990. In the future, it can be expected that the rationale, as well as the degree of "risk" inherent in a TLV® will be more explicitly described in its *Documentation*. This degree of transparency in the *Documentation* is necessary since the definition of "virtually safe" or "insignificant risk" with respect to workplace exposure will change as the values of society evolve regarding the definition of "safe" (Paustenbach et al., 1990a).

The degree of reduction in TLVs® or other OELs that will undoubtedly occur in the coming years will vary depending on the type of adverse health effect to be prevented, e.g., central nervous system depression, acute toxicity, odor, irritation, developmental effects, or others. It is unclear to what degree the TLV® Committee will rely on various predictive toxicity models or the risk criteria they will adopt as we enter the next century. However, one thing is clear, transparency in the approach used to set any OEL will be expected by those who are going to apply these guidance values.

During the past 15 years, the ACGIH® TLV® Committee has worked diligently to make the Committee's work more transparent. The benefits of this effort are generally evident in the *Documentation* for the various TLVs® that have been published recently. For example, the 1996 *Documentation* of the TLV® for Formaldehyde is 24 pages in length and cites more than 200 references. The version published in 1990 was two pages long and cited about 20 papers. Regrettably, the fact that the *Documentation* are much more thorough than in the past does not necessarily mean that all of the key papers have been carefully read and thoroughly understood. Often, the truly important or definitive papers have been weighed equally with the many more of lesser quality.

The availability of funds or professional resources is not the only reason that many current TLVs® are not sufficiently well documented to satisfy every health professional. The other reason is that "the bar has been set much higher than its historical placement." In short, the quality of analysis that health professionals, lawyers, and the courts have come to expect since about 1985 has frequently gone beyond what can be provided by volunteers. The compilation of information and the analyses that were invested in the NIOSH criteria documents and the proposed OSHA standards of the 1980s and early 1990s were much more comprehensive than that envisioned as necessary by those who originally established the TLV® setting process. One reason the TLV® *Documentation* fell short was that it was not the original intent of ACGIH® to publish "permanent" values that would be "cast out in stone." Rather, the goal was to disseminate information to health professionals to assist them in helping to protect workers from harm or discomfort. To perform this service, more than 50 years ago ACGIH® asked the most reputable and knowledgeable persons in the field to analyze what was known about a chemical and to suggest a value that was expected to be protective. After the value was proposed, then various parties could submit comments and documentation for the next two years so that a new TLV® could be proposed, if appropriate. In short, the TLVs® did not have to be scientifically "bulletproof" since they were to be dynamic values that could be changed in rather short periods of time.

ACGIH® anticipated that those who worked with chemicals each day would have an interest in providing the best possible guidance to fellow health professionals, so they expected any shortcomings in the documentation to be brought to their attention by their colleagues. One must remember that the TLVs® came about as a direct result of a concern by health professionals that insufficient information was being shared within the occupational health community to help hygienists protect workers. They were never intended to be dissertation quality analyses. The Committee has always encouraged those who were truly experts, including professionals in firms who used various chemicals, to submit information that would help identify the best possible limits.

Does this discussion suggest that the TLV® setting process is in need of change? The answer to this question resides within ACGIH®. If it feels obligated to issue "notice of intended changes" or new TLVs® that are virtually bulletproof, then the current approach probably cannot provide that service. There is simply too much information, and the scientific expertise needed to interpret many of these studies is so specialized, that unassailable work products (e.g., *Documentation of TLVs® and BEIs®*) cannot be reasonably provided by a group of volunteers who have a very limited support staff. If, however, ACGIH® continues to make their *Documentation of the TLVs® and BEIs®* more transparent, if it encourages interested parties to perform the vast amount of tedious work needed to document the rationale for the suggested OEL, and if it is willing to provide modest stipends to the committee members to support graduate students who can do much of the background work, then the present system should be able to satisfy critics. During this period of introspection about how to improve the TLV® setting process, we should not forget that thousands of lives have been saved as a result of these values and that, to the best of our knowledge, as stated by Stokinger more than 20 years ago, few if any workers "have been shown to have sustained serious adverse effects on his health as a result of exposure to these concentrations of an industrial chemical" (Stokinger, 1981).

CORPORATE OELs

Although exposure limits or guides like TLVs® or WEELs for most large volume chemicals have been established, and the vast majority of workers are exposed to processes for which these guidance values are applicable, the majority of the 5,000 or so chemicals to which workers are routinely exposed in industry do not have PELs, RELs, WEELs, or MAKs. As a result, about 50 companies in the United States have chosen to establish internal or corporate limits to protect their employees, as well as those who purchase the chemicals.

The need for internal limits is generally identified by the manufacturing divisions within large companies; although the Health, Safety and Environmental Affairs Department may also initiate the process (Paustenbach and Langner, 1986). A panel of toxicologists, hygienists, physicians, and epidemiologists usually gather the scientific data and make the technical assessment, much like the ACGIH® TLV® Committee. The data considered by the group are similar to those considered by other OEL setting bodies to account for uncertainty and other variability when identifying an applicable OEL (see Table 12-6). Their deliberations are often reviewed by

Table 12-6
Criteria and guidelines for the application of UFs

Type	Magnitude	Comments
Interindividual variation	3-5	Intended to account for variation in susceptibility among the human population (i.e., high risk groups).
Interspecies variation	3-5	Intended to account for uncertainty in extrapolating results obtained in animals to the general human population. Applied in all cases where the NOAEL and LOAEL were derived from an animal study.
Subchronic to chronic or	5 or 10	Intended to account for the uncertainty in extrapolating from less-than-lifetime to lifetime exposure or subchronic to chronic exposure. UF of 10 applied whenever study duration is 90 days or less in rodent species or approximately 1/10 to 1/5 life span in other species.
LOAEL to NOAEL	5 or 10	Intended to account for uncertainty in extrapolating downward from an LOAEL to an NOAEL.

Source: Paustenbach (1997)

an oversight group, which integrates the scientific input with information provided by the manufacturing, legal, law, regulatory, and other groups in the company to establish, as appropriate, internal exposure levels, including in some instances, maximum exposure levels and short-term exposure levels.

Most firms who have established OELs believe that the management of occupational exposure requires limits or criteria much like a manufacturing group needs quality control criteria. Some companies have found that manufacturing groups use OELs to define acceptable versus unacceptable manufacturing conditions. Without these limits as guides, operations managers claim that they would not know when conditions are unhealthy, when personnel need to be protected and, if monitoring is performed, how the results should be interpreted. In short, the experience of the past 30 years indicates that corporate or internal OELs serve a useful purpose.

The question could be asked, "If internal OELs are so beneficial, why don't all companies set them?" The answer to the question is complex. First, the cost of establishing limits through committees is substantial. Based on the author's experience, to establish and document a corporate OEL, about 150–400 professional hours are invested in:

1. Identifying the proper studies,
2. Reading and interpreting them,
3. Selecting a preliminary OEL,
4. Writing the documentation,
5. Having committee meetings,
6. Revising the documentation and OEL based on the committee's suggestions, and
7. Obtaining reviews of corporate management and the legal department.

At a cost of about $175 per hour, this equates to an investment of about $25,000 to $60,000 per OEL. Second, many firms believe that setting an internal OEL establishes a legal responsibility to meet this limit at all times and that it generates a liability that is otherwise unnecessary. Although these are the two most important issues, others have been mentioned (Paustenbach and Langner, 1986).

Nearly all the firms who set OELs have found that perhaps the most difficult and controversial aspects are the legal ramifications. For example, lawyers have noted that if a company develops internal standards on its chemicals or chooses to adopt values for a chemical that are more conservative than those of a regulatory agency, the firm had better plan to comply with them. On the other hand, many lawyers believe that perhaps an equal legal exposure exists with those firms who know a great deal about the potential hazards of a chemical but do not set internal limits. Admittedly, such a scenario puts manufacturers between a rock and a hard place. For example, some firms may feel that the workmen's compensation immunity does not encourage them to set internal limits on their own chemicals. It is, however, worth bearing in mind that the manufacturer of a chemical could be sued by someone else's employee who, if injured, could claim that the manufacturer did not supply enough data. Although internal limits have been set for nearly 40 years by various firms, there remains controversy about these complex legal issues.

MODELS FOR ADJUSTING OELs

Several researchers have proposed mathematical formulae or models for adjusting OELs (PELs, TLVs®, etc.) for use during unusual work schedules, and these have received a good deal of interest in the industrial and regulatory arenas (Brief and Scala, 1975; Mason and Dershin, 1976; Hickey, 1977, 1980, 1983; Hickey and Reist, 1977, 1979; Roach, 1977, 1978; Paustenbach, 1994; Andersen et al., 1987a). Although OSHA has not officially promulgated specific exposure limits applicable to unusual work shifts, it has published guidelines for use by OSHA compliance officers for adjusting exposure limits (OSHA, 1989). These generally apply to work shifts longer than eight hours per day.

Brief and Scala Model

In the early 1970s, due to the increasingly large number of workers who had begun working unusual schedules, the Exxon Corporation began investigating approaches to modifying the OELs for their employees on 12-hour shifts. In 1975, the first recommendations for modifying TLVs® and OSHA PELs were published by Brief and Scala (1975), wherein they suggested that TLVs® and PELs should be modified for individuals exposed to chemicals during novel or unusual work schedules.

They called attention to the fact that, for example, in a 12-hour workday the period of exposure to toxicants was 50% greater than in the 8-hour workday, and that the period of recovery between exposures was shortened by 25%, or from 16 to 12 hours. Brief and Scala noted that repeated exposure during longer workdays might, in some cases, stress the detoxication mechanisms to a point that accumulation of a toxicant might occur in target tissues, and that alternate pathways of metabolism might be initiated. It has been generally held that due to the margin of safety in most of the TLVs®, there was little potential for frank toxicity to occur due to unusually long work schedules. Based on recent evaluations of the TLVs®, the margins of safety are probably not as great as was believed prior to 1990 (Roach and Rappaport, 1990).

Brief and Scala's approach was simple but important since it emphasized that unless worker exposure to systemic toxicants was lowered, the daily dose would be greater, and due to the lesser time for recovery between exposures, peak tissue levels might be higher during unusual shifts than during normal shifts. This outcome (Figure 12-4) was to be prevented with the following equation for a 5-day work week.

$$\text{TLV Reduction Factor (RF)} = \frac{8}{h} \times \frac{24-h}{16} \qquad \text{[Equation 12-5]}$$

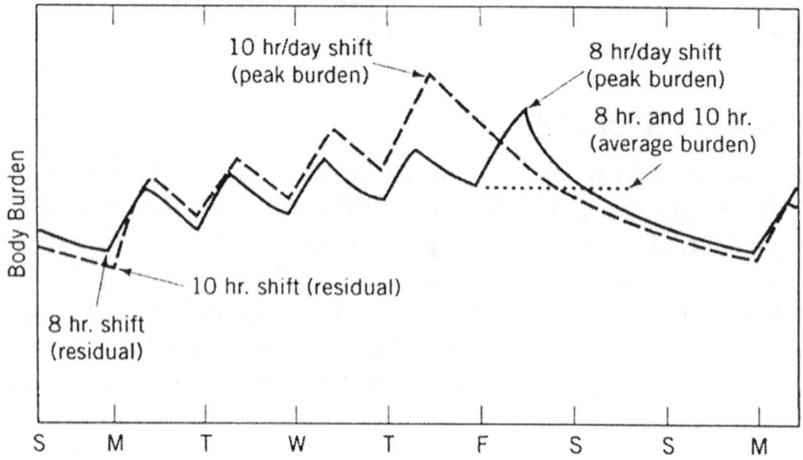

Figure 12-4. Comparison of the peak, average, and residual body burdens of an air contaminant following exposure during a standard (8-hr/day) and unusual (10-hr/day) workweek. In this case, the weekly average body burdens are the same for both schedules since each involved a 40-hr/week. The residual (Monday morning) body burden of the 8-hr shift worker, however, is greater than the 10-hr shift worker and the peak body burden of the person who worked the 10-hr shift is higher than the 8-hr worker. Based on Hickey and Reist (1979).

Where: h = hours worked per day

For a 7-day workweek, the authors suggested that the formula be driven by the 40-hour exposure period; consequently, they developed the following formula, which accounts for both the period of exposure and period of recovery.

$$\text{TLV}^{\circledR}\ \text{Reduction Factor (RF)} = \frac{40}{h} \times \frac{168 - h}{128} \qquad \text{[Equation 12-6]}$$

Where: h = hours exposed per week

One advantage of these formulas is that the biologic half-life of the chemical and the mechanism of action are not needed to calculate a modified TLV®. Such a simplification has shortcomings since the reduction factor for a given work schedule is the same for all chemicals even though the biologic half-lives of different chemicals vary widely. Consequently, the Brief and Scala approach should overestimate the degree to which the limit should be lowered.

Brief and Scala (1975) were cautious in describing the strength of their proposal and offered the following guidelines for its use. Their caveats should be considered when applying this model and other approaches:

$$EF = (EF_8 - 1)RF + 1$$

"(l) Where the TLV® is based on systemic effect (acute or chronic), the TLV® reduction factor will be applied and the reduced TLV® will be considered as a time-weighted average (TWA). Acute responses are viewed as falling into two categories: (a) rapid with immediate onset and (b) manifest with time during a single exposure. The former are guarded by the C notation and the latter are presumed time and concentration dependent, and hence, are amenable to the modifications proposed. Number of days worked per week is not considered, except for a 7-day workweek discussed later."

"(2) Excursion factors for TWA limits (Appendix D of the 1974 TLV® publication) will be reduced according to the following equation:
Where:

EF	=	desired excursion
EF_8	=	value in Appendix D for 8-hour TWA
TW	=	TLV® Reduction Factor"

"(3) Special case of 7-day workweek. Determine the TLV® Reduction Factor based on exposure hours per week and exposure-free hours per week."

"(4) When the novel work schedule involves 24-hour continuous exposure, such as in a submissive or other totally enclosed environment designed for living and working, the TLV® reduction technique cannot be used. In such cases, the 90-day continuous exposure limits of the National Academy of Science should be considered, where applicable limits apply."

"(5) The techniques are not applicable to work schedules less than seven to eight hours per day or $\leq$ 40 hours per week."

In short, the Brief and Scala approach is dependent solely on the number of hours worked per day or week and the period of time between exposures. For example, for any systemic toxin, this approach recommends that persons who are employed on a 12 hours per day, three or four day workweek, should not be exposed to air concentrations of a toxicant greater than one-half that of workers who work on an eight hours per day, 5 day schedule.

Brief and Scala acknowledged the importance of a chemical's biologic half-life when adjusting exposure limits, but because they believed this information was rarely available, they were comfortable with their proposal. They noted that a reduction in an occupational exposure limit is probably not necessary for chemicals whose primary untoward effect is irritation since the threshold for irritation response is not likely to be altered downward by an increase in the number of hours worked each day (Brief and Scala, 1975); that is, irritation is concentration rather than time dependent. Although this appears to be a reasonable assumption, some researchers believe that it may not be entirely justified since duration of exposure could possibly be a factor in causing irritation in susceptible individuals who are not otherwise irritated during normal eight hours per day exposure periods (Alarie and Nielsen, 1982).

Example 12-4: ***Brief and Scala Model***

Refinery operators often work a 6-week schedule of three 12-hour workdays for three weeks, followed by four 12-hour workdays for three weeks. What is the adjusted TLV® for methanol (1993 TLV® = 200 ppm) for these workers? Note that the weekly average exposure (over 9 weeks) is only slightly greater than that of a normal work schedule.

$$\begin{aligned} \text{Adjusted TLV}^{®} &= \text{RF} \times \text{TLV}^{®} \\ &= 0.5 \times 200 \text{ ppm} \\ &= 100 \text{ ppm} \end{aligned}$$

Note: The TLV® Reduction Factor of 0.5 applies to the 12-hour workday, whether exposure is for three, four, or five days per week.

Example 12-5: ***Brief and Scala Model***

What is the modified TLV® for tetrachloroethylene (1993 TLV® = 25 ppm) for a 10 hours per day, four days per week work schedule if the biologic half-life in humans is 144 hours?

$$\text{RF} = \frac{8}{10} \times \frac{24 - 10}{16} = 0.7$$

$$\begin{aligned} \text{Adjusted TLV}^{®} &= 0.7 \text{ x } 25 \\ &= 15 \text{ ppm} \end{aligned}$$

Note: This model and the one used by OSHA *do not consider* the pharmacokinetics (biologic half-life) of the chemical when deriving a modified TLV®. Other models to be discussed later *do* take this into account.

Haber's Law Model

Most toxicologists believe that, in general, the intensity and likelihood of a toxic response is a function of the mass that reaches the site of action per unit time (Fiserova-Bergerova et al., 1980; Andersen et al., 1987b). This "delivered dose" is usually expressed as the concentration of the chemical in the blood for systematic agents. This principle is simplistic and, while it may not apply to irritants and sensitizers, it is clearly true for the systemic toxicants. This assumption is the basis for the OSHA model for modifying PELs for unusual shifts (OSHA, 1979), which is based on Haber's Law. The originators of the model assumed that for chemicals that cause an acute response, if the daily uptake (concentration × time) during a long workday was limited to the amount that would be absorbed during a standard workday, then the same degree of protection would be given to workers on the longer shifts. For chemicals with cumulative effects (i.e., those with a long half-life), the adjustment model was based on the dose imparted through exposure during the normal workweek (40 hours) rather than the normal workday (eight hours).

OSHA recognized that the rationale for the OELs for the various chemicals was based on different types of toxic effects. After OSHA adopted the ~500 TLVs® of 1968 (OSHA, 1970) as PELs, and attempted to do the same in 1989, they placed each of the chemicals into different toxicity categories to assure that an appropriate adjustment model would be used by their hygienists. As can be seen in the following Examples, the degree to which an exposure limit is to be adjusted, if at all, is based to a large degree, on the primary toxic effect of the chemical.

Irrespective of the model that is used to make the adjustments, including the pharmacokinetic models to be discussed, the table in the Federal Register (OSHA, 1989) should be consulted before the hygienist begins the task of modifying an exposure limit. The use of the OSHA tables or the designations of primary adverse effect shown in the most recent ACGIH® TLV® Book (ACGIH®, 1998), combined with some professional judgment, will prevent hygienists from requiring control measures when they are unnecessarily restrictive, as well as minimize the risk of injury or discomfort from overexposure during an unusual exposure schedule.

The objective of OSHA's approach for acute toxicants is to modify the limit for the unusually long shift to a level that would produce a dose (mg) no greater than that obtained during eight hours of exposure at the PEL. Examples of chemicals with exclusively acute effects include carbon monoxide or phosphine. The following equation is recommended by OSHA for calculating an adjustment limit (Equivalent PEL) for these types of chemicals:

$$8 \text{ hour PEL} \times \frac{8 \text{ hours}}{\text{hours exposure in 1 day}} \qquad \text{[Equation 12-7]}$$

The other formula recommended by OSHA applies to chemicals for which the PEL is intended to prevent the cumulative effects of repeated exposure, e.g., the chronic toxicants. For example, PCBs, polybrominated biphenyls (PBB), mercury, lead, and 1,1,1-trichloro-2,2-bis(p-chlorophenyl)ethane (DDT) are considered cumulative toxins because repeated exposure is usually required to cause an adverse effect and the overall biologic half-life is clearly in excess of 10 hours. The goal of PELs in this category is to prevent excessive cumulation in the body following many days or even years of exposure. Accordingly, Equation 12-8 is offered to OSHA compliance officers as a viable approach for calculating a modified limit for chemicals whose half-life would suggest that not all of the chemical will be eliminated before returning to work the following day. Its intent is to ensure that workers exposed more than 40 hours per week will not eventually develop a body burden of that substance in excess of persons who work normal eight hours per day, 40 hours per week schedules.

$$\text{Equivalent PEL} = 8 \text{ hour PEL} \times \frac{40 \text{ hours}}{\text{hours of exposure in one week}} \qquad \text{[Equation 12-8]}$$

The OSHA models, although less rigorous than the pharmacokinetic models which will be discussed, have certain advantages since they do account for the kind of toxic effect to be avoided, require no pharmacokinetic data, and tend to be more conservative than the pharmacokinetic models.

Example 12-6: *OSHA Model*

An occupational exposure limit of 1 microgram per cubic meter ($\mu g/m^3$) has been suggested by NIOSH for PCBs. In animal tests, it has been found that the biologic half-life of PCBs could be as long as several years and that may cause cancer. What adjustment to the occupational exposure limit might be suggested by NIOSH for workers on a 12-hour work shift involving four days of work per week if they adopted the simple OSHA formula?

$$\text{Recommended Limit} = 8 \text{ hr PEL} \times 40 \text{ hr}/48 \text{ hr}$$

$$\text{Recommended Limit} = 1\ \mu g/m^3 \times 0.833 = 0.833\ \mu g/m^3$$

Exposure limits for chemical irritants are currently thought not to require adjustment. Until more is known about response to irritants during unusually long durations of exposure, the physician, nurse, and hygienist should make note of the employee tolerance to the presence of irritants at levels at or near the TLV®. Eventually, human experience will identify those classes of chemicals for which irritation could be a time dependent phenomenon.

Pharmacokinetic Models

Pharmacokinetic models for adjusting occupational limits have been proposed by several researchers (Mason and Dershin, 1976; Hickey, 1977, 1980, 1983; Hickey and Reist, 1977;

1979; Roach, 1966, 1977, 1978; Veng-Pederson et al., 1987; Andersen et al., 1987a). These models acknowledge that the maximum body burden arising from a particular work schedule is a function of the biological half-life of the substance. Pharmacokinetic models generate a correction factor that is based on the elimination half-life of the substance as well as the number of hours worked each day and week. This is applied to the OEL to determine a modified limit. Unlike the OSHA, as well as the Brief and Scala approaches, by accounting for a chemical's pharmacokinetics, these models can also identify those exposure schedules where a reduction in the limit is not necessary.

The rationale for a pharmacokinetic approach to modifying limits is that, during exposure to the TLV® for a normal workweek, the body burden rises and falls by amounts governed by the biologic half-life of the substance. A general formula provides a modified limit for exposure during unusual work shifts so that the peak body burden accumulated during the unusual schedule is no greater than the body burden accumulated during the normal schedule. This is the goal of all of the pharmacokinetic models which have, thus far, been developed.

It is worthwhile to note that the maximum body burden arising from continuous uniform exposure under the standard eight hours per day work schedule nearly always occurs at the end of the last work shift before the 2-day weekend. The only exception is for some of the agents whose biologic half-life is quite long (greater than 16 hours). On the other hand, the maximum body burden under an extraordinary work schedule may not occur at the end of the last shift of that schedule. This is especially true when the duration and spacing of work shifts that precede the last shift differ markedly from the standard week. Because unusual work schedules can be based on a 2-week, 3-week, 4-week, or even 11-week cycle and the work shift may be 10, 12, 16, or even 24 hours in duration, no generalization regarding the time of peak body burden can be offered. The time of peak tissue burden for unusual schedules must, therefore, be calculated for each specific schedule.

Hickey and Reist Model (1977)

In 1977, Hickey and Reist published a paper describing a general formula approach to modifying exposure limits that was fundamentally equivalent to that of Mason and Dershin (1976). The benefits of their work were manifold. First, they validated their approach to some extent by comparing the results with published biological data. Second, they proposed broader uses of the pharmacokinetic approach to modifying limits and presented a number of graphs that could be used to adjust exposure limits for a wide number of exposure schedules. The graphs were based on (a) the biologic half-life of the material, (b) hours worked each day, and (c) hours worked per week.

Over the next three years, Hickey and Reist wrote publications which illustrated how their model could be used to set limits for persons on overtime (Hickey and Reist, 1979) and for seasonal workers (Hickey, 1980). Hickey's treatment of the topic of adjusting exposure limits is quite thorough, and his publications are primarily responsible for most of the interest and research activity in this area.

As discussed, it is clear that, for any schedule, the degree of toxicant accumulation in tissue is a function of the biologic half-life of the substance. Figure 12-4 illustrates how a toxicant might behave in a biologic system or a tissue following repeated exposure to a particular average air concentration during a typical work schedule. Note that the peak body burden, rather than the average or residual body burden, is the parameter of interest. The biologic half-life not only dictates the level to which a chemical accumulates with repeated exposure, it dictates the time at which steady-state will be reached for any given exposure regimen (normal, unusual, or continuous). For example, for moderately volatile substances (e.g., solvents), which have half-lives in the range of 12–60 hours, and for most work schedules, the steady-state tissue burden will be reached after approximately two to six weeks of repeated exposure. For most volatile chemicals (low molecular weight solvents) with shorter half-lives, the steady-state blood levels will be reached after about two to four workdays. Under conditions of continuous uniform exposure, most chemicals will be within 10% of the steady-state levels following about four times the biologic half-life of the chemical, and after seven half-lives, concentrations will be within 1% of the plateau (steady-state) levels (Gibaldi and Perrier, 1975).

Several indices of body or tissue burden could have been chosen as the basis for predicting equal protection for any two different exposure regimens. These indices are the peak, residual, and average body or tissue burden of a substance. As in Mason and Dershin's model (1976), Hickey and Reist selected the peak body burden as the criterion since it is more likely to predict the occurrence of a toxic effect than either the average or residual tissue concentrations. A thorough discussion of the rationale for selecting the peak burden rather than the residual or average for building the models can be found in Hickey's dissertation (Hickey, 1977).

In short, other choices for modeling are problematic. For most chemicals, the residual body burden goes to virtually zero for most chemicals after a weekend away from exposure. Consequently, modeling to control this criterion would not prevent excessive peak burdens. The use of the average burden reduces the model to Haber's Law (David et al., 1981). This, of course, would allow high tissue burdens to occur for long periods even though the time-weighted average burden might be acceptable. Peak burden, therefore, is generally the best criterion.

The Hickey and Reist model can be used to determine a special exposure limit for workers on extraordinary schedules, which will prevent peak tissue or body burdens from being greater than that observed during standard shifts. This special limit is expressed as a decimal adjustment factor (F_p) which, when multiplied by the appropriate exposure limit, would yield the "modified" limit. It is worthwhile to note that all of the researchers have been careful not to assert that currently prescribed or recommended occupational health limits are safe, but only that the special limit that can be predicted from their models should yield "equal protection" during a special exposure situation! Many example problems and a longer description of how to adjust OELs can be found in Paustenbach (1994).

The general equation for regular repetitive schedules is:

$$F_p = \frac{[1 - e^{-kt_{1n}}][1 - e^{-k(t_{1n} + t_{2n})n}][1 - e^{-kT_s}][1 - e^{-k(t_{1s} + t_{2s})}]}{[1 - e^{-kt_{1s}}][1 - e^{-k(t_{1s} + t_{2s})m}][1 - e^{-kT_n}][1 - e^{-k(t_{1n} + t_{2n})}]} \qquad \text{[Equation 12-9]}$$

In which, using hours as the time unit,

t_{1n} = length of normal daily work shift (8 hours),

t_{2n} = length of normal daily nonexposure periods (16 hours),

$t_{1n} + t_{2n}$ = length of normal day (24 hours),

T_n = length of normal week (168 hours),

n = number of workdays per normal week (5),

t_{1s} = length of special "daily" work shift, hours,

t_{2s} = length of special "daily" work shift, hours,

$t_{1s} + t_{2s}$ = length of basic work cycle, analogous to the "day," hours,

m = number of work "days" per work "week" in the special schedule.

Where the special or extraordinary work cycle uses normal days and weeks, the following can be used:

$$F_p = \frac{(1 - e^{-8k})(1 - e^{-120k})(1 - e^{-840k})}{(1 - e^{kH})(1 - e^{-24kDH \text{ or } kHD})(1 - e^{168kW})} \qquad \text{[Equation 12-10]}$$

In which, using hours as the time unit,

F_p = TLV® or PEL reduction factor

k = uptake and excretion rate of the substance in the body ($k = \ln 2/t_{1/2}$)

H = length of special daily work shift

D = number of workdays per "workweek" in the special schedule

W = number of weeks in the special work season (weeks/year)

$t_{1/2}$ = biological half-life

The model may be used to predict the permissible level and duration of exposure necessary to avoid exceeding the normal peak body burden during intrashift, short, high-level exposures. The model does this by establishing excursion limits that will provide equal protection for these situations.

$$F_p = \frac{1 - e^{-kt_n}}{1 - e^{-kt_e}} \qquad \text{[Equation 12-11]}$$

Where:

$k = \ln 2/t_{1/2}$

t_e = exposure time (hrs)

t_n = normal shift length (8 hrs)

Hickey and Reist have noted that for substances with short biologic half-lives, less than three hours, no adjustment needs to be applied for workers on most extraordinary work shifts since there is no opportunity for accumulation. Figure 12-5 presents possible body burdens for the normal workweek compared to workweeks from one to seven 8-hour days. It can be seen that exposure limits may not be increased, even if exposure is only for one day per week, unless the substance half-life is greater than six hours. Similarly, limits need not be decreased for six or seven day workweeks involving exposures of eight hours per day unless the substance half-life is greater than about 16 hours. For substances with very long half-lives, those in excess of 400 hours, F_p is simply proportional to the number of hours worked per week, as compared to 40 hours.

Hickey offered sound advice when he noted that one need not resort to the conservative approaches of Brief and Scala, OSHA, or others when the biologic half-life of the substance is not known. By assuming that the chemical has a half-life that would cause the greatest degree of day-to-day accumulation for that particular work schedule, the worst-case F_p can be calculated for any exposure schedule. Some of these worst case values of F_p for selected schedules are

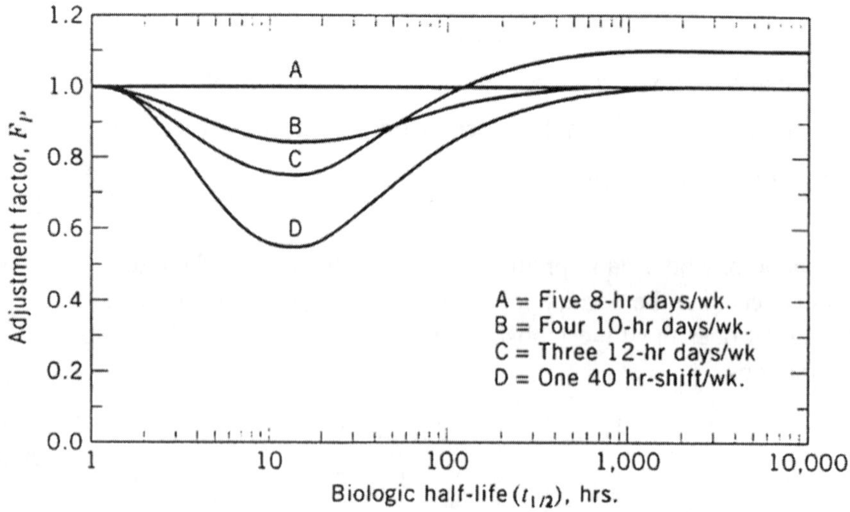

Figure 12-5. Adjustment factor (Fp) as a function of substance half-life ($t_{1/2}$) for various exposure regimens (shift schedules). (From Hickey and Reist, 1977).

shown in Figure 12-5. For example, since F_p varies as a function of the half-life, the worst case condition is 0.84 for four 10-hour days, 0.75 for three 12-hour days, and 0.54 for the consecutive 40-hour shift. Where the half-life is not known, the worst case F_p can be used. Consequently, the pharmacokinetic models can be used to accurately protect workers on any schedule even when the pharmacokinetic behavior of the specific chemical is not known.

A Physiologically-based Pharmacokinetic Approach to Adjusting OELs

The previous section described how one can adjust the OEL using mathematics to insure that peak tissue levels are not exceeded during long shifts. Unfortunately, these models cannot account for biological factors like enzyme induction or approximate target tissue concentrations. The optimal method for adjusting exposure limits to account for peak blood levels for either long work days, or even continuous exposure, is to base it on a pharmacokinetic approach that accounts for the differences between animals and humans, as well as other biologic factors which are not dealt with through the use of a single term like biologic half-life. The most sophisticated approach for incorporating these factors involves the use of a PB-PK model. At this time, about 40 papers have been published that describe evaluations of at least that many chemicals (see Table 12-7).

At present, the PB-PK approach is recommended only for exposure periods of four to 16 hours per day. Pharmacokinetic approaches alone should not be relied on for exposure periods greater than 16 hours per day or less than four hours per day because the mechanisms of toxicity for some chemicals may vary for very short- or very long-term exposure. For these altered schedules, biological information on recovery, rest periods, and mechanisms of toxicity is considered before any adjustment should be attempted. As noted by Andersen et al. (1987b), when pharmacokinetic data are not available, a simple inverse formula may be sufficient for adjustment in most instances. Their paper illustrated the use of the PB-PK approach on two industrially important chemicals: styrene and methylene chloride.

OCCUPATIONAL EXPOSURE LIMITS OUTSIDE THE UNITED STATES

OELs have been adopted or established in a number of countries outside the United States. Some information about these various limits is presented in this section. Most of this information was obtained from Cook (1986), and it represents only a fraction of the information on limits used in other countries. A compilation of virtually all of the OELs used by various countries has been published by WHO (1995).

Argentina

Under ANNEX III, Article 60 of the regulation approved for Decree No. 351/79 lists OELs for Argentina that are essentially the same as those of the 1978 ACGIH® TLVs®. These

Table 12-7
PB-PK Models for toxic substances

Benzene	Lead
Benzo(a)pyrene	Methanol
Butoxyethanol	Methoxyethanol
Carbon Tetrachloride	Methylethylketone
Chlorfenvinphos	Nickel
Chloralkanes	Nichotine
Chloroform	Parathion
Chlropentafluorobenzene	Physostigmine
cis-Dichlorodiammine Platinum	PBBs
Dichloroethane	PCBs
Dichloroethylene	Styrene
Dichloromethane	Toluene
Dieldrin	TCDF
Diisopropylfluorophosphate	TCDD
Dimethyloxazolidine dione	Tetrachloroethylene
Dioxane	Trichloroethane
Ethylene oxide	Trichloroethylene
Hexane	Trichlorotribluoroethane
Kepone	Vinylidene Fluoride
	Xylene

Source: Leung and Paustenbach, 1995

OELs continue in force. The principal difference from the ACGIH® list is that, for the 144 substances (of the total of 630) for which no STELs are listed by ACGIH®, the values used for the Argentina TWAs are also entered under this heading.

Commonwealth of Australia

The National Health and Medical Research Council of the Commonwealth of Australia adopted a revised edition of the "Occupational Health Guide Threshold Limit Values (1990-91)" in 1992. The OELs have no legal status in Australia, except where specifically incorporated into law by reference. The ACGIH® TLVs® are published in Australia as an appendix to the occupational health guides, revised with ACGIH® revisions in odd-numbered years.

Austria

The acronym MAC or maximal acceptable (or allowable) concentration was a term used in the United States during the years before ACGIH® introduced the expression "TLV®." The MAC was translated by both Austria and Germany to MAK, with the same pronunciation, for Maximale Arbeitsplatzkonzentration.

The Austrian list of values, recommended by the Expert Committee of the Worker Protection Commission for Appraisal of MAC Values in cooperation with the General Accident Prevention Institute of the Chemical Workers Trade Union, is used by the Federal Ministry for Social Administration. The MAC values were declared as obligatory by the Federal Ministry on December 23, 1982. Number 61,710/24-4/82. They are applied by the Labor Inspectorate under the authority of the Labor Protection Law, Section 6, Abs. 2 (Cook, 1986).

Belgium

The Administration of Hygiene and Occupational Medicine of the Ministry of Employment and of Labor uses the TLVs® of ACGIH® as a guideline in control of occupational health hazards. As of July 13, 1983, the periodically issued Belgian TLV® publication was being brought up-to-date with the 1982 ACGIH® TLVs® (Cook, 1986).

Brazil

The TLVs® of ACGIH® have been used as the basis for the occupational health legislation of Brazil since June 8, 1978 through the 3214/78 Edict published by the Ministry of Labor. As the Brazilian work week is usually 48 hours, the values of ACGIH® (1977 TLVs® with changes recommended for 1978) were adjusted in conformity with a formula developed for this purpose. The ACGIH® list was not adopted in its entirety, but only for those air contaminants that, at the time, had nationwide application. The Ministry of Labor has brought the limits up-to-date with establishment of values for additional contaminants in accordance with recommendations from the Fundacentro Foundation of Occupational Safety and Medicine (Cook, 1986).

Canada

For each of the various Provinces in Canada, there can be different OELs. In the Province of Alberta, OELs are under the direction of Alberta Regulation 8/82 of the Occupational Health and Safety Act, Chemical Hazard Regulation with amendments up to and including Regulation 242/83. Section 2 of this Act requires the employer to ensure that workers are not exposed above the limits. They are under the direction of the Standards & Projects Section, Occupational Hygiene Branch, Occupational Health and Safety Division of the Workers' Health, Safety, and Compensation Department (Cook, 1986).

In the Province of British Columbia, the Industrial Hygiene Department, Industrial Health and Safety Division of the Workers' Compensation Board, administers the Industrial Health and Safety Regulations that are legal requirements with which most of British Columbia industry must comply. These requirements refer to the current schedule of TLVs® for atmospheric contaminants published by ACGIH® (Cook, 1986).

In the province of Manitoba, the Department of Environment and Workplace Safety and Health is responsible for legislation and its administration concerning the OELs. Manitoba relies on the provisions of the Workplace Safety and Health Act (Chapter W210), ("assented to" June 11, 1976) for protection of workers. Section 4 provides that "every employer shall provide and maintain a workplace ... this is safe and without the risks of health, as far as is reasonably practicable" together with other sections on enforcement by the Department of Safety and Health Officer. The guidelines currently used to interpret risk to health are the ACGIH® TLVs®, with the exception that carcinogens are given a zero exposure level "so far as is reasonably practicable" (Cook, 1986).

The New Brunswick Occupational Health and Safety Commission is the authority responsible for the health and safety of workers in this province. Under the Occupational Health and Safety Act, New Brunswick has adopted the ACGIH® TLVs®. The applicable standards are those published in the latest ACGIH® issue and, in case of an infraction, it is the issue in publication at the time of infraction that dictates compliance.

The Northwest Territories (NWT) Safety Division of the Justice and Service Department regulates workplace safety for non-federal employees in the NWT. The latest edition of the ACGIH® TLVs® has been adopted by the Division and these values have the force of law (Cook, 1986).

The Department of Labour and Manpower administers the Public Health Act of Nova Scotia under which the list of OELs are a legal requirement. This list is the same as that of the ACGIH® as published in 1976 and its subsequent amendments and revisions. These permissible exposure levels have been adopted and constituted as regulations under both the Industrial Safety Act and Regulations and the Construction Safety Act and Regulations (Cook, 1986).

In the Province of Ontario, regulations for a number of hazardous substances are enforced under the Occupational Health and Safety Act, Revised Statutes of Ontario, 1980, Chapter 321. These regulations are administered by the Occupational Hygiene Services, Occupational Health Branch, Ministry of Labour. Each Regulation is published each in a separate booklet that includes the permissible exposure level and codes for respiratory equipment, techniques for measuring airborne concentrations, and medical surveillance approaches.

In the Province of Quebec, the Act is administered by the Commission of Occupational Health and Safety. Permissible exposure levels are similar to the ACGIH® TLVs®, and compliance with the permissible exposure levels for workplace air contaminants is a legal requirement in Quebec.

Chile

The term "Concentraciones Ambientalis Maximas Permissibles" (CAMP) is used for maximum permissible atmospheric concentrations in Chile. This refers to an average exposure during eight hours daily with a weekly exposure of 48 hours, which can be exceeded only momentarily. In the list of permissible exposure limits for Chile, the concentration of 11 of the substances, indicated by an asterisk, cannot be exceeded for even a moment, these having the capacity of causing acute, severe, or fatal effects.

The CAMP Standards are legally enforceable under Title III on environmental contamination of workplaces of Decreto No. 78 in conformity with the "Sanitary Code." The administration of the Regulation is by the Department of Occupational Health and Environmental Contamination, Institute of Public Health of Chile, Ministry of Health. The values in the Chile standard are those of the ACGIH® TLVs®, to which a factor of 0.8 is applied in view of the 48-hour week.

Denmark

The Danish OELs of April 1984 include values for 542 chemical substances and 20 particulates. It is legally required that these not be exceeded as time-weighted averages. Data from ACGIH® are used in the preparation of the Danish standards. About ¼ of the values are different from those of ACGIH® with nearly all of these being somewhat more stringent.

Ecuador

The Division of Occupational Hazards of the Ecuadorian Institute of Social Security is responsible for control of occupational health hazards but at this time Ecuador does not have a list of permissible exposure levels incorporated into its legislation. The TLVs® of ACGIH® are used as a guide for good industrial hygiene practice.

Finland

The OELs for Finland are defined as concentrations that are deemed to be hazardous to at least some workers on long term exposure. In the establishment of these values, possible effects to especially sensitive persons such as those with allergies are not taken into consideration, nor are those exposures where the possibility of deleterious effect is very improbable.

Whereas ACGIH® has as their philosophy that nearly all workers may be reportedly exposed to substances below the threshold limit value without adverse effect, the viewpoint in Finland is just the opposite in considering that where exposures are above the limiting value, deleterious effects on health may occur.

Germany

The definition of the MAC value is "the maximum permissible concentration of a chemical compound present in the air within a working area (as gas, vapor, particulate matter) which, according to current knowledge, generally does not impair the health of the employee nor cause undue annoyance. Under these conditions, exposure can be repeated and of long duration over a daily period of eight hours, constituting an average work week of 40 hours (42 hours per week as averaged over four successive weeks for firms having four work shifts)...Scientifically based criteria for health protection, rather than their technical or economical feasibility, are employed."

Ireland

Workplace air contaminants are regulated by Section 20 of the Safety in Industry Act of 1980 under the direction of the Industrial Inspectorate of the Ministry of Labor. For the purpose of enforcing and interpreting this Section, the latest TLVs® of ACGIH® are normally used. However, the ACGIH® list is not incorporated in the national laws or regulations.

Japan

The process for setting limits in Japan has been described by Cook (1986). It is not significantly different from that used in most other countries. In recent years, the process for setting limits for certain classes of chemicals has changed from the traditional approach that relied on safety factors to a greater emphasis on more complex approaches. For example, when setting OELs for carcinogens, low-dose extrapolation models and risk criterion like 1 in 1,000 have been adopted. Examples of how this approach has been applied were recently described by Kaneko et al. (Kaneko et al., 1998).

Netherlands

The OELs are used as a guide by the Labour Inspectorate administered by the Director General of Labour. The levels listed for the Netherlands are those published as the "National MAC-list 1985 Arbeidsinspectie P No. 145." Most desirably, the substances in this list include the Chemical Abstracts Service numbers. These values are a revision of the 1981 and 1982/83 lists based on the advice of the National MAC-Commission.

These MAC values are taken largely from the list of ACGIH®. A number of values are based on those of the Federal Republic of Germany's Senate Commission for the Investigation of Health Hazards of Work Materials; others are taken from the recommendations of the United States National Institute for Occupational Safety and Health (NIOSH).

The "Maximal Accepted Concentration" is the term used in the Netherlands with emphasis on the word "accepted" rather than acceptable to point to the fact that the MAC value has been accepted by the authorities. The MAC of a gas, vapor, fume, or dust of a substance is defined as "that concentration in the workplace air which, according to present knowledge, after repeated long-term exposure even up to a whole working life, in general does not harm the health of workers or their offspring." It is to be noted that the words "or their offspring" in this definition are not included in the definition of ACGIH®, and that the TLVs® are based on that definition. In the MAC list, two types of values are used:

1. "Maximum Aanaarde Concentratie—tijdgewogen gemiddelde (MAC-TGG)." This is the maximal accepted concentration averaged over an exposure period up to eight hours per day and 40 hours per week.
2. "Maximale Aanvaarde Concentratie—Ceiling (MAC-C)." This concentration may not be exceeded in any case. A MAC-C is based on short-term toxic action.

Philippines

The Bureau of Working Conditions has adopted the entire list of the 1970 TLVs® of ACGIH®. This list was incorporated into the Occupational Safety and Health Standards promulgated in 1978. In 1986, the only deviations from the 1970 list were 50 ppm for vinyl chloride and 0.15 mg/m^3 for lead, inorganic compounds, fume, and dust (Cook, 1986).

Organization of Russian States (Former USSR)

The former USSR established many of its limits with the goal of eliminating any possibility for even reversible effects, such as those involving subtle changes in behavioral response, irritation, or discomfort. The philosophical differences between limits set in the USSR and in the United States have been discussed by Letavet (1961), a Russian toxicologist, who stated that:

> "The method of conditioned reflexes, provided it is used with due care and patience, is highly sensitive and therefore it is a highly valuable method for the determination of threshold concentrations of toxic substances.
>
> At times disagreement is voiced with Soviet MACs for toxic substances, and the argument is that these standards are founded on a method which is excessively sensitive, namely the method of conditioned reflexes. Unfortunately, science suffers not a surplus of excessively sensitive methods, but their lack. This is particularly true with regard to medicine and biology (Letavet, 1961).
>
> Although the methods of examination of the higher nervous activity are very sensitive, they cannot be considered to always be the most sensitive indicator of an

> adverse response and to enable us always to discover the harmful after-effects of being exposed to a poison at the earliest time."

Such subclinical and fully reversible responses to workplace exposures have, thus far, been considered too restrictive to be useful in the United States and in most other countries. In fact, due to the economic and engineering difficulties in achieving such low levels of air contaminants in the workplace, there is little indication that these limits have actually been achieved in countries that have adopted them. Instead, the limits appear to serve more as idealized goals rather than limits that manufacturers are legally bound or morally committed to achieve (Elkins, 1961; Stokinger, 1963).

Vincent (1998) provides an analysis of the process used in Russia. One of the most informative portions of his report summarizes the Russian MAC setting process:

> "The MAC development process is entirely toxicological, without reference to occupational hygiene or epidemiology. It is carried out under the auspices of the Ministry of Health, acting through the Russian Federation Department of Sanitary and Epidemiological Surveillance.
>
> In preparing the MACs, the material considered is derived mostly from Russian sources. There is no discussion about prevailing exposure levels in industry, technical feasibility, or economical implications. The development of a MAC for a given substance is based entirely on its potential impact on the health of the worker, to the exclusion of all other considerations. But, even at the level of basic health effects, the criterion for setting a MAC is generally more stringent than that adopted by other standards setting bodies."

Of all countries, the OELs used by Russia or the other countries formerly part of the USSR are the most difficult to interpret due to the factors mentioned here. Their applicability to other countries is, therefore, limited.

CONCLUSIONS

Although it is not possible for any single book chapter to discuss how each of the various biological issues that need to be considered when establishing an OEL should be quantitatively accounted for, most of them have at least been qualitatively addressed here. It should be clear from the discussion that the process for setting OELs remains remarkably similar to those that were used in the late 1940s, but that the data used to set these limits and the methodology has evolved as science has evolved. It is also clear that our better understanding of toxicology and medicine, our ability to quantitatively account for pharmacokinetic differences among chemicals, and our better knowledge of the mechanisms of action of toxicants require that more refined approaches be continually incorporated into the procedures. Hopefully, the end result will be that future occupational exposure limits will be based on better scientific information and, therefore, our confidence that workers will be protected at these limits will be even greater than it is today.

QUESTIONS

12.1. Who were the first groups to establish occupational exposure limits (OELs) and for what purpose?

12.2. List at least four organizations in the United States who recommend occupational exposure limits.

12.3. What is the purpose of uncertainty (or safety) factors as they relate to occupational or environmental health standards?

12.4. When establishing an OEL, it is best to apply different uncertainty factors to the different classes of chemicals which produce toxic effects. List five or more categories of toxic effects which OELs attempt to prevent.

12.5. Assume that you are presented the following information about a chemical. Using what is described in this chapter, what might be a reasonable occupational exposure limit?

<u>Chemical A</u>

(1) The primary adverse effect observed in the rat is liver toxicity.

(2) NOEL in an 8 hour/day, 20 day inhalation study with rats is 400 ppm.

(3) Test results in the mouse are similar to the rat (NOEL was 300 ppm).

(4) The biologic half-life in rats is 3 hours and in humans it is 6 hours.

(5) Eye irritation has been reported in laboratory workers exposed to 50 ppm.

(6) In the Ames Test (mutagenicity) and in the mouse lymphoma test (reverse mutations), the chemical was negative (no activity).

(7) UF_1 = 5 to 10 (to account for animal to human differences).
UF_2 = 3 to 10 (to account for differences in sensitivity among humans).
UF_3 = 3 to 10 (to account for conducting a less-than-chronic study).

12.6. Virtually all limits for exposure to carcinogens in our food, water, ambient air, pharmaceuticals, and soil are established based on the results of low-dose extrapolation models. Why don't the ACGIH® TLV® Committee or the AIHA WEEL Committee rely upon these models to identify safe levels of exposure?

12.7. List four different occupations or job descriptions that you think might require adjustments to their OEL as a result of unusual work schedules.

12.8. The use of physiologically-based pharmacokinetic (PB-PK) models has become more common in recent years when attempting to identify appropriate OELs. Please describe the models and why they are important.

REFERENCES

Abdel-Rahman, M.S.: The Third Annual Workshop on Evaluation of Default Safety Factors in Health Risk Assessment. HERA 5(5):963 (1999).

Abdel-Rahman, M.S.: The Fourth Annual Workshop on Evaluation of Default Safety Factors in Health Risk Assessment. HERA 7(1):5 (2001).

Abraham, M.H.; Whiting, G.S.; Alarie, Y.; et al.: Hydrogen Bonding 12. A New QSAR for Upper Respiratory Tract Irritation by Airborne Chemicals in Mice. Quant Struct Relat 9:6-10 (1990).

Abraham, M.H.; Andonian-Haftvan, J.; Cometto-Muniz, J.E.; et al.: An Analysis of Nasal Irritation Thresholds using a New Solvation Equation. Funda Appl Toxicol 31: 71-76 (1996).

Abraham, M.H.; Kumarsingh, R.; Cometto-Muniz, J.E.; et al.: An Algorithm for Nasal Pungency Thresholds in Man. Arch Toxicol 72: 227-232 (1998).

Adkins, L.E.; et al.: Letter to the Editor. Appl Occup Environ Hyg 5(11):748-750 (1990).

Alarie, Y.: Dose Response Analysis in Animal Studies: Prediction of Human Responses. Environ Health Perspect 42:9-13 (1981).

Alarie, Y.; Nielsen, G.D.: Sensory Irritation, Pulmonary Irritation, and Respiratory Stimulation by Airborne Benzene and Alkylbenzenes: Prediction of Safe Industrial Exposure Levels and Correlation with their Thermodynamic Properties. Toxicol Appl Pharmacol 65:459-477 (1982).

Alavanja, M.C.R.; Brown, C.; Spirtas, R.; et al.: Risk Assessment of Carcinogens: A Comparison of the ACGIH and the EPA. Appl Occup Environ Hyg 5(8):510-517 (1990).

Albert, R.E.: Carcinogen Risk Assessment in the U.S. Crit Rev Toxicol 24:75-95 (1994).

Allen, B.C.; Crump, K.S.; Shipp, A.M.: Correlation Between Carcinogenic Potency of Chemicals in Animals and Humans. Risk Anal 8:531-544 (1988).

Allen, B.; Kimmel, C.; Kavlock, R.; et al.: Dose-response Assessment for Development Toxicity. II. Comparison of Generic Benchmark Dose Estimates with NOAELs. Fund Appl Toxicol (in press).

American Conference of Governmental Industrial Hygienists: Documentation of Threshold Limit Values. American Conference of Governmental Industrial Hygienists, Cincinnati, OH (1994).

American Conference of Governmental Industrial Hygienists: Threshold Limit Values: A More Balanced Appraisal. Appl Occup Environ Hygiene 5:340-344 (1990).

American Conference of Governmental Industrial Hygienists: 2002 Threshold Limit Values for Chemical Substances and Physical Agents and Biological Exposure Indices. American Conference of Governmental Industrial Hygienists, Cincinnati, OH (2002).

American Industrial Hygiene Association: An International Review of Procedures for Establishing Occupational Exposure Limits. American Industrial Hygiene Association, Fairfax, VA (1996).

American Public Health Association: Health Based Exposure Limits and Lowest National Occupational Exposure Limits. Draft #5. November 6, 1991. American Public Health Association. Washington DC (1991).

American Industrial Hygiene Association: Workplace Environment Exposure Limits (WEELs). AIHA, Fairfax, VA (2000).

Ames, B.N.: Six Common Errors Relating to Environmental Pollution. Regul Toxicol Pharmacol 7:379-383 (1987).

Ames, B.N.; Gold, L.S.: Environmental Pollution and Cancer: Some Misconceptions in Phantom Risk. Scientific Interference and the Law, pp. 153-181 (1993).

Ames, B.N.; Gold, L.S.: Too Many Rodent Carcinogens: Mitogenesis Increases Mutagenesis. Science 249:970-971 (1990).

Ames, B.N.; Gold, L.S.; Willett, W.C.: The Causes and Prevention of Cancer. Proc Natl Acad Sci USA 92: 5258-5265 (1995).

Ames, B.N.; Gold, L.S.; Shigenaga, M.K.: Cancer Prevention, Rodent High-dose Cancer Tests, and Risk Assessment. Risk Anal 16: 613-617 (1996).

Andersen, M.E.: Pharmacokinetics of Inhaled Gases and Vapors. Neurobehavioral Toxicol Teratol 3:383-389 (1981).

Andersen, M.E.; Clewell, H.J.; Krishnan, K.: Tissue Dosimetry, Pharmacokinetics Modeling, and Interspecies Scaling Factors. Risk Anal 15-533-537 (1995).

Andersen, M.E.: Quantitative Risk Assessment and Chemical Carcinogens in Occupational Environments. Appl Occup Env Hyg 3: 267-273 (1991).

Andersen, M.E.; MacNaughton, M.G.; Clewell, H.J.; et al.: Adjusting Exposure Limits for Long and Short Exposure Periods Using a Physiological Pharmacokinetic Model. Am Ind Hyg Assoc J 48:335-343 (1987a).

Andersen, M.E.; Clewell, H.J.; Gargas, M.L.; et al.: Physiologically Based Pharmacokinetics and the Risk Assessment Process for Methylene Chloride. Toxicol Appl Pharmacol 87:185-205 (1987b).

Anger, W.K.: Worksite Behavioral Research: Results, Sensitive Methods, Test Batteries and the Transition from Laboratory Data to Human Health. Neurotoxicology 11:629-720 (1990).

Astrand, I.: Uptake of Solvents in the Blood and Tissues of Man–A Review. Scand J Work Enviorn Health 1:199-218 (1975).

Astrand, I.; Gamberale, F.: Effects on Humans of Solvents in the Inspiratory Air: A Method of Estimation of Uptake. Environ Res 15:1-4 (1978).

Astrand, I.; Ehrner-Sanuel, H.; Kilbom, A.; et al.: Toluene Exposure. I. Concentration in Alveolar Air and Blood at Rest and During Exercise. Work Environ Health 9:119-130 (1972).

Baetjer, A.M.: The Early Days of Industrial Hygiene: Their Contribution to Current Problems. Am Ind Hyg Assoc J 41:773-777 (1980).

Bailer, J.C.; Crouch, E.A.C.; Shaikh, R.; et al.: One-Hit Models of Carcinogenesis: Conservative or Not? Risk Anal 8:485-497 (1988).

Ball, W.L.: The Toxicological Basis of Threshold Limit Values: Paper #6 Report of Prague symposium on international threshold limit values. Am Ind Hyg Assn J 20:370-373 (1959).

Barnes, D.G.; Dourson, M.: Reference Dose (RfD): Description and Use in Health Risk Assessments. Regul Toxicol Pharmacol 8:471-486 (1988).

Barnes D.G.; Daston, G.P.; Evans, J.S.; et al.: Benchmark Dose Workshop: Criteria for Use of a Benchmark Dose to Estimate a Reference Dose. Reg Toxicol Pharmacol 21: 296-306 (1995).

Beck, B.D.; Clewell, H.J.: Uncertainty/Safety Factors in Health Risk Assessment: Opportunities for Improvement. Human Ecol Risk Assess 7(1):203 (2001).

Beck, B.D.; Conolly, R.B.; Dourson, M.L.; et al.: Symposium Overview: Improvements in Quantitative Non-cancer Risk Assessments. Fundam Appl Toxicol 20:1-14 (1993).

Bierbaum, P.J.; Dormeny, L.J.; Smith, J.P.; et al.: U.S. Approach to Air Sampling of Workplace Contaminants. Current Base and Future Options. Appl Occup Environ Hygiene 8:247-250 (1993).

Bjerner, B.; Holm, A.; Swensson, A.: Diurnal Variation in Mental Performance: A Study of 3-Shift Workers. Brit J Ind Med 12:103-110 (1955).

Bjerner, B.; Holm, A.; Swensson, A.: Studies on Night and Shiftwork. In: Shiftwork and Health. Scandinavian University Books, Oslo, Norway (1964).

Botzum, G.D.; Lucas, R.L.: Slide Shift Evaluation–A Practical Look At Rapid Rotation Theory. Proceedings of Human Factors Society pp. 207-211 (1980).

Bowditch, M.; Drinker, D.K.; Drinker, P.; et al.: Code for Safe Concentrations of Certain Common Toxic Substances Used in Industry. J Ind Hyg Tox 22:251 (1940).

Briatico-Vangosa, G.; Braun, C.L.; Cookman, G.; et al.: Respiratory Allergy: Hazard Identification and Risk Assessment. Review. Fundam Appl Toxicol 23:145-158 (1994).

Brandt, A.: On the Influence of Various Shift Systems on the Health of the Workers. XVI Int Congr Occup Health, Tokyo, p.106 (1969).
Brief, R.S.; Scala, R.A.: Occupational Exposure Limits for Novel Work Schedules. Am Ind Hyg Assoc J 36:467-471 (1975).
Bus, J.S.; Gibson, J.E.: Body Defense Mechanisms to Toxicant Exposure. In: Patty's Industrial Hygiene and Toxicology, 3rd Edition, Volume 3B, Harris; Cralley; Cralley, Eds. John Wiley & Sons, New York (1994).
Butterworth, B.E.; Conolly, R.B.; Morgan, K.T.: A Strategy for Establishing Mode of Action of Chemical Carcinogens as a Guide for Approaches to Risk Assessments. Cancer Lett 93:129-146 (1995).
Butterworth, B.E.; Slaga, T.: Nongenotoxic Mechanisms in Carcinogenesis: Banbury Report 25. Cold Spring Harbor Laboratory, Cold Spring Harbor, NY (1987).
Butterworth, B.E.; Eldridge, S.R.: A Decision Tree Approach for Carcinogen Risk Assessment. In: Growth Factors and Tumor Promotion, M. McClain; B.E. Butterworth, Eds., Progress in Clinical and Biological Research. Wiley-Liss, New York (1996).
Calabrese, E.J.: Methodological Approaches to Deriving Environmental and Occupational Health Standards. John Wiley and Sons, New York (1978).
Calabrese, E.J.: Principles of Animal Extrapolation. John Wiley and Sons, New York (1983).
Calabrese, E.J.: Further Comments on Novel Schedule TLVs®. Am Ind Hyg Assoc J 38:443-446 (1977).
Carlsson, A.: Exposure to Toluene: Uptake, Distribution, and Elimination in Man. Scand J Work Environ Health 8:43-56 (1982).
Castleman, B.I.; Ziem, G.E.: Corporate Influence on Threshold Limit Values. Am J Ind Med 13:531-559 (1988).
Castleman, B.I.; Ziem, G.E.: American Conference of Governmental Industrial Hygienists: Low Threshold Credibility. Am J Ind Med 26:133-143 (1994).
Cherrie, J.W.; Robertson, A.: Biologically Relevant Assessment of Dermal Exposure. Ann Occup Hyg 39:387-92 (1995).
Coffin, D.L.; Gardner, D.E.; Sidorenko, G.I.; et al.: Role of Time as a Factor in the Toxicity of Chemical Compounds in Intermittent and Continuous Exposures, Part II. Effects of intermittent exposure. J Toxicol Environ Health 3:821-828 (1977).
Colburn, W.A.; Matthews, H.B.: Pharmacokinetics in the Interpretation of Chronic Toxicity Tests: The Last-In, First-Out Phenomena. Toxicol Appl Pharm 48:387-395 (1979).
Colquhoun, W.P.; Blake, M.J.F.; Edwards, R.S.: Experimental Studies of Shiftwork. I: A Comparison of Rotating and Stabilized 4-Hour Shift System. Ergonomics 11:437-447 (1968a).
Colquhoun, W.P.; Blake, M.J.F.; Edwards, R.S.: Experimental Studies of Shiftwork. II: Stabilized 8-Hour Shift Systems. Ergonomics 11:527-537 (1968b).
Colquhoun, W.P.: Circadian Variations in Mental Efficiency. In: Biological Rhythms and Human Performance, W.P. Colquhoun, Ed. Academic Press, London and New York (1971).
Colquhoun, W.P.; Blake, M.J.F.; Edwards, R.S.: Experimental Studies of Shiftwork. III: Stabilized 12-Hour Shift Systems. Ergonomics 12:865-875 (1969).
Cometto-Muniz, J.E.; Cain, W.S.: Sensory irritation. Relation to Indoor Air Pollution. Ann NY Acad Sci 641: 137-151 (1992).
Cometto-Muniz, J.E.; Cain, W.S.: Sensory Reactions of Nasal Pungency and Odor to Volatile Organic Compounds: The Alkyl Benzenes. Am Ind Hyg Assoc J 55(9): 811-817 (1994).
Cometto, J.E.; Cain, W.S.; Abraham, M.A.: Nasal Pungency and Odor of Homologous Aldehydes and Carboxylic Acids. Exp Brain Res 118: 180-188 (1998).
Commonwealth of Pennsylvania: Threshold Limit Values and Short-term Limits, Title 25, Part 1, Subpart D, Article IV, Chapter 201, Subchapter A, Threshold Limits. Rules and Regulations, 1 Pa B 1985 (November 1971).
Cook, W.A.: Maximum Allowable Concentrations of Industrial Contaminants. Ind Med 14(11):936-946 (1945).
Cook, W.A.: Occupational Exposure Limits-Worldwide. Am Ind Hyg Assoc, Akron, OH (1986).
Cooper, W. Clark: Indicators of Susceptibility to Industrial Chemicals. J Occ Med 15(4):355-359 (1973).
Corn, M.; Esman, N.A.: Workplace Exposure Zones for Classification of Employee Exposures to Physical and Chemical Agents. Am Ind Hyg Asso J 40:47-57 (1979).

Corn, M.: Historical Perspective on Approaches to Estimation of Inhalation Risks by Air Sampling. Am J Ind Med 21:113-123 (1992).

Crump, K.S.; Hoel, D.G.; Langley, G.H.; et al.: Carcinogenic Processes and their Implications for Low Dose Risk Assessment. Cancer Res 36:2973-2979 (1976).

Crump, K.: A New Method for Determining Allowable Daily Intakes. Fund Appl Toxicol 4:854-871 (1984).

Crump, K.: Calculation of benchmark doses from continuous data. Risk Anal 15: 79-85 (1995).

Dalton, P.: Evaluating the Human Response to Sensory Irritation: Implications for Setting Occupational Exposure Limits. Am Ind Hyg Assoc J 62:723-729 (2001).

David, A.; Frantik, E.; Holvsa, R.; et al.: Role of Time and Concentration on Carbon Tetrachloride Toxicity in Rats. Inter Arch Occup Environ Health 48:49-60 (1981).

de la Cruz, P. L.; Sarvadi, D.G.: OSHA PELs: Where Do We Go From Here? Am Ind Hyg Assoc J 55:894-900 (1994).

De Carvalho, M.; Falson-Rieg, F.; Eynard, I.; et al.: Changes in Vehicle Composition During Skin Permeation Studies. In: Prediction of Percutaneous Penetration, Volume 3b, K.R. Brain; V.J. James; K.A.Walters, Eds., pp. 251-254. STS Publishing Ltd., Cardiff, England (1993).

DeRosa, C.T.; Dourson, M.L.; Osborne, R.: Risk Assessment Initiatives for Noncancer Endpoints: Implications for Risk Characterizations of Chemical Mixtures. Tox Ind Health 5(5):805-824 (1989).

Deutsche Forschungsgemeinschaft Commission for the Investigation of Health Hazards of Chemical Compounds in the Work Area: List of MAK and BAT Values, pp.185. Wiley-VCH, Weinheim, Federal Republic of Germany (1997).

DiVincenzo, G.D.; Yanno, F.J.; Astill, B.D.: Human and Canine Exposure to Methylene Chloride Vapor. Am Ind Hyg Assoc J 33:125-135 (1972).

Doull, J.: The ACGIH® Approach and Practice. Appl Occup Environ Hyg 9(1):23-24 (1994).

Dourson, M.L.; Stara, J.F.: Regulatory History and Experimental Support of Uncertainty (Safety) Factors. Regulatory Toxicol Pharm 3:224-238 (1983).

Dourson, M.L.: Toxicity-based Methodology, Thresholds and Possible Approaches, and Uncertainty Factors. In: Approaches to Risk Assessment for Multiple Chemical Exposures. EPA 600/9-84-008, pp. 2-42 (1984).

Dourson, M.L.; Hertzberg, R.C.; Hartung, R.; et al.: Novel Methods for the Estimation of Acceptable Daily Intake. Toxicol Ind Health 1:23-33 (1985).

Dourson, M.L.: New Approaches in the Derivation of the Acceptable Daily Intake (ADI). Comments Toxicol 1: 35-48 (1986).

Dourson, M.L.; DeRosa, C.T.: The Use of Uncertainty Factors in Establishing Safe Levels of Exposure. In: Statistics in Toxicology, D. Krewski and C. Franklin, Eds.. Gordon and Breach Science Publisher, New York (1991).

Dourson, M.L.: Modifying Uncertainty Factors for Noncancer Endpoints. In: Advanced Topics in Risk Assessment. Society of Toxicology continuing education Course #10. 32nd Annual Meeting, New Orleans, LA (March 1993).

Dourson, M.L.: Methods for Establishing Oral Reference Doses (RfDs). In: Risk Assessment of Essential Elements, W. Mertz; C.O. Abernathy; S.S. Olin, Eds., pp.51-61. ILSI Press, Washington, DC (1994).

Dourson, M.L.; Felter, S.P.; Robinson, D.: Evolution of Science-based Uncertainty Factors in Noncancer Risk Assessment. Reg Tox Pharmacol 24:108-120 (1996).

Dourson, M.L.; Felter, S.P.: Route-to-route Extrapolation of the Toxic Potency of MTBE. Risk Anal 25: 43-57 (1997).

Droz, P.O.; Fernandez, J.G.: Effect of Physical Workload on Retention and Metabolism of Inhaled Organic Solvents–A Comparative Theoretical Approach and its Applications with Regards to Exposure Monitoring. Int Arch Occup Environ Health 38:231-240 (1997).

Durnin, J.V.G.; Passmore, R.: Energy, Work, and Leisure. Heinemann Educational Books, Ltd London (1967).

ECETOC: Percutaneous Absorption. Monograph No. 20, pp.1-80. European Centre for Ecotoxicology and Toxicology of Chemicals, Brussels (1993).

ECETOC: Strategy for Assigning a Skin Notation. Revised ECETOC Document No. 31. European Centre for Ecotoxicology and Toxicology of Chemicals, Brussels (1994).

Ehrenberg, L.; Hieschke, K.D.; Osteman-Golkar, S.; et al.: Evaluation of Genetic Risks of Alkylating Agents: Tissue Doses in the Mouse from Air Contaminated with Ethylene Oxide. Mutation Research 24:83-103 (1974).

Fenske, R.: Dermal Exposure Assessment Techniques. Ann Occup Hyg 37:687-706 (1993).
Fieldner, A.C.; Katz, S.H.; Kenney, S.P.: Gas Masks for Gases Met in Fighting Fires. USA Bureau of Mines Bulletin 248. Pittsburgh, PA (1921).
Finkel, A.M.: Commentary: Who's Exaggerating? Discover, pp. 48-54 (May 1996).
Finklea, J.A.: Threshold Limit Values: A Timely Look. Am J Ind Med 14:211-212 (1988).
Finley, B.; Proctor, D.; Paustenbach, D.J.: An Alternative to the U.S. EPA's Proposed Inhalation Reference Concentration for Hexavalent and Trivalent Chromium. Regul Toxicol Pharm 16:161-176 (1992).
Fiserova-Bergerova, V.: Simulation of Uptake, Distribution, Metabolism, and Excretion of Lipid Soluble Solvents in Man. Aerospace Medical Research Laboratory Report No. AMRL-TR-72-130 (Paper No. 4). Wright-Patterson Air Force Base, Dayton, OH (1972).
Fiserova-Bergerova, V.; Holaday, D.A.: Uptake and Clearance of Inhalation Anesthetics in Man. Drug Metab Reviews 9(1):43-60 (1979).
Fiserova-Bergerova, V.; Vlach, J.; Cassady, J.C.: Predictable Individual Differences in Uptake and Excretion of Gases and Lipid Soluble Vapors Simulation Study. Br J Ind Med 37:42-49 (1980).
Fiserova-Bergerova, V.; Pierce, J.T.; Droz, P.O.: Dermal Absorption of Industrial Chemicals: Criteria for Skin Notation. Am J Ind Med 17:617-35 (1990).
Fiserova-Bergerova, V.: Relevance of Occupational Skin Exposure. Ann Occup Hyg 37:673-85 (1993).
Flury, F.; Zernik, F.: Schadliche Gase, Dampfe, Nebel, Rauch-und Staubarten. Julius Springer Verlag, Berlin (1931).
Flury, F.; Heubner, W.: Uber Wirkung and Eingiftung eingeatmeter Blausaure. Biochem Z 95:249-256 (1919).
Fottler, M.D.: Employee Acceptance of Four Day Workweek. Acad Mangt J 20:100 (December 1977).
Gargas, M.L.; Burgess, R.J.; Voisard, D.E.; et al.: Partition Coefficients of Low-molecular-weight Volatile Chemicals in Various Liquids and Tissues. Toxicol Appl Pharmacol 98(1):87-99 (1989).
Gaylor, D.W.: Quantitative Risk Analysis for Quantal Reproductive and Developmental Effects. Environ Health Perspect 79:243-246 (1989).
Gaylor, D.W.; Slikker, W.: Risk Assessment for Neurotoxic Effeects. Neuro Toxicology 11:211-218 (1990).
Gibaldi, M.; Perrier, D.: Pharmacokinetics. Marcel Dekker, Inc., New York (1975).
Gibaldi, M.; Weintraub, H.: Some Considerations as to the Determination and Significance of Biologic Half-life. J Pharm Sci 60:624-626 (1971).
Gold, L.S.; Ames, B.N.; Sloan, T.H.: Misconceptions About the Cause of Cancer. In: Human and Ecological Risk Assessment, Theory and Practice, pp. 1415-1460. D.J. Paustenbach, Ed. John Wiley & Sons, New York, NY (2002).
Graham, C.; Rosenkrantz, H.S.; Karol, M.H.: Structure-Activity Model of Chemicals that Cause Human Respiratory Sensitization. Regul Toxicol Pharm 26:296-306 (1977).
Graham, J.; Hartwell, J. (Eds.): The Greening of Industry. Harvard Univ Press, Cambridge, MA (1997).
Grandjean, P.; Berlin, A.; Gilbert, M.; et al.: Preventing percutaneous absorption of industrial chemicals: The skin denotation. Am J Ind Med 14:97-107 (1988).
Guberan, E.; Fernandez, J.: Control of Industrial Exposure to Tetrachloroethylene by Measuring Alveolar Concentrations: Theoretical Approach Using a Mathematical Model. Br J Ind Med 31:159-167 (1974).
Handy, R.; Schindler, A.: Estimation of Permissible Concentrations of Pollutants for Continuous Exposure. U.S. Environmental Protection Agency, Pub. No. EPA-600/2-76-155. Washington, DC (1976).
Hart, W.L.; Reynolds, R.C.; Drasavage, W.J.; et al.: Evaluation of Developmental Toxicity Data: A Discussion of Some Pertinent Factors and a Proposal. Risk Anal 8:59-70 (1988).
Hayes, A.W.: Principles and Methods of Toxicology, 4th Edition. Raven Press, New York (2000).
Henderson, Y.; Haggard, H.H.: Noxious Gases and the Principles of Respiration Influencing Their Action. Reinhold Publishing Corp., New York (1943).
Henschler, D.: Exposure Limits: History, Philosophy, Future Developments. Ann Occup Hyg 28:79-92 (1984).
Henschler, D.: Development of Occupational Limits in Europe. Ann Am Conf Govt Ind Hyg 12:37-40 (1985).
Hertzberg, R.C.; Patterson, J.: Approaches to the Quantitative Estimation of Health Risk from Exposure to Chemical Mixtures. In: Health & Environmental Reserach on Complex Organic Mixtures, R.H. Gray; E.K. Chess; P.J. Mellinger, et al., Eds., pp. 747-757. Twenty-Fourth Handford Life Sciences Symposium, October 20-24, 1985 (1987).

Elkins, H.B.: Maximum Acceptable Concentrations, A Comparison in Russia and the United States. AMA Arch Environ Health 2:45-50 (1961).
Federal Focus: Epidemiology and Risk Assessment. Washington, DC (1996).
Federal Register: Proposed Occupational Exposure Limits, Volume 54, pp.2435-2443. Washington, DC (1989).
Felter, S.P.; Dollarhide, J.S.: Acrylonitrile: A Reevaluation of the Database to Support an Inhalation Cancer Risk Assessment. Reg Tox Pharmacol 26: 281-287 (1997).
Hewitt, P.: Interpretation and Use of Occupational Exposure Limits for Chronic Disease Agents. In: Occupational Medicine State-of-the-Art Reviews. Hanley and Belfus Publishing, Inc., Philadelphia, PA 11(3):561-590 (1996).
Hickey, J.L.S.: Application of Occupational Exposure Limits to Unusual Work Schedules and Excursions. Ph.D. dissertation. University of North Carolina at Chapel Hill, Chapel Hill, NC (1977).
Hickey, J.L.S.; Reist, P.C.: Application of Occupational Exposure Limits to Unusual Work Schedules. Am Ind Hyg Assoc J 38:613-621 (1977).
Hickey, J.L.S.; Reist, P.C.: Adjusting Occupational Exposure Limits for Moonlighting, Overtime, and Environmental Exposures. Am Ind Hyg Assoc J 40:727-734 (1979).
Hickey, J.L.S.: Adjustment of Occupational Exposure Limits for Seasonal Occupations. Am Ind Hyg Assoc J 41:261-263 (1980).
Hickey, J.L.S.: The ATWAP@ in the Lead Standard. Am Ind Hyg Assoc J 44(4):310-311 (1983).
Holland, C.D.; Sielken, Jr., R.L.: Quantitative Cancer Modeling and Risk Assessment. Prentice Hall, Englewood Cliffs, NJ (1993).
Hunter, D.: Occupational Diseases, 6th Edition, Little Brown and Co., Boston, MA (1978).
Ikeda, M.; Immamura, T.; Hayashi, M.; et al.: Biological Half-Life of Styrene in Human Subjects. Int Arch Arbeitsmed 32:93-100 (1974).
International Life Sciences Institute (ILSI): Variability in Susceptibility. Washington, DC (1992).
Iuliucci, R.L.: 12-Hour TLVs®. Pollution Eng, pp. 25-57 (November 1982).
Integrated Risk Information System 2002. Environmental Protection Agency, IRIS (2002).
Jacobi, W.: Basic Concepts of Radiation Protection. J Ecotox Environ Safety 4(4):434-443 (1980).
Jarabek, A.M.; Menache, M.G.; et al.: The U.S. Environmental Protection Agency's Inhalation RfD Methodology: Risk Assessment for Air Toxics. Tox Ind Health 6(5):279-301 (1990).
Johanson, G.; Boman, A.: Percutaneous Absorption of 2-Butoxyethanol Vapour in Human Subjects. Brit J Ind Med 48:788-92 (1991).
Johnson, E.M.; Christian, M.S.; Dansky, L.; et al.: Use of the Adult Developmental Relationship in Pre-screening for Developmental Hazards. Terat Carc Mut 7:273-285 (1987).
Johnson, E.M.: Cross-Species Extrapolations and the Biologic Basis for Safety Factor Determinations in Developmental Toxicology. Regul Toxicol Pharmacol 8:22-36 (1988).
Johnson, E.M.: A Case Study of Developmental Toxicity Risk Estimation Based on Animal Data. The Drug Bendectin, In: The Risk Assessment of Environmental Hazard: A Textbook of Case Studies, D.J. Paustenbach, Ed., pp. 771-724. John Wiley & Sons, New York (1989).
Kane, L.E.; Alarie, Y.: Sensory Irritation to Formaldehyde and Acrolein during Single and Repeated Exposures in Mills. Am Ind Hyg Assoc J 38:509-522 (1977).
Kaneko, T.; Wang, P.Y.; Sato, A.: Development of Occupational Exposure Limits in Japan. Int J Occup Med Environ Health 11(1): 81-98 (1998).
Karnovsky, D.A.: Chapter 8. In: Teratology: Principle and Techniques, J.G. Wilson; J. Warrany, Eds. University of Chicago Press, Chicago, IL (1965).
Karol, M.H.: Animal Models of Occupational Asthma. Eur Respir J 7(3): 555-68 (1994).
Karol, M.H.; Graham, C.; et al.: Structure-activity Relationships and Computer-assisted Analysis of Respiratory Sensitization Potential. Toxicol Lett 86(2-3): 187-91 (1996).
Kazmina, N.P.: Study of the Adaptation Processes of the Liver to Monotonous and Intermittent Exposures to Carbon Tetrachloride (in Russian). Gig Tr Prof Zabol 3:39-45 (1976).
Kezic, S.; Mahieu, K.; Monster, A.C.; et al.: Dermal Absorption of Vaporous and Liquid 2-Methoxyethanol and 2-Ethoxyethanol in Volunteers. Occup Environ Med 54:38-43 (1997).

Kim, Y.; Carlson, G.P.: The Effect of an Unusual Worshift on Chemical Toxicity: I: Studies on the Exposure of Rats and Mice to Dichloromethane. Fund Appl Toxicol 6:162-171 (1986a).
Kim, Y.C.; Carlson, G.P.: The Effect of an Unusual Workshift on Chemical Toxicity: II. Studies on the Exposure of Rates to Aniline. Fund Appl Toxicol 7:144-152 (1986b).
Kimber, I.: The Role of Skin in the Development of Chemical Respiratory Hypersensitivity. Toxicol Lett 86:89-92 (1992).
Kimber, I.; Dearman, R.: The Mechanisms and Evaluation of Chemically Induced Allergy Toxicol Lett 64-65, 79-84 (1992).
Kimmel, C.A.; Kavlock, R.J.; Allen, B.C.; et al.: Benchmark Dose Concept Applied to Data from Conventional Development Toxicity Studies. Toxicol Lett 82/88:549-554 (1995).
Kimmerle G.; Eben, A.: Metabolism, Excretion and Toxicology of Trichoroethylene After Inhalation: II. Experimental Human Exposure. Arch Toxicol 30:127-138 (1973).
Klaassen, C. D.: Casarett and Doull's Toxicology: The Basic Science of Poisons, 6th Edition. McGraw-Hill, New York (2001).
Kobert, R.: The Smallest Amounts of Noxious Industrial Gases which are Toxic and the Amounts which May Perhaps be Endured. In: Compendium of Practical Toxicology 5:45 (1912).
Kodell, R.L.; Howe, R.B.; Chen, J.J.; et al.: Mathematical Modeling of Reproductive and Developmental Toxic Effects for Quantitative Risk Assessment. Risk Anal 11:583-590 (1991).
Krewski, D.; Murdoch, D.; Witney, J.: Recent Developments in Carcinogenic Risk Assessment. Health Physics 57:313-325 (1989).
Krewski, D.; Zhu, Y.: A Simple Data Transformation for Estimating Benchmark Doses in Developing Toxicity Experiment. Risk Anal 15:29-39 (1995).
Krewski, D.; Brown, C.; Murdoch, D.: Determining "Safe" Levels of Exposure: Safety Factors or Mathematical Models. Fund Appl Toxicol 4:383-394 (1984).
Krishman, K.: and Andersen, M.: Physiologically-based Pharmacokinetic Modeling in Toxicology. Principles and Methods in Toxicology. Chp. 5 A.W. Hayes, Ed. Taylor and Francis, Philadelphia, PA (2001).
Krivanek, N.; et al.: Monomethylformamide Levels in Human Urine after Repetitive Exposure to Dimethylformamide Vapor. J Occup Med 20:179-187 (1978).
LaNier, M.E.: Threshold Limit Values: Discussion and 35 year index with recommendations (TLVs®: 1946-81). American Conference of Governmental Industrial Hygienists, Cincinnati, OH (1984).
Lauwerys, R. R.: Occupational Toxicology in Casarett and Doull's Toxicology, Klaassen, Ed., pp. 987-1011. McGraw-Hill, New York (1996).
Lehmann, K.B.: Experimentelle Studien uber den Einfluss Technisch und Hygienisch Wichtiger Gase und Dampfe auf Organismus: Ammoniak und Salzsauregas. Arch Hyg 5:1-12 (1886).
Lehmann, K.B.; Schmidt-Kehl, L.: Die 13 Wichtigsten Chlorkohlenwasserstoffe der Fettreihe vom Standpunkt der Gewerbehygiene. Arch Hyg Barkt 116:131-268 (1936).
Lehmann, K.B.; Flury, F.: Toxikologie und Hygiene der technischen Losungsmittel. Julius Springer Verlag, Berlin (1938).
Lehman, A.; Fitzhugh, O.G.: 100-fold Margin of Safety. Q Bull-Assoc Food Drug Off 18:33-35 (1954).
Lehnert, G.; Ladendorf, R.D.; Szadkowski, D.: The Relevance of the Accumulation of Organic Solvents for Organization of Screening Tests in Occupational Medicine. Results of Toxicological Analyses of More than 6000 Samples. Int Arch Occup Environ Health 41:95-102 (1978).
Letavet, A.A.: Scientific Principles for the Establishment of the Maximum Allowable Concentrations of Toxic Substances in the USSR. In: Proceedings of the 13th Annual Congress on Occupational Health July 25-29, 1960. Book Craftsmen Assoc, New York (1961).
Leung, H.W.; Paustenbach, D.J.: Application of Pharmacokinetics to Derive Biological Exposure Indexes from Threshold Limit Values. Am Ind Hyg Assoc Journal 49:445-450 (1988a).
Leung, H.W.; Paustenbach, D.J.: Setting Occupational Exposure Limits for Irritant Organic Acids and Bases Based on Their Equilibrium Dissociation Constants. Appl Occup Environ Hygiene 3:115-118 (1988b).
Leung, H.W.; Paustenbach, D.J.; Murray, F.J.: A Proposed Occupational Exposure Limit for 2,3,7,8-TCDD. Am Ind Hyg Assoc J 49(9):466-474 (1988).

Leung, H.W.; Paustenbach, D.J.; Andersen, M.E.: A Physiologically-Based Pharmacokinetic Model for 2,3,7,8-Tetrachlorodibenzo-p-dioxin. Chemosphere 18:659-664 (1989).

Leung, H.W.; Paustenbach, D.J.: Organic Acids and Bases: A Review of Toxicological Studies. Am J Ind Med 18:717-735 (1990).

Leung, H.W.; Paustenbach, D.J.: Techniques for Estimating the Percutaneous Absorption of Chemicals Due to Occupational and Environmental Exposure. Appl Occup Environ Hyg 9(3):187-197 (1994).

Leung, H.W.: Techniques for Setting Occupational Exposure Limits. In: Human and Ecological Risk Assessment: Theory and Practice, pp. 647-671. D.J. Paustenbach, Ed. John Wiley and Sons, New York (2002).

Lippman, M.: Exposure data needs in Risk assessment and risk management; database information needs. Appl Occup Environ Hygiene 10:238-243 (1995).

Lu, F.C.; Dourson, M.L.: Safety Risk Assessment of Chemicals: Principles, Procedures and Examples. J Occup Med Tox 1(4): 321-335 (1992).

Lundberg, P.: National and International Approaches to Occupational Standard Setting Within Europe. Appl Occup Environ Hyg 9:25-27 (1994).

Lundberg, P.: General Concepts, In: Occupational Hazards in the Health Professions, D.K. Brune; C. Edling, Eds., pp.1-10. CRC Press, Boca Raton, FL (1989).

Lyublina, E.I.: Some Methods Used in Establishing the Maximum Allowable Concentrations. MAC of Toxic Substances in Industry. IUPAC pp.109-112 (1962).

Magnuson, H.L.: Industrial Toxicology in the Soviet Union Theoretical and Applied. Amer Ind Hyg Assoc J 25:185-190 (1964).

Manson, J.M.; Kang, Y.J.: Test Methods for Assessing Female Reproductive and Development Toxicology. In: Principles and Methods in Toxicology, Hayes, Ed. Raven Press, New York (1994).

Mason, J.W.; Hughes, J.: A Proposed Approach to Adjusting TLVs® for Carcinogenic Chemicals (unpublished report). Univ of Alabama, Birmingham (1985).

Mason, J.W.; Dershin, H.: Limits to Occupational Exposure in Chemical Environments Under Novel Work Schedules. J Occup Med 18:603-607 (1976).

Meldrum, M.: Setting OELs for Sensory Irritants. The Approach in the European Union. Am Ind Hyg Assn J (2001).

Monster, A.C.: Difference in Uptake, Elimination and Metabolism in Exposure to Trichlorethylene, 1,1,1-Trichloroethane and Tetrachloroethylene. Int Arch Occup Environ Health 42:311-317 (1979).

Moolenaar, R.J.: Overhauling Carcinogen Classification. Environ Sci Tech 8:70-75 (1992).

Moolgavkar, S.H.; Dewangi, A.; Venson, D.J.: A Stochastic Two-stage Model for Cancer Risk Assessment: The Hazard Function and the Probability of Tumor. Risk Anal 8:383-392 (1988).

National Safety Council: Final Report of the Committee of the Chemical and Rubber Sector on Benzene. National Bureau of Casualty and Surety Underwriters (May 1926).

National Institute of Occupational Safety and Health (NIOSH): Recommended Exposure Limit (REL) Policy. NIOSH, Cincinnati, OH (May 1996).

Naumann, B.D.; Sargent, E.V.; Starkman, B.S.; et al.: Performance-based exposure control limits for pharmaceutical active ingredients. Am Ind Hyg Assoc J 57:33-42 (1996).

NCRP: Maximum Permissible Body Burdens and Maximum Permissible Concentrations of Radionuclides in Air and in Water for Occupational Exposure. NBS Handbook No 69. US Government Printing Office, Washington, DC (1959).

Neumann, D.A.; Kimmel, C.A.: Human Variability in Response to Chemical Exposures: Measures, Modeling and Risk Assessment. CRC Press, Boca Raton (1998).

Nielsen, G.D.: Mechanisms of Activation of the Sensory Irritant Receptor by Airborne Chemicals. Crit Reviews in Toxicol 21:183-208 (1991).

Nollen, S.D.; Martin, V.H.: Alternative Work Schedules Part 3: The Compressed Workweek. AMACOM, New York (1978).

Nollen, S.D.: The Compressed Workweek: Is it Worth the Effort? Ind Eng pp. 58-63 (1981).

Notari, R.E.: Biopharmaceutics and Pharmacokinetics, 3rd Edition. Marcel Dekker, Inc., New York (1985).

O'Flaherty, E.J.; Hammond, P.B.; Lerner, S.I.: Dependence of Apparent Blood Lead Half-Life on the Length of Previous Lead Exposure in Humans. Fund Appl Toxicol 2:49-54 (1982).

O'Reilly, W.J.: Pharmacokinetic in Drug Metabolism and Toxicology. Canadian J Pharm Sci 7:66-77 (1972).
Occupational Safety and Health Administration (OSHA): Air Contaminants: Final Rule. Federal Register 54(12):2332-2983 (January 1989).
Occupational Safety and Health Administration: OSHA Compliance Officers Field Manual. Department of Labor, Washington, DC (1979).
Occupational Safety and Health Administration (OSHA), U.S. Congress (91st) S. 2193, Public Law 91-596, Washington, DC (1970).
Ottobani, A.: The Dose Makes the Poison. (1993).
Park, C.N.; Hawkins, N.C.: Cancer risk assessment. In: Patty's Industrial Hygiene and Toxicology, 3rd Edition, Volume 3, L.J. Cralley; L.V. Cralley; J. Bus, Eds. John Wiley & Sons, New York (1995).
Park, C.; Snee, R.: Quantitative Risk Assessment: State of the Art for Carcinogenesis. Fund Appl Toxicol 3:320-333 (1983).
Patterson, J.; Schoeny, R.; Daunt, P.: The U.S. Environmental Protection Agency's Integrated Risk Information System (IRIS). In: Regulating Drinking Water Quality, C. Gilbert; E. Calabrese, Eds., pp. 273-281. Lewis Publishers, Inc., Chelsea, MI (1991).
Paull, J.M.: The Origin and Basis of Threshold Limit Values. Am J Ind Med 5:227-238 (1984).
Paustenbach, D.J.; Langner, R.R.: Setting Corporate Exposure Limits: State of the Art. Am Ind Hyg Assoc J 47:809-818 (1986).
Paustenbach, D.J.; Carlson, G.P.; Christian, J.E.; et al.: The Effect of the 11.5 Hr/Day Exposure Schedule on the Distribution and Toxicity of Inhaled Carbon Tetrachloride in the Rat. Fund Appl Toxicol 6:472-483 (1986a).
Paustenbach, D.J.; Carlson, G.P.; Christian, J.E.; et al.: A Comparative Study of the Pharmacokinetics of Carbon Tetrachloride in the Rat Following Repeated Inhalation Exposure of 8 hr/day and 11.5 hr/day. Fund Appl Toxicol 6:484-497 (1986b).
Paustenbach, D.J.: Assessment of the Developmental Risks Resulting from Occupational Exposure to Select Glycol Ethers within the semi-conductor Industry. J Toxicol Env Health 23:53-96 (1988).
Paustenbach, D.J.: Health Risk Assessments: Opportunities and Pitfalls. Columbia J Environ Law 14(2):379-410 (1989).
Paustenbach, D.J.: Occupational Exposure Limits: Their Critical Role in Preventive Medicine and Risk Management (editorial). Am Ind Hyg Assoc J 51:A332-a336 (1990a).
Paustenbach, D.J.: Health Risk Assessment and the Practice of Industrial Hygiene. Am Ind Hyg Assoc J 51:339-351 (1990b).
Paustenbach, D.J.: The History and Biological Basis of Occupational Exposure Limits for Chemical Agents. Chapter 41. In: Patty's Industrial Hygiene, 5th Edition, Volume 3, pp. 1903-2000. R. Harris, Ed. John Wiley and Sons, New York, NY (2000a).
Paustenbach, D.J.; Jernigan, J.D.; Finley, B.L.; et al.: The Current Practice of Health Risk Assessment: Potential Impact and Standards for Toxic Air Contaminants. J Air Pollution Control Association 40(12):1620-1630 (1990).
Paustenbach, D.J.; Price, P.; Ollison, W.; et al.: A Reevaluation of Benzene Exposure for the Pliofilm (Rubber Worker) Cohort (1936-1976). J Toxicol Environ Health 36:177-231 (1992).
Paustenbach, D.J.: Occupational Exposure Limits, Pharmacokinetics, and Unusual Work Schedules. In: Patty's Industrial Hygiene and Toxicology, 3rd Edition, Volume 3(A), R.L. Harris; L.J. Cralley; L.V. Cralley, Eds., pp.191-301. The Work Environment. John Wiley & Sons, New York (1994).
Paustenbach, D.J.; Alarie, Y.; Kulle, T.; et al.: A Recommended Occupational Exposure Limit For Formaldehyde Based On Irritation. J Toxicol Environ Health 50:217-263 (1997).
Paustenbach, D. J.: Approaches and Considerations for Setting Occupaitonal Exposure Limits (OELs) for Sensory Irritants: Report of a Recent Symposia. Appl Occ Env Hygiene (2001).
Paustenbach, D.J.: Putting Politics Aside, Updating the PELs. Amer Ind Hyg Assn J 58:845-849 (1997).
Paustenbach, D.J.: The Practice of Exposure Assessment: A State of the Art Review. J Toxicol Environ Health Part B 3:179-291 (2000b).
Paustenbach, D.J.: The History and Biological Basis of Occupational Exposure Limits. In Patty's Industrial Hygiene, Chapter 41, 5th Ed., Volume 3. R. Harris, Ed. John Wiley & Sons, New York, NY (2000c).

Peto, R.: Carcinogenic Effects of Chronic Exposure to Very Low Levels of Toxic Substances. Environ Health Perspect 22:155-159 (1978).

Piotrowski, J.K.: Exposure Tests for Organic Compounds in Industrial Toxicology. Department of Health, Education and Welfare, pp. 77-144, (NIOSH) Cincinnati, OH (1977).

Price, P.; Keenan, R.; Swartout, J.; et al.: An Approach for Modeling Noncancer Dose Responses with an Emphasis on Uncertainty. Risk Anal 17 (4):427-437 (1997).

Pitot, H. C.; Dragan, Y. P.: Chemical carcinogenesis. In: Casarett and Doull's Toxicology, 5th Edition, pp. 201-268. McGraw-Hill, New York (1996).

Powell, C. J.; Berry, C. L.: Non-genotoxic or Epigenetic Carcinogenesis. In: General and Applied Toxicology, 2nd Edition, B. Ballantyne; T. Marrs; T. Syversen, Eds., pp. 1119-1137. Macmillian Reference Ltd., London (1999).

Ramazinni, B.: De Morbis Atrificum Diatriba (Diseases of Workers) (Translated by W.C. Wright in 1940). The University of Chicago Press, Chicago, IL (1700).

Rappaport, S.M.: The Rules of the Game: An Analysis of OSHA's Enforcement Strategy. Am J Ind Med 6:291-303 (1981).

Rappaport, S.M.: Threshold Limit Values, Permissible Exposure Limits, and Feasibility: The Bases for Exposure Limits in the United States. Am J Ind Med 23(5): 683-694 (1993).

Rappaport, S.M.; Kromhout, H.; Symanski, E.: Variation of Exposure Between Workers in Homogeneous Exposure Groups. Am Ind Hyg Assoc J 54:654-662 (1993).

Reitz, R.H.; Andersen, M.; Gargas, M.: A PB-PK Model for Vinyl Chloride. Toxicol Appl Pharmacol (1995).

Renwick, A. G.: The Use of an Additional Safety or Uncertainty Factor for Nature of Toxicity in the Estimation of Acceptable Daily Intake and Tolerable Daily Intake Values. Regul Toxicol Pharmacol 22:250-261 (1995).

Renwick, A.G.; Lazarus, N.R.: Human Variability and Non-Cancer Risk Assessment: An Analysis of Default Uncertainty Factor. Regul Toxicol Pharm 27:3-120 (1998).

Report of Miners': Phthisis Prevention Committee. Johannesburg, Union of South Africa (1916).

Riihimakj, V.; Pfaffli, P.: Percutaneous Absorption of Solvent Vaports in Man. Scand J Work Environ Health 4:73-85 (1978).

Ripple, S.R.: The Use of Retrospective Health Risk Assessment: Looking Back at Nuclear Weapons Facilities. Environ Sci Tech 26:1270-1276 (1992).

Roach, S.A.: A More Rational Basis for Air Sampling Programs. Am Ind Hyg Assoc J 27:1-19 (1966).

Roach, S.A.: A Most Rational Basis for Air Sampling Programs. Ann Occup Hyg 20:65-84 (1977).

Roach, S.A.: Threshold Limit Values for Extraordinary Work Schedules. Amer Ind Hyg Assoc J 39:345-364 (1978).

Roach, S.A.; Rappaport, S.M.: But They are not Thresholds: A Critical Analysis of the Documentation of Threshold Limit Values. Amer J Ind Med 17:727-753 (1990).

Robinson, J.C.; Paxman, D.G.; Rappaport, S.M.: Implications of OSHA's Reliance on TLVs® in Developing the Air Contaminants Standard. Am J Ind Med 19:9-13 (1991).

Rodricks, J.V.; Brett, S.; Wrenn, G.: Significant Risk Decisions in Federal Regulatory Agencies. Regul Toxicol Pharmacol 7:307-320 (1987).

Rodricks, J.V.; Taylor, M.: Application of Risk Assessment to Food Safety Decision Making. Regul Toxicol Pharmacol 3:275-307 (1983).

Rowe, V.K.; Wolf, M.A.; Weil, C.S.; et al.: The Toxicological Basis of Threshold Limit Values: 2. Pathological and Biochemical Criteria. Am Ind Hyg Assoc J 20:350-356 (1959).

Rupp, E.M.; Parzychk, D.C.; Booth, R.S.; et al.: Composite Hazard Index for Assessing Limiting Exposures to Environmental Pollutants: Application Through a Case Study. Environ Sci Technol 12(7):802-807 (1978).

Ruzic, A.: Pharmacokinetic Modelling of Various Theoretical Systems. J Pharm Sci 11:110-150 (1970).

Saffioti, U.: Identification and Definition of Chemical Carcinogens: Review of Criteria and Research Needs. J Toxicol Environ Health 6(5):1029-1058 (1980).

Sarlo, K.; Karol, M.H.: Guinea Pig Productive Tests for Respiratory Allergy. In: Immunotoxicology and Immunopharmacology, Dean; Luster; Munson; Kimber, Eds. Raven Press, New York (1994).

Sayers, R.R.: Toxicology of Gases and Vapors. International Critical Tables of Numerical Data, Physics, Chemistry and Toxicology. McGraw-Hill, New York 2:318-321 (1927).

Sayers, R.R.; DalleValle, J.M.: Prevention of Occupational Diseases other than those that are Caused by Toxic Dust. Mech Eng 57:230-234 (1935).

Scansetti, G.; Piolatto, G.; Rubino, G.: Skin Notation in the Context of Workplace Exposure Standards. Am J Ind Med 14:725-32 (1988).

Schaper, M.: Development of a Database for Sensory Irritants and its Use in Establishing Occupational Exposure Limits. Am Ind Hyg Assoc J 54(9): 488-544 (1993).

Schardein, J.L.: Teratogenic Risk Assessment. In: Issues and Reviews of Toxicology, Volume I, pp.181-214. Plenum Press, New York (1983).

Schrenk, H.H.: Interpretation of Permissible Limits. Am Ind Hyg Assoc Qtrly 8:55-60 (1947).

Schumann, A.M.; Quast, J.F.; Watanbe, P.G.: The Pharmacokinetics and Macromolecular Interactions of Perchloroethylene in Mice and Rats as Related to Oncogenicity. Toxicol Appl Pharm 55:207-219 (1980).

Seiler, J.P.: Apparent and Real Thresholds: A Study of Two Mutagens. In: Progress in Genetic Toxicology, D. Scott; B.A. Bridges; F.H. Sobels, Eds. Elsevier Biomedical Press, New York (1977).

Silva, P.: TLVs® to Protect "Nearly All Workers." Appl Occup Environ Hyg 1:49-53 (1986).

Smith, R.G.; Olishifski, J.B.: Industrial Toxicology. In: Fundamentals of Industrial Hygiene, J.Olishifski, Ed., pp.354-386. National Safety Council, Chicago, IL (1988).

Smolensky, M.H.: Human Biological Rhythms and Their Pertinence to Shift Work and Occupational Health. Chronobiologia 7:378-390 (1980).

Smolensky, M.H.; Paustenbach, D.J.; Schering, L.E.: Biological Rhythms, Shiftwork and Occupational Health. In: Patty's Industrial Hygiene and Toxicology 2nd Edition, Volume 3b, Biological Responses, Cralley; Cralley, Eds., pp. 175-312. John Wiley & Sons, New York (1985).

Smyth, H.F.: Improved Communication; Hygienic Standard for Daily Inhalation. Am Ind Hyg Assoc Qtrly 17:129-185 (1956).

Spirtas, R.; Steinberg, M.; Wands. R.C.; et al.: Identification and Classification of Carcinogens: Procedures for the Chemical Substances Threshold Limit Value Committee, ACGIH. Am J Public Health 10:1232-5 (1986).

Stara, J.F.; Bruins, R.J.F.; Dourson, M.L.; et al.: Risk Assessment is a Developing Science: Approaches to Improve Evaluation of Single Chemicals and Chemical Mixtures. In: Methods for Assessing the Effects of Mixtures of Chemicals, V.B. Vouk; G.C. Butler; A.C. Upton; et al., Eds., pp. 719-743. SCOPE (1987).

Stokinger, H.E.: Threshold Limits and Maximum Acceptable Concentrations; Their Definition and Interpretation. Am Ind Hyg Assoc J 23:45-47 (1962).

Stokinger, H.E.: International Threshold Limits Values. Am Ind Hyg Assoc J 24:469 (1963).

Stokinger, H.E.: Modus Operandi of Threshold Limits Committee of ACGIH®. In: Transactions of Twenty-Sixth Annual Meeting of the American Conference of Governmental Industrial Hygienists. 26:23-2 (1964).

Stokinger, H.E.: Current Problems of Setting Occupational Exposure Standards. Arch Environ Health 19:277-280 (1969).

Stokinger, H.E.: Criteria and Procedures for Assessing the Toxic Responses to Industrial Chemicals. In: Permissible Levels of Toxic Substances in the Working Environment. ILO, World Health Organization, Geneva (1970).

Stokinger, H.E.: Intended Use and Application of the TLVs®. In: Transactions of the Thirty-third Annual Meeting of the American Conference of Governmental Industrial Hygienists. 33:113-116 (1971).

Stokinger, H.E.: The Case for Carcinogen TLVs® Continues Strong. Occup Health and Safety (March-April) 46:54-58 (1977).

Stokinger, H.E.: Threshold Limit Values: Part I. In: Dangerous Properties of Industrial Materials Report, pp. 8-13 (May and June 1981).

Stott, W.T.; Reitz, R.H.; Schumann, A.M.; et al.: Genetic and Nongenetic Events in Neoplasia. Food Cosmet Tox 19:567-576 (1981).

Swartout, J.; Price, P.; Dourson, M.; et al.: A Probabilistic Framework for the Reference Dose. Risk Anal (1997).

Tarlau, E.S.: Industrial Hygiene With No Limits. A Guest Editorial. Amer Ind Hyg Assoc J 51:A9-A10 (1990).

Taylor, P.J.: The Problems of Shift Work. Proceedings of an International Symposium on Night and Shiftwork, Oslo, Sweden (1969).

Tengs, T.O.; Adams, M.E.; Pliskin, J.S.; et al.: Five-hundred Life-saving Interventions and Their Cost-effectiveness. Risk Anal 15:369-390 (1995).

Thiis-Evensen, E.: Shift Work and Health. Ind Med Surg 27:493-513 (1958).

Thran, C.D.: Linearity and Superposition in Pharmacokinetics. Pharmacological Reviews 26:3-31 (1974).

Travis, C.C.; Richter, S.A.; Crouch, E.A.; et al.: Cancer Risk Management: A Review of 132 Federal Regulatory Decisions. Env Sci Tech 21(5):415-420 (1987).

Travis, C.C.; Hattemer-Frey, H.A.: Determining an Acceptable Level of Risk. Environ Sci Technol 22:873-876 (1988).

Tuggie, R.M.: The NIOSH Decision Scheme. Am Ind Hyg Assoc J 42:493-498 (1981).

U.S. Environmental Protection Agency: Approaches to risk Assessment for multiple Chemical exposures. EPA 600/9-84-008, EPA, Washington, DC (1984).

U.S. Environmental Protection Agency: Proposed Guidelines for Health Assessment of Suspect Developmental Toxicants. Federal Register, EPA, Washington, DC (1986).

U.S. Environmental Protection Agency: Proposed Guidelines for Assessing Female Reproductive Risk. Federal Register 53126:14834-24847, June 30, EPA, Washington, DC (1988).

U.S. Environmental Protection Agency: Dermal Exposure Assessments: Principals and Applications. Exposure Assessment Group. Office of Health and Environmental Assessment, EPA, Washington, DC (1992).

U.S. Environmental Protection Agency: The Proposed Carcinogen Guidelines. EPA, Washington, DC (1996).

U.S Environmental Protection Agency: Interim Methods for Development of Inhalation Reference Doses. Office of Health and Environmental Assessment. U.S. Environmental Protection Agency. EPA/600/8-90/066A, Washington, DC (1990).

U.S. Environmental Protection Agency: Methods for Derivation of Inhalation Reference Concentrations and Application of Inhalation Dosimetry. Office of Health and Environmental Assessment, Office of Research and Development, U.S. Environmental Protection Agency, EPA/600/8-90/066F, Research Triangle Park, NC (1994).

U.S. Environmental Protection Agency: Reducing Risk: A Comparative Assessment of Environmental Problems. EPA Science Advisory Board, Washington, DC (1990).

U.S. Environmental Protection Agency: Final report. Principles of Neurotoxicity Risk Assessment. Federal Register 59(158): 42360-42404. EPA, Washington, DC (1994).

U.S. Environmental Protection Agency: List of Available Toxicological Reviews and Other Support Documents. Office of Research and Development. http://www.epa.gov/iris (2000).

Velazquez, S.F.; Schoeny, R.; Rice, G.; et al.: Cancer Risk Assessment: Historical Perspectives, Current Issues and Future Directions. Drug Chemical Toxicol 19(3):161-185 (1996).

Van Hemmen, J.J.; Van Goldstein Brouwers, Y.G.C.; Brouwer, D.H.: Pesticide Exposure and Re-entry in Agriculture. In: Methods of Pesticide Exposure Assessment, P.B. Curry; S. Iyengar; P.A. Maloney; M. Maroni, Eds. London (1995).

Van Hemmen, J.J.; Brouwer, D.H.: Assessment of Dermal Exposure to Chemicals. Sci Total Environ 168.131-41 (1995).

Veng-Pederson, P.; Paustenbach, D.J.; Carlson, G.P.; et al.: A Linear Systems Approach to Analyzing the Pharmacokinetics of Carbon Tetrachloride in the Rat Following Exposure to an 8-hour/day and 12-hour/day Simulated Workweek. Arch Toxicol 60:355-364 (1987).

Vincent, J.H.: International Occupational Exposure Standards: A Review and Commentary. Am Ind Hyg Assoc J 59:729-742 (1998).

Wang, R.; Schwertz, B.A.: An Evaluation System for Ranking Chemicals with Teratogenic Potential. Terat Carc Mut 7:133-139 (1987).

Watanabe, P.G.;Young, J.D.; Gehring, P.J.: The Importance of Non-Linear (Dose-Dependent) Pharmacokinetics In Hazard Assessment. J Environ Path Toxicol 1:147-159 (1977).

Watanabe, P.G.; Reitz, R.H.; Schumann, A.M.; et al.: Implications of the Mechanisms of Tumorigenicity for Risk Assessment. In: The Scientific Basis of Toxicity Assessment, M. Witschi, Ed., pp. 69-88. Elsevier/North-Holland Press (1980).

Weil, C.S.: Statistics Versus Safety Factors and Scientific Judgment in the Evaluation of Safety for Man. Toxicol Appl Pharmicol 21:454-463 (1972).

Wheeler, K.; Gurman, R.; Tarnowieski, D.: The Four-Day Week. AMACOM, New York (1972).

Whipple, C.: De Minimus Risk. Plenum Pub, New York (1987).

Wildavsky, A.: Risk in The Workplace. University of Columbia Press, New York (1990).

Wilkinson, C.F.: Being More Realistic about Chemical Carcinogenesis. Environ Sci Technol 9:843-848 (1988).
Willhite, C.C.; Oseas, N.N.: Reconciliation of American TLVs® and German MAKs. Occup Hyg 4:1-16 (1997).
Willhite, C.C.: Paper presented at the 1998 AIHC meeting in Atlanta. TLV/MAK Collaboration and Excursion (peak) Limitations (1998).
Williams, G.M.; Weisburger, J.H.: Chemical Carcinogenesis. In: Casarett and Doull's Toxicology: The Basic Science of Poisons 4th Edition, C. Klaassen, Ed. McGraw-Hill, New York (1991).
Williams, R.T.: Inter-species Variations in the Metabolism of Xenobiotics. Biochem Soc Trans 2:359-377 (1974).
Williams, G.M.: Epigenetic Mechanisms of Action of Carcinogenic Organochlorine Pesticides, ACS, Symp Series: ISS Pest Chem Mod Toxicol 160:45-56 (1981).
Wilschut, A.; Ten Berg, W.F.; Robinson, P.J.; et al.: Estimating Skin Permeation. The Validation of Five Mathematical Skin Permeation Models. Chemosphere 30:1275-96 (1995).
Wilson, J.T.; Rose, K.M.: The Twelve-Hour Shift in the Petroleum and Chemical Industries of the United States and Canada: A Study of Current Experience. Wharton Business School, University of Pennsylvania, Philadelphia, PA (1978).
Wilson, H.K.: Recent Policy and Technical Developments in Biological Monitoring in the United Kingdon. Sci Total Environ 199:101-5 (1997).
World Health Organization (WHO): Methods Used in Establishing Permissible Levels in Occupational Exposure to Harmful Agents. Tech Report 601. International Labor Office, WHO, Geneva (1977).
World Health Organization (WHO): Occupational Exposure Limits for Airborne Toxic Substances, 4th Edition, Occ Safety and Health Series, No. 37. International Labor Office, WHO, Geneva (1995).
Zapp, J.A.: The Toxicological Basis of Threshold Limit Values: Physiological Criteria. Am Ind Hyg Assn J 20:346-349 (1959).
Zenick, H.; Clegg, E.D.; Perreault, S.D.; et al.: Assessment of Male Reproductive Toxicity: A Risk Assessment Approach. In: Principle and Methods of Toxicology, A.W. Hays, Ed. Raven Press, New York (1994).
Zenz, C.; Berg, B.A.: Influence of Submaximal Work on Solvent Uptake. J Occup Med 12:367-369 (1970).
Zielhuis, R.L.: Permissible Limits for Occupational Exposure to Toxic Agents: A Discussion of Differences in Approach Between U.S. and USSR. Int Arch Arbeitsmed 33:1 (1974).
Zielhuis, R.L.; Van Der Kreek, F.: Calculations of a Safety Factor in Setting Health Based Permissible Levels for Occupational Exposure. A Proposal I. Int Arch Occup Environ Health 42:191-201 (1979a).
Zielhuis, R.L.; Van Der Kreek, F.W.: Calculations of a Safety Factor in Setting Health Based Permissible Levels for Occupational Exposure. A Proposal II. Comparison of Extrapolated and Published Permissible Levels. Int Arch Occup Environ. Health 42:203-215 (1979b).
Ziem, G.E.; Castleman, B.I.: Threshold Limit Values: Historical Perspective and Current Practice. J Occup Med 13:910-918 (1989).

SUPPLEMENTAL REFERENCES

American Industrial Hygiene Association: A Guide to Product Health and Safety and the Right to Know. Ch III Product Liability: The Right to Know, Ch V Risk Assessment, and Ch VI Communicating Product Safety and Health Information. Akron, OH (1986).
American National Standards Institute, Inc: American National Standard for the Precautionary Labeling of Hazardous Industrial Chemicals. January 15. Sponsored by the Manufacturing Chemists Association, Inc (1976).
Anderson, R.C.; Henderson, F.G.; Chen, K.K.: The Rat as a Suitable Animal for the Study of Prolonged Medication. J Am Pharm A (Scient Ed) 32(8):204-208 (1943).
Association of Food & Drug Officials of the United States: 100-fold Margin of Safety. Q Bull Assoc Food & Drug Officials U.S. January: 33-35 (1954).
Association of Food & Drug Officials of the United States: A Procedure for Determining the Leachability of Uncertified Pigments for Printing Food Wraps and Coloring Plastic Food Containers. 61st Annual Conference, Louisville, KY, May 6-11, pp. 113-115 (1957).

Baetjer, A.M.: The Early Days of Industrial Hygiene - Their Contribution to Current Problems. Transactions of the 42th Annual Meeting of the American Conference of Governmental Industrial Hygienists. pp. 10-17 (1981).

Ball, W.L.: Threshold Limits for Pesticides. AMA Arch Indust Health. 14(2):178-185 (1956).

Ball, W.L.: The Toxicological Basis of Threshold Limit Values: 4.Theoretical Approach to Prediction of Toxicity of Mixtures. Am Indust Hyg Assoc J 20(5):357-63 (1959).

Bar, F.: The Toxicological Evaluation (margin of safety, tolerances) for Food Legislation (food additives and pesticide residues). Arch Toxicol 32:51-62 (1974).

Bardodej, Z.: Occupational Exposure Limits for Protection of Working People in Modern Society. Ann Am Conf Ind Hyg 12:73-78 (1985).

Barkley, J.F.: Accepted Limit Values of Air Pollutants. Information Circular 7682. United States Dept of the Interior, Bureau of Mines (1954).

Barnes, E.C.: Keeping Score on Toxic Materials and Processes in The Plant. Industrial Hygiene Foundation of America, Inc, Seventh Annual Meeting of Members, November 10-11, Pittsburgh, PA (1942).

Barnes, J.M.: The Use of Experimental Pathological Techniques in Assessing the Toxicity of Chemical Compounds. In: The Application of Scientific Methods to Industrial and Service Medicine. Medical Research Council. London: His Mafesty's Stationery Office. pp. 49-55 (1951).

Barnes, J.M.: Toxic Hazards of Certain Pesticides to Man. Bull World Health Org 8: 419-490 (1953).

Barnes, J.M.; Denz, F.A.: Experimental Methods Used in Determining Chronic Toxicity. Pharmacol Rev 6(2):191-242 (1954).

Barnes, J.M.: The Potential Toxicity of Chemicals Used in Food Technology. Proc Nutrition Soc 15(2):148-154 (1956).

Bateman, A.J.: Testing Chemicals for Mutagenicity in a Mammal. Nature 210(5032): 205-206 (1966).

Bates, R.R.: Regulation of Carcinogenic Food Additives and Drugs in the U.S. In: Carcinogenic Risks. Strategies for Intervention, W. Davis; C. Rosenfeld, Eds. International Agency for Research on Cancer, Lyon (1979).

Behrend, B.; Jaeger, E.: Drug Evaluation in Industrial Medicine - Purposes and Procedures. Ind Med Surg 29(7):319-323 (1960).

Bellingham, E.F.; Bloomfield, J.J.; Dreessen, W.C.: Bibliography of Industrial Hygiene, 1900-1943. Publ Health Bull 289:1-95. United States Public Health Service (1945).

Berliner, V.R.: U.S. Food and Drug Administration Requirements for Toxicity Testing of Contraceptive Products. In: Pharmacological Models in Contraceptive Development, M.H. Briggs; E. Diczfalusy, Eds. WHO, Geneva (1974).

Berry, C.M.: Safety in the Use of Economic Poisons. Transactions of the 26th Annual Meeting of the American Conference of Governmental Industrial Hygienists. Philadelphia, PA, April 25-28, pp. 82-86 (1964).

Bing, F.C.: Re: Perspective Versus Caprice in Evaluating Toxicity of Chemicals in Man. JAMA 154(11):939 (1954).

Bingham, E.; Falk, H.L.: Environmental Carcinogens. The Modifying Effect of Cocarcinogens on the Threshold Response. Arch Environ Health 19:779-783 (1969).

Bliss, C.I.: The Determination of the Dosage-Mortality Curve from Small Numbers. J Pharm Pharmacol 11:192-216 (1938).

Bloomfield, J.J.: What the ACGIH Has Done for Industrial Hygiene. Am Ind Hyg Assoc J 19:338-344 (1958).

Bloomfield, J.J.: Fragmentation in Industrial Hygiene, an Opportunity to Strengthen an Expanding Field. Transactions of the 30th Annual Meeting of the American Conference of Governmental Industrial Hygienists, St. Louis, Missouri, May 12-14, pp. 16-19 (1968).

Bock, F.G.: Dose Response: Experimental Carcinogenesis. In: Toward a Less Harmful Cigarette. Nat Cancer Inst Monograph 28, E.L. Wynder, Ed, pp. 57-63 (1968).

Bolt, H.M.: Short-term Exposure Limits. Ann Am Conf Ind Hyg 12:85-87 (1985).

Borasi, M.: Standardization of Toxicity Tests. Farmaco (fasc straord) 6(7):949-953 (1951).

Boughton, L.L; Stoland, O.O.: The Effects of Drugs Administered Daily in Therapeutic Doses Throughout the Life Cycle of Albino Rats. Univ Kansas Sci Bull 27(3):27-60 (1941).

Bowditch, M.; Elkins, H.B.: Chronic Exposure to Benzene (benzol). I. The Industrial Aspects. J Ind Hyg Toxicol 21(8):321-330 (1939).

Bowditch, M.: In Setting Threshold Limits. Transactions of the Seventh Annual Meeting of the National Conference of Governmental Industrial Hygienists. St. Louis, Missouri, May 9, pp. 29-32 (1944).

Boyd, E.M.: Toxicological Studies. J New Drugs 1(3):104-9 (1961).
Boyd, E.M.: Predictive Drug Toxicity: Assessment of Drug Safety Before Human Use. Canad Med Assoc J 98:278-293 (1968).
Brandt, A.D.: Engineering and Chemical Application of Standards. Am Ind Hyg Assoc Q 17:286-292 (1956).
Bratton, A.C.: A Short-term Chronic Toxicity Test Employing Mice. J Pharmocol Exp Therap 85:111-118 (1945).
Brewer, J.H.; Bryant, H.H.: The Toxicity and Safety Testing of Disposable Medical and Pharmaceutical Materials. J Am Pharm Assoc 49(10):652-656 (1960).
Brinkley, P.C.: Industry's Responsibility in the Toxicity Testing, Manufacture, Compounding, and Use of Economic Poisons. American Ind Hyg Assoc J 26(6): 611-614, November-December (1965).
British Medical Association: Labelling of Poisons. Brit Med J, No 5204, October 1 (1960).
Brown, J.B.; Fryer, M.P.; Randall, P.; et al.: Silicon Compounds in Plastic Surgery. Laboratory and Clinical Investigations, a Preliminary Report. Plast and Reconst Surg 12(5):374-376 (1953).
Brown, J.B.; Ohlwiler, D.A.; Fryer, M.P.: Investigation of and Use of Dimethyl Siloxanes, Halogenated Carbons, and Polyvinyl Alcohol as Subcutaneous Prostheses. Ann Surg 152(3):534-47 (1960).
Brown, J.B.: Studies of Silicones and Teflon as Subcutaneous Prostheses. Plast Reconstr Surg 28:86-7 (1961).
Brown, J.B.; Fryer, M.P.; Ohlwiler, D.A.; et al.: Dimethylsiloxane and Halogenated Carbons as Subcutaneous Prosthesis. Am Surg 28(3):146-8 (1962).
Brues, A.M.: Critique of the Linear Theory of Carcinogenesis. Science 128(3326):693-699 (1958).
Burchell, H.B.: A Plea for the Labeling of Drugs. Postgrad Med 21(6):644-645 (1957).
Calvery, H.O.: Safeguarding Foods and Drugs in Wartime. Am Sci 32(2):103-119 (1944).
Camp, W.J.R.: The Toxicologist and Industrial Toxicology. Ind Med Surg 19(7):321-322 (1950).
Campbell, W.G.: Progress in Food, Drug and Cosmetic Control Under the New Federal Food, Drug and Cosmetic Act. Assoc Food Drug Officials of U.S. 5(1):15-22 (1941).
Carbide & Carbon Chemicals Corp: Manual of Hazards to Health from Chemicals Used at the Plants and Laboratories of Carbide and Carbon Chemicals Corp, 1st Edition. May 1 (1949).
Carpenter, C.P.; Smyth, H.F.; Pozzani, U.C.: Assay of Acute Vapor Toxicity, and the Grading and Interpretation of Results on 96 Chemical Compounds. J Ind Hyg and Toxicol 31(6):343-346 (1949).
Carter, V.L.: Development of Hygiene Guides. Ann Am Conf Govt Ind Hyg 3:113-115 (1982).
Carter, V.L.: Modus Operandi of Committee on Threshold Limit Values for Chemical Substances. Ann Am Conf Ind Hyg 12:11-13 (1985).
Chemical Manufacturers Association: Legislative Action Alert, Pennsylvania Right to Know. State Affairs Report (1984).
Chemical Manufacturers Association: An International Review of Procedures for Establishing Occupational Exposure Limits (1992).
Child, G.P.; Paquin, H.O.; Deichmann, W.B.: Chronic Toxicity of Methylpolysiloxane? DC Antifoam A? in dogs. AMA Arch Ind Hyg 3(5):479-482 (1951).
Clayton Environmental Consultants: OSHA's Hazard Communication Standard. Issued. No 15: January (1984).
Coleman, A.L.: Report of Committee on Threshold Limits. Transactions of the 16th Annual Meeting of the American Conference of Governmental Industrial Hygienists. Chicago, IL, April 24-27, pp. 22-26 (1954).
Coleman, A.L.: Threshold Limits of Organic Vapors. Transactions of the 16th Annual Meeting of the American Conference of Governmental Industrial Hygienists. Chicago, IL, April 24-27, pp. 50-52 (1954).
Coleman, L.A.: The American Medical Association proposed act for labeling hazardous substances; legal implications for industry. AMA Arch Ind Health 19(3): 271-273 (1959).
Conley, B.E.: Principles for Precautionary Labeling of Hazardous Chemicals. JAMA 166(17):2154-2157 (1958).
Cook, W.A.: Maximum Allowable Concentrations of Industrial Atmospheric Contaminants. Ind Med 14(11):936-946 (1945).
Cook, W.A.: Present Trends in MACs. Am Ind Hyg Assoc Q 17(3):273-274 (1956).
Cook, W.A.: Problems of Setting Occupational Exposure Standards - Background. Arch Env Health 19:272-276 (1969).

Cook, W.A.: History of ACGIH TLVs. Ann Am Conf Ind Hyg 12:3-9 (1985).
Cordle, F.; Kolbye, A.C.: Food Safety and Public Health. Interaction of Science and Law in the Federal Regulatory Process. Cancer 43:2143-2150 (1979).
Couchman, C.E.: Echoes from the 1959 American Conference of Governmental Industrial Hygienists. Transactions of the 30th Annual Meeting of the American Conference of Governmental Industrial Hygienists. St. Louis, Missouri, May 12-14, pp. 23-25 (1968).
Coulston, F.: Some Principles Involved in Assessment of Drug Safety. Canad Med Assoc J 98:276-277 (1968).
Cralley, L.J.: Industrial hygiene in the U.S. Public Health Service (1914-1968). Appl Occup Environ Hyg 11(3):147-155 (1996).
Cramer, G.M.; Ford, R.A.: Estimation of Toxic Hazard - a Decision Tree Approach. Fd Cosmet Toxicol 16:255-276 (1978).
Crawford, C.W.: Evaluation of Toxicity of Chemicals. JAMA 154(11):938-939 (1954).
Curry, A.S.: Toxicological Analysis. J Pharm Pharmacol 12(6):321-39 (1960).
Cutting, W.C.: Toxicity of Silicones. Stanford Med Bull 10(1):23-26 (1952).
Darby, T.D.: Pharmacologic Considerations in the Design of Toxicology Experiments. Clin Toxicol 12(2):229-238 (1978).
Darby, T.D.: Safety Evaluation of Polymer Materials. Ann Rev Pharmacol Toxicol 27:157-167 (1987).
Daughters, G.T.: The Food Additive Amendment to the Food, Drug, & Cosmetic Act. Assoc Food Drug Officials of U.S. 23(2):73-76 (1959).
Davey, D.G.: The Study of the Toxicity of a Potential Drug? Basic Principles. Supplement to Experimental Studies and Clincal Experience: The Assessment of Risk. Proceedings of the European Society for the Study of Drug Toxicity. Vol 6:3-13. Amsterdam: Excerpta Medica Foundation (1965).
Day, P.L.: The Food and Drug Administration Faces New Responsibilities. Nutr Rev18(1) 1-5 (1960).
Deere & Co.: Labeling Code Category Classification Guide List. Department of Industrial Health & Hygiene, Moline, IL (1952).
Deichmann, W.B.; LeBlanc, T.J.: Determination of the Approximate Lethal Dose With About Six Animals. J Ind Hyg Toxicol 25(9):415-417 (1943).
Deichmann, W.B.; Mergard, E.G.: Comparative Evaluation of Methods Employed to Express the Degree of Toxicity of a Compound. J Ind Hyg and Toxicol 30(6):373-378 (1948).
Deuel, Jr., H.J.:Toxicologic Determination of Suitability of Food Additives. Food Res 20(3):215-220 (1955).
Devignat, R.: Estimation of Virulent or Toxic Doses or Their Antagonists by Means of Groups of Six Mice. Rev Immunol 17(4-5):211-223 (1953).
Doull: The ACGIH TLVs: Past, Present, and Future. Appl Occup Environ Hyg 6(2):89-90 (1991).
Doyle, H.N.; Smith, R.G.; Silverman, L.; et al.:Recent Industrial Hygiene Developments –A Symposium. AIHA Q 17(3):330-344 (1956).
Doyle, H.N.: The Federal Industrial Hygiene Agency. A History of the Division of Occupational Health.United States Public Health Service (1977).
Draize, J.H.; Woodard, G.; Calvery, H.O.: Methods for the Study of Irritation and Toxicity of Substances Applied Topically to the Skin and Mucous Membranes. J Pharmacol Exp Ther 82:377-390 (1944).
DuBois, K.P.; Geiling, E.M.K.: Textbook of Toxicology (Ch. 2, General Principles of Toxicology). Oxford University Press, New York (1959).
Duggan, J.J.: A Progress Report on In-Plant Hazard Identification Systems. Arch of Environ Health 2:269-277, January through June (1961).
Eckardt, R.E.: In Defense of TLVs. J Occup Med 33(9):945-948 (1991).
Edson, E.F.: Pesticide Toxicology - Experimental, Applied and Regulatory.World Rev Pest Control 10:24-30 (1971).
Ehrlich,G.E.: Guidelines for Antiinflammatory Drug Research. J Clin Pharmacol 17(11-12):697-703 (1977).
Eisler, M.: Pesticides: Toxicology and Safety Evaluation. Trans NY Academy Sci Ser II, 31(6):720-730 (1969).
Elder, R.L.: The Cosmetic Ingredient Review - a Safety Evaluation Program. Cosm Ingredient Rev 11(6):1168-1174 (1984).
Electronic Industries Association: Labeling of Hazardous Materials. Safety and Health Committee, Industrial Relations Department, Data Sheet No. 101. Washington, D.C. January (1966).

Elkins, H.B.: The Case for Maximum Allowable Concentrations. Am Ind Hyg Assoc Q 9(1):22-25 (1948).
Elkins, H.B.: Labeling Requirements for Toxic Substances. AMA Arch of Ind Health 18:451-456, July through December (1958).
Elkins, H.B.: The Chemistry of Industrial Toxicology, 2nd Edition. Chapter 15: Maximum Allowable Concentrations. John Wiley & Sons, New York (1959).
Elkins, H.B.: Maximum Acceptable Concentrations. Arch Environ Health 2:45-49 (1961).
Elkins, H.B.: Maximum Allowable Concentrations of Mixtures. Am Ind Hyg Assoc J 23(2):132-6 (1962).
Elkins, H.B.: Threshold Limit Values and Their Significance. Transactions of the 28th Annual Meeting of the American Conference of Governmental Industrial Hygienists. Pittsburgh, PA, May 15-17, pp. 116-122 (1966).
Elkins, H.B.: Excretory and Biologic Threshold Limits. Am Ind Hyg Assoc J 28:305-314 (1967).
Elkins, H.B.: The Real World or Science Fiction. Ann Am Conf Ind Hyg 4:5-11 (1983).
Fairchild, E.J.: Occupational Exposure Limits and the Sensitive Worker: the Dilemma of International Standards. Ann Am Conf Govt Ind Hyg 3:83-89 (1982).
Fairhall, L.T.: The Relative Toxicity of Lead and Some of its Common Compounds. Transactions of the Third Annual Meeting of the National Conference of Governmental Industrial Hygienists. Bethesda, MD, April 30-May 2, pp. 155-164 (1940).
Fairhall, L.T.: Inorganic Industrial Hazards. Physiol Rev 25(1):182-202 (1945).
Fairhall, L.T.: Toxicology (1951).
Fassett, D.W.; Roudabush, R.L.: Short-term Intraperitoneal Toxicity Tests. AMA Arch Indust Hyg Occup Med 6(6):525-529 (1952).
Fassett, D.W.: Industrial Toxicology. In: Kirk-Othmer: Encyclopedia of Chemical Technology, 2nd Edition 11:595-610 (1966).
FDA: Hazardous Substances: Definitions and Procedural and Interpretative Regulations. U.S. Dept of Health, Education, and Welfare, Food and Drug Administration (1951).
FDA: Read the Label on Foods, Drugs, Devices, and Cosmetics. Miscellaneous Publication No 3. Food and Drug Administration. U.S. Dept of Health, Education, and Welfare (1957).
FDA: Appraisal of the Safety of Chemicals in Foods, Drugs and Cosmetics. The Association of Food and Drug Officials of the United States. Food and Drug Administration (1959).
FDA: Notices of Judgment under the Federal Hazardous Substances Labeling Act. U.S. Dept of Health, Education, and Welfare, Food and Drug Administration (1963).
FDA: Food and Drug Administration Advisory Committee on Protocols for Safety Evaluations: Panel on Reproduction Report on Reproduction Studies in the Safety Evaluation of Food Additives and Pesticide Residues. Toxicol Appl Pharmacol 16:264-296 (1970).
Finney, D.J.: The Statistical Treatment of Toxicological Data Relating to More Than One Dosage Factor. Ann Appl Biol 30(1):71-79 (1943).
Food Safety Council: Proposed System for Food Safety Assessment. Food Cosmet Toxicol 16: Supplement 2 (1978).
Fraser, D.A.: Re: Corporate Influence on Threshold Limit Values. Am J Ind Med 15:235-236 (1989).
Frawley, J.P.: Scientific Evidence and Common Sense as a Basis for Food-packaging Regulations. Food Cosmet Toxicol 5:293-308 (1967).
Frazer, A.C.: Synthetic Chemicals and the Food Industry. J Sci Food Agric 2(1):1-7 (1951).
Frazer, A.C.: Pharmacological Aspects of Chemicals in Food. Endeavour 12(45):43-47 (1953).
Fredrick, W.G.: The Birth of the ACGIH Threshold Limit Values Committee and its Influence on the Development of Industrial Hygiene. Transactions of the 30th Annual Meeting of the American Conference of Governmental Industrial Hygienists. St. Louis, MO, May 12-14, pp. 40-43 (1968).
Friedman, L.: Symposium on the Evaluation of the Safety of Food Additives and Chical Residues: II. The Role of the Laboratory Animal Study of Intermediate Duration for Evaluation of Safety (1969).
Friend, D.G.; Hoskins, R.G.: Reactions to Drugs. Med Clin N Am 44(5):1381-92 (1960).
Frohberg, H.: Tasks and Possibilities of Toxicology in the Pharmaceutical Industry. Drugs Made in Germany 13:1-44 (1970).
Fuess, J.T.: The American Medical Association Proposed Act for Labeling Hazardous Substances. AMA Arch of Ind Health 19:274-277, January through June (1959).

Gad, S.C.: Defining the Objective: Product Safety Assessment Program Design and Scheduling. In: Product Safety Evaluation Handbook. Marcel Dekker, Inc. New York (1988).

Gaddum, J.H.: The Estimation of Safe Dose. Brit J Pharmacol 11:156-160 (1956).

Gelzer, J.: Governmental Toxicology Regulations: an Encumbrance to Drug Research? Arch Toxicol 43:19-26 (1979).

Gidel, R.D.: Providing Hazardous Substances Information. Transactions of the 31th Annual Meeting of the American Conference of Governmental Industrial Hygienists. Denver, CO, May 11-13, pp. 137-144 (1969).

Giovacchini, R.P.: Premarket Testing Procedures of a Cosmetics Manufacturer. Toxicol Appl Pharmacol Supplement No 3:13-18 (1969).

Giovacchini, R.P.: Toxicological Evaluation of Product Safety. Pediatric Clinics N Am 17(3):645-652 (1970).

Giovacchini, R.P.: Old and New Issues in the Safety Evaluation of Cosmetics and Toiletries. CRC Crit Rev in Toxicol1(4):361-378 (1972).

Glaser, E.M.: Experiments on the Side Effects of Drugs. Brit J Pharmacol 8:187-192 (1953). Gobinet, G.: The Council of Europe Approach to Toxicity Testing and Toxicological Evaluation. Arch Toxicol Suppl 5:45-47 (1982).

Golberg, L.: The Assessment of Safety-in-use: Just How Much is Contributed by Feeding Studies in Animals. J Soc Cosmetic Chemists 15:177-194 (1964).

Goldberg, L.: Chemical and Biochemical Implications of Human and Animal Exposure to Toxic Substances in Food. Pure Appl Chem 21:309-330 (1970).

Goldenthal, E.I.; Aguanno, W.: Evaluation of Drugs. In: Appraisal of the Safety of Chemicals in Foods, Drugs and Cosmetics. The Association of Food and Drug Officials of the U.S., pp. 60-67 (1959).

Goldwater, L.: Toxicology. In: Dangerous Properties of Industrial Materials. By NI Sax Reinhold Publishing Corp, New York (1994).

Goodwin, L.G.; Rose, F.L.: The Evaluation of New Drugs. J Pharmacy & Pharmacol London 10(Suppl.):24-39 (1958).

Grasso, P.: Bioactiviation, Toxicity and Safety. Food Cosmet Toxicol 15(4):355-356 (1977).

Green S.: Present and Future Uses of Mutagenicity Tests for Assessment of the Safety of Food Additives. J Environ Path Toxicol 1:49-54 (1977).

Grice, H.C.: The Changing Role of Pathology in Modern Safety Evaluation. CRC Critical Rev in Toxicol 1(2):119-152 (1972).

Guess, W.L.: Safety Evaluation of Medical Plastics. Clin Toxicol 12(1):77-95 (1978).

Gulf Oil Corporation: Summary, Your Rights to Hazard Communication Information as a Result of the Issuance by OSHA of the "Hazard Communications" Standard: 29CFR Part 1910.1200 (1984).

Hagan, E.C.: Acute Toxicity. In: Appraisal of the Safety of Chemicals in Foods, Drugs and Cosmetics. The Association of Food and Drug Officials of the U.S., pp. 17-25 (1959).

Hamilton, A.: The Toxicity of the Chlorinated Hydrocarbons. Yale J Biol and Med 15:787-801 (1943).

Hamilton, A.: Forty Years in the Poisonous Trades. American Ind Hyg Assoc Q 9(1):5-16, March (1948).

Hanley, T.; Udall, V.; Weatherall, M.:An Industrial View of Current Practice in Predicting Drug Toxicity. Brit Med Bull 26(3):203-211 (1970).

Hanzlik, P.J.; Newman, H.W.; Van Winkle, W.; et al.: Toxicity, Fats and Excretion of Propylene Glycol and Some Other Glycols. J Pharmacol Exp Therap 67(1):101-113 (1939).

Hart, F.L.: Controlling Labelling of Common Hazardous Substances. Connecticut State Med J 20(12):962-966 (1956).

Haskins, A.L.: Caveat emptor. Bull School Med Univ Maryland 43(3):51-53 (1958).

Hatch, T.F.: Significant Dimensions of the Dose-response Relationship in Industrial Toxicology. Transactions of the 29th Annual Meeting of the American Conference of Governmental Industrial Hygienists. Chicago, IL, May 1-2, pp. 39-50 (1967).

Hayes, W.J.: Toxicological Evaluation of Some of the Newer Pesticides. Transactions of the Fourteenth Annual Meeting of the American Conference of Governmental Industrial Hygienists. Cincinnati, Ohio, April 19-22, pp. 17-21 (1952).

Hewett, P.: Interpretation and Use of Occupational Exposure Limits for Chronic Disease Agents. Occup Med: State of the Art Reviews 11(3):561-590 (1996).

Hill, S.J.: The Manufacturing Chemists' Association Labeling Program. AMA Arch of Ind Health 12:378-382 (1955).
Hine, C.H.; Jacobsen, N.W.: Safe Handling Procedures for Compounds Developed by the Petro-Chemical Industry. Am Ind Hyg Assoc J 15:141-144, June (1954).
Hine, C.H.; Dunlap, M.K.; Kodama, J.K.: Industrial Toxicology. I. General Principles and New Developments. AMA Arch Intern Med 104(5):816-26 (1959).
Hine, C.H.: Toxicology and Occupational Health. J Occup Med 4(9):457-464 (1962).
Hodge, H.C.; Sterner, J.H.: Tabulation of Toxicity Classes. Am Ind Hyg Assoc Q 10(4):93-96 (1949).
Hodge, H.C.; Downs, W.L.: The Approximate Oral Toxicity in Rats of Selected Household Products. Toxicol Appl Pharmacol 3:689-95 (1961).
Hogan, E.J.: Basis Principles for Precautionary Labeling. Ind Med Surg 29(11):530-533 (1960).
Holmberg, B.;Lundberg, P.: Exposure Limits for Mixtures. Ann Am Conf Ind Hyg12:111-118 (1985).
Homrowski, S.: Current Problems in Safety Evaluation of Plastics (introductory lecture). Pol J Pharmacol Pharm 32:65-75 (1980).
Hopkins, H.: Food Additives: Double Check on Safety. FDA Consumer 11(5):8-13 (1977).
Horner, G.H. (editor): The Solubility of Silica. Brit Med J 1:931-932 (1939).
Horner, G.H. (editor): Industrial Solvents. Brit Med J 2:177 (1939).
Hosey: Labeling of Hazardous Substances Used in Industry. Transactions of the 31th Annual Meeting of the American Conference of Governmental Industrial Hygienists. Denver, CO, May 11-13, pp. 129-132 (1969).
Hunter, O.B., Jr.: Statement of the A.M.A. to the Subcommittee on Health and Safety, Committee on Interstate and Foreign Commerce, House of Representatives. Federal Hazardous Substances Labelling Act. March 14, 1960. JAMA 173:263-5 (1960).
Hutt, P.B.: A History of Government Regulation of Adulteration and Misbranding of Medical Devices. Food Drug Cosm etLaw J 44(2):99-118 (1989).
Industrial Health Foundation: The Voluntary Approach to Worker Health Since 1935. American Ind Hyg Assoc J 47(11):667-669 (1986).
Industrial Hygiene Foundation of America: Summary of Chemical-Toxicological Conference. 18th Annual Meeting Conferences (1953).
Industrial Medical Association: Report of an Investigation of Threshold Limit Values and Their Usage. J Occup Med 8(5):280-283 (1966).
Industrial Medicine: Toxic logic (editorial). Ind Med Ind Hyg Sec 1(4):53-54 (1940).
Irish, D.D.: The Value to the Industrial Physician of Toxicological Information. Ind Hyg Q 9(1):25-28 March (1948).
Ivy, R.H.: Circumstances Leading to Organization of the American Board of Plastic Surgery. Plast Reconstruct Surg 16(2):77-87 (1955).
Jolly, D.W.: The History and Function of Toxicity Testing. In: How Safe is Safe? The Design of Policy on Drugs and Food Additives, D.W. Jolly, Ed, National Academy of Sciences, Washington, DC (1974).
Kantner, L.M.: Drug Control in Maryland. Assoc of Food & Drug Officials of the US 9(4):141-148 (1945).
Katz, D.: Psychology of Margin of Safety. Nervenarzt 22:375-376 (1951).
Keane, W.T.: OSHA - Interpretation of the Industrial Hygienist. Am Ind Hyg Assoc J33:547-557 (1972).
Keith, L.H.; Walters, D.B.: The Compendium of Safety Data Sheets for Research and Industrial Chemicals. VCH Publishers (1986).
Keplinger, M.L.: Use of Humans to Evaluate Safety of Chemicals. Arch Environ Health 6:342-349 (1963).
Kerlan, I.; Molinas, S.: Current Status of the Federal Hazardous Substances Labeling Act. New Physician 11:72-6 (1962).
Keysser, C.H.: Preclinical Safety Testing of New Drugs. Ann Clin Lab Sci 6(2):197-205 (1976).
Kittelton, J.D.: Legal Considerations in Drafting Warning Labels. Arch of Environ Health 2:263-268, January through June (1961).
Klumpp, T.G.: The Philosophy of the Administration of the Drug Sections of the Food, Drug, and Cosmetic Act. Presented at the Forty-Fifth Annual Conference of Food and Drug Officials. St. Paul, Minnesota, June. Pages 83-89 (1941).
Klumpp, T.G.: The Food and Drug Administration and the Pharmaceutical Industry. J Indiana Med Assoc 54(11):1680-90 (1961).

Kohlstaedt, K.G.: Developing and Testing of New Drugs by the Pharmaceutical Industry. Clin Pharmacol Ther 1(2):192-201 (1960).

Lakey, J.F.: Food and Drug Laws; Interpretation and Enforcement. Pediatrics 1:534-537 (1948).

Landsteiner, K.; Jacobs, J.J.: Studies on the Sensitization of Animals with Simple Chemical Compounds. J Exp Med 61:643-657 (1935).

Larrick, G.P.: The Evolution of a Drug Control Program. Presented at the 49th Annual Conference. Buffalo, NY, June. Pages 42-48 (1945).

Lasagna, L.: Congress, the FDA, and New Drug Development: Before and after 1962. Perspect Biol Med 32(3):322-343 (1989).

Lazarus, J.; Cooper, J.:Absorption, Testing, and Clinical Evaluation of Oral Prolonged-action Drugs. J Pharmaceutical Sci 50(9):715-732 (1961).

Leeper, P.: Finding the Bad Actors in a World of Chemicals. New Report, pp. 4-10. March (1984).

Lehman, A.J.: The Toxicology of the Newer Agricultural Chemicals. Assoc of Food Drug Officials of US 12(3):82-89 (1948).

Lehman, A.J.: Pharmacological Considerations of Insecticides. Assoc of Food Drug Officials of the U.S. 13(2):65-70 (1949).

Lehman, A.J.; Laug, G.W.; Draize, J.H.; et al.: Procedures for the Appraisal of the Toxicity of Chemicals in Foods. Food Drug Cosmet Law Q September: 412-434 (1949).

Lehman, A.J.; Hartzell, A.; Ward, J.C.: Effects on Beneficial Forms of Life, Crops and Soil and Residue Hazards. JAMA Sept 9:108 (1950).

Lehman, A.J.: Proof of Safety: Some Interpretations. J Am Pharm Assoc (Scient Ed) 40(7):305-308 (1951).

Lehman, A.J.: Chemicals in Foods: A Report to the Association of Food and Drug Officials on Current Developments. Assoc of Food Drug Officials of U.S. 15(3):82-89 (1951).

Lehman, A.J.: Chemicals in Foods: A Report to the Association of Food and Drug Officials on Current Developments. Part II. Pesticides. Assoc of Food Drug Officials of U.S. 15(4):122-133 (1951).

Lehman, A.J.; Patterson, W.I.: F&DA Acceptance Criteria. Modern Packaging January:115-174 (1955).

Lehman, A.J.; Patterson, W.I.; Davidow, B.; et al.: Procedures for the Appraisal of the Toxicity of Chemicals in Foods, Drugs and Cosmetics. Food Drug Cosmet Law J October: 679-748 (1955).

Lehman, A.J.: The Food and Drug Administration and Drug Safety. Minnesota Med 41(8):574-6 (1958).

Lewis, C.E.: A Method for Reporting Toxicological Information. AMA Arch of Ind Health 18:457-459 July through December (1958).

Lewis, T.R.: Identification of Sensitive Subjects Not Adequately Protected by TLVs. Ann Am Conf Ind Hyg 12:167-172 (1985).

Ligon, E.W.: Federal Hazardous Substances Labeling Act. Arch Environ Health 10(4):596-598 April (1965).

Liljestrand, A.: Safety Aspects on Long-term Medication. Acta Pharm Cuecica 10:371-380 (1973).

Litchfield, J.T.; Wicoxon, F.: A Simplified Method of Evaluating Dose-effect Experiments. J Pharmacol Exp Ther 96(2):99-113 (1949).

Litchfield, J.T., Jr.: Evaluation of the Safety of New Drugs by Means of Tests in Animals. Clinc Pharmacol Therap 3(5):665-672 (1962).

Lu, F.C.: International Activities in the Field of Food Additives, with Particular Reference to Carcinogenicity. Ecotox Environ Safety 3:301-309 (1979).

Luckens, M.M.: The Interpretation of Toxicity Data. Safety Maintenance 123(6):44-47 (1962).

Ludwig, H.F.: Chemicals and Environmental Health. Am J Pub Health 45(7):874-879 (1955).

Luther, H.G.: Where Do We Go From Here? pp 139-161 (19??).

MacDougall, J.D.B.: Toxicity Studies on Silicone Rubber and Other Substances. Nature 172 (4368):124-125 (1953).

Magnuson, H.J.: Occupational Health in Official Agencies Retrospect and Prospect. Transactions of the 30th Annual Meeting of the American Conference of Governmental Industrial Hygienists. St. Louis, Missouri, May 12-14, pp. 44-49 (1968).

Mancuso, T.F.: Forces and Trends Which Developed the Present State of the Art. Transactions of the 30th Annual Meeting of the American Conference of Governmental Industrial Hygienists, St. Louis, MO, May 12-14, pp. 20-22 (1968).

Mantel, N.; Heston,W.E.;Gurian, J.M.: Thresholds in Linear Dose-response Models for Carcinogenesis. J Nat Cancer Inst 27(1):203-215 (1961).

Mantel, N.; Bryan,W.R.: "Safety" Testing of Carcinogenic Agents. J Natl Cancer Inst 27(2):455-470 (1961).

Mantel, N.: Part IV. The Concept of Threshold in Carcinogenesis. Clin Pharm Ther 4(1):104-109 (1963).

Mantel, N.; Bohidar, N.R.; Brown, C.C.; et al.: An Improved Mante-Bryan Procedure for "Safety" Testing of Carcinogens. Cancer Res 35:865-872 (1975).

Manufacturing Chemists' Association, Inc., Washington, D.C., Chemical Safety Data Sheets

Acetaldehyde, 1952, Manual Sheet SD-43.
Acetylene, 1947, Manual Sheet SD-7.
Ammonium Dichromate, 1952, Manual Sheet SD-45.
Arsenic Trioxide, 1956, Manual Sheet SD-60.
Benzene (third revision), 1960, Manual Sheet SD-2.
Betanaphthylamine, 1949, Manual Sheet SD-32.
Butyraldehydes, 1960, March, Manual Sheet SD-78.
Carbon Tetrachloride (revised), 1963, Manual Sheet SD-3.
Chloroform, 1974, Manual Sheet SD-89.
Cresol, 1952, Manual Sheet SD-48.
Chromic Acid (Chromium Trioxide), 1952, Manual Sheet SD-44.
Diethylenetriamine, 1959, September, Manual Sheet SD-76.
Dimethyl Sulfate, 1947, Manual Sheet SD-19.
Ethyl Chloride, 1953, Manual Sheet SD-50.
Formaldehyde (revised), 1960, April, Manual Sheet SD-1.
Hydrocyanic Acid, 1961, Manual Sheet SD-67.
Lead Oxides, 1956, August, Manual Sheet SD-64.
Methyl Acrylate and Ethyl Acrylate, 1960, April, Manual Sheet SD-79.
Methylamines, 1955, Manual Sheet SD-57.
Methyl Ethyl Ketone, 1961, Manual Sheet SD-83.
Naphthalene, 1956, Manual Sheet SD-58.
Nitrobenzene (revised), 1967, August, Manual Sheet SD-21.
Phthalic Anhydride (commercial), 1956, Manual Sheet SD-61.
Sodium and Potassium Dichromates and Chromates, 1952, Manual Sheet SD-46.
Styrene Monomer, 1950, Manual Sheet SD-37.
Tetrachloroethane, 1949, Manual Sheet SD-34.
Toluene, 1956, Manual Sheet SD-63.
1,1,1-Trichloroethane, 1965, Manual Sheet SD-90.
Trichloroethylene, 1947, Manual Sheet SD-14.
Vinyl Chloride (revised), 1972, Manual Sheet SD-56.

Manufacturing Chemists' Association: A Guide for the Preparation of Warning Labels for Hazardous Chemicals. Manual L-1 (1945).

Manufacturing Chemists' Association: Product Caution Labels. Manual L-2 May (1945).

Manufacturing Chemists' Association: Warning Labels, A Guide for the Preparation of Warning Labels for Hazardous Chemicals. Manual L-1 (1949).

Manufacturing Chemists' Association: Warning Labels, A Guide for the Preparation of Warning Labels for Hazardous Chemicals. Manual L-1 (1956).

Manufacturing Chemists' Association: Guide to Precautionary Labeling of Hazardous Chemicals. Manual L-1 (1961).

Manufacturing Chemists' Association: Guide to Precautionary Labeling of Hazardous Chemicals, 7th Edition. Manual L-1 (1970).

Markuson, K.E.: Plant Conditions. To What Extent Should Official Findings Regarding Them be Made Available to Workers? Transactions of the Seventh Annual Meeting of the National Conference of Governmental Industrial Hygienists. St. Louis, Missouri, May 9, pp. 22-25 (1944).

Marzoni, F.A.; Upchurch, S.E.; Lambert, C.J.: An Experimental Study of Silicone as a Soft Tissue Substitute. Plast Reconstr Surg Transplantation Bull 24(6):600-8 (1959).

Masek, E.J.: Uniform Principles for Precautionary Labeling. Transactions of the 31th Annual Meeting of the American Conference of Governmental Industrial Hygienists. Denver, CO, May 11-13, pp. 133-136 (1969).

Mastromatteo, E.: Presented to the Mining Section, National Safety Council, October 26, 1971, Chicago, IL In: Ann Am Conf Ind Hyg 9:207-213 (1984) (1971).

Mastromatteo, E.: On the Concept of Threshold. Presented at the Am Ind Hyg Conference Portland, OR, May 24-29. In: Ann Am Conf Ind Hyg 9:331-340 (1984) (1981).

Matter, B.E.: Problems of Testing Drugs for Potential Mutagenicity. Mutation Res 38:243-258 (1976).

Mayer, R.L.: Shortcomings of the Experimental Methods for Detection of Sensitizing Properties of New Drugs. J Allergy 26(2):133-140 (1955).

Maynard, E.A.: Toxicity Testing of Chemical Additives. Food Technol 6(9):351-353 (1952).Mayo, C.W.; Fishbein, M.; Covet, S.: A Plea for the Labeling of Drugs. Postgraduate Med June: 644-645 (1957).

McCarl, G.W.: Present Status of Labeling of Materials in Industry. Am Ind Hyg Assoc Q 17(1):510-513 (1956).

McGee, L.C.: Chlorinated insecticides: Toxicity for Man. Ind Med Surg 24(3):101-109 (1955).

McGowan, J.C.: Physically Toxic Chemicals and Industrial Hygiene. AMA Arch Ind Health 11(4):315-323 (1955).

McLean, A.E.M.: Testing of Industrial Chemicals. The Lancet Nov 19:1070-1071 (1977).

McLean, A.E.M.: Symposium on “Safety in Man’s Food.” Risk and Benefit in Food and Additives. Proc Nutr Soc 36:85-90 (1977).

McNamara, B.P.: Toxicological Test Methods. Assoc Food Drug Off U.S. Q Bull 38(1):33-50 (1974).

McNamara, B.P.: Concepts in Health Evaluation of Commercial and Industrial Chemicals. In: Advances in Modern Toxicology, Volume 1, Part 1: New Concepts in Safety Evaluation, M.A. Mehlman; R.E. Shapiro; H. Blumenthal, Eds. Hemisphere Publishing Corp, Washington (1976).

Medical Research Council: Assessment of Toxicity. Memorandum of Toxicology Committee. Month Bull Min Health Publ Health Lab Serv16:2-5 (1957).

Mellan, I.; Mellan, E.: Encyclopedia of Chemical Labeling. Chemical Publishing Co Inc, New York (1961).

Merrill, R.A.: Regulation of Drugs and Devices: an Evolution. Health Affairs 13(3):47-69 (1994).

Modell, W.: Problems in the Evaluation of Drugs in Man. J Pharm Pharmacol 11(10):577-594 (1959).

Morgan, J.F.: Summary of Chemical and Toxicological Conference. Transactions of Thirteenth Annual Meeting of Industrial Hygiene Foundation of America Inc, November 18. Pages 38-41 (1948).

Morse, K.M.: Industrial Hygiene - Progress and Inertia. Transactions of the 30th Annual Meeting of the American Conference of Governmental Industrial Hygienists. St. Louis, Missouri, May 12-14, pp. 12-15 (1968).

Nale, T.W.: Current Method of Determining Safety in the Application of New Chemicals. Indt Med 20(11):501-506 (1951).

Nale, T.W.: The Chemical Industry and Precautionary Labeling. Transactions of the 20th Annual Meeting of the American Conference of Governmental Industrial Hygienists. Atlantic City, NJ, April 19-22 (1958).

Nale, T.W.: The Federal Hazardous Substances Labeling Act. Arch of Environ Health 4:239-246, January through June (1962).

National Academy of Sciences: Principles and Procedures for Evaluating the Safety of Food Additives. Publication 750. National Research Council, Washington, DC (1959).

National Academy of Sciences: Basis for Establishing Emergency Inhalation Exposure Limits Applicable to Military and Space Chemicals. National Research Council, Washington, DC (1964).

National Academy of Sciences: Principles and Procedures for Evaluating the Toxicity of Household Substances. National Research Council, Washington, DC (1964).

National Academy of Sciences: Some Considerations in the Use of Human Subjects in Safety Evaluation of Pesticides and Food Chemicals. Publication 1270. National Research Council, Washington, DC (1965).

National Academy of Sciences: Principles and Procedures for Evaluating the Toxicity of Household Substances. National Research Council, Washington, DC (1977).

National Fire Protection Association: A Table of Common Hazardous Chemicals, 7th Edition. Committee on Hazardous Chemicals and Explosives, Boston (1944).

National Safety Council: Chemical Safety References. Data Sheet 486, Revision A (extensive), Chicago (1968).

Nelson, E.E.: Biological Assays and Enforcement Activities. Presented at the Forty-Fourth Annual conference of Food and Drug Officials, New Orleans, LA Pages 22-28 (1940).

Newberne, P.M.: Pathology: Studies of Chronic Toxicity and Carcinogenicity. J of the AOAC 58(4):650-656 (1975).

New Jersey State Department of Health: Threshold Limit Values. Occupational Health Bulletin 5(3):1-7 (1964).

New York State Department of Health: Right to Know Bill - Resources for Training Materials. Bureau of Toxic Substance Assessment, January (1981).

New York State Department of Health: Chemical Fact Sheet, Trichloroethylene. Bureau of Toxic Substance Assessment, July (1981).

New York State Department of Health: Right to Know Bill - Implementation and Work Plan. Bureau of Toxic Substance Assessment, May (1982).

NIOSH: A Recommended Standard for an Identification System for Occupationally Hazardous Materials. United States Department of Health, Education, and Welfare, Public Health Service, Centers for Disease Control, National Institute for Occupational Safety and Health (1974).

NIOSH: The Toxic Substances List, 1974 Edition, H.E.Christensen; T.T. Luginbyhl; B.S.Carroll, Eds. National Institute for Occupational Safety and Health, U.S. Dept of Health, Education, and Welfare (1974).

NIOSH: Registry of Toxic Effects of Chemical Substances,1985-86 Edition. User's Guide, D.V.Sweet, Ed. National Institute for Occupational Safety and Health, U.S. Dept of Health, Education, and Welfare, April (1987).

Olishifski, J.B.: The Elements of Industrial Toxicology. National Safety News, October (1967).

Oser, B.L.: Gaging the Toxicity of Chemicals in Food. Chem Eng News 29(28):2808-2812 (1951).

Oser, B.L.: What are the Goals of a Safety Evaluation Program? Toxicol Appl Pharmacol Suppl No 3:126-130 (1969).

OSHA: Hazard Materials Labeling; Notice of Proposed Rulemaking and Public Hearings. Federal Register 42(19):5372-5374, January 28. OSHA (1977).

OSHA: Hazard Identification; Notice of Proposed Rulemaking and Public Hearings. Federal Register, 46(11):4412-4453, January 16. OSHA (1982).

OSHA: Hazard Communication; Notice of Proposed Rulemaking and Public Hearings. Federal Register, 47(54):12092-12101, March 19. OSHA (1982).

OSHA: Chemical Hazard Communication. OSHA (1983).

OSHA: Technical Note # 19, OSHA Clarifies Warning Labels Required by Hazard Communication Standard. OSHA (1986).

Paget, G.E.: Toxicity tests: A Guide for Clinicians. J New Drugs 2(2):78-83 (1962).

Paget, G.E.: Standards for the Laboratory Evaluation of the Toxicity of a Drug? Viewpoint of the Expert Committee on Drug Toxicity of the Association of the British Pharmaceutical Industry. Proc Eur Soc Study Drug Toxicol 2:7-13 (1963).

Pangman, W.J.; Wallace, R.M.:Use of Plastic Prosthesis in Breast Plastic and Other Soft Tissue Surgery. West J Surg Obstetrics Gyn 63:503-512 (1955).

Parmer, L.G.: Toxicological Aspects of Common Food and Drugs. Ind Med and Surg 27(6):285-6 (1958).

Patton, D.E., (1980). Legal Aspects of Pesticides and Toxic Substances Testing Requirements. J Environ Sci Health B15 (6):645-663 (1980).

Paustenbach, D.J.: Methods for Setting Limits for Acute and Chronic Toxic Ambient Air Contaminants. Appl Occup Environ Hyg 12(6):418-428 (1997).

Peck, H.M.: An Appraisal of Drug Safety Evaluation in Animals and the Extrapolation of Results to <Man. In: Importance of Fundamental Principles of Drug Evaluation, D.H. Tedeschi; R.E. Tedeschi, Eds, pp. 449-471. Raven Press, New York (1968).

Pennsylvania Department of Health: Short Term Limits for Exposure to Airborne Contaminants. A Documentation. Division of Occupational Health (1969).

The General Assembly of Pennsylvania: House Bill No 1236, Section 6. Labeling. pp. 26-31 (1983).

The General Assembly of Pennsylvania: Worker and Community Right-to-Know Act. Act 1984-159, pp.734-757 (1984).

Pier, S.M.; Allison, R.C.; Cunningham, E.M.; et al.: Methods for Categorization of Hazardous Materials. In: Proc of the 1978 Nat'l Conf on Control of Hazardous Material Spills. April 11-13, Miami Beach, FL (1978).

Polson, C.J.; Tattersall, R.N.: Advances in Clinical Toxicology. The Practitioner 187:549-56 (1961).

Poulsen, E.: Toxicological Aspects of Food Safety. Introduction to the Symposium. Arch Toxicol Suppl 1:15-21 (1978).

Pozzani, U.C.; Weil, C.S.; Carpenter, C.P.: The Toxicological Basis of TLVs: 5. The Experimental Inhalation of Vapor Mixtures by Rats, with Notes upon the Relationship between Single Dose Inhalation and Single Dose Oral Data. Am Ind Hyg Assoc J 20(5):364-9 (1959).

Price, P.E.: Coming to Grips with the "Right to Know." Metal Producing, pp. 36-37, June (1986).

Prusak, L.P.: The Requirement for Proof of Utility in Patent Applications for Therapeutic Products. Am J Pharm 125(7):240-249 (1953).

Queen, W.A.: Progress in Drug Control. Presented at the Conference of Pharmaceutical Law Enforcement Officials, Cleveland, OH, September 7. Pp. 49-64 (1944).

Randolph, T.G.: Unlabeled Allergenic Constituents of Commercial Food and Drugs: A Critique of Food, Drug, and Cosmetic Act Annals of Allergy. 9:151-165 (1951).

Rankin, W.B.: Criminal Evidence in Misbranded Drug and Device Cases. Presented at the 32nd Annual Conference of the Central Atlantic States Association, Baltimore, MD, May 26-27. pp. 105-111 (1948).

Ratney, R.S.: The Development of Workplace Exposure Limits for Toxic Substances. Appl Ind Hyg Special Issue, December: 47-49 (1989).

Reindollar, W.F.: Laboratory Program for Drug Control. Presented at the Forty-fourth Annual Conference, New Orleans, LA, October. Pages 118-120 (1940).

Reznikoff, P.; Rose, H.M.; Stern, M.; et al.: Toxic Effects of Therapeutic Agents; Transcription of a Panel Meeting on Therapeutics. Bull NY Acad M 32(11), pp. 796-818 (1956).

Rhode Island Department of Labor: Hazardous Substance Right-to-Know Act. Chapter 28-21 (1984).

Rinehart, W.E.; Kaschak, M.; Pfitzer, E.A.: Range-Finding Toxicity Data for 43 Compounds. Industrial Hygiene Foundation of America, Chemical-Toxicological Series, Bulletin No 6 (1967).

Rowe, V.K.; Spencer, H.C.; Bass, S.L.: Toxicologic Studies on Certain Commercial Silicones and Hydrolyzable Silane Intermediates. J Ind Hyg Toxicol 30(6):332-352 (1948).

Rowe, V.K.; Spencer, H.C.; Bass, S.L.: Toxicologic Studies on Certain Commercial Silicones. Arch Ind Hyg Occup Med 1(5):539-544 (1950).

Rowe, V.K.; Spencer, H.C.; McCollister, D.D.; et al.: Toxicity of Ethylene Dibromide Determined on Experimental Animals. AMA Arch Ind Hyg Occup Med 6(2):158-173 (1952).

Rowe, V.K.; McCollister, D.D.; Spencer, H.C.; et al.: Toxicology of Mono-, Di-, and Tri-propylene Glycol Methyl Ethers. AMA Arch Ind Hyg Occup Med 9(6):509-525 (1954).

Rowe, V.K.; Adams, E.M.: Problems of Health in the Marketing of Chemicals. AMA Arch Ind Hyg Occup Med 10(1):50-53 (1954).

Rowe, V.K.; Wolf, M.A.; Weil, C.S.; et al.: The Toxicological Basis of Threshold Limit Values: 2. Pathological and Biochemical Criteria. Am Ind Hyg Assoc J 20(5):346-349, October (1959).

Rowe, V.K.: The Significance and Application of Threshold Limit Data. Nat Safety Cong 12:33-36 (1963).

Sachs, M.: The Need for Threshold Limits. Am Ind Hyg Assoc Q 17(3):274-281, September (1956).

Schmutz, J.F.: Chronic Health Hazards, A National Challenge. Presentation to the National Symposium on Chronic Hazards. November 29 (1977).

Schneiderman, M.A.; Mantel, N.: The Delaney Clause and a Scheme for Rewarding Good Experimentation. Prev Med 2:165-170 (1973).

Schrenk, H.H.: Toxic Logic. Ind Med 1(4):53-54 (1940).

Schrenk, H.H.: Interpretations of Permissible Limits. Am Ind Hyg Assoc Q 8(3):55-60 (1947).

Schrenk, H.H.: Interpretation of Permissible Limits in the Breathing of Toxic Substances in Air. United States Department of the Interior, Bureau of Mines. May (1948).

Schrenk, H.H.: Development of New Products Includes Responsibility for Toxicological Data for Safe Manufacture and Use. Ind Eng Chem 42:81A October (1950).

Schrenk, H.H.: Toxicity of Carbon Tetrachloride and Mercury Hazards in Laboratories are Subjects of Recent Reports. Ind Eng Chem 44:131A October (1952).

Schrenk, H.H.: Pitfalls in Using Maximum Allowable Concentrations in Air Pollution. Am Ind Hyg Assoc Q 16(3):230-234 September (1955).

Schrenk, H.H.: Growth and Progress in Industrial Hygiene. Ind Hyg Q June:113-118 (1957).

Sebrell, W.H.; Cannon, P.R.; Davidson, C.S.; et al.: Some Considerations in the Use of Human Subjects in Safety Evaluation of Pesticides and Food Chemicals: A Report of the Ad Hoc Subcommittee on Use of Human Subjects in Safety Evaluation of the Food Protection Committee, Food and Nutrition Board, National Academy of Sciences - National Research Council (1965).

Seevers, M.H.: Perspective Versus Caprice in Evaluating Toxicity of Chemicals in Man. J Am Med Assoc 153(15):1329-1333 December 12 (1953).

Sellers, E.A.: Unexpected Hazards Associated with Drugs and Chemicals. Canadian Med Assoc J 86(16):721-724 (1962).

Shaffer, C.B.: Health Aspects of Production and Marketing of Chemicals. AMA Arch Ind Health 19(3):298-301 (1959).

Shenoy, K.G.; Grice, H.C.; Campbell, J.A.: Acute Toxicity as a Method of Assessing Sustained Release Preparations. Toxicol Appl Pharmacol 2:100-110 (1960).

Shubik, P.; Sice, J.: Chemical Carcinogenesis as a Chronic Toxicity Test, a review. Cancer Res 16(8):728-742 (1956).

Shubik, P.: Symposium on the Evaluation of the Safety of Food Additives and Chemical Residues: III. The Role of the Chronic Study in the Laboratory Animal for Evaluation of Safety. Toxicol Appl Pharmacol 16:507-512 (1970).

Silson, J.E.: The Significance of Maximum Allowable Concentrations. Monthly Review of the Div of Ind Hyg & Safety Standards. NYS Dept of Labor 28(2):5-8 (1949).

Simon, I.: Suggestion of a Method for the Determination of the Minimal Lethal Dose of Drugs Which Could Form the Basis of an International Convention on this Subject. Sci Med Ital 2(3-4):653-661 (1951-1952).

Slocum, G.G.: Pure foods- safe drugs: The Food and Drug Administration's Role in Public Health. Am J Pub Health 46(8):973-977 (1956).

Smith, A.E.: Needed Medical Research Developments in Industry. J Am Pharm Assoc Sci 34(4):123-126 (1945).

Smith, C.: A Short Term Chronic Toxicity Test. J Pharmocol Exp Therap 100(4):408-420 (1950).

Smith, R.G.: Problems Raised by New Drugs. Assoc of Food Drug Officials of U.S. 19(4):144-150. Assoc Food Drug Officials of U.S. 23(1):44-49 (1955).

Smith, R.G.: Evaluation of Safety of New Drugs by the Food and Drug Administration. J New Drugs 1(2):59-64 (1961).

Smith, S.E.: The Safety of Drugs. Nursing Times March 31:423-424 (1967).

Smyth, H.F.: Report of the Committee on Volatile Solvents to the Industrial Hygiene Section of the American Public Health Association. October 17 (1939).

Smyth, H.F.; Carpenter, C.P.: The Place of the Range Finding Test in the Industrial toxicology Laboratory. J Ind Hyg Toxicol 26(8):269-273 (1944).

Smyth, H.F.: Solving the Problem of the Toxicity of New Chemicals in Industry. The W Virginia Med J 42(7):9-13 (1946).

Smyth, H.F.; Holden, F.R.: Summary of Conference on Chemistry and Toxicology. Transactions of Eleventh Annual Meeting of Industrial Hygiene Foundation of America, Inc, November 7 (1946).

Smyth, H.F.; Morgan, J.F.: Summary of Chemical and Toxicological Conference. Transactions of Twelfth Annual Meeting of Industrial Hygiene Foundation of America, Inc, November 20 (1947).

Smyth, H.F.; Carpenter, C.P.: Further Experience with the Range Finding Test in the Industrial Toxicology Laboratory. J Ind Hyg Toxicol 30(1):63-68 (1948).

Smyth, H.F.; Carpenter, C.P.; Weil, C.S.: Range-finding Toxicity Data, List III. J Ind Hyg and Toxicol 31(1):60-62 (1949).

Smyth, H.F.; Carpenter, C.P.; Weil, C.S.: Range-finding Toxicity Data: List IV. AMA Arch Ind Hyg Occup Med 4(2):119-122 (1951).
Smyth, H.F.; Weil, C.S.; Adams, E.M.; et al.: Efficiency of Criteria of Stress in Toxicological Tests. AMA Arch Ind Hyg Occup Med 6(1):32-36 (1952).
Smyth, H.F.: Toxicological Data - Sources of Information and Future Needs. Am Ind Hyg Assoc Q 15:203-205 (1954).
Smyth, H.F.: Toxicology. In: Annual Review of Medicine. W.C. Cutting; H.W. Newman, Eds. Annual Review, Inc, Stanford, CA (1954).
Smyth, H.F.; Carpenter, C.P.; Weil, C.S.; et al.: Range-finding Toxicity Data. List V. AMA Arch Ind Hyg Occup Med 10(1):61-68 (1954).
Smyth, H.F.: Improved Communication - Hygienic Standards for Daily Inhalation. Am Ind Hyg Assoc Q 17:129-185 (1956).
Smyth, H.F., Jr.: The Toxicological Basis of Threshold Limit Values:1. Experience with Threshold Limit Values Based on Animal Data. Am Ind Hyg Assoc J 20(5):341-5 (1959).
Smyth, H.F.; Carpenter, C.P.; Weil, C.S.; et al.: Range-finding Toxicity Data: List VI. Am Ind Hyg Assoc J 23(2):95-107 (1962).
Smyth, H.F.: A Toxicologist's View of Threshold Limits. Am Ind Hyg Assoc J 23:37-43 (1962).
Smyth, H.F.: Range-finding Toxicity Data: List VII. Am Ind Hyg Assoc J September-October: 470-476 (1951).
Smyth, H.F.: Current Confidence in Occupational Health. Presented at the Am Ind Hyg Conf, Chicago, IL, May 27-June1. In: Ann Am Conf Ind Hyg 9:323-329 (1984) (1979).
Starr, I.: The Testing of New Drugs and Other Therapeutic Agents. JAMA 177(1):84-92 (1961).
Steinberg, M.: ACGIH TLVs and the Sensitive Worker. Ann Am Conf Govt Ind Hyg 3:77-81 (1982).
Sterner, J.H.: A Program for Detecting Possible Toxic Responses to Varied Organic Chemical Exposures. NY State J Med 41(6):594-599 (1941).
Sterner, J.H.: Determining Margins of Safety - Criteria for Defining a "Harmful" Exposure. Ind Med 12(8):514-518 (1943).
Sterner, J.H.: Threshold Limits - A Panel Discussion. Am Ind Hyg Assoc Q 16(1):27-39 March (1955).
Sterner, J.H.: Methods of Establishing Threshold Limits. Am Ind Hyg Assoc Q 17(3):280-286 (1956).
Stewart, C.P.; Stolman, A. (editors): Toxicology - Mechanisms and Analytical Methods. Volume 1 (Ch. 1). Academic Press, New York (1960).
Stokinger, H.E.: Present Status of MAC of Gases.Transactions of the 16th Annual Meeting of the American Conference of Governmental Industrial Hygienists. Chicago, IL, April 24-27, pp. 53-55 (1954).
Stokinger, H.E.: Standards for Safeguarding the Health of the Industrial Worker. Public Health Reports 70(1):1-11 (1955).
Stokinger, H.E.: Advances in Industrial Toxicology for the Year 1955. AMA Arch Ind Health 14:206-212 (1956).
Stokinger, H.E.: Toxicologic Aspects of Occupational Hazards. Ann Rev Med 7:177-194 (1956).
Stokinger, H.E.: Threshold Limits and Maximal Acceptable Concentrations: Definition and Interpretation,1961. Arch Env Health 4:121-123 (1962).
Stokinger, H.E.: Pharmacodynamic, Biochemical, and Toxicologic Methods as Bases for Air Quality Standards. Transactions of the 25th Annual Meeting of the American Conference of Governmental Industrial Hygienists, Cincinnati, Ohio, May 6-10, pp. 25-40 (1963).
Stokinger, H.E.: Industrial Contribution to Threshold Limit Values. Arch Env Health 10:609-611 (1965).
Stokinger, H.E.: Criteria and Procedures for Assessing the Toxic Responses to Industrial Chemicals. Presented at the First ILO/WHO Meeting on International Limits, June. In: Ann Am Conf Ind Hyg 9:155-163 (1984) (1968).
Stokinger, H.E.: Suggested Principles and Procedures for Developing Data for Threshold Limit Values for Air. Industrial Hygiene Foundation of America, Chemical-Toxicological Series, Bulletin No. 8-69 (1969).
Stokinger, H.E.: Sanity in Research and Evaluation of Environmental Health: How to Achieve a Realistic Evaluation (in Seven Commandments). Science 174:662-665 (1971).
Stokinger, H.E.: Toxicity of Airborne Chemicals: Air Quality Standards - a National and International View. Annual Rev Pharmacol Toxicol 12:407-422 (1972).

Stokinger, H.E.: Concepts of Thresholds in Standards Setting: an Analysis of the Concept and its Application to Industrial Air Limits (TLVs). Arch Env Health 25:153-157 (1972).
Stokinger, H.E.: Industrial Air Standards - Theory and Practice. J Occup Med 15(5):429-431 (1973).
Stokinger, H.E.: The Case for Carcinogen TLVs Continues Strong. Presented at the ACGIH Symposium on Workplace Control of Carcinogens, October 25-26, 1976, Kansas City, MO. In Ann Am Conf Ind Hyg 9:257-264 (1984) (1976).
Stokinger, H.E.: Stokinger Lecture. Courtesy of Bill Wagner, American Conference of Governmental Industrial Hygienists (1980).
Stokinger, H.E.: Historic Aspects of Occupational Health Standards and the Sensitive Worker. Ann Am Conf Gov Ind Hyg 3:65-69 (1982).
Stokinger, H.E.: Suggested Principles and Procedures for Developing Experimental Animal Data for Threshold Limit Values for Air. Ann Am Conf Ind Hyg 9:177-186 (1984).
Stokinger, H.E.: Occupational Carcinogenesis. In: Patty's, Pages 2879-2931 (19??).
Stolz, D.R.; Poirier, L.A.; Irving, C.C.; et al.: Evaluation of Short-term Tests for Carcinogenicity. Toxicol Appl Pharmacol 29:157-180 (1974).
Stone, G.B.: The New Product Challenge. Am J Pharmacy 131(9):327-336 (1959).
Stormont, R.T.: New Program of Operation for Evaluation of Drugs. JAMA 158(13):1170-1171 (1955).
Teleky, L.: Toxic limits. Ind Med Ind Hyg Sec 4:68-71 (1940).
Thomas, M.J.; Majors, P.A.: Animal, Human, and Microbiological Safety Testing of Cosmetic Products. J Soc Cosmet Chem 24:135-146 (1973).
Tice, L.F.: The Expansion of Drug Certification by the FDA. Am J Pharmacy 132(12):431-2 (1960).
Tice, L.F.: Package Inserts for Prescription Products. Am J Pharmacy 133(7):240-2 (1961).
Truhaut, R.: The Problem of Thresholds for Chemical Carcinogens - Its Importance in Industrial Hygiene, Especially in the Field of Permissible Limits for Occupational Exposure. Am Ind Hyg Assoc J 41:685-692 (1980).
Turfitt, G.E.: Recent Advances in Toxicologic Analysis. J Pharm Pharmacol 3(6):321-337 (1951).
United States of America Standards Institute: USA Standard Acceptable Concentrations of Formaldehyde. September 12. Sponsored by the American Industrial Hygiene Association (1967).
United States of America Standards Institute: USA Standard Acceptable Concentrations of Carbon Tetrachloride. October 9. Sponsored by the American Industrial Hygiene Association (1967).
United States Congress: Federal Hazardous Substances Labeling Act. Pages 1-10. July 12 (1960).
United States Congress: Chemical Dangers in the Workplace. Thirty-fourth Report by the Committee on Government Operations. Washington, DC (1976).
Unknown: Table of Common Hazardous Chemicals. Chem Eng News 23(14):1249-1256 (1945).
Unknown: Chronological History of Occupational Safety and Health Programs (1986).
Vainio, H.; Tomatis, L.: Exposure to Carcinogens: Scientific and Regulatory Aspects. Ann Am Conf Ind Hyg 12:135-143 (1985).
Van Arsdell, P.M.: Health Hazards of Common Laboratory Reagents. Chem Engin News 26(5):304-309, February 2 (1948).
Von Oettingen, W.F.: Variations in the Toxicity of Chemical Compounds under Different Conditions. Annual Safety Congress Transactions (1935).
Von Oettingen, W.F.: The Toxicity and Potential Dangers of Aliphatic and Aromatic Hydrocarbons. Yale J Biol Med 15(2):167-184 (1942).
Von Oettingen, W.F.; Neal, P.A.; Donahue, D.D.; et al.: The Toxicity and Potential Dangers of Toluene with Special Reference to its Maximal Permissible Concentration. Publ Health Bull 279: 1-50. United States Public Health Service (1942).
Von Oettingen, W.F.: The Aliphatic Alcohols: Their Toxicity and Potential Dangers in Relation to Their Chemical Constitution and Their Fate in Metabolism. U.S. Public Health Service, p. 253. Public Health Bull 281 (1943).
Vorhes, F.A.: Requirements of Analytical Data. Ag Food Chem 4(5):415-416 (1956).
Wands, R.C.: The Unpublished Volumes of Toxicology Literature. Am Ind Hyg Assoc J July-August: 30(4)344-346 (1969).

Wands, R.C.: Industrial Health in 1978: a Perspective. Presented at the Am Ind Hyg Conf Los Angeles, CA, May 7-12. In: Ann Am Conf Ind Hyg 9:317-321 (1984) (1978).

Wands, R.C.; Steinber, M.; Weisburger, E.K.; et al.: Threshold Limit Values for Carcinogens - Current Status. Ann Am Conf Ind Hyg 12:263-265 (1985).

University of Washington: A Toxicity Evaluation of Several Trade-Name Solvents. In: Occupational Health Newsletter. Environmental Research Laboratory. August (1954).

Weil, C.S.: Tables for Convenient Calculations of Median-effective Dose (LD50 or ED50) and Instructions in Their Use. Biometrics 8(3):249-263 (1952).

Weil, C.S.; Carpenter, C.P.; Smyth, H.F.: Specifications for Calculating Median Effective Dose. Am Ind Hyg Assoc Quart 14(3):200-206 (1953).

Weil, C.S.: Application of Methods of Statistical Analysis to Efficient Repeated-dose Toxicological Tests. I. General Considerations and Problems Involved. Sex Differences in Rat Liver and Kidney Weights. Toxic Appl Pharmacol 4:561-71 (1962).

Weil, C.S.; McCollister, D.D.: Relationships Between Short- and Long-term Feeding Studies in Designing an Effective Toxicity Test. J Agr Food Chem 11:486-491(1963).

Weil, C.S.; Woodside, M.D.; Bernard, J.R.; et al.: Relationship Between Single-peroral, One-week, and Ninety-day Rat Feeding Studies. Toxicol Appl Pharmacol 14:426-431 (1969).

Weil, C.S.; Carpenter, C.P.: Abnormal Values in Control Groups During Repeated Dose Toxicologic Studies. Toxicol Appl Pharmacol 14:335-339 (1969).

Weil, C.S.: Editorial Toxicol Appl Pharmacol 17(2):i-ii (1970).

Weil, C.S.: Guidelines for Experiments to Predict the Degree of Safety of a Material for Man. Toxicol Appl Pharmacol 21:194-199 (1972).

Weil, C.S.: Statistics vs Safety Factors and Scientific Judgment in the Evaluation of Safety for Man. Toxicol Appl Pharmacol 21:454-463 (1972).

Weisburger, J.H.; Weisburger, E.K.: Food Additives and Chemical Carcinogens: On the Concept of Zero Tolerance. Food Cosmet Toxicol 6:235-242 (1968).

Wexler, P.(19xx):Information Resources in Toxicology, 2nd edition. Elsevier, New York.

Wheatly, G.M.: The Federal Hazardous Substances Labeling Act. Pediatrics 28(3):499-500 (1961).

WHO: Principles for Pre-Clinical Testing of Drug Safety. WHO Tech Rep Ser No 341, pp. 3-22. World Health Organization, Geneva (1966).

WHO: Principles for the Clinical Evaluation of Drugs. Tech Rep Ser No 403, pp. 9-10. World Health Organization, Geneva (1968).

WHO: WHO on drug safety. Fd Cosmetics Toxicol 7(6):674-675. World Health Organization (1969).

WHO: Permissible Levels of Occupational Exposure to Airborne Toxic Substances. Sixth Report of the Joint ILO/WHO Committee on Occupational Health. World Health Organization Geneva (1969).

WHO: Principles and Methods for Evaluating the Toxicity of Chemicals. Part 1.World Health Organization, Geneva (1978).

WHO: Evaluation of Certain Food Additives and Contaminants. Technical Report Series 631. Twenty-second Report of the Joint FAO/WHO Expert Committee on Food Additives. World Health Organization (1978).

White, W.B.: The Addition of Chemicals to Food. Food Drug Cosmetic Law Q 2(4):475-481 (1947).

Wilson, R.H.; DeEds, F.: Importance of Diet in Studies of Chronic Toxicity. Arch Ind Hyg Occup Med 1(1):73-80 (1950).

Wilson, R.H.; McCormick, W.E.: Toxicology of Plastics and Rubber? Plastomers and Monomers. Indust Med 23(11):479-486 (1954).

Winnell, M.: An international Comparison of Hygienic Standards for Chemicals in the Work Environment. AMBIO 4(1):34-36 (1975).

Wodicka, V.O.: Food Ingredient Safety Criteria. Food Tech 31(1):84-88 (1977).

Wodicka, V.O.: Risk and Responsibility. Nutrition Rev 38(1):45-52 (1980).

Wodicka, V.O.: Evaluating the Safety of Food Constituents. J Environ Pathol Toxicol 3:139-147 (1980).

Woodard, G.; Calvery, H.O.: Acute and Chronic Toxicity. Ind Med 12(1):55-59 (1943).

Woolmer, R.: Clinical Tests of New Drugs. Proc Royal Soc Med 52(2):98-100 (1959).
Worden, A.N.: Toxicological Methods. Toxicol 2:359-370 (1974).
Yant, W.P.: Industrial Hygiene Codes and Regulations. Transactions of Thirteenth Annual Meeting of Industrial Hygiene Foundation of America, Inc, November 18. Pages 48-61 (1948).
Zapp, J.A.: An Acceptable Level of Exposure. Presented at the Am Ind Hyg Conf, New Orleans, LA, May 22-27. In: Ann Am Conf Ind Hyg 9:309-315 (1984).
Zapp, J.A.; Doull, J.: Industrial Toxicology: Retrospect and Prospect. In: Patty's Industrial Hygiene and Toxicology, 4th Edition, 2A, G.D. Clayton; F.E. Clayton, Eds (1993).
Zbinden, G.: The Problem of the Toxicological Examination of Drugs in Animals and Their Safety in Man. Clin Pharmacol Therap 5(5):537-545 (1964).
Zbinden, G.: Drug Safety: Experimental Programs. Problems and Solutions of the Past 10 Years are Critically Reviewed. Science 164(3880):643-647 (1969).

Appendix

ANSWERS TO STUDY QUESTIONS

Chapter 1. The Respiratory System

1.1. gas exchange; acid-base regulation; metabolism of drugs, toxins, and endogenous substances; and cardiovascular pressure regulation.

1.2.

a. Irritative symptoms of the upper and large airways, i.e. any of the following: cough, central burning chest pain, possibly wheeze, dry, scratchy or sore throat, hoarseness, headache, stuffy or runny nose, disruption of smell sensation, nosebleeds, or blood-streaked sputum.

b. Upper and lower airways (primarily the latter). The respiratory zone would be spared by inability of the particles to penetrate to that depth.

c. History of asthma or respiratory allergy (atopic status) seems to be related to a higher rate of development of long-term effects. Status as a smoker may also contribute.

1.3. exposure intensity; physical and chemical characteristics of the agent; host defense mechanisms; anatomical site of contact; and mechanism of injury.

1.4.

a. Asbestosis is a fibrotic lung disease characterized by reduced lung air volumes, decreased diffusing capacity, and increased work of breathing. Spirometry and lung volumes would be valuable to assess and quantify reduction in air volumes (usually present in asbestosis) and the presence or absence of airway obstruction (not usually present in asbestosis). D_LCO is useful to quantify reduction in diffusing capacity for oxygen. Chest x-ray may show characteristic changes associated with pneumoconiosis. Cardiopulmonary stress test can quantify exercise limitations due to reduced maximum oxygen utilization. Peak flow, and bronchoprovocation testing are not particularly useful because airway hyperresponsiveness is not usually a factor in asbestosis.

b. Any variety of asthma is characterized by episodic and reversible periods of airway obstruction due to bronchospasm. Spirometry and lung volumes would be valuable to assess and quantify changes in air volumes (air trapping and hyperinflation may occur in chronic obstructive disease) and the presence or absence of airway obstruction (usually present in asthma). Peak flow self-tests can document airway obstruction that occurs episodically and help identify triggering environmental circumstances. Bronchoprovocation testing can quantify the airway hyperresponsiveness (always present in asthma). Cardiopulmonary stress test, D_LCO, and chest x-ray are not particularly useful except to rule out other processes.

c. Heart disease may mimic a respiratory problem because heart function is necessary for normal oxygen utilization. It can produce cardiogenic pulmonary edema (congestive heart failure). The cardiopulmonary stress test would be expected to show limitations due to impaired oxygen transport, but all of the other tests would be of value only to rule out concurrent respiratory disease.

Chapter 2. Respiratory Tract Deposition and Penetration

2.1. d. b and c are correct

2.2. c. particle solubility

2.3. impaction, sedimentation, diffusion, interception; impaction is associated with the deposition of very large particles, diffusion is associated with the deposition of very small (nanometer size) particles.

Chapter 3. Occupational Dermatotoxicology: Significance of Skin Exposure in the Workplace

3.1. Skin exposures in the past have certainly been responsible for their share of occupational illness, but in many cases may have gone by unsuspected and undocumented. During the past decades, air exposure criteria have been revised downward as better knowledge of the toxic effects and the societal desire and economic need to reduce the risk of illness increased. The recent attempt by OSHA to reduce the standard for some glycol ethers from 25 ppm to 0.1 ppm is an example of this downward trend.

However, given the low vapor pressure of 2-methoxyethanol acetate of 2 mm Hg, for example, it is unlikely that elevated air concentrations will occur if basic engineering controls are present. If inhalation at the proposed PEL occurred for 8 hours, only about 5 milligrams would be inhaled. On the other hand, the absorbed dose from a skin contact surface of only 2 cm^2 exposed for one hour would exceed the exposure at the PEL. Thus, the standard, without adequate attention to skin protection, would have been ineffective in preventing illness. The most efficient use of economic resources would be to control the hazard according to the predominant route of exposure. The trend toward reducing the inhalation risk by substituting less volatile compounds that remain on surfaces longer has often resulted in an imbalanced apportionment of resources to control exposures.

3.2. First try to determine if the cause is actually work related. This requires an interview of the afflicted employee and an attempt to identify patterns of the cause or the worsening of the condition. The work environment may be the cause or may exacerbate the condition. If this seems possible, carefully observing the workplace and activities may provide clues. A complete inventory of chemical agents potentially present in the employee's surroundings should be developed. Unfortunately, complete identification of the components of product mixes is difficult if the chemical is present in less than 1% concentration or is an in-situ reaction product; but drawing an association to a product might still be possible from the history of occur rence of the condition. This information should be provided and sent along with the employee to the dermatologist or occupational physician.

3.3. It is a common misconception that a skin condition is of minor consequence and may simply be part of the job. This attitude can be a recipe for disaster. Both irritant and allergic contact dermatitis tend to worsen with continuing exposure to the offending agent. Self-treatment by the employee can result in sensitization to the medication. By far, the best course of action is for all employees to report the skin condition to the appropriate employer representative immediately so that corrective action can be taken. If mitigation occurs early, the long-term prognosis is much more favorable, reducing both future suffering by the employee as well as possible workmen's compensation and other costs incurred by the employer.

3.4. Sensitization by LMW compounds is clearly a concentration-dependent event. In addition, sufficient time of contact must also occur. Furthermore, not all allergic compounds are highly potent. Finally, strong sensitization resulting in a clinically apparent response is more likely to occur with repeated contacts. Thus, it may require highly concentrated exposures that are left on the skin to sensitize. HMW compounds, like natural rubber latex proteins, cannot readily diffuse through intact skin, but may penetrate damaged skin quite effectively. These requirements for sensitization make sensitization a fairly rare event, which, in effect, provides multiple opportunities for prevention.

3.5. First, try to estimate is reasonable area of skin surface exposure that might become contaminated by this material. You might assume from your observations that the average worker might touch contaminated parts 10 times during the work day. Since the material is not readily volatile, you might also assume that it remains on the skin until washed off. Even if wiped, significant amounts would remain on the

3.6. Skin absorption of chemicals is dependent upon many influences that could affect rate of absorption. Even if the calculation above makes a few conservative worst-case assumptions, it also ignores possible damaged skin conditions, enhancement of penetration by other components of the mixture, elevated temperatures in the factory, and the effect of occlusion if workers periodically donned gloves over contaminated skin. If in doubt regarding the adequacy of your initial risk estimate, actual workplace skin exposure sampling could be warranted.

3.7. To identify the cause of the problem, first try to establish whether the job causes or exacerbates the condition. Does the condition improve when away from work? How rapidly does it reoccur when returning to work? Determine if the employee is self-medicating and discourage all interventions outside of using mild soaps and hypo-allergenic emollient creams on clean skin. If the condition appears to be caused or worsened by the job, the next step is to identify all materials the employee may come in contact with. Do not overlook chemical protective clothing he may wear, or exposure to physical agents. This information should be provided with referral of the worker to a qualified dermatologist or occupational physician for evaluation. The physician should preferably assess the dermatological condition while it is still occurring. The evaluation might include possible patch testing to identify allergic reactions to any of the potentially allergenic substances that were identified. If an allergic response to a workplace substance is determined, the employee will not likely ever be able to tolerate even minute exposures, and should be transferred to another department, or preferably a substitute for the material should be found. It may not be feasible to adequately protect highly sensitized persons from further exposure. If irritation is indicated, the physician may recommend a skin restorative cream or lotion to assist healing. If other commercial products are considered, demand that the vendor provide evaluation data on safety and efficacy before distributing any product to your workforce. In the workplace, modified work practices and process technology to reduce chemical contact should be explored. Other workers should be encouraged to report dermatological conditions as soon as they occur so that remedial steps can be taken early.

Chapter 4. The Practice and Management of Occupational Ergonomics

4.1. First is the prevention of occupational injuries to people at work. The second is the organization of work (both physical set-up and work methods) to improve worker effectiveness to ensure operator comfort and safety.

4.2. When demand exceeds capability, physical injury and process errors are more likely to occur. By making jobs fit within the capabilities of people, the overall stresses are reduced and the jobs are less taxing on the individual. This makes the person less fatigued at the end of the work shift, and makes the job more attractive. This leads to higher job satisfaction and ultimately to lower turnover and absenteeism. While this economic impact is challenging to measure, it is significant where it has been measured.

4.3. Ergonomic job improvements frequently result in improvements to worker effectiveness with benefits in productivity, product quality, lower operating costs, increased uptime, and other similar operational benefits.

The loss of an experienced worker can be quite costly considering the time and expense needed to replace this individual (i.e., recruitment, hiring, and training time/cost for both management and new employees) plus the loss of work process knowledge developed through years of job familiarity. Unlike a new worker, an experienced worker can identify potential problems early in the process and take steps to mitigate the issue before serious consequences occur.

The economic value of ergonomic improvements can be 3%-6% of labor costs. When a 6% increase in productivity is possible in addition to controlling occupational injuries, ergonomics becomes a powerful tool indeed. Similarly, including ergonomics at the design phase of a new facility can easily save 5%-25% of development costs.

4.4. Ergonomics influences the design of hand tools, machines, and facilities layout to meet the physical limits of the individual, limiting fatigue and injury potential. It includes the physiological response to physical work load, response to environmental conditions (e.g., heat, noise, lighting, and vibration), ability to utilize hand-eye coordination in complex psychomotor tasks, and visual monitoring of the work process (e.g., control panel design and layout).

Human factors, on the other hand, focus on reducing processing error during interaction with these devices. It relies on complying with common stereotypes (e.g. flip the toggle switch up to turn the light on or down to turn the light off and using a green light to indicate that a machine is energized and a red light when de-energized), placing the control switch below or to the side of the corresponding display, ensuring that the display is not visually obstructed when using the control, and color coding to facilitate the association between groups of controls.

The gap between ergonomics and human factors engineering narrows as we are better able to remove unnecessary effort and demand. We do this through improved information transfer between people and/or product (inspection). As a result, we affect how people accomplish tasks and influence improvements in productivity and profit.

4.5. "Ergos" means work and "logos" means laws, in other words ergonomics means "the laws (or study) of work." This discipline incorporates statistics and management, manufacturing technology, mechanical engineering, anatomy and physiology, and psychology. These are the building blocks for development of task analysis and work methods, occupational biomechanics, human factors, and industrial psychology. The whole work process can be divided into individual tasks, then analyzed for issues related to posture, force, repetition, mechanical stress, temperature, vibration, and the influence of frequency and duration.

4.6. The client may assess the industrial hygienist's competency based on knowledge of its service or product development process, knowledge of the industry language, or familiarity with the current business climate.

Identifying if the company's business has a specialty, dying, or competitive market and understanding its product life may provide an insight into potential funding issues and bandwidth for proposed improvements.

A company that is rapidly changing, uses expensive state-of-the-art equipment, and runs high profit margins is more likely to spend money on improving next-generation machine and machine-interface designs (e.g., semiconductor and computer industries). A company that does not change significantly in technology and product may be more interested in improving the way employees interface with the machines.

4.7. Recommendations that improve cycle time, improve quality, minimize rework, eliminate downtime, minimize staffing, keep experienced employees injury-free and on the job, and improve business performance.

4.8. First, funding must be budgeted in advance for modifications costing more than $500 (capital budget). This is particularly important when changes must be made on more than one machine or location. It will be important to test improvements at a single site before making changes at other locations. Smaller and less costly changes can be made at a single workstation in the development phase. These improvements can then be incorporated for all other locations to ensure final success. Well-planned improvements are more likely to have process or equipment changes that pay for themselves through productivity improvements.

Second, ensuring the health and safety of the experienced employee has long- term benefits. An employee who is more comfortable and who feels that the company cares about each employee is more likely to have better attendance, better attitude, and become more of an advocate for company goals and processes.

4.9. Cycle time (the time is takes to complete a task one time) improvements get products out the door faster and allow more product to be made in a given period of time. This can be accomplished by eliminating unnecessary steps, minimizing the number and types of fasteners needed, and providing efficient hand tools.

A well-designed process flow system allows the entire component of each work process to be completed before that unit of work is passed along to the next step in the process. The flow should be linear and should not require steps to be repeated out of sequence. If a change in any part of the production process is recommended, it should first be evaluated for potential impact on product quality, cycle time, and physical demands both upstream and downstream of this step in the process to prevent product damage or bottle neck situations.

4.10. If a machine is installed next to a wall or other obstruction, leaving insufficient space for the maintenance technician or hand tools, then it may take longer than otherwise necessary to complete the repair or maintenance process. When a machine is not operational for any reason, it cannot be used for production, and therefore, cannot provide revenue.

4.11. Work is generally categorized as piece-rate, machine-paced, or self-paced. With piece-rate operations, the worker is paid per unit of completed product. These individuals often deny themselves necessary breaks to appropriately recover from the physical demands of the job because the more they make the more they are paid. With machine-paced operations, the worker may not have adequate recovery time between tasks before the next unit of work is presented to the worker to complete. With self-paced operations, the worker can take small breaks as needed between tasks so long as productivity needs are met.

4.12. Arm length of the smallest female (5th percentile female) is the data point used for maximal reach to supplies at a workstation. This includes forward, overhead, and downward reach. If the smallest female can reach supplies at these locations, then certainly taller people should be able to reach with relative ease.

Hand, shoulder, and trunk girth of the largest male (95th percentile male) are the data points used for minimum clearance for openings. If the largest hand can fit between two machine parts to access controls, then certainly smaller hands should be able to fit with relative ease. The same theory holds true for aisle way clearance, overhead clearance, and leg clearance beneath the tabletop.

When possible workstation design incorporates adjustability ranging from smallest female to largest male anthropometry needs.

4.13. Horizontal distance of the load from the center of the body has the greatest influence on the trunk muscles, particularly muscles and other structures of the low back. Injuries to the low back are the leading cause of industrial musculoskeletal injuries in the U.S. today according to the Bureau of Labor Statistics.

4.14. As the wrist moves away from the straight or neutral posture, strength diminishes. The flexor muscles of the forearm and wrist are forced into a shortened length (i.e., a mechanical disadvantage) and cannot generate adequate force to squeeze the trigger for any significant duration. The flexor muscles fatigue quickly and this action becomes rather uncomfortable.

4.15. In general, people respond to familiar signs and signals with little effort and tend to make fewer judgment errors as compared to new or "non-stereotype" input. The cycle time for familiar and stereotypic tasks is usually short. In a production setting, reducing cycle time and error are two important components of productivity measures.

4.16. Inadequate and non-adjustable workplace design that requires people to work in awkward postures. Fatigue and muscle/joint discomfort are likely outcomes.Work organization with little opportunity for task variation and inadequate recovery.

Task variation would otherwise allow for different muscle groups to operate and allow overused muscles to physiologically recover from fatigue.

Work methods, machines, and hand tools that require excessive manual force, frequent and repetitive motions, static holding (i.e., constant muscle contraction), and vibration.

Lifting heavy or awkward loads, especially on a frequent basis. This may often be the result of a poor load or layout design.

4.17. Although off the job activities cannot be mandated or controlled by the employer, it is important to understand if symptom development is partially or primarily the result of off the job hobbies, sports, or other activities. The bucket model of symptom development demonstrates that each person has a limited capacity or tolerance for physical activity. If the primary requirement for physical demand comes from work-related activities, then the employer should take steps to keep exposure to a reasonable limit. However, if the employee participates in activities outside of work that push physical demands to or beyond reasonable limits, then the employee should be informed about the cumulative impact of muscle or body overuse to prevent further physical injury. In these cases, modifying the work environment will have little to no benefit for injury reduction.

4.18. The primary reach envelope is defined by the arch of motion about the elbow allowing the elbow to stay close to the trunk. The secondary reach envelope is defined by the arch of motion about the shoulder where the elbow moves away from the trunk. Supplies and parts that are handled frequently should be placed in the primary reach zone to minimize time, effort, and muscle utilization. Supplies and parts accessed infrequently may be placed in the secondary reach zone. Supplies and parts that are rarely used should be placed at sufficient distance to require the worker to walk over to the storage area. This discourages trunk bending and twisting that may occur if objects are located just beyond fingertip reach.

4.19. Static work involves holding muscles still in a constant contraction. Muscles fatigue and use up available energy (oxygen) supply quickly. Recovery takes significantly longer than dynamic work that involves frequent muscle movement. This muscle movement "pumps" oxygen through the body to provide an adequate energy source for the work required. Therefore, dynamic work is the preferred method of design for long duration activities.

4.20. For long duration, precision.

4.21. Boxing and packing is best performed at a standing workstation because the tasks involve constant movement and walking around the warehouse. The worker would probably choose to stand even if a high chair is provided because of the short duration cycle time. However, a "lean-to" chair may be a good supplemental device because the worker can lean against it from a standing position for short duration weight shifting. A foot rail or footrest should be provided for all standing workstations to facilitate weight shifting and posture changes.

Repairing watches is a precision activity and is best performed at a sitting workstation where the forearms can be stabilized by the tabletop.

Telemarketing requires a long duration of being in one area, however, all work does not need to be performed in a sitting position. A sit/stand workstation allows the worker to use the computer from both a sitting and standing posture. This option allows the worker to stand at will, changing postures often and protecting the low back from problems associated with prolonged sitting. A telephone headset allows hands free communication and further improves trunk posture.

4.22. Some common quick fixes include a standing platform to raise the worker, a spacer or platform on top of the workstation to raise the work height, padding for sharp or hard workstation edges, lean-to or sit/stand chairs, foot rests or rails, providing knee and toe clearance, providing long handled tools for reaching, padding on hand tools, specialty hand tools, spring loaded hand tools, suspending hand tools, clips or vices to hold parts in place, and reorganizing the location of supplies on the work surface.

4.23. The first design principle meets the anthropometry needs of the smallest female to the largest male, usually relying on adjustable features with quick fix options, and without the need for additional accommodations or adaptive devices.

The second design principle suggests the use of anthropometry tables to identify maximum reach limits for product or supplies handled at the workstation. This is one of the least expensive design techniques and it is widely used.

The third design principle requires adjustability to fit the necessary range of workers. This method is typically among the most costly, yet it is widely used. Adjustability is one of the offset expenses of having multiple shifts and/or multiple operators use one piece of equipment.

The fourth design principle requires a design for the end user population. For example, if designing for people in the Pacific Rim, use anthropometry tables for that population as opposed to North America.

The fifth design principle states that if adjustability is not possible, then design for the average. The result is inconvenience for many people, but when adjustment is not possible, it allows one design to serve the greatest number of people.

4.24. First, many ergonomics related symptoms and injuries are not reported because people may be concerned about job security or they may not recognize the association between symptoms and job activities. A good symptom awareness and early reporting program will help to minimize under reporting. Second, requesting employee input about difficult tasks often reveals the presence of ergonomics risk factors that may not have been reported. Often workers have modified the workspace themselves to make the job easier. Finally, using your own "ergonomics eye" to identify potential problems such as awkward postures and equipment handling problems may pinpoint unrecognized ergonomics risk factors.

4.25. Complete a job overview to understand the entire process and how the task(s) in question relate to the whole process. This information will help with both problem identification and problem solving upstream and downstream of the jobs under investigation.

Review injury data for important information related to body parts injured, how the injury occurred, and what may have caused the symptoms.

Talk with people performing the job and look for a pattern of symptoms or complaints of discomfort. A symptom survey may be used to solicit input from a larger population of employees, then analyzed for trends.

Observe and analyze the job using a variety of evaluation tools. First, a general survey for high/moderate risk problems can be performed to establish a need to investigate further. More in-depth analysis methods may include the NIOSH Lifting Guideline, motion analysis studies, psychophysical tables, and others. Also break the job down into its various tasks and the ergonomics risk factors for each task, then identify the root cause for each risk factor.

4.26. The chosen solution may not be practical in terms of installation, technology, or funding, therefore, a list of alternatives may be useful for future use. Engineering solutions may be more expensive initially, however, the benefits are usually longer lasting, more effective, and do not depend on the employee to remember specific rules or work methods.

4.27. Develop a list of who, what, when, where, and why to hold people accountable and to list specific responsibilities and due dates for project completion. Next, test the modification at only one workstation initially to "work out the bugs." Solicit user feedback, then use any design improvements for similar operations.

4.28. Horizontal distance has the greater influence in this example. The horizontal multiplier has the greater influence on reducing the recommended weight limit than the other multipliers because of the biomechanical influence on the lower part of the spine.

4.29. No, a Lifting Index (LI) of 1.5 means that the actual weight of the load is 1.5 times the recommended weight. An LI between 1 and 2 indicates the need for administrative controls or quick fix options, whereas an LI above 2 indicates the need for engineering controls.

4.30. The body has a greater biomechanical advantage when pushing rather than pulling a cart. The operator can lock the elbows by the trunk and push with the legs rather than arms alone. When pulling, the body is twisting (putting the back at risk for injury) and the arms work harder than necessary. The operator may prefer pulling to pushing when unable to see over the top of the cart, when the wheels are hard to maneuver in a direct path, or if there is no steering mechanism on the wheels.

4.31. Certification in ergonomics has several key benefits. It is a first step to help an employer immediately recognize the basic qualifications of potential employees. It helps companies to distinguish between qualified ergonomics consultants and individuals who simply sell products with ergonomic features, but who do not possess an acceptable level of technical competency in ergonomics.

4.32. The Certified Ergonomics Associate (CEA). This certification was developed to provide a technical level of certification to meet the growing need for certified ergonomists who use commonly accepted tools and techniques for analysis and enhancement of human performance in existing systems, but are not required to solve complex and unique problems or develop advanced measurement tools.

Chapter 5. Biological Agents – Recognition

5.1. Biological agents are defined as agents from plant or animal matter or from microorganisms, and can thus include a wide range of agents such as toxins from microorganisms and plants; plant, animal and microbial allergens; and carcinogens. The most important diseases associated with these exposures are 1) Infectious diseases; 2) Respiratory diseases; and 3) Cancer.

5.2. Health care workers: They may develop diseases such as Hepatitis B and Aids due to contact with blood products. In addition, they may develop a large number of other viral, bacterial, and parasitic diseases as listed in Table 5-2.

Forestry workers (foresters, hunters, trappers, ranchers): They may develop Lyme disease due to tick bites. In addition, they may develop several other infectious diseases listed in Table 5-2.

Aid workers (and other workers that travel to remote tropical areas such as soldiers, field biologists, etc): They may develop malaria, yellow fever, and Dengue fever due to mosquito bites. In addition, they may develop several other infectious diseases listed in Table 5-2.

5.3. Allergic diseases are caused by allergen exposure, resulting in 1) sensitization, thus requiring a specific antibody response (e.g., IgE or IgG), and 2) symptoms after being sensitized to the particular allergen. The process of sensitization and subsequent development of an allergy can take up to months and often years. Allergic respiratory diseases can develop only in hypersensitive individuals. Examples of allergic respiratory conditions include allergic asthma, allergic rhinitis, and hypersensitivity pneumonitis.

Non-allergic respiratory diseases associated with exposure to biological agents are not caused by allergens, do not require prior sensitization, and thus, do not involve an antibody mediated inflammatory response. Instead, a non-specific immune response is involved resulting in inflammation without the involvement of specific antibodies. Symptoms can develop the first day on the job, and with sufficiently high exposures most workers can develop symptoms. Examples of non-allergic respiratory conditions include non-allergic asthma, ODTS, MMI, chronic bronchitis, and chronic airflow obstruction.

5.4. Sawmill workers are exposed to a large variety of biological agents, and can potentially develop most of the diseases described in this chapter. However, infectious diseases (perhaps with the exception of aspergillosis that may occur incidentally) are not likely to occur. Some exposures are highly dependent on specific conditions such as microbial exposure; elevated exposures may only occur in case of soft wood (e.g., pine, cedar, spruce, etc.) that is not appropriately stored or dried, and in moderate climates.

Asthma: both non-allergic and allergic asthma may occur. Depending on the specific conditions and the type of wood being processed (see above) exposures may include microbial agents such as bacterial endotoxin, ß(1,3)-glucans, mycotoxins, etc. (potentially causing non-allergic asthma). Also, specific components in the dust (e.g., abietic acid in pine, and plicatic acid in western red cedar) have been described as potential allergens, thus potentially causing allergic asthma in workers exposed to pine and western red cedar. Similar agents (i.e., other resin acids) in other tree species may be involved in asthma observed in workers processing other tree species. Finally, various volatile monoterpenes may be involved.

HP: hypersensitivity pneumonitis has been described in sawmill workers caused by the sometimes high exposure levels of certain fungi.

ODTS: Organic dust toxic syndrome has been described in sawmill workers, and this is most likely caused by high exposures to microorganisms (both fungi and bacteria) and their specific agents.

MMI: mucous membrane irritations have been described in sawmill workers and may be caused by high levels of fine dust, volatile monoterpenes and endotoxins.

Bronchitis and chronically reduced lung function have been described in the literature and should be considered. Specific causal exposures are unknown but sawdust and microbial agents may be involved.

Cancer, and particularly sinonasal cancer: The risk appears to be highest among workers who have been exposed to dust from hardwoods. Currently, it is not clear whether the excess cancer risks are due to the wood dust itself, the various chemicals applied to the wood, or other carcinogenic agents involved in working with wood.

5.5. Laboratory animal workers are generally only exposed to type I allergens from rats and/or mice, and allergic asthma and rhinitis are the most likely occupational diseases to occur in this group of workers. However, the possibility of other diseases can never be entirely excluded and is dependent on the specific situation. Infectious diseases are unlikely to occur since most laboratory animals are free of infectious diseases.

5.6. Bacterial endotoxins (from gram negative bacteria): This group of agents can be found in a wide variety of environments including agricultural and related industries, industries using recycled process water, waste processing industry, metal cutting industry, buildings with contaminated ventilation systems and humidifiers, and various other industries listed in Tables 5-12 and 5-13.

5.7. Bacterial peptidoglycans, fungal ß(1,3)-glucans and mycotoxins. However, although these agents have been suggested to be potentially relevant, only limited evidence for a causal role is currently available.

5.8. Farming, slaughterhouses, wood-based industries, boot and shoe manufacturing, rubber industry.

Chapter 6. Biological Agents – Monitoring and Evaluation of Bioaerosol Exposure

6.1. Viable and non-viable sampling techniques differ in many aspects.The limitations of viable sampling are mainly associated with the viability of micro-organisms. Longer sampling times reduce the viability, treatment of the sample (dissolving media, gelatine filters, dilution series) also potentially reduces the viability leading to underestimation of the true concentration. Growth media, even broad spectrum ones are, by definition, also selective. This can also be considered an advantage; there is a wide range of commercially available specific media for many micro-organisms. The issue of growth media leads to biased results, especially in complex populations of micro-organisms where little is known about the direction and magnitude of the bias. Samplers with agar media as impaction plates have practicalconstraints such an upper limit of detection that is usually occurs at low levels of exposure. Few samplers are available for personal sampling of viable particles (impingers). Impactors with growth media plates are still popular because sample treatment is relatively simple compared to many other methods. Viable sampling leads to considerable lower counts per m^3 relative to non-viable sampling. The relevance of this difference depends completely on the aims of the exposure assessment study (focused on infectious organisms?).

The great advantage of non-viable sampling is that most (modern) personal dust sampling equipment can be used and sampling times can be as long as a full shift. The most complex part is usually the detection of specific micro-organisms by, for instance, staining techniques or microscopic evaluation of morphology.

Simple total counts can usually be performed by laboratory technicians or hygienists. Identification of micro-organisms either by specific plating, microscopic evaluation, or organoleptic approaches requires input from experienced and qualified personnel.

6.2. Classical sampling chaacteristics such as sampling efficiency, wall losses, loss of viability in the sampler and on the medium, duration of sampling, problem of multiple hits, sampling medium, incubation temperature after sampling, incubation time, etc., all play a role and can lead to underestimated biased results.

6.3. PCR-based technology can be used best when one is interested in exposure to a specific *infectious* species or a limited number of sero-types, such as Mycobacterium sp. (in case of tuberculosis), Legionella sp., or when bio-terrorism is suspect with species such as Anthrax, and rapid analysis is required. A trade off of the technique is that probes need to be available that are specific for the organism of interest. If little is known about potential exposures, or if one is interested in exposure profiles to complex populations of micro-organisms, classical methods are usually more appropriate. PCR-based approaches are also less effective if one is specifically interested in specific toxins or allergens from micro-organisms because often production of this agent depends on many aspects such as temperature, growth phase, substrate, etc., and not only on the presence of the organism.

6.4. Systematic differences between assays occur. Although this issue has not been studied in great detail so far, differences are known to be associated with assay type (inhibition, sandwich), antibody source, (monoclonal, polyclonal), filter extraction medium, and capturing filter type.

6.5 One approach is the development of pragmatic exposure standards that correspond with an excess risk as low as possible or as reasonable as possible in comparison with non-occupationally exposed populations. Another approach might be a trial and error type of approach. Exposure reduction to an arbitrary level has to be followed by an evaluation of the effectiveness in terms of reduced health risk (sensitization or allergy prevalence or incidence).

Chapter 7. Biological Agents – Control in the Occupational Environment

7.1. Substitution with a less hazardous material is the preferred approach to providing a safe work environment. Research activities in microbiology laboratories can be conducted on attenuated or avirulent strains that reduce or eliminate the risk of worker infection. As with many other situations involving microbiological agents, this option may not be available since the agents are usually not an intentional process material. However, the nutrients upon which the organisms feed may be amenable to substitution with substrates that are less conducive to growth. When resolving microbiological contamination of building materials in an indoor environment, materials that contain a high proportion of cellulose may be replaced with other composites.

7.2. To elicit a response (infectious or immunologic) in a susceptible individual, the microorganism must be present (reservoir), must be capable of propagation to concentrations necessary to induce a response (amplification), and must be dispersed to the susceptible individual (dissemination).

7.3. Infection is proportional to the virulence multiplied by the pathogen dose and divided by the susceptibility or resistance state of the host.

7.4. Acceptable levels of airborne microorganisms have not been established for a) total culturable or countable bioaerosols; b) specific culturable or countable bioaerosols other than infectious agents; c) infectious agents; or d) assayable biological contaminants for the reasons identified below. Total culturable or countable bioaerosols exist as a complex mixture of different microbial, plant, and animal particles which produce widely varying human responses, depending on the susceptibilities of the individual. This fact, and the lack of a single sampling method that can adequately characterize all bioaerosol components, limits the ability to produce sufficient information to describe the exposure-response relationships that can predict human health effects. The scientific data which are available consist of case reports and qualitative exposure assessments which are generally insufficient as well. The absence of good epidemiologic data on exposure-response relationships related to specific culturable or countable bioaerosols have resulted from the derivation of data from indicator measurements as opposed to actual effector agents. Also, the variations of bioaerosol components and concentrations among differing occupational and environmental settings, combined with the limitations of sampling methodologies, have resulted in the inaccurate representation of workplace exposures. Human dose-response relationships are available for some infectious bioaerosols. However, facilities associated with increased risks for airborne disease transmission rely on the application of engineering and administrative controls to minimize the potential for workplace exposure. In the future, some assayable, biologically derived contaminants (e.g., endotoxin) may lend themselves to the development of numeric criteria due to advances in assay methods.

7.5. Tuberculosis, viral hepatitis from hepatitis B virus, and Lyme disease.

7.6. Preventive measures tend to have the greatest efficacy in settings where workers are concentrated, where the workforce is well structured, where the hazards are recognized, and where there are sufficient resources to provide physical controls, training programs, and surveillance.

7.7. Air should be introduced and, subsequently exhausted, in a manner that moves the uncontaminated airstream from the health care worker's (HCW) breathing zone through the patient's breathing zone. Therefore, supply/exhaust diffuser placement becomes critical to the mixing and general movement of the air in the room. Poor mixing may result in short-circuiting (between supply and exhaust diffusers) and/or stagnation zones that do not allow contaminants to be efficiently removed. Perfect mixing usually does not occur (mixing factor = 1); for very poor air distribution the mixing factor can range to 10.

7.8. Universal precautions are defined as the treatment of all human blood and certain other human body fluids as if known to be infected with HIV, HBV, and other bloodborne pathogens. Under the OSHA regulation, blood is defined as human blood, human blood components, and products made from human blood. Bloodborne pathogens include any pathogenic agents that are present in human blood and that can cause disease in humans. The definition of other potentially infectious materials are inclusive of human body fluids; any unfixed tissue or organ from a human; HIV-containing cell or tissue cultures, organ cultures, and HIV- or HBV-containing culture medium or other solutions; and blood, organs, or other tissues from experimental animals infected with HIV and HBV.

7.9. Class II BSCs are most often recommended as a primary containment device. They are designed to protect the inner working environment of the cabinet (i.e., providing a clean air stream for contamination sensitive laboratory activities) and significantly minimize the escape of microbial contaminants into the surrounding laboratory thereby exposing personnel. These cabinets are classified into Type A or B as defined by the cabinet construction, air flow velocities and patterns, and exhaust system design. Both types utilize a downward laminar flow of HEPA-filtered air that separates at the working surface to slots in the back and the front of the cabinet.

Type A cabinets use an internal system of ducts, HEPA filters, and a fan to recirculate a substantial portion (up to 70%) of HEPA filtered air back to the cabinet workspace and to exhaust the remaining 30% back into the laboratory (Figure 6). Most Class II, Type A cabinets have dampers to modulate this 30/70 division of airflow. This recirculation makes these cabinets suitable for microbiological activities but not for work with volatile or toxic chemicals and radionuclides. The face velocities of these systems should be maintained at a minimum of 75 lfpm (0.4 m/sec). It is possible to duct the exhaust from a Type A cabinet out of the building. However, it must be done in a manner that does not alter the balance of the exhaust system, which could result in a disturbance of the internal cabinet air flow or other cabinets attached to the system. The usual method is to use a "thimble", or canopy hood, which maintains a small opening around the cabinet exhaust filter housing. Type B cabinets are similar in design to Type A, however, this cabinet allows for separate exhaust of up to 100% of HEPA filtered air to the building dedicated exhaust system. Additionally, the plenum is maintained under negative pressure. These modifications ensure that contaminated materials (including toxic chemicals and radionuclides) are contained in the cabinet assembly until exiting the exhaust system. For all Type B cabinets, face velocities should be maintained at a minimum of 100 lfpm (0.5 m/sec).

7.10. For microbiological proliferation, the necessary elements include the presence of the biocontaminant; the susceptibility of a material to contamination by microorganisms, nutrients, and moisture; and the appropriate environmental conditions (i.e., temperature and humidity) for growth of a specific microorganism.

To reflect the degree of exhaust to a dedicated, external system, Type B cabinets are further divided into three sub-categories: B1, B2, and B3. Type B1 systems recirculate 30% of the air through a HEPA filter back into the cabinet work space with the remaining 70% being exhausted to a dedicated system. Type B2 is a total exhaust cabinet that is, no air is recirculated. All air from the cabinet work space is ducted to a dedicated exhaust system. This type of cabinet is expensive to operate since it exhausts large quantities of conditioned room air. If the building or cabinet exhaust fail, this type of cabinet will be pressurized, resulting in a flow of air from the work area back into the laboratory. Cabinets built since the early 1980s are incorporated with an interlock system to prevent the supply blower from operating whenever the exhaust flow is insufficient. Presence of such an interlock system should be verified; existing systems can be retrofitted if necessary. Exhaust air movement should be monitored by a pressure-independent device. Type B3 cabinets are identical to a Class II, Type A with the exception that the 30% component is ducted to a dedicated exhaust system. All positive pressure contaminated plenums within the cabinet are surrounded by a negative air pressure plenum. Thus, leakage in a contaminated plenum will be into the cabinet and not into the environment.

7.11. Moisture can pass through a building envelope by one or more of four transport mechanisms; liquid flow, capillary action, air movement, and vapor diffusion.

7.12. Inherent in the definition of remediation is a *removal* of the microbiological contaminant. That removal will result in the disruption of microbiological reservoirs. The airborne dissemination of these bioaerosols can pose a significant exposure concern for the remediation workers. Additionally, these aerosols can be spread to uncontaminated areas of a building, increasing the hazard for the remaining occupants and adding to the difficulty of clean-up. Therefore, it is important that all remediation activities be conducted with an awareness of the potential bioaerosol exposures and with minimal disturbance of contaminated materials. Specifically, controls must be instituted that protect both the *worker* and the *adjacent environment.*

7.13. Various grain saprophytes including thermoactinomycetes, thermotolerant fungi (e.g., *Aspergillus* sp.), and mites.

7.14. The work practices of the operators can be a determining factor in the potential for occupational exposure. For example, during the collection of fermentor tank sample volumes in the enzyme manufacturing study, operators were observed purging the sample port with a "blast" of pressurized steam prior to the collection of a broth sample. The steam served to clean and decontaminate the interior surfaces of the pipes to ensure a pure sample for subsequent analysis. However, the contact time between the steam and residual microbial populations in the sample line were not adequate to kill the microorganisms and therefore resulted in a dissemination of viable aerosols into the surrounding environment. Work practices are most reliable when used in combination with effective engineering measures such as isolation of the machinery from the operator or automation of the process. For example, microbial exposures during the separation of solids can be reduced by limiting operator interaction with those processes or, if this is not possible, the observance of proper and safe work practices.

Chapter 8. Biological Monitoring in Occupational Health

8.1. At 50 watts, V= 16 LPM and Q=9 LPM, so 89% and 73% for the two compounds, respectively.

8.2. When significant exposure occurs via the dermal and/or oral routes.

8.3. Biological monitoring is not useful for local irritants that do not have systemic effects.

8.4. Estimating that the sample was taken 20 hours after the spill and cleanup (when the worst potential exposure probably took place), and using the exponential decay equation (below), you estimate that the peak level in the urine at the time of the incident was 3.5 g/g creatinine and clearly an overexposure took place at the time.

$C_o = C_t e^{(0.693/12)(20)}$ with 12 and 20 being the half-life and elapsed time, respectively

C_0 =(1.1g/g creatinine)$e^{1.16}$=(1.1g/g creatinine)(3.17) = 3.5g/g creatinine

8.5 Since the company was not paying the worker over the weekend, the only recourse is to advise the worker of the potential risks of overexposure and remind her of the steps she should take (PPE, etc.) to reduce exposure to an ALARA level (as low as reasonably achievable).

Chapter 9. Dermal Exposure Assessment

9.1. Direct contact with bulk contaminants — unloading a bag of chemicals; Immersion in liquids — dipping parts into a solvent tank; Deposition of spray on skin — paint spray; Contact with contaminated surfaces — contaminated tools or workbench

9.2. Activities that involve contact with contaminated surfaces can result in skin exposure. Ways to reduce exposure: Proper use of chemical protective clothing, washing, especially before meals, maintenance of a clean workplace.

9.3. Surrogate skin technique: captures all of the contaminant that reaches the skin; may overestimate by adsorbing more material than would the skin.

Chemical removal technique: measures material that has actually been deposited on the skin; almost certainly underestimates exposure, so removal efficiency studies need to be done in advance.

Fluorescent tracer technique: visualizes patterns of exposure across the entire body; complex quantitation process raises questions about the accuracy of the measurements.

9.4. Contact rate; exposed skin surface area; duration of contact.

9.5. "An occupational exposure limit needs clear definitions. In the case of dermal occupational exposure limits there is a need to define "potential dermal exposure" as it pertains to particular work tasks. The application of such limits will also require a reliable and well-accepted method for measurement of dermal exposure. Dermal occupational exposure limits could provide importance guidance to industrial hygienists if such definitions and methods can be developed."

Chapter 10. Epidemiology

10.1. Epidemiology is the study of determinants of disease in human populations. In the occupational setting, epidemiology is used:

- to identify occupational causes of disease.
- to develop interventions designed to prevent disease caused by occupational factors.
- to set work place and general environmental regulatory standards.
- to carry out disease surveillance.
- to develop risk assessments.
- to evaluate interventions.
- to determine directions for future research.

10.2. Four measures of disease frequency are:

- incidence rate, the number of cases of a disease, divided by the total person-time experience giving rise to the cases;
- mortality rate, the number of deaths from a disease, divided by the total person-time experience giving rise to the cases;
- incidence proportion, the proportion of a closed population at risk that develops a disease during a specified study period; and
- prevalence proportion, the proportion of a population or study group that has a disease or other medical condition at a particular point in time.

10.3. Four relative measures of association used in epidemiology, and the corresponding study design in which each is used, are:

- incidence rate ratio - retrospective and prospective follow-up studies of open study groups;
- mortality rate ratio - retrospective and prospective follow-up studies of open study groups;
- incidence proportion ratio - retrospective and prospective follow-up studies of closed study groups; and
- odds ratio - case-control studies.

10.4. Three problems that threaten validity, and standard procedures for minimizing these problems, are:

- selection bias – systematic error that arises because of procedures used to select subjects for inclusion in a study; minimized by identifying and selecting 100% of eligible subjects, by selecting a random sample of all eligible subjects, and by maximizing participation if participation is voluntary;

- information bias – systematic error in classifying subjects with regard to exposure or disease; minimized by using objective sources of data on exposure and disease, using comparable data collection and development procedures for the subject groups to be compared in a study, and using blind exposure and disease assessment techniques; and
- confounding – an error that occurs when the groups being compared in a study differ with respect to disease determinants other than the factor whose possible effect is being assessed and when such a difference is ignored in the analysis of the study; minimized by identifying, obtaining data on, and analyzing data on established and potential disease determinants or by restricting the study to subjects in one category of a potential confounder.

10.5. The 95% confidence interval of 0.9 to 2.4 means that, if the data from the study are valid, it is 95% likely that the true value of the odds ratio lies in the range of 0.9 (a protective effect) to 2.4.

10.6. An "open study group" is a study group in which the date of entry and the date of leaving observation vary from subject to subject. In studies of open study groups, a unit of person-time is the unit of observation, and the number of person-years accrued during the overall study period. This may vary markedly from subject to subject. Open study groups are typically used in occupational retrospective follow-up studies that have long (decades) study periods. A "closed study group" is a study group in which the date of entry or age at entry and the length of observation are approximately the same for each subject. The unit of observation is typically a person, rather than person-time. Closed study groups are typically used in follow-up studies, particularly prospective follow-up studies, having relatively short study periods.

10.7. The first step in conducting a follow-up study entails specifying hypotheses of interest and study objectives. In occupational epidemiology, the objectives may be specific (e.g., to evaluate the hypothesis that exposure to butadiene causes leukemia) or may be quite broad (e.g., to determine if an occupational group has particularly high or low mortality rates when compared to a general population). Formulation of objectives requires an understanding of the occupational setting to be investigated (potential exposures, concerns of employees and management), and of the pertinent epidemiologic and toxicologic literature.

The next step is to identify subjects in the occupational setting to be investigated and to specify subgroups on the basis of any exposure to be evaluated. Subject identification requires review of employee personnel, benefits, and other types of records kept by management (and sometimes by unions) of the facilities under investigation. Identification of subgroups of subjects having exposures or other work characteristics of interest often requires an in-depth review of subjects' employment records to reconstruct job histories. Collection, review, and categorization of industrial hygiene and engineering information, with the construction of job-exposure matrices, may also be needed to develop estimates of exposure to specific agents in the occupational setting.

The next step is the identification of comparison groups. The investigator must consider using both external (general population) and internal (unexposed subjects) comparison groups and must weigh the advantages and disadvantages of each.

Follow-up to determine subjects' vital status, disease status, and person-years of observation is the next step. Vital status determination typically involves checking employers' benefits and employment records on study group members, linking identifying data on study group members with the National Death Index and with records kept by the Internal Revenue Service and the Social Security Administration. Individual contact with subjects or their family members may also be required. Disease status may be based entirely on cause of death data obtained from death certificates, on information from other disease registries such as population-based state cancer registries, or on self-reports of illnesses obtained from questionnaires.

The next step is the epidemiologic analysis. This usually involves comparing the disease-specific mortality or incidence rates of the study group with the rates of a general population or comparing the mortality or incidence rates of one subgroup of the subjects with those of another subgroup (e.g., comparing the rates of the workers exposed to an agent with the rates of the unexposed workers in the study group). The analysis should control for confounding, assess induction time effects, and evaluate dose-response. It should also determine whether factors such as age, race, period of employment, and the like have any impact on the results of the analysis.

The last step is interpretation of results. This entails assessing the internal validity of the study: that is, considering the potential impact on the validity of the results of systematic errors such as selection bias, information bias, and residual confounding. Interpretation also requires considering the results of the study in the broader context of all available epidemiologic studies and other pertinent information.

10.8. The main advantage of using an external, rather than an internal, comparison group in a follow-up study is the enhancement of precision. The main disadvantages are an increased potential for uncontrolled confounding, largely due to differences between occupational study groups and the general population in correlates of socioeconomic status, and an increased potential for information bias because of differences between the study subjects and the general population in procedures used to identify and classify deaths or disease cases.

10.9. Data sources commonly used to determine subjects' vital status are employers' benefits and personnel records, the Social Security Administration's Death Master

File, the National Death Index, records of the Internal Revenue Service, and records of state divisions of motor vehicles.

10.10. The overall lung cancer incidence rate is:

100 cases/200,000 person-years or 50/100,000 years.

The incidence rate among the exposed is 50 cases/50,000 person-years. The incidence rate among the unexposed subjects is 50 cases/150,000 person-years. The crude IRR is:

(50/50,000 y) / (50/150,000 y) or 3.0

10.11. Two procedures used to obtain confounder-adjusted summary rate ratios in follow-up studies of open study groups are stratification and modeling.

10.12. Prospective follow-up studies are not often done in evaluating the relationship between an occupational factor and cancer because of the potentially long induction time required to observe an effect, if any exists. A long induction time would require the investigator to follow-up subjects into the future for 20 years or more.

10.13. The two main types of case-control study are the nested case-control study and the registry-based case-control study. Nested case-control studies are most often done after a preliminary follow-up study of a particular occupational setting has been completed. The ensuing nested case-control study provides an efficient approach to obtaining detailed information on occupational and nonoccupational exposures – information that was not feasible to obtaine in the parent follow-up study because of the much larger size of the follow-up study. In nested case-control studies, both cases and controls are selected from a specific study group roster that lists all subjects in the parent follow-up study. Nested case-control studies often have a decades-long study period.

Registry-based case-control studies are not done within the context of a prior follow-up study. Cases usually come from a general population registry, and controls are obtained by recruiting persons using random digit dialing within the geographic region covered by the registry, by selecting persons from other disease categories included in the registry, or by selecting persons from population lists such as voters registration lists or birth certificates. Registry-based case-control studies often evaluate a variety of, rather than a single, occupational settings or jobs. Registry-based case-control studies typically have a study period of three to five years.

10.14. In nested case-control studies, controls for a case are usually selected randomly from a risk set, consisting of all subjects who match with the case on gender, race and year of birth, who were alive at the time of the case's death of diagnosis with the disease under study, and who started employment or exposure at the study facility before the case died or was diagnosed. Typically, two to 10 controls are chosen for each case.

10.15. The data are from a case-control study of ethylene glycol ether (EGE) and cleft palate and are summarized as follows:

	EGE-exposed	EGE-unexposed	Total
Cases	30	70	100
Controls	60	340	400

The odds ratio for the association between EGE exposure and disease is computed as the exposure odds among the cases (30/70), divided by the exposure odds among the controls (60/340):

$$(30/70) \,/\, (60/340) = 2.429 = 2.4.$$

10.16. Follow-up study advantages:

- minimal potential for selection bias;
- the ability to assess relationships between the occupation or exposure under study and many different diseases;
- high frequency of exposure in the study group;
- use of objective procedures to assess exposure;
- prospective follow-up study allows for assessment of confounding by nonoccupational factors; and
- prospective follow-up study can be used to study nonfatal diseases.

Follow-up study disadvantages:

- a long time period or a very large study group is required to obtain precise data on rare diseases;
- often uninformative for very rare diseases;
- high expense of prospective follow-up studies; and
- frequent losses to follow-up in prospective follow-up studies.

Nested case-control study advantages:

- relatively low expense compared to follow-up studies;
- relatively good control of confounding by nonoccupational disease risk factors compared to follow-up studies;
- less potential for selection bias than in registry-based studies;
- use of objective procedures to determine exposure; and
- relative high frequency of occupational exposure than in registry-based case-control studies and better precision.

Nested case-control study disadvantages:

- greater potential for selection bias than in follow-up study;
- less precision than in follow-up study;
- poorer information on nonoccupational confounders than in registry-based case-control study; and
- limited informativeness for rare diseases.

Registry-based case-control study advantages:

- relatively good for studying rare diseases;
- useful for studying occupations and diseases that cannot be assessed using retrospective follow-up studies because of unavailability of information on groups of workers or unavailability of data on the medical conditions of interest;
- relatively good potential for high participation; and
- relatively good information on nonoccupational confounders.

Registry-based case-control study disadvantages:

- low frequency of the occupational exposure or job of interest and, hence, low precision;
- relatively high potential for information bias, due to use of unvalidated self-reports of exposure; and
- relatively high potential for selection bias.

10.17. Validity is assessed by evaluating the adequacy of procedures used in a study to control for confounding and to minimize selection bias and information bias. A highly valid study will have identified, obtained information for, and analyzed data for all important potential confounders. It will have used objective procedures to recruit and enroll subjects, and will have achieved high participation among eligible subjects. It will have used objective, reasonably accurate procedures to assess exposure status and disease status. It will have found a consistent dose-response relationship and will have found a temporal pattern consistent with any expected induction time effect. The association reported in the study will be biologically plausible and will be strong enough to render bias or confounding unlikely as explanations. In order to be convincing, a study will need to be reasonably precise, as well as valid.

10.18. Causality is evaluated by assessing the consistency of the results of a set of epidemiologic studies judged to be relevant to the hypothesis of interest, to be valid and to be informative (i.e., at least some of the studies should be reasonably precise), and by considering all other pertinent biologic information.

Chapter 11. Health Risk Assessment and the Practice of Industrial Hygiene

11.1. The four components of a health risk assessment are: (1) hazard identification, (2) dose-response assessment, (3) exposure assessment, and (4) risk characterization.

11.2. In the hazard identification component, one should to find toxicological, epidemiological, biological and/or structural analogy data on the types of health injury or disease that may be produced in humans by a chemical at any dose. The hazard identification should ask the question, what adverse effects can the chemical in question cause, and what is the likelihood that they will occur in humans? A good discussion would note whether the health effects were likely to occur at plausible levels of exposure.

The dose-response assessment attempts to define the quantitative relationship between the chemical dose and biological response. In the dose-response assessment, the reader should become familiar with the results of various models for estimating the cancer risk at plausible doses, including low-dose extrapolation models. In addition, he/she should learn about the results of applying extrapolation methods to estimate non-cancer effects. In this section, one should learn of the results of the benchmark dose and safety factor approaches.

In the exposure assessment, all of the calculations and models used to predict exposure and system uptake of the chemical(s) should be described. Exposure assessment estimates the magnitude and probability of intake of chemicals from various sources through different routes of exposure (oral, dermal, or inhalation). Uncertainty around the exposure calculations and results should be quantitatively evaluated.

In the risk characterization, all of the dose-response and exposure assessment data from the previous steps are integrated, summarized, and interpreted in an attempt to predict the likely cancer and non-cancer risks posed by a specific level of exposure to one or more chemicals. A quantitative analysis of uncertainty using a Monte Carlo analysis should accompany the risk characterization, and uncertainties in the risk assessment should be identified. Any regulatory criteria should be presented and its significance discussed.

11.3. A routine activity for industrial hygienists is consulting the TLV® *Documentation* for chemicals of interest. When an industrial hygienist observes that persons are exposed to a chemical and then reads the *Documentation of the TLVs® and BEIs®*, he/she is conducting an abbreviated form of hazard identification.

11.4. For most industrial hygienists, the OEL-setting group has conducted the dose-response portion of the risk assessment. It is generally assumed that the expert panel has considered the slope of the dose-response curve and various methods of estimating safe doses based on that information. Based on that information, the group has concluded that the OEL set is "safe" for nearly everyone.

11.5. The industrial hygienist has exceptional skills at being able to measure the concentration of chemicals in workplace air. This has been the forte of the profession for more than 50 years. Thus, exposure assessment is an integral part of the profession. Due to the information collected by industrial hygienists over the past decades, the uncertainty associated with risk assessments has been improved. For example, the range of plausible values for the inhalation of air by a worker during an 8-hour workday extends from just 5 m^3 to 10 m^3, a factor of 2. It is also known that the body weight for a typical male worker will normally be between 165 and 250 lbs, a factor of only 1.5. Other exposure factors will generally be less than two-fold between the most different of persons. This level of uncertainty is much less than that which is addressed in the dose-response assessment where a rat might be able to metabolize a chemical to a harmless metabolite while a human may not be able to change it at all after many years. At the same time, some uncertainties associated with risk assessments have yet to be improved upon. For example, hygienists have been aware that dermal exposure can be a significant avenue for absorbing chemicals. Hygienists can improve the quality and thoroughness of their work by adding an assessment of dermal exposure to their analysis thereby reducing some of the uncertainty in the risk assessment itself.

11.6. Dose-response assessments, by their very nature, rely upon results of animal testing and, rarely, human epidemiology sources. Thus, there are rarely enough studies of the various possible adverse effects to understand all of the possible adverse health effects. To complicate matters, when attempting to estimate the risk of exposure to cancer-causing chemicals, the models for estimating the hazard at low doses are quite uncertain. For example, equally plausible dose extrapolation models can have dose estimates that differ by a factor of 1,000 at a risk level of 1 in 1,000,000. This range adds to the uncertainty associated with dose-response assessments.

11.7. Risk characterizations are the written description of the results of the work in the prior three sections of the risk assessment. They present calculations which describe the likely degree of risk, both cancer and non-cancer, to the exposed population. To write a good risk characterization, one must compare the predicted exposure estimates to the various "known" risk estimates, which include both consensus and regulatory criteria. The risk characterization should show a number of charts and graphs that illustrate the results of the analysis in a way that is understandable to any observer. Such analysis may include cost effectiveness evaluation wherein the cost of achieving every 5 or 10 fold reduction in risk is presented. All in all, to write a risk characterization, one must integrate scientific analysis, regulatory analysis, qualitative uncertainty analysis, and cost/benefit analysis. To do this, one must be proficient in several fields.

11.8. PB-PK models are mathematical descriptions of how a chemical enters a test animal, is distributed and metabolized within that animal, and is eliminated from that animal. PB-PK models also help determine forms to which a chemical it is converted once administered. The PB-PK model, which is actually a series of differential equations, attempts to predict how the chemical is handled by the test animal (usually a rodent) at a given dose, and then attempts to predict how the chemical will be handled at other, usually lower, doses. The proof of the validity and usefulness of a PB-PK model is when it is able to successfully predict the absorption, distribution, metabolism, and elimination of a chemical in humans based on the test animal data. When this occurs, it is theoretically possible to use animal data to predict the actual risk from chemicals to humans over a wide range of doses and for a number of toxic endpoints, allowing for much more data to be obtained than that obtained using more traditional approaches.

Chapter 12. The History and Biological Basis of Occupational Exposure Limits

12.1. The first groups to establish OELs include the Hygienic Institute at Munich (in 1883 and 1886), the U.S. Bureau of Mines (in 1921), the National Safety Council (in 1926), The USSR Ministry of Labor (in 1930), the American Standards Association (ASA) (in 1939 and 1940). The ASA and a number of industrial hygienists combined to form ACGIH® in 1938.

12.2. Examples of organizations in the United States who recommend OELs include:

- American Conference of Governmental Industrial Hygienists (ACGIH®) – Threshold Limit Values (TLVs®)
- National Institute for Occupational Safety and Health (NIOSH), part of the U.S. Department of Human Health Services – Recommended Exposure Limits (RELs)
- Occupational Safety and Health Administration (OSHA) – Permissible Exposure Limits (PELs)
- American Industrial Hygiene Association (AIHA) – Workplace Environmental Exposure Limits (WEELs)

12.3. Uncertainty factors are used to account for the lack of complete knowledge about various components of risk assessment. Uncertainty factors provide a margin of safety that can be used to establish safe OELs for almost all adverse effects for most persons.

12.4. Examples of categories of toxic effects which OELs attempt to prevent include:

- Systemic effects
- Irritant effects (e.g., eye, nose, and throat)
- Neurological effects
- Reproductive/developmental effects
- Cancer

12.5. To identify an appropriate OEL, it is necessary to weigh all of the available human and animal toxicology information provided. First, the primary adverse effect is a systemic toxic effect (e.g., liver toxicity), rather than irritation or other chronic effect such as carcinogenicity. Second, the NOEL for liver effects is similar between species; however, it is appropriate to choose the lower of the two NOELs for deriving the OEL. Third, the rate of elimination of the chemical is not markedly different between human and rodent, and the faster the metabolism of the smaller animal is consistent with what one would expect; therefore, no UF for unexpected pharmacokinetic differences is required. Fourth, although the data are only from a 20-day inhalation study, since the adverse effect is liver toxicity, a large UF is probably not necessary to predict the chronic toxicity (although a 6- or 12-month study would be preferable). Fifth, since it is an eye irritant in employees at 50 ppm, the OEL will need to be at least 50% less than the effect level (e.g., at least as low as 25 ppm). Sixth, the lack of mutagenic activity lends some support for a lack of cancer hazard. If there is going to be wide exposure to this chemical, especially to downstream consumers, then longer term testing is necessary.

Based on the above, it is reasonable to select an UF of 5 to account for differences in susceptibility between rodents and humans. This is reasonable since the difference in results between two different animal species was small. Another UF of 3 could be applied to account for differences in susceptibility among humans. Lastly, it would be reasonable to apply an additional UF of 5 to account for the less than lifetime exposure.

Given the following information provided for Chemical A, to solve for a reasonable OEL:

$$OEL = \frac{NOEL}{(UF_1)(UF_2)(UF_3)} \quad \text{(provided in text)}$$

where:

$NOEL = 300$ ppm

$UF_1 = 5$

$$UF_2 = 3$$

$$UF3 = 5$$

$$OEL = \frac{300}{(5)(3)(5)}$$

$$= 4 \text{ ppm}$$

12.6. OEL-setting bodies do not generally rely upon low-dose extrapolation models for a number of reasons. First, existing low dose models, such as Weibull and multistage models, show a high level in variability among risk estimates making it difficult to determine which model is the best one for setting OELs. Second, the extrapolation from high dose data to lower doses is often erroneously applied, since the high dose data usually involves test animals exposed to levels of contaminants 1000-fold or more above those to which humans might be exposed. These doses are so high that they often produce effects that would not occur at the doses to which people would be exposed. Third, the mechanisms of action for various chemicals are not fully understood, making it difficult to estimate how humans will respond at low doses. In addition, toxicity may vary among species, even with the same target dose, making extrapolation efforts more complicated. Differences in metabolism, absorption, and excretion among species should be taken into account. Fourth, the goodness of fit for dose-response relationships seen for experimental, high dose data may vary significantly at lower doses, thereby affecting the predicted risks at low doses. Finally, and most important, low dose models do not quantitatively incorporate biological protective mechanisms.

12.7 Occupations and/or job descriptions that might require adjustments to their OEL as a result of unusual work schedules include:

- Bakers or other such persons who work the night shift
- Workers in process operations, such as those found in oil refineries, chemical plants, steel and aluminum mills, pharmaceutical manufacturing, glass plants, and paper mills, that require round-the-clock manning
- Workers in the military or railroad industry where a workday of 16-24 hours in not uncommon
- Craftsmen and equipment repair workers who work overtime to accommodate routine equipment repair, fill in for absent workers, or deal with seasonal fluctuations in demand

12.8. PB-PK models are mathematical descriptions of how a chemical enters a test animal, is distributed and metabolized within that animal, and is eliminated from that animal. PB-PK models also help determine forms to which a chemical it is converted once administered. The PB-PK model, which is actually a series of differential equations, attempts to predict how the chemical is handled by the test animal (usually a rodent) at a given dose, and then attempts to predict how the chemical will be handled at other, usually lower, doses. PB-PK models are extremely useful in the field of risk assessment because, based on the validity of a particular PB-PK model, it is theoretically possible to use animal data to predict the actual risk from chemicals to humans over a wide range of doses and for a number of toxic end-points, allowing for much more data to be obtained than that obtained using more traditional approaches.

INDEX

2-Butoxyethanol, 444
2-Ethoxyacetic acid, 394
2-Ethoxyethanol, 394
2-Ethoxymethanol, 394
2,5-Hexanedione, 394
1-Hydroxypyrene (1HP) , 393
2-Methoxyacetic acid, 394
4,4'-Methylene-bis-2-chloroaniline (MOCA), 439

A

α-Antitrypsin, 433
Abatement, 368
Absolute measure of association, 503
Absorbed dose, 569
Absorbent gloves, 478
Acetylcholinesterase (AchE) inhibition, 396
ACGIH®, 393, 560, 633
ACH, 343, 345
Acrylate, 51
  delayed-irritant, 53
  irritants, 53
  seasonal sensitivity, 54
  irritant susceptibility, 54
Activity median aerodynamic diameter (AMAD) , 31
Acute toxicity, 647
ADD, 588
Additive interactions, 413
Adducts, 395, 441, 442
Adjustability of workspace, 162, 194
Administrative controls, 194, 195
Administrative solutions, 177, 205, 206
Adverse effects, 641, 643
Aero-allergens, 303
Aerodynamic diameter, 28
Aerosols, 27, 351
Age and skin, 84
Aging, 34
AIHA, see American Industrial Hygiene Association
Air changes per hour (ACH), 343, 345
Air filtration methods, 297
Air-locks, 360
Air sampling, 369, 384, 389
  advantages, 384
  disadvantages, 389
Airway narrowing, 34
Allergen, 114

Allergenic agents, 264
Allergic, 335
Allergic (Type I or IgE-mediated) asthma, 236, 255
Allergic contact dermatitis (ACD), 62, 589
  chromate, 63
  colophony, 63
  epoxy resins, 62
  formaldehyde, 62
  isothiazolinones, 63
  nickel, 62
  rubber, 62
  thiuram sulfides, 62
Allergic inflammation, 238
Allergic reactions, 663
Allergic respiratory diseases, 236
Allergic rhinitis, 236
Allergy, 238
American Conference of Governmental Industrial Hygienists (ACGIH®), 393, 560, 633
American Industrial Hygiene Association (AIHA), 640
American National Standards Institute, 636
Analytic epidemiology, 498
Anatomy, skin, 74
  cellular constituents, 76
  dermis, 76
  stratum corneum, 76
  viable epidermis, 76
Animals, 266
  confinement buildings, 371
  feed delivery systems, 371
  studies, 555, 638
ANSI, see American National Standards Institute
Antagonism, 413
  chemical, 413
Anthropometry, 150, 164
Appendageal route, 66
Aspect ratio, 30
Aspergillosis, 341
Aspergillus, 340, 370, 372
Assay, 306
Assay method, 308
Asthma
  occupational, 237
Asthma-like syndrome, 253, 256
Asthmatics, 34
Atopy, 61, 238, 239
Attenuated total reflectance fourier transform infrared (ATR-FTIR), 485
Autoclaves, 357, 359, 360, 361

B

$\beta(1\rightarrow3)$-glucan, 306
Bacillus thuringiensis, 330
Bacteria, 221
Bacterial endotoxins, 267, 268, 305
Bacterial peptidoglycans, 267
Barrier function, skin, 112
BEIs®, see Biological Exposure Indices
Biological Exposure Indices (BEIs®), 393, 423, 637, 651
  Documentation, 419
Benchmark dose method, 563, 660
Benefits and costs, ergonomics, 213
Benzidine, 394
  and adducts, 394
Berylliosis, 438

Bifurcations, bronchial, in between, 28, 39, 40
Bio-allergens, 312
Bioaerosols, 293, 330, 336, 361, 363, 366, 373
practical evaluation of levels, 318
Bioavailability, 598, 599
Biological agents, 219
isolation and control of, 332, 341, 345, 359, 366, 368, 370
plants as, 266, 275
Biological basis, 633
Biological monitoring, 383, 414, 488, 577
Biological safety cabinets, (BSC) 351, 356, 361
Biological significance, 650
Biomarkers, 392
early effect, 396
effective dose, 394
Biosafety level (BSL), 351, 359
Biosafety officers, 340, 341
Blood as the central compartment, 407
Bloodborne pathogens, 347
Body burden, 686
Boltzman equilibrium, 30
Boltzmann constant, 29
Branching angle, 32
Branching pattern, 32, 33
Breathing mode, 33
Brief and Scala Model, 682
Brownian diffusion, 27
Buehler test, 105
Bulk samples, 336
Bulk, surface, and/or air samples, 369
Byssinosis, 256

C

Cancer, 276, 556, 639
Cancer risk, 583
Carbon monoxide, 396
Carboxyhemoglobin, 396
Carcinogenic, 639
Carcinogens, 559, 665
Carina, 40
Carinal ridges, 39
Case-control study, 499
CDC, see Centers for Disease Control and Preven tion
Centers for Disease Control and Prevention (CDC), 330, 337, 340, 341, 345, 346, 347, 350, 361
Centrifuges, 372
Certification, 214
Chitin, 306
Chromosomal damage assays, 443
Chromosome aberration assay, 443
Chronic airflow obstruction, 263
Chronic bronchitis, 34, 261
Chronic obstructive airway disease, 34
COPD, 34
Cladosporium, 363, 371
Clearance, 150, 162, 164, 193
Closed population, 500
Coding and cognitive errors, 159
Colony forming units, 298
Comparison group, 522
Competitive organisms, 302
Confidence interval, 516

Confirmed Animal Carcinogen with Unknown Relevance to Humans, 667
Confirmed Human Carcinogen, 667
Confounding, 507
Contact allergens, 62
Contact dermatitis, 63
  cutting oils, 63
  gloves, 63
  soap, 63
  water, 63
Contaminated surfaces, 471
Corneosurfametry, 115
Corporate OELs, 678
Cost benefit analyses, 586
Creatinine, 411
Critical fluxes, 593
Cross-sectional area, 29
Cross-sectional study, 499
Culture-based methods, 294
Cumulative incidence, 500
Cumulative toxins, 685
Cycle time, 143

D

Decontamination efficacy, 104
Decontamination of the skin, 100
  hot water, 101
Decontamination treatments, 103
Density, 31
Deposit, 27
Deposition, 471
Dermal contact, 574
Dermal exposure, 387, 467
Dermal exposure assessment, 467
Dermal exposure control strategies, 489
Dermal exposure pathways, 469
Dermal fluorescence, 484
Dermal occupational exposure limits, 490
Dermal transfer coefficients, 472
Dermatitis, 50
  causes, 59
  contact, 63
  delayed-type, 50
  induction, 50, 51
  locations, 51
  susceptibility, 51
  urticaria, 51
Dermatoses, 57, 87
Dermis, 76
Descriptive epidemiology, 498
Detergents, 58
Developmental toxicants, 658
Diagnosis of, 56
  allergic contact dermatitis, 56
  cytokine profiles, 56
  patch testing, 55
  urticaria, 56
Differential misclassification, 510
Diffusion, 29, 66, 77
Diffusion coefficient, 29
Dlution ventilation, 332
Direct count methods, 299
Disease status, determination of, 523
Dispositional antagonism, 413
Documentation of TLVs® and BEIs®, Chapters 11 and 12, 419

Dodecahedral lighting system, 484
Dose, 639
Dose-extrapolation models, 555
Dose-response, 557, 648
  assessment, 556
Duration of contact, 473

E

EAA, 249
Eccrine, 76
Ecologic or correlational studies, 547
Economics, 136, 142
Electrostatic precipitation, 27
Elimination, 400
Emphysema, 34
Encapsulating suit, 361
Endotoxins, 306
Endurance, 156
Engineering measures, 331
Engineering solutions, 176, 205, 206
Environment, 157, 386
Environmental Protection Agency (EPA), 559
Enzymes, 265, 372, 373
Epidemiologic studies, 666
Epidemiology, 573
  occupational, 498
Ergosterol, 306, 311
Excursion factors, 683
Exhalation, 400
Exposure, 556
  assessment, 556
  by all routes, 387
  limits, 555
  reconstruction, 395
  repeated, 94, 365
  standards, 316
Extracellular polysaccharides (EPS), 305, 310
Extraction method, 308
Extremophiles, 329

F

Face velocity, 353, 354, 355
Farmer's lung, 249, 370
Fate, 399
Fermentor, 372, 373
Fibrotic disease, 34
Fick's First Law of Diffusion, 79, 591
Filter press, 372
FISH, 303
Flow cytometry, 298
Fluorescent In Situ Hybridization (FISH) Techniques, 303, 304
Fluorescent tracer techniques, 482
Fluorescent whitening agent, 482
Follow-up study, 499
Formaldehyde, 63
Frequency of skin contact, 93
Functional antagonism, 413
Fungal ß(1→3)-glucans, 267, 272
Fungal volatile components, 311

G

Gases and MACs, 634
Gender, 84
Gender and skin, 84
General ventilation, 343

Genetic screening, 431
Genotoxic, 639
Genotype, 429
Geometric standard deviation, 31
Gloves, 59, 63, 91
Glucans, 309
Glucose-6-phosphate dehydrogenase (G6PD), 433
Glycophorin A assay (GPA), 441
Grain fever, 259
Gram negative bacteria, 335, 363, 371
Growth media, 298
Guinea pig maximization (GPM), 105

H

Haber's Law, 685
Hair follicles, 76
Half-lives, 661
Handwash, 481
Hazard identification, 556
Hazard index (HI), 580
Hazard quotient, 579
HBV, 348
Health hazard evaluations, 313
Health risk assessment, 555
Helminths, 222
HEPA, 343, 353, 354, 355, 357, 360, 366, 368
Hepatitis, 332, 337, 339, 347
Hickey and Reist Model, 687
High molecular weight sensitizers, 312
Histoplasma capsulatum, 369
Historical exposures, 422
HIV, 348
HLA-DPb1 Glu 69, 438
Human factors, 137
Humidifier fever, 259
Hydration, 101
Hydrophilic, 362
Hygroscopicity, 32
Hyperresponsiveness, bronchial, 238, 239
Hypersensitivity, 235
Hypersensitivity pneumonitis (HP), 249, 257, 335
Hypoxanthine guanine phosphoribosoyl transferase, 441

I

ICRP, see International Commission on Radiological Protection
IgG, 370
Image charges, 30
Immersion of skin, 471
Immunoassays, 303, 312
Immunocompromised, 339
Immunology, 234
Impaction, 27, 28
Impactor methods, 296
Incidence, 57, 500
Incidence proportion, 500
    ratio, 504
Incidence rate ratio, 504
Incinerator, 357
Independent effects, 412
Individual medical records, 421
Individual susceptibility, 428, 637
Industrial hygiene, 330, 555, 638
Industrial hygienists, 340, 341, 361, 421
Inertia, 28
Infectious agents, 64

Infectious diseases, 64, 220
Inhalability , 35
Inhalation fever, 259
Injury data, 196
Integrated Risk Information System (IRIS), 579, 646
Intended use of TLVs®, 637
Inter-individual differences, 84
Inter-individual variability, 568
Interception, 27, 30
Interlock, 355, 361
Internal comparison groups, 522
Internal consistency, 549
Internal dose, 392
International Commission on Radiological Protection (ICRP), 40
Interspecies adjustments, 561
Interspecies extrapolation, 568
Ionization, 81
IRIS, see Integrated Risk Information Systems
Irritants, 62, 101, 114
  common, 62
  soaps, 102
Irritant rhinitis, 261
Irritation, 87, 88
Isocyanate compounds, 73
Isolation and control of biological agents, 332, 341, 345, 359, 366, 368, 370
Isolation rooms, 345

J

Job demands, 136
Job-exposure matrix, 520

L

Labeling , 159
  microorganisms, 299
Laboratory-associated infections, 347
Langerhan's cells, 76
Larynx, 35, 39
Laser scanning confocal microscopy (LSCM), 299
Lead, 71
Legionella pneumophila, 298, 330
Legionellosis, 330, 337, 339, 362, 368
Local exhaust ventilation (LEV), 342, 370
Lifetime average daily dose (LADD), 582
Lifetime cancer risk, 573
Lifting, 152, 154, 160
Lifting Index (LI), 181, 183
Limulus amebocyte lysate (LAL), 306
  interferences, 308
Lipopolysaccharide (LPS) , 268
Liquid impinger methods, 297
Local lymph node assays, 105
Lognormal distribution, 31
Low dose models, 666
Low-dose carcinogenic risks, 569
Low-dose extrapolation, 561, 568
Low-dose extrapolation models, 571
Lowest Observable Adverse Effect Level (LOAEL), 564, 645

M

Manual materials handling, 152, 154, 160, 172, 174, 178
  mechanical aids, 175
Maple bark stripper's disease, 249

Margin of safety, 642
Margin of exposure, 563
Markers for assessment of mold concentrations, 305
Mass median aerodynamic diameter(MMAD), 31
Mathematical modeling of skin permeation, 77
Matrices, biological monitoring, 401
Maximum allowable concentrations (MAC), 635
Maximum exposure levels, 679
Mean free path, 28
Median diameter, 31
Mesophilic, 362
Meta-analysis, 547
Metabolism, 400
Metal fume fever, 259
Metals, 59, 68
Methemoglobinemia (MetHb), 396
Methylene diphenyldiisocyanate, 49
Micronucleus assay, 443
Microscopy, 298
Mill fever, 259
Misclassification, 437, 509
Mixing factor, 345
Mixtures, 671
  interactions, 414
Models, 566
Modifying factors (MF), 642, 643
Moisture, 363, 364
Molecular biological approaches, 303
Monodisperse aerosol, 31
Monte Carlo model, 573-575
Moonlighting, 415
Mortality rate, 500
  ratio, 504
Mouse ear sensitivity test, 105
MSDS, see Musculoskeletal disorders
Mucous membrane irritations, 261
Multiple chemical exposure, 580
Musculoskeletal disorders (MSDs), 155, 159, 185
Mushroom grower's lung, 249
Mycobacterium tuberculosis, 340, 342, 347, 350
Mycotoxins, 267, 274, 276, 278, 310

N

N-Acetyltransferase, 430, 433
Nasal turbinates, 39
Nasopharynx, 35
National Council on Radiation Protection (NCRP), 40
National Institute for Occupational Safety and Health , see NIOSH
National Sanitation Foundation (NSF), 357
  certification, 358
Natural rubber latex allergens, 266
NCRP, see National Council on Radiation Protection
Negative pressure, 345, 353, 355, 357, 360, 361, 366
Nested case-control study, 532
Neurotoxic agents, 660
New York City Department of Health (NYCDH), 366
NIOSH, 333, 336, 340, 347, 368, 370
  Lifting Guidelines, 153, 178, 181, 182, 184
N,N-dimethylformamide, 444
No Observed Adverse Effect Level (NOAEL), 563, 645

No Observed Effect Level (NOEL), 644
No-threshold assumption, 569
Non-allergic asthma, 253, 255, 256
Non-allergic respiratory diseases, 252
Non-carcinogens and risk assessment, 560, 577
Non-culture methods, 294
Nondifferential misclassification, 510
Nose, 35
Nosocomial infections, 333, 334, 340
Not Classifiable as a Human Carcinogen, 667
Not Suspected as a Human Carcinogen, 667
NSF, see National Sanitation Foundation, 358
Null value, 504
Nutrient sources, 362

O

Obstruction, 34
Occlusion, 88
Occupational asthma, 237
Occupational epidemiology, 498
Occupational exposure limits (OELs), Chapter 12
 abroad, 690-696
Occupational Safety and Health Act (OSHA), 332, 333, 348, 639
Occupational Safety and Health Administration, see OSHA
Octanol-water partition coefficient, 79
Odor threshold, 660
Oils, 58
Open population, 500
Opportunistic pathogens, 335
Oral cavity, 35
Oral ingestion, 70
Organic acids and bases, 654
Organic dust toxic syndrome (ODTS), 251, 258
Organo-metallic compounds, 68
Oronasal breathing, 33
OSHA, 332, 333, 348

P

P-value, 516
Parasites, 222
Particle, 31
 size, 31
 surface area, 31
 terminal velocity, 27, 29
Partitioning, 80, 96
Patch samplers, 477
Path lengths, 32
Pathway, 331
PELs, see Permissible Exposure Limits
Penetrate, 27
Penicillin, 372
Penicillium, 363, 371
Peptidoglycans, 272
Percutaneous absorption, 110, 465, 604
Permeability, 77
 coefficient, 591
Permeation, 96
 rate, 96
 testing, 109
Permissible exposure limits (PELs), 640
Persistent chemicals, 661
Person-time, 500
Personal monitoring, 474
Personal protective equipment (PPE), 331, 333,

339, 348, 360, 366, 368, 374, 473
Pharmaceutical, 372
Pharmacokinetics, 638, 666
Pharynx, 35
Phenotype, 429
Physiological response, 635
Physiologically-based pharmacokinetic (PB-PK) model, 555, 642
Pigeon breeder's lung, 249
pKa, 654
Plants as biological agents, 266, 275
Plausibility, biologic, 549
Pollens, 266
Polydisperse aerosol, 31
Polymerase Chain Reaction techniques (PCR), 303
Population at risk, 500
Porous materials, 362
Positive hole correction, 296
Posture, 167
Potentiation, 413, 672
Power of test, 517
PPE, see Personal protective equipment
Practical evaluation of bioaerosol levels, 318
Practical threshold, 665
Prevalence, 57, 500
Prevention, 65
Pro-inflammatory and toxic agents, 267
Probabilistic analysis, 575
Process flow, 142
Productivity and ergonomics, 210
Prognosis, 59, 65
Proportional mortality ratio (PMR) studies, 546
Prospective follow-up study, 530
Protective clothing, 489
Protein and DNA adducts, 441, 442
Protozoa, 222
Psychosocial conditions, 147
Psychrophilic, 362
Pulmonary region, 35, 38
Push and pull limits, 184
Pushing tasks, 175

Q

Quantitative exposure assessments, 555
Quantitative risk assessment, 556

R

Race and skin, 84
Radon progeny, 40
RD_{50} approach, 653
Re-certification, 358
Reach, 150, 162, 164, 193
  envelopes, 167, 168, 169
Redistribution, 399, 400
Reference concentrations (RfCs), 579, 644
Reference doses (RfD), 563, 644
Registry-based (population-based) case-control study, 537
Regulatory decisions, 556
Relative humidity, 91, 362, 363
Relative measure of association, 503
Removal efficiency, 478
Repeated exposures, 94, 635
Repetitive irritancy test (RIT), 115
Reproductive toxicants, 655
Residual body burden, 688
Respirable fraction, 35
Respirators, 346, 371

Respiratory, 232
diseases, 232
protection programs, 333
sensitization, 72
sensitizers, 663
Retrospective follow-up study, 525
Reynolds number, 33
Rhinitis, irritant, 261
Risk, 555
assessment, 555
characterization, 556
factors, 61, 155, 202, 203, 204
Routes of administration, 568

S

Safe doses, 644
Safety factor approach, 563
Sample matrix, 308
Sampling, 331
strategies, 486
Saprophytes, 370
Saprophytic, 329
Scientific judgment, 560
Sebaceous glands, 76
Sedimentation, 27,29
Segment II test, 659
Selection bias, 508
Sensory irritants, 650
Serial bolus aerosol, 33
Shape, 31
Sharps, 349
Short-term exposure levels, 679
Silo unloaders syndrome, 259
Sitting, 171, 172
Skills checklist, 141
Skin
barrier function, 87
bioengineering, 112
exposure, 467
notation, 592
physical damage, 81, 85
protective product (SPP), 114
surface area, 599
ultraviolet irradiation, 86
Skin protecting, 65
emollients, 66
gloves, 66
Slip correction, 28
Small airways disease (SAD), 34
Smokers, 34
Soil, 97, 99, 100
Solubility, 76
Source control, 331
Soy allergens, 266
Specific enzyme immunoassays, 305
Specific gravity alignment, 410
Spore count methods, 299
Spray contact, 471
Stachybotrys chartarum, 363, 366
Standing, 171, 172, 173
Stokes number, 28
Stratified analysis, 524
Strength of the evidence, 559
Structure activity relationship (SAR), 647, 662
Study base, 500
Subchronic, 574
Substitution, 332
Sugar beet, 371

Suppressed immune systems, 334, 335
Surgical masks, 347
Surrogate, 395
biomarker, 395
skin techniques, 474
Surveillance, 339
Susceptibility, 84
Suspected Human Carcinogen, 667
Symptom development, 162
Synergism, 413
Synergistic action, 672
Systemic effects, 643
Systemic exposure, 66

T

T-lymphocytes, 76
Target organs, 395
Task analysis, 187
Temperature, 91, 362, 363
Test, 105
Test methods, 104
Buehler test, 105
Corrositex®, 108
Draize guinea pig test, 105, 107
Finn Chambers®, 107
Guinea Pig Maximization (GPM) test, 105
human repeat insult patch test (HRIPT), 106
local lymph node assays (LLNA), 105
mouse ear sensitivity test (MEST), 105
prick test, 106
sensitization potential, 105
TRUE test®, 107
Thermoactinomycetes, 370
Thermophilic, 362
Thermotolerant, 362
Threshold carcinogen, 560
Threshold dose, 639
Threshold Limit Values, *see* TLVs®
Health protectiveness of, 672
Time-weighted average (TWA), 651
TLVs®, Chapter 12
TLV® reduction factor, 683
TLV® or PEL reduction factor, 689
Total risk, 582
Toxic, 259
alveolitis, 259
effects, 558
response, 685
Toxicokinetics and toxicodynamics, 397
Toxigenic fungi, 331
Tracers, 482
Tracheobronchial, 35
region, 35
tree, 37
Transepidermal water loss (TEWL), 112
Translocations, 443
Trauma, 302
Tryptase soy agar, 298
Tuberculosis, 335, 337, 339, 340, 346
Mycobacterium, 340, 342, 347, 350

U

Ultrafine particles, 39, 32, 38
Ultraviolet illumination, 484
Uncertainty factor (UFs), 644

Universal precautions, 332, 337, 348
Unusual work schedules, 681
Upper respiratory tract, 35
Urinary mutagenicity, 441
Urticaria, 64, 77

V

Validity, 507, 548
Vapors, 92, 634
    diffusion, 364
    penetration, 92
    pressure, 49
    retarding systems, 365
Vehicle and permeation, 94
Vehicle effects, 94
Ventilation system, 362, 363, 368
Viable epidermis, 76
Video imaging technique for assessing exposure, 482
Viruses, 221
Viscosity, 27, 28
Viscous resistance, 27
VITAE system, 482
Vital status tracing, 522
Volatility, 93

W

Walk-through investigation, 331
Washing, 102,473, 479
WEELs, see Workplace Environment Exposure Limits
Weight of evidence, 559, 666
Wet work, 58
Wheat allergens, 266
Whole-body garment samplers, 478
Wipe samples, 608
Wiping, 479
Work, 160
    organization, 160
    practices, 331, 332, 348, 374
    schedules, staffing, breaks, and overtime, 145
Workplace, 160
    design, 160, 161, 162, 164, 167, 189
    layout, 144, 164
Workplace Environment Exposure Limits (WEELs), 640

X

Xerophilic, 362
Xerotolerant, 362